1. PERIODIC CHART OF THE ELEMENTS

1A	2A		3B	4B	5B	6B	7B		8B		1B	2B	3A	4A	5A	6A	7A	
1 **H** 1.0079																		2 **He** 4.00260
3 **Li** 6.941	4 **Be** 9.01218												5 **B** 10.81	6 **C** 12.011	7 **N** 14.0067	8 **O** 15.9994	9 **F** 18.99840	10 **Ne** 20.179
11 **Na** 22.98977	12 **Mg** 24.305												13 **Al** 26.98154	14 **Si** 28.0855	15 **P** 30.97376	16 **S** 32.06	17 **Cl** 35.453	18 **Ar** 39.948
19 **K** 39.0983	20 **Ca** 40.08	21 **Sc** 44.9559	22 **Ti** 47.90	23 **V** 50.9415	24 **Cr** 51.996	25 **Mn** 54.9380	26 **Fe** 55.847	27 **Co** 58.9332	28 **Ni** 58.70	29 **Cu** 63.546	30 **Zn** 65.38	31 **Ga** 69.72	32 **Ge** 72.59	33 **As** 74.9216	34 **Se** 78.96	35 **Br** 79.904	36 **Kr** 83.80
37 **Rb** 85.4678	38 **Sr** 87.62	39 **Y** 88.9059	40 **Zr** 91.22	41 **Nb** 92.9064	42 **Mo** 95.94	43 **Tc** (98)	44 **Ru** 101.07	45 **Rh** 102.9055	46 **Pd** 106.4	47 **Ag** 107.868	48 **Cd** 112.41	49 **In** 114.82	50 **Sn** 118.69	51 **Sb** 121.75	52 **Te** 127.60	53 **I** 126.9045	54 **Xe** 131.30
55 **Cs** 132.9054	56 **Ba** 137.33	57– see La series	72 **Hf** 178.49	73 **Ta** 180.9479	74 **W** 183.85	75 **Re** 186.207	76 **Os** 190.2	77 **Ir** 192.22	78 **Pt** 195.09	79 **Au** 196.9665	80 **Hg** 200.59	81 **Tl** 204.37	82 **Pb** 207.2	83 **Bi** 208.9804	84 **Po** (209)	85 **At** (210)	86 **Rn** (222)
87 **Fr** (223)	88 **Ra** 226.0254	89– see Ac series	104	105													

La Series

57 **La** 138.9055	58 **Ce** 140.12	59 **Pr** 140.9077	60 **Nd** 144.24	61 **Pm** (145)	62 **Sm** 150.4	63 **Eu** 151.96	64 **Gd** 157.25	65 **Tb** 158.9254	66 **Dy** 162.50	67 **Ho** 164.9304	68 **Er** 167.26	69 **Tm** 168.9342	70 **Yb** 173.04	71 **Lu** 174.967

Ac Series

89 **Ac** 227.0278	90 **Th** 232.0381	91 **Pa** 231.0359	92 **U** 238.029	93 **Np** 237.0482	94 **Pu** (244)	95 **Am** (243)	96 **Cm** (247)	97 **Bk** (247)	98 **Cf** (251)	99 **Es** (252)	100 **Fm** (257)	101 **Md** (258)	102 **No** (259)	103 **Lr** (260)

For B, C, H, Pb, Sm, and S, the precision of the atomic weight is limited by variations in the isotopic abundances in terrestrial samples. For B, Li, and U, variations in atomic weights in processed material may occur because of commercial isotopic separations. Values are reliable to ±1 in the last digit, or ±3 if the last digit is smaller. Mass numbers in brackets are for most stable or best known isotope.

WALTER E. HARRIS

Professor of Analytical Chemistry
University of Alberta

BYRON KRATOCHVIL

Professor of Analytical Chemistry
University of Alberta

AN

INTRODUCTION

TO

CHEMICAL ANALYSIS

SAUNDERS GOLDEN SUNBURST SERIES

SAUNDERS COLLEGE PUBLISHING
Philadelphia

SAUNDERS COLLEGE PUBLISHING/CBS
Educational and Professional
Publishing, A Division of CBS Inc.
Address Orders to: 383 Madison Ave.,
New York, NY 10017
Address Editorial correspondence to:
West Washington Square
Philadelphia, PA 19105

The cover. One of the rewards of analytical chemistry is the opportunity it affords to work with a variety of colors. The items displayed suggest both the instrumental and noninstrumental sides of analysis. The visible region of the spectrum on the spine represents the important role of spectral methods in analysis. The other items displayed are: a molecular model of the magnesium–calmagite indicator complex; mixing coils from an automatic analyzer; an inductively coupled plasma flame during the measurement of a 1000-ppm solution of barium chloride; bromocresol green acid-base indicator at 0.2 pH intervals from pH 3.0 to 5.8; a liquid-liquid extraction separation of lead dithizone; titration of iron(II) with dichromate using diphenylaminesulfonate indicator; and a gas-chromatographic column.

An Introduction to Chemical Analysis   ISBN 0-03-056882-X

CBS COLLEGE PUBLISHING

PREFACE

Our goals in writing this textbook are several: (1) to present analytical chemistry that is both accurate and current; (2) to provide laboratory procedures that emphasize workable, student-tested experiments; and (3) to present a readable, consistent level of theory with the student in mind.

We have aimed for a logical and concise style of presentation, illustrated often by examples that employ real data. The level of the material is designed for students who have a background in general chemistry and have sufficient appreciation of the need for high-quality experimental work to benefit from an intensive course in chemical measurements. Our overall objective is to develop in students a justified confidence in the quality of their experimental work, while providing them with the related theoretical background. The questions and problems at the end of each chapter are intended to serve as a systemic review of the material. For convenience in use, the examples and experiments have been set in a different format.

To reflect the practice of teaching a combination of noninstrumental and instrumental topics in introductory analytical courses, this book treats the two areas with about equal emphasis. It is designed to provide both the theoretical basis for analysis and a set of tested experiments on a level consistent with the description of the principles. No prior analytical experience is assumed; the book begins with basic analytical operations and builds progressively to current instrumental techniques. Our aim throughout is to give the reader a sense of the practical and its relation to theory. In explaining the fundamental basis of titrimetry we illustrate the calculated theoretical curves with data obtained in the laboratory. Such quantities as formal potentials and conditional equilibrium constants are of greater practical use than the more fundamental concepts of standard potential and thermodynamic equilibrium constants. Similarly, distribution ratios and partition ratios are more useful to the working analyst than distribution constants. The theory of sampling error is not only discussed, but its measurement is illustrated by an experiment.

Our experience indicates that student satisfaction and competence are enhanced when appropriate attention is paid initially to the fundamental techniques of measurement of weight and volume. Accordingly, more than usual detail on these topics is included. Also, we have emphasized polarography over more advanced electroanalytical methods in the belief that a student who understands direct-current polarography can quickly understand the more sophisticated techniques.

The detailed instructions for the procedures have been chosen, tested, and refined over a period of years. Unknown samples are used throughout; there is objective evidence that they increase both teaching effectiveness and student interest. We have attempted to recognize the realities of the marketplace by deleting a well-worked-out experiment because of the escalating cost of silver nitrate; in other procedures more dilute solutions of silver are used with, we trust, minimal loss of instructional benefit.

As far as practical, SI units, with the exceptions listed in *Analytical Chemistry*, and the nomenclature, symbols, and abbreviations recommended by IUPAC have been adopted. In certain cases awkward situations arise. For instance, in chromatography the symbol D_m is widely used to denote diffusion coefficient in the mobile phase, while the same symbol is proposed for the mass distribution ratio. To reduce confusion we use the symbol k for mass distribution ratio (partition ratio).

The accompanying instructors' manual gives both philosophy and practical suggestions on the teaching of introductory analytical chemistry. It provides information on techniques for the development of workable procedures, valid samples, and sound grading scales.

Colleagues and associates have been more than liberal with their assistance. Extraordinary appreciation must be expressed to Lu Ziola and Phyllis Harris. Ms. Ziola's knowledge of TEXTFORM programming and computer typesetting carried the book through to the completion of pages. This allowed us to make changes up to the last minute, with a substantial saving in publication time.* Phyllis Harris enhanced the quality in matters of style, clarity, and consistency and edited the copy.

Professor Fred Cantwell of the University of Alberta not only taught from the draft manuscript for a year but also willingly reviewed the manuscript twice. In both contexts he provided innumerable incisive comments and suggestions that we deeply appreciate. Dorothy Cox made many contributions, particularly in the laboratory experiments and discussions, and in writing the problems.

Appreciation is also expressed to those who reviewed the manuscript, Gary Asleson of the College of Charleston, James Anderson of the University of Georgia, and Stanley Pons of the University of Alberta; to Cathy Johnson for assisting with manuscript preparation, particularly with the figures and problems; and to Faye Nagle for drafting services. We also wish to thank John Vondeling and Lee Walters of Saunders for their encouragement and assistance in allowing us total responsibility for the book including typesetting at the University of Alberta.

Finally, we acknowledge the contributions of many others, including the teaching assistants and students with whom we have had the pleasure of working over the years. We welcome comments, suggestions, and criticisms.

W. E. HARRIS
B. KRATOCHVIL

August, 1980

*Writing and entry of material was begun in July, 1979, class material was produced as required throughout the following academic year, the first draft was completed in May, 1980, the revised manuscript on August 18, and page composition on September 30.

CONTENTS

LIST OF EXPERIMENTS

1

INTRODUCTION

When it has once been given a man to do some sensible things, afterwards his life is a little different.

A. Einstein

1-1 THE NATURE OF ANALYTICAL CHEMISTRY

Analytical chemistry deals with the identification, characterization, and measurement of the chemical species present in a sample. It may also be involved in the characterization of a sample through measurement of its physical properties, such as the viscosity of an oil or the crystal structure of a mineral. The scope of modern analytical chemistry is thus extremely broad, and it continues to expand as society and technology impose new demands on it. These demands extend from the identification and measurement (typically at the part-per-billion level) of environmental contaminants in water, air, and soil, or of impurities in materials for semiconductors in electronics, to continuous monitoring of the levels of critical substances in the bloodstream of hospital patients. To meet such demands has required the development of a host of new analytical techniques and instruments.

Analytical chemistry is a dynamic, challenging field called upon to solve many kinds of problems with technology that constantly provides new measurement tools. For example, more and more substances are being found whose concentration levels in the human body serve as indicators of health or of specific disorders. In pharmacology, accurate adjustment of a drug dose for an individual may necessitate the determination of its levels in the bloodstream. Because these measurements must be made rapidly and in large numbers, the development of efficient automated methods is important. Thus the role of an analytical chemist is essential not only to chemical science but also to allied areas in biology, food science, medicine, biochemistry, and engineering.

That part of analytical chemistry dealing with the measurement of the quantity of a substance present in a material is termed *quantitative analysis*, and is the area considered in this book. Training in this area includes more than an introduction to various methods of analysis. It involves evaluation of problems from the standpoints of proper acquisition of samples, selection of

1

analytical methods, consideration of possible interfering substances, and sound interpretation of results. It also fosters an appreciation of the care necessary in laboratory operations and provides experience in logic and planning that is valuable in all fields.

1-2 ANALYTICAL CHEMISTRY AND PROBLEMS IN SOCIETY. PUBLIC PERCEPTIONS OF TOXICITY

> Analytical chemistry sits at the core of all regulations designed to protect the public from chemical exposure that may present a significant health risk.
>
> J. G. Rodricks

Analytical chemistry, because it permeates almost every aspect of society, has profound implications for human welfare. An example is the analysis of materials for chemicals of possible toxicity to man. Threats to health, whether perceived or real, are of intense concern to the public. In the setting, monitoring, and enforcing of regulations regarding toxic substances, analytical chemistry plays a key role.

When foods were tested in past decades for toxic materials, they were presumed safe if such materials were not detected. Today, all foods can be shown to contain detectable amounts of a host of toxic materials. Such information is unsettling to the public. Nevertheless, food now is probably generally safer and of better quality than ever before. What has happened? To clarify this seeming contradiction, two terms relating to toxicity and one concerning the limit of analytical measurement should be described. The *fatality limit*[1] is the level of chemical exposure that when maintained for any length of time results in death. This limit depends on the nature of the chemical, the type of exposure, and on individual human factors such as size, age, general health, and smoking habits. The *toxicity limit* is the limit where well-being is first noticeably affected. This limit also depends on the chemical, the type of exposure, and individual human factors. It cannot be sharply defined. For benzene the toxicity limit is about 25 ppm, for the refrigerant gas freon about 1000 ppm, and for the fungal metabolite aflatoxin about 0.03 ppm.

Between the fatality limit and the toxicity limit an individual's health is affected. For the many chemicals essential to normal health (including water, oxygen, potassium, fluoride, selenium, and chromium) intake must not be reduced too far or health is affected by deficiency, which in the limit is fatal also.

The *detection limit* is the minimum amount of a substance that can be determined by a laboratory test procedure. This limit depends on the state of the art in analytical laboratories. The detection limit is completely independent of the other two and is continually being lowered as knowledge

---

[1]Both fatality and toxicity limits may be either acute or chronic. The acute limit is that for an exposure of short duration; the chronic limit is for exposures of long duration. Chronic limits are lower than acute limits. Types of exposure include inhalation, ingestion, and skin contact.

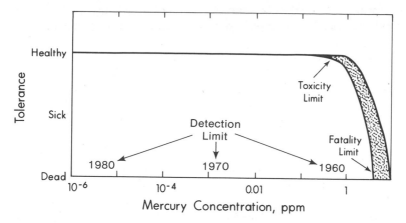

Figure 1-1. Schematic diagram illustrating the relation between the fatality and toxicity limits and the changing detection limits with time for mercury.

of chemistry improves and more sensitive instrumentation is developed.

The difference between detection limit and fatality or toxicity limit is not significant as long as the detection limit is below the toxicity limit. One of the greatest differences exists for the detection of radioactive substances, where the detection limit is billions of times lower than the toxicity limit. The levels of radioactivity normally present in our bodies from such atoms as carbon-14 and potassium-40 are easily measurable by present techniques.

Two or three decades ago, detection limits and toxicity limits for many materials were in the same range. But now detection limits have dropped far below toxicity limits. Unfortunately, misinterpretation of these greatly decreased detection limits has resulted in serious abuses in the use and interpretation of modern analytical information.

EXAMPLE 1-1

Twenty years ago detection and toxicity limits for mercury were about the same, and the detection of mercury in food was cause for justifiable concern. Since then detection limits have been dramatically lowered. Figure 1-1 depicts schematically the relation between the fatality limit, toxicity limit, and changing detection limits for mercury. A decade ago calls for the banning of food products were made because mercury had been detected in them — even though the amount was well below the toxicity limit.

Toxicity limits are established from the evaluation of numerous medical records. As indicated in Figure 1-1, the toxicity curve in the region of the toxicity limit shows a gradual rather than a sharp change, indicating the limit to be somewhat arbitrary. This applies to all substances. The value of this limit is unlikely to change substantially in the future. On the other hand, as analytical knowledge and techniques improve, detection limits will undoubtedly continue to fall, but never reach zero.

The general public and even many scientists have been slow to recognize the significance of the dramatic changes in detection limits brought about by improved analytical methods. Instead of being reassured by more precise information on levels of potentially hazardous substances, they have often responded negatively to reports that toxic substances have been detected in our food and general environment, without realizing the difference between what is detectable and what is harmful. Although alarm is certainly warranted when a substance is present at an unsafe level, it is not warranted when the substance is detectable but at far below toxic levels. Thus, public reaction to reports that polychlorinated biphenyls (PCB's) have been detected in chicken, DDT in milk, and arsenic in vegetables, has not taken into account that their detection may simply be the result of improved analytical sensitivity. Every mouthful of food that we or our ancestors have ever eaten is radioactive at far above present detection limits, and all foods contain numerous toxic substances well above these limits. In short, it is no exaggeration to say that detectable amounts of almost everything are present in almost anything.[2] The pursuit of lower and lower detection limits along with more rapid, efficient analytical methods of high specificity increases our ability to improve our environment. Misuse of the information is a major problem, one which analytically trained scientists must assume some responsibility in solving.

1–3 STEPS IN ANALYSIS

An introductory course in quantitative analysis cannot turn out an analytical chemist prepared to attack every problem that may come his way. But it can provide a basis for the most important attributes an analyst can possess — sound knowledge of the fundamentals of the field and confidence in his ability to gather accurate, reproducible laboratory data. Such a course requires an integrated combination of theory and experimental work.

The analytical operation comprises up to five steps — (1) definition of the problem, (2) sampling and preliminary treatment, (3) separation, (4) measurement, and (5) calculation and evaluation of data. All are important, and this book provides an introduction to each. More comprehensive coverage of the various steps is available in subsequent courses that build on the foundation provided here.

The level of background assumed is a one-year course in general chemistry, including a brief introduction to organic chemistry. A second prerequisite is an appreciation of the need for carefully planned and executed experimental work in the solution of problems of interest to modern society.

This section offers brief comment on the key steps involved in chemical analysis and is designed to put them into perspective.

Sampling

Before an analysis can be performed, a sample of the material to be analyzed must be obtained that is representative of the whole. This part of

---

[2]Absolute safety, zero exposure, and completely nontoxic substances exist only in the imagination. The first could not be proven in less than infinite time. Should the second be achieved, it could not be confirmed experimentally. As for the last, since chemicals include everything we see, eat, or walk on, excessive exposure to any substance can have adverse effects. The chemical H_2O, for instance, has doubtless caused more fatalities than any other.

the operation is critical, since the quality of the analysis can be no better than the degree to which the sample represents the body of material under study. The study of proper sampling procedures is therefore of major importance.

Preliminary Operations

Once a representative sample is received, a decision must be made as to the analytical method to be applied. This decision is based on a number of considerations: the methods available for the substance to be determined, possible interferences and their effects on the various methods, the equipment available, the precision and accuracy with which the result needs to be known, and the time and money available. An additional consideration is the percentage of sought-for substance in the sample; a *major constituent* constitutes at least 1%; a *minor constituent*, 0.01 to 1%; and a *trace constituent*, less than 0.01%.

The method selected may require a variety of preliminary operations before the actual measurement. These may include dissolution of the sample, separation of the sought-for component from substances that would interfere in the measurement, and adjustment of conditions for the measurement itself. Sometimes preliminary operations are unnecessary, as in measuring the free hydrogen ion concentration in a sample of lake or river water. At other times they may constitute the major part of the analytical process, as in determining the distribution of various hydrocarbons in a gasoline sample by gas chromatography, where the measurement is simple once separation is achieved. Separations, required for many samples, usually involve the formation of two phases, with the wanted material concentrated in one and the unwanted in the other. The two phases may be a combination of gases, liquids, or solids, such as gas and liquid, gas and solid, or liquid and solid.

Measurement

Procedures for the measurement of a sought-for substance may be classified according to several general types. (1) *Chemical* methods involve adding to the sample a reagent that reacts in a defined way with the component to be determined. Either the amount of reagent needed to complete the reaction or the amount of product obtained can then be measured. (2) *Electrical* methods involve measuring potential, current, conductivity, or other electrical properties that can be related to the amount of sought-for substance, either directly or upon addition of a reagent. (3) *Spectroscopic* methods involve measuring the quantity of electromagnetic radiation absorbed or emitted by a substance. Radiation from the visible, ultraviolet, and infrared regions of the spectrum is most often measured for this purpose; high energy (gamma rays or X-rays) or low energy (radio frequency) radiation is also measured, though less often. (4) *Other physical* methods include the measurement of properties such as density, refractive index, optical rotation, emission of alpha or beta particles from atomic nuclei, magnetism, and rates of chemical reaction.

Calculation and Evaluation of Results

Following the measurement step the analyst must relate the data (amount of reagent required, instrument reading, or other) to the amount of sought-for substance. This relation, which may be simple or complex, is a frequent source of grief for students. Of equal importance is the assessment of the data in terms of *precision*, or reproducibility, and *accuracy*, or closeness to the true value. A variety of statistical tests are available to answer questions about the reliability of a set of measurements, rejection of a suspect measurement, and so on. These tests also help one decide how many analyses of a sample are necessary to give the required level of confidence in the results.

1–4 MANAGEMENT OF EXPERIMENTAL DATA

Statistical thinking will one day be as necessary for efficient citizenship as the ability to read and write.

H. G. Wells

Significant Figures

To indicate the uncertainty in a quantity, the correct number of *significant figures* must be selected. For instance, if a certain volume of water were measured in a graduated cylinder, the reported result might be 41 mL, a number having two significant figures. If the water were measured in a buret, the result might be 41.21 mL, four significant figures. If the water were weighed on a top-loading balance, the result might be 41.206 mL — here five figures are significant. In each case no fewer digits would normally be appropriate and no more would be justifiable.

Confusion may arise in the selection of significant figures where zeros are involved. To avoid ambiguity, remember that any zero needed only to locate the decimal point is not a significant figure. Thus, 0.005 has one significant figure, but 2.0400 has five. If confusion appears likely, numbers should be written in exponential form. For instance, 6.00×10^{10} is unambiguous; it has three significant figures. To indicate that the number 40 000 has but one significant figure, write it in exponential form, 4×10^{4}, and to indicate it has three, 4.00×10^{4}.

When *rounding* numbers, if the quantity being rounded is less than 5, drop it; if it is greater than 5, increase the last significant figure by 1; if it is 5, round to the nearest even number.[3] For example, 4.25, 6.35, and 1.05 are rounded to 4.2, 6.4, and 1.0. During calculations at least one digit more than the number of significant figures to be used in the final answer should be carried along for purposes of eventual rounding. If possible, avoid rounding until the final answer in a calculation is obtained. With electronic calculators this practice requires little or no additional effort.

The last digit of a result is generally assumed to be uncertain by 1. But in analytical work the uncertainty in the last digit may be greater than 1; this is

---

[3]Rounding to the nearest even number ensures that results of a series of calculations will not be biased up (if all fives are rounded up) or down (if all fives are dropped).

sometimes indicated by lowering the digit a half space. Thus 25.8_5 means that the uncertainty in the 5 is more than 1. In this instance, rounding to 25.8 implies an uncertainty of four parts per thousand, which may be greater than is warranted. (See the footnote to the periodic table at the back of the book for examples.)

Use of Calculators

Electronic calculators range from simple four-function models — which add, subtract, multiply, and divide — to sophisticated programmable units. All types allow calculations to be completed quickly and accurately. Experience is necessary, however, before any calculator can be operated efficiently. To obtain experience, calculate the value of each of the following, applying the appropriate instructions for the calculator available. Report the correct number of significant figures in each case.

a. $8 \times 6/(3 \times 4)$

b. $3.260 \times 8.264 \times 0.001\ 264/(0.1376 \times 2.165)$
(Answer: 0.1143)

c. $40.32 \times 3.636 \times 0.2942/(7.319 \times 17.94 \times 14.44)$
(Answer: 0.02275)
$40.32 \times 3.636 \times 0.2942/(7.319 \times 17.94 \times 14.52)$
$40.32 \times 3.636 \times 0.2942/(7.319 \times 17.94 \times 14.47)$
$40.32 \times 3.636 \times 0.2942/(7.319 \times 17.94 \times 14.41)$
$40.32 \times 3.636 \times 0.2942/(7.319 \times 17.94 \times 14.93)$

d. $0.01234/88\ 888$ (Answer: 1.388×10^{-7})

e. $1\ 234\ 567 \times 9\ 999\ 999$ (Answer: $1.234\ 567 \times 10^{13}$)

f. $123\ 456.789/999.999\ 999$ (Answer: 123.456 789)

Electronic calculators lacking the ability to operate in exponential notation may balk at calculations involving many digits.

Errors in Chemical Measurements

In quantitative chemical measurements it is essential to reduce to acceptable levels the errors to which all measurements are subject. First, possible sources of error must be identified, and second, the effort required to reduce them to an acceptable level must be assessed. Limitations of time, cost, and technique usually keep a method from being so accurate and precise that a single measurement can give an answer that corresponds to the true value with a high degree of confidence. For this reason a measurement normally is made in replicate, and the results are treated statistically to provide an estimate of the error of the method. Evaluation of experimental results therefore comprises two facets: (1) an assessment of the extent and direction of errors in individual measurements and (2) a statistical treatment of the resulting set of accumulated measurements to give an estimate of the best answer and the degree of confidence warranted in it. Some definitions and discussion of terms in common use are appropriate here.

The *error* of a measurement is the difference between the value of that measurement and the true value:

$$\text{error} = \text{observed value} - \text{true value}$$

A measurement is termed *accurate* when its error is small.

Error may be classified as indeterminate (random) or determinate (systematic). *Indeterminate errors* vary in a nonreproducible way around the true value. For a set of measurements they always show two characteristics: large errors occur less frequently than small ones, and the errors are distributed evenly above and below the average value. These errors can be treated statistically by use of the laws of probability. *Determinate errors* tend to be reproducible from analysis to analysis and generally bias a result in one direction. They can be assigned to definite causes, at least in principle, even though a cause may not have been located. In general, statistical concepts should not be applied to sets of data that contain determinate errors.

Most errors are partly indeterminate and partly determinate. A search for and reduction of determinate errors in analysis is generally worthwhile. Indeterminate errors are not so readily discovered and eliminated, and consequently they are more often treated statistically and taken into account in reporting an analytical result.

The *precision* of a group of measurements refers to how closely individual measurements agree. When the values of a set of measurements are close to each other, they are said to have high precision, and when scattered, they are said to have low precision. High precision does not, however, necessarily mean high accuracy because a constant error may be present that causes bias in all the results. Low precision indicates errors large enough to produce untrustworthy results unless the central value of a large number of values is taken.

The central value of a set of chemical measurements is most often expressed quantitatively by either the average or the median. The *average*, or *mean*, $\bar{x}$ is defined by

$$\bar{x} = (x_1 + x_2 + \cdots + x_n)/n$$

where x_1, x_2 $\cdots$ are the values for individual measurements and n is the total number of measurements. The average is generally the most valid statistical estimate of the true value provided that determinate errors are absent.

EXAMPLE 1–2

The values for a set of measurements of the density of a solution are 1.012, 1.007, 1.022, 1.020, 1.017, and 1.014. The average is

$$(1.012 + 1.007 + 1.022 + 1.020 + 1.017 + 1.014)/6 = 1.015$$

To obtain the *median*, arrange the values of a set of measurements in numerical sequence from the largest to the smallest. The median is the middle value in the set if the number of measurements is odd and the average of the central pair if the number is even. The median is especially useful for small numbers of measurements having considerable scatter. It has the advantage of being quickly determinable and being less sensitive than the average to deviant values in a set. In the foregoing. example the median is 1.016.

EXAMPLE 1–3

Thirteen students carefully and independently read and recorded the volume on a buret with the following results: 33.26, 33.22, 33.20, 34.21, 33.22, 33.20, 33.21, 33.20, 33.22, 33.22, 33.19, 33.22, and 33.21. What is the average and the median?

The average, which is the sum of the numbers divided by 13, is 33.29. If the numbers are ranked in order of size, the value of the seventh number is 33.22, which is the median and in this example is clearly more reliable than the average.

The scatter in a set of measurements may be estimated by the *average deviation*, the average of the absolute individual deviations from the mean. In Example 1-2 the average deviation is 0.004. When the median is used, the scatter is better expressed by the *probable deviation*, the median of the absolute individual deviations from the median for a set. For Example 1-2 the probable deviation is also 0.004.

A more statistically useful measure of the scatter, particularly for indeterminate errors, is the *standard deviation s*:

$$s = \sqrt{\frac{d_1^2 + d_2^2 + d_3^2 + \cdots + d_n^2}{n - 1}}$$

where $d_1 = x_1 - x$, $d_2 = x_2 - \bar{x}$, and so forth.[4] The standard deviation is used widely in mathematical treatments of data. If the scatter is known to be caused by indeterminate (random) error only, calculations based on the standard deviation can give a measure of the degree of confidence justifiable in a set of results.

Errors may be reported in either absolute or relative terms. An *absolute error*, defined as the difference between the observed and the true value, has the dimensions of the measurement. The absolute magnitude of an error is not useful, however, unless the observed value also is given. For example, an error of 0.2 mg would not be significant if 10 g of a salt were being weighed for a synthesis, but would be if a 10-mg organic sample were being weighed for microanalytical carbon and hydrogen determinations. To relate the absolute error to the size of the measured quantity, a relative value often is used. *Relative error* is defined as the absolute error divided by the true value:

$$\text{relative error} = \text{absolute error/true value}$$

---

[4]A convenient form of this equation for use with calculators is

$$s = \sqrt{\frac{\Sigma x^2 - (\Sigma x)^2/n}{n - 1}}$$

Many scientific calculators include built-in programs for calculation of standard deviation.

Similarly, *relative standard deviation* is given by standard deviation/average $s/\bar{x}$.

Relative quantities have no units and are normally expressed as a decimal fraction or in parts per hundred (pph), parts per thousand (ppt), parts per million (ppm), and so forth — whichever is most suitable. They can be expressed in terms of percentage, but because absolute values for analyses are often given in that form, the "parts per" nomenclature is recommended to avoid confusion.

EXAMPLE 1–4

Calculation of deviations for the measurements of density in Example 1-2.

Value, x	Deviation from Average, $d = |\bar{x} - x|$	Deviation ^{2}, d^2
1.012	0.003	0.000009
1.007	0.008	0.000064
1.022	0.007	0.000049
1.020	0.005	0.000025
1.017	0.002	0.000004
1.014	0.001	0.000001
$\bar{x} = 1.015$	Sum = 0.026	Sum = 0.000152

Average deviation = 0.026/6 = 0.004
Relative average deviation = 0.004/1.015 = 0.004, or 4 ppt
Standard deviation = $\sqrt{0.000152/(6-1)}$ = $\sqrt{0.000030}$ = 0.0055
Relative standard deviation = 0.0055/1.015 = 0.0054, or 5 ppt

In addition or subtraction the absolute error of a result is at least as large as that of the number with the largest absolute error. For example, if 0.34 is subtracted from 44.6436, the result will have only four significant figures and should be reported as 44.30, not 44.3036. In multiplication and division the relative uncertainty of the result should be of the same order as that of the least accurately known factor. Thus, if 121.463 is multiplied by 0.026, the result should be reported as 3.2, not 3 or 3.16.[5]

For logarithmic terms, such as pH, retain as many digits to the right of the decimal point as there are significant figures in the number written in nonexponential form. Digits to the left of the decimal point in a logarithm serve only to locate the decimal point in the number. By this rule, log 186.3

[5]Estimation of the error of a computed result from the errors of the terms used to calculate the result can be done by several paths. Which to choose depends on the error (determinate or random) and on the computation (addition-subtraction, multiplication-division, or other). For the computation $R = A + B - C$, the determinate error ϵ_R is given by $\epsilon_R = \epsilon_A + \epsilon_B - \epsilon_C$ and the random error by $s_R = (s_A^2 + s_B^2 + s_C^2)^{0.5}$, where s_R corresponds to the standard deviation. For the computation $R = AB/C$, the *relative* determinate error is $\epsilon_R/R = \epsilon_A/A + \epsilon_B/B - \epsilon_C/C$, and the random error, $(s_R/R) = ((s_A/A)^2 + (s_B/B)^2 + (s_C/C)^2)^{0.5}$, where s_R/R is the *relative* standard deviation. A general expression covering all mathematical operations has been derived; see the second reference at the end of the chapter.

$= 2.2702$, $\log 1.863 = 0.2702$, $\log 2.67 \times 10^{-10} = -9.573$, antilog $0.157 = 1.44$, and antilog $0.057 = 1.1$.

Determinate Error

Because most of the statistical tests described in the next section cannot be applied to determinate errors, it is important to identify and then eliminate or minimize them in analytical procedures. Several kinds of determinate errors occur. Classified according to source, they include errors inherent in the analytical method chosen, such as an incomplete chemical reaction, side reactions, and interferences; errors resulting from the analytical system employed, such as malfunctioning or improperly calibrated equipment and faulty reagents; and errors caused by the operator, such as faulty color perception, incorrect handling of equipment, nonquantitative technique, calculation errors, improper sampling, and bias. Once identified, these errors can often be reduced significantly. Method errors may be reduced by modifying the method, by performing blank or standard control runs, or by standard addition procedures. System errors may be reduced by calibration and proper use of apparatus or by obtaining higher-quality reagents. Operator errors may be reduced by increased awareness of potential sources of error or by greater care in critical operations.

Determinate errors may also be classified as constant, proportional, or variable. A reactive impurity in a fixed volume of solvent used to dissolve a sample would give a constant error, a wrong concentration of a standard solution used as a titrant would give a proportional error, and gain or loss of water by a sample before being weighed for analysis would give a variable error. How much time and effort should be expended on reducing these errors to a minimum depends on what they are, how significant they are judged to be, and how expensive their reduction would be.

1–5 STATISTICS FOR SMALL SETS OF NUMBERS

Dealing with the small sets of numbers common to most chemical analyses raises several questions. Foremost is what to select as the best value when one of the three or four determinations run on a sample seems to disagree with the others. The first point to consider is the possibility of determinate error. Objective judgment is required in deciding the probability of such an error and whether rejection of the suspect value is reasonable. When this suspect value cannot justifiably be discarded on the basis of a known determinate error, the difficulty may be caused by unknown determinate error or simply random error. With a small number of observations ($n < 10$), statistical methods applicable to large numbers become less useful or even misleading. Therefore, methods have been developed especially for cases involving few measurements. When only three or four values are available and both indeterminate and variable determinate errors are present, the median is often more reliable than the average in that it is less sensitive to widely divergent values.

Criterion for Rejection of Suspect Values: The Q Test

When the number of values is large, confidence limits can be calculated that enable a suspect value to be judged in relation to its position on the Gaussian-distribution curve (Section 22-1). When, however, the number of values is small, conventional statistical methods cannot be used.

The best test available for objectively handling data rejection when dealing with small numbers of observations is the Q *test*:

$$Q_{obs} = \left| \frac{\text{suspect value} - \text{nearest value}}{\text{largest value} - \text{smallest value}} \right|$$

where "nearest value" refers to the value numerically closest to the suspect value. In calculations of averages and standard deviations the suspect number should be rejected if Q_{obs} exceeds the tabulated value Q_{tab}. Table 1-1 gives Q_{tab} values for small numbers of measurements at the 90% confidence level. The Q test is applicable only once to a set of data. If two or more values are suspect, either make more measurements (preferable), or use the median.

For three values the suspect one usually must be highly divergent before it can be rejected. When results are divergent, additional measurements to provide a more reliable average should be considered. If determinate errors are absent, there is safety in numbers.

EXAMPLE 1-5

Four determinations of chloride in a sample give values of 44.28, 44.56, 44.37, and 44.33%. Can the value 44.56% be rejected? For these measurements

$$Q_{obs} = \left| \frac{44.56 - 44.37}{44.56 - 44.28} \right| = \frac{0.19}{0.28} = 0.68$$

Since 0.68 does not exceed the $Q_{90\%}$ value of 0.76 for four measurements, the value 44.56% cannot be discarded. The best average to report for this set of results is 44.38%. If the suspect value were 44.87% instead of 44.56%, Q_{obs} would be 0.85. Because the suspect value 44.87% would cause Q_{obs} to exceed $Q_{90\%}$ for four measurements, it would not be included in the average. The best average would then be 44.33%.

Table 1-1. Q-Test Values for Three to Ten Measurements at the 90% Confidence Level

n	3	4	5	6	7	8	9	10
$Q_{90\%}$	0.94	0.76	0.64	0.56	0.51	0.47	0.44	0.41

Criteria for data rejection such as the Q test are not applied to medians. Medians are taken directly from the entire set of data. In the foregoing example the median would remain 44.35%, whether the suspect value were 44.56 or 44.87%.

Many defective results are caused by errors in calculation. Common sources of such mistakes include the use of incorrect formulas, use of incorrect molecular weights, incorrect digits, misplaced decimal points, omission of factors in calculation, and too few significant figures. Inadequate record keeping is another frequent cause of poor results. A positive correlation exists between the quality of records and the quality of experimental work. If specific causes of poor results for several experiments cannot be located, the work probably is being performed without adequate comprehension of the background. Poor agreement among replicates often serves as a warning that results will be unsatisfactory.[6]

PROBLEMS

1-1. Explain what is meant by fatality limit, toxicity limit, and detection limit. How is each established? What is the effect of decreasing detection limits on public perceptions of health hazards?

1-2. What are the five key steps in a chemical analysis?

1-3. How many significant figures do the following quantities contain: 1.56×10^{-4}, 40.02, 400.2, 0.000 156, 156 000.0, the sum of 40.02 and 402, the product of 402 and 156 000.0?

1-4. The mass of the sun is 330 000 times that of the earth, and its radius is 700 000 km. Express these numbers to indicate unambiguously that they are accurate to three significant figures.

1-5. Round to four significant figures: 43.6424; 0.777 77; 4.426×10^{-7}; 30 000.24.

1-6. Express the molecular weight of the following compounds to the maximum allowable number of significant figures (using the atomic weights from the periodic table at back of book): Ag_2O, B_2O_3, $EuCl_3$, PbS, and H_2O.

1-7. Using the correct number of significant figures, express the result of $0.957\ 28 - (0.1132 \times 1.37)$.

1-8. Explain what is meant by each of the following terms: determinate error, indeterminate error, absolute error, relative error, precision, accuracy.

1-9. Define and illustrate the following: average, median, mean, average deviation, relative average deviation, probable deviation, relative probable deviation, standard deviation.

1-10. What is the relative error in percentage in a result that reflects a difference of 0.02 mL of titrant when the total volume of titrant required is (a) 2 mL, (b) 20 mL, (c) 40 mL?

---

[6] A recent survey of follow-up reports at the University of Alberta, based on combined student and instructor assessment of poor results, showed that of the approximately 16% of analyses that earned a grade of 1 (out of 5) 56% revealed errors in calculation, 12% appeared to involve errors in procedure or technique, and 18% had poor agreement of replicate analyses. In 14% of the cases (2% of the total analyses reported) the cause could not be determined. For grades of 2 out of 5 (12% of all the analyses reported) 25% had 1 or more errors in calculation, 20% had errors in procedure or technique, 23% had poor agreement of replicates, and 32% (about 4% of the total analyses reported) had no ascertainable source of error. When specific problems are encountered it is recommended that you discuss with your laboratory instructor the possible causes of poor results.

1-11.   A sample of silver nitrate containing 75.52% silver is analyzed. One analyst obtains 74.88, 75.32, and 75.59% for three independent determinations, and reports the mean value. (a) How many parts per thousand does each value differ from the mean? (b) What is the absolute error of the mean? (c) What is the relative error of the mean and of the median?

1-12.   Express the uncertainty of the following measured quantities in parts per thousand: 0.1000 g; 5.12 g; 2.45×10^4 mL; 0.014 cm.

1-13.   For the following sets of numbers calculate the average deviation and the relative average deviation: (a) 40.00 and 40.13; (b) 0.511, 0.515, and 0.512; (c) 1234, 1247, and 1238.

1-14.   Nine students independently analyzed a sample of iron ore using dichromate as the titrant and obtained the following results for the percentage of iron: 33.50, 32.98, 33.23, 33.35, 33.43, 33.25, 33.12, 33.30, 33.64. Calculate the average, average deviation, standard deviation, relative standard deviation, median, probable deviation, and relative probable deviation for these results. If the correct result is 33.37, what is the accuracy of the average?

1-15.   Nine different students independently analyzed another sample of iron ore and obtained the following results: 35.15, 34.87, 34.77, 35.16, 35.27, 34.67, 34.54, 34.96, and 35.25. Is the precision of their average result as high as that for the preceding problem?

1-16.   Eight students analyzed a sample containing uranium for thorium-234. Their independent results in picograms were: 3.34, 3.47, 3.60, 3.55, 3.57, 3.27, 3.55, and 3.53. For this set of results calculate the average, median, and relative standard deviation.

1-17.   Eight students independently analyzed a sample of nickel for trace copper by atomic absorption and obtained the following results for the percentage of copper: 0.343, 0.283, 0.300, 0.236, 0.272, 0.333, 0.322, and 0.308. What is the average, median, relative standard deviation, and relative probable deviation for this set of results.

1-18.   A sample of pure potassium chloride was analyzed for chloride with the following results: 47.24, 47.08, 47.31, 47.42, 47.29, and 47.38%. What is the precision in terms of standard deviation and relative standard deviation in percentage, and what is the absolute and relative accuracy of the average analytical result?

1-19.   In the preceding problem, can any of the results be rejected on the basis of the Q test?

1-20.   The maple leaf coin is of pure gold, nominally weighing one troy ounce (31.1035 g). If the weights of 10 such coins were found to be 31.2104, 31.1991, 31.1938, 31.2127, 31.2054, 31.2092, 31.1673, 31.1713, 31.1839, and 31.1721 g, what is the average and the standard deviation? With gold at $500 per ounce, what is the standard deviation expressed as dollars per coin? What is the average accuracy in dollars per coin?

1-21.   After testing any questionable result by the Q test, calculate the average and median of the following results for percentage of copper in an alloy: 10.00, 10.10, 10.18, 10.00. What value should be reported?

1-22.   Repeat the calculations of the preceding question on the values 10.00, 10.10, 10.25, and 11.00. In your judgment, what is the best value to report in this case?

1-23.   Determination of the percentage of calcium in a sample of pure calcium carbonate yielded the following results: 39.77, 39.87, 39.63, 39.68, 39.71, 39.92, and 39.55. What is the average and the standard deviation? Can any of the results be rejected? Is there a determinate error? If so, how large is it?

1-24.   Consider the following set of experimental data: 40.00, 40.15, 40.11, 40.02, 40.07, 40.01, 40.32. What is the median, the probable deviation, the best value for the average, the average deviation, and the standard deviation?

1-25.   Repeat the calculations of the preceding problem for the following: (*a*) 12.34, 12.35, 12.34, 12.37, 12.39, 12.45, 12.37; (*b*) 0.437, 0.432, 0.429, 0.431; (*c*) 122.8, 120.2, 121.9, 122.2, 122.6; (*d*) 87.34, 87.21, 87.46, 87.74.

REFERENCES

International Union of Pure and Applied Chemistry, *Compendium of Analytical Nomenclature*, Pergamon Press, Oxford, 1978, pp 8–11. A compilation of recommended terms and symbols for reporting analytical data.

D. E. Jones, *J. Chem. Ed.* **1972**, *19*, 753. A discussion of significant digits in logarithm-antilogarithm interconversions.

H. A. Laitinen and W. E. Harris, *Chemical Analysis*, McGraw-Hill, New York, 1975, Chap. 26. ✓ Presents more advanced methods of calculating the error of a computed result.

J. C. Martin, *Anal. Chem.* **1980**, *52*, 253A. A critical assessment of the concept of absolute zero risk by a chemist turned U.S. Congressman.

J. G. Rodricks, *Anal. Chem.* **1980**, *52*, 515A. A discussion of the major role of analytical chemistry in the regulatory responsibilities of the U.S. Food and Drug Administration.

2

BASIC LABORATORY OPERATIONS

If you can measure that of which you speak, and can express it by a number, you know something of your subject; but if you cannot measure it, your knowledge is meager and unsatisfactory.

Lord Kelvin

2-1 MEASUREMENT OF WEIGHT [1]

Weighing[2] provides, either directly or indirectly, the foundation on which exact chemical knowledge rests. The precision with which analytical measurements can be made, and therefore the quality of the experimental results, depends ultimately on proper operation of the correct kind of balance. In this section the design and operation of the *analytical balance*, the *top-loading balance*, and the *triple-beam balance* are described.

The Analytical Balance: Background

The user of an analytical balance should know how to determine whether it is operating properly and how to make the simple adjustments that are occasionally necessary. After instruction, adjustments in optical-scale sensitivity and zero position of a single-pan balance can be made easily in 5 to 10 min.

Weighing usually involves application of the lever principle. The weight of an object is determined by (1) addition or removal of weights from one arm of the lever until it is brought back to its initial position (the null-balance principle, illustrated in Figure 2-1) or (2) measurement of the extent to which the lever is deflected from its initial position by the weight of the object (the direct-deflection principle, Figure 2-2). Most usual is a

---

[1]Probably the first quantitative analyses were checks on the qualilty of goods traded by early man. Stone weights used in Babylon 4600 years ago have survived to present times, indicating that some sort of balance must have been in use in those cultures. The oldest quantitative analysis was likely the fire-assay method for measuring the purity of gold. This procedure requires the weighing of the material under assay before and after the heating operation.

[2]We use the term *weight* rather than *mass* throughout this book, while recognizing that a distinction can be made between the two. *Mass* is a measure of the quantity of matter in an object, and *weight* is the force of gravitational attraction between an object and its surroundings. Gravitational attraction varies with locality, being less at higher altitudes on earth, for example, whereas mass is unaffected by locality. The relation between the two is $W = M g$, where g is the acceleration due to gravity. Analytical balances measure mass and so are unaffected by locality.

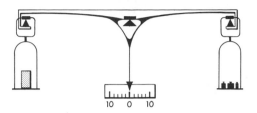

Figure 2-1. Schematic illustration of the null-balance principle in weighing with a two-pan balance. Weights are added to opposite arm of lever until beam returns to zero position, as indicated by attached pointer. Load on beam varies with weight of object.

combination of the two in which the direct-deflection principle is employed for the last fraction of the weight.

Early analytical balances of the two-pan type were highly precise but tedious to operate. They also were subject to decreasing sensitivity with load; that is, the larger the load on the beam, the smaller the beam deflection per unit of weight. This variation in sensitivity made use of the direct-deflection principle in two-pan balances unreliable for the determination of any but a minute fraction, typically 1 to 3 mg, of the weight of an object.

In the single-pan balance of the constant-load type a large counterweight is permanently attached to one end of the beam, making a knife-edge unnecessary at this end (Figures 2-2 and 2-3). The analytical weights and the pan holding objects to be weighed are suspended by a knife-edge from the opposite end. The object is placed on the pan, and weights are removed until they correspond to slightly less than the weight of the object. The last fraction of the weight of the object is obtained by use of the direct-deflection principle and is read from an optical scale. Because the beam has a constant load upon it, the sensitivity of the balance is independent of the weight of the object, and therefore a large fraction of the weight may be obtained by direct deflection. The full optical scale is typically 1 g (sometimes 100 mg) in a balance of 100- to 160-g capacity. For a 1-g scale the deflection is read from an image of a 0- to 1000-mg scale attached to the beam and projected onto a frosted glass or plastic panel near the front of the balance. This optical scale

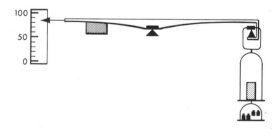

Figure 2-2. Schematic illustration of the direct-deflection and constant-load principles in a single-pan balance. Weights are removed from lever arm supporting object until beam returns to near zero position. Last portion of weight is read from optical scale. Beam is under constant load at all times.

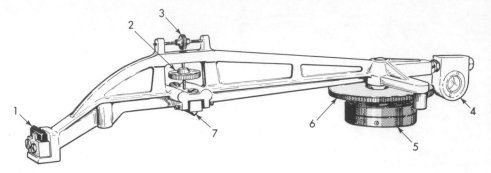

Figure 2-3. Beam from a constant-load balance: (*1*) terminal knife-edge; (*2*) nut for adjustment of optical-scale sensitivity; (*3*) coarse zero-adjustment knob; (*4*) reticle for optical-scale readout; (*5*) counterweight; (*6*) piston portion of air damper and part of counterweight; (*7*) central knife-edge.

can be read to the nearest 0.1 mg. The optical-scale sensitivity can be adjusted by changing the center of gravity of the balance beam through raising or lowering the nut provided for this purpose (Figure 2-3, *2*). Raising the nut increases the sensitivity. The single-pan balance has important advantages over the classical two-pan balance: (1) weighing is more rapid and convenient owing to the use of direct deflection for the last portion of the weight and to the inclusion of built-in weights operated by exterior knobs and (2) accuracy and precision can be maintained over long periods because only two knife-edges are required instead of three and because the use of internal weights minimizes the chances of weight changes from corrosion or mishandling.

An analytical balance must be treated with the care due any precision instrument. It should be mounted on a base that cushions the effects of vibration. To reduce wear on the knife-edges, the beam should be arrested between readings. The zero setting should be checked frequently and the optical scale periodically. Each should be adjusted as needed according to the

Figure 2-4. Single-pan analytical balances.

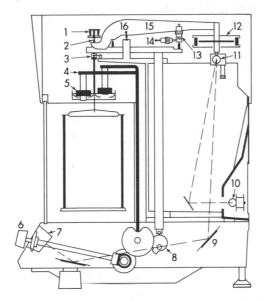

Figure 2-5. Schematic drawing of components of a single-pan analytical balance: (*1*) pan-support stirrup; (*2*) front knife-edge; (*3*) pan brake; (*4*) weight carriage; (*5*) built-in weights; (*6*) weight control mechanism; (*7*) optical readout; (*8*) beam-arrest shaft; (*9*) mirror (adjustable for zero setting); (*10*) bulb; (*11*) scale and objective; (*12*) damping and counterweight; (*13*) sensitivity adjustment; (*14*) coarse zero adjustment; (*15*) beam; (*16*) center knife-edge.

procedure outlined in the experimental work that follows. Two typical analytical balances are shown in Figure 2-4. A schematic diagram of the internal components is shown in Figure 2-5, and the external controls of a different model in Figure 2-6.

Increasing numbers of analytical balances are now appearing that operate on the principle of electromagnetic compensation. Lever arms and

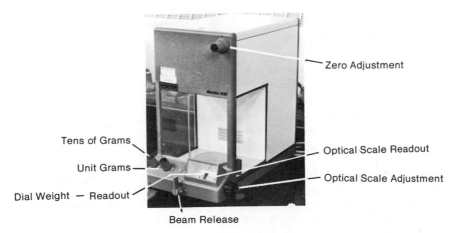

Figure 2-6. External components of a single-pan analytical balance.

knife-edges are not required in this system. When an object to be weighed is placed on the pan, the pan moves downward, displacing a metal core located within a coil of wire. An electric current is passed through the coil, creating an electromagnetic force that draws the core upward. The quantity of current required to restore the core and the pan assembly to the original position is proportional to the weight of the object on the pan. The initial, or null, position is typically determined by an optical device in which a light beam passes through the pan assembly to strike a photodetector device at null, but not in any other position. An optical-scale range of up to 30 g with a sensitivity of 0.1 mg is feasible with these systems at the present time; therefore the balance must be designed so that internal weights can be removed from the pan assembly when larger objects are to be weighed.

The Analytical Balance: Routine Weighing

Direct Weighing

Selection of the best method for weighing depends on the number and nature of the samples. When only one or two are required, *direct weighing* is preferable.

Weigh an empty vessel and record the value. Set the balance to read an additional amount equal to the weight of sample to be obtained. Add sample in small amounts until approximate balance is achieved. Weigh the sample and vessel and record the value. The difference in readings is the sample weight. With this technique each sample requires two weighings. Sample may be added while the vessel is on the balance pan provided that the sample is not corrosive and no material is spilled in the weighing compartment. When moving a vessel between weighings, do not handle it directly with the fingers; use tongs, fingercots, or other means to avoid any change in weight from oils and other contaminants from the skin.

Some balances are equipped with a built-in compensator (taring mechanism) that enables the operator to set the balance to read zero with an empty vessel on the pan. This method reduces the likelihood of major errors in recording or subtraction.

Weighing by Difference

When several samples are required, *weighing by difference* is highly recommended. After the first sample each succeeding one requires only one weighing. An additional advantage of this method is that it enables hygroscopic or volatile samples to be weighed in closed containers with little exposure to the atmosphere.

Put sufficient material for all the samples into a vessel. Weigh the vessel and contents and record the reading. Set the balance to read less by the approximate weight of sample desired. Using a camel's-hair brush and a spatula, quantitatively transfer small amounts of material to a clean beaker or flask until the approximate weight is reached. Weigh the vessel again. The difference between the two weights is the weight of the first sample. The weight of the second sample is obtained in the same manner, but since the initial weight is already known, only one additional weighing is needed. Thus three samples require only four weighings.

Weighing Corrosive Material

For accurate weighing of corrosive materials, approximate amounts are weighed first on a triple-beam or other auxiliary balance to protect the analytical balance from spillage and corrosive vapors. Silver nitrate or iodine, for instance, are weighed in this manner.

Weigh an empty vessel on an analytical balance (using a cover if the material is volatile). Record the weight. Take the vessel to a rough balance and weigh into it the approximate amount of material to be taken. Accurately reweigh the vessel plus material on the analytical balance.

The Analytical Balance: Sources of Error in Weighing

Two potential sources of error in weighing with an analytical balance, inaccurate weights and shifts in balance zero or sensitivity, have already been discussed. Two other important sources of error, weight loss and buoyancy, are considered here. Errors from changes in weight of the sample include absorption of moisture (especially serious when a hygroscopic material is being weighed), volatilization, air currents from hot sample or container, and static charges on containers. Errors from air buoyancy occur because an object immersed in a gas or a liquid is buoyed up by a force equal to the volume of fluid displaced. This gives the weight of an object weighed in air an apparent weight that is less than the true weight by the weight of the displaced air. The same effect, of course, operates also on the weights used. If the density of the weights and of the object are equal or nearly so, buoyancy errors are zero or negligible. Stainless-steel weights, the most common type in current use, have a density of 7.8. With such weights a correction for buoyancy is necessary at the level of part-per-thousand accuracy only when the density of the object being weighed is 1.0 or less. This is encountered primarily when liquids are weighed. The buoyant effect of air on an object B_{obj} is given by

$$B_{obj} = \frac{\text{apparent weight}}{\text{density of object}} \times \text{density of air}$$

The weight of an object divided by its density gives the volume of the object and therefore the volume of air displaced. Multiplying this volume by the density of air[3] gives the weight of air displaced and therefore the buoyant force on the object. Similarly, the buoyant effect of air on stainless-steel weights is

---

[3]The density of dry air depends on barometric pressure and temperature. At 25°C a density of 0.0011 g/mL can be used for barometric pressures in the range 670 to 740 mm of mercury, and 0.0012 g/mL for pressures above 740 mm. An average pressure of 740 mm corresponds to an altitude of about 300 m. At the standard sea-level pressure of 760 mm, 0.0012 g/mL can be used over the temperature range 10 to 33°C.

$$B_{\text{wts}} = \frac{\text{apparent weight}}{7.8} \times \text{density of air}$$

The net buoyant effect[4] is the difference between these two values, $B_{\text{obj}} - B_{\text{wts}}$; this difference is added algebraically to the apparent weight to give the true weight. The use of buoyancy corrections in weighing is illustrated in the procedure for pipet calibration in Section 2-9.

The Top-Loading Balance

A top-loading balance is used for rapid weighings with a precision between that of a triple-beam balance and an analytical balance. Two mechanical top-loading balances are shown in Figure 2-7, and a schematic drawing of the internal components in Figure 2-8. This type of balance operates under constant load in the same manner as an analytical balance. The range in the optical scale is large; four significant figures can normally be obtained from it alone.

After an object is placed on the pan, the balance comes to rest with the aid of magnetic damping in 3 to 4 s. Often provided is a taring device that minimizes weighing errors by eliminating the need for recording and subtracting the weight of the sample container. A taring knob controls a small spring that exerts a force on the beam to bring the optical scale back to zero. This permits the weight to be read directly from the optical scale when sample is added to the container. Because the pan of a top-loading balance is not ordinarily enclosed in a case, it must be protected from air currents by some other means. A variety of shields are available. Correct adjustment of the optical-scale sensitivity of a top-loading balance must be verified periodically in the same way as for an analytical balance. In those models having no internal weights an accurate external weight must be used for this operation.

---

[4] The weight reading of the optical scale, if the scale is calibrated with an internal weight of the balance (as outlined in the experimental procedure), is included in the apparent weight without error. See F. F. Cantwell, B. Kratochvil, and W. E. Harris, *Anal. Chem.* **1978**, *50*, 1010 for background on air buoyancy as applied to optical-scale readings.

Figure 2-7. Top-loading balances.

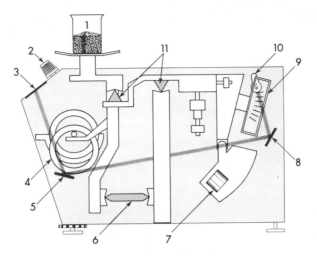

Figure 2-8. Schematic drawing of components of a top-loading balance: (*1*) object weighed; (*2*) weight knob; (*3*) optical readout; (*4*) weights; (*5*) mirror; (*6*) guide plate; (*7*) magnet; (*8*) mirror; (*9*) optical scale; (*10*) lamp; (*11*) knife-edges.

The capacities of the balances of one manufacturer (Mettler) range from 2 mg to 10 kg. The model with 2-mg capacity can weigh objects to the nearest 0.1 or 0.2 μg, and the 10-kg industrial model to the nearest 0.5 g. In both, the optical scale extends over the full weighing range, and no beam arrest is used.

For semiquantitative laboratory work a top-loading balance with a capacity of about 150 g and a sensitivity of about 1 mg is practical.[5]

Electronic top-loading balances are now available that weigh on the principle of electronically controlled force compensation. The components of this system are shown schematically in Figure 2-9. When a load is placed on

[5]The Mettler P-163 model has a capacity of 160 g, an optical-scale range of 10 g, a tare capacity of 10 g, a sensitivity of 1 mg, and a set of weights up to 150 g in 10-g intervals that are operated by an external knob. The beam bearings are of synthetic sapphire.

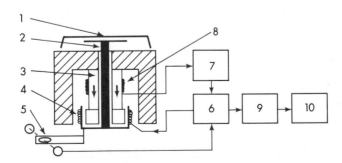

Figure 2-9. Schematic drawing of components of a top-loading electronic balance: (*1*) weighing pan; (*2*) pan holder; (*3*) magnet; (*4*) coil; (*5*) optical scanning system; (*6*) regulator/converter; (*7*) constant current source; (*8*) temperature sensor; (*9*) tare calculator; (*10*) digital display.

the weighing pan, an optical scanning system detects the vertical movement of the pan holder. This information is sent to a regulator unit, which produces an electric current. The current passes through a coil that operates as an electromagnet, drawing the pan holder back to its original position. The current required to restore the initial position is proportional to the load on the pan; its value is converted into units of weight and shown on a digital display. With this system containers can be tared readily; the tare value can be stored in a memory and subtracted from the total weight or can be recalled as necessary. Weight information can also be transferred to other systems such as electronic calculators, industrial process controllers, or computers.

The Triple-Beam Balance

Triple-beam balances are designed for rapid weighings with an accuracy of 0.01 to 1 g, depending on the capacity. Capacities range from about 100 g to several kilograms. Three graduated scales are used, each with a weight attached. The beam typically has a zero-adjustment nut, two knife-edges of hard metal, and bearings of agate or metal. The instrument usually incorporates magnetic damping, and sometimes a beam-arrest mechanism.

2–2 MEASUREMENT OF VOLUME

> Do not descend into a well on an old rope.
>
> Turkish proverb

For making accurate measurements in analytical procedures, next in importance to the balance is equipment for the measurement of volume. In this section burets, pipets, volumetric flasks, and graduated cylinders are discussed. The experimental procedures include gravimetric calibration of burets, pipets, and volumetric flasks to provide checks on the accuracy of the markings as well as to aid in development of proper technique. The few minutes required to learn correct use of this equipment will in the long run save time and give better results. For example, a typical experienced person using incorrect technique takes about 10 s to read a buret and will have a standard deviation in the readings of 0.02 mL or more. A person of the same experience using correct technique takes about 6 s and will have a standard deviation in the readings of less than 0.01 mL.

Burets

Burets are designed to accurately deliver measurable, variable volumes of liquid, particularly for titrations. A 50-mL buret, the most common size, has 0.1-mL graduations along its length and can be read by interpolation to the nearest 0.01 mL. Parallax errors in reading are reduced by extension of every tenth graduation around the tube. For reproducible delivery, proper design of the tip is important. Any change affecting the orifice will also affect reproducibility; a buret with a chipped or fire-polished tip should not be

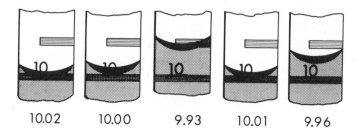

10.02   10.00   9.93   10.01   9.96

Figure 2-10. Enlarged sections of a 50-mL buret showing the meniscus at several positions; correct readings are given below each section. (For clarity the vertical scale is exaggerated over the horizontal.)

employed for accurate work. Drainage errors are usually minimized if the tip is constricted so that the meniscus falls at a rate not exceeding 0.5 cm/s.

The accuracy of the graduation marks on volumetric burets depends on the uniformity of the bore. The most convenient way to determine actual volumes is to weigh the amount of water delivered. Burets can be purchased already calibrated and accompanied by a calibration certificate.[6] However, burets are best calibrated by the user; the procedure is simple and also is an excellent means of learning to use burets correctly.

Several characteristics of a buret should be recognized when the analyst chooses it to measure volumes. Because graduations on a 50-mL buret are 0.1 mL apart, volumes between the marks must be estimated. In this estimation the width of the lines must be taken into account. The thickness of a line on a 50-mL buret takes up about one fifth of the distance from one mark to the next and therefore is usually equivalent to about 0.02 mL, as illustrated in Figure 2-10. Generally it is preferable to read the bottom of the meniscus and to take as the value for a given line the point where the meniscus bottom just touches the top of the line. For most people this edge-to-edge technique gives consistent results. Figure 2-10 shows readings of various meniscus positions with this technique. Parallax, another source of error in reading a buret, occurs if the eye is above or below the level of the meniscus. This error can be minimized by use of the encircling markings on the buret as guides to keep the eye level with the meniscus.

The apparent position of the meniscus is significantly affected by the way it is illuminated. Lighting errors are minimized by use of a reading card consisting of a dull-black strip of paper on a white background. As in Figure 2-11, the card is placed behind and against the buret so that the top of the black portion is almost flush with, or no more than a millimeter below, the bottom edge of the meniscus. For some solutions, such as permanganate and iodine, the bottom of the meniscus may be difficult to see; in such cases the top may have to be read.

To provide effective control of the delivery rate when titrating, operate the stopcock with the left hand around the barrel (if right-handed) as in Figure 2-12. The other hand is free to swirl the titration flask.

---

[6]Such certificates are of little value if the technique being used differs from that used in calibration.

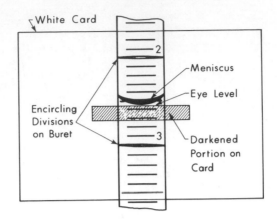

Figure 2-11. Reading card in correct position.

Pipets

Graduated (Mohr) pipets are used to deliver small volumes of liquid with an accuracy of about 1%.

Several manufacturers now produce micropipets that deliver from 1 μL to 1 mL with ± 1% accuracy and reproducibility. These pipets consist of a cylinder with a thumb-operated air-tight plunger. A disposable plastic tip is placed on the end of the cylinder, the plunger is depressed, and the tip is immersed in the sample solution. The plunger is released, allowing sample to enter the tip. The pipet is moved to the delivery vessel, and the plunger depressed again to deliver the sample. Sample solution touches only the tip, never the plunger. Volumes are usually fixed, though some models have provision for more than one volume setting. These models are especially

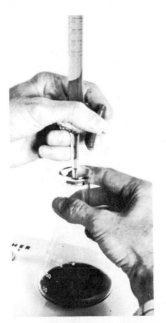

Figure 2-12. How to hold buret stopcock and titration flask during a titration.

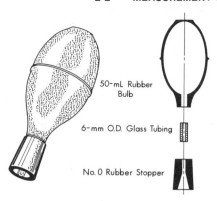

50-mL Rubber Bulb

6-mm O.D. Glass Tubing

No. 0 Rubber Stopper

Figure 2-13. Inexpensive pipetting bulb, prepared from a rubber stopper, a 2-cm length of 6-mm o.d. capillary glass tubing, and a 50-mL rubber bulb.

useful in clinical and biochemical laboratories where errors of ± 1% are acceptable and where problems of contamination make disposable units desirable.

Volumetric (transfer) pipets are used to deliver precisely a single, definite volume of liquid. The tip must meet stringent requirements[7] because drainage time is controlled by the internal tip diameter. Provided that pipets are used in a reproducible manner, accuracies of one part per thousand in the amount of liquid delivered can be attained readily.

Procedure for Pipet Use

First clean the pipet so that water drains smoothly from the interior surface. Rinse the interior by drawing a portion of the liquid to be pipetted into the pipet with the aid of a suction bulb (Figure 2-13), and then tilt and turn the pipet until all the inner surface has been wetted. Discard this portion of solution and repeat the operation twice. Then draw solution above the mark, wipe the tip and stem of the pipet carefully with a lintless tissue or a clean towel to remove external droplets, and allow the solution to drain until the bottom of the meniscus is even with the calibration mark. Keep the pipet vertical, with the tip in contact with a tilted beaker or flask wall, during this operation. Move the pipet to the receiving vessel and allow the solution to drain freely. During drainage, hold the pipet vertically, with the tip in contact with the container as illustrated in Figure 2-14. Keep the tip in contact for about 5 s after free solution flow has stopped, and then remove it. Leave any remaining liquid in the tip; do not blow out this portion.

Figure 2-14 shows the proper way to hold a pipet during use. If difficulty is experienced at first with a leaky index finger, try moistening it slightly; too much moisture, however, makes fine control of the outflow rate difficult. After a little practice pipetting will become both rapid and accurate.

A rubber suction bulb is recommended for all pipetting (Figure 2-13). Particularly, do not pipet poisonous or corrosive solutions by mouth. Rinse

---

[7]According to *Circular 602* of the National Bureau of Standards, Washington, D.C.: "The tip should be made with a gradual taper of from 2 to 3 cm, the taper at the extreme end being slight. A sudden constriction at the orifice is not permitted, and the end of the tip must be ground perpendicular to the axis of the tube. The outside edge should be beveled slightly ...."

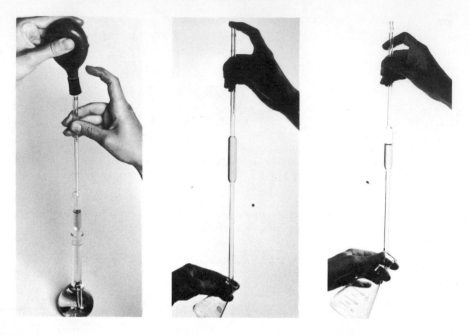

Figure 2-14. How to fill, hold, and drain a pipet. During adjustment of meniscus the end is covered by the index finger, not the thumb.

the pipet thoroughly after use. Avoid drawing liquid into the bulb. If the bulb is contaminated accidentally, thoroughly rinse and dry it before reuse.

Pipets are marked to deliver a known volume of water at a defined temperature, usually 20 or 25°C, when used in a specified manner. When pipets are used correctly the manufacturer's markings are seldom in error by more than a part or two per thousand. Since the volume of liquid delivered depends on how the pipet is manipulated, the technique needs to be carefully specified. The calibration exercise outlined in Section 2-9 is recommended to compensate for individual variations in handling.

Volumetric Flasks

Volumetric flasks are designed to contain a specified volume of liquid. Since accuracy in their use is not highly dependent on technique, calibration is generally unnecessary for work at the level of a part per thousand. The principal source of error is variation in temperature, which causes enough expansion or contraction of aqueous solutions to give errors on the order of 0.1%/5°C.

Before use a volumetric flask should be cleaned thoroughly. The solution to be diluted or the solid to be dissolved is then transferred to it with the aid of a funnel. Often it is more convenient for a solid to be dissolved in a beaker or a conical flask first, and then the resulting solution transferred quantitatively to the flask. In the most accurate work the flask is filled to just below the mark and immersed in a water bath at the calibration temperature before final volume adjustment. Alternatively, the temperature of the solution may be taken and a correction made. In careful work droplets of solvent above the meniscus may be removed with a lintless towel after the meniscus has been adjusted but before mixing.

The solution in a volumetric flask is mixed thoroughly by inverting, shaking, and turning upright at least 10 times.

Graduated Cylinders

Graduated cylinders, or *graduates*, are used for only approximate measurements. Numerous sizes are available; by use of the smallest size that will hold the volume being measured, an accuracy within 1 to 5% can be attained. A graduated cylinder is not suitable for measurements of part-per-thousand accuracy.

Calibration of Volumetric Equipment

Calibration of volumetric equipment is carried out by weighing the amount of water contained or delivered. Since the density, or weight per unit volume, of water changes about 0.03%/°C at 25°C, the temperature of the water used in the calibration must be known. The actual volume delivered can be obtained from a table relating density and temperature, such as Table 2-1. Because the density of water is relatively low, a correction must be made for buoyancy (Section 2-9). The correction takes into account the upward force exerted on the water and flask by the air they displace. This force corresponds to the net weight of the air displaced by the water, flask, and weights. If the density of dry air is 0.0011 g/mL under the conditions of calibration and that of stainless-steel weights 7.8 g/mL, the net correction amounts to about 0.0010 g/mL of water weighed.

Although errors in placement of buret and pipet markings by the manufacturer are usually small, they cannot be neglected in careful work. For example, if a 10-mL pipet actually delivers 9.990 mL, an error of 1 part per 1000 will be introduced unless the true value is employed. Errors in volumetric-flask markings are also small, and calibration is not often necessary because their use is not so dependent on technique as is that of burets and pipets. The two main sources of error in the use of volumetric flasks are differences between the temperature of the flask contents and that at which dilution was made, and failure to mix the contents adequately before withdrawing portions.

The major sources of error in volumetric work arise from nonadditivity of

Table 2-1. Density of Water at Various Temperatures

Temp,°C	Density, g/mL	Temp,°C	Density, g/mL
15	0.9991	26	0.9968
20	0.9982	27	0.9965
22	0.9978	28	0.9963
24	0.9973	29	0.9960
25	0.9971	30	0.9957

volumes, variation in solution properties (drainage errors from change of viscosity, for example) with the solute, errors in volume measurements through calibration errors, operator error, and temperature effects.

A change in temperature causes expansion of the glass and of the solution in the container. With an increase in temperature a pipet or buret delivers a larger volume, and a volumetric flask holds a larger volume. Fortunately, the cubic coefficient of expansion of borosilicate glass is only about 0.00001/°C. Thus a flask whose capacity is 1000.0 mL at 20°C will expand by $10 \times 1000 \times 0.00001$, or 0.1 mL, when the temperature is raised to 30°C (or about 0.001%/°C). For experimental work precise to 1 part per 1000 such a change is negligible. The expansion of water with temperature is another matter. The magnitude of the effect can be calculated from the change in density with temperature (Table 2-1) and is of the order of 0.02%/°C.

2–3 BACKGROUND FOR BEGINNING LABORATORY WORK
Planning Laboratory Work

Before starting an experiment, read carefully and understand thoroughly the background and procedure. Knowing each step of the procedure before starting the work and why it is included will save time and reduce mistakes. Even as minor an item as dilution of a solution before titrating may be important. Should a question arise during study of the experiment, ask the laboratory instructor for an explanation.

Come to each laboratory period with a work plan. Record the sample identification in your laboratory notebook immediately after each sample is obtained. The data page in the notebook should be prepared before starting each experiment. Schedule your work to allow for lengthy procedures such as drying or dissolution of samples. For example, the experiment in Section 5-10 calls for dry sodium carbonate in the standardization of a hydrochloric acid solution. Drying can be completed ahead of time while other work is proceeding, and the dry material stored in a tightly closed, labeled container until needed. With effective planning, two or even three experiments may be underway simultaneously.

A minimum of three valid results usually constitutes the basis for selection of a value to report. A sound practice, therefore, is to carry four samples through a determination whenever time and equipment permit. Four results give sufficiently improved statistical validity over three (Section 22-1) that the extra work required is generally worthwhile. A fourth sample also allows for accidental loss of a replicate without unduly jeopardizing the validity of the final result.

It is possible to be overly cautious in making measurements. Time should not be wasted in using a buret when a graduated cylinder is precise enough, or a graduated cylinder when a rough estimation of volume in a beaker is adequate. In measuring solids, do not use an analytical balance when a triple-beam balance is adequate or a triple-beam balance when visual estimation is adequate. The reverse, of course, is also true. Never use a graduated cylinder or a triple-beam balance for an operation requiring the precision of a pipet or an analytical balance. Save your patience for situations where exceptional accuracy is essential, and then be prepared to use care.

Drying and Storing Materials

Some materials pick up varying amounts of water from the atmosphere and so need to be dried before samples are weighed for analysis. Such treatment is especially necessary for finely powdered chemicals that have large surface areas. Metal powders and nonhygroscopic materials that do not require drying can be weighed as received. Drying is usually carried out in an oven at atmospheric pressure, although some substances that tend to decompose on heating can be effectively dried at lower temperatures under reduced pressure.

Once dried, samples must be kept dry until weighing is completed. Uptake of atmospheric moisture or carbon dioxide is reduced if samples are placed in a well stoppered bottle. A *weighing bottle* is a container with a ground-glass stopper, designed to hold samples during drying, storage, and weighing, though any well stoppered container that is convenient to use on an analytical balance is satisfactory. For added protection, dry samples are often kept in a larger container called a *desiccator*. Examples of weighing bottles and desiccators are shown in Figure 2-15. Desiccators are designed to contain a drying agent such as magnesium perchlorate, phosphorus(V) oxide, calcium sulfate, or calcium chloride. They are generally of glass or aluminum; coffee or jam cans with snugly fitting plastic lids are often satisfactory. A desiccator maintains a dry atmosphere provided that the lid is not removed frequently and the desiccant is fresh. Normally it is not used to dry materials, but is convenient for storage and transport in the laboratory of such objects as dried samples or ignited crucibles. In a laboratory where the relative humidity is low, a desiccator can be dispensed with for most samples.

The effectiveness of various drying agents for use in desiccators has been studied. The most effective is anhydrous magnesium perchlorate, which at equilibrium leaves less than a microgram of water per liter of air. Phosphorus(V) oxide also is highly effective, but it tends to form a surface film of phosphoric acid that limits its rate of water uptake.

Figure 2-15. Examples of weighing bottles and desiccators.

Handling Chemicals

Take care to avoid contamination of reagent chemicals or solutions. Do not place spatulas or individual eyedroppers into reagent containers of either solids or solutions; instead, transfer the approximate amount needed into a beaker or graduated cylinder and measure the quantity required from that portion. In many cases a beaker can be placed on a triple-beam balance, tared, and the reagent poured directly into the beaker until the desired weight is obtained; in this method pouring must be done carefully so as to avoid spillage. The flow of a powdered chemical can often be controlled by gently rolling the bottle during pouring. Avoid taking excessive amounts, which results in costly waste. Also avoid returning the remaining portion to the original container and thus possibly contaminating the entire contents. Instead, discard it in a safe manner.

Laboratory Safety and Cleanliness

General safety practices should be observed at all times. Although this text contains no specifically hazardous experiments, danger is inherent in all chemical experimentation. Carelessness leads to accidents, and warnings cannot replace common sense and reasonable caution. A list of basic precautions follows:

1. Learn the location of the safety shower and when it should be used.

2. Report every accident to the laboratory instructor. First aid will be provided. If necessary you will be taken for emergency treatment; do not try to go alone.

3. Wear eye protection in the laboratory. Safety goggles are best, but safety glasses with sideshields are adequate for most analytical work.[8] Even in a procedure as innocuous as washing glassware, protection may be needed from flying glass or from chemicals inadvertently splashed by nearby workers.

4. Protect both hands with several thicknesses of cloth toweling while inserting glass tubing into rubber stoppers. First fire-polish the ends of the tubing. Then wet the tubing with soapy water or dilute aqueous ammonia and apply pressure slowly with a twisting motion.

5. Beware of burns from forgotten, still ignited burners and from hot glass.

6. Organic solvents are generally flammable. Do not use them near an open flame.

7. Some chemicals, such as mercury(II) chloride and sodium cyanide, are poisonous. Handle them with care; wipe up spilled material promptly and completely.

8. Handle concentrated acids and bases with care. Quickly flood spills on skin or in eyes with large volumes of water. Continue flooding until the laboratory instructor advises otherwise. Neutralize spills on bench or floor with sodium bicarbonate.

Keep all equipment in good condition. Glassware should be cleaned with a brush and detergent as soon as possible after use. Not only is it easier to clean, but it is then ready for later use, and possible contamination of reagents and samples is avoided. Volumetric glassware must be sufficiently free of grease to drain smoothly and uniformly without drops of solution adhering to the interior

---

[8]Ordinary prescription glasses are considered adequate in some laboratories; others require goggles over glasses.

walls. Burets should be stored filled with distilled water. At the end of each laboratory period, clean and put in order your work space and any community equipment or space you used.

2–4 THE LABORATORY NOTEBOOK

A laboratory notebook should be designed to provide a permanent record of firsthand observations. A basic criterion for sound record keeping is that someone else can readily locate pertinent data and results for an experiment. Although reasonable legibility and neatness are desirable, the usefulness of a record is determined largely by whether it is original, systematic, and complete — not by whether it is a work of art.

The notebook should be hard-covered and bound. A 7- by 9-in. size is convenient; 8- by 10 1/2-in. is also common. Number the pages consecutively, leaving two or three pages at the front for an index.

To keep an efficient experimental record the following is recommended:

1. Enter data directly in ink as soon as taken. Never recopy numbers or use loose sheets of paper. Cancel errors and rejected data by drawing a single line through them; cancel a complete page if necessary by a single diagonal line across it. Do not erase data or remove pages. The notebook should be a permanent record of the original laboratory work.

2. Enter data only on the right page. Use the left page for recording observations that may be useful in evaluating results, for calculations, and so on.

3. Clearly label all entries. To facilitate direct entry of experimental work, set up a data page for each experiment before starting. Align data for each sample in a single column; align data for replicates in parallel columns. Examples are shown in Figures 2-16 and 2-17. Leave extra pages for each experiment so that a repeat determination or recalculation can be accommodated.

4. Make the record of experiments complete but concise. Date each entry. Give each experiment a title and list it in the index. Securely attach all graphs and plots next to the corresponding calculation or data entries. Affix the graph for buret calibrations inside the front cover.

5. Record all calculations, including formulas with molecular weights where applicable. Calculate a separate. result for each sample and label it clearly; also clearly identify the value to be reported for the analysis.

6. Include with calculations all relevant statistical information such as Q-test calculation, average and average deviation, and median and probable deviation.

2–5 PRELIMINARY OPERATIONS

This section lists background exercises to provide familiarity with proper use of the analytical balance and volumetric glassware and gives directions for preparation and checking of equipment for the experiments that follow. Your instructor will indicate any changes necessary, including the order in which the items are to be done.

Determination of Chloride by Fajans Method

Sample Code Number *A2837*   Date: September 28, 1981

Preparation of Standard AgNO₃ Solution

Wt beaker and AgNO₃, g	29.4775
Wt beaker, g	28.6503
Wt AgNO₃, g	0.8272
Molarity of AgNO₃ solution	0.009739

Preparation of Standard KCl Solution

Wt beaker and KCl, g	26.8684
Wt beaker, g	26.2405
Wt KCl, g	0.6279
Molarity of KCl solution	0.03369

Preparation of Sample Solution

Wt beaker and sample, g	28.7288
Wt beaker, g	28.1941
Wt chloride sample, g	0.5347

Titration of KCl	1	2	3
Final buret reading	35.47	34.58	35.16
Initial buret reading	0.94	0.03	0.58
Difference	34.53	34.55	34.58
Net buret correction	+0.04	+0.04	+0.04
Corrected volume, mL	34.57	34.59	34.62
Molarity of AgNO₃ solution	0.009745	0.009740	0.009731
Average molarity		0.009739	
Median molarity		0.009740	

Titration of Unknown Sample	1	2	3
Final buret reading	36.10	35.04	38.96
Initial buret reading	1.52	0.48	4.35
Difference	34.58	34.56	34.61
Net buret correction	+0.04	+0.04	+0.04
Corrected volume, mL	34.62	34.60	34.65
% chloride (from average AgNO₃ molarity)	55.88	55.85	55.93
Average % chloride	55.89		
Relative average deviation	0.5 ppt		
% chloride (from median AgNO₃ molarity)	55.89	55.86	55.94
Median % chloride	55.88		
Relative probable deviation	0.5 ppt		
Result Reported	55.89%	Grade Obtained: 5	

Figure 2-16. Sample of a summary data page for the volumetric determination of chloride by Fajans method.

Determination of Carbonate

Sample Code Number *B23ll*   Date: October 10, 1981

Standardization of HCl against Na$_2$CO$_3$	1	2	3	4
Wt weighing bottle + sample, g	19.2750	18.8346	18.4640	18.0229
Wt weighing bottle, g	18.8346	18.4640	18.0229	17.6120
Wt Na$_2$CO$_3$ sample, g	0.4404	0.3706	0.4411	0.4109
Final buret reading	40.22		43.41	40.97
Initial buret reading	0.14	1.03	3.23	3.48
Difference	40.08	End	40.18	37.49
Net buret correction	+0.02	point	+0.01	+0.02
Volume HCl(corrected), mL	40.10	missed	40.19	37.51
Molarity of HCl	0.2072		0.2071	0.2067

Average molarity of HCl   0.2070
Relative average deviation   1 ppt
Median molarity   0.2071
Relative probable deviation   0.5 ppt

Determination of Na$_2$CO$_3$

Sample *B2311*	1	2	3	4
Wt weighing bottle + sample, g	16.4361	16.0632	15.6321	15.2207
Wt weighing bottle, g	16.0632	15.6321	15.2207	14.8100
Wt sample, g	0.3729	0.4311	0.4114	0.4107
Final buret reading	33.96	43.90	38.31	36.44
Initial buret reading	1.23	5.84	1.86	0.16
Difference	32.73	38.06	36.45	36.28
Net buret correction	+0.04	+0.01	+0.03	+0.03
Volume HCl(corrected), mL	32.77	38.07	36.48	36.31
% Na$_2$CO$_3$ (from average HCl molarity)	96.40	96.87	97.27	96.99
% Na$_2$CO$_3$ (from median HCl molarity)	96.45	96.92	97.32	97.03

Average % Na$_2$CO$_3$ 96.88%
Relative average deviation   3.1 ppt
Median % Na$_2$CO$_3$ 96.98%
Relative probable deviation   2.1 ppt
Result Reported   96.9$_8$%   Grade Obtained: 5

Figure 2-17. Sample of a summary data page for the volumetric determination of carbonate by titration with hydrochloric acid.

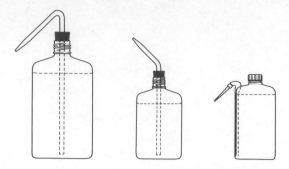

Figure 2-18. Three types of polyethylene wash bottles.

1. Check in locker equipment. Consult the next subsection for criteria to be applied. If in doubt about the acceptability of an item, consult your instructor.

2. Carry out the balance assignment described in Section 2-6. Record all observations and conclusions directly in your laboratory notebook.

3. Obtain a buret and calibrate it (Section 2-8). Record all data directly in your laboratory notebook.

4. Calculate buret corrections and draw a buret calibration graph.

5. Calibrate the pipets provided, following the procedure outlined in Section 2-9. Again, record all data directly in the laboratory notebook. Calculate pipet calibration volumes.

6. Prepare or obtain a 1-L polyethylene wash bottle with the delivery tip pointing downward. In addition, prepare or obtain a 500-mL polyethylene wash bottle with the delivery tip pointing upward. (See Figure 2-18 for illustrations.) An upward pointing tip is useful in the quantitative transfer of solutions and solids.

7. Examine crucibles for individual distinguishing marks. If necessary, mark with *small* dots of 10% cobalt(II) chloride solution on the outer surface near the top, and heat strongly for 10 to 15 min over a Bunsen burner to fuse the markings into the surface. Cool, and then scrub the marked areas with a brush and water to remove unfused cobalt(II) chloride.

8. After reading Section 2-10, perform the aliquot exercise as directed.

9. Read Section 2-4 and the procedure portion of the first experiment assigned (such as the determination of chloride in Section 4-4), and prepare a blank summary data page in your notebook for that experiment.

10. Read Section 1-4 on management of experimental data, and solve the problems dealing with this topic at the end of the chapter. Check all answers for the correct number of significant figures.

11. As each item in this list is satisfactorily completed, ask the laboratory instructor to check it and initial the notebook if satisfactory. Record values for the calibrated volumes of the pipets inside the front cover, along with the buret calibration graph, for convenient reference during calculations.

12. On completion of this list a practical test may be scheduled, consisting of items such as a short written quiz, reading several burets, calculations and significant figures, taking an aliquot of a solution, and weighing with an analytical balance. Review of Items 2, 3, 5, 8, and 10 in the above list is helpful before a test of this kind.

Criteria for Glassware on Check-in and Check-out

Inspect all glassware during check-in to ensure that it is in first-class condition. Top-quality work requires top-quality glassware. The following criteria, which are also applied at the end of the term, should be applied by you at check-in:

1. Burets and pipets: No chips on tips. Even small chips affect reproducible delivery of volume.

2. Beakers: No excessive scratches on inner surfaces, which can interfere with quantitative transfer of precipitates.

3. In general: No chips, cracks, etching, or deep heavy scratches that can interfere with normal operation. Stirring rods fire-polished at both ends to minimize scratching of glassware.

Taking the following precautions will make glassware last longer and give better results.

1. Replace worn brushes that might scratch glassware during cleaning.

2. To prevent scratching, avoid striking the sides and bottoms of beakers with stirring rods while mixing solutions.

3. Use care in nesting glassware for storage.

2–6   USE OF THE ANALYTICAL BALANCE

Before beginning work, read the background information on the balance and obtain instruction on its design and correct operation. A balance will be assigned to you for use during the course. After an instructional session, complete the operations assigned using that balance.

Operating Instructions

Before every weighing, ensure that the balance is level, the pan compartment clean, and the door to the compartment closed. Turn the beam-arrest knob to full release, and with the fine zero knob adjust the balance to read zero. Arrest the beam, place the object to be weighed on the pan, and carefully turn the beam-arrest knob to the partial-release position.[9] Remove weights by first rotating the knob for tens of grams so that the numbers appear in the order 9, 8, 7, 6, ... until the lighted portion of the optical scale changes from a position less than zero to one greater than zero. Next, remove smaller weights by rotating the knob for grams and, on some balances, for tenths of a gram until the optical scale reads greater than zero. Then carefully turn the beam-arrest knob to full release, and rotate the micrometer knob to align the scale pointer with the next lower division on the moving scale. This allows direct readout of the last fraction of the weight. Record the weight. Arrest the beam, release it again, and

---

[9]Some balances incorporate a *preweigh* feature, which replaces the partial-release position with a preweigh position. The object to be weighed is placed on the pan, and the beam-arrest knob turned to preweigh. The weight of the object is read to the nearest gram from the optical scale; this weight is dialed directly (not the reverse as on a balance with partial release), the beam-arrest knob turned to the full-release position, and the final weight read in the usual way.

double-check the weighing. Again arrest the beam. Remove the object, dial the weights to zero, and check the zero reading. If it has changed, reset the zero and reweigh the object.

Some balances provide digital readout of the weight; in certain models fractions of grams are read directly and appear automatically when the correct gram weights have been dialed.

When finished, leave the balance clean and in the arrest position. Dial the weights to zero, and close the door to the pan compartment. Chemicals must always be weighed in a container such as a beaker or a flask, never directly on the pan. Whenever possible, transfer chemicals outside the balance compartment. Wipe up any spilled material immediately.

Procedure for Familiarization with the Analytical Balance

1. *Reproducibility of Readings with and without Pan Load.* Zero the balance by releasing the beam and adjusting the fine zero knob. Arrest the beam. With forceps add a crude 100-g weight[10] to the pan. Carefully dial the appropriate knob to remove 100 g of balance weights, and release the beam. Record the reading. (If the actual weight is less than 100 g, it may be necessary to dial 99 g to obtain a reading.) Arrest the beam, remove the weight, dial zero grams, and record the new reading. Do not reset the zero. Reload the 100-g weight onto the pan, dial 100 g, and record the reading; again remove the 100-g weight, dial zero grams, and determine the reading. Repeat these operations three or four times. Calculate the average zero reading, the average reading for the 100-g weight (taking into consideration any change in the preceding zero reading), and the average deviation for all the readings for the 100-g weight.

2. *Test of Optical-Scale Sensitivity.* Add a crude 1-g weight to the pan and dial 1 g. Release the beam and zero the optical scale with the zero adjustment. Now dial 0.0 g. (This step adds the 1-g balance weight to the front end of the beam.) Record the reading of the optical scale. Repeat the above operations to obtain a duplicate reading of the optical scale. Does the scale read correctly? If not, adjust it by the procedure in Item 5 before proceeding to Item 3.

3. *Weighing.*[11]

   a. The responsiveness of a balance to small differences in weight may be illustrated as follows: Weigh a 50-mL beaker containing a few milliliters of water. Determine the rate of evaporation of the water by observing the reading at 1-min intervals for 5 min. What is the rate of evaporation in milligrams per minute?

   b. Weigh an object supplied by the laboratory instructor. Report to him the value obtained.

   c. Samples for analysis are weighed either by addition or by difference. To practice weighing by addition, proceed as follows: Weigh a clean, dry 50-mL beaker on the analytical balance. Record the weight. With finger cots or a folded paper strip wrapped around the beaker to prevent direct contact with

---

[10]Substitute a smaller weight on balances with capacities of less than 100 g.

[11]When performing weighings, make a habit of the following series of operations: check level, check zero, place object on pan, record weight, remove object, check zero. On a less regular basis check the sensitivity.

the fingers, remove the beaker from the balance case and measure into it approximately 2 g of clean sand. (To estimate the quantity required, weigh 2 g of sand on the triple-beam balance.) Replace beaker and sand on the balance pan, close the pan-compartment door, and weigh to the nearest 0.1 mg. Record the weight. This sample has been weighed by addition; the weight of the sample is the difference between the two weighings. Note that two weighings are required to arrive at the weight of one sample by this method.

To practice weighing by difference, proceed as follows: Obtain two additional small beakers, a spatula, and a camel's-hair brush. The weight of the initial 50-mL beaker plus sand is already known to you. Now weigh each of the two additional small beakers. Remove the beaker containing sand from the balance case, and quantitatively transfer with the spatula about 0.5 g of sand to one of the small beakers. Avoid spillage. Use the camel's-hair brush to transfer the last traces into the receiving beaker. Now return the original beaker and sand to the balance pan, close the door, and reweigh. Remove the beaker from the balance case, and in the same way transfer an approximately 0.5-g sample into the second beaker. Again return the stock sample to the balance pan and reweigh. Note that you have weighed two samples with three weighings and that each subsequent sample requires only one weighing.[12] Bear in mind that transfers must be made quantitatively. This highly efficient procedure should be used whenever possible. For this exercise confirm the weights of samples transferred by weighing each of the two beakers containing the 0.5-g samples and comparing the total weight of sand removed from the original beaker with the sum of the two small samples.

Before proceeding to Items 4 and 5, ask the laboratory instructor to put the balance significantly out of zero and sensitivity adjustment.

4. *Adjustment of Balance Zero.* Release the beam. Note that the balance cannot be set at zero with the fine zero knob. Arrest the beam, remove the balance cover, and give the coarse zero nut (Figure 2-3) one or two turns in the required direction.[13] Replace the cover, release the beam, and observe the deflection of the optical scale. Continue as above until the balance reads zero with the fine zero knob approximately centered.

5. *Adjustment of Optical-Scale Sensitivity.* Adjust the optical scale to be internally consistent with the balance weights as follows: Add a crude 1-g weight to the pan. Dial 1 g. Release the beam and set the optical scale to read zero.[14] Dial the balance back to 0.0 g and read the optical scale. Arrest the beam and remove the cover. Turn the sensitivity nut in the proper direction and replace the cover. Repeat the above operations until the optical scale reads correctly within 0.1 mg.

---

[12]Weighing the empty small beakers is only for the purpose of confirming the weights of the samples transferred in this exercise; it is unnecessary in routine weighing by difference.

[13]If the beam moves too far forward, giving high readings on the optical scale, the coarse zero nut must be moved toward the rear of the balance. If the beam tilts too far back, the nut must be moved toward the front.

[14]Setting the optical scale to read zero will usually displace it from the original zero of the balance because the crude weight is not exactly 1 g. This displacement is unimportant, since the only purpose of this weight is to allow the balance to be zeroed with the 1-g weight dialed. In this way the optical scale can be checked against the 1-g internal weight simply by dialing 0.0 and observing the optical-scale reading. Using a crude weight along with setting the optical scale to zero with 1 g on the dial eliminates the need for a high-quality external 1-g weight by substituting for it the one in the balance.

The scale is now consistent with the 1-g internal balance weight. Ask the instructor to confirm that the balance is in correct adjustment. The optical-scale adjustment should be inspected every two or three weeks, and adjustments made as necessary.

Experiments Illustrating Balance Characteristics

The following items illustrate (1) the effect of sample weight on the sensitivity of a single-pan balance and (2) a method of checking the internal consistency of the built-in set of weights. They provide supplementary information on the behavior of the single-pan balance and additional practice in its operation.

Sensitivity as a Function of Pan Load. Add crude 1-g and 100-g weights to the pan. Dial 101 g. Release the beam and set the optical scale to read zero. Now dial 100 g and read the optical scale. Arrest the beam. Remove the 100-g weight and dial 1 g. Set the optical scale to read zero. Now dial 0.0 g and read the optical scale. Repeat the above operations.

The difference between the optical-scale deflection with and without the 100-g sample weight is a measure of sensitivity error. Does the sensitivity vary with sample weight on the pan (sensitivity error)? Since the balance operates under constant load, should the optical-scale reading be affected by the sample weight on the pan?

Internal Consistency of Weights. Add a crude 100-g weight to the pan, dial 100 g, release the beam, and zero the optical scale with the fine zero adjustment. Arrest the beam. Now dial (carefully) 99 g, release the beam, and again record the reading. Repeat these operations. If the weights are internally consistent, the difference between the zero and the full-scale optical-scale readings will be the same as that observed in the test of optical-scale sensitivity.

Compensating errors are not detected by these operations. Generally speaking, the weights built into constant-load balances are internally consistent; that is, a large weight is the sum of the smaller weights plus the full optical-scale reading.

Show your data for this section to the laboratory instructor, who will initial your laboratory notebook if the data are satisfactory.

Comments on Experimental Work with the Balance

When a balance is operating properly, the beam-release mechanism operates smoothly, the beam comes to a stable rest point in a few seconds, the zero point is the same before and after a weighing, and consecutive readings for an object are the same. A test of the reproducibility of the readings with and without load gives an impression of what to expect of a balance and a justified feeling (or lack) of confidence in its performance.

A change in temperature of several degrees in the vicinity of a balance may cause the zero point to shift as much as a milligram within a few minutes. In such a situation the zero point should be observed frequently.

If a balance in which part of the weight is obtained by the

direct-deflection principle is properly designed, the sensitivity will not change with load on the pan. In the test of sensitivity in Item 2 a constant-load balance should give the same optical-scale readings with or without a load on the pan.

The procedure for optical-scale adjustment is strongly recommended over the more obvious one of calibration against an external standard weight. First, the internal 100-mg weight supplied with the balance is of high quality and, because it is not handled, remains so. Second, adjustment of the optical scale can be completed more quickly and with less confusion because the crude 1-g weight is left on the pan throughout the operations. The procedure is not only faster, but more nearly foolproof, particularly against unawareness that turning the sensitivity nut usually results in a change in the zero.

The recommended check on internal consistency can be carried out at levels other than 100 g, such as 5, 10, or 50. It gives an idea of the quality of the weights, but because compensating errors can occur, it is in no sense a calibration. Remember also that discordant results may be obtained if the optical scale is not in adjustment. The optical scale will read full scale only when the weights are internally consistent and the optical scale is correctly adjusted.

Because the quality of the weights supplied with constant-load balances is usually equal or superior to Class-S tolerances,[15] weight calibrations of the bad old days are unnecessary.

Under conditions of low humidity static electrical charge can cause serious errors in weighings if precautions are not taken to dissipate the charges.

2–7   USE OF THE TOP-LOADING BALANCE [16]

1. Verify that the balance is level. If not, adjust the leveling feet at the front of the balance until the bubble (located at the left of readout display) is centered in the circle.

2. Periodically verify that the optical scale is in proper adjustment according to the following procedure: Place a crude weight that is approximately equal to the full optical range (here 10 g) on the balance pan. Dial 10 g. Set the optical scale to read zero with the zero-adjustment knob. Dial the weight knob to read zero. Read the optical scale and adjust internal sensitivity if the reading is less or more than full scale.

3. Turn the tare knob (located to right of readout display) to zero.

4. Turn on the lamp switch. Adjust the readout display to zero with the fine zero-adjustment knob (located at lower right side of the balance behind the leveling foot).

---

[15]The tolerances for various classes of weights are defined by the National Bureau of Standards, Washington, D.C., *Circular 500*. Class-S tolerances for 100-mg, 1-g, and 10-g weights are 0.025 mg, 0.054 mg, and 0.074 mg.

[16]These instructions are for one top-loading balance in common use (Mettler P-163). Although details of component locations may vary, the basic sequence of operations is similar for most mechanical top-loading balances on the market.

5.   Place an object to be weighed on the pan. A readout display of $+++$ indicates that the sample weighs more than 10 g (the limit of the optical scale). Dial the 10-g weight knob slowly clockwise until the $+++$ display disappears.

6.   To obtain the last fraction of the weight, turn the fine weight-adjustment knob until the guide line is centered in the point of the arrow.

7.   Read the weight directly to the nearest milligram. The reading for tens of grams is obtained from the small window at the left, the remainder from the lighted optical scale. The comma in the display is read as a decimal point.

8.   Dial weights back to zero and remove the object from the pan.

9.   Ensure that the zero has not drifted during the weighing. If it has, rezero and then reweigh the object.

10.   The tare knob may be used to cancel up to 10 g of the weight of a container or sample vessel. This feature is especially convenient when repetitive weighings are to be made into or out of a container. Consult the laboratory instructor before attempting to use the tare, however, to ensure that you understand how to operate it properly.

2-8 CALIBRATION OF BURETS

Before using a buret, clean it thoroughly with soap and a buret brush, taking care not to scratch the interior surface. Rinse it well with distilled water; the buret is clean when water drains from the inside surface uniformly without the formation of droplets. Store the buret filled with distilled water and with the Teflon stopcock nut loosened slightly. Before use rinse the buret at least twice with titrant solution, tightening the Teflon stopcock nut only enough to prevent leakage of solution. After filling the buret with titrant, make certain no air bubbles are present in the tip. Bring the solution level to or slightly below the zero mark, remove the drop adhering to the tip, and wait a few seconds for solution above the meniscus to drain before taking the initial reading.

Procedure

Because water evaporates at an appreciable rate at room temperature (recall the evaporation study in the balance exercise), complete the calibration procedure before starting the calculations. Before calibration ask the laboratory instructor to check your buret; if it is adequately clean, you will be assigned a top-loading balance. Before going to the balance, prepare a data page in your laboratory notebook with charts for two calibrations (see example in Table 2-2). Also assemble the following equipment: clean buret, buret clamp, 50-mL conical flask with dry stopper, funnel, 400-mL beaker of distilled water, and buret reading card.

At the top-loading balance carry out the following procedure:

1.   Place a thermometer in the beaker of distilled water.

2.   Mount the buret in a buret clamp beside the balance.

Table 2-2. Data and Calculations for Calibration of 50-mL Buret*

Approx. Interval	Buret Readings	Apparent Volume	Weight	Apparent Weight	True Weight	True Volume	Correction	Cumulative Correction
Initial	0.03B†		36.450E					
0–10	10.02	9.99	46.420	9.970	9.980	10.01	0.02	0.02
10–20	20.01	9.99	56.381	9.961	9.971	10.00	0.01	0.03
20–30	30.01	10.00	66.362	9.981	9.991	10.02	0.02	0.05
30–40	39.98	9.97	76.264	9.902	9.912	9.94	-0.03	0.02
40–50	49.99A	10.01	86.205D	9.941F	9.951	9.98G	-0.03	-0.01C

*At 25°C.
†The letters after some entries refer to the arithmetic check of Equation (2-1).

3.   Check the level and zero of the balance. Ensure that the zero-adjustment knob, not the tare knob or the leveling feet, is used for the zeroing operation.

4.   Read and record the temperature of the distilled water.

5.   Fill the buret with distilled water, using the buret funnel. Run out water so that the meniscus is just below zero.

6.   Ensure that no bubbles are trapped near the stopcock or in the tip.

7.   Weigh the conical flask and dry stopper. Record the weight to the nearest milligram in the appropriate chart space in the notebook.

8.   Read (Section 2-2) the initial buret volume and record it on the chart.

9.   Drain approximately 10 mL of water into the conical flask. Touch off the final drop on the inner wall of the flask, below the farthest point of insertion of the stopper. Restopper the flask.

10. Recheck the zero. Weigh the flask plus water and record the weight.

11. Read the buret and record the reading.

12. Repeat items 9 through 11 until 50 mL of water has been delivered and weighed. Keep the flask stoppered except when adding liquid. No more than 10 to 15 min should be needed to acquire a set of calibration data.

13. Empty the flask, wipe the neck, refill the buret, and repeat the calibration, recording the data on the second chart.

Calculations. Do not attempt to make calculations at the balance. Calculate the buret corrections, following the outline of Table 2-2. Apply the arithmetic check on *each* calibration. If agreement between the duplicate runs is not within the designated acceptable limits, repeat the calibration.

Check the calculations for the buret calibration by substituting the corresponding values from your table for the letters in the equation

$$A - B + C = D - E + 5(G - F) \tag{2-1}$$

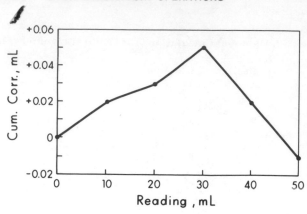

Figure 2-19. Example of a calibration chart for a buret.

If the arithmetic is correct, the two sides of the equation should agree within 0.01 mL. After completing this check, ask the laboratory instructor to initial your notebook beside the table.

Duplicate calibration corrections should agree within 0.02 mL, and duplicate cumulative corrections within 0.04 mL at all volumes. If either of these conditions is not met, the technique for reading the buret probably needs improvement, and an additional calibration should be performed. Plot the average cumulative corrections against buret volume on a graph as in Figure 2-19, and attach it to the inside front cover of the laboratory notebook.

All subsequent buret readings at titration end points should be corrected by the use of this graph. The net correction for a titration volume is equal to the correction for the final reading minus the correction for the initial reading. This net correction is then added to the net reading. For instance, if the final reading is 40.81 mL and the initial reading is 1.03 mL, the net correction interpolated from Figure 2-19 is +0.02 – 0.00, or +0.02 mL. The correct titration volume is therefore 40.81 – 1.03 + 0.02 = 39.80 mL.

2–9 CALIBRATION OF PIPETS

Ask the laboratory instructor for any supplementary instructions, and then clean (with pipe cleaners if necessary) and calibrate the pipets provided, following the directions for pipet use given in Section 2-2.

Procedure

Before carrying out the pipet calibration ask the laboratory instructor to verify that your pipet is clean.

Prepare a data page in your laboratory notebook for two calibrations (see Example 2-1). Assemble the following equipment: clean 10- or 20-mL pipet, pipetting bulb, 50-mL conical flask with dry stopper, 400-mL beaker of distilled water, and lintless tissue or clean towel.

At the top-loading balance carry out the following procedure:

1.   Place a thermometer in the beaker of distilled water.

2. Check the level and zero of the balance. Ensure that the zero-adjustment knob, not the tare knob or the leveling feet, is used for the zeroing operation.

3. Weigh the conical flask and stopper, recording in your notebook the weight to the nearest milligram.

4. Read and record the temperature of the distilled water. Remove the thermometer.

5. Pipet a portion of the distilled water into the flask. Remove the final drop by touching the pipet tip to the inner wall of the flask below the farthest point of insertion of the stopper. Restopper the flask.

6. Recheck the balance zero. Weigh and record the weight of the flask and water.

7. Add a second portion, removing the stopper just before the addition. Restopper, recheck the zero of the balance, weigh, and record the weight.

The three weighings for the set of two portions should not take longer than 15 min.

Calculations. Leave the balance and calculate the delivered volumes as in Example 2-1. If agreement between duplicates is not within 0.005 mL, carry out another set of calibrations. Calibrate all the pipets provided.

Use the average value of the duplicate calibrations for all subsequent measurements with the pipets.

EXAMPLE 2–1

Temperature of water	26 °C
Weight of flask + water	24.678 g
Weight of flask	14.713 g
Apparent weight of water delivered	9.965 g
True weight of water delivered*	9.975 g
True volume of pipet**	10.007 mL

*The weight of air displaced by the water is 9.965 g $\times$ 0.0011 = 0.011 g (with the assumption that the density of water is unity). The weight of air displaced by the stainless-steel weights used to weigh the water is (9.965/7.8)0.0011 = 0.0014 g. The net buoyant effect of air on the water weighed is 0.011 – 0.0014 = 0.010 g. The true weight of water delivered is 9.965 + 0.010 = 9.975 g.

**At 26°C, the density of water is 0.9968 (Table 2-1). The true volume of 9.975 g of water at 26°C is 9.975/0.9968 = 10.007 mL.

2–10 USE OF PIPETS AND VOLUMETRIC FLASKS

Preparation of a Solution in a Volumetric Flask

The following experiment is designed to provide experience in correct use of the volumetric flask: Weigh a 50-mL beaker on the triple-beam balance. Set the weights for an additional 2 g, and add solid $KMnO_4$ to the beaker until the beam is again level. Dissolve the potassium permanganate in about 20 mL of distilled water, stirring gently to avoid loss. Quantitatively transfer the solution to a 100-mL volumetric flask fitted with a small funnel. To prevent solution running down the outside of the beaker, pour the solution down the stirring rod, and then touch the rod to the spout of the beaker to remove the last drop. Add more water to the beaker, stir, and repeat the procedure. Note the amount of washing required to quantitatively transfer the permanganate from the beaker to the flask. Finally, rinse the last portions of solution from the stirring rod into the volumetric flask with a stream of water from the wash bottle. Rinse the funnel and remove it. Dilute the solution in the flask until the bottom of the meniscus is even with the graduation mark. Stopper, invert, and shake the flask. Return it to the upright position, and allow the air bubble to return all the way to the top of the neck. Repeat until the solution is completely homogeneous; about 10 inversions and shakings are required. Save the solution for later use.

Taking an Aliquot

Whenever a buret or pipet is used to deliver a measured volume of solution, the liquid it contains before measurement should have the same composition as the solution to be dispensed. The following operations are designed to illustrate the minimum effort needed to ensure this:

Fill a pipet with a 2% solution of potassium permanganate and let it drain. Draw some distilled water from a 50-mL beaker into the pipet, rinse, and discard the rinse solution. Determine the minimum *number* of such rinsings required to remove completely the permanganate color from the pipet. If the technique is efficient, three rinsings will suffice. Again fill the pipet with permanganate and proceed as before. This time determine the minimum *volume* of rinse water required to remove the color by collecting the rinsings in a graduated cylinder (less than 5 mL is enough with efficient technique). In the rinsing operations was the water in the 50-mL beaker contaminated with permanganate? If a pink color shows that it was; repeat the exercise with more care.

As a test of your aliquoting technique, ask the laboratory instructor to observe and comment on the following operation. Rinse a 10-mL pipet several times with the 2% solution of potassium permanganate you prepared. Pipet 10 mL of the permanganate solution into a 250-mL volumetric flask. Dilute the solution to volume. Mix by repeatedly inverting and shaking the flask. (Note the effort required to disperse the permanganate color uniformly through the solution.) Rinse the pipet with the solution in the volumetric flask. Pipet a 10-mL aliquot of the solution into a conical flask.

If the laboratory instructor considers your technique satisfactory, ask him to initial your notebook to that effect.

Calibration of Volumetric Flasks

Weigh a clean, dry, stoppered flask to the nearest milligram on an analytical balance or a top-loading balance. (The capacity of the balance determines the maximum size of flask that can be calibrated.) Insert a clean, dry funnel into the flask so that the stem extends below the calibration mark. Fill the flask with water at room temperature to just below the mark. Carefully remove the funnel and add water with a medicine dropper until the bottom of the meniscus coincides with the calibration line. Remove droplets of water present above the line with a lintless towel or a strip of filter paper. Stopper the flask and reweigh

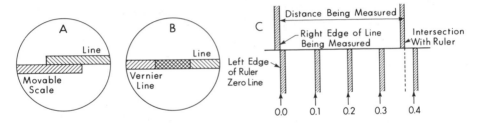

Figure 2-20. Correct methods for reading scales: (*A*) meters, ruler, and so forth; (*B*) verniers; (*C*) placement and reading of a ruler (line thicknesses and spacing magnified for clarity).

it. Immediately after weighing, measure and record the temperature of the water in the flask.

Calculate the true volume of the flask in the same way as for the transfer pipet. Obtain a duplicate calibration by emptying the flask and drying it. Air drying is best if time permits; it can be hastened by drawing air through the flask with the aid of an aspirator and a piece of glass tubing. Alternatively, *filtered* (clean, grease-free) compressed air can be used. Duplicate measurements should agree to about 5 parts in 10 000.

2-11 LINE WIDTHS AND MEASUREMENT TECHNIQUES

In Section 2-2 the instructions for reading a buret point out that the finite width of the lines printed on a buret must be taken into account in careful work. Similarly, recorder tracings, lines on rulers, and needles on meters all have finite width, and more precise results are obtained when these line widths are correctly taken into account (Figure 2-20). When measuring with a ruler work consistently from line edges, which are more easily located than line centers. Bring the left edge of the zero line on the rule up to the right edge of the line edge being measured, then read the distance on the ruler corresponding to the point where the right edge of the second line meets the ruler. In Figure 2-20C the distance is between 0.3 and 0.4 ruler units; the correct reading is 0.37.

An exception to this method is with vernier scales, which are more conveniently read at the position where the lines superimpose, as in Figure 2-20B.

PROBLEMS [17]

2-1. Explain why the modern single-pan (constant-load) analytical balance yields accurate weights even though the balance arms are of unequal length.

2-2. What is meant by sensitivity error in a balance? How can it be minimized?

2-3. Identify the main sources of potential error to be considered in obtaining a weight with a balance. Explain briefly how they are minimized in modern analytical balances.

2-4. Explain the differences between direct weighing and weighing by difference. What are the advantages of each method?

---

[17]The more difficult or challenging problems throughout the book are indicated by a dagger.

2-5.   What is meant by the buoyancy correction in weighing? When does a buoyancy correction need to be made to an observed weight?

2-6.   A student finds that, although his analytical balance is level, it cannot be zeroed with the fine zero adjustment. If the zero reads 0.0153 g, should the zero-adjustment nut be moved forward or to the rear?

2-7.   A balance was tested for variation in sensitivity with pan load as described in Section 2-6. An optical-scale reading of 97.3 mg was obtained with 100.0 g of weights dialed; an optical-scale reading of 97.2 mg was obtained with 0.0 g dialed. Is there a sensitivity error? (Recall that an analytical balance weighs to a precision of about 0.1 mg.) What corrective action should be taken before the balance is used?

2-8.   The optical-scale setting of a balance is tested as follows: Add a rough 1-g weight to the pan. Dial 1 g. Release the beam and set the optical scale to zero. Dial 0.0 g. Read the optical scale. Arrest the beam. In this procedure the rough weight need not be exactly 1 g. Why?

2-9.   Why is no sensitivity error with pan load expected in a constant-load balance?

2-10.   What is the true weight of an object having an apparent weight of 20.0000 g when weighed with stainless-steel weights if the object is pure aluminum (density 2.7); platinum (density 21.4); stainless steel (density 7.8); borosilicate glass (density 2.7)?

2-11.   Devise a set of seven individual weights that will provide any combination of weights from 0 to 100 g at 1-g intervals. If one more weight were permitted, what set might be selected?

2-12.   The optical scale of a balance reads 97.5 mg when it should read 100.0 mg. If two revolutions of the nut for adjustment of optical-scale sensitivity in Figure 2-3 are required to change the sensitivity by 1 mg, in what direction and how much should it be turned for the optical scale to read correctly?

2-13.   A sample of NaCl (density 2.16) had an apparent weight of 5.0000 g when weighed on a constant-load analytical balance (stainless-steel weights). What is the true weight of the NaCl?

2–14.†   When making buoyancy corrections, one *must include* that portion of the weight obtained from the optical-scale reading. Why?

2-15.   What is the true weight of (a) diethyl ether (density 0.714), which weighs 0.5000 g in air, when weighed with stainless-steel weights; (b) mercury (density 13.55), which weighs 3.0000 g in air.

2–16.†   A solid glass paperweight weighs 60.4231 g. When suspended by a wire in water at 25°C, it weighs 43.0462 g. The wire suspended in water to the same depth weighs 0.2345 g. All weighings were made with stainless-steel weights of density 7.8. What is the true weight, volume, and density of the glass?

2-17.   What is the correct volume of a 25-mL pipet that delivers 24.912 g of water (apparent weight) at 26°C?

2-18.   In the calibration of a 50-mL pipet by weighing the water delivered, how accurately should the weighings be made if the calibration error from weighing is not to exceed 1 part per 1000?

2-19.   In the determination of the volume of a 20-mL pipet, how accurately must the temperature of the water be measured to obtain a calibration accurate to 0.002 mL?

2-20.   A sample is measured in an uncalibrated 10-mL pipet. If the true pipet volume is 10.046 mL, will the results be high or low? By how much?

2-21.   If a pipet calibrated in the normal way was used to deliver a highly concentrated salt solution, would the amount of liquid delivered be more or less than the calibration volume? Why?

2-22.   If the inside diameter of a 10-mL pipet stem is 3.5 mm and the error in adjusting the meniscus is 0.2 mm, what is the corresponding error in the volume delivered? If the overall delivery error is 0.005 mL, what is the most likely source of error?

2-23.   A 10-mL pipet calibrated at 25°C is known to deliver 9.973 mL. What error is introduced if it is used at (a) 22°C; (b) at 30°C? Is this acceptable?

2-24.   In the calibration of a buret at 27°C for buret readings of 0.02, 10.01, 19.99, 29.98, 39.90, and 49.99 mL, the weight of the flask plus water was 36.2513, 46.2520, 56.2518, 66.2510, 76.2508, and 86.2509 g. Calculate the corrections which must be made to observed volumes, do the arithmetic check, and sketch a calibration graph for this buret. Did the weighings need to be made this precisely?

2-25.   Why should a 20-mL transfer pipet be calibrated to the nearest 0.002 mL, whereas a 50-mL buret need not be calibrated to better than 0.01 mL?

2-26.   A buret is read initially with the eye below the level of the meniscus, and after delivery with the eye above that level. Will the apparent volume be more or less than that actually delivered?

2-27.   A student prepares a sample solution in a volumetric flask but fails to mix the solution after dilution to the mark. What will be the effect on the results if a portion of this solution is withdrawn and used in a determination?

2-28.   While a clean buret is being calibrated, a bubble of air, which was present in the tip, escapes. Will this occurrence affect the results obtained? If so, will the apparent volume delivered be high or low?

2-29.   A calibration run was made on a buret at 22°C. For buret readings of 0.00, 9.98, 20.01, 30.09, 40.09, and 49.97 the weights were 36.209, 46.186, 56.225, 66.127, 76.130, and 86.028. Complete the calculations for the calibration, and then use the arithmetic check to ensure that no calculation errors were made. Construct a buret calibration graph.

2-30.   When using a buret calibration graph to correct the volume of a titration a student applied a correction of +0.003 mL. Is a correction at this level justified? If so, why? If not, what should the correction have been?

2-31.   The initial volume in a titration was 15.83 mL, and the final volume 37.21 mL. The buret calibration chart is shown in Figure 2-19. What is the corrected volume of titrant delivered?

2-32.   An unknown solution is prepared on a warm fall day and diluted to volume in a 100-mL volumetric flask. The next day the level of the meniscus in the flask is well below the mark on the neck. How should you proceed to take the required aliquots?

2-33.   In the preparation of 0.02 M standard potassium dichromate a 1-L volumetric flask is filled at 15°C. The temperature 24 h later has risen to 25°C. If the neck of the flask has an inner diameter of 20 mm, how far above the mark does the solution now stand?

2-34.   A 1-L volumetric flask has a volume of 1000.25 mL at 25°C. What is its volume at 20°C?

2-35.   A sample requires addition of the equivalent of 0.1 mL of 3 M H_2SO_4 with an error not exceeding about 3%. How can this be accomplished with a graduated cylinder?

2-36.†   Exactly 0.1000 mol of potassium chloride is dissolved in water, and the resulting solution diluted to the mark at 22°C in a flask having a volume of 499.86 mL. What would be the molarity of the solution if the temperature of the solution were changed to (a) 20°C; (b) 25°C; (c) 30°C? What is the maximum allowable difference between the temperatures of preparation and use if the error is to be kept below 1 part per 1000?

2-37.   A sample in a 200-mL volumetric flask is accidentally filled above the graduation mark. A second sample is not available. How can an analysis precise to within 1 part per 1000 still be obtained without resorting to another volumetric flask?

2-38.   What is the effect on the amount of solute delivered if, when an aliquot is taken as directed in the check on aliquoting procedures, (a) the freshly washed pipet is not rinsed with the concentrated solution; (b) the solution is inadequately mixed after dilution; (c) the same pipet is used throughout, but inadequately rinsed with the diluted solution before the second pipetting?

2-39.   A 0.01136-mol sample was dissolved in water, and the resulting solution diluted to the mark in a 100-mL volumetric flask. A series of 10-mL aliquots was pipetted into flasks. How many moles of sample were present in each of the flasks if the calibration volumes of the volumetric flask and pipet were (a) 100.00 and 10.020 mL; (b) 100.20 and 10.000 mL; (c) 100.20 and 10.020 mL?

2-40.   A solution is 0.2046 M when prepared at 30°C. What is its concentration when its temperature is lowered to 20°C?

2-41.   Several readings for a balance with vernier readout are shown below. What are the readings in grams?

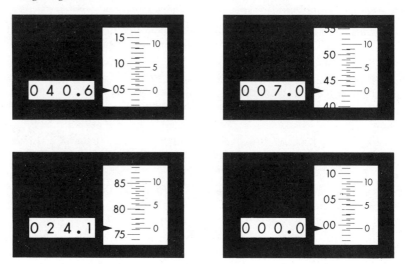

REFERENCES

W. J. Blaedel and V. W. Meloche, *Elementary Quantitative Analysis*, 2nd ed., Harper & Row, New York, 1963. Thorough discussion of calibration of volumetric glassware and of fundamentals of analytical weighing.

R. O. Leonard, *Anal. Chem.* **1976**, *48*, 879A. Discussion of the variety of electronic balances now being manufactured and how they work.

3

CHEMICAL EQUILIBRIUM.
INTRODUCTION TO TITRIMETRY

Theory guides, experiment decides.

I. M. Kolthoff

3-1 REVIEW OF CHEMICAL EQUILIBRIUM

The rate of reaction is proportional to the active concentration of the reacting substances, each raised to the power numerically equal to the number of molecules appearing in the balanced equation.

C. Guldberg and P. Waage, *J. Prakt. Chem. (2)* **1879**, *19*, 69.

For a reaction to be quantitative, the position of equilibrium and the rate of reaction must be favorable. For reversible reactions, therefore, concepts involving equilibria must be applied to the requirements for making chemical measurements (Section 3-2). In this section some of the main concepts are reviewed.

In a chemical system at *equilibrium*, no change in composition is detectable with time. Moreover, this same composition can be attained by carrying out the reaction in either the forward or the reverse direction. For the reaction

$$I_2 + I^- \rightleftarrows I_3^- \tag{3-1}$$

the concentrations of all three chemical species will be the same, whether one starts with iodide and iodine or with the triiodide ion (provided that proper attention was paid to the initial concentrations). By itself, the absence of change with time is not a criterion for equilibrium. For example, hydrogen and oxygen gases may be mixed essentially indefinitely without appreciable reaction to form water. The same mixture of hydrogen and oxygen is not produced, however, from water on standing. When reactions are slow, it may be difficult to know what the equilibrium state would be.

The position at which equilibrium is attained depends on the rates of the forward and reverse reactions. Thus, for the generalized reaction

$$A + B \rightleftharpoons C + D \qquad (3\text{-}2)$$

the rate of reaction of A and B to form the products C and D depends on the number of collisions between A and B and on how fruitful these collisions are in producing a reaction. In turn, the fraction of fruitful collisions depends on factors such as temperature, presence of catalysts, and the nature of A and of B. When conditions are held constant, this fraction is a constant. The reaction rate, then, is simply proportional to the number of collisions between A and B per unit time. Should the number of collisions double, then the rate would also double.

In an ideal system containing A and B the number of collisions between the two is directly proportional to the concentration of A. Should A be doubled, the number of collisions and the rate would likewise be doubled. Similarly, the rate is directly proportional to the concentration of B. In such an ideal system the rate of the forward reaction is then equal to

$$\text{forward rate} = k_f[A][B]$$

where k_f is a proportionality constant called the *rate constant* for the forward reaction. It, too, depends on the proportion of fruitful collisions. Although this is the expression for the rate, it says nothing about the actual chemical mechanism for the reaction.

Most, but not all, reactions of value in analytical chemistry are reversible. As indicated by Equation (3-2), reversibility means that the products can react to regenerate the reactants:

$$C + D \rightarrow A + B \qquad (3\text{-}3)$$

As with (3-2), the rate depends on the number of collisions between C and D and the fraction that are fruitful. In an ideal system the rate of this reverse reaction is

$$\text{reverse rate} = k_r[C][D]$$

If the mixture contains only the reactants A and B, the rate of the forward reaction initially is some value that decreases according to changes in the concentrations of A and B. Since the concentrations of C and D would initially be zero, the reverse rate would be zero. But, as C and D are formed from the reaction of A and B, the rate would become higher and higher. Thus one reaction goes slower and slower and the other faster and faster. After a time, C and D will react as fast as they are formed, and the same will be true for A and B. The concentrations will now not change with time, and the system is in a state of equilibrium. Both reactions proceed at equal rates (Figure 3-1), a state characteristic of *dynamic equilibrium*. Thus

$$k_f[A][B] = k_r[C][D]$$

On rearrangement we obtain

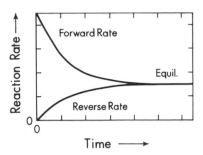

Figure 3-1. Forward and reverse reaction rates as a function of time. At equilibrium both reactions occur at equal rates.

$$\frac{[C][D]}{[A][B]} = \frac{k_f}{k_r} = K_{eq}$$

The ratio of the forward and the reverse reaction rate constants is called the *equilibrium constant*. This constant contains the equilibrium values of the concentrations of the reactants and products.

By convention the concentrations of the products are in the numerator of the expression, and those of the reactants in the denominator. When K_{eq} is large, the reaction has a strong tendency to proceed to the right, that is, A and B react to a significant extent. When K_{eq} is small, A and B have reacted to only a slight extent by the time equilibrium is reached, and there is no further net change.

The equilibrium constant tells nothing about how long it takes for equilibrium to be reached, whether microseconds or years. It also tells nothing about details of the mechanism of the reaction. For the equilibrium involving iodine in Equation 3-1 the expression for the equilibrium constant is

$$[I_3^-]/[I_2][I^-] = K_{eq} = 710$$

For a reaction such as

$$Ag^+ + 2CN^- \rightleftharpoons Ag(CN)_2^-$$

where one of the species reacts in a 2:1 ratio with the other, the expression[1] is

$$K_{eq} = \frac{[Ag(CN)_2^-]}{[Ag^+][CN^-]^2}$$

---

[1]For the general reaction aA + bB $\rightleftharpoons$ cC + dD,

$$K_{eq} = \frac{[C]^c[D]^d}{[A]^a[B]^b}.$$

When a system is at equilibrium and conditions are changed, either the system or the position of equilibrium must change in some way to relieve the stress (Le Châtelier's principle).   Suppose that Reaction (3-2) involves the generation of heat and the system is at equilibrium. Now, if the temperature is raised, the equilibrium will shift to form more A and B to absorb some of the heat.[2] If the reactants are gases and the products a condensed phase, solid or liquid, then increasing pressure on the system at equilibrium will cause it to shift to form more of the condensed phase.

When an equilibrium constant is small, significant amounts of the reactants may still be present at equilibrium. In this circumstance addition of a large excess of A may reduce the amount of B to a small value, making the reaction complete.

EXAMPLE 3-1

For a solution initially containing 0.01 M I_2 and 0.01 M I^- we can calculate from the equilibrium constant of 710 that at equilibrium $[I_3^-] = 0.0069$ M, $[I_2] = 0.0031$ M, and $[I^-] = 0.0031$ M. In many iodimetric procedures I_2 is produced. Because molecular iodine is volatile, adding a substantial excess of iodide reduces error that may be introduced by loss through volatilization. In a solution initially 0.3 M in KI and 0.01 M in I_2, at equilibrium $[I_3^-] = 0.01$ M, $[I_2] = 5 \times 10^{-5}$ M, and $[I^-] = 0.3$ M. Thus the addition of excess KI reduces the I_2 concentration about 100-fold and diminishes volatility losses proportionately.

Strictly speaking, equilibrium constants are valid only when activities are used, rather than concentrations. The *activity* a of an ion or molecule is sometimes called the *effective concentration* and is expressed in the same units as concentration. The activity of a substance is equal to the product of an *activity coefficient* γ and concentration C,

$$a = \gamma C \tag{3-4}$$

The absolute activity of an individual species is an unmeasurable quantity; only relative changes can be determined. The changes are measured with respect to a completely *arbitrarily selected standard state*, where the substance is *assigned* an activity of unity. For solutes in solution the standard state is usually defined so that, as the concentration approaches zero, the activity coefficient approaches unity; activity then approaches concentration as a limit. For solutes at moderate concentration the activity coefficient as defined is usually less than unity. Values for activity coefficients depend on total ionic strength, that is, on the concentrations of all electrolytes in solution.

The theory of solute activity coefficients developed by Debye and Hückel provided a much clearer understanding of solutions. With the simplifying assumptions that ions are point charges of negligible size, the simplest form of the Debye-Hückel equation, called the *Debye-Hückel limiting law*, was

---

[2]But note that, in addition to shifting the position of equilibrium, a change in the temperature of the system may also result in a change in the magnitude of the equilibrium constant. See, for example, G. M. Bodner, *J. Chem. Ed.* **1980**, *57*, 117.

obtained. According to this law, for highly dilute solutions of a strong electrolyte

$$-\log \gamma_i = A\, Z_i^2 \sqrt{\mu} \tag{3-5}$$

where γ_i is the activity coefficient of ion i, A is a constant with a value of 0.511 at 25°C in water, Z_i is the charge of the ion, and μ is the ionic strength, defined by

$$\mu = 0.5\, \Sigma C_i Z_i^2 \tag{3-6}$$

In (3-6) C_i is the concentration of ion i in the solution.

EXAMPLE 3-2

Calculate the ionic strength of a 0.1 M solution of Na_2SO_4.
From Equation (3-6), $\mu = 0.5\,([Na^+](1)^2 + [SO_4^{2-}](2)^2) = 0.5\,((0.2) + (0.1)(4))$ = 0.3.

Calculate the ionic strength of a 0.02 M solution of $MgCrO_4$.
$\mu = 0.5((0.02)(2)^2 + (0.02)(2)^2) = 0.08$.
Note that the ionic strength depends on ion concentration and charge.[3]

EXAMPLE 3-3

The mean activity coefficients of potassium chloride (the mean of the activity coefficients of the chloride and potassium ions) in 1.0 M, 0.1 M, 0.01 M, and 0.001 M solution are 0.606, 0.769, 0.901, and 0.965. The activity of potassium chloride in 1 M solution is therefore 0.606.

In dilute solutions, activity coefficients as defined are fortunately near unity, and therefore activities and concentrations are nearly the same. Except for some special cases, concentrations will be applied here in most calculations concerning equilibria, thereby simplifying the arithmetic.

For solvents and pure substances generally, concentration is conventionally expressed as mole fraction, where the mole fraction is unity for the pure material. For such substances, changes in activity are measured with respect to a standard state defined so that the activity coefficient is unity for the pure material. Thus water as solvent is assigned an activity value of unity, solid silver chloride an activity of unity, and so on. For gases the concentration is given in terms of pressure; an *ideal gas* is defined as one having an activity of unity at a partial pressure of one atmosphere.

3-2 TITRIMETRIC METHODS

> With the aid of titrimetric analysis, analytical chemistry could be introduced into practical life. I would be contented if even to a small extent I could open the door through which science could enter into the life of industry and technology.
>
> *K. H. Schwarz,* 1853
> Author of the first textbook of titrimetry.
> (Quoted in F. Szabadvary, *History of Analytical Chemistry*)

---

[3]The relation between ionic strength μ and concentration C for salts of various charge types is as follows: A^+B^-, $\mu = C$; $A^{2+}B_2^-$ or $A_2^+B^{2-}$, $\mu = 3C$; $A^{2+}B^{2-}$, $\mu = 4C$; $A^{3+}B_3^-$ or $A_3^-B^{3-}$, $\mu = 6C$.

Probably the most broadly applicable approach to making chemical measurements is *titrimetry*, or *titrimetric analysis* (sometimes called *volumetric analysis*). Titrimetric methods are illustrated in several of the ensuing chapters in the discussions of oxidation-reduction, acid-base, and complexation reactions and of such end-point detection techniques as amperometric, potentiometric, and photometric. In this chapter the terminology of titrimetric analysis is explained, and the fundamentals are illustrated by precipitation reactions.

Titrimetric analysis involves a reaction between the substance to be measured and a solution of a reagent of known concentration:

$$\text{substance} + \text{reagent} \longrightarrow \text{products}$$

The reagent solution of known concentration, normally a *standard solution*, is called the *titrant* (the titrating agent) and the substance is called the *analyte*. The process of bringing about the reaction is called a *titration* . When the amount of reagent added is equal chemically to the substance being determined, the *equivalence point*, or *theoretical end point* (or *stoichiometric end point*), has been reached. At this point the weight of the substance can be calculated, usually from the volume and concentration of standard solution that was necessary to complete the reaction. The completion of the titration is generally noted by auxiliary means. A reagent used to reveal the completion of the titration process by a visual signal such as a color change is called an *indicator*. The *end point* is the experimentally obtained estimate of the equivalence point in the titration; it may be determined with the aid of a visual indicator or by instrumental means. In an ideal titration, the end point coincides with the equivalence point. In actuality a finite *titration error* is almost always present. This error is defined as the amount of reagent corresponding to the difference between the two points. Experimental conditions should always be sought that will ensure only small titration error. A *titration curve* is a record of some function of concentration of a titration species such as pH or pCl as a function of the amount of reactant added.

If a reaction between a substance and a reagent is to result in precise and accurate quantitative measurements, it must meet four fundamental conditions:

1. The reaction must be quantitatively complete (usually 99.9 + %). This condition can normally be predicted from equilibrium data.

2. The reaction must be sufficiently rapid that the measurement can be completed in a reasonable time. This condition must be determined experimentally. Many oxidation-reduction reactions, for example, are unsuitably slow.

3. So that the amount of substance can be calculated from the amount of reagent used, the reaction should occur in a single, well defined process.

4. A method must be available to indicate when equivalent amounts of reactants have been added, that is, a method of determining the probable

equivalence point.

Although only a small fraction of the thousands of possible chemical reactions meet these four requirements, titrimetry can still be applied to the measurement of an enormous variety of substances.

The wide adoption of titrimetric methods is a result of their several practical advantages. They are highly precise and accurate, are rapid, and often require only simple and inexpensive equipment.

Titrimetric methods can be classified according to the type of chemical reaction involved:

1. *Precipitation titrations*, in which a slightly soluble product is formed (Chapter 4)

2. *Acid-base titrations*, in which a slightly dissociated product such as water, a weak acid, or a weak base is formed (Chapters 5 and 6)

3. *Complexation titrations*, in which a slightly dissociated product of another type is formed (Chapter 7)

4. *Oxidation-reduction titrations*, in which oxidation states are changed (Chapter 10)

Titrimetric methods can also be classified according to method of end-point detection:

1. *Visual-indicator titrations*, which exploit a change in color or fluorescence, appearance of a precipitate, and so on

2. *Potentiometric titrations*, in which the change in concentration of a substance, reagent, or product is monitored through measurements of potential (Section 11-8)

3. *Amperometric titrations*, in which changes are monitored through measurements of current (Section 12-3)

4. *Conductance titrations*, in which changes are monitored through measurements of conductivity (Section 12-5)

5. *Photometric titrations*, in which changes are monitored through measurements of radiant energy (Section 14-10).

Other titrimetric methods in addition to those in which the volumes of solutions are measured include *gravimetric titrations,* in which the weight of solution is measured, and *coulometric titrations,* in which the product of current and time determines the quantity of reagent added.

Direct titrations are concerned with the straightforward reaction of the titrant and substance of interest; *indirect* titrations are concerned with the reaction of the titrant with an intermediary substance that has a known relation to the substance of interest; and *back* titrations are based on the addition of a measured amount of a reagent of known concentration in excess of the amount to react with a substance followed by titration of the excess with a second reagent of known concentration.

Primary Standards

In the preparation of solutions of known concentration there must be some means of calibration of the solution in terms of a known amount of the substance to be measured. The ultimate standard, carbon-12, is unsuitable as a working standard for chemical analysis. Pure metallic silver and the coulomb are two important operational standards for evaluating the purity of other reagents and for standardizing solutions. For practical work a variety of primary standards have been developed. A *primary standard* is defined as a material that (1) is obtainable in pure form (usually 99.99% pure or better); (2) is stable so that it can be stored indefinitely without decomposing; (3) has properties that permit it to be weighed accurately without special precautions (for example, it must be nonhygroscopic); (4) reacts stoichiometrically with the substance to be titrated; (5) is soluble in appropriate solvents; and (6) has a high equivalent weight to minimize relative weighing errors.

A *primary standard solution* is one that can be prepared directly by weighing a quantity of a primary standard and diluting it to a known (measurable) volume. For example, a primary standard solution can be prepared by weighing a specified amount of a suitable primary standard, such as pure potassium chloride, and diluting it to a definite volume in a volumetric flask.

A *secondary standard solution* is one whose concentration is determined indirectly by standardization. *Standardization* is the process of causing a solution to react with a suitable primary standard, usually by titration, to determine its precise concentration. The standardization procedure should, whenever possible, be the same as or similar to the procedure for which it is to be used in subsequent analyses. For example, sulfuric acid is not a primary standard; a sulfuric acid solution must be standardized by reaction with a primary standard such as pure sodium carbonate.

Alternation

Alternation consists in carrying out a series of laboratory operations on standards and samples in alternating sequence. In this way errors resulting from changes in operator technique, solution concentrations, instrument drift, end-point judgment, and so forth, tend to be canceled. The technique is especially advantageous in titrimetry, but competent analytical chemists use it whenever possible, even though it is not written into the procedures. For example, in Section 5-10, where sodium carbonate is determined by titration with hydrochloric acid, alternation involves preparation for titration of both the sodium carbonate standards and the samples. The titrations are then carried out in the order: standard, sample, standard, sample, and so on. Procedural modifications needed to incorporate alternation are generally obvious and usually are not included in the directions. This technique requires only slightly more organization and adds polish to experimental work.

Weight Titrations

A *weight*, or *gravimetric, titration* is one in which the weight rather than the volume of titrant is measured. The advantages are that (1) weight can be determined more accurately and precisely than volume; (2) procedures necessitating use of volumetric ware (burets, pipets, and flasks), with the necessity for calibration and removing traces of grease for efficient drainage, are not required; (3) small amounts of titrant and sample do not cause loss of precision or accuracy when reagents are expensive or sample is limited; and (4) errors in volume measurement due to expansion or contraction of solvent with temperature are avoided. The technique is particularly applicable in photometric titrations, where small volumes are more convenient and give greater precision in that they permit titration directly in a spectrophotometric cell without removal of the cell from the instrument.

Despite these advantages, weight titrations have not been popular, principally because several weighings may be required, a tedious process with old-style balances. But with the advent of top-loading balances capable of weighing quickly and accurately to within a milligram or better, weight titrations are now feasible, and often may be preferable to volumetric methods. The experiment in Section 6-5 illustrates a situation for which weight titration is recommended — titration with an organic solvent. Acetic acid, like most organic liquids, has a cubic coefficient of thermal expansion almost five times that of water; the change in volume is about 1 ppt/°C. Thus, small changes in temperature affect the volume of titrant markedly.

3-3 CALCULATIONS IN TITRIMETRIC ANALYSIS

Solution Concentration

The most common ways of expressing solution concentration in titrimetric analysis are as molarity, normality, and titer. All express the amount of a solute present per unit volume of solution. *Molar concentration (molarity)* may be defined as the number of gram molecular weights (moles) of a substance dissolved in 1 L of solution.[4] A one-molar (1 M) solution of silver nitrate contains one mole of silver nitrate (169.87 g) per liter of solution, or one millimole per milliliter of solution. *Normal concentration (normality)* may be defined as the number of gram equivalent weights (equivalents) of a substance dissolved in one liter of solution, or the number of milliequivalents per milliliter. The equivalent weight of a substance depends upon the type of reaction it undergoes. In an acid-base neutralization reaction the *equivalent weight* is the weight that either contributes or reacts with one molecular weight of hydrogen ion in the reaction. Thus for an acid or base containing only one reactive hydrogen or hydroxyl group, as in HCl or NaOH, the equivalent weight is equal to the molecular weight; if two reactive hydrogen or hydroxyl groups are present, as

---

[4]The term *formality* is sometimes used in place of molarity to emphasize that the nature of the species present in solution is not defined. The term is not an acceptable exception to SI usage (*Anal. Chem.* **1980**, *52*, 222) and is not used in this book.

in H_2SO_4 or $Ba(OH)_2$, the equivalent weight is equal to the molecular weight divided by two. For acids or bases containing two or more reactive groups of differing strengths, stoichiometry according to two or more reactions is possible. Thus, the base CO_3^{2-} may be neutralized only to the bicarbonate ion,

$$H^+ + CO_3^{2-} \rightleftarrows HCO_3^-$$

or all the way to carbonic acid,

$$2H^+ + CO_3^{2-} \rightleftarrows H_2CO_3.$$

In the first instance the gram equivalent weight of carbonate is equal to the gram molecular weight; in the second it is equal to the gram molecular weight divided by two, since carbonate in this case is reacting with two protons. The equivalent weight of a substance is always based on its behavior in a specified chemical reaction. A substance may have several equivalent weights, each corresponding to a different reaction stoichiometry. Since the gram equivalent weight of a substance varies with the reaction in which it is involved, a single solution may have several different normalities. Because of possible ambiguity, normalities are not used in this book.[5]

The *titer* of a solution is the weight of a substance that is equivalent to, or reacts with, a unit volume of the solution. Titer values simplify calculations, and are used frequently in routine analysis. The chloride titer of a 0.01000 M silver nitrate solution is (0.01000 M) (mol wt Cl/1000), or (0.01000)(0.03545). The weight of chloride in a sample can be determined simply by multiplying the milliliters of silver nitrate required to titrate the sample by the value of the chloride titer.

Solution concentrations may also be expressed in terms of weight, that is, as the amount present per unit weight of solution or solvent. Units include *molality*, defined as the number of gram molecular weights of a substance dissolved in one kilogram of solvent; *weight percent*, defined as (wt solute/wt solution)(100); and *weight molarity*, defined as the number of gram molecular weights of a substance per kilogram of solution. Both molality and weight percent are generally inconvenient to apply to titrimetric analysis. Weight molarity is useful when titrations are carried out by measurement of the weight of solutions rather than volume (Section 6-5). Weight percent is often used to express the concentrations of commercial aqueous reagents (see Appendix for examples).

Trace concentrations are sometimes expressed in units of parts per million (ppm) or parts per billion (ppb) by weight. Thus 1 mg of calcium in 1 L of water corresponds to a calcium concentration of 1 ppm (with the density of water assumed to be 1).

---

[5]Normality, like formality, is not listed among the acceptable exceptions to SI usage (*Anal. Chem.*, **1980**, 5, 222).

Calculation of Percentages

Volumetric calculations most often involve conversion of experimentally obtained data — generally volumes, concentrations, and weights — to the percentage of a substance present in a sample. A useful relation is that the grams of material present is equal to the number of moles of material in a sample multiplied by the molecular weight w:

$$g = (mol)(w)$$

The number of moles of material in a given volume of solution is obtained by multiplying the volume V by the molarity M. Therefore, the weight of material present in a dissolved sample is given by

$$g = (V)(M)(w)$$

where V is in liters. In titrations, V is more conveniently expressed in milliliters; then

$$g = V_{mL}Mw/1000$$

In stoichiometric calculations, cultivate the habit of leaving the arithmetic until last. When a numerical answer is not required, report the answer in factorial form. Thus the weight of sodium sulfate in 45 mL of $0.3\ M$ solution would be reported as

$$g\ Na_2SO_4 = (45\ mL)(1\ L/1000\ mL)(0.3\ mol/L)(142.04\ g\ Na_2SO_4/mol)$$

The following examples illustrate calculations often encountered in titrimetric work.[6]

EXAMPLE 3–4

Preparation of a standard solution. How much silver nitrate is required to prepare 250.0 mL of a 0.1000 M solution?

$$g\ AgNO_3 = \frac{250.0\ mL \times 0.1000\ mol/L \times 169.87\ g/mol}{1000\ mL/L} = 4.25\ g$$

EXAMPLE 3–5

Titration of a chloride sample. The chloride in a 0.5185-g sample required 44.20 mL of 0.1000 M AgNO$_3$ for titration. What is the percentage of chloride in the sample?

---

[6]In this section units have been included with all the terms in the arithmetic equations. Inclusion of units when setting up problems often assists correct handling of terms, and provides a verification when they are combined to give the units of the result. Although units are not included in every sample calculation in succeeding chapters, their inclusion is recommended whenever uncertainty is present in the setup of a calculation.

$$\% \ Cl = g \ Cl(100)/(g \ sample)$$

$$g \ Cl = (moles \ Cl)(g \ Cl/mol)$$

One mole of silver nitrate reacts quantitatively with one mole of chloride in the balanced equation,

$$Ag^+ + Cl^- \rightarrow AgCl(s)$$

and so the number of moles of chloride present in the sample equals the number of moles of silver used in the titration. Since

$$mol \ AgNO_3 = (mL \ AgNO_3)(mol \ AgNO_3)/L)(1 \ L/1000 \ mL)$$

the foregoing three relations can be combined to give

$$\% \ Cl = \frac{(mL \ AgNO_3)(M \ AgNO_3)(g \ Cl/mol)(1 \ L/1000 \ mL)}{(g \ sample)} \quad (100)$$

$$\% \ Cl = \frac{(44.20 \ mL)(0.1000 \ mol/L)(35.45 \ g/mol)(100)}{(0.5185 \ g)(1000 \ mL/L)} = 30.22$$

If a solution of silver nitrate is to be standardized against a pure salt such as sodium chloride, the same expression can be used to determine the molarity of the silver nitrate by substituting for % Cl the correct value for pure NaCl, 60.66%, and solving for $M \ AgNO_3$.

EXAMPLE 3-6

Calculation of titrant volume needed for a sample. How many milliliters of 0.1000 M AgNO$_3$ are required to react with 0.4250 g of BaCl$_2\cdot$2H$_2$O? Since one mole of BaCl$_2$ reacts with two moles of AgNO$_3$ in the balanced equation,

$$BaCl_2 + 2AgNO_3 \rightarrow 2AgCl(s) + Ba(NO_3)_2$$

$$mol \ AgNO_3 = (2)(mol \ BaCl_2\cdot2H_2O)$$

$$= (2) \ \frac{g \ BaCl_2\cdot2H_2O}{mol \ wt \ BaCl_2\cdot2H_2O}$$

$$mL \ AgNO_3 = \frac{(mol \ AgNO_3)(1000 \ mL/L)}{M \ AgNO_3}$$

$$mL \ AgNO_3 = \frac{(2) \ g \ BaCl_2\cdot2H_2O \ (1000 \ mL/L)}{(mol \ wt \ BaCl_2\cdot2H_2O)(M \ AgNO_3)}$$

$$mL \ AgNO_3 = \frac{(2)(0.4250 \ g)(1000 \ mL/L)}{(244.28 \ g/mol)(0.1000 \ mol/L)} = 34.80$$

The uncertainty of a result includes both indeterminate and determinate errors in the composition of the sample and in the procedure, reagents, and technique. Since the precision of the result is the statistical combination of these separate factors, the number of significant figures should reflect a reasonable estimate of the combined uncertainty.

PROBLEMS

3-1. Define each of the following terms: equilibrium, equilibrium constant, Le Châtelier's principle, activity, activity coefficient, standard state, ideal gas.

3-2. Explain how one can tell whether a system is at equilibrium.

3-3. How are the forward and reverse rate constants for a reaction related to the equilibrium constant of the reaction?

3-4. What is the relation between the equilibrium constant for a reaction and the relative concentrations of products and reactants?

3-5. Write the equilibrium-constant expressions for each of the following reactions:
(a) $Cl^- + Ag^+ \rightleftarrows AgCl(s)$
(b) $Ba^{2+} + SO_4^{2-} \rightleftarrows BaSO_4(s)$
(c) $H^+ + OH^- \rightleftarrows H_2O$
(d) $HOAc + OH^- \rightleftarrows H_2O + OAc^-$
(e) $H^+ + CO_3^{2-} \rightleftarrows HCO_3^-$
(f) $2H^+ + CO_3^{2-} \rightleftarrows H_2CO_3$
(g) $MnO_4^- + 8H^+ + 5Fe^{2+} \rightleftarrows Mn^{2+} + 5Fe^{3+} + 4H_2O$

3-6. Calculate the ionic strength of each of the following: (a) 0.05 M HCl; (b) 0.2 M NaOH; (c) 0.01 M H_2SO_4; (d) 0.25 M $Ca(NO_3)_2$; (e) 0.1 M $NaNO_3$ + 0.12 M KCl; (f) 0.04 M Li_2SO_4 + 0.15 M $MgBr_2$; (g) 0.3 M NH_4Cl + 0.075 M $COCl_2$; (h) 0.14 M $AgNO_3$ + 0.006 M $Zn(ClO_4)_2$.

3-7. Calculate the activity coefficient for each of the following species: (a) H^+ in 0.5 M HCl; (b) OH^- in 0.2 M NaOH; (c) SO_4^{2-} in 0.01 M H_2SO_4; (d) Ca^{2+} in 0.25 M $Ca(NO_3)_2$.

3-8. Define the following types of titration, and give an example of each: (a) direct, (b) indirect, (c) back.

3-9. What four conditions must be met for a method of titrimetric analysis?

3-10. Clearly distinguish between end point and equivalence point, giving an example.

3-11. Define or explain the meaning of titration, indicator, molarity, titration error, and the letter "p" in the term pCl.

3-12. What is a primary standard? List its requirements.

3-13. List the advantages of gravimetric over volumetric titrations.

3-14. In a titrimetric procedure for the determination of iron, 6 M phosphoric acid is required. Describe how to prepare 500 mL of 6 M H_3PO_4 by starting with concentrated phosphoric acid, which has a density of 1.69 and contains 85% H_3PO_4.

3-15. How many moles of magnesium chloride are in each of the following: (a) 5 mL of a 1% solution, (b) 25 mL of a 0.4 M solution, (c) 500 mL of a 2.0 M solution, (d) 50 mL of a solution containing 25 g/L.

3-16. If sea water contains 10^{-4} ppm of silver by weight, calculate its molar concentration, assuming the density of sea water is unity. How much silver would be in a cubic kilometer?

3-17.   On the average the earth's crust is 46.60% oxygen, 27.72% silicon, 8.13% aluminum, 5.00% iron, 2.09% magnesium, and 3.63% calcium, in addition to smaller amounts of all the other elements. Assuming the density of the earth's crust to be 3.0, calculate the molar concentration of these six most abundant elements.

3-18.   The overall concentration of mercury in the earth's crust is 0.5 ppm by weight. What is its molar concentration, assuming the density of the earth's crust to be 3.0? What is its weight molarity?

3-19.   The average igneous rock is 59.1% silica. If the average density is 2.6, what is the molar concentration of silica in igneous rock?

3-20.   Polychlorinated biphenyls (PCB's) have been known since about 1880 and have been widely used as electrical insulating liquids since 1929. They are stable and unreactive and are now found even in the polar ice caps. A limit of 2 ppm in fish for human consumption has been set. If a person eats 225 g of fish containing 0.8 ppm of PCB, how many grams of PCB are ingested? If the average molecular weight is 288, how many moles are ingested?

3-21.   To neutralize 1.23 moles of sulfuric acid, how many moles of sodium hydroxide are required? What volume of 0.24 M NaOH and what volume of 50% NaOH of density 1.53 are required? What is the weight molarity of the sodium hydroxide?

3-22.   What volume of 38% hydrochloric acid of density 1.19 is required to prepare 2 L of 0.2 M hydrochloric acid? What volume of 6 M hydrochloric acid would be required?

3-23.   The normal range in concentrations for calcium ion in human urine is from 0.2 to 0.5 g/L. Express this concentration range in units of (a) molarity; (b) parts per million.

3-24.   Sea water has an average density of about 1.024 g/mL. It has an average magnesium concentration of 1270 ppm, and bromine (as Br⁻), of 65 ppm. Calculate the concentration of each of these elements in (a) molarity, (b) weight molarity, (c) weight percent, and (d) parts per thousand.

3-25.   Calculate the molarity of (a) concentrated hydrochloric acid, density 1.19 and containing 38 weight percent HCl; (b) concentrated ammonia, density 0.90 and 28 weight percent NH_3; and (c) concentrated sulfuric acid, density 1.83 and 94 weight percent H_2SO_4.

3-26.   What weight of pure silver nitrate is required to prepare 500 mL of 0.03 M solution?

3-27.   Calculate the weight of (a) NaOH in 36.47 mL of 0.1135 M sodium hydroxide, (b) $K_2Cr_2O_7$ in 34.29 mL of 0.1085 M potassium dichromate, and (c) $AgNO_3$ in 40.12 mL of 0.1247 M silver nitrate.

3-28.   Calculate the volume of (a) 6 M HCl required to prepare 1 L of 0.1 M hydrochloric acid, (b) 3 M H_2SO_4 required to prepare 500 mL of 0.2 M sulfuric acid, and (c) 6 M NH required to prepare 250 mL of 0.15 M ammonia.

REFERENCES

H. A. Laitinen and W. E. Harris, *Chemical Analysis*, 2nd ed., McGraw-Hill, New York, 1974. Chapter 2 deals with equilibrium and activity. Chapter 5 includes a discussion of ultimate, operational, and primary standards.

F. Szabadvary, *History of Analytical Chemistry*, Pergamon Press, New York, 1966, Chapter 8. Describes the rocky road that titrimetric methods had to travel before being accepted. Szabadvary notes: "The famous analytical chemists of this period were contemptuous of the new methods of titrimetric analysis." Gay-Lussac had substantial influence on titrimetry becoming respectable when his new titrimetric methods gave more accurate results for the silver in French coinage than the methods then in use.

PRECIPITATION EQUILIBRIA AND TITRATIONS

I think it was in a course in quantitative chemical analysis that an appreciation of the scientific method ... began really to take hold of me.... There were no shortcuts to beat clear thinking and careful technique.

R. S. Mulliken, Nobel Laureate, 1966

4-1 EQUILIBRIA IN PRECIPITATION TITRATIONS

The reagent most commonly used in precipitation titrations is silver nitrate, particularly for the determination of chloride by the reaction

$$Ag^+ + Cl^- \rightleftarrows AgCl(s)$$

where s stands for solid precipitate. This system is used here, not because it is more important than other types, but because it conveniently demonstrates several main concepts of titration theory.

Section 3-2 notes that not every chemical reaction can serve as the basis for a titration. The silver chloride precipitation reaction qualifies. It is rapid, equilibrium is attained quickly at all stages of the titration, the stoichiometry is definite, and side reactions are negligible. Also, both chemical and physical means have been developed for end-point detection. The remaining questions are the completeness of the reaction at the equivalence point and the precision with which the equivalence point can be detected. These can best be answered by examining a specific titration curve. For the case of silver chloride precipitation let us calculate the chloride ion concentration at several points during a titration of 50.00 mL of 0.1 M NaCl with 0.1 M AgNO$_3$. Activity coefficients of unity and complete dissociation of dissolved salts are assumed.

The initial concentration of chloride ion is 0.1 M. The first point on the titration curve, then, will correspond to this value.

If 5 mL of silver nitrate is added, then essentially 5 mL $\times$ 0.1 mmol/mL or 0.5 mmol of AgCl precipitate will have formed. (The amount of AgCl present in solution owing to dissolved precipitate can be assumed negligible.) Since we started with 5 mmol of chloride, then 4.5 mmol must remain in solution. The volume of solution is 50 + 5, or 55 mL. Therefore [Cl$^-$] is 4.5 mmol/55 mL = 8.2 $\times$ 10^{-2} M.

Similar calculations at 25, 40, 45, and 49 mL of added $AgNO_3$ give chloride ion concentrations of 3.3×10^{-2}, 1.1×10^{-2}, 5.3×10^{-3}, and 1.0×10^{-3} M.

At the equivalence point, when 50 mL of silver nitrate has been added, by definition neither chloride nor silver ions can be present in excess. The amount of each in solution is, then, determined solely by the solubility of silver chloride. This can be calculated from the equilibrium

$$AgCl(s) \rightleftarrows Ag^+ + Cl^-$$

for which we can write the expression for the equilibrium constant, here the solubility product K_{sp},

$$K_{sp} = [Ag^+][Cl^-] = 10^{-10}$$

At the equivalence point $[Ag^+]$ must be equal to $[Cl^-]$. Substituting $[Cl^-]$ for $[Ag^+]$ in the solubility-product expression gives

$$[Cl^-][Cl^-] = 10^{-10}$$

$$[Cl^-] = \sqrt{10^{-10}} = 10^{-5} \ M$$

Beyond the equivalence point an excess of silver ion is present. Thus at 51 mL the first 50 mL has undergone reaction to form 5.0 mmol of AgCl, and 1 mL of 0.1 M (0.1 mmol) $AgNO_3$ is present in excess. The volume of solution is $50 + 51$, or 101 mL. Therefore, $[Ag^+] = 0.1$ mmol/101 mL $= 10^{-3}$ M. The value for $[Cl^-]$ can be calculated from the solubility-product expression

$$[Cl^-] = 10^{-10}/[Ag^+] = 10^{-10}/10^{-3} = 10^{-7} \ M.$$

By similar calculations we find at 55 and 60 mL the chloride ion concentration to be 2.1×10^{-8} and 1.1×10^{-8} M.

The first observation to be made is that the chloride ion concentration decreases continuously throughout the titration and dramatically so in the region of the equivalence point.

Calculations in the region near the equivalence point are more complicated because the simplifying assumption that the amount of dissolved silver chloride is negligible compared with the total ion concentrations no longer holds. Strictly speaking, at all volumes before the equivalence point, chloride ions come from two sources: the excess in solution and the slight solubility of silver chloride. In the foregoing calculations we considered only the excess chloride. This is acceptable whenever the chloride ion concentration is appreciable, because the small amount of dissolved silver chloride can be neglected. Near the equivalence point, however, it must be taken into account. Consider, for example, that 49.95 mL of silver nitrate had been added. The rigorous expression, valid at all points during titration, is

$$[Cl^-] = \frac{\text{mmol excess chloride}}{\text{total volume soln}} + \frac{K_{sp}}{[Cl^-]}$$

At 49.95 mL of silver nitrate the concentration of excess chloride is $(0.05 \text{ mL})(0.1 \text{ mmol/mL})/99.95 \text{ mL} = 5 \times 10^{-5} M$. From dissolved silver chloride we must obtain equal numbers of silver and chloride ions. Therefore $[Ag^+]$ from dissolved $AgCl = K_{sp}/[Cl^-]$. Then

$$[Cl^-] = 0.00005 + K_{sp}/[Cl^-]$$

Rearranging, and substituting 10^{-10} for K_{sp}, we have

$$[Cl^-]^2 - 0.00005[Cl^-] - 10^{-10} = 0$$

Solving this quadratic[1] equation for $[Cl^-]$ yields $5.2 \times 10^{-5} M$. Had the chloride ion from the dissolved silver chloride been neglected, the result would have been $5.0 \times 10^{-5} M$. The resulting error of some 4% is not serious in this type of calculation. The nearer one is to the equivalence point, the more important is the inclusion of dissolved silver chloride in the calculation. In general, before the equivalence point

$$[Cl^-] = \frac{(\text{volume excess Cl}^-, \text{mL})(M_{Cl})}{\text{volume soln, mL}} + \frac{K_{sp}}{[Cl^-]}$$

Immediately after the equivalence point a similar exact calculation must be made. The difference now is that the only source of chloride ion is dissolved silver chloride, whereas silver ion comes from two sources: excess titrant and dissolved silver chloride. Accordingly, after the equivalence point

$$[Ag^+] = \frac{(\text{volume excess Ag}^+, \text{mL})(M_{AgNO_3})}{\text{volume soln, mL}} + \frac{K_{sp}}{[Ag^+]}$$

Near the equivalence point the second term cannot be neglected. As before, a quadratic equation can be set up and solved for $[Ag^+]$. From this value for $[Ag^+]$, $[Cl^-]$ can be calculated. For example, at 50.05 mL, we obtain $[Cl^-] = 1.9 \times 10^{-6} M$.

The first two columns of Table 4-1 show the relation between $[Cl^-]$ and the volume of silver nitrate for points from the beginning of the titration to past the equivalence point.

In any scientific investigation a key step is obtaining a valid set of data. With high-quality data, the necessary interpretations and decisions can be made with more reliability and less effort. Much can be learned by inspection

---

[1]The solution for the quadratic expression $ax^2 + bx + c = 0$ is

$$x = \frac{-b \pm \sqrt{b^2 - 4ac}}{2a}$$

Table 4-1. Data for Titration of 50.00 mL of 0.1 M NaCl with 0.1 M AgNO$_3$

Volume AgNO$_3$, mL	[Cl$^-$], M	$-$log [Cl$^-$], pCl	$-$log [Ag$^+$], pAg
0	0.1	1.00	9.00
5	8.2×10^{-2}	1.09	8.91
25	3.3×10^{-2}	1.48	8.52
40	1.1×10^{-2}	1.96	8.04
45	5.3×10^{-3}	2.28	7.72
49	1.0×10^{-3}	3.00	7.00
49.9	1.0×10^{-4}	4.00	6.00
49.95	5.2×10^{-5}	4.28	5.72
49.99	1.6×10^{-5}	4.80	5.20
50.00	1.0×10^{-5}	5.00	5.00
50.01	6.0×10^{-6}	5.22	4.78
50.05	1.9×10^{-6}	5.72	4.28
50.1	1.0×10^{-6}	6.00	4.00
51	1.0×10^{-7}	7.00	3.00
55	2.1×10^{-8}	7.68	2.32
60	1.1×10^{-8}	7.96	2.04

of the data in the first two columns of Table 4-1. It is usually easier, however, to visualize trends after plotting data on a graph. Plotting concentration of chloride against volume of silver nitrate yields a slightly curved line approaching the base line and becoming indistinguishable from it at about 49 mL. No further change would be seen past 49 mL because the chloride ion concentration would be low at that point and remain so. Changes in concentration can be visualized more clearly over wide concentrations by use of an exponential scale on the vertical axis. The simplest way to obtain data for an exponential type of plot is to convert the exponential numbers in the table to numbers analogous to pH. Thus, pCl may be defined as the negative logarithm of [Cl$^-$]: pCl = $-$log [Cl$^-$]. At 5 mL in Table 4-1, for example, pCl = $-$log 8.2×10^{-2} = $-(0.91 - 2)$ = $-(-1.09)$ = 1.09. Similarly, pCl values can be obtained for each of the chloride ion concentrations.

Changes in silver ion concentration may also be conveniently calculated during the titration. The changes are as major as with chloride. Values of $-$log [Ag$^+$], or pAg, can be calculated readily from pCl and the solubility product. We see, then, that the relation between [Ag$^+$] and [Cl$^-$] can be written in several forms, all equivalent:

$$[Ag^+][Cl^-] = 10^{-10}$$

$$\log [Ag^+] + \log [Cl^-] = \log 10^{-10}$$

$$-\log [Ag^+] - \log[Cl^-] = -\log 10^{-10} = 10 = pK_{sp}$$

$$pAg + pCl = 10 = pK_{sp}$$

$$pAg = 10 - pCl = pK_{sp} - pCl$$

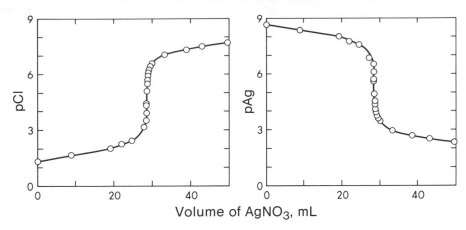

Figure 4-1. Titration curves for 30.00 mL of 0.1 M NaCl with 0.1 M AgNO$_3$, showing pCl against volume and pAg against volume. Data by B. Findlay.

Thus, if pCl = 1.09, pAg = 8.91.

Either pCl or pAg can be plotted on a graph to show the changes that occur during titration. Figure 4-1 shows the titration curves for both. The figure and also Table 4-1 confirm that the most dramatic changes occur at or near the equivalence point. The [Cl$^-$] decreases continuously throughout the titration, while pCl increases. Conversely, pAg decreases continuously throughout the titration. Note also that both [Ag$^+$] and [Cl$^-$] are small but significant at the equivalence point. The kind of calculations illustrated here, and the corresponding graphs, are valuable aids in selecting conditions for detection of the end point and in estimating the errors likely to accompany a particular choice.

With a perfect indicator the end point would appear precisely at a pAg of 5. Since the equivalence-point and end-point volumes would be identical, the volume of titrant at this point would be 50.0 mL and the titration error would be zero. An end point at pAg = 6 would be seen when 49.9 mL of titrant had been added. The titration error now would be 49.9 – 50.0 mL = –0.1 mL, the negative sign denoting an end point appearing before the equivalence point. In a 50-mL volume, 0.1 mL is 2 parts per 1000, or 0.2% error.

Real indicators change color or appearance only gradually. For example, when we attempt to observe an end point at a pAg of 5.0 with a suitable indicator, we might well detect end points beginning before 5.0 and continuing to change after 5.0. The average of several titrations may still be near a pAg of 5, but the titration error would rarely be zero. Although by refinement of technique the pAg range over which end points actually are selected can be restricted, the titration error would still almost never be zero.

Should a titration error of more than 1 part in 1000 be unacceptable, it is necessary to select end points at pAg values neither lower than 4.3 (50.05 mL) nor higher than 5.7 (49.95 mL). Ensuring that end points fall within this pAg range allows little leeway.

The titration curve for chloride varies with the concentration of the

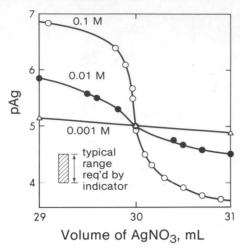

Figure 4-2. Experimental titration curves in the region of the equivalence point for titration of sodium chloride with silver nitrate. Concentration of reactants is 0.1, 0.01, or 0.001 M as indicated. Data by B. Findlay.

reactants. If a 0.01 M chloride solution were titrated with 0.01 M silver nitrate, the initial pCl value would be 2 instead of 1, the equivalence-point value would be 5 as before, and the values after the equivalence point would be lower. Changes near the equivalence point therefore are smaller and more gradual. In general, the more dilute the solutions, the less will be the change in pCl or pAg near the equivalence point. In Figure 4-2 the regions 2% on either side of the equivalence point are shown. Keep in mind that the precision with which an end point can be selected depends strongly on the slope of the titration curve in the region of the equivalence point. When the curve has a steep slope, as for 0.1 M solutions, the precision can be high. The slope becomes less steep as the concentration of the reactants decreases. For 0.001 M solutions, where the slope is extremely gradual, no known indicator gives an end point precise to within 1 part per 1000 of the equivalence point.

Nature of the Reactants

In a precipitation titration the value of the solubility product is an important variable. For the reaction of iodide with silver ion (K_{sp} for AgI equals 10^{-16}) the pI value at the equivalence point is 8 for the titration of 50.00 mL of 0.1 M NaI with 0.1 M AgNO$_3$. At 49.95 mL (1 part per 1000 before the equivalence point) the pI value is 4, and at 50.05 mL (1 part per 1000 past the equivalence point) it is 12. This is a change of 8 units. The comparable change for chloride over this same range is less than 2 units. Figure 4-3 compares titration curves for 0.1 M solutions of silver chloride, bromide, and iodide in the region of the equivalence point. A larger change in pAg (or pBr or pI) means a wider range over which accurate end points can be taken, a broader choice of indicators, and the possibility of titrating more dilute solutions.

Other conditions being equal, the steepness of the slope of the titration curve increases as the solubility product decreases. The calculations become somewhat more complex when other than simple 1:1 reactions are involved. With calcium fluoride, for example, the solubility-product constant involves

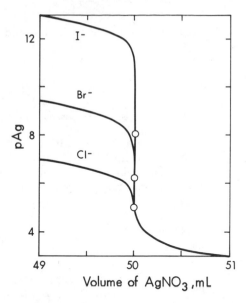

Figure 4-3. Titration curves near the equivalence point for 0.1 M NaI, NaBr, and NaCl with 0.1 M AgNO$_3$.

the square of the fluoride ion concentration. Solubility, and therefore the completeness of a titration, may also depend on pH or on the presence of complex-forming reagents. Thus, if the precipitate involves the anion of a weak acid, then the solubility depends on acid-base equilibria (Chapter 5) in addition to the precipitation equilibrium. Similarly, if the precipitate includes a cation which can form a complex with a substance in solution, then the solubility will depend partly on complexation equilibria (Chapter 7).

4–2 END–POINT DETECTION: METHODS AND INDICATORS

The end point in a precipitation titration can be detected in a variety of ways. In this section we consider the most important methods for the silver halides.

Fractional Precipitation. The Mohr Method

The *fractional precipitation* method of end-point detection, first proposed by Mohr in 1856, is one of the oldest still in use. It is based on the formation of an orange-red precipitate of silver chromate at the end point. When a mixture of chloride and chromate is titrated with silver nitrate, silver chloride precipitates first, being the less soluble. After the chloride in solution has precipitated, the first excess of silver ion precipitates silver chromate to mark the end point.

EXAMPLE 4–1

If a standard solution of silver nitrate is added to a solution 0.01 M in chloride ion and 0.002 M in chromate ion, silver chloride will begin to precipitate when

$$[Ag^+] = 10^{-10}/[Cl^-] = 10^{-10}/10^{-2} = 10^{-8} \ M$$

The expression for the solubility product for silver chromate, Ag_2CrO_4, is

$$[Ag^+]^2[CrO_4^{2-}] = K_{sp} = 2 \times 10^{-12}$$

Silver chromate therefore will begin to precipitate when

$$[Ag^+] = \sqrt{2 \times 10^{-12}/[CrO_4^{2-}]} = \sqrt{2 \times 10^{-12}/0.002} = 3 \times 10^{-5} \ M$$

Thus, addition of silver nitrate to a solution containing chloride and chromate results in precipitation of silver chloride long before silver chromate. At the equivalence point for silver chloride the silver ion concentration is 10^{-5} M. The start of silver chromate precipitation, then, is beyond the equivalence point. In addition, the first traces of silver chromate formed are not visible; a moderate concentration must be present before a red color is discernible. Together, these two factors make the volume at the end point larger than that at the equivalence point.

The concentration of chromate needed for the precipitation of silver chromate to just begin at the equivalence point in a chloride titration can be calculated from the solubility-product expression to be $2 \times 10^{-12}/(10^{-5})^2$, or 0.02 M. Chromate solutions, however, are of a sufficiently intense yellow to reduce the precision with which the first color of silver chromate can be detected. Common practice is therefore to use a chromate concentration of about 0.002 M and to determine an *indicator blank* to take into account the difference between the end-point and equivalence- point volumes. A *blank* titration is one in which all the reagents are added but the sample is not. It compensates for errors due to titratable impurities in the reagents and to late end points, but seldom for those due to early end points. The best way to correct for errors resulting from early end points and from some titratable impurities in reagents is to standardize the titrant against a pure sample of the same material.

The indicator blank in the Mohr titration is determined by measuring the volume of silver nitrate required to form the first visible silver chromate in a solution of the same volume as in a titration, but where the silver chloride precipitate is replaced by a suspension of a solid (usually calcium carbonate) having similar properties.

The Mohr end point is taken in a solution having a pH of about 8 or 9. At lower pH values protonation of CrO_4^{2-} occurs:

$$H^+ + CrO_4^{2-} \rightleftarrows HCrO_4^-$$

The ionization constant for this reaction is about 10^{-7}. Hence at low pH the chromate ion concentration is reduced substantially, and a larger silver ion concentration will be required to initiate precipitation. If the acidity is too high, part of the chromate is converted to dichromate, and the end point is late. If the solution is too basic, silver oxide begins to form. The pH therefore must be adjusted to between 6 and 10, typically by addition of sodium bicarbonate.

The end-point error in the Mohr method is appreciable, depending in part on the amount of silver chloride precipitate. Increasing amounts of the precipitate obscure the first traces of silver chromate. Errors can be reduced by standardizing the silver nitrate against primary-standard sodium chloride. If the amounts of chloride present in the standard and sample titrations are about equal, end-point errors will cancel each other. When silver nitrate is used directly as a standard, the error can be reduced by titration of a blank containing a white insoluble compound such as calcium carbonate to simulate the silver chloride precipitate.

Adsorption Indicators. The Fajans Method

Adsorption-indicator methods were first proposed by Fajans and coworkers in 1923. A visible color change is produced when an appropriate dye is adsorbed on either a positively or negatively charged surface of a precipitate. The fluorescein anion, for example, will be adsorbed on positively charged silver chloride, and the rhodamine 6G cation on negatively charged silver bromide. The color of the indicator adsorbed on the surface is different from that in solution.

Dichlorofluorescein, a weak organic acid, may be used for the detection of the end point in the titration of chloride with silver. The

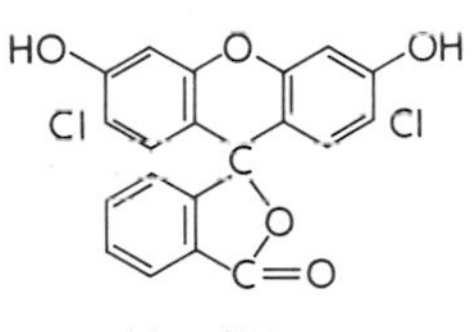

Dichlorofluorescein 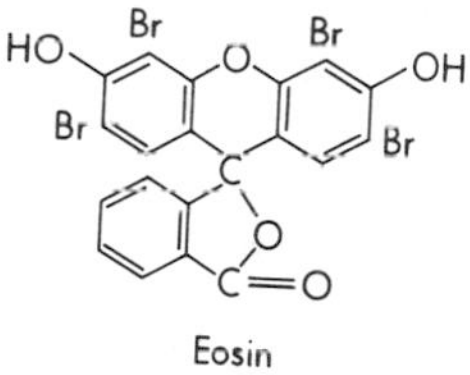   Eosin

dichlorofluoresceinate anion tends to be adsorbed on the surface of precipitated silver chloride, but less strongly than chloride. In the first stages of a titration, chloride ions in solution are adsorbed preferentially on the surface of the precipitate, as shown in Figure 4-4. As the titration proceeds the chloride ion concentration diminishes until it approaches a small value at the equivalence point. After the equivalence point the slight excess of silver ion tends to be strongly adsorbed on the surface of the precipitate. The

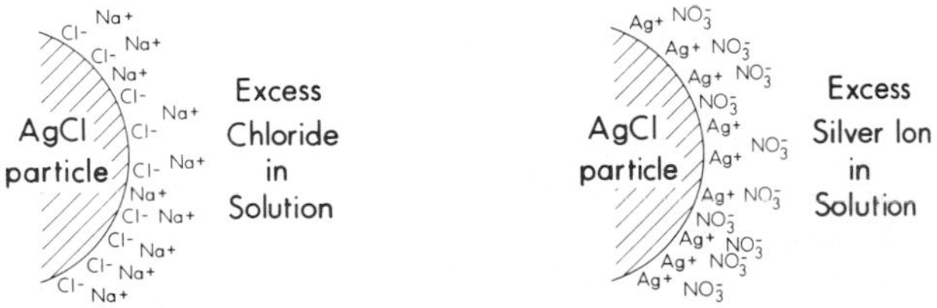

Figure 4-4. Schematic diagram to illustrate adsorption giving the surface of silver chloride either a negative charge, or a positive charge.

attraction between adsorbed silver ions and dichlorofluoresceinate anions causes formation of a pink adsorbed layer of silver dichlorofluoresceinate on the precipitate, which is taken to be the end point. Silver ions do not interact with the dye in the bulk of the solution. Because the color change occurs on the surface of the precipitate, it is of course more distinct when the surface area is large. Certain compounds, such as dextrin, prevent precipitates of silver halides from coagulating near the equivalence point and provide the needed surface area.

Before the equivalence point the excess of chloride present will be adsorbed to give the surface of the precipitate a negative charge. After the equivalence point adsorbed silver ions give the surface a positive charge. In passing through the equivalence-point region during a titration the surface charge switches from negative to positive, and an abrupt color change is observed on the precipitate when the dichlorofluoresceinate ions are adsorbed by the process of *secondary adsorption*.

Since dichlorofluorescein is a weak acid, at low pH values only a small proportion is present as the anion; the protonated species does not function as an indicator. An acetate buffer is normally added to control the pH during titration. The upper pH limit, about 10, is determined by the precipitation of AgOH.

The sensitivity of silver halides to light is enhanced by adsorbed indicator:

$$2AgCl \rightarrow 2Ag(s) + Cl_2(g)$$

Exposure to light, especially direct sunlight, during absorption- indicator titrations should be kept low.

Tetrabromofluorescein (eosin), a stronger acid than either fluorescein or dichlorofluorescein, is useful in solutions as acidic as pH 2. Eosin anions are so strongly adsorbed that they displace chloride ions from silver chloride precipitates, and no end point is observed. Tetrabromofluorescein is a useful adsorption indicator for bromide, iodide, and thiocyanate titrations.

Detection by Complex Formation. The Volhard Method

In the method of *detection by complex formation*, proposed by Volhard in 1874, the amount of silver in solution can be determined accurately by titration with potassium thiocyanate as the precipitant. For the determination of halides the procedure involves the addition of a measured excess of standard silver nitrate solution followed by titration with thiocyanate. The indicator is iron(III) ammonium sulfate. For the analytical reaction

$$Ag^+ + SCN^- \rightleftarrows AgSCN(s)$$

the first excess of thiocyanate ion is detected when the indicator reaction is

$$Fe^{3+} + SCN^- \rightleftarrows FeSCN^{2+}$$

This deep-red complex becomes visible in solution at about 10^{-5} M. The titration error is therefore usually negligible. Since iron(III) readily hydrolyzes, the titration must be carried out at low pH.

The necessity for the thiocyanate ion concentration to be about 10^{-5} M before the color change appears causes some difficulty. The solubility product for silver thiocyanate is about 10^{-12}, which means that at the end point the silver ion concentration must be about $10^{-12}/10^{-5} = 10^{-7}$ M. If, however, chloride ion were being determined by the indirect Volhard method, the solution would be saturated with silver chloride, and the chloride ion concentration would be $10^{-10}/10^{-7} = 10^{-3}$ M at equilibrium. Since for silver chloride the equivalence-point concentration of chloride is 10^{-5} M, considerable error in the chloride determination would result from the back titration of excess silver. This error can be reduced by the addition of nitrobenzene to coat the silver chloride particles and prevent the rapid establishment of equilibrium. With nitrobenzene as an aid, accurate results for chloride are obtained when the concentration of chloride is above about 0.02 M.

Other Methods of End-Point Detection

A wide variety of physical methods can be utilized for the detection of end points in precipitation titrations, including potentiometric (Section 11-8), conductometric (Section 12-5), amperometric (Section 12-3) , and thermometric. Additional discussions of these topics are available in advanced textbooks.

4-3 APPLICATIONS

Since chemical analysis is a dynamic field, the relative importance of methods changes with time. Many titrimetric precipitation methods, largely developed more than a century ago, have been displaced by faster or better procedures. As an example, the precipitation titration of zinc with ferrocyanide has been replaced by a complexation titration with EDTA. Nevertheless, the precipitation titration of chloride continues to play a significant role in organic elemental analysis and for the determination of inorganic chloride and silver. Other examples are the determination of bromide and iodide with silver nitrate, of silver with potassium thiocyanate, of sulfate with barium chloride, and of fluoride with thorium nitrate.[2]

4-4 ARGENTIMETRIC DETERMINATION OF CHLORIDE

Procedure (time 5 to 6 h)

Preparation of 0.01 M AgNO_3

Weigh accurately on an analytical balance a clean, dry 50-mL beaker. Transfer it to a triple-beam balance, and weigh into it enough $AgNO_3$ to prepare

---

[2]Escalation of the cost of silver-containing reagents may restrict the application of some excellent methods of analysis.

500 mL of 0.01 M solution.[3] Reweigh the beaker plus compound on the analytical balance. Dissolve the $AgNO_3$ in a small portion of chloride-free distilled water.[4] Transfer this solution quantitatively to a clean 500-mL volumetric flask with the aid of a funnel, and dilute to volume with distilled water. Mix well and store protected from light. Check the concentration of this solution by titration of aliquots of a standard solution of potassium chloride, as directed in the following procedure.

Preparation of Sample Solution

Weigh, to the nearest 0.1 mg in a 50-mL beaker, from 0.5 to 0.6 g of a dry[5] sample, and dissolve it in a small volume of distilled water. Transfer the solution quantitatively to a 250-mL volumetric flask, dilute to volume, and mix well. Pipet 10-mL aliquots (Section 2-10) of this solution into each of several 200-mL conical (Erlenmeyer) flasks, using a calibrated pipet. Add about 10 mL of distilled water to each flask.

Preparation of 0.030 M Potassium Chloride

Weigh, to the nearest 0.1 mg in a 50-mL beaker, from 0.6 to 0.7 g of primary-standard potassium chloride, and dissolve it in a few milliliters of distilled water. Transfer the solution quantitatively to a 250-mL volumetric flask, dilute to volume with distilled water, and mix well. Pipet 10-mL aliquots of the standard potassium chloride solution into a set of 200-mL conical flasks, using a calibrated pipet. Add about 10 mL of water to each flask.

Titrate the standards and samples with silver nitrate solution according to either the Mohr or the Fajans procedure as directed by the laboratory instructor. We recommend that titrations of aliquots of standard and unknown samples be carried out in alternation (Section 3-2).

Mohr Titration Procedure

To obtain experience with the Mohr end point, first determine an indicator blank value. Place about 0.5 g of powdered $CaCO_3$ and about 0.5 g of $NaHCO_3$ in 10 mL of distilled water.[6] Add 1 mL of 5% K_2CrO_4 solution, and then titrate dropwise with standard silver nitrate solution to the first permanent appearance of red-orange in the yellow chromate solution.[7] Titrate two or three blanks and record the values.

To each of the aliquots of potassium chloride standard and unknown sample

---

[3]The weight of $AgNO_3$ taken should be within 5% of the amount required to make 500 mL of 0.01 M solution. Since $AgNO_3$ is corrosive, it must not be handled near an analytical balance. If the solution is standardized against KCl as described later, precise preparation is unnecessary. For beginning workers, however, it affords a useful check on technique.

[4]Chloride-free distilled water should be used throughout this experiment. Anionic impurities in water are most readily removed by passage through a column containing an anion-exchange resin in the hydroxide form. If desired, both anionic and cationic impurities can be removed in one step by use of a mixture of cation-exchange resin in the hydrogen form and anion-exchange resin in the hydroxide form.

[5]Consult the laboratory instructor concerning drying of samples.

[6]The $CaCO_3$ simulates the white AgCl precipitate. If no precipitate is present in the blank titration, the end point appears from 0.02 to 0.10 mL early.

[7]On the first blank, practice identifying the end point by titrating to the apparent end point and then adding a drop or two more titrant to observe the effect. Next add a crystal or two of NaCl to the flask, swirl, and titrate again. Go back and forth over the end point several times in this way until the color change is familiar.

add about 0.5 g of $NaHCO_3$ and 1 mL of 5% K_2CrO_4 solution.[8] Titrate with standard $AgNO_3$ solution, swirling the flask continuously to prevent a local excess of Ag_2CrO_4 from being occluded by the AgCl precipitate. The end point is the first permanent appearance of red-orange in the yellow chromate solution.[9]

Fajans Titration Procedure

To each of the aliquots of potassium chloride standard and unknown sample add about 1 mL of 0.5 M acetic acid–0.5 M sodium acetate buffer and 0.1 g of dextrin. Titrate with $AgNO_3$ solution to within about 1 mL of the end point, and then add 5 drops of 0.1% dichlorofluorescein solution.[10] Immediately continue the titration until the precipitate becomes pale pink, swirling the flask constantly. No blank correction is necessary.

Calculate the percentage of chloride in the sample, using (a) the molarity of the silver nitrate based on the titrations of the standard potassium chloride aliquots and (b) the molarity based on the weight of silver nitrate taken. On the basis of your results submit a report of the analysis.

Calculations

Because a large number of poor results are due to errors in the calculations, all should be checked. Many hours of careful laboratory work can be forfeited because of a simple mistake in arithmetic or an incorrect setup for the calculations. In this and other early experiments the necessary calculations are outlined in detail. The directions in later experiments become less detailed in the expectation that the calculations can be performed with only an occasional suggestion. It is important to understand the reasons behind every step in every calculation.

EXAMPLE 4-2. (See also Section 3-3.)

$$\% \; Cl = \frac{wt \; Cl}{wt \; sample \; in \; aliquot} (100)$$

$$wt \; Cl = (mol \; Cl)(g \; Cl/mol)$$

Since one mole of Ag^+ reacts with one mole of Cl^-,

$$wt \; Cl = \frac{(mL \; AgNO_3)(M \; AgNO_3)(35.45 \; g/mol)}{1000 \; mL/L}$$

---

[8]If the samples are neutral, pH adjustment is unnecessary. Because H_2CO_3 may be present in the solution through absorption of CO_2, $NaHCO_3$ is added to raise the pH above 6.

[9]If a gradual indicator change makes end-point selection uncertain, record the buret reading. Then add another drop and note and record any change; repeat this procedure. Select as the end point the reading corresponding to the drop that causes the greatest change.

[10]Dichlorofluorescein promotes photoreduction of silver in the AgCl precipitate. The resulting finely divided dark silver obscures the end point. Addition of the indicator near the equivalence point reduces this difficulty. If the approximate location of the end point is unknown, indicator must be added early in the first titration, and the approximate end-point volume determined. To lessen photodecomposition, which may cause error even when indicator is added late in the titration, avoid direct sunlight on the titration flask.

$$\% \ Cl = \frac{(mL \ AgNO_3)(M \ AgNO_3)(35.45 \ g/mol)}{(1000 \ mL/L)(wt \ sample \ in \ aliquot)}(100)$$

The weight of sample in the aliquot taken for each titration is given by

$$(initial \ wt \ sample) \ \frac{calibration \ vol \ of \ 10\text{-}mL \ pipet}{250 \ mL}$$

Note that the molarity is expressed in moles per liter rather than millimoles per milliliter, and the formula weight of chloride in grams per mole rather than milligrams per millimole. Accordingly, the volume must be converted to liters by dividing by 1000. This convention is used in this example; other combinations of units will be seen elsewhere. No single, exclusive method must be followed — use the one you find most convenient. A check of units at the end will frequently reveal errors. Such a check in the first part of this example gives

$$\% \ Cl = \frac{(mL)(mol/L)(g/mol)}{(mL/L)(g)}(100)$$

The molarity of the $AgNO_3$ solution based on the weight taken is calculated from the moles $AgNO_3$ and volume of solution:

$$mol \ AgNO_3 = \frac{g \ AgNO_3}{g \ AgNO_3/mol} = \frac{wt \ AgNO_3}{169.87}$$

$$L \ AgNO_3 = 500.0 \ mL/(1000 \ mL/L)$$

$$M \ AgNO_3 = \frac{g \ AgNO_3}{(169.87 \ g/mol)(0.5000 \ L)}$$

The molarity of the $AgNO_3$ solution based on the titration of aliquots of standard KCl solution is obtained from

$$M \ AgNO_3 = \frac{(M \ KCl)(mL \ KCl)}{(mL \ AgNO_3)}$$

The molarity of the KCl solution based on the weight of salt taken is calculated in the same way as that for the $AgNO_3$ solution:

$$M \ KCl = \frac{g \ KCl}{(g \ KCl/mol)(0.250 \ L)}$$

An estimate should be made of the combined uncertainties associated with the sample, procedure, reagents, and technique to guide your choice of the appropriate number of significant figures to report in the result. In this experiment, uncertainties of about 1 part per 1000 may reasonably be ascribed to the procedure and reagents. Although the uncertainty in the composition of the sample is unknown, samples for instructional purposes are usually highly uniform, and an uncertainty of less than 1 part per 1000 is reasonable (say 0.5/1000).

Considering the uncertainties in the standardization, we have the uncertainty in the weight of 0.85 g of $AgNO_3$, estimated to be 0.3 mg, or 0.3/0.85 ppt; the uncertainty in the weighing of the 0.5-g sample, estimated to be 0.3/0.5 ppt; and the uncertainty in the taking of an aliquot, estimated to be about 1 ppt. The uncertainty in the end point for this titration is estimated to be 0.8 ppt. The uncertainty in the final result due to random error is then the square root of

$$\sigma^2_{sample} + \sigma^2_{wt\ AgNO_3} + \sigma^2_{wt\ sample} + \sigma^2_{aliquot} + \sigma^2_{end\ point}$$

$$= (0.5)^2 + (0.3/0.85)^2 + (0.3/0.5)^2 + (1)^2 + (0.8)^2 = 2.4$$

The square root of 2.4 is 1.5 ppt.

4–5   TITRIMETRIC DETERMINATION OF SULFATE WITH BARIUM USING AN ADSORPTION INDICATOR

Background

The determination of sulfate is required in countless situations. Unfortunately, few rapid and accurate methods are available. Although the classic gravimetric method involving precipitation of barium sulfate is slow and subject to errors from coprecipitation, few reactions other than with the barium ion have been exploited. One widely used titrimetric method for small amounts of sulfate is the precipitation titration with barium perchlorate in an alcohol–water mixture with the use of an adsorption indicator. The optimum conditions are a pH of 3 to 4 in 80% 2-propanol–20% water, with dilute standard barium perchlorate as titrant in the same solvent. In contrast with the dense crystals precipitated from water, the barium sulfate precipitated from this solvent mixture is flocculent and has a large surface area suitable for proper functioning of the adsorption indicator thorin.

Thorin, 2-(2-dihydroxy-3,6-disulfo-1-naphthylazo) benzenearsonic acid, has the structure

In the pH region 3 to 4 both of the sulfonic acid protons and one of the arsonic acid protons are dissociated from the molecule, and the anion is adsorbed onto the barium sulfate precipitate when an excess of barium ions is present. This behavior is analogous to that of dichlorofluorescein in the chloride titration with silver nitrate. The color change is from yellow in solution to pink after adsorption.

Coprecipitation of ions other than sulfate, a serious problem in the gravimetric determination of sulfate, also causes inaccuracies in this titration (Section 8-2). Most cations coprecipitate with sulfate, yielding low results;

potassium ions are particularly troublesome. Sodium and ammonium ions coprecipitate to a smaller extent, but still on the order of 2 to 3% when the concentrations of the interfering ions are comparable to that of sulfate. Magnesium ions coprecipitate little, and therefore magnesium acetate may be used to adjust the pH of acidic samples prior to titration. Some metal cations also interfere, forming colored complexes with thorin that block proper indicator action.

Interfering cations can be removed if necessary by passage of a solution of the sample through a column containing a cation-exchange resin in the hydrogen form (Section 19-1). Anions in the sample may also cause errors in the determination of sulfate by coprecipitation as their barium salts. Chloride, bromide, and perchlorate cause only slight error, whereas phosphate, fluoride, and nitrate cause large error.

A method recommended for the accurate determination of sulfur dioxide or sulfur trioxide in the atmosphere involves drawing a measured volume of air through a solution of dilute hydrogen peroxide to produce sulfate, followed by a microtitration of the sulfate with dilute barium perchlorate incorporating thorin indicator. Sulfur in organic compounds may be determined by burning the compounds in a flask containing an atmosphere of oxygen over a solution of hydrogen peroxide (Schöniger oxygen-flask method). The sulfur dioxide produced during combustion reacts with peroxide to form sulfate ions, which can then be titrated.

Procedure (time 3 to 6 h)

Preparation of 0.01 M Barium Perchlorate Solution

On a triple-beam balance weigh 2 ± 0.1 g of barium perchlorate trihydrate into a clean weighing bottle. Place the bottle with contents in an oven at 150 to 160°C for at least 4 h.[11] Allow it to cool to near room temperature, in a desiccator if necessary, and then stopper it. Weigh the bottle and contents to the nearest 0.1 mg, and then pour the contents into a clean 250-mL beaker. Do not attempt to transfer the last traces. Immediately restopper the bottle and reweigh it. Add 10 to 20 mL of distilled water to the beaker, dissolve the salt, and transfer quantitatively to a 500-mL volumetric flask; up to 50 mL of distilled water may be used to aid in the transfer. Add 400 mL of 2-propanol, swirl gently to mix, and then ensure that the pH[12] is between 3.5 and 4. If not, adjust it to within this range by dropwise addition of 1 M HClO$_4$ or 1 M Mg(OAc)$_2$.[13] Dilute to volume with distilled water and mix.

---

[11]Anhydrous barium perchlorate takes up water readily to form a series of hydrates. The commercial material is usually the trihydrate, Ba(ClO$_4$)$_2\cdot$3H$_2$O, from which the water must be removed by drying over the range 100 to 250°C. A drying temperature of 160°C is recommended here, though 110°C will suffice. The dried material can be used as a standard if an uncertainty of 0.5% can be tolerated; for microanalyses this is normally acceptable. Should greater accuracy be needed, the barium perchlorate titrant should be standardized against a standard sulfuric acid solution or against weighed portions of pure, dry sodium sulfate that have been dissolved in water and passed through a cation-exchange column to produce sulfuric acid.

[12]The pH of a solution containing water plus another solvent cannot be related directly to values in water alone. The values used here are relative, and are measured in the water–2-propanol mixture with a pH meter–electrode system previously calibrated with aqueous buffers.

[13]Magnesium acetate rather than sodium hydroxide is used here to neutralize excess acidity because the magnesium ion is the less strongly coprecipitated.

Preparation of Sample

When a sample contains metal sulfate salts, removal of metal ions by passage through a cation-exchange resin may be necessary. For this purpose, perform the procedure outlined in Section 19-7 prior to titration. When the sample contains no interfering metals, dilute it to volume in a volumetric flask so that a 10-mL aliquot will require about 30 to 40 mL of barium perchlorate titrant (0.3 to 0.4 mmol per aliquot).

Titration Procedure

Transfer a 10-mL aliquot of unknown sample to a 200-mL conical flask. Add 50 mL of 2-propanol and 1 or 2 drops of 0.2% thorin indicator. Adjust the pH to 3.5 to 4 by dropwise addition of 1 M perchloric acid or magnesium acetate. Titrate with 0.01 M barium perchlorate solution, swirling the flask vigorously. The end point is the first permanent change from yellow or yellow-orange to pink. Titrate moderately rapidly until near the end point, indicated by the appearance of pink where the titrant enters the solution. Then proceed dropwise, giving the system time to equilibrate between increments.

Calculate the grams of sulfate present in the sample, and submit a report of the analysis.

Calculations

The molar concentration of the barium perchlorate titrant is obtained from

$$M \, Ba(ClO_4)_2 = \frac{g \, Ba(ClO_4)_2}{[\text{mol wt } Ba(ClO_4)_2](0.500 \text{ L})}$$

The molecular weight of $Ba(ClO_4)_2$ is 336.24.

Barium ion reacts with sulfate ion in a 1:1 ratio. Therefore, the total weight in grams of sulfate in a solution is given by

$$g \text{ sulfate} = \frac{[\text{mL } Ba(ClO_4)_2] \, (M_{Ba(ClO_4)_2})(96.06 \text{ g/mol})(\text{total vol soln})}{(1000 \text{ mL/L})(\text{vol aliquot})}$$

PROBLEMS (See Appendix for solubility-product constants)

4-1. In the chloride titration with standard silver nitrate by the Mohr (chromate indicator) method, why is it desirable to standardize the silver nitrate with primary-standard potassium chloride even though pure silver nitrate is itself a primary standard?

4-2. A 0.3297-g sample of a mixture of potassium nitrate and sodium chloride required 44.25 mL of 0.0995 M silver nitrate for titration. What was the percentage of chloride present?

4-3. What concentration of silver nitrate would be required to give a solution equivalent to 1 mg Cl/mL; 1 mg Br/mL; 1 mg I/mL?

4-4. What volume of 0.1009 M silver nitrate is required to titrate a 0.2212-g sample of pure potassium bromide?

4-5. A sample of pure lithium iodide required 42.89 mL of 0.1014 M silver nitrate for titration. What was the weight of the sample?

4-6.   In the titration of 50 mL of 0.01 M NaCl with 0.01 M AgNO$_3$, what is pCl after 49 and 49.99 mL of titrant? What is pCl at the equivalence point?

4-7.   A 25.00-mL sample of 0.01000 M sodium chloride solution is titrated with 0.0200 M silver nitrate.
a. Calculate the concentration of chloride ion in solution — in molarity and in pCl — after 5, 10, 15, and 25 mL of silver nitrate solution has been added.
b. Calculate [Cl$^-$] and the pCl at the equivalence point of the titration.
c. Plot pCl against volume of silver nitrate for the calculated points and draw the titration curve.

4-8.   In a titration of 25 mL of a 0.07634 M solution of potassium chloride with 0.0500 M silver nitrate, what is the pAg after the addition of 1, 10, 25, 35, 36, 37, 38, and 40 mL of the standard solution? What is the pAg at the equivalence point? Plot pAg against volume of silver nitrate, and sketch the titration curve.

4-9.   In a titration of 0.1 M sodium bromide with 0.1 M silver nitrate, what is the pAg after 1, 10, 50, 90, 99, 99.9, 100, 100.1, 101, and 110% of the stoichiometric amounts of silver nitrate?

4–10.†   The known chloride content of a sample is in the range 30 to 40%. What are the minimum and maximum sample sizes that may be taken if the volume of 0.1000 M silver nitrate used in the titration is to be in the range 35 to 50 mL?

4-11.   In the titration of 25 mL of 0.05 M sodium bromide with standard 0.05 M silver nitrate, calculate pBr after the addition of 1, 5, 20, 24, 25, 26, and 30 mL of the silver nitrate solution. Draw the titration curve.

4-12.   In the titration of 80 mL of 0.03 M silver nitrate with 0.1 M potassium thiocyanate, calculate the pAg after the addition of 1, 5, 20, 23, 24, 25, and 30 mL of the potassium thiocyanate solution. Draw the titration curve.

4-13.   What concentration of silver nitrate would be required for 1 mL of titrant to represent 0.1% of chloride in 10-mL aliquots of a salt solution?

4-14.   A 0.2246-g sample of pure sodium chloride was dissolved and diluted to 100 mL. A 10.006-mL aliquot required 36.74 mL of silver nitrate titrant to reach the end point. What was the concentration of the silver nitrate?

4–15.†   Biphenyl, C$_6$H$_5$C$_6$H$_5$, is chlorinated to produce a PCB (polychlorinated biphenyl). A 0.2436-g sample is decomposed, and the chloride titrated with silver nitrate. In the titration, 44.06 mL of 0.06432 M solution is required. What is the molecular weight of the PCB? What is its formula?

4-16.   Would the results of chloride analysis be high or low in a Mohr titration if an indicator blank were not applied? Explain.

4-17.   A chemist wished to find the volume of a large barrel and had no equipment for measuring the amount of water needed to fill it. He added 440 g of sodium chloride to the barrel and filled it with water. After the solution was mixed, a 100.0-mL sample required 36.66 mL of 0.0505 M silver nitrate for titration. What was the volume of the barrel?

4-18.   Sea water contains about 2.9% sodium chloride. For a 5-mL sample, what should be the molarity of a silver nitrate solution if the titration volume is to be about 35 mL?

4-19.   A 10.006-mL sample of a salt solution requires 25.67 mL of 0.1006 M silver nitrate for titration, with chromate as indicator. The indicator blank is 0.06 mL. What is the percentage of sodium chloride in the solution if its density is 1.024 g/mL?

4-20.   A mixture contains only sodium chloride and potassium bromide. Titration of the total halide in a 0.3456-g sample of the mixture required 36.66 mL of 0.1111 M silver nitrate.

What is the composition of the mixture?

4-21.   What is the chloride ion concentration in a 0.002 M potassium chromate solution that contains both solid silver chloride and solid silver chromate?

4-22.   What is the bromide ion concentration in a 0.002 M solution of potassium chromate that contains both solid silver bromide and solid silver chromate? What would be the iodide concentration if solid silver iodide were also present?

4-23.   If the pH of a 0.002 M solution of silver nitrate is raised by the addition of sodium hydroxide, at what pH will a precipitate of silver hydroxide first form? What is the maximum pH for a titration of chloride by the chromate indicator method? Explain.

4-24.   Calculate the theoretical concentration of chromate ion needed to make the end point and the equivalence point coincide in the titration of a 0.05 M solution of potassium bromide with 0.05 M silver nitrate.

4-25.   A solid sample is known to contain only sodium hydroxide, sodium chloride, and water. A 2.6835-g portion is dissolved in water and diluted to volume in a 100-mL volumetric flask. A 9.993-mL aliquot of this solution requires 33.56 mL of 0.02132 M hydrochloric acid when titrated to the phenolphthalein end point. A second aliquot, titrated with 0.05000 M silver nitrate by the adsorption-indicator method requires 49.18 mL to reach the end point. What is the composition of the original sample?

4-26.   Within what pH range must the titration of chloride by the Fajans (adsorption indicator) method be performed? Would an ammonia–ammonium chloride buffer be suitable in this titration? Explain.

4-27.   Explain why, with an adsorption indicator, a large surface area favors a sharp end point.

4-28.   Calculate the theoretical concentration of chromate ion needed to make the end point and the equivalence point coincide in the titration of a 0.05 M solution of potassium thiocyanate with 0.05 M silver nitrate.

4-29.   A 0.4560-g sample was analyzed for chloride by back titration; 50.06 mL of 0.05236 M silver nitrate was added, and 12.22 mL of 0.06212 M thiocyanate was required for titration of the excess. Calculate the percentage of chloride present.

4-30.   A 10.47-L sample of air was passed through a hydrogen peroxide solution. The sulfuric acid formed was titrated with 0.01072 M barium perchlorate, requiring 8.44 mL to reach the end point. Calculate the weight of SO_2 in the air sample. Calculate the milligrams per liter of SO_2 in the air sample.

4-31.   A solution containing sulfate is diluted to volume in a 100-mL volumetric flask. A 10.00-mL aliquot of the solution requires 8.56 mL of 0.01032 M barium perchlorate to reach the end point. How much sodium sulfate was in the sample?

4-32.   Sulfur in coal is determined by combustion of the sample, absorption in water of the sulfur trioxide formed, and titration of the sulfate with standard barium chloride solution. If a 0.7224-g sample of coal is analyzed for sulfur and 14.21 mL of 0.0632 M barium chloride is required to titrate the sulfate produced, what is the percentage of sulfur in the coal?

4-33.   Zinc can be titrated by ferrocyanide according to the reaction
$$3Zn^{2+} + 2K^+ + 2Fe(CN)_6^{4-} \rightarrow K_2Zn_3(Fe(CN)_6)_2(s)$$
If 28.61 mL of 0.1111 M $K_4Fe(CN)_6$ is required to titrate the zinc in a 0.5522-g sample of smithsonite ore ($ZnCO_3$), what is the percentage of zinc in the ore?

4-34.   Calculate the percentage of fluoride in a 0.2436-g sample that requires 25.55 mL of a 0.02112 M solution of thorium nitrate when the reaction is
$$Th^{4+} + 4F^- \rightarrow ThF_4(s)$$

REFERENCES

K. Fajans and O. Hassel, *Z. Elektrochem.* **1923,** *29,* 495. Original paper on the use of fluorescein as adsorption indicator for the titration of silver chloride. Dichlorofluorescein was suggested a few years later by Kolthoff, Lauer, and Sunde as being somewhat preferable to fluorescein in solutions of pH less than 4 because it is a stronger acid.

J. S. Fritz and S. S. Yamamura, *Anal. Chem.* **1955,** *27,* 1461. The original paper describing the method for sulfate using thorin indicator, applying it as a microtitration for the determination of sulfate in water samples.

M. Katz, Ed., *Methods of Air Sampling and Analysis,* 2nd ed., Am. Public Health Assoc., Washington, D.C., 1977, Method 711, p 737. Describes the application of the microtitration procedure of Fritz and Yamamura to the determination of sulfur dioxide and sulfur trioxide in air.

I. M. Kolthoff, E. B. Sandell, E. J. Meehan, and S. Bruckenstein, *Textbook of Quantitative Inorganic Analysis,* 4th ed., Macmillan, New York, 1969, pp 719 and 796.

H. A. Laitinen and W. E. Harris, *Chemical Analysis,* 2nd ed., McGraw-Hill, 1975, Chapter 7. Treats in detail the solubility of precipitates.

F. Mohr, *Annalen* **1856,** *97,* 335. Original paper on chromate indicator for chloride titration.

R. M. Ramette, *J. Chem. Educ.* **1960,** *37,* 348. The experimental determination of equilibrium constants for the silver chloride system, including the formation of $AgCl_2^-$ at high ratios of chloride to silver.

5

ACID-BASE EQUILIBRIA
AND TITRATIONS

Having extracted the whole of the saline matter from the several Potashes, it was judged that the quantity of true alkali in the salts might be discovered by their power of saturating acids, compared with that of an alkali of known purity; and this method succeeded so well, that it is hereafter proposed for the assaying of Potashes.

From *Experiments and Observations on American Potashes,*
with an Easy Method of Determining their Respective Qualities,
by William Lewis, London, 1767.
(Quoted in F. Szabadvary, *Histsory of Analytical Chemistry,* p. 202)

5-1 REVIEW OF FUNDAMENTAL CONCEPTS

Water as Solvent

Acids and bases were defined and described by a number of early chemists, including Boyle, Lavoisier, Davy, Berzelius, Liebig, and Arrhenius. Because most acid and base reactions take place in water as solvent, some of the structural and solvent properties of this extraordinary liquid are discussed here.

A single water molecule is almost spherical in shape, with a radius (0.138 nm) only slightly larger than that of the oxygen atom (0.135 nm).[1] The two hydrogen atoms are positioned at an angle of 105° from each other, and in the vicinity of their nuclei the water molecule is positively charged. On the opposite side of the molecule the two pairs of electrons that are not involved in the formation of hydrogen–oxygen bonds produce a region of negative charge. The result is that a water molecule has a positive and a negative end. Such a species is termed a *dipole*. Attraction between hydrogen atoms on one water molecule and lone pairs of electrons on another leads to a special kind of association called *hydrogen bonding*. The intense hydrogen bonding between neighboring water molecules is the cause of many of the unusual properties of water, including unexpectedly high boiling and melting points and expansion rather than contraction upon freezing. In liquid water the

---

[1]The recommendation for expressing length is in meters or 10^3 fractions or multiples thereof, rather than in units such as angstroms (1 nm = 10^{-9} m = 10Å).

numerous hydrogen bonds are continually breaking and re-forming with nearby water molecules. No wonder the structure of water is sometimes described as "flickering clusters"!

The dipolar character of water molecules and the ability to form hydrogen bonds with many solutes makes it a powerful solvent. Ionic substances generally dissolve in water through solvation (hydration) of the ions by interaction with the water dipoles; covalent or molecular substances and many anions may interact through hydrogen bonding. Because water is the most important medium in chemistry and biochemistry, reactions in it merit major attention. Other solvents also find special use in analytical work; Chapter 6 considers nonaqueous systems.

Equilibrium in Pure Water

Pure water dissociates only slightly to form hydrated H^+ ions and OH^- ions. This ionization process can be written

$$2H_2O \rightleftarrows H_3O^+ + OH^-$$

or conventionally,

$$H_2O \rightleftarrows H^+ + OH^- \tag{5-1}$$

Here it is understood that H^+ represents a proton solvated by water; it is more correctly written H_3O^+.

The expression for the equilibrium constant for Equation (5-1) is

$$K_{eq} = \frac{a_{H^+}a_{OH^-}}{a_{H_2O}} \tag{5-2}$$

where a is activity. Because the amount of H^+ and OH^- formed at equilibrium under normal conditions is so small, a_{H_2O} is little changed from the value it would have if no dissociation took place. Since pure water by definition has an activity of 1, Equation (5-2) is usually written

$$K_w = a_{H^+} a_{OH^-} \tag{5-3}$$

where K_w is sometimes termed the *dissociation*, or *ionization*, *constant* but more often the *ion-product*, or *autoprotolysis*, *constant* of water. It has a value of 1.0×10^{-14} at 24°C.[2] Hence a_{H^+} and a_{OH^-} are equal to 1×10^{-7} M in pure water at this temperature.

For most analytical equilibrium calculations the use of concentrations rather than activities yields sufficient accuracy and is more convenient. On the assumption of activity coefficients of unity, Equation (5-3) becomes

$$K_w = [H^+][OH^-] \tag{5-4}$$

---

[2]We use this value of K_w throughout the book. Values range from 0.11×10^{-14} at 0°C to 2.09×10^{-14} at 35°C; at higher temperatures and pressures K_w becomes larger.

The pH Scale

Because calculations with the small numbers often encountered in hydrogen and hydroxyl ion concentrations are inconvenient, a simplified system was devised in 1909 by S. P. L. Sørensen, a Danish biochemist. He proposed use of the term pH, defined[3] by $pH = -\log[H^+]$. Thus, if $[H^+] = 1 \times 10^{-6}\ M$, pH is 6, whereas if $[H^+] = 6 \times 10^{-9}\ M$, pH is $-\log(6 \times 10^{-9}) = -\log 6 - \log (10^{-9}) = -0.78 + 9 = 8.22$. The Sørensen system can be applied to any quantity. For example, $pOH = -\log [OH^-]$ is often employed in studies of solution equilibrium, and terms such as pAg, pCl, pCu, and pK_{eq} are common.

The relation between pH and pOH can be obtained by taking the negative logarithm of both sides of Equation (5-4):

$$-\log K_w = -\log [H^+] + (-\log [OH^-])$$

$$pK_w = pH + pOH$$

$$14 = pH + pOH$$

This relation is especially useful in many calculations involving aqueous systems.

Electrolytes

Water does not conduct electricity to a significant extent in the absence of dissolved electrolytes. *Electrolytes* are substances that produce ions on dissolving in water. If dissociation into ions is complete, as in the dissolution of a crystal of NaCl to give the ions Na^+ and Cl^-, or HCl to give the ions H^+ and Cl^-, the substance is called a *strong electrolyte*; if ionization is incomplete, as in a solution of acetic acid[4] to give a mixture of undissociated HOAc molecules and H^+ and OAc^- ions, the substance is called a *weak electrolyte*. Most salts ionize almost completely in water and therefore are strong electrolytes:

$$NaCl \rightarrow Na^+ + Cl^-$$

$$KOAc \rightarrow K^+ + OAc^-$$

$$Mg(NO_3)_2 \rightarrow Mg^{2+} + 2NO_3^-$$

Although only a few acids and bases are strong electrolytes, these few are especially significant. The strong acids (100% dissociated) of analytical importance are hydrochloric (HCl), nitric (HNO_3), perchloric ($HClO_4$), and

---

[3]Alternative forms of the expression are $pH = \log (1/[H^+])$ and $[H^+] = 10^{-pH}$. In the modern definition (Section 11-7) of the pH scale, pH represents $-\log a_{H^+}$.

[4]Acetic acid, the principal component of vinegar and one of the most common organic acids, has the formula CH_3COOH. The formula also may be written $HC_2H_3O_2$ or HOAc, where Ac refers to the acetyl group, CH_3CO.

sulfuric (H_2SO_4). Other strong acids are hydrobromic (HBr) and hydroiodic (HI). The strong bases of analytical importance are sodium hydroxide (NaOH), potassium hydroxide (KOH), and tetraethylammonium hydroxide [$(C_2H_5)_4NOH$].

The term strong must not be confused with simply a high concentration of a solute. Whereas a strong electrolyte is essentially completely dissociated into ions, a *concentrated* solution contains a large amount of solute per unit volume, regardless of the degree of dissociation.

Conjugate Acid-Base Systems

Of the reactions occurring in water, those involving acids, bases, and their neutralization are of special significance. This is true not only in chemical analysis, but in almost every natural process, from the formation and dissolution of rocks to the digestion and distribution of food in the human body. Of the several definitions available, that of Brønsted and Lowry is the most useful in analytical work: an *acid* is a substance having a tendency to donate a proton, and a *base* a substance having a tendency to accept a proton. Thus, for the general case of an acid HA a corresponding base A^- exists, and the two are related according to

$$HA \rightleftarrows H^+ + A^-$$

HA and A^- constitute a *conjugate* acid-base pair, with HA the conjugate acid of A^-, and A^- the conjugate base of HA. The proton H^+ does not exist independently in solution, always being attached to some base. Therefore, the dissociation of a proton from A^- is accompanied by its uptake by another base.

A variety of conjugate acid-base systems of varying charge type exist. For example, NH_4^+ is the conjugate acid of NH_3, H_3O^+ that of H_2O, H_2O that of OH^-, and HCO_3^- that of CO_3^{2-}. Two examples of the transfer of a proton from the acid of one conjugate acid-base pair to the base of another are

$$HCl + H_2O \rightleftarrows H_3O^+ + Cl^-$$

where HCl and Cl^- constitute one conjugate pair, and H_2O and H_3O^+ the other, and

$$NH_3 + H_2O \rightleftarrows NH_4^+ + OH^-$$

where the two conjugate pairs are NH_3–NH_4^+ and H_2O–OH^-. Note that H_2O can function as either an acid or a base, depending on the other conjugate acid-base pair involved in the reaction. Such a substance is said to be *amphiprotic*. Amphiprotic properties are common to many species, including the hydrogen phosphate ion, HPO_4^{2-},

$$HPO_4^{2-} + OH^- \rightleftarrows PO_4^{3-} + H_2O$$

$$HPO_4^{2-} + H_3O^+ \rightleftarrows H_2PO_4^- + H_2O$$

and the amino acid glycine,

$$H_2NCH_2COOH + H_3O^+ \rightleftharpoons {}^+H_3NCH_2COOH + H_2O$$

$$H_2NCH_2COOH + OH^- \rightleftharpoons H_2NCH_2COO^- + H_2O$$

5–2   CALCULATION OF pH IN SOLUTIONS OF ACIDS, BASES, AND THEIR SALTS. CALCULATION OF CONCENTRATIONS OF SPECIES PRESENT AT A GIVEN pH

Solutions of Strong Acids and Bases

A *strong acid* when added to water ionizes completely to liberate an equivalent amount of hydrogen ion. Similarly, a *strong base* when added to water liberates an equivalent amount of hydroxide ion. The pH can be calculated from either $[H^+]$ or $[OH^-]$, and is straightforward[5] unless the concentration of acid is so low that the hydrogen ion contribution from dissociation of water must be taken into account. This does not occur above 10^{-6} M or so.

EXAMPLE 5–1
Calculate the pH of a 0.0546 M solution of perchloric acid.

Since $HClO_4$ is a strong acid, $[H^+] = 0.0546$, and pH $= -\log 0.0546 = 1.26$.

EXAMPLE 5–2
Calculate the pH of a 0.0224 M solution of sodium hydroxide.

Since NaOH is a strong base, $[OH^-] = 0.0224$, and pOH $= -\log 0.0224 = 1.65$. Then pH $= 14 - $ pOH $= 12.35$.

EXAMPLE 5–3
Calculate the pH of a 1×10^{-7} M solution of hydrochloric acid.

Even though hydrochloric acid is a strong acid, at this low concentration the contribution of H^+ from water dissociation is significant and must be taken into account. The *total* hydrogen ion concentration at equilibrium is equal to the sum of the contributions from the HCl, 1×10^{-7} M, and from the dissociation of water:

$$[H^+]_{total} = [H^+]_{HCl} + [H^+]_{H_2O}$$

For every H^+ ion contributed by water, an OH^- ion must be formed. Therefore $[OH^-]$, and correspondingly $[H^+]_{H_2O}$, can be obtained from the expression for the dissociation of water:

---

[5]The contribution to $[H^+]$ and $[OH^-]$ from the dissociation of water is ignored except in those cases where important. Since the inclusion of these quantities is critical only when concentrations are extremely low, their omission causes little error. A general, rigorous equation that makes no assumptions concerning water has been described (W. J. Felty, *J. Chem. Educ.* **1978,** *55,* 576).

$$[OH^-]_{H_2O} = [H^+]_{H_2O} = K_w/[H^+]_{total}$$

$$[H^+]_{total} = [H^+]_{HCl} + K_w/[H^+]_{total} = 1 \times 10^{-7} + 1 \times 10^{14}/[H^+]_{total}$$

$$[H^+]_{total}^2 - (1 \times 10^{-7})[H^+]_{total} - (1 \times 10^{-14}) = 0$$

Solution of this quadratic equation gives

$$[H^+] = \frac{(1 \times 10^{-7}) \pm \sqrt{(1 \times 10^{-7})^2 + 4(1 \times 10^{-14})}}{2} = 1.6 \times 10^{-7} \qquad pH = 6.79.$$

Solutions of Weak Acids and Bases

Ionization Constants

A *weak acid* is a substance that ionizes to form H^+ to only a limited extent. The dissociation reaction for a weak acid in water is

$$HA \rightleftharpoons H^+ + A^-$$

The equilibrium constant for the dissociation is termed the *dissociation, ionization,* or *acidity constant* of the acid, and is given the symbol K_a:

$$K_a = \frac{[H^+][A^-]}{[HA]} \tag{5-5}$$

Values of acid dissociation constants have been compiled; a partial list can be found in the Appendix.

EXAMPLE 5-4

Write the expression for the acid dissociation constant for acetic acid, and calculate the pK_a. Do the same for the ammonium ion.

For acetic acid, HOAc,

$$HOAc \rightleftharpoons H^+ + OAc^-$$

$$K_a = \frac{[H^+][OAc^-]}{[HOAc]} = 1.9 \times 10^{-5}$$

$$pK_a = -\log(1.9 \times 10^{-5}) = 4.72$$

For ammonium ion, NH_4^+,

$$NH_4^+ \rightleftharpoons NH_3 + H^+$$

$$K_a = \frac{[H^+][NH_3]}{[NH_4^+]} = 5.5 \times 10^{-10} \qquad pK_a = 9.26$$

Note that ammonium ion can exist in water only as a salt such as $(NH_4)_2SO_4$ or NH_4Cl. For our purposes most salts can be considered completely dissociated in water; therefore, the nature of the anion will usually not affect equilibria involving the cation.

Some weak acids can lose more than one proton. In such cases, expressions for the chemical equations and acid dissociation constants are written stepwise. Thus for the first dissociation step of oxalic acid, HOOCCOOH or H_2Ox,

$$H_2Ox \rightleftharpoons H^+ + HOx^-$$

$$K_1 = \frac{[H^+][HOx^-]}{[H_2Ox]} = 8.8 \times 10^{-2} \qquad pK_1 = 1.06$$

For the second step

$$HOx^- \rightleftharpoons H^+ + Ox^{2-}$$

$$K_2 = \frac{[H^+][Ox^{2-}]}{[HOx^-]} = 5.1 \times 10^{-5} \qquad pK_2 = 4.29$$

It is more difficult to remove a proton from a species with greater negative charge; therefore the first ionization constant is larger than the second.

A *weak base* is a substance that accepts protons to only a limited extent. The equilibrium concepts outlined for weak acids apply in the same way to bases. Two charge types of weak bases are common: anions (such as acetate) and neutral molecules (such as ammonia). Reactions of these bases in water and the corresponding equilibrium constants are in general written

$$A^- + H_2O \rightleftharpoons HA + OH^- \qquad K_b = \frac{[HA][OH^-]}{[A^-]}$$

$$B + H_2O \rightleftharpoons BH^+ + OH^- \qquad K_b = \frac{[BH^+][OH^-]}{[B]}$$

Values for K_b may be obtained from tables in the same way as those for K_a.

EXAMPLE 5-5

Write the expressions for the base dissociation constant for acetate ion and for ammonia, and calculate the pK_b for each.

For acetate ion,

$$OAc^- + H_2O \rightleftharpoons HOAc + OH^-$$

$$K_b = \frac{[HOAc][OH^-]}{[OAc^-]} = 5.3 \times 10^{-10} \qquad pK_b = 9.28$$

For ammonia,

$$NH_3 + H_2O \rightleftharpoons NH_4^+ + OH^-$$

$$K_b = \frac{[NH_4^+][OH^-]}{[NH_3]} = 1.8 \times 10^{-5} \qquad pK_b = 4.74$$

A useful relation is that, for a conjugate acid-base pair in water, $K_a K_b = K_w$, and therefore $pK_a + pK_b = pK_w = 14$. Thus if K_a for any acid is available, K_b for the conjugate base can readily be obtained. For example, for phenol, C_6H_5OH, $K_a = 1 \times 10^{-10}$, and so K_b and pK_b for the phenolate anion are calculated according to

$$C_6H_5O^- + H_2O \rightleftharpoons C_6H_5OH + OH^-$$

$$K_b = K_w/K_a = (1 \times 10^{-14})/(1 \times 10^{-10}) = 1 \times 10^{-4}$$

$$pK_b = -\log K_b = -\log (1 \times 10^{-4}) = 4$$

pK_b could also be calculated directly:

$$pK_b = 14 - pK_a = 14 - 10 = 4.$$

Calculation of pH in Solutions of Weak Acids or Bases

In a solution containing only a weak acid HA, ionization produces H^+ and A^- in equal amounts. The pH of the solution can be calculated from the expression for the acid dissociation constant provided the value of the constant and the total concentration of acid are known.

EXAMPLE 5-6

Calculate the pH of a 0.4 M solution of acetic acid in water.

$$HOAc \rightleftharpoons H^+ + OAc^-$$

$$K_a = \frac{[H^+][OAc^-]}{[HOAc]} = 1.9 \times 10^{-5}$$

The chemical equation shows that one H^+ is formed for every OAc^- produced, and so their concentrations must be equal. The contribution of $[H^+]$ from the dissociation of water [Equation (5-1)] is not considered. This neglect is valid as long as the acid is neither too weak nor its initial concentration too small. The error introduced will be less than 5% if the product of K_a and the initial concentration of the weak acid is greater than about 1×10^{-8}.

The concentration of HOAc at equilibrium must be equal to the total concentration,[6] 0.4 M, minus the amount dissociated, which is equal to $[H^+]$ (or $[OAc^-]$):

---

[6]The total concentration of a species added to a solution is sometimes called the *analytical concentration* because on analysis this is the amount that would be determined. In this example the analytical concentration of acetic acid C_{HOAc} is 0.4 M even though two forms, undissociated acetic acid and acetate ion, are present in solution.

$$[HOAc] = 0.4 - [H^+]$$

Substitution in the K_a expression gives

$$1.9 \times 10^{-5} = \frac{[H^+][H^+]}{C_{HOAc} - [H^+]} = \frac{[H^+]^2}{0.4 - [H^+]}$$

On rearrangement

$$[H^+]^2 + (1.9 \times 10^{-5})\,[H^+] - (7.6 \times 10^{-6}) = 0$$

This is another example of a quadratic equation, which can be solved by the quadratic formula to give the exact value of $[H^+]$:

$$[H^+] = 2.75 \times 10^{-3} \qquad\qquad pH = 2.56$$

Note that [HOAc] in the denominator at equilibrium is $0.4 - 0.00275$, or 0.397. This is sufficiently close to the total concentration, 0.4, that nearly the same answer would be obtained if $[H^+]$ were neglected. With 0.4 in the denominator as an estimate, or approximation, for [HOAc]

$$1.9 \times 10^{-5} = [H^+]^2/0.4$$
$$[H^+] = 2.76 \times 10^{-3} \qquad\qquad pH = 2.56$$

The calculation is simplified, and the answer within 0.01 pH unit of that calculated rigorously. This method is applicable when the ionization constant is less than about 10^{-4} and the initial concentration greater than 10^{-3} M. Always check the result to ensure that the error introduced by the approximation is negligible. If the term neglected is less than 5 to 10% of the total, the result is acceptable. If larger, it is necessary to apply either the quadratic equation or the method of successive approximations.[7]

EXAMPLE 5-7

Calculate the pH of a 0.0096 M solution of ammonium chloride.

Ammonium chloride dissociates completely (as we assume all salts do) in water to give NH_4^+ and Cl^- ions. The NH_4^+ ion can undergo ionization according to

$$NH_4^+ \rightleftharpoons H^+ + NH_3 \qquad\qquad K_a = 5.5 \times 10^{-10}$$

Substituting in the K_a expression gives

$$5.5 \times 10^{-10} = \frac{[H^+]\,[NH_3]}{0.0096 - [H^+]}$$

---

[7]In this method, for Example 5-6 the value for $[H^+]$ calculated in the first approximation is substituted for $[H^+]$ in the denominator, and the calculation repeated. This process of substituting the previous result into the next calculation is repeated until successive answers agree within about 5%.

Since the solution initially contained only NH_4^+, the only source of NH_3 and H^+ may be assumed to be from dissociation. Then $[H^+] = [NH_3]$, and with the assumption that $[H^+]$ is small enough compared with 0.0096 that it can be neglected in the denominator,

$$5.5 \times 10^{-10} = [H^+]^2/0.0096$$

$$[H^+]^2 = 5.3 \times 10^{-12}$$

$$[H^+] = 2.3 \text{ x } 10^{-6} \qquad pH = 5.64$$

Since 2.3×10^{-6} is only 2% of 0.0096, the assumption is valid.

The calculations for a solution containing only a weak base are handled similarly, but replacing the dissociation constant K_a with K_b.

EXAMPLE 5–8

Calculate the pH of a 0.12 M solution of methylamine.

$$CH_3NH_2 + H_2O \rightleftarrows CH_3NH_3^+ + OH^-$$

$$K_b = \frac{[CH_3NH_3^+] [OH^-]}{[CH_3NH_2]} = 4 \times 10^{-4}$$

To reduce the equation to one unknown, two relations are used: (1) $[CH_3NH_3^+]$ is equal to $[OH^-]$ when no other source of either species, such as a methylammonium salt or other base, is present and (2) $[CH_3NH_2]$ at equilibrium is equal to the initial concentration of methylamine minus the concentration of $CH_3NH_3^+$ (or OH^-) formed. Then

$$4 \times 10^{-4} = \frac{[OH^-] [OH^-]}{0.12 - [OH^-]}$$

With the assumption that $[OH^-] \ll 0.12$ and therefore negligible in the denominator, $4 \times 10^{-4} = [OH^-]^2/0.12$, $[OH^-]^2 = 0.48 \times 10^{-4}$, or 48×10^{-6}, and $[OH^-] = 6.9 \times 10^{-3}$ M. $pOH = -\log (6.9 \times 10^{-3}) = 2.16$, and $pH = 14 - 2.16 = 11.84$.

As a test of the assumption, determine the magnitude of $[OH^-]$ relative to the analytical concentration of methylamine, 0.12 M, that is, 0.0069/0.12, or 0.058. This value of 5.8% is sufficiently low to be acceptable. For comparison, calculation of $[OH^-]$ using the quadratic equation gives 6.7×10^{-3} M, or pOH 2.17. The difference from 6.9×10^{-3} M is about 3%, or 0.01 pH unit. Since the precision of pH measurements is no better than a few hundredths of a pH unit without rigorous precautions, the difference can be considered negligible.

EXAMPLE 5–9

Calculate the pH of a 0.024 M solution of ammonia, NH_3.

$$NH_3 + H_2O \rightleftarrows NH_4^+ + OH^-$$

$$K_b = \frac{[NH_4^+] [OH^-]}{[NH_3]} = 1.8 \times 10^{-5}$$

Because the only source of NH_4^+ is from the reaction of NH_3 with water, $[NH_4^+]$ = $[OH^-]$. Substituting $[OH^-]$ for $[NH_4^+]$ in the numerator of the K_b equation, and replacing $[NH_3]$ with the original concentration minus the amount that reacted with water, gives

$$1.8 \times 10^{-5} = \frac{[OH^-] \, [OH^-]}{0.024 - [OH^-]}$$

Assuming $[OH^-] \ll 0.024$ and rearranging, we have

$$(1.8 \times 10^{-5})(0.024) = [OH^-]^2$$

$$[OH^-] = 6.57 \times 10^{-4}, \text{ or pOH} = 3.18 \qquad \text{pH} = 10.82$$

Since $[OH^-]$ is only 3% of 0.024, the assumption is valid.

Calculation of pH in Mixtures of Weak Acids or Bases and their Salts

The calculations that follow are for solutions that contain initially both the acid and base form of a conjugate pair.

EXAMPLE 5–10

Calculate the pH of a solution prepared by placing 0.1 mol of acetic acid and 0.02 mol of sodium acetate in a flask and diluting to 1 L.

Initially, $[HOAc] = 0.1 \, M$. Also, since NaOAc is completely ionized, $[OAc^-]$ = $0.02 \, M$. Substituting these values in the K_a expression for acetic acid gives

$$1.9 \times 10^{-5} = \frac{[H^+] \, (0.02)}{0.1}$$

$$[H^+] = 9.5 \times 10^{-5} \qquad \text{pH} = 4.02$$

When both the acid and the conjugate base of an acid-base pair are present initially in significant amounts, the reaction of either species with water will be negligible and hence need not be included in calculations of equilibrium concentrations.

EXAMPLE 5–11

Calculate the pH of a solution that initially contains $0.01066 \, M$ ammonia and $0.00533 \, M$ ammonium nitrate.
In the equilibrium expression substitute $[NH_3] = 0.01066$ and $[NH_4^+]$ = 0.00533:

$$K_b = \frac{[NH_4^+][OH^-]}{[NH_3]}$$

$$1.8 \times 10^{-5} = \frac{(0.00533) \, [OH^-]}{(0.01066)}$$

Then $[OH^-] = 3.6 \times 10^{-5}$, pOH = 4.44, and pH = 14 − 4.44 = 9.56.

In the preceding examples the concentrations of species are expressed in units of molarity. To prevent confusion, other units are better converted to molarity or millimolarity before substitution into the equations of these examples. When necessary, concentrations may be converted to other units after the calculations are completed.

Calculation of Concentrations of Species
Present at a Given pH

Often the pH of a solution is known, and it is of interest to calculate the concentrations of the other species at equilibrium. For example, knowledge of the concentration of an anion that forms metal complexes may be needed (Section 7-2). These concentrations are most conveniently calculated by defining a series of alpha (α) terms in the following way.

We consider first the simple case of a single acid-base system:

$$HA \rightleftharpoons H^+ + A^- \qquad K_a = \frac{[H^+][A^-]}{[HA]}$$

If we let α_{HA} be the fraction of the total concentration ($[HA] + [A^-]$) in the form HA, then

$$\alpha_{HA} = \frac{[HA]}{[HA] + [A^-]} \qquad (5\text{-}6)$$

Substituting $[HA] = [H^+][A^-]/K_a$ in (5-6), and multiplying the right side by $K_2/[A^-]$, we have

$$\alpha_{HA} = \frac{[H^+][A^-]/K_a}{([H^+][A^-]/K_a) + [A^-]} = \frac{[H^+]}{[H^+] + K_a} \qquad (5\text{-}7)$$

From (5-7) we see that α_{HA} depends on only the pH and the acid dissociation constant and not on the amount of HA present. With C_{HA} defined as equal to $[HA] + [A^-]$, (5-6) can be written

$$\alpha_{HA} = [HA]/C_{HA}$$

If the total concentration of the acid, C_{HA}, is known, [HA] can be calculated.

Similarly, if we let α_{A^-} be the fraction of the total concentration in the form A^-,

$$\alpha_{A^-} = \frac{[A^-]}{C_{HA}} = \frac{[A^-]}{[HA] + [A^-]} = \frac{K_a}{[H^+] + K_a} \qquad (5\text{-}8)$$

EXAMPLE 5-12

Calculate [HOAc] and [OAc⁻] in a solution having $C_{HOAc} = 0.1\ M$ and a pH of (a) 3 and (b) 6.

(a) $$\alpha_{HOAc} = \frac{[H^+]}{[H^+] + K_a} = \frac{1 \times 10^{-3}}{(1 \times 10^{-3}) + (1.9 \times 10^{-5})} = 0.98$$

$$\alpha_{HOAc} = \frac{[HOAc]}{C_{HOAc}}$$

$$[HOAc] = \alpha_{HOAc} C_{HOAc} = (0.98)(0.1) = 0.098$$

$$\alpha_{OAc^-} = \frac{K_a}{[H^+] + K_a} = \frac{1.9 \times 10^{-5}}{(1 \times 10^{-3}) + (1.9 \times 10^{-5})} = 0.019$$

$$\alpha_{OAc^-} = \frac{[OAc^-]}{C_{HOAc}}$$

$$[OAc^-] = \alpha_{OAc^-} C_{HOAc} = (0.019)(0.1) = 1.9 \times 10^{-3}$$

(b) At pH 6, α_{HOAc} is 0.05, $[HOAc] = 0.005$, and $[OAc] = 0.095$.

We now consider a dibasic acid H_2A and its ions. Here the total concentration is

$$C_{H_2A} = [H_2A] + [HA^-] + [A^{2-}] \tag{5-9}$$

and we calculate the fractions α_{H_2A}, α_{HA^-}, and $\alpha_{A^{2-}}$ in the forms H_2A, HA^-, and A^{2-}. Substituting the acid dissociation constants K_1 and K_2 for H_2A in (5-9), we obtain

$$C_{H_2A} = [H_2A] + \frac{K_1[H_2A]}{[H^+]} + \frac{K_1K_2[H_2A]}{[H^+]^2}$$

$$= [H_2A] \frac{[H^+]^2 + K_1[H^+] + K_1K_2}{[H^+]^2}$$

$$\alpha_{H_2A} = \frac{[H_2A]}{C_{H_2A}} = \frac{[H^+]^2}{[H^+]^2 + K_1[H^+] + K_1K_2} \tag{5-10}$$

In a similar way

$$\alpha_{HA^-} = \frac{[HA^-]}{C_{H_2A}} = \frac{K_1[H^+]}{[H^+]^2 + K_1[H^+] + K_1K_2} \tag{5-11}$$

$$\alpha_{A^{2-}} = \frac{[A^{2-}]}{C_{H_2A}} = \frac{K_1K_2}{[H^+]^2 + K_1[H^+] + K_1K_2} \tag{5-12}$$

Expressions of this type are especially useful for the determination of various ionic concentrations in buffers for purposes of ionic-strength calculations, for the estimation of the ratio of acid to base forms of

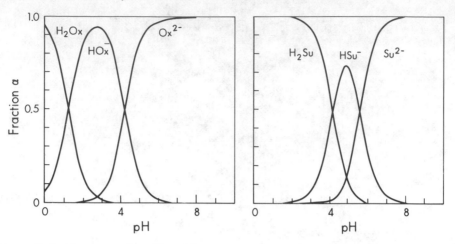

Figure 5-1. Fraction of oxalic acid and succinic acid present in solution in various forms as a function of pH.

indicators, and so on.[8] They are also convenient in calculations involving metal-ion complexation reactions (Chapter 7).

EXAMPLE 5–13

Calculate $[Ox^{2-}]$ for a 0.1 M solution of oxalic acid, H_2Ox, at pH 3.0.

For H_2Ox, $K_1 = 6 \times 10^{-2}$ and $K_2 = 6 \times 10^{-5}$. Then

$$\alpha_{Ox^{2-}} = \frac{(6 \times 10^{-2})\,(6 \times 10^{-5})}{(1 \times 10^{-3})^2 + (6 \times 10^{-2})(1 \times 10^{-3}) + (6 \times 10^{-2})(6 \times 10^{-5})}$$

$$\alpha_{Ox^{2-}} = \frac{3.6 \times 10^{-6}}{(1 \times 10^{-6}) + (6 \times 10^{-5}) + (3.6 \times 10^{-6})} = 0.056$$

$$[Ox^{2-}] = (0.1)(0.056) = 0.0056\ M$$

A graph of the distribution of various species as a function of pH can often clarify the overall view of the effect of pH on a system. Figure 5-1 illustrates the distribution of species for oxalic acid, where K_1 and K_2 are well separated, and for succinic acid, where they are close together.[9]

---

[8]The general case is

$$\alpha_{A^{n-}} = \frac{K_1 K_2 \cdots K_n}{[H^+]^n + K_1[H^+]^{n-1} + K_1 K_2[H^+]^{n-2} + \cdots + K_1 K_2 \cdots K_n}$$

and

$$\alpha_{H_2A} = \frac{[H^+]^n}{[H^+]^n + K_1[H^+]^{n-1} + K_1 K_2[H^+]^{n-2} + \cdots + K_1 K_2 \cdots K_n}$$

The numerator in each of the expressions for the intermediate species corresponds to one of the individual terms in the denominator, the second denominator term appearing in the numerator for calculation of $\alpha_{H_{n-1}A^-}$, the third for $\alpha_{H_{n-2}A^{2-}}$, and so on.

[9]Equation (5-12) was used to calculate α as a function of pH for the unprotonated species Ox^{2-} and Su^{2-} in Figure 5-1.

EXAMPLE 5-14

From Figure 5-1 estimate the fraction of succinic acid present as the unprotonated anion at pH 6.

At pH 6, α for Su^{2-} is 0.71, that is, 71% of the total succinic acid is present in the unprotonated form. About 28% of the total is present in the monoprotonated form, and about 1% in the diprotonated form.

5-3 BUFFER SOLUTIONS

A *buffer* solution is one that resists change in pH on dilution or on addition of small amounts of either a strong acid or a strong base.[10] Usually a buffer solution consists of a weak acid HA and its conjugate base A^- (often as the sodium or potassium salt) or a weak base B and its conjugate acid BH^+ (often as the chloride or nitrate salt). An example is a solution containing ammonia and ammonium chloride, or one containing acetic acid and sodium acetate. On addition of a strong acid the hydrogen ions of the acid react with the conjugate base of the buffer; on addition of a strong base the hydroxyl ions of the base react with the conjugate acid of the buffer.

Buffers are important in many areas. Sea water is buffered at a pH of about 8 by the carbonate–bicarbonate system formed through the uptake of carbon dioxide from the atmosphere. Fluids in and around the cells of living organisms are held within narrow limits of pH by conjugate acid-base pairs. The blood of human beings must be held within a few tenths of a pH unit of 7.4, or the catalytic activity of enzymes may be affected and cause irreparable damage to cells. The major buffer for extracellular fluids is H_2CO_3–HCO_3^- ($pK_a = 6.5$), and for intracellular fluids, $H_2PO_4^-$–HPO_4^{2-} ($pK_a = 7.2$). The ability of a buffer to tolerate the addition of quantities of strong acid or base is called *buffer capacity*.[11] Blood plasma has a high buffer capacity. The addition of 10 mL of 1 M HCl to a liter of water will result in a pH drop of five units, from about 7 to 2; addition of the same quantity of HCl to a liter of blood plasma causes a pH drop of only 0.2 pH unit, from 7.4 to 7.2.

For maximum buffering effectiveness both species of a conjugate acid-base pair should be present in appreciable concentration. When the two concentrations are equal, or [HA] = [A⁻] for a weak-acid buffer system, then from Equation (5-5) $K_a = [H^+]$, or $pK_a = pH$. Thus, the most effective buffering region for a weak acid and its conjugate base is at a pH corresponding to the pK_a for the acid. In practice a buffer may be considered effective over a range of one pH unit above or below the pK_a for the acid. In this range the ratio of [HA] to [A⁻] varies from 10 to 0.1. The ratio of the two conjugate species is called the *buffer ratio*.

EXAMPLE 5-15

What is the pH range over which the acetic acid–sodium acetate buffer system is useful?

---

[10]Occasionally the term is employed with species other than hydrogen ion, such as metal ions or anions. Thus a metal ion buffer resists change in pM.

[11]The quantitative definition of buffer capacity is the number of moles of hydrogen or hydroxyl ion required to cause 1 L of buffer to undergo a one-unit change in pH.

For acetic acid, $K_a = 1.9 \times 10^{-5}$, and $pK_a = 4.72$. The acetic acid–acetate buffer system is therefore most effective over a pH range of 3.7 to 5.7.

Similarly, for a base the optimum buffering range is one pOH unit above and below the pK_b value for the base.

EXAMPLE 5-16

What is the pH range over which the ammonia–ammonium chloride buffer system is useful?

For ammonia, $K_b = 1.8 \times 10^{-5}$, and $pK_b = 4.74$; so this system would be most effective at a pOH of 4.74 and useful over the range 3.7 to 5.7. This corresponds to a pH range of 10.3 to 8.3. Note that this calculation could also have been made by considering ammonium ion a weak acid, with ammonia its conjugate base, and following the same approach as for the acetic acid system. The value of K_a for ammonium ion can be obtained either from tables of acid dissociation constants or from K_b for ammonia by the relation $K_a = K_w/K_b$.

EXAMPLE 5-17

Calculate the pH shift that would result on addition of 10 mL of 1 M HCl to 500 mL of a pH 4.72 buffer consisting of 0.5 M HOAc and 0.5 M NaOAc. Compare the result with the addition of the same quantity of 1 M HCl to a solution of dilute acid of pH 4.72.

The addition of 10 mmol of HCl converts 10 mmol of OAc^- to HOAc. Then the new [HOAc] is given by the original amount present, (500 mL)(0.5 mmol/mL) = 250 mmol, plus an additional 10 mmol, divided by the new volume of 510 mL:

$$[HOAc] = \frac{(250 + 10) \text{ mmol}}{(500 + 10) \text{ mL}} = 0.51 \ M$$

Similarly, the new [OAc^-] is given by

$$\frac{(500 \text{ mL})(0.5 \text{ mmol/mL}) - (10 \text{ mL})(1 \text{ mmol/mL})}{500 + 10 \text{ mL}} = 0.47 \ M$$

Then

$$[H^+] = K_a [HOAc]/[OAc^-] = (1.9 \times 10^{-5})(0.51 \ M)/(0.47 \ M) = 2.06 \times 10^{-5}$$

pH = 4.69, a shift of only 0.03 pH unit. If only dilute acid were present initially, on addition of the 10 mL of acid

$$[H^+] = \frac{(500 \text{ mL})(1.9 \times 10^{-5} \text{ mmol/mL}) + (10 \text{ mL}) (1 \text{ mmol/mL})}{(500 + 10) \text{ mL}} = 1.96 \times 10^{-2}$$

pH = 1.71, a shift of *three* pH units from the original value.

EXAMPLE 5-18

Calculate the pH shift that would result on the dilution of 50 mL of the HOAc–OAc^- buffer in Example 5-17 to 500 mL with pure water. Compare the result with the addition of the same quantity of water to a solution of hydrochloric acid of pH 4.72.

On dilution of 50 mL of the buffer to 500 mL, [HOAc] and [OAc⁻] become 0.05 M. Then

$$[H^+] = \frac{(1.9 \times 10^{-5})[HOAc]}{[OAc^-]} = \frac{(1.9 \times 10^{-5})(0.05)}{(0.05)} = 1.9 \times 10^{-5}$$

pH = 4.72, unchanged from the original value.

If only dilute hydrochloric acid were present initially, $[H^+]$ = (50 mL)(1.9 × 10⁻⁵ mmol/mL)/500 mL = 1.9 × 10⁻⁶ M.

pH = 5.72, a change of *one* pH unit from the original value.

Preparation of Buffer Solutions

To prepare a buffer solution of a specified pH, three preliminary steps are required: (1) selection of the appropriate conjugate acid-base pair, (2) calculation of the appropriate ratio of acid form to base form, and (3) calculation of the amounts of materials required. The procedure is as follows. First pick a conjugate acid-base pair with a pK_a within one unit of the desired pH value. Then fix the concentration of one species in the conjugate pair at an appropriate level, typically 0.1 to 1 M, and calculate from the K_a expression the concentration required for the other species. Finally, calculate the weight (or volume of solution if a liquid) of each of the required substances.

EXAMPLE 5–19

Describe the preparation of a liter of pH 9 buffer.

From a table of values of acid-base dissociation constants, the ammonium ion–ammonia pair (pK_b = 4.74, pK_a = 9.26) has a pK value in the appropriate range.

Arbitrarily take the concentration of NH_3 to be 0.1 M. The required concentration of NH_4^+ is given by

$$K_a = \frac{[H^+][NH_3]}{[NH_4^+]} = 10^{-9.26}$$

If pH = 9,

$$10^{-9.26} = \frac{(10^{-9})(0.1)}{[NH_4^+]}$$

$$[NH_4^+] = \frac{(10^{-9})(10^{-1})}{10^{-9.26}} = 10^{-0.74} = 0.18 \ M.$$

Therefore, a solution 0.1 M in NH_3 and 0.18 M in the ammonium salt NH_4Cl would provide a buffer of pH 9. This solution could be prepared from concentrated ammonia (commercial concentrated NH_3 is 15 M) and ammonium chloride. Add 6.7 mL of 15 M ammonia to $(53.49)(0.18)$, or 9.6, g of NH_4Cl and dilute to 1 L.

The resulting pH will be approximate because of factors such as activity-coefficient effects on the acid dissociation constants; measurement with a carefully standardized pH meter is necessary when a more accurate value is needed.

5-4 ACID-BASE TITRATIONS

Analytical acid-base titrations are almost always performed with a strong acid or base as titrant because they produce the largest change at the end point. In an acid-base titration a plot of some function of the concentration of a reactant or product, such as pH, against the quantity of titrant is helpful in understanding the course of the reaction, selecting an appropriate indicator, and estimating the titration error. Points of particular importance in acid-base titrations are the midpoint and the equivalence point.

In Section 3-2 the equivalence point was defined generally; in an acid-base titration it is that point where exactly stoichiometric amounts of the reacting acid and base are present. The equivalence point can be estimated by several methods, including the use of colored indicators or measurements with a pH meter. The pH at the end point, the point obtained by the indicating system, ideally should coincide with the pH at the equivalence point. The *midpoint* is that point in the titration half way to the equivalence point. At the midpoint the concentration of the species being titrated and its conjugate acid or base are usually the same, and the pH is equal to the pK_a of the acid of the conjugate pair. Thus the midpoint pH value in a titration of a weak acid or base can be obtained directly from the pK_a or pK_b for the species being titrated. Calculations of the pH at the beginning, midpoint, equivalence point, and past the equivalence point in the titration of strong and weak acids and bases are illustrated in the examples that follow.

Titrations Involving Strong Acids and Strong Bases

In the titration of a strong acid with a strong base such as sodium hydroxide, the pH before the equivalence point is determined by the amount of acid not neutralized by hydroxide. At the equivalence point the only source of H^+ or OH^- is the dissociation of water, and the pH will be 7.00 (at 24°C). The pH after the equivalence point is determined by the amount of sodium hydroxide added in excess of that required to neutralize the acid.

EXAMPLE 5-20

For the titration of 30 mL of 0.1 M sodium hydroxide with 0.1 M hydrochloric acid, calculate the pH at several points during the neutralization.

At start. Determine the pH from the concentration of NaOH present, 0.1 M. $[OH^-] = 0.1$ M, pOH = 1, and pH = $14 - 1 = 13$.

At other points. Determine the pH at other points on the titration curve prior to the equivalence point by calculating for each point the millimoles of OH^- remaining from the relation mmol NaOH taken initially – mmol HCl added = mmol OH$^-$ remaining

Then divide the millimoles of OH$^-$ by the total volume of solution at that point to obtain [OH$^-$].

At equivalence point.[12] The acid is completely neutralized, [H$^+$] = [OH$^-$], and so from K_w, [H$^+$] = 1 × 10^{-7} and pH = 7.

Past equivalence point. Determine [H$^+$] by calculating for each point the millimoles of HCl added in excess over the millimoles of NaOH initially taken and dividing by the total solution volume. Then convert to pH.

Curves for this titration and for the reverse titration of 30 mL of 0.1 M HCl with 0.1 M NaOH are shown in Figure 5-2.

EXAMPLE 5-21

For the titration of 20.0 mL of 0.16 M hydrochloric acid with 0.16 M sodium hydroxide, calculate the pH at the start, midpoint, equivalence point, and 50% past the equivalence point. Sketch the titration curve.

At start. Solution is 0.16 M in HCl. Since hydrochloric acid is a strong acid, [H$^+$] = 0.16 M and pH = 0.80.

At midpoint. The total quantity of acid initially is (0.020 L)(0.24 mol/L) = 0.0032 mol. Half this total acid is neutralized by hydroxyl ion, leaving 0.0016 mol of acid in a volume of (0.020 + 0.020) L, since 20 mL (0.020 L) of NaOH solution has been added for neutralization to this stage. Then the molarity = 0.0016 mol/0.040 L = 0.04 M, and pH = 1.40.

At equivalence point. Just sufficient base has been added to neutralize the acid, giving a solution of sodium chloride. Since the only source of hydrogen or hydroxyl ions is the dissociation of water, [H$^+$] = [OH$^-$] = 1 × 10^{-7} M, and pH = 7.

At 50% past equivalence point. Here 0.0032 mol of acid × 0.5 = 0.0016 mol of base has been added in excess. Therefore 0.0016 mol of OH$^-$ is present in a volume of (0.020 + 0.060) L, giving [OH$^-$] = 0.0016/(0.020 + 0.060) = 0.02 M. pOH = 1.70, and pH = 14 – 1.70 = 12.30.

See Figure 5-3 for a sketch of the titration, along with titrations performed when both the sample and titrant are diluted by factors of 10 and 100.

EXAMPLE 5-22

Calculate the pH at the start, midpoint, equivalence point, and 50% past the equivalence point in the titration of 20.0 mL of 0.016 M sodium hydroxide with 0.016 M hydrochloric acid. Sketch the titration curve.

At start. Solution is 0.016 M in sodium hydroxide, and so [OH$^-$] = 0.016 M. pOH = 1.80, and pH = 12.20.

At midpoint. The moles of hydroxide present equals the initial moles of NaOH, (0.020 L)(0.0240 M), minus the moles of HCl added, (0.010 L)(0.016 M), or 1.6 × 10^{-4}. The total volume of solution is (0.020 + 0.010) L, or 0.030 L. [OH$^-$] = (1.6 × 10^{-4} mol)/(0.030 L) = 0.0053 M. This corresponds to a pOH of 2.27 or a pH of 11.73.

At equivalence point. Since the only hydrogen or hydroxyl ions present are from dissociation of water, pH = 7.0.

50% past equivalence point. The moles of excess hydrogen ion equals total hydrogen ion added minus the base initially present, (0.030 L)(0.0160 M) –

---

[12]At or near the equivalence point the hydrogen ion concentration contributed by the dissociation of water becomes significant.

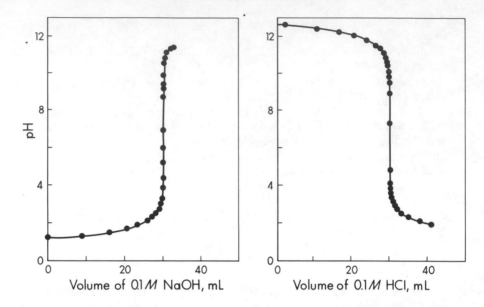

Figure 5-2. Titration of 30 mL of 0.1 M sodium hydroxide with 0.1 M hydrochloric acid, and of 0.1 M hydrochloric acid with 0.1 M sodium hydroxide. Data by L. Riley and C. Johnson.

(0.020 L)(0.016 M), or 1.6 $\times$ 10^{-4}. The total volume is (0.030 + 0.020 L), or 0.050 L. Therefore [H^+] = (1.6 $\times$ 10^{-4} mol)/(0.050 L) = 3.2 $\times$ 10^{-3} M, and pH = 2.49.

See Figure 5-3 for a sketch of the titration and of titrations of solutions a factor of 10 more concentrated and a factor of 10 more dilute.

Titration of a Weak Base with a Strong Acid

In the titration of weak bases with a strong acid the pH prior to the equivalence point is determined by the equilibrium amounts of weak base and its conjugate acid that are present. After the equivalence point it is determined by the excess of strong acid added over that required to react with the weak base.

At start. In the beginning the solution contains only weak base, and the pH is determined by the extent of dissociation of the base, as in the examples earlier for ammonia and for sodium acetate.

At points before equivalence point. During the titration both B and BH^+ are present. If the initial amount of B and the amount of strong acid added are known, [B] and [BH^+] can be calculated. From K_b and knowledge of [B] and [BH^+], [OH^-] can be obtained, and thence pH. At the midpoint the special condition of [B] = [BH^+] occurs, and pOH = pK_b. The pH, then, is equal to 14 – pK_b.

At equivalence point. At the equivalence point an amount of hydrogen ion equivalent to the amount of base present has been added, and the base has been converted to its conjugate acid BH^+:

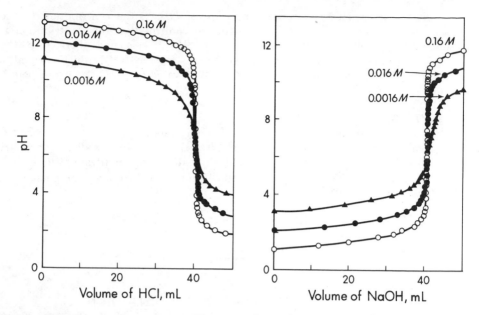

Figure 5-3. Curves for strong acid-strong base titrations. *A*, plots of pH against volume of sodium hydroxide for titration of 20 mL of 0.16, 0.016, and 0.0016 *M* hydrochloric acid with 0.16, 0.016, and 0.0016 *M* sodium hydroxide. *B*, pH against volume of hydrochloric acid for titration of 20 mL of 0.16, 0.016, and 0.0016 *M* sodium hydroxide with 0.16, 0.016, and 0.0016 *M* hydrochloric acid. Data by R. Smillie and C. Johnson.

$$H^+ + B \rightleftharpoons BH^+$$

Therefore, the system can be considered a solution of the weak acid BH^+ in water and treated as in Example 5-6.

At 50% past equivalence point. Once past the equivalence point by 10% or more, the excess acid suppresses dissociation of the salt BH^+, and the pH may be calculated from the concentration of excess acid added in the same way as pOH was calculated from the concentration of excess base in the titration of a solution of a strong acid.

EXAMPLE 5-23

For the titration of 20 mL of 0.024 *M* ammonia with 0.016 *M* hydrochloric acid, calculate the pH at the start, midpoint, equivalence point, and 50% past the equivalence point. Sketch the titration curve.

     The pH at the start of the titration is the same as that of 0.024 *M* ammonia and is calculated in Example 5-9 to be 10.82. At the midpoint, $pOH = pK_b$, and so $-\log (1.8 \times 10^{-5}) = 4.74$, and pH $= 9.26$. At the equivalence point the concentration of NH_4Cl is 0.48 mmol/50 mL, or 0.0096 *M*, and the pH is determined by the dissociation of the ammonium ion. This calculation is handled as in Example 5-7 to give a pH of 5.64. At 50% past the equivalence point the pH is determined by the excess HCl present, $(15 \text{ mL})(0.016 \text{ } M) = 0.24$ mmol in 65 mL. This gives a hydrochloric acid concentration in the solution at this point of $3.7 \times 10^{-3} M$, and so $[H^+] = 3.7 \times 10^{-3}$, and pH $= 2.43$. Note that, since H^+

from the excess HCl suppresses ionization of NH_4^+, the contribution to the acidity from this source can be ignored. To confirm that this contribution is indeed negligible, substitute the value of $[H^+]$ calculated here into the ionization-constant expression for $[NH_4^+]$ and solve for $[NH_3]$, which is equal to the $[H^+]$ provided by NH_4^+ dissociation. The value of $[NH_3]$ from this source should be found to be negligible.

EXAMPLE 5-24

Calculate the pH at the start, midpoint, equivalence point, and 50% past the equivalence point for the titration of 20 mL of 0.024 M sodium glycinate, $NaOOCCH_2NH_2$, with 0.016 M hydrochloric acid. Sketch the titration curve.[13]

At start. For glycine, $K_a = 2.5 \times 10^{-10}$, and so K_b for the glycinate ion, Gly^-, is K_w/K_a, or 4.0×10^{-5}.

$$Gly^- + H_2O \rightleftarrows HGly + OH^-$$

$$K_b = \frac{[HGly][OH^-]}{[Gly^-]} = 4.0 \times 10^{-5}$$

$[HGly] = [OH^-]$, and $[Gly^-] = 0.024 - [OH^-]$

$$4.0 \times 10^{-5} = \frac{[OH^-]^2}{0.024 - [OH^-]}$$

Assuming $[OH^-] \ll 0.024$ and solving gives $[OH^-] = 9.8 \times 10^{-4}$. Since $[OH^-] = 4\%$ of 0.024, the assumption is valid. Then $[H^+] = K_w/[OH^-] = 1.02 \times 10^{-11}$, and pH = 10.99.

At midpoint. Since half the Gly^- has been converted to HGly, $[Gly^-] = [HGly]$, and $K_b = [OH^-]$. Therefore $[OH^-] = 4.0 \times 10^{-5}$, pOH = 4.40, and pH = 9.60.

At equivalence point. With the assumption that all Gly^- has been converted to HGly,

$$[HGly] = \frac{(0.020 \text{ L})(0.024 \text{ M})}{(0.020 + 0.030)\text{L}} = 9.60 \times 10^{-3} \text{ M}$$

Only slight dissociation of the weak acid HGly can be expected:

$$HGly \rightleftarrows H^+ + Gly^-$$

$$K_a = \frac{[H^+][Gly^-]}{[HGly]}$$

$[H^+] = [Gly^-]$, and $[HGly] = (9.6 \times 10^{-3}) - [H^+]$

With the assumption that $[H^+] \ll 9.6 \times 10^{-3}$,

---

[13]In this calculation protonation of the amino group is omitted for simplicity. The properties of amino acids are discussed in Section 5-5.

$$K_a = \frac{[H^+]^2}{9.6 \times 10^{-3}}$$

$$[H^+] = 1.5 \times 10^{-6} \qquad\qquad pH = 5.81$$

Since 1.5×10^{-6} is $< 1\%$ of 9.6×10^{-3}, the assumption is valid.

At 50% past equivalence point. Since 30 mL of HCl was required to reach the equivalence point, 50% past it requires $(30)(1.5) = 45$ mL. Total moles of acid added minus moles of sodium glycinate initially present $=$ $(0.045\ L)(0.016\ M) - (0.020\ L)(0.024\ M) = 2.4 \times 10^{-4}$ mol hydrogen ion present in excess. Then

$$[H^+] = \frac{2.4 \times 10^{-4}\ \text{mol}}{(0.045 + 0.020)\ L} = 3.69 \times 10^{-3} \qquad\qquad pH = 2.43$$

See Figure 5-4 for a similar titration curve.

Titration of a Mixture of Two Acids or Two Bases

Mixtures of acids or bases can be of several types. Three classes are considered here.

1. The component acids in mixtures of strong acids such as hydrochloric and nitric react simultaneously with a strong base titrant because both are completely dissociated. The pH remains low until both have reacted and then changes rapidly. The titration curve is indistinguishable from that of a single strong acid (Figure 5-3) at the same total concentration. Thus acid-base titrimetry in water cannot resolve such a mixture.

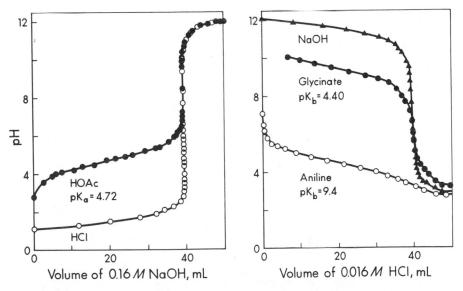

Figure 5-4. Curves for the titration of hydrochloric acid and acetic acid with sodium hydroxide and for the titration of sodium hydroxide, sodium glycinate, and aniline with hydrochloric acid. Data by D. Knudsen, R. Smillie, D. Hemmings, and C. Johnson.

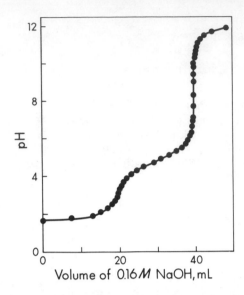

Figure 5-5. Titration with sodium hydroxide of a mixture of hydrochloric and acetic acids. Data by R. Smillie.

A mixture of strong bases such as sodium and potassium hydroxide, when titrated with a strong acid, similarly yields a titration curve indistinguishable from that of a single strong base at the same total concentration.

2. A mixture of a strong and a weak acid, when titrated with standard strong base, will always yield one, and sometimes two, analytically useful end points. The strong acid in the mixture will be titrated first and give a sufficient change in pH at the equivalence point for the amount present to be determined if the weak acid is indeed weak and the solution is not too dilute. If the weak acid is not so weak and has a pK_a on the order of 2 to 4, a single end point is seen and the total acidity is obtained. But if the pK_a of the weak acid is in the range 4 to 8, a second end point can be located. Figure 5-5 illustrates the titration of a mixture of hydrochloric and acetic acids with sodium hydroxide.

Should the pK_a of the weak acid be 8 or greater, the change in pH at the equivalence point will be too small to be of value. The reason is that the equilibrium constant for the reaction becomes too small for the reaction to be driven to completion at the equivalence point, and appreciable concentrations of the reactants remain in the solution.

Under these conditions only one good end point is obtained, corresponding to the amount of the strong acid in the mixture.

A mixture of a strong and a weak base when titrated with a standard strong acid will similarly provide one or two end points, depending on the ionization constant of the weak base and the concentrations of the two bases.

3. A mixture of two weak acids may provide one useful end point provided either that both pK_a values are larger than about 8 or that one is greater than about 8 and the other three to four pK units smaller. In the first instance the total acidity can be obtained; in the second only the stronger

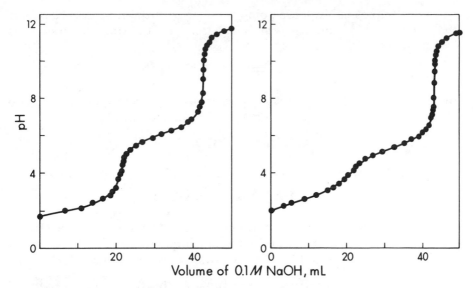

Figure 5-6. **Titration of 2 mmol each of malonic and maleic acid with 0.1 M NaOH.**
Data by L. Creagh.

of the two is determined. An analogous statement applies to mixtures of two weak bases.

The pH during titrations of mixtures of acids or bases can be calculated as in the ensuing discussion of the titration of polyprotic acid-base systems. A titration in which two breaks are sufficiently separated to allow the estimation of two equivalence points is called a *differentiating titration*. When the dissociation constants of two acids or bases are not greatly different, but both are strong enough to be titratable when present separately, only a single break corresponding to the sum of the two is seen.

Titration of Polyprotic Acids and Bases

Some acids and bases have sites for donation or acceptance of more than one proton. Acids having two dissociable protons, such as H_2SO_4, H_2CO_3, and oxalic acid (HOOCCOOH), are called *dibasic acids*; those with three, such as H_3PO_4, are termed *tribasic acids*. Similarly, there exist *diacidic bases* such as CO_3^{2-} and ethylenediamine ($H_2NCH_2CH_2NH_2$) and *triacidic bases* such as PO_4^{3-}. Often pK_a or pK_b values for successive proton transfers in a given polyprotic species are sufficiently different that a separate pH break is seen at each equivalence point. Thus the titration of an acid having two titratable protons whose acid dissociation constants differ by 10^3 to 10^4 will provide two analytically useful pH breaks, just as does the titration of an equimolar mixture of two separate acids of differing pK_a values. Figure 5-6 shows the titration of malonic acid ($pK_1 = 2.65$, $pK_2 = 5.28$) and of maleic acid ($pK_1 = 1.75$, $pK_2 = 5.83$) with sodium hydroxide.

The pH at various stages in the titration of a dibasic acid H_2A whose pK_a values differ by two or more units can be calculated in stepwise fashion. The procedure is as follows:

1. Prior to the first equivalence point consider only the first dissociation of the stronger acid, and perform calculations by treating the system as containing only that acid:

$$H_2A \rightleftharpoons H^+ + HA^-$$

2. At the first equivalence point H_2A has been converted quantitatively to HA^-. This species can react as a base with water to form H_2A,

$$HA^- + H_2O \rightleftharpoons H_2A + OH^-$$

    or as an acid to form A^{2-},

$$HA^- \rightleftharpoons H^+ + A^{2-}$$

    Upon combination of the equilibrium constants for these two reactions, and with the assumption that $[HA^-]$ is much larger than K_1, the relation between the K_a values and $[H^+]$ is to a first approximation,

$$[H^+] \approx \sqrt{K_1K_2}, \text{ or } pH = (pK_1 + pK_2)/2 \qquad (5\text{-}13)$$

3. After the first equivalence point the major species present are HA^- and A^{2-}, and only K_2 is needed in the calculations. Thus the system can be treated as though only the acid HA^-, having a dissociation constant of K_2, and its conjugate base A^{2-} were present.

EXAMPLE 5–25

Calculate the pH at the start, the two midpoints, and the two equivalence points in the titration of 20.0 mL of 0.1 M maleic acid with 0.1 M NaOH. For maleic acid, $K_1 = 1.8 \times 10^{-2}$, and $K_2 = 1.5 \times 10^{-6}$.

At start. With $[A^{2-}]$ neglected because K_1 is much larger than K_2,

$$K_1 = \frac{[H^+][HA^-]}{[H_2A]} = \frac{[H^+]^2}{0.1 - [H^+]} = 1.8 \times 10^{-2}$$

Assuming $[H^+] \ll 0.1$ and solving gives $[H^+] = 0.042$. Since $[H^+]$ is about 40% of 0.1, the assumption is not valid. Retaining $[H^+]$ in the denominator and rearranging gives the quadratic equation

$$[H^+]^2 + 0.018 [H^+] - 0.0018 = 0$$

Solving by the quadratic formula, we obtain

$$[H^+] = 0.0344 \qquad\qquad pH = 2.46.$$

At first midpoint. With neglect of K_2 as at the beginning, $pH = pK_1 = 1.74$.

At first equivalence point. With the approximation given in Equation (5-13),

$$[H^+] = \sqrt{K_1K_2} = \sqrt{(1.8 \times 10^{-2})(1.5 \times 10^{-6})} = 1.6 \times 10^{-4}$$

$pH = 3.78$.

At second midpoint. With neglect of K_1 (a valid assumption), $pH = pK_2 = 5.82$.

At second equivalence point. With only K_2 considered to be important and A^{2-} to be the major species, the reaction of interest is

$$A^{2-} + H_2O \rightleftarrows OH^- + HA^-$$

$$K_{eq} = \frac{[OH^-][HA^-]}{[A^{2-}]} = K_b = \frac{K_w}{K_2} = \frac{1 \times 10^{-14}}{1.5 \times 10^{-6}} = 6.7 \times 10^{-9}$$

The initial concentration of A^{2-} before reaction with H_2O is

$$\frac{(20.0 \text{ mL})(0.1 \ M)}{(20.0 + 40.0) \text{ mL}} = \frac{2.0 \text{ mmol}}{60.0 \text{ mL}} = 0.0333 \ M$$

$$\frac{[OH]^{-2}}{0.0333 - [OH^-]} = 6.7 \times 10^{-9}$$

Assuming $[OH^-] \ll 0.0333$ and solving, we obtain $[OH^-] = 1.5 \times 10^{-5}$, and pH = 9.17. Since $[OH^-] = 0.04\%$ of 0.033, the assumption is valid.

5–5 MOLECULES WITH BOTH ACIDIC AND BASIC PROPERTIES. AMINO ACIDS

Many molecules encountered in analytical work contain both acidic and basic groups, and often display ionic character in solution owing to internal ionization. Among these are compounds important in biochemistry and medicine, such as the amino acids and many drugs. The naturally occurring amino acids contain an acidic carboxyl group, -COOH, and on the carbon atom adjacent to the carboxyl group, a basic amino, $-NH_2$, group. With the simplest amino acid, glycine, H_2NCH_2COOH, the following reaction takes place:

$$H_2NCH_2COOH \rightleftarrows {}^+H_3NCH_2COO^-$$

Acid-base reactions for such molecules are often best handled as for a dibasic acid. Protonated glycine can be considered a dibasic acid H_2A^+ possessing two dissociable protons. The pH of a solution of any glycine species can then be calculated as outlined previously for a dibasic acid. The internally ionized form is called a *zwitterion*.

5–6 ACID-BASE INDICATORS

The end point in acid-base titrations is ordinarily determined by visual indicators or by potentiometric methods employing a glass electrode for sensing hydrogen ions. The potentiometric measurement of pH is treated in Chapter 11. Visual indicators are important in many practical situations because of their simplicity and convenience. An introduction to the use of indicators is given in Section 4-2.

Almost all acid-base indicators in use today are organic dyes that are weak acids or bases in which the acid form HIn and the base form In⁻ differ in color.[14] A small quantity of indicator is added to the solution to be titrated. Near the equivalence point the pH of the solution changes abruptly, converting the indicator from one form to the other with a corresponding change in color. Although some titrant is consumed in the indicator reaction, the amount required is small because so little of the intensely colored indicator is used.

Indicators change from the acidic to the basic color over a range of one to two pH units. The range of color change can be calculated by treating an indicator like other weak acid or base systems. Thus, for an indicator HIn the dissociation reaction in water is

$$HIn \rightleftharpoons H^+ + In^-$$

and the expression for the acid dissociation constant is

$$K_{In} = \frac{[H^+][In^-]}{[HIn]}$$

where K_{In} denotes K_a for the indicator. Rearranged,

$$[H^+] = \frac{K_{In}[HIn]}{[In^-]}$$

Taking the negative logarithm of each side, we have

$$-\log[H^+] = -\log K_{In} - \log \frac{[HIn]}{[In^-]}$$

$$pH = pK_{In} + \log \frac{[In^-]}{[HIn]} \tag{5-14}$$

When $[In^-] = [HIn]$, the term $\log [In^-]/[HIn]$ becomes zero and $pH = pK_{In}$. Thus the pH at which the indicator changes color is determined by the value of pK_{In}.

At pH values above or below pK_{In} the change in the ratio of $[In^-]$ to $[HIn]$ causes visible changes in the color of the solutions (Section 4-2). If the intensities[15] of the colors of HIn and In⁻ are about the same, then the eye will see a mixture of the colors until the concentration of one species is about ten times that of the other. An increase in the ratio above 10:1 does not

---

[14]For purposes of this discussion the acid indicator form is designated HIn and the base form In⁻. The completely analogous charge forms HIn⁺ and In, or HIn⁻ and In²⁻, could also be used; indicators of all three charge types are known.

[15]The relation between concentration and color intensity is discussed in Section 14-3.

Table 5-1. pH Transition Ranges of Selected Indicators

A Amber	C Colorless	P Purple	Y Yellow
B Blue	O Orange	R Red	

Chart of pH transition ranges (pH scale 0 to 14):

- Picric acid: C ▒ Y
- Thymol blue: R ▒ Y — Y ▒ B
- Cresol red: O ▒ A — A ▒ R
- Bromphenol blue: Y ▒ B
- Methyl orange: O ▒ Y
- Bromcresol green: Y ▒ B
- Methyl red: R ▒ Y
- Bromcresol purple: Y ▒ P
- Bromthymol blue: Y ▒ B
- Phenol red: Y ▒ R
- Thymol blue: R ▒ Y — Y ▒ B
- Phenolphthalein: C ▒ R
- Thymolphthalein: C ▒ B
- Tropaeolin: Y ▒ O

change the appearance of the solution; only the color of the more concentrated species is seen. Therefore the indicator changes color over the range of [In⁻]/[HIn] from 10:1 to 1:10. This corresponds to a change in log ([In⁻])/[HIn]) from 1 to –1. Inserting these values into Equation (14-5) gives a pH transition range of $pK_{In} \pm 1$. Table 5-1 depicts the pH transition ranges of a number of indicators.

EXAMPLE 5-26

What is the pH range over which the indicator phenol red would be expected to change? From the Appendix, pK_a for phenol red is 7.4.

Since for phenol red $pK_a = 7.4$, the pH transition range is predicted from Equation (5-14) to be 7.4 ± 1, or 6.4 to 8.4. According to Table 5-1 the actual range is about 6.8 to 8.6. The reason for the slightly higher pH values reported in the table for the visual range is the greater sensitivity of the human eye to the red basic form than to the yellow acidic form.

In theory, selection of an indicator for a given acid-base titration involves first calculation of the pH at the equivalence point and then selection of an indicator whose pK_{In} corresponds to that pH. In practice it is preferable to select an indicator whose transition range is centered on the appropriate equivalence-point pH from a table listing the transition ranges of indicators.

Some indicators, such as thymol blue and cresol red, have two protonation sites whose pK_a values differ greatly, and they undergo two color changes. Thus, thymol blue is yellow in neutral solution, changing to blue between pH 8 and 9.5 and to red between pH 1 and 3.

Occasionally an indicator cannot be found that changes at the equivalence point in a titration. This lack of coincidence of color change with the equivalence point will obviously lead to end-point error in the titration.

The error can often be reduced substantially by standardizing the titrant against a pure portion of the compound being titrated. In this way the end-point error in the titrations of the sample, and accordingly in the analytical results, will be reduced. If the color change is gradual, precision can be improved by comparing the color to a closely spaced set of standards or by matching it to a particular reference color.

The color of an indicator in some instances varies with temperature. This characteristic can arise from the effect of temperature either on the pH of the solution or directly on the response of the indicator. Although the pH of a solution of 0.01 M hydrochloric acid is not sensitive to temperature, the pOH is sensitive because the ion product of water changes markedly with temperature. An indicator such as phenolphthalein which responds to pOH would, then, have a color dependent on temperature. Bromocresol green is another example of an indicator whose color is strongly temperature-dependent.

5-7 TITRANTS AND PRIMARY STANDARDS

Acid-base titrations are important because they are simple, rapid, and applicable to a large number of substances, either directly or after conversion to a titratable acid or base by ion exchange (Chapter 19) or chemical reaction. The most widely used titrants in water are sodium hydroxide for titrating acids and hydrochloric acid for titrating bases. Other strong acids and bases such as sulfuric acid and potassium hydroxide are less common titrants. In nonaqueous solvents several other titrant systems are applicable (Chapter 6).

Although end points in acid-base titrations are usually determined by visual indicators, a pH meter with a glass indicator electrode and a suitable reference electrode can be used to follow the pH during a titration, allowing the equivalence point to be estimated from a plot of pH against volume of titrant (Section 11-8).

Preparation and Standardization of Titrants

Hydrochloric and Other Acid Titrants

Hydrochloric acid is available commercially as a concentrated solution in water. Solutions for use as titrants must be prepared by dilution to the concentration desired and then standardized. In solutions of the concentrations used in analytical work the volatility of hydrogen chloride is negligible even at elevated temperatures and so causes no error. Solutions of hydrochloric acid can be stored for years without change in concentration if the container remains tightly closed. Nevertheless, in careful work it is good practice to standardize any titrant solution within a day or so of use.

Other suitable titrants are sulfuric, nitric, and perchloric acids. All are stable and react rapidly with bases. Each has certain drawbacks, however. The oxidizing properties of nitric acid may be unacceptable in some applications. Perchloric acid may cause precipitation of potassium or ammonium perchlorate, and sulfuric acid precipitation of barium sulfate, in solutions containing appreciable concentrations of these cations.

Primary Standard Bases

The primary standards most used for the standardization of strong-acid titrants are sodium carbonate, tris(hydroxymethyl)amino methane, and 4-aminopyridine.

Sodium carbonate (formula wt 106.00) is available in highly pure form except for the presence of small amounts of sodium bicarbonate. This contamination is readily removed by heating the salt at 160°C or higher for several hours. This converts the $NaHCO_3$ to Na_2CO_3:

$$2NaHCO_3 \rightleftarrows Na_2CO_3 + H_2O + CO_2$$

When sodium carbonate is titrated with a strong acid, it is converted quantitatively to carbonic acid. This reaction is discussed in Section 5-10.

4-Aminopyridine (formula wt 94.12) is obtainable in high purity and is easily handled. It does not pick up water or carbon dioxide from the air, so that drying is normally unnecessary. If desired, it can be dried at 110°C to remove traces of surface water, though it will slowly sublime if held at this temperature for extended periods. Since it is not a strong base ($K_b = 1.3 \times 10^{-5}$, about the strength of ammonia), an indicator changing over the pH range 4 to 6, such as methyl red, should be employed.

Tris(hydroxymethyl)amino methane (tris, THAM) (formula wt 121.14) is similar to 4-aminopyridine in character, though somewhat weaker as a base ($K_b = 1.2 \times 10^{-6}$). Drying is recommended at 100 rather than 110°C, where some decomposition occurs. The purity of the commercial material is not always so dependable as that of the other two primary standards mentioned.

Sodium Hydroxide as Titrant

A solution of sodium hydroxide is the most common titrant for acids in water. Potassium and barium hydroxides have no advantage over sodium hydroxide and are somewhat more expensive. Since none of the strong bases are suitable as primary standards because of their affinity for carbon dioxide and water from the air, they are prepared and standardized against a standard acid.

Solutions of strong bases such as sodium hydroxide react with carbon dioxide from the air according to

$$2OH^- + CO_2 \rightleftarrows CO_3^{2-} + H_2O$$

Therefore, a solution of sodium hydroxide containing carbonate ion will have a different molarity when used as a titrant for a strong acid than it will for a weak one; a strong acid will react with both OH^- and CO_3^{2-}, but a weak one with only OH^-. One way to avoid this complication is to take every precaution in the preparation and storage of sodium hydroxide solutions to minimize carbon dioxide contamination; another is to standardize the solution in a manner that takes into account the extent of carbonate reaction with the sample.

Carbonate-free solutions of sodium hydroxide are usually prepared by dilution of a saturated solution of sodium hydroxide that has been allowed to stand for a time. Sodium carbonate is insoluble in concentrated solutions of sodium hydroxide and slowly settles out. Portions of the clear supernatant solution can be poured or siphoned off and diluted with CO_2-free water for preparation of titrant. Carbon dioxide is most easily removed from distilled water by boiling for a few minutes. This precipitation method of removal of carbonate is not applicable to solutions of potassium hydroxide, since potassium carbonate is soluble in concentrated solutions of the hydroxide. Another excellent method of removing carbonate is to pass the sodium hydroxide solution through a column of anion-exchange resin in the hydroxide form, so that carbonate ions are quantitatively replaced by hydroxyl ions:

$$2R^+OH^- + CO_3{}^{2-} \rightleftarrows (R^+)_2CO_3{}^{2-} + 2OH^-$$

Here each R^+ represents a fixed, positively charged site in the resin (Section 19-1). Once prepared, sodium hydroxide solutions should be protected from atmospheric carbon dioxide. A tightly capped polyethylene bottle is normally satisfactory.

Since solutions of NaOH attack glass they should not be allowed to stand for extended periods in a buret or for more than a few days in a glass storage bottle.

Solutions of sodium and potassium hydroxide are usually standardized against potassium hydrogen phthalate (KHP), but occasionally against sulfamic or benzoic acid.

Potassium Hydrogen Phthalate as Primary Standard Acid

KHP (formula wt 204.23) is a white crystalline solid that has most of the properties sought in a primary standard. It is readily available in high purity, is anhydrous and nonhygroscopic, does not absorb carbon dioxide from the atmosphere, has a high equivalent weight so that weighing errors are minimized, and can be dried at 110°C for the removal of traces of adsorbed water. The only minor disadvantage is that the acid dissociation constant for the second hydrogen ion in KHP is small, 3.1×10^{-6} (the first hydrogen ion is already replaced by a potassium ion). Accordingly, the pH change at the equivalence point is somewhat less than desired. For titration by strong bases this characteristic is of minor importance. Phenolphthalein serves as a workable indicator.

5-8 APPLICATIONS

Acids and bases may be classified as strong ($K_a = 10^{-3}$ or greater), weak ($K_a = 10^{-3}$ to 10^{-8}), and very weak ($K_a = 10^{-8}$ or smaller). Strong acids are titratable even at high dilution by solutions of strong bases, and any of several indicators with transition ranges near pH 7 are suitable. Weak acids can be titrated visually at moderate concentrations or higher; an indicator with a pH transition range on the alkaline side (7 to 10) must be chosen. Very weak acids do not provide enough change in pH at the equivalence point for a visual indicator to be satisfactory. With some very weak acids a useful end point can be obtained from a graph of data from a potentiometric titration of pH against volume of sodium hydroxide titrant.

The same situation prevails in the titration of bases with a strong acid. Strong bases can be titrated at low concentrations; weak bases ($K_b = 10^{-3}$ to 10^{-8}) can be titrated only at moderate or high concentrations, with an indicator having a transition range on the acid side of neutrality; and very weak bases ($K_b < 10^{-8}$) cannot be titrated with visual indicators at any concentration.

As noted in Section 5-4, many salts are titratable as weak acids or bases. One example is the titration of carbonate.

Titration of Carbonate

The titration of a solution of a carbonate salt such as sodium carbonate with a strong acid such as HCl leads to two inflection points, two protons being added to the carbonate ion stepwise:

$$CO_3^{2-} + H^+ \rightleftarrows HCO_3^-$$

$$HCO_3^- + H^+ \rightleftarrows H_2CO_3$$

The titration curve shows two inflection points as in Figure 5-7. Either the first or second equivalence point (C or E in Figure 5-7) can be used for the determination of carbonate. The pH change is not large at either point. An uncertainty of 0.1 pH unit at either end point results in an uncertainty of about 1% in the amount of hydrochloric acid required. The error can be reduced if the titration is carried to a preselected indicator color. When a solution is titrated to the second equivalence point, a more satisfactory approach is to take advantage of the equilibrium that exists between carbonic

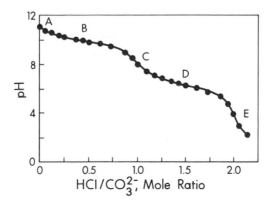

Figure 5-7. **Curve for the titration of 0.1 M sodium carbonate with 0.1 M hydrochloric acid.** Data by L. Riley.

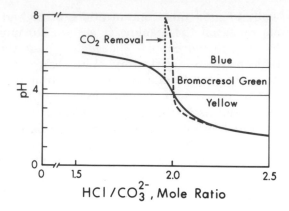

Figure 5-8. Effect of removal of carbon dioxide on pH at the second equivalence point in a titration of carbonate with hydrochloric acid. Band indicates region of change of indicator color. (*Solid line*) titration without boiling; (*dotted line*) pH change on removal of carbon dioxide by boiling; (*dashed line*) titration after removal of carbon dioxide.

acid and solutions of carbon dioxide in water. Shaking or boiling a solution of carbonic acid causes the equilibrium

$$H_2CO_3 \rightleftarrows H_2O + CO_2(g)$$

to be driven to the right through loss of carbon dioxide. If a carbonate or bicarbonate solution is titrated to just before the equivalence point at pH 4 and then shaken or boiled,[16] the pH will rise to about 8 as the concentration of carbonic acid drops (dotted line in Figure 5-8). The pH is no longer controlled by dissociation of a relatively large concentration of carbonic acid but by a small concentration of bicarbonate. When the titration is continued, the pH drops sharply because the amount of carbonic acid formed is small and the buffering effect is therefore negligible (dashed line in Figure 5-8). Bromocresol green gives a sharp color change under these conditions; methyl red, although having a transition range at higher pH values, may also be used.

Titration of Mixtures of Carbonate and Bicarbonate

In a mixture of carbonate and bicarbonate both can be determined in a single titration with hydrochloric acid. The first equivalence point at about pH 8.4 corresponds to conversion of the carbonate to bicarbonate; phenolphthalein is the indicator of choice. The bicarbonate present at this point is from two sources — the original bicarbonate plus that formed from the carbonate. Continuation of the titration to the second equivalence point converts the bicarbonate to carbonic acid, with bromocresol green or methyl

---

[16]In mammals the CO_2 produced through biological oxidation is carried by the blood to the lungs, where it is exchanged for oxygen. Part of the CO_2 is present in the blood as H_2CO_3. Since the time available in the lungs for exchange is short, the dissociation of H_2CO_3 to CO_2 and H_2O is accelerated by the enzyme carbonic acid anhydrase, a zinc-containing protein of high molecular weight. Thus nature need not resort to either boiling or shaking.

red serving as indicator. The volume required to the first end point allows calculation of the amount of carbonate originally present in the sample. The volume required from the first to the second end point is greater than that required to the first by an amount equivalent to the amount of bicarbonate in the sample. Consequently, both carbonate and bicarbonate can be determined in a single sample by titration to two end points.

Both components in mixtures of hydroxide and carbonate can be determined similarly. Here titration to the phenolphthalein end point yields the sum of the hydroxide plus carbonate according to

$$OH^- + H^+ \rightleftharpoons H_2O$$

$$CO_3^{2-} + H^+ \rightleftharpoons HCO_3^-$$

Further titration of the solution to the bromocresol green end point yields the amount of bicarbonate ion produced in the previous step:

$$HCO_3^- + H^+ \rightleftharpoons H_2CO_3$$

The second titration volume corresponds to the amount of carbonate originally present, and the difference between the first and second titration volumes corresponds to the original amount of hydroxide, since the same volume was required to convert HCO_3^- to H_2CO_3 as was required to convert CO_3^{2-} to HCO_3^-.

Because hydroxide and bicarbonate react to form carbonate, they cannot coexist in significant amounts.

Kjeldahl Determination of Nitrogen in Organic Material

An important application of acid-base chemistry is the determination of nitrogen in plant and animal materials by conversion to ammonia and titration with standard hydrochloric acid. Protein contains considerable nitrogen, mostly in the form of the amide group, and so a measure of the amount of nitrogen present serves as an indication of protein content. Knowledge of the percentage of nitrogen in fertilizers and many other organic compounds is also important. The procedure developed by Kjeldahl[17] involves heating of the organic matter with concentrated sulfuric acid to convert the carbon to carbon dioxide and the nitrogen to ammonium acid sulfate. The solution is next made basic by addition of excess sodium hydroxide to convert the ammonium ion to free ammonia. This is then separated by distillation into a receiving flask containing a saturated solution of boric acid to neutralize the ammonia and thereby prevent loss through volatilization. The borate produced is titrated with standard hydrochloric acid. Figure 5-9 illustrates the apparatus now used for the distillation operation.

---

[17]Johan Kjeldahl directed the chemical section of the Carlsberg Laboratory in Copenhagen from 1876 to 1900. When he found his studies of the changes in protein content of grain during germination and fermentation hindered by the poor methods then available for determining nitrogen in organic material, he stopped his other work and developed the method described here.

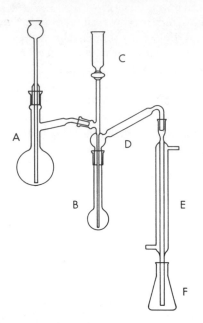

Figure 5-9. Diagram of distillation apparatus for micro Kjeldahl determination of nitrogen. (*A*) Steam generator; (*B*) Kjeldahl flask containing digested sample; (*C*) reservoir containing 40% solution of NaOH; (*D*) spray trap to block carryover of NaOH; (*E*) water-cooled condenser; (*F*) receiving flask containing saturated solution of boric acid.

The destruction of organic matter by oxidation of the carbon present to carbon dioxide without oxidation of the nitrogen to N_2 or higher oxidation states is difficult. Sulfuric acid accomplishes this by a combination of dehydration (to form carbon) and oxidation. Although these reactions are often slow, they may be accelerated either by addition of sodium or potassium sulfate to raise the boiling point of the digestion mixture or by addition of catalysts such as copper, mercury, or selenium. Copper selenite, $CuSeO_3$, is especially effective and widely used.

Nitrogen in higher oxidation states than ammonia must first be reduced to that state before the distillation step. A variety of reductants have been found applicable to different compounds. For example, nitrate ion can be reduced to ammonia with Devarda's alloy (50% Cu, 45% Al, 5% Zn) in basic solution.

The flask used for the digestion and distillation steps has an elongated neck to reduce loss of sample during digestion. The long neck also reduces carry-over of sodium hydroxide from the flask to the receiver by spray or bumping during distillation. Carry-over may be reduced further still by incorporation of a spray trap between the flask and the receiver (Figure 5-9).

The amount of boric acid used in the receiving flask is not critical[18] as long as it is in excess of the ammonia to be distilled. The basis of the method is the reaction of boric acid with ammonia to form ammonium borate:

---

[18]Kjeldahl originally placed a measured excess of a standard solution of hydrochloric acid in the receiving flask to convert the ammonia to ammonium ion and thereby avoid loss by volatilization. The acid remaining after distillation of the ammonia was titrated with a standard solution of sodium hydroxide, and the amount of ammonia calculated from the difference between the millimoles of HCl originally present in the receiving flask and the millimoles of NaOH required in the titration. The modern procedure replaces the HCl with a saturated solution of boric acid, H_3BO_3, thereby eliminating the preparation and standardization of a solution of NaOH in addition to HCl.

$$NH_3 + H_3BO_3 \rightleftharpoons NH_4^+ + H_2BO_3^-$$

Although the equilibrium constant for this reaction is only about unity, the reaction is driven to the right by the high concentration of boric acid. In effect, the volatile weak base NH_3 is replaced by the nonvolatile weak base $H_2BO_3^-$, which can be titrated directly with standard HCl in the same way as NH_3.

The Kjeldahl method remains important despite the empirical nature of the sulfuric acid digestion step. Intensive study has yet to produce a satisfactory alternative procedure. Digestion conditions must be carefully controlled to ensure complete decomposition of organic material without loss of nitrogen as N_2. Both the digestion and the distillation steps are time- and labor-intensive.

5-9 TITRATION OF POTASSIUM ACID PHTHALATE

This experiment introduces the basic operations in acid-base titrimetry. A solution of approximately 0.1 M sodium hydroxide is prepared and standardized against primary-standard potassium acid phthalate (KHP). (See Section 5-7 for structure and properties of KHP.) An unknown sample containing KHP is then titrated with it. The ionic reaction, with Na^+ and K^+ ions neglected, is

$$OH^- + C_8H_4O_4H^- \rightleftharpoons C_8H_4O_4^{2-} + H_2O$$

Procedure (time 3 to 4 h)

Preparation of 0.1 M NaOH Solution

Gently boil about 800 mL of distilled water for 4 or 5 min to remove dissolved carbon dioxide. Cover the container with a watch glass, and allow the water to cool to room temperature. Transfer about 500 mL to a clean polyethylene or borosilicate-glass bottle.[19] Add sufficient NaOH to give an approximately 0.1 M solution and mix.[20] Keep tightly covered to minimize absorption of CO_2.

Standardization of NaOH solution against KHP

Weigh, to the nearest 0.1 mg, samples of 0.8 to 0.9 g of primary standard KHP into conical (Erlenmeyer) flasks.[21] Use the technique of weighing by difference. Add 50 to 75 mL of distilled water to each flask, swirl to dissolve, and add 2 to 3 drops of 0.1% phenolphthalein indicator solution. Titrate with NaOH

---

[19]Avoid glass-stoppered bottles, since strongly alkaline solutions may cause the stopper to freeze.

[20]Because reagent-grade NaOH may contain considerable Na_2CO_3, it is frequently dispensed as a saturated solution, in which Na_2CO_3 is sparingly soluble. To prepare a saturated solution, add 50 g of NaOH pellets to 50 mL of distilled water, and shake until dissolved. Let stand until the insoluble Na_2CO_3 has settled (several days may be required). Then remove the required quantity with a Mohr pipet and rubber bulb (about 3.5 mL for 500 mL of 0.1 M solution). This solution may be provided in the laboratory.

[21]KHP may have some surface moisture, which can be removed if desired by drying the material at 110°C for 2 to 3 h. Because the amount of moisture is a few hundredths of a percent at most, this step may generally be omitted in all but the most precise work.

solution to the first permanent pink, swirling the flask continuously. Add titrant rapidly at first, then dropwise as the pink persists. Near the end point rinse the walls of the flask once or twice with a small amount of distilled water from a wash bottle. For greater precision, add fractions of a drop when within the last one or two tenths of a milliliter.[22] The pink fades on standing owing to absorption of CO_2 from the atmosphere. Record the final buret reading, including the applicable calibration corrections.

Determination of KHP in an Unknown Sample

Dry the sample for 2 to 3 h at 110°C. Allow to cool 20 to 30 min before weighing. Then weigh by difference (to the nearest 0.1 mg) three or four portions of 1.0 to 1.3 g each into 200-mL conical flasks. Dissolve the sample in 50 to 75 mL of distilled water, and add 2 to 3 drops of 0.1% phenolphthalein solution. Titrate with a standard solution of sodium hydroxide to the first permanent pink as in the standardization.

Calculate and report the percentage of KHP in the sample.

Calculations

To calculate the percentage of KHP in the unknown sample, first calculate the molarity of the NaOH solution from the standardization titrations. Then, with the best value of the NaOH molarity, calculate the percentage of KHP from the titrations of the unknown samples. Use either averages or medians throughout the calculations. If averages are chosen, use the Q test (Section 1-5) as the criterion for the rejection of suspect experimental data.

Molarity of NaOH

Since KHP and NaOH react in a one-to-one mole ratio according to the balanced equation, the molarity of the NaOH titrant solution may be obtained from the following relations:

$$M \text{ NaOH} = \frac{\text{mmol NaOH}}{\text{mL NaOH}} = \frac{\text{mmol KHP}}{\text{mL NaOH}} = \frac{\text{g KHP}}{(\text{mmol wt KHP})(\text{mL NaOH})}$$

Percentage of KHP in Sample

From the average or median molarity of the NaOH solution, the percentage of KHP in each sample can be calculated as follows:

$$\% \text{ KHP} = \frac{\text{g KHP in sample}}{\text{g sample}} \times 100 = \frac{(\text{mmol KHP in sample})(\text{g KHP/mmol})(100)}{\text{g sample}}$$

$$\% \text{ KHP} = \frac{(\text{mL NaOH})(\text{molarity NaOH})(\text{g KHP/mmol})(100)}{\text{g sample}}$$

---

[22]To deliver amounts less than 1 drop from a buret, first let a droplet form on the tip, and then touch the tip momentarily to the inside wall of the flask. Rinse the wall with a small amount of distilled water from a wash bottle to ensure that the titrant is washed into the solution. Do not rinse the tip of the buret.

Calculate the average or median of the set of titrations and report the value. *Recheck all calculations.* Errors in calculation may affect analytical results as much as errors in experimental procedure. Do not throw away hours of careful laboratory work by a careless mistake in equations or arithmetic.

5-10 DETERMINATION OF SODIUM CARBONATE

In this experiment a solution of hydrochloric acid is prepared, standardized against pure sodium carbonate, and used to determine the percentage of sodium carbonate in a sample. The characteristics of this reaction have been discussed in Section 5-8, which should be read carefully in preparation for the experimental work.

Pure anhydrous sodium carbonate, in addition to possessing most of the properties of a suitable primary-standard base, has the added advantage in this instance of being the same compound as the substance being determined. This tends to compensate for determinate errors in end-point selection. Because of this tendency, pure forms of the material being determined, where available, are almost always recommended for the standardization of titrants.

Procedure (time 4 to 5 h)

Preparation of 0.2 M HCl

Put a little less than 1 L of distilled water into a clean 1-L bottle. Calculate the volume of 6 *M* HCl required to prepare 1 L of 0.2 *M* HCl, and measure this quantity into a graduated cylinder. Transfer it to the bottle and mix thoroughly. Label.

Weighing of Primary Standard Na₂CO₃ and Sample

Dry 1.5 to 2.0 g of pure Na_2CO_3 in a glass weighing bottle at 150 to 160°C for at least 4 h.[23,24] Allow to cool, in a desiccator if necessary, and then weigh by difference (to the nearest 0.1 mg) three or four 0.35- to 0.45-g portions of the dry material into clean 200-mL conical (Erlenmeyer) flasks.

Dry the sample in a weighing bottle or vial at least 4 h at 150 to 160°C. Weigh into clean 200-mL flasks, to the nearest 0.1 mg, samples of 0.35 to 0.45 g of the dry material.

Titration of Standards and Samples

Add about 50 mL of distilled water to each flask and swirl gently to dissolve the salt. Add 4 drops of bromocresol green indicator to one of the flasks, and titrate with the HCl solution to an intermediate green. At this point stop the titration and boil the solution gently for 2 to 3 min, taking care that no solution is

---

[23]Na_2CO_3 tends to absorb H_2O from the air to form $Na_2CO_3 \cdot H_2O$, and CO_2 to form $NaHCO_3$. At least several hours of drying at 160°C is necessary to remove all H_2O and CO_2.

[24]Use a pencil or felt marking pen to label the container with the name or sample number of the contents and with your locker number. The container may be placed inside a small glass beaker, and a watch glass, raised with several bent portions of glass rod, placed on top for protection. Avoid leaving chemicals or equipment in the drying oven longer than necessary; this not only causes crowding, but increases the chance of equipment being broken or samples contaminated by spilled chemicals.

lost during the process.[25] Cool the solution to room temperature, wash the flask walls with distilled water from a wash bottle, and then continue the titration to the first appearance of yellow.[26]Just before the end point the titrant is best added in fractions of a drop. Record the buret reading and add to it the buret calibration correction. Complete one titration before beginning the next. Alternate titration of standards and samples.

Calculate and report the percentage of Na_2CO_3 in the sample.

Calculations

Calculate the molarity of the HCl solution. The procedure outlined in the discussion below may be used as a guide. Relative deviations of individual values from the average should not exceed about 2 parts per 1000.

When calculating the percentage of Na_2CO_3 in the sample, either the median or the average may be reported. If the average is chosen, use the Q test (Section 1-5) as the criterion for rejection of suspect experimental data. If the median is chosen, the median value for the molarity of the HCl should be used in the calculations rather than the average value. See Section 1-4 for a discussion of medians and averages.

The percentage of Na_2CO_3 in a sample can be calculated in two steps: (1) determination of the molarity of the HCl titrant from the standardization and (2) calculation of the percentage of Na_2CO_3 from titrations of the sample.

1. *Molarity of HCl.* In titrations of Na_2CO_3 with HCl to the pH 4 end point, two moles of HCl are added for each mole of Na_2CO_3:

$$2HCl + Na_2CO_3 \rightleftarrows H_2CO_3 + 2NaCl$$

The HCl molarity is obtained from the following relations:[27]

$$M\ HCl = \frac{mmol\ HCl}{mL\ HCl} = \frac{(mmol\ Na_2CO_3)(2)}{(mL\ HCl)}$$

$$= \frac{(g\ Na_2CO_3)(2)}{(mmol\ wt\ Na_2CO_3)(mL\ HCl)}$$

The factor 2 is required because each mole of Na_2CO_3 reacts quantitatively with two moles of HCl.

2. *Percentage of Na_2CO_3 in Sample.* Calculate the percentage of Na_2CO_3 in the sample as follows:

$$\%\ Na_2CO_3 = \frac{g\ Na_2CO_3\ in\ sample\ (100)}{g\ sample}$$

---

[25]To drive out CO_2 quantitatively, place the flask on a wire gauze and iron ring about 10 cm above a moderate Bunsen flame, and bring the solution to a rolling boil. Continue boiling for 2 to 3 min. If the solution tends to bump, insert a stirring rod and reduce the flame.

[26]Bromocresol green is an example of an indicator whose color depends strongly on temperature (Section 5-6).

[27]In these calculations millimolecular weight, equal to molecular weight/1000, is used because it is convenient to work with volumes in milliliters. If desired, molecular weights and volumes in liters may be employed.

$$= \frac{(\text{mmols Na}_2\text{CO}_3)(\text{g Na}_2\text{CO}_3/\text{mmol})(100)}{\text{g sample}}$$

$$= \frac{(\text{mL HCl})(\text{molarity HCl})(\text{g Na}_2\text{CO}_3/\text{mmol})(100)}{(2)(\text{g sample})}$$

Remember: Poor results are often caused by errors in calculation rather than by faulty laboratory technique. Check all calculations before reporting results. Calculate the average or median of the set of titrations and report the value.

EXAMPLE 5-27

A 0.3729-g sample of a mixture of Na_2CO_3 and inert material required 32.77 mL of HCl for titration. A 0.4404-g sample of pure Na_2CO_3 required 40.12 mL of the same HCl solution. What is the percentage of Na_2CO_3 in the mixture?

First calculate the HCl molarity:

$$M \text{ HCl} = \frac{(0.4404 \text{ g})(2)}{(40.12 \text{ mL})(0.10599 \text{ g/mmol})} = 0.2071 \text{ mmol/mL}.$$

Then calculate the percentage of Na_2CO_3 in the sample:

$$\% \text{ Na}_2\text{CO}_3 = \frac{(32.77 \text{ mL})(0.2071 \text{mmol/mL})(0.10599 \text{ g/mmol})(100)}{(2)(0.3729 \text{ g})} = 96.46\%$$

Notice that the units cancel each other.

PROBLEMS

5-1.   The melting and boiling points of water are unexpectedly high. What is the explanation? Why is it such an effective solvent?

5-2.   Calculate the pH corresponding to a hydrogen ion concentration of (a) 2.3×10^{-4} M, (b) 4.6×10^{-7} M, (c) 9.6×10^{-2} M; corresponding to a hydroxyl ion concentration (at 24°C) of (d) 4.2×10^{-10} M, (e) 3.1×10^{-6} M, and (f) 3.9×10^{-3} M.

5-3.   Convert the following pH values to hydrogen ion concentration: 2.16, 5.38, 7.15, 8.09, 11.12, 4.68. Convert the following pH values to hydroxyl ion concentration (24°C): 4.26, 4.77, 10.14, 7.62, 12.11.

5-4.   Clearly distinguish between a concentrated solution of an acid and a solution of a strong acid, and give examples.

5-5.   Define a conjugate acid-base pair and give an example. What is the relation between the ionization constant of an acid and the ionization constant of its conjugate base?

5-6.   Write the expressions for the equilibrium constants for the following acids and bases, giving the value of the constant for each (see Appendix): formic acid, hydrofluoric acid, ammonium ion, phenol, aniline, hydroxylamine, cyanide ion.

5-7.   Write the formula for the conjugate acid of each of the following bases: dihydrogen phosphate, carbonate, pyridine, glycinate, aniline.

5-8.   Write the formula for the conjugate base of each of the following acids: hydrofluoric acid, dihydrogen phosphate, ammonium, glycine hydrochloride, bicarbonate.

5-9.   Calculate the pH of a 0.050 M solution of a weak acid with a $K_a = 5.0 \times 10^{-6}$.

5-10.†   (a) Calculate the pH of a 2×10^{-8} M solution of sodium hydroxide; of a 10^{-10} M solution of sodium hydroxide. (b) Calculate the pH of a 1×10^{-8} M solution of HCl.

5-11.   Calculate the pH of a solution that initially contained 1.22 g/L of benzoic acid and 2.88 g/L of sodium benzoate.

5-12.   What is the pH of a solution (a) containing 2.5 g of sodium hydroxide in 150 mL, (b) containing 5 mL of 6 M hydrochloric acid in 500 mL, (c) 0.2 M in acetic acid, (d) 0.5 M in ammonium chloride, (e) containing 0.35 mol of sodium acetate in 100 mL, (f) containing 4.7 g of NaCl and 0.01 mol of nitric acid in 200 mL, (g) containing equal weights of sodium carbonate and sodium bicarbonate?

5-13.   (a) Calculate the fractions of carbonic acid present as H_2CO_3, HCO_3^-, and CO_3^{2-} at pH 5. If the total concentration of all species is 0.05 M, what is the concentration of each species at this pH? (b) Carry out the calculations of (a) at pH 10.

5-14.†   A phosphate buffer solution is prepared by addition of concentrated sodium hydroxide solution to a 0.1 M solution of phosphoric acid until the pH is 7.5. (a) Neglecting dilution, calculate the concentration of each of the various phosphate species in the solution. (b) What is the ionic strength of the resulting solution?

5-15.   What is the pH of a solution that is 0.3 M in sodium acetate and 0.3 M in acetic acid? What is the pH of a solution 0.01 M in sodium acetate and 0.01 M in acetic acid? Which solution would be the preferred buffer for most applications? Explain.

5-16.   In what ratio would acetic acid and sodium acetate need to be mixed to prepare a buffer with a pH of 5? Give specific directions for the preparation of 500 mL of an acetate buffer, using solid sodium acetate and glacial acetic acid (100%) of density 1.05. How would the buffer be prepared from acetic acid and solid sodium hydroxide?

5-17.   In what ratio would ammonia and ammonium chloride need to be mixed to prepare a buffer with a pH of 8.7? Give specific directions for its preparation, using appropriate starting materials. What happens to the pH if the buffer is added to a solution containing some ammonium sulfate?

5-18.   How much solid sodium acetate should be added to 300 mL of 0.8 M acetic acid to produce a buffer with a pH of 4.3?

5-19.   How much phosphoric acid should be added to 300 mL of 0.4 M sodium dihydrogen phosphate solution to produce a buffer with a pH of 4.3?

5-20.   What ratio of phosphoric acid to sodium dihydrogen phosphate would be needed in a solution to produce a buffer with a pH of 5?

5-21.   Give detailed directions for the preparation of 250 mL of a phosphate buffer of pH 5, using 85% phosphoric acid of density 1.69 and $NaH_2PO_4 \cdot H_2O$.

5-22.   What is the ratio of concentrations of sodium acetate and acetic acid required to produce a buffer with a pH of 4? What ratio of sodium benzoate and benzoic acid? If the two buffers are 0.1 M in the appropriate acid, which is the preferred buffer? Explain.

5-23.   To prepare a buffer of pH 6, would you choose acetic acid–sodium acetate or ammonia–ammonium chloride? Explain.

5-24.   Of the acids and bases listed in the Appendix, which one, along with one of its salts, would be the most appropriate for preparing a buffer of pH 7.0?

5-25.   You have available in the laboratory 6 M HCl, 6 M NH$_3$, and solid NH$_4$Cl. Explain how to prepare 1 L of a buffer solution of pH 10.3. Calculate the required amount of each reagent.

5-26.   A sample contains pyridine hydrochloride, a salt of pyridine. Could a satisfactory acid-base titration of this material be carried out with a visual indicator? If so, explain why, naming a suitable titrant and indicator.

5-27.   Why can borax, a sodium salt of boric acid, be titrated precisely by an acid-base reaction, whereas boric acid cannot?

5-28.   What volume of 37% hydrochloric acid of density 1.18 would be required to prepare 2 L of 0.10 M solution?

5-29.   Calculate the pH at the start, midpoint, equivalence point, and 50% past the equivalence point for the titration of 0.1 M ammonia with 0.1 M hydrochloric acid. Sketch the titration curve expected, and select a suitable indicator from the list in Table 5-1.

5-30.   Calculate the pH at the start, midpoint, and equivalence point for the titration of 0.1 M trimethylamine with 0.1 M hydrochloric acid. On the basis of the calculations, sketch the titration curve expected. Select a suitable indicator from the list in Table 5-1.

5-31.   Sketch the approximate titration curves expected (calculations not required) for the titration with 0.1 M hydrochloric acid of (a) a sample of pure sodium bicarbonate; (b) an equimolar mixture of sodium bicarbonate and sodium carbonate; (c) an equimolar mixture of sodium hydroxide and sodium carbonate. Label the axes clearly and completely.

5-32.   Why does the titration curve for phosphoric acid show two but not three analytically useful end points?

5-33.   A 1.146-g sample of sausage is analyzed for nitrogen (protein) content by the Kjeldahl method. The ammonia produced is collected in a boric acid solution. The borate thus produced requires 23.12 mL of 0.1018 M standard hydrochloric acid to reach the methyl red end point. What is the nitrogen content of the sausage? (The practical problem of determination of the sample weight owing to variable water content is a major source of uncertainty in this analysis.)

5-34.   A 50.00-mL sample of a household cleaner is diluted to 250 mL in a volumetric flask. A 20.00-mL aliquot requires 37.62 mL of 0.2222 M hydrochloric acid to neutralize it to a bromocresol green end point. Calculate the percentage of NH$_3$ in the cleaner, assuming its density is 0.974. Would cresol purple be a suitable indicator for the titration?

5-35.   What precautions are necessary in the use of a standard solution of sodium hydroxide? Why is it not a suitable primary standard? What happens to its strength as a base if carbon dioxide is absorbed?

5-36.   A 0.3752-g sample of pure sodium carbonate requires 38.21 mL of a hydrochloric acid solution to reach the bromocresol green end point. What is the molarity of the hydrochloric acid?

5-37.   A 0.5063-g sample of a mixture of sodium and potassium carbonates requires 41.84 mL of a 0.2023 M hydrochloric acid solution to reach the bromocresol green end point. What is the percentage of carbonate as CO_3^{2-} in the sample?

5-38.   Which pairs of the four compounds sodium hydroxide, sodium bicarbonate, sodium carbonate, and carbonic acid can exist in significant concentrations in solution?

5-39.   A slurry contains two of the compounds sodium hydroxide, sodium bicarbonate, and sodium carbonate. A 0.2116-g sample requires 36.55 mL of 0.1020 M hydrochloric acid to reach a phenolphthalein end point (pH 8.4) and an additional 6.34 mL to reach the bromocresol green end point. Which compounds are present, and what is the percentage of each?

5-40.   What fraction of the acid-base indicator phenol red is in the acidic form at pH 6; at pH 8?

5-41.   (a) Calculate the equilibrium constant for $NH_3 + H_3BO_3 \rightleftharpoons NH_4^+ + H_2BO_3^-$. (b) If 3.00 mmol of ammonia is distilled into a receiver flask containing 75 mL of saturated H_3BO_3, what is the concentration of NH_3 in the flask at equilibrium? (c) What fraction of the total ammonia is present as NH_3? The solubility of H_3BO_3 in water is about 60 g/L.

5-42.   Using data obtained with a pH meter during a titration explain how the ionization constant of an acid may be determined experimentally; the ionization constant of a base.

5-43.   What weight of sample that contains mainly sodium carbonate should be taken for its determination with a standard solution of 0.23 M hydrochloric acid?

5-44.   Outline a method for the standardization of a solution of barium hydroxide. What precautions should be taken to ensure that its composition does not change?

5–45.†   (a) Why are indicators used only in the form of dilute solutions such as 0.1%? (b) Suppose that 0.1% methyl red (mol wt 269) is used as indicator in a titration to determine the amount of bicarbonate in a water supply. Five drops (0.25 mL) of methyl red is added to a 100-mL sample of water, and 4.74 mL of 0.01072 M hydrochloric acid is required to change the indicator to the midpoint in its transition range. On the assumption of no error from the indicator, what is the percentage of bicarbonate as calcium bicarbonate in the sample? If the indicator was initially in its acid form, what is the magnitude of the indicator error? What is the correct value for the bicarbonate analysis?

5-46.   Sketch the titration curve to be expected for the titration of a 0.1 M solution of hydroxylamine with a suitable analytical titrant. What titrant might be used; what indicator? Would the end point be analytically useful? What is the pH at the equivalence point?

5-47.   Calculate the pH in the titration of 20 mL of 0.1 M phosphoric acid after the addition of the following volumes of 0.1 M NaOH: (a) 0 mL, (b) 10 mL, (c) 20 mL, (d) 30 mL, (e) 40 mL. Sketch the titration curve.

5-48.   Calculate the pH of a 1×10^{-8} M solution of HCl.

5-49.   A 10 mL sample of vinegear (acetic acid, $HC_2H_3O_2$) was pipetted into a flask, 2 drops of phenolphthalein indicator added, and the acid titrated with 0.1008 M sodium hydroxide. If 45.62 mL was required for the titration, what was the molar concentration of acetic acid in the sample? If the density of the pipetted acetic acid solution was 1.004 g/mL, what was the percentage of acetic acid in the sample?

5–50.†   How many grams of sodium acetate must be added to 500 mL of 0.0217 M H_2SO_4 to give a solution of pH 4.03?

5–51.†   How many grams of trisodium phosphate (Na_3PO_4) must be added to 500 mL of 0.0192 M H_2SO_4 to give a solution of pH 7.5? If the calculated weight were added, the solution mixed, and the pH measured, would the value likely be exactly 7.5? Explain.

5-52.   In a photometric titration of iron a chloroacetate buffer is required. If 2 g of NaOH is added to 24 g of monochloroacetic acid, $CH_2ClCOOH$, how many moles of the chloroacetate salt are formed and how many moles of the chloroacetic acid remain? What is their ratio?

5-53.   To adjust the pH in the Fajans (adsorption indicator) titration of Section 4-4, a buffer of 0.5 M acetic acid and 0.5 M sodium acetate is required. Describe how to prepare 250 mL of this buffer from 6 M acetic acid and solid sodium acetate.

REFERENCES

P. A. Giguere, *J. Chem. Educ.* **1979,** *56,* 571. Reviews the evidence concerning the nature of the hydronium ion and presents the case for formulating it as H_3O^+.

H. A. Laitinen and W. E. Harris, *Chemical Analysis*, 2nd ed., McGraw-Hill, N.Y., 1974, Chapter 3. Concise treatment of acid-base equilibria in water at an advanced level.

A. E. Martell and R. M. Smith, *Critical Stability Constants*, Plenum, N.Y., 1974–76. In four volumes. A compilation of selected acid and base equilibrium constants.

A. L. Underwood, *Anal. Chem.* **1961,** *33,* 955. The enzyme carbonic acid anhydrase accelerates the hydration of carbon dioxide dissolved in water; the resulting carbonic acid can be titrated rapidly and accurately.

6

ACID-BASE TITRATIONS IN
NONAQUEOUS SOLVENTS

Our titration method represents, we believe, a new departure in alkalimetry and is not only interesting from a theoretical standpoint but practically should become applicable to a large number of organic acids which hitherto it has not been possible to determine by titration.

O. Folin and H. Wentworth,
J. Biol. Chem. **1910** 7, 421.

Water is so important and ubiquitous that it is easy to overlook the role it plays as solvent in analytical reactions. In fact, water as solvent is directly involved in all the acid-base and precipitation reactions considered thus far. Is it ever advantageous to use a solvent other than water as a reaction medium for analysis? In many circumstances the answer is "yes" for precipitation, acid-base, complexation, and oxidation-reduction titrations, as well as for gravimetry and the determination of organic functional groups. The most widespread applications have been in the area of acid-base titrations. For this reason, and because proton-transfer reactions effectively illustrate solvent involvement in the reaction process, we limit our discussion to this area. More detailed information is available in the references listed at the end of the chapter.

Solvents other than water are more expensive and complicated to use. Interfering impurities must be removed. (A common one is water itself, which is difficult both to remove and to keep removed.) Therefore, water is replaced by another solvent only when necessary, the principal reasons having to do with solubility and extent of reaction. Many compounds insoluble in water are soluble in organic solvents. Also, acids up to billions of times stronger than hydrochloric or sulfuric acid can exist in nonaqueous solvents. These "superacids" permit the titration of many organic compounds that are too weakly basic to be titrated in water. Similarly, "superbases" can be prepared in nonaqueous solvents for the titration of exceedingly weak acids.

Acid-base character and dielectric constant are the two main properties of solvents that affect acid-base reactions. In this chapter we consider these properties, along with the range of applicability of acid-base titrations in nonaqueous solvents and some practical aspects of working with these systems.

6-1   PROPERTIES OF SOLVENTS

Acid-Base Character

Consider a base B with a K_b in water of 1×10^{-10}. This base cannot be titrated quantitatively in water because H_3O^+ is not a sufficiently strong acid to protonate B completely when an equimolar amount of H_3O^+ is added to B. That is, the equilibrium for the reaction

$$B + H_3O^+ \rightleftharpoons BH^+ + H_2O \qquad (6\text{-}1)$$

does not lie far enough to the right to make titration feasible. The necessary acid strength can be obtained only by replacing H_2O with another solvent S that has less affinity for protons. In other words, the solvent S must be a weaker base than H_2O, so that its corresponding conjugate acid, HS^+, is stronger than H_3O^+. With this stronger acid it is now reasonable that equilibrium for the reaction

$$B + HS^+ \rightleftharpoons BH^+ + S$$

will lie farther to the right than that for (6-1), and the reaction may be analytically useful.

In this example a base B can be titrated in the solvent S because S, being a much weaker base than water, does not compete for protons. Examples of such solvents are acetonitrile (CH_3CN) and glacial acetic acid (CH_3COOH).[1]

Selection of the acid to be used as titrant with solvent S is important. Acids considered strong in water do not necessarily function as strong acids in weakly basic solvents. In acetic acid as solvent, perchloric acid has been found experimentally to be much stronger than hydrochloric. The equilibrium constant for the reaction

$$HClO_4 + HOAc \rightleftharpoons H_2OAc^+ + ClO_4^-$$

is much larger than for

$$HCl + HOAc \rightleftharpoons H_2OAc^+ + Cl^-$$

We conclude, then, that ClO_4^- attracts protons less strongly than does Cl^- and therefore is a weaker base. Both $HClO_4$ and HCl are strong acids in water, which is sufficiently basic that proton transfer to H_2O from $HClO_4$ or HCl to form H_3O^+ is quantitative at all concentrations. No acid stronger

---

[1]Glacial acetic acid refers to 100% acetic acid, the common form of the commercial reagent. The name derives from the fact that the material is concentrated by cooling until pure acetic acid freezes out at 16.7°C.

than H_3O^+ can exist in water. The acid strengths of $HClO_4$ and HCl are both reduced to that of H_3O^+ and are said to have been *leveled* to that of H_3O^+. This is referred to as the *solvent-leveling effect*.

The way to determine a base too weak to be titrated in water is then to carry out the titration in a solvent having as little basic character as possible, utilizing as titrant an acid that is strong in that solvent. The most widely used titrant-solvent system for the titration of weak bases is perchloric acid in glacial acetic acid.

For the determination of acids too weak to be titrated in water, corresponding conditions hold. The strongest base that can exist in water is aquated OH^-; any stronger base reacts with water to form hydroxide ion:

$$B + H_2O \rightleftharpoons BH^+ + OH^-$$

Thus, water levels all strong bases to the strength of the hydroxide ion. By substituting for water a solvent that has minimal acid properties, we can titrate weaker bases. No single strong base–solvent combination is used so universally as the strong acid–solvent system of perchloric acid in acetic acid; the most important combinations are tetrabutylammonium hydroxide in a toluene–methanol mixture or 2-propanol, and potassium methoxide in toluene–methanol.

The key property of acetic acid that makes it so useful for the titration of weak bases is its weak basicity. That it is moderately acidic is unimportant. Other solvents, such as ethylenediamine ($H_2NCH_2CH_2NH_2$), may be weakly acidic but moderately basic. Solvents that possess both acidic and basic properties are sometimes termed *amphiprotic*. Amphiprotic solvents undergo self-dissociation, or *autoprotolysis*, according to

$$2SH \rightleftharpoons SH_2^+ + S^-$$

Examples of amphiprotic solvents include water, acetic acid, ammonia, ethylenediamine, and methanol:

$$2H_2O \rightleftharpoons H_3O^+ + OH^-$$

$$2HOAc \rightleftharpoons H_2OAc^+ + OAc^-$$

$$2NH_3 \rightleftharpoons NH_4^+ + NH_2^-$$

$$2H_2NCH_2CH_2NH_2 \rightleftharpoons H_2NCH_2CH_2NH_3^+ + H_2NCH_2CH_2NH^-$$

$$2CH_3OH \rightleftharpoons CH_3OH_2^+ + CH_3O^-$$

The equilibrium constant for any one of these reactions, called the *autoprotolysis constant*, is generally written

$$K_S = [SH_2^+][S^-]$$

Table 6-1.   Properties of Several Nonaqueous Solvents

Solvent	pK_S	Dielectric constant
Acetic acid	14.5	6.1
Acetone	---	20.7
Acetonitrile	32	36.0
Ammonia	29.8	16.9
Benzene	---	2.3
Carbon tetrachloride	---	2.2
Chloroform	---	4.8
Dimethylformamide	18.0	36.7
Dimethylsulfoxide	33	46.6
Ethanol	18.9	24.3
Ethylenediamine	15.2	12.5
Formic Acid	6.2	57
Methanol	16.5	32.6
Methylisobutyl ketone	---	13
2-Propanol	20.6	18.0
Pyridine	---	12.0
Toluene	---	2.4
Water	14.0	78.5

Values for the negative logarithms of K_S, pK_S, are listed for a number of solvents in Table 6-1.

Solvents that have little or no acidic or basic character also have little or no tendency to undergo self-dissociation; these solvents are termed *aprotic*. Aprotic solvents having slight acid-base properties include dimethylsulfoxide $((CH_3)_2SO)$, acetonitrile (CH_3CN) and dimethylformamide $((CH_3)_2NCHO)$. Those having no appreciable acid-base properties include carbon tetrachloride (CCl_4), benzene (C_6H_6), and hexane (C_6H_{14}); these are often termed *inert* because they show so little tendency to undergo reaction with substances dissolved in them.

A few solvents, including pyridine, the ethers, and several ketones, do not fit readily into either the amphiprotic or aprotic classes. They possess basic properties, being able to accept protons on basic nitrogen or oxygen atoms, but their acid properties are negligible, and so they are often useful solvents for the titration of weak acids.

A solvent with a small K_S interacts little with solute species, a desirable characteristic for acid-base titrations. When K_S is large, protonation or deprotonation of solvent is likely, resulting in leveling of the strength of the acid or base and a decrease in the equilibrium constant for the sample-titrant reaction. Acetic acid has about the same K_S value as water. It is a stronger acid than water, but a weaker base. Being itself a very weak base, it is useful in titrations of weak bases. Ethylenediamine also has about the same K_S as water, but being a very weak acid, can be used in the titration of weak acids. Solvents such as carbon tetrachloride show no measurable acid or base properties and so are ideal from the standpoint of noninterference with proton transfer between the titrant and sample molecules. At the other extreme, formic acid, HFm, has so large a K_S value that even acids of only

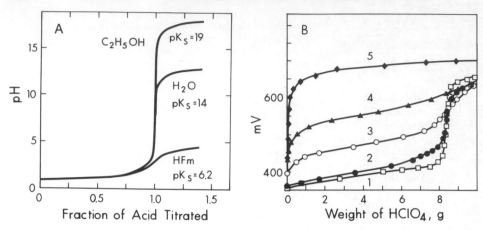

Figure 6-1. (A) Calculated titrations of 0.1 M solutions of the strongest possible acid with the strongest possible base in ethanol, water, and formic acid. (B) Titrations of amines of varying base strength with perchloric acid in glacial acetic acid. (The ordinate is a millivolt scale instead of pH to indicate that actual pH is unknown in this system.) pK_b values for H_2O as solvent: (1) Aniline, 9.4; (2) p-bromoaniline, 10.1; (3) o-chloroaniline, 11.2; (4) p-nitroaniline, 12.1; (5) blank. Data by D. Leung.

moderate strength are leveled to H_2Fm^+, the protonated solvent, and little change in H_2Fm^+ can occur at the equivalence point.

EXAMPLE 6–1

For a titration of 0.1 M $H_2Fm^+Cl^-$ with 0.1 M Na^+Fm^- in pure formic acid as solvent, calculate the pH at the start, the equivalence point, and 100% past the equivalence point. Sketch the approximate titration curve.

For this titration of a strong acid with a base the calculations are similar to those in Chapter 5 for water, except that a value for K_S of 6×10^{-7} replaces K_w and pH is defined as $-\log [H_2Fm^+]$.

At the start: $[H_2Fm^+] = 0.1$, and pH = 1.

At the equivalence point: $[H_2Fm^+] = [Fm^-]$, and so $K_S = [H_2Fm^+]^2$, $[H_2Fm^+]$ = the square root of 6×10^{-7}, or 7.7×10^{-4}, and pH = 3.1.

At 100% past the equivalence point: On the assumption that reaction is complete, $[Fm^-] = 0.1/3$, or 0.033 (the 3 taking into account dilution caused by the addition of two volumes of Na^+Fm^- to each volume of $H_2Fm^+Cl^-$ originally present). Then $[H_2Fm^+] = K_S/[Fm^-] = (6 \times 10^{-7})/0.033 = 1.8 \times 10^{-5}$, and pH = 4.7. Thus the equivalence-point break in the titration is slight.

Figure 6-1A includes a curve for the titration in the preceding example, along with curves for comparable titrations of strong acid with strong base in water and ethanol. Note that pH values well above 14 are feasible in solvents of large pK_S. Precise measurement of pH values larger than about 16 is difficult.

Dielectric Constant

The acid or base properties of a solute are strongly affected by interactions of solute molecules or ions with solvent molecules. Dissociation of an uncharged acid HA, when dissolved in a solvent SH, takes place in two steps. The first is formation of an *ion pair,* and the second dissociation of the ion pair to form free ions:

$$HA + SH \rightleftharpoons SH_2{}^+A^- \rightleftharpoons SH_2{}^+ + A^-$$

For a base B a similar formation and separation of ions may occur:

$$B + SH \rightleftharpoons BH^+S \rightleftharpoons BH^+ + S^-$$

A measure of the capacity of a solvent to separate ions of opposite charge is given by a property called the *dielectric constant.* With a solvent of high dielectric constant, solute ions tend to be surrounded by solvent molecules through ion-molecule interaction; this process facilitates separation of oppositely charged ions of the solute. The higher the dielectric constant of a solvent, the more effective it is in promoting ion separation. Since the value for water is relatively high (78.5), most ionic solutes are completely dissociated in it. At the other end of the scale, few dissociated ions are observed in benzene (dielectric constant 2.3). Ion formation may still occur in benzene, but the ions produced will tend to stay together through electrostatic attraction as ion pairs. Thus acid-base titrations may be carried out quantitatively in solvents such as benzene even though free $SH_2{}^+$, and therefore the pH, may change little throughout the titration. Indicators must be used to follow such titrations; a pH meter and glass indicator electrode system responds only to free $SH_2{}^+$.

6-2   TITRANTS AND SOLVENTS

Titrants

Strong Acids

Perchloric acid, because it is the strongest of the common acids, is attractive as a titrant for the determination of bases too weak to be titrated in water. Solutions in glacial acetic acid as solvent are prepared by adding the required amount of concentrated (60 or 70 to 72%) perchloric acid. Commercial-reagent perchloric acid[2] (72%) has the composition $HClO_4 \cdot 2H_2O$; the water introduced with the titrant usually does not cause difficulty except when very weak bases are to be titrated. In these instances it can be removed

---

[2]Perchloric acid is widely used in several areas of analytical chemistry, but must be handled with care. Cold or dilute, it is not an oxidizing agent, but hot and concentrated, it is an extremely powerful oxidizing and dehydrating agent that can cause explosions if improperly handled. Aqueous solutions, marketed as 72% or 60% perchloric acid, are completely stable, but they should of course be handled with the care accorded all concentrated strong acids. In addition, contact with oxidizable organic material should be avoided; on standing perchlorate esters may form, which may become shock-sensitive when dry. Metal ions, especially of the transition metals, often serve as catalysts for explosive reactions between perchlorates and oxidizable material. Good laboratory practice requires immediate cleanup of all spills. Rinse all materials such as paper towels used in the cleanup of perchloric acid or perchlorate compounds with copious volumes of water. The use of perchloric acid as an oxidant in sample preparation is discussed in Section 21-3.

by adding an equivalent amount of acetic anhydride and allowing the mixture to stand about a half hour. The acetic anhydride reacts with water to form acetic acid:

$$CH_3-C{\overset{O}{\underset{O}{\diagdown}}}\ \ CH_3-C{\overset{O}{\diagdown}} + H_2O \longrightarrow 2CH_3COOH$$

An excess of acetic anhydride is acceptable unless primary ($R-NH_2$) or secondary (R_2NH) amines are to be titrated; these compounds react with the excess to form amides. Solutions of 0.01 to 0.1 M perchloric acid in glacial acetic acid are stable for months.

When the leveling effect of acetic acid is undesirable in titrations of mixtures of bases, 1,4-dioxane sometimes serves as a replacement. Although with dioxane the water introduced into the perchloric acid cannot be removed, the effect on the results is often slight. Commercial dioxane may contain impurities that cause perchloric acid solutions to turn brown but do not cause titration error. The impurities may be removed if desired by distillation of the dioxane before the addition of perchloric acid.

1,4-Dioxane

Perchloric acid solutions in glacial acetic acid are usually standardized against potassium acid phthalate (KHP). In this titration the KHP acts as a *base*, the HP^- ion accepting a proton from the acid $CH_3COOH_2^+ClO_4^-$ to form H_2P:

$$HClO_4 + \underset{CO_2H}{\overset{CO_2K}{\bigcirc}} \longrightarrow \underset{CO_2H}{\overset{CO_2H}{\bigcirc}} + KClO_4$$

This reaction is driven to the right in part by the strength of the titrant acid and in part by the tendency of HP^- to accept a proton in a solvent system such as glacial acetic acid. (From the first ionization constant as an acid of 1.1×10^{-3}, the ionization constant as a base is calculated to be 9.1×10^{-12}.) This is the reverse of the reaction of KHP in water, where it serves as a primary-standard *acid* for the standardization of solutions of strong bases. The sluggish dissolution of KHP in glacial acetic acid may be speeded by warming.

In addition to its superior acid strength, perchloric acid has the advantage that most perchlorate salts of organic bases are soluble in it and so do not interfere in the titration. Also, end points can be detected either by visual indicator or potentiometrically.

Certain aromatic sulfonic acids, particularly *p*-toluenesulfonic acid in chloroform or acetonitrile, though not so strong as perchloric, have the advantage that their salts are often more soluble than the perchlorates and so less likely to interfere in potentiometric titrations.

Strong Bases

Tetrabutylammonium hydroxide, $(C_4H_9)_4N^+OH^-$, in toluene–methanol or in 2-propanol is the most important titrant for weak acids in nonaqueous systems. A small amount of methanol provides the required solubility; the dilution with toluene or 2-propanol yields sharper end points than are observed in high concentrations of methanol.[3] The base is available commercially as a concentrated (1 M, 25%) solution in methanol that is readily diluted to 0.1 M with toluene or 2-propanol. These titrant solutions undergo slow decomposition and so must be standardized at the time of use. Decomposition is inhibited by storage at lower temperatures. Absorption of carbon dioxide leads to errors from carbonate, as in aqueous solutions of sodium hydroxide (Section 5-7). Precautions to minimize carbonate contamination during and after preparation of all strongly basic titrants are essential.

Potassium methoxide, $KOCH_3$, also is an effective titrant for a variety of weak acids. It is prepared by allowing potassium metal to react with methanol:

$$2K + 2HOCH_3 \rightarrow 2KOCH_3 + H_2(g)$$

The resulting potassium methoxide–methanol mixture is diluted with toluene to give an approximately 0.1 M solution of $KOCH_3$ in 90% toluene–10% methanol. Sodium may be substituted for potassium with the same result.

All the above titrants are prepared to an approximate concentration and then standardized, ordinarily against pure benzoic acid. This acid, in addition to having all the qualities of a good primary standard, is soluble in most

Solvents

The solvents often selected for dissolution of samples may be unsatisfactory for the preparation of titrants because of their slow reaction with strong acid or base reagents.

An ideal solvent provides adequate solubility for, but is unreactive with, reagent, sample, and reaction products. For the titration of either weak acids or weak bases it is either aprotic or inert. Among the solvents frequently used are acetonitrile, acetone, methylisobutyl ketone, and the lower alcohols (methanol, ethanol, and 2-propanol). Sometimes mixtures of two solvents are preferred for special samples. For example, mixtures of a glycol, such as ethylene glycol, and a solvent for hydrocarbons, such as 2-propanol, constitute the so-called G–H solvents for dissolution and titration of carboxylic acid salts and other weak bases.

For titration of weak acids, solvents that are also used include, in addition to the ones previously mentioned, dimethylformamide, dimethylsulfoxide, ethylenediamine, and t-butyl alcohol. For the titration of weak bases, other useful solvents are acetic acid, 1,4-dioxane, and nitromethane. Solvent mixtures of acetic acid and acetic anhydride may provide sharper end points in the titration of some very weak bases.

[3]Toluene is recommended as replacement for the more toxic benzene used in earlier work. Titration precision is reported to be unaffected by the change. (W. Selig, *Microchem. J.* **1978**, *23*, 466.)

When the titration of a mixture of acids of differing base strengths is to be followed with a pH meter, the solvent should be selected with thought for possible leveling effects. For example, in the titration of a mixture of butylamine and pyridine in glacial acetic acid, the butylamine reacts with the acetic acid solvent to give acetate ion:

$$C_4H_9NH_2 + HOAc \rightleftharpoons C_4H_9NH_3^+ + OAc^-$$

Since the base strength of OAc^- is about the same as that of pyridine, only one inflection in the titration curve is seen. If the titration is carried out in a solvent such as acetonitrile, however, two breaks are seen because acetonitrile, having little acidic character, exerts no leveling effect. In general, aprotic solvents such as acetonitrile are preferred for differentiating titrations.

6-3 DETECTION OF THE END POINT

The end point in nonaqueous acid-base titrations, as in aqueous systems, may be detected by indicator or by potentiometric titration with a pH meter and reference electrode–indicator electrode pair. For titrations of weak bases with perchloric acid in glacial acetic acid, the indicators of choice are usually crystal violet, the closely related methyl violet, or α-naphtholbenzein. The change with crystal violet is from violet to blue, then to green, and finally to yellow. The violet-to-blue transition usually yields the most satisfactory end point. The change for α-naphtholbenzein is from green to yellow or olive.

For the titration of weak acids a variety of indicators including bromophenol blue and thymol blue have been recommended, depending on the strength of the acid. Greatest precision and accuracy are achieved by performing a preliminary potentiometric titration and then picking an indicator whose color transition coincides most nearly with the potentiometric end point. For potentiometric titrations the procedure is the same as in water, except for the electrode system. The glass electrode behaves acceptably in most solvents as a hydrogen-ion indicator electrode, and usually yields satisfactory titration curves in solvents of moderate to high dielectric constant. The tendency of the potential of this electrode to vary from day to day in many solvents is not a drawback if the electrode is needed only to determine the end point in a potentiometric titration. The response of the electrode is often sharpened by storage between titrations in the organic solvent in which it is used. The sensitivity of the glass electrode to change in hydrogen ion activity is about the same in nonaqueous solvents as in water.

Reference electrodes for measurements in nonaqueous solutions usually must be modified to avoid the presence of potassium chloride and water at the junction. The low solubility of potassium chloride in many organic solvents often produces plugged junctions at the water-solvent interface, and water may be undesirable because of its acid-base properties. The simplest modification is to use a conventional reference electrode, such as the saturated calomel electrode, and make contact between it and the solution being titrated with a salt bridge containing a solution of lithium or tetrabutylammonium nitrate or perchlorate in acetonitrile or alcohol. Alternatively, the aqueous potassium chloride solution in a calomel or silver–silver chloride reference electrode can be replaced by a nonaqueous solvent-solute mixture.

Other methods of end-point detection include conductometry (Section 12-5) and spectrophotometry (Section 14-10).

6-4 APPLICATIONS

Titration of Bases

Many organic compounds can be titrated as bases with perchloric acid. A large proportion of these contain nitrogen as primary (RNH$_2$), secondary (R$_2$NH), and tertiary (R$_3$N) aliphatic amines; aromatic amines (ArNH$_2$), amino acids; and compounds in which the nitrogen is incorporated into a ring structure, such as pyridine (pK_b in H$_2$O = 8.8) and pyrazole (pK_b in H$_2$O = 11.5).

Pyridine   Pyrazole

In general, amines having a pK_b in water of 12 or less can be titrated potentiometrically in glacial acetic acid (Figure 6-1B); a pK_b of 10 or less is required if an indicator is to be used.

Another group of compounds that can be titrated by this method includes the salts of weak acids. The reaction is

$$H_2OAc^+ + A^- \rightleftarrows HA + HOAc$$

Recall from Section 5-8 that in aqueous solution only the salt of a very weak acid, that is, an acid with a pK_a of approximately 8 or greater, can be titrated in this way. In glacial acetic acid the salts of carboxylic acids, of nitric acid, and of the first proton of sulfuric acid react quantitatively:

$$H_2OAc^+ + C_6H_5COO^- \rightleftarrows C_6H_5COOH + HOAc$$

$$H_2OAc^+ + NO_3^- \rightleftarrows HNO_3 + HOAc$$

$$H_2OAc^+ + SO_4^{2-} \rightleftarrows HSO_4^- + HOAc$$

The titration of KHP for the standardization of solutions of perchloric acid in glacial acetic acid is another example of this class of reaction. The titration of salts is of considerable importance in many areas of chemistry, particularly the pharmaceutical field.

Titration of Acids

An acid with a pK_a in water of approximately 10 or less is titratable with reasonable precision and accuracy in suitable nonaqueous solvents. The usually preferred titrant, tetrabutylammonium hydroxide, must be used with precautions to avoid contamination from carbonate and tertiary amines formed as decomposition products. End points can be determined either with a pH meter and a glass indicator electrode–reference electrode pair or with a visual indicator.

Although many of the simpler carboxylic acids are sufficiently acidic to be titratable in water, some of the substituted carboxylic acids such as amino acids are not. Glycine cannot be titrated in water, but in pyridine it can be titrated with tetrabutylammonium hydroxide. Other important compounds that can be titrated as weak acids include phenol and many substituted phenols. A typical reaction is

$$C_6H_5OH + OH^- \rightleftharpoons C_6H_5O^- + H_2O$$

The sulfonamide group present in sulfa drugs can also be titrated as an acid in nonaqueous solvents. The reaction can be written (Ar indicates an aryl group)

$$Ar-SO_2NHR + OH^- \rightleftharpoons Ar-SO_2NR^- + H_2O$$

A number of salts can be titrated as weak acids. Of particular importance are the salts of amines. The general reaction is

$$BH^+ + OH^- \rightleftharpoons B + H_2O$$

Many drugs used in medicine contain basic sites and are administered as salts of the bases. For instance, amine hydrochlorides are common. Most of these compounds are titratable with tetrabutylammonium hydroxide in solvents such as ethylenediamine, dimethylsulfoxide, or dimethylformamide. Salts containing small anions of high charge density, such as chloride, are only slightly soluble in organic solvents. Sometimes addition of a little water may provide the needed solubility; a small amount does not ordinarily affect the sharpness of the end point or the accuracy of the results.

6–5 NONAQUEOUS TITRATION OF OXINE

Background

In this experiment the amount of a weak base, 8-hydroxyquinoline (oxine), is determined by titration with a strong acid, perchloric acid, in the organic solvent acetic acid. The perchloric acid titrant also is dissolved in glacial acetic acid; it is standardized against primary-standard potassium acid phthalate. The indicator is α-naphtholbenzein. The technique of weight titration is applicable here (Section 3-2).

Oxine is so weak a base in water ($pK_b = 9.8$) that it could not be titrated accurately even if it were soluble enough to attain a useful concentration. In contrast, a solvent such as acetic acid provides both solubility and the requisite acid-base properties.

In the procedure described here the titrant is delivered from an ordinary 10-mL syringe with a stainless-steel needle. The syringe serves as a convenient vessel for weighing and delivering the titrant and reduces losses from evaporation. It is weighed to the nearest milligram on a top-loading balance before and after each titration. In this way a titration requiring, say, 2 g of titrant to reach the end point can be carried out with a precision of better than 1 ppt.

Procedure (time 2½ to 3 h)

Preparation of Titrant, Unknown Sample, and Standard Solutions

Prepare an approximately 0.1 M solution of perchloric acid in glacial acetic acid as follows. Measure with a Mohr pipet approximately 0.85 mL of 70 to 72% perchloric acid into 10 to 15 mL of reagent-grade glacial acetic acid. Add 2 mL of acetic anhydride and allow the solution to stand for about 30 min. Dilute the solution to 100 mL with glacial acetic acid and store in a well-stoppered glass container.[4]

Prepare a solution of the unknown sample by weighing about 0.3 g of sample to the nearest 0.1 mg and transferring into a clean, dry, glass-stoppered 50-mL flask previously weighed on a top-loading balance to the nearest milligram. Dilute the sample to about 50 mL with acetic acid, stopper the flask, and weigh again.

Prepare solutions of primary-standard potassium acid phthalate by weighing, to the nearest 0.1 mg, samples of approximately 0.15 g into 200-mL conical flasks. Add 15 to 20 mL of glacial acetic acid and swirl to dissolve.[5] Add 2 to 3 drops of a 0.1% solution of α-naphtholbenzein in glacial acetic acid as indicator.

Titration Procedure

Standardization of Perchloric Acid Titrant. Obtain a 10-mL hypodermic syringe.[6] Draw 5 to 10 mL of perchloric acid titrant solution into the syringe.

---

[4]Both glacial acetic acid and perchloric acid are corrosive to bench top, skin, and clothing. *Handle with care!* Wipe up all spills promptly, and remove all traces of perchloric acid from the material used for wiping by rinsing with cold water before discarding into waste containers.

[5]Potassium acid phthalate is only slowly soluble in glacial acetic acid; warming on a hot plate will accelerate the process. Cool to room temperature before proceeding with the titration.

[6]The syringe may be fitted either with a 22-gauge 1-in. stainless-steel or Teflon needle or with a glass or polyethylene tip having a fine orifice at the end. A thin film of silicone grease on the plunger allows control of titrant addition and prevents titrant loss along the ground-glass surface. The syringe may be supported during weighing by a cradle constructed from a 16-oz polyethylene bottle as shown in Figure 6-2.

Wipe the needle, and then weigh the syringe to the nearest milligram and record the weight. Titrate one of the solutions of KHP with the perchloric acid to a color change from olive to green.[7] When the end point is reached, retract the plunger slightly to avoid loss or evaporation of the drop that would otherwise remain on the delivery tip, and obtain and record the weight of the syringe and remaining contents.

Calculate the weight molarity of the perchloric acid titrant in units of millimoles of perchloric acid per gram of titrant solution.

Titration of Unknown Sample. Weigh a 200-mL conical flask on a top-loading balance to the nearest milligram. Add 10 to 12 g of sample solution and weigh again.[8] Add 2 to 3 drops of 0.1% α-naphtholbenzein indicator, and titrate as for the standardization.

Calculate and report the percentage of oxine in the unknown sample.

6–6   NONAQUEOUS PHOTOMETRIC TITRATION OF OXINE

Background

The use of a spectrophotometer to locate the end point in a photometric titration is discussed in Section 14-10. In the present experiment this method of end-point detection is used to determine the amount of a weak base, 8-hydroxyquinoline (oxine), by titration with a strong acid, p-toluenesulfonic acid (p-TSA), in the organic solvent acetonitrile. In addition, the technique of weight titration is employed. The advantages of weight titrations are discussed in Section 3-2.

The weak basicity of oxine ($pK_b = 9.8$) and its low solubility in water precludes water as the titration medium. In contrast, a solvent such as acetonitrile, CH_3CN, provides a medium in which oxine is readily soluble and can be titrated with an acid. p-TSA is used as the titrant because it is a moderately strong acid, is stable in acetonitrile, and being available in highly pure form, can be weighed directly as a primary standard. The p-TSA–acetonitrile combination is also less corrosive than perchloric acid–acetic acid mixtures. Photometric end-point detection is employed because oxine is too weak a base to give a sharp visual color change with p-TSA titrant. A plot of absorbance against volume of titrant is necessary in this method.

In the procedure the titrant is delivered from a syringe. After each delivery of titrant the syringe is weighed to the nearest milligram on a top-loading balance. To accommodate the moderately large volumes of sample, a test tube serves as the spectrophotometer cell.

---

[7]Since both hands are often needed to operate the syringe, a magnetic stirrer is convenient. Alternatively, the syringe may be mounted with a clamp over the titration flask, and the plunger operated with one hand while the flask is swirled with the other.

[8]If the solvent is volatile it may be necessary to stopper the flask to diminish solvent loss by evaporation prior to completion of the weighing. For acetic acid (bp 127°C) evaporation is not a problem because weighing is done immediately after addition of sample solution to the titration flask.

Procedure (time 3 to 4 h)

Preparation of p-TSA Titrant

Prepare a standard solution of p-toluenesulfonic acid monohydrate (p-TSA, formula weight 190.22) in acetonitrile by weighing to the nearest 0.1 mg about 0.5 g of pure p-TSA[9] into a small beaker. Weigh a clean, dry, glass-stoppered 50-mL flask to the nearest milligram on a top-loading balance.[10] Dissolve the p-TSA in a small amount of acetonitrile, and transfer the solution quantitatively to the flask. Rinse the contents of the beaker into the flask with several small portions of acetonitrile. Dilute to about 50 mL with acetonitrile, stopper tightly, mix, and weigh the flask plus contents to the nearest milligram.[11]

Preparation of Unknown Sample

Weigh a clean, dry, glass-stoppered 100-mL flask on a top-loading balance to the nearest milligram. Weigh about 0.3 g of sample to the nearest 0.1 mg and transfer it to the flask. Dilute the sample to about 100 mL with acetonitrile, stopper the flask, and weigh again to the nearest milligram.

Titration Procedure

Obtain a spectrophotometer, a 10- to 15-mL cell (verify that the cell compartment on the instrument will accommodate the cell provided), a stirrer, and a 10-mL syringe.[6] (A Bausch and Lomb Spectronic 20 spectrophotometer with a sample compartment accommodating a 25-mm test tube is convenient. Operating instructions are given in Section 14-12.) Set the wavelength control at 540 nm.

Weigh the cell (Figure 6-2) on a top-loading balance to the nearest milligram, add 10 to 15 g of sample solution, and weigh again. Set the spectrophotometer to zero on the percent-transmittance scale. Insert the cell, and add 10 drops of a 0.1% solution of 4-phenylazodiphenylamine indicator in acetonitrile. Insert a stirrer, [12] mix, cover the cell, and set the instrument to zero absorbance (100% transmittance).

Draw 5 to 10 mL of p-TSA titrant solution into the syringe. Weigh the syringe to the nearest milligram, record the weight, and add 0.4 to 0.5 g of titrant

---

[9]Reagent-grade p-TSA, obtained as the monohydrate, typically assays 99.8% or higher. It is stable and nonhygroscopic. Should the purity of a particular lot of the acid be unknown or doubtful, test it by titration of a pure sample of 8-hydroxyquinoline (best prepared by sublimation of a few tenths of a gram under vacuum) according to the procedure of this experiment.

[10]The concentration of titrant is defined as grams of p-TSA per gram of solution. The flask must be dry, not only because excess water would reduce the strength of the acid, but also because any acetonitrile or water initially in the flask is not included as part of the solution weight.

[11]Stock solutions of p-TSA and oxine must be kept tightly stoppered to prevent evaporation. Once a portion of oxine solution has been weighed into the spectrophotometer cell, however, evaporation of solvent does not affect the amount of oxine present. The situation is analogous to pipetting an aliquot of sample into a flask for titration. Changes in total volume of solution after pipetting, either from evaporation or from addition of solvent when washing the sides of the flask, do not affect the amount of sample being titrated.

[12]The stirrer, conveniently a polyethylene rod or a 16-gauge copper wire sealed in polyethylene tubing (Figure 6-2), must be positioned in the cell so that it does not obstruct the light path during measurements. Do not remove it, or solution may be lost.

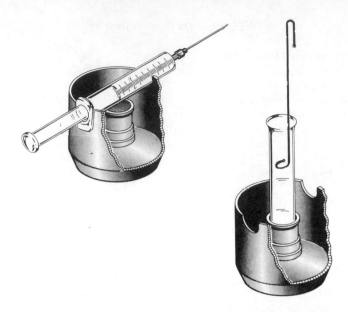

Figure 6-2. Support for syringe and titration cell during weighing. Stirrer is placed in cell after weighing and left there throughout titration.

solution to the cell. Retract the plunger slightly after each portion of titrant is added to minimize evaporation of the drop that would otherwise remain on the delivery tip. Stir the cell solution, cover the cell, and record the absorbance. Record the weight of the syringe, add another portion of titrant, and continue as before. When the absorbance reaches 0.08 to 0.1, take readings after each 0.1 or 0.2 g of titrant up to an absorbance of about 0.45, when the titration may be terminated.

To determine the weight of titrant required at the end point, plot weight of titrant against absorbance, and draw straight lines through the points before and after the break point (Figure 6-3). A major source of error in this experiment is

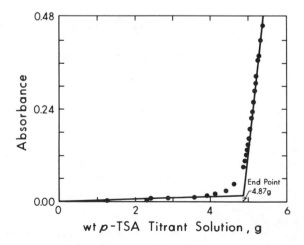

Figure 6-3. Plot of absorbance against weight of _p_-toluenesulfonic acid solution for photometric titration of oxine. Data by E. Findlay.

insufficient care in drawing the lines used to select the weight at the end point. The line after the break point should be drawn through the steepest portion of the curve.

Calculations

From the weight of p-TSA and the weight of solvent plus p-TSA, calculate the concentration of p-TSA titrant in grams of p-TSA per gram of solution. From the weights of titrant (obtained from the graph) and sample, calculate the concentration of oxine in the sample. The formula weight of oxine is 145.16. The percentage of oxine is then

$$\% \text{ oxine} = \frac{(\text{g } p\text{-TSA/g titrant})(\text{wt titrant})(\text{formula wt oxine})}{(\text{mol wt } p\text{-TSA})(\text{wt sample soln})(\text{g sample/g soln})} \times 100$$

6–7   TITRATION OF ACETYLSALICYLIC ACID

Background

Acetylsalicylic acid (ASA, aspirin) in tablet form may be determined by titration with sodium hydroxide to a phenolphthalein end point. Because ASA is readily hydrolyzed to form acetic acid and salicylic acid,[13] water is not a feasible solvent.

acetylsalicylic acid   acetic acid   salicylic acid
$pK_a = 3.5$ $pK_a = 4.72$ $pK_1 = 3.0$ (carboxyl)
                                           $pK_2 = 13.45$ (hydroxyl)

Since the carboxyl protons on both acetic and salicylic acids are neutralized at the phenolphthalein end point, hydrolysis of ASA in the titration solution will produce high results. This hydrolysis reaction occurs more slowly at lower temperatures; it is further reduced by initially dissolving the sample in a nonaqueous solvent such as ethanol. If the titration is completed reasonably quickly in a cool solution, the rate of reaction of ASA with water is sufficiently slow that an aqueous solution of sodium hydroxide is acceptable as titrant.

---

[13]The "whiskers" occasionally found growing on old ASA tablets are crystals of salicylic acid formed by the reaction of ASA with moisture from the air. Such tablets often have the odor of acetic acid.

Procedure (time 2½ to 3 h)

Preparation of 0.1 M NaOH Titrant

Prepare a solution of NaOH titrant as in Section 5-9. Keep solution well stoppered when not in use to minimize the absorption of carbon dioxide.

Preparation of Standard and Sample Solutions

Preparation of Standards. Weigh, to the nearest 0.1 mg, samples of approximately 0.45 g of primary-standard benzoic acid into 125-mL conical flasks. Add 25 mL of ethanol, swirl to dissolve, and then add 3 drops of 0.1% phenolphthalein indicator to each flask.

Preparation of Unknown Samples. Weigh, to the nearest milligram, 10 ASA tablets selected from the sample provided.[14] Place the tablets in a clean, dry mortar and grind thoroughly. Weigh, to the nearest 0.1 mg, approximately 0.8-g portions of the powdered material into 125-mL conical flasks.

Titration Procedure

Determine an indicator blank by placing 25 mL of ethanol and 25 mL of water into two or three 125-mL conical flasks, add 3 drops of 0.1% phenolphthalein indicator, and titrate with 0.1 M NaOH to the first permanent pink.

To standardize the NaOH solution, add 25 mL of ethanol to each of the benzoic acid samples, swirl to dissolve, add 3 drops of 0.1% phenolphthalein indicator, and titrate with 0.1 M NaOH to the first permanent pink.

To titrate the samples, add 25 mL of cold ethanol to one of the flasks, swirl to dissolve the ASA,[15] add 3 drops of 0.1% phenolphthalein indicator, and titrate immediately to the first permanent pink.[16] Rinse the nner walls of the flask once or twice near the end point with ethanol.

Report the percentage of ASA in the material. Also report the amount present in milligrams per tablet and in grains per tablet, based on the average weight of the tablets taken (1 grain = 0.0648 g). The formula weight of ASA is 180.16; that of benzoic acid is 122.13.

PROBLEMS

6-1.   Explain how the solvent plays a direct role in acid-base reactions. What are the principal reasons for substituting another solvent for water?

6-2.   What is meant by the solvent-leveling effect? Give an example of a differentiating solvent.

6-3.   What is an amphiprotic solvent? Is acetic acid such a solvent? Explain.

6-4.   Are large or small values of K_S preferred in nonaqueous titrations? Explain.

---

[14]This weight will be used to estimate the amount of ASA present per tablet.

[15]Inert material such as starch is often added to ASA tablets to minimize crumbling. This material may be insoluble, but the suspended particles will not affect the titration and can be ignored.

[16]It is convenient to place a 500-mL flask of ethanol, along with a wash bottle of ethanol, in an ice bath to cool well ahead of time. The same bath may be used to cool the titration flask during titrations as necessary.

6-5.   Write expressions for the self-dissociation of the following solvents: (a) H_2O, (b) CH_3OH, (c) CH_3COOH, (d) NH_3, (e) C_2H_5OH, (f) HCOOH. Also write expressions for the autoprotolysis constant of each.

6-6.   For the solvent ethylenediamine (Table 6-1) what would be the expected pH of a solution of 0.01 M strong acid; of 0.01 M strong base? Assume complete dissociation in both cases.

6-7.   For the solvent dimethylsulfoxide (Table 6-1) what would be the expected pH of a solution of 0.01 M strong acid; of 0.01 M strong base? Assume complete dissociation of the acid or base.

6-8.   If the acid dissociation constant of acetic acid in ethanol is 5×10^{-11}, what is the pH of a 0.01 M solution of acetic acid in ethanol? What would be the pH of an equimolar buffer of sodium acetate and acetic acid in ethanol?

6-9.   Explain the importance of the dielectric constant in the formation of ion pairs.

6-10.   In anhydrous methanol ($K_S = 10^{-16.7}$) an acid HA has a K_a value of $10^{-6.0}$. What would be the K_b value for the conjugate base A⁻ in methanol? What would be the pH of the pure alcohol?

6-11.   Write the chemical equation for the self-ionization of ethanol, C_2H_5OH. If K_s for ethanol is $10^{-19.1}$, what would be the pH of pure ethanol?

6-12.   Explain why perchloric acid is no stronger in water as solvent than is hydrochloric acid. Which acid is stronger in glacial acetic acid as solvent? Explain. How could this point be demonstrated experimentally?

6-13.   In liquid ammonia as solvent, would the extent of dissociation of acetic, hydrochloric, and perchloric acids be expected to differ from that in water? Explain.

6-14.   What are the principal factors that determine the strength of an acid such as acetic acid in a nonaqueous solvent? Explain how these factors operate to determine the strength of an acid.

6-15   Which of the solvents in Table 6-1 may be classified as aprotic and which as inert?

6-16.   The strongest possible acid in water is the hydrated proton. What is the strongest possible acid in acetic acid; in ammonia; in ethanol? What is the strongest possible base in each of these solvents?

6-17.   From Table 6-1 calculate the pH of each of the following pure solvents: (a) formic acid, (b) acetic acid, (c) ammonia, (d) acetonitrile.

6-18.   For the titration in pure ethanol of 25 mL of 0.100 M sodium ethoxide (C_2H_5ONa) with 0.100 M $HClO_4$, calculate the pH (-log $[C_2H_5OH_2^+]$) after the addition of 0, 12.5, 24, 25, 26, and 30 mL of titrant. Assume that both the perchloric acid titrant and the sodium ethoxide are completely dissociated in the solvent. Plot the curve. Compare the change in pH from 24 to 26 mL with the pH change for the same range if water were the solvent and sodium hydroxide the base.

6-19.   Sketch the curve for the titration in water of 0.1 M urea ($K_b = 1.5 \times 10^{-14}$) with 0.1 M hydrochloric acid and compare it with the curve for 0.1 M pyridine ($K_b = 1 \times 10^{-9}$) and for 0.1 M methylamine ($K_b = 4 \times 10^{-4}$). Which of these substances could be determined precisely by acid-base titration in water as solvent? What solvent might be chosen to improve the precision for the others?

6-20.   Why is glacial acetic acid not a suitable solvent for the titration of a weak acid such as boric acid or ammonium ion? Give an example of one that is suitable.

6-21.   For which of the following would perchloric acid and glacial acetic acid as titrant and solvent be a suitable combination: sodium formate, pyridine, trimethylamine, methyl ammonium chloride, phenol?

6-22. For which of the following would tetrabutylammonium hydroxide and ethylenediamine as titrant and solvent be a suitable combination: hydrocyanic acid, ammonium chloride, sodium benzoate, aniline, boric acid, pyridinium chloride, hydroxylamine?

6-23. Sketch and compare the titration curves of benzoic acid with 0.1 M strong base in water as solvent ($K_a = 6.7 \times 10^{-5}$) and in ethanol as solvent ($K_a = 1 \times 10^{-10}$).

6-24. Suggest a solvent and titrant for the determination of each of the following compounds in 0.1 M solutions: (a) pyridine, (b) phenol, (c) benzoic acid, (d) sodium acetate, (e) pyridinium hydrochloride, (f) potassium phenolate. Suggest a method of end-point detection in each case.

6-25. Name several methods for the detection of the equivalence point in a nonaqueous titration.

6-26. In potentiometric titrations in nonaqueous solvents, what problems can be encountered when an aqueous saturated calomel reference electrode is employed?

6-27. What is meant by a gravimetric titration?

6-28. Explain why potassium acid phthalate can be used as only a primary-standard acid in water, but as either a primary-standard acid or base in some solvents.

6-29. A 0.2957-g sample of impure oxine was dissolved in acetonitrile to give a total weight of solution of 77.876 g. A 12.144-g portion was titrated by the procedure of Section 6-5; 3.220 g of titrant was required. If the weight molarity of the perchloric acid titrant was 0.0975 mmol/g, what was the percentage of oxine in the sample?

6-30. Give three reasons why oxine cannot be determined precisely by titration with p-toluenesulfonic acid in water.

6-31. Why is a photometric end point employed in the determination of oxine with p-TSA titrant?

6-32. A 0.3101-g sample of impure oxine was dissolved in acetonitrile, the total weighing 77.938 g. An 11.200-g portion was titrated by the procedure of Section 6-6. The p-TSA titrant solution was prepared by dissolving 0.9623 g of p-TSA in acetonitrile to give 39.035 g of solution. The following data were obtained.

Wt Titrant	Absorbance
0.483	0.007
0.967	0.010
1.293	0.012
1.543	0.014
1.852	0.081
2.110	0.021
2.361	0.093
2.506	0.201
2.592	0.264
2.731	0.362
2.814	0.415
2.947	0.498

What was the percentage of oxine in the sample?

REFERENCES

J. S. Fritz, *Acid-Base Titrations in Nonaqueous Solvents*, Allyn and Bacon, Boston, 1973. A concise, practical treatment of nonaqueous acid-base titrations.

T. Higuchi and E. Brochmann-Hanssen, *Pharmaceutical Analysis*, Interscience Publishers, N.Y., 1971. A discussion of methods for the determination of acetylsalicylic acid by acid-base titration.

H. A. Laitinen and W. E. Harris, *Chemical Analysis*, 2nd ed., McGraw-Hill, N.Y., 1974, Chapter 4. A systematic discussion at an advanced level of the theory of nonaqueous acid-base systems.

Mettler Instrument Corp., *Gravimetric Titrimetry, Technical Information Bulletin* No. 1014, Princeton, N.J., 1967. An introduction to the concept of weight titrations and their advantages.

S. R. Palit, *Anal. Chem.* **1946**, *18*, 246. Introduced G-H solvent mixtures for the titration of weak bases.

7

COMPLEXATION EQUILIBRIA AND TITRATIONS

There is little exaggeration in the statement that in an aqueous solution complex ions are the rule and simple ions the exception.

A. Ringbom and E. Wänninen

A *complex* is an ion or molecule formed by reaction between two or more ions or molecules capable of independent existence. The most important complexation reactions from an analytical point of view are those in which one reactant is a metal ion and the other an ion or molecule possessing an atom with an unshared pair of electrons. The electron-pair donor ion or molecule, called a *ligand*, is usually either uncharged or an anion. Examples of ligands are H_2O, NH_3, Cl^-, and the amino acid glycine (H_2NCH_2COOH).

A metal ion can usually form a bond with more than one donor atom. The maximum depends on the *coordination number*, the number of electron pairs the metal ion can accept; it depends also on the number and location of the unshared electron pairs on the ligand. Coordination numbers and geometries for some metal ions in aqueous solution are listed in Table 7-1.

Ligands that provide one pair of electrons to a bond are called *unidentate*, or *monodentate* (one-toothed); examples are NH_3, Br^-, SCN^-, and CH_3CN. Those that are capable of bonding to a metal ion by two or more electron pairs are *multidentate*, or *polydentate*. The terms bidentate, tridentate, tetradentate, pentadentate, or hexadentate may be used to specify the number of binding groups on the ligand. Multidentate ligands are also called *chelates* (from the Greek "chelos," meaning claw). Examples of chelate ligands are

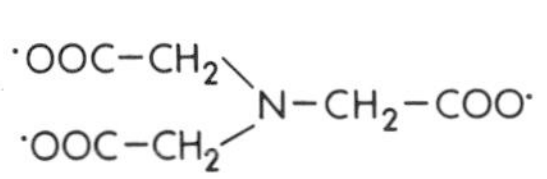

nitrilotriacetate
(NTA)

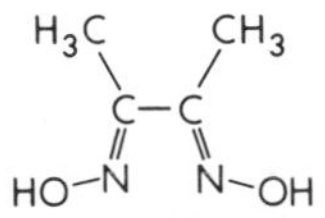

dimethylglyoxime
(DMG)

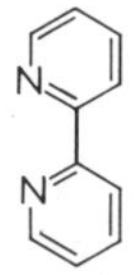

bipyridine
(Bipy)

Table 7-1. Coordination Numbers and Geometries of Metal Ions in Water

Metal Ion	Usual Coordination Number in Aqueous Solution	Usual Geometry	
Ag(I)	2	Linear	$-\overset{+}{\text{Ag}}-$
Cu(II), Ni(II) Co(II), Cd(II)	4 or 6, depending on ligand	4 Planar	
		or	
		4 Tetrahedral	
Fe(II),(III) Co(III)	6	Octahedral*	

*The 6 atoms are at the corners of an octahedron.

In these compounds the sites of coordination to metal ions are either nitrogen or oxygen; for organic ligands the coordinating groups generally contain nitrogen, oxygen, or sulfur atoms. The groups may be either acidic, in which a proton is lost before complex formation with a metal can occur, or neutral, in which complex formation can occur directly. Among the acidic groups are $-COOH, -OH$ (as enol or phenol), $-SH, -PO_3H_2, -C=NOH$ (oxime), and R_3NH^+. Among the neutral groups are $R_3N, -C=O$ (as ketone or aldehyde), $-COR$ (as ether or ester), $-CONR_2$ (amide), $-OH$ (as alcohol), $-S-$ (thioether), $-C=S$ (thioketone), $-C\equiv N$ (nitrile), $-C\equiv C-$ (acetylenic), $-PR_2$, and $-AsR_2$.

Inorganic coordinating groups include the halides (F^-, Cl^-, Br^-, I^-) and the pseudohalides (CN^-, SCN^-, N_3^-) as well as a number of low-molecular-weight compounds (H_2O, NH_3, H_2S, NO, CO) and ions (PO_4^{3-}, SO_4^{2-}).

In water most metal ions are hydrated, $M(H_2O)_x^{n+}$, and a ligand must displace a molecule of water if it is to form a complex. The overall reaction of a metal ion of charge n with a singly charged ligand can be written

$$M(H_2O)_x^{n+} + xL^- \rightleftarrows ML_x^{n-x} + xH_2O$$

For simplicity, water usually is omitted in the equilibrium expressions; however, the rate at which the reactions occur usually depends on the rate at which water leaves a metal coordination site. When the rate of ligand exchange is fast, the complex is termed *labile*; when slow, it is termed *inert*.

Rates of ligand exchange depend on the ligand and on the metal and its oxidation state. For example, rates with Cr(III) and Co(III) are so slow that hours or days may be required for a reaction to achieve equilibrium. On the other hand, rates with Cu(II), Zn(II), and Fe(III) and (II) are usually rapid (but slow with certain complexes of these metals such as $Fe(CN)_6^{3-}$). Rates with Al(III) and Ni(II) are intermediate, though generally fast enough to allow direct titration.

7–1 EQUILIBRIA WITH UNIDENTATE LIGANDS

Complexes form stepwise. Thus, the complexation of Al^{3+} ions with fluoride ions involves a series of equilibria:

$$Al^{3+} + F^- \rightleftharpoons AlF^{2+}$$
$$AlF^{2+} + F^- \rightleftharpoons AlF_2^+$$
$$AlF_2^+ + F^- \rightleftharpoons AlF_3$$
$$AlF_3 + F^- \rightleftharpoons AlF_4^-$$
$$AlF_4^- + F^- \rightleftharpoons AlF_5^{2-}$$
$$AlF_5^{2-} + F^- \rightleftharpoons AlF_6^{3-}$$

for which corresponding expressions for equilibrium constants can be written (water associated with the metal ion omitted for simplicity). In contrast with the practice of expressing reactions as dissociations in acid-base and solubility equilibria, complexation reactions are usually expressed as complex formation. Therefore, complexation equilibrium constants are expressed as *stability*, or *formation*, constants. For the formation of the first two aluminum fluoride complexes the equilibrium expressions are

$$\frac{[AlF^{2+}]}{[Al^{3+}][F^-]} = K_1 = 1.3 \times 10^6$$

$$\frac{[AlF_2^+]}{[AlF^{2+}][F^-]} = K_2 = 1.0 \times 10^5$$

Similar expressions for K_3, K_4, and K_5 have values of 7600, 1000, and 25. For the last stage,

$$\frac{[AlF_6^{3-}]}{[AlF_5^{2-}][F^-]} = K_6 = 2.5$$

K_1, K_2, $\cdots$, K_6 are *stepwise formation constants*. (The numerical values quoted here are for complexation in solutions with an ionic strength of 0.5.) The *overall formation constant*, which is the equilibrium expression[1] for the overall formation reaction $Al^{3+} + 6F^- \rightleftharpoons AlF_6^{3-}$, is written

---

[1]K_{eq} is sometimes written K_f or K_S.

$$K_{eq} = \frac{[AlF_6^{3-}]}{[Al^{3+}][F^-]^6} = \beta_6 = 6.3 \times 10^{19}$$

and given the symbol β. An overall formation constant can be written for each species. The relation between stepwise and overall formation constants is $\beta_1 = K_1$, $\beta_2 = K_1K_2$, $\beta_3 = K_1K_2K_3$, and so on. Thus $\beta_6 = K_1K_2 ... K_6$. The term β is useful in the calculation of equilibria involving complex formation, as shown in the next section.

EXAMPLE 7-1
Determine the values of β_x for the tetraammine complex of Zn^{2+} and the diammine complex of Ag^+.

For Zn^{2+} the values of K_1, K_2, K_3, and K_4 are 210, 310, 200, and 160, and so $\beta_4 = K_1K_2K_3K_4 = 2.1 \times 10^9$. Similarly, $\beta_2 = K_1K_2 = 6.5 \times 10^4$, and $\beta_3 = K_1K_2K_3 = 1.3 \times 10^7$.

For Ag^+ the values of K_1 and K_2 are 2×10^3 and 7.9×10^3, $\beta_2 = K_1K_2 = 1.6 \times 10^7$, and $\beta_1 = K_1 = 2 \times 10^3$.

Analytical Applications of Unidentate Ligands

The analytical use of unidentate ligands is currently limited to only a few titrations. The most important are the titration of cyanide with silver to form $Ag(CN)_2^-$; the titration of Hg^{2+}, Cu^{2+}, and Ni^{2+} with cyanide to form $Hg(CN)_2$, $Cu(CN)_4^{2-}$, and $Ni(CN)_4^{2-}$; and the titration of Hg^{2+} with SCN^-, Br^-, or thiourea. Because all steps of these complexes are highly stable, stepwise formation does not affect their analytical usefulness.

7-2 CALCULATION OF CONCENTRATIONS OF SPECIES IN METAL-LIGAND SYSTEMS

In titrations and other complexation reactions it is often important to determine how much of a metal ion or ligand in a mixture is present in the free or hydrated form, that is, complexed only by the solvent. For example, free metal ions in a solution may catalyze unwanted reactions, or they may poison plant or animal life by forming complexes with enzymes and thereby deactivating them. However, the same metal ions may be inactive when bound in stable complexes. Thus, measurement of the total metal ion concentration in a solution may not tell the whole story. The concentrations of species present at equilibrium are most conveniently calculated by application of the method described in Section 5-2 for acid-base systems.

Consider a system at fixed pH, with a concentration of ligand higher than that of the metal ion. The fraction of the total metal ion in the free form is calculated in the following way. The fraction α_M of the metal present as free ion is the ratio

$$\alpha_M = [M]/C_M \tag{7-1}$$

where C_M is the concentration of M in solution in all forms. This concentration is referred to as the *analytical concentration*. For a solution of silver(I) in ammonia, the expression becomes

$$\alpha_{Ag^+} = \frac{[Ag^+]}{C_{Ag}} = \frac{[Ag^+]}{[Ag^+] + [Ag(NH_3)^+] + [Ag(NH_3)_2^+]} \tag{7-2}$$

Substituting the formation-constant expressions gives

$$\alpha_{Ag^+} = \frac{[Ag^+]}{[Ag^+] + [Ag^+]K_1[NH_3] + [Ag^+]K_1K_2[NH_3]^2}$$

$$= \frac{1}{1 + K_1[NH_3] + K_1K_2[NH_3]^2}$$

Substituting the β terms gives

$$\alpha_{Ag^+} = \frac{1}{1 + \beta_1[NH_3] + \beta_2[NH_3]^2}$$

The general case for the reaction

$$M + xL \rightleftarrows ML_x$$

is

$$\alpha_M = \frac{1}{1 + \beta_1[L] + \beta_2[L]^2 + \cdots + \beta_x[L]^x} \tag{7-3}$$

EXAMPLE 7-2

(a) Calculate α_{Ag^+} for a solution in which the total concentration of silver(I) is $1 \times 10^{-4}\ M$, and $[NH_3]$ is $0.1\ M$.

$$\alpha_{Ag^+} = \frac{1}{1 + (2 \times 10^3)(0.1) + (2 \times 10^3)(7.9 \times 10^3)(0.1)^2}$$

$$= \frac{1}{1 + 200 + 158\ 000} = 6.3 \times 10^{-6}$$

Thus the fraction of free, or uncomplexed, silver is small. Note that the terms in the denominator indicate the relative amounts of each of the species present in solution. Here most of the silver, 158 000/158 201, is present as the $Ag(NH_3)_2^+$ complex, with minor amounts, 200/158 201 and 1/158 201, as $Ag(NH_3)^+$ and Ag^+. Note also that knowledge of the total silver ion concentration is not required.

(b) Calculate $[Ag^+]$, the concentration of free silver ion, present under the conditions of part (a).

From (a), $\alpha_{Ag^+} = 6.3 \times 10^{-6}$. On rearrangement of Equation (7-2),

$[Ag^+] = \alpha_{Ag^+} C_{Ag} = (6.3 \times 10^{-6})(1 \times 10^{-4}) = 6.3 \times 10^{-10}\ M$.

Effect of pH on Metal Ion–Ligand Equilibria. Conditional Formation Constants

The pH of a solution affects the ability of ligands to complex metals because sites that coordinate to metals, such as $-COO^-$ and $-NH_2$, also form stable bonds with protons. In effect, metal ions and protons compete for ligands. Formation constants listed in tables are normally for reactions written without taking the effect of hydrogen ions into account, whereas in real situations it often cannot be ignored.

To deal with the effect of ligand protonation upon complex formation, a system similar to the one described previously for calculation of free [M] is set up as follows. Suppose only the species L^{2-} is capable of coordination to a metal ion M. If protonation of L^{2-} to form HL^- or H_2L occurs, $[L^{2-}]$ must first be calculated if the usual formation-constant expression is to be used.

$$M^{n+} + L^{2-} \rightleftharpoons ML^{n-2}$$

$$K_f = \frac{[ML^{n-2}]}{[M^{n+}][L^{2-}]}$$

This calculation involves use of the α expression (Section 5-2), which here is

$$\alpha_{L^2} = \frac{[L^{2-}]}{[H_2L] + [HL^-] + [L^{2-}]} = \frac{K_1K_2}{[H^+]^2 + K_1[H^+] + K_1K_2}$$

Expressions for species other than the completely unprotonated anion can be calculated but are of lesser importance in complexation reactions.

When a metal ion is to be titrated with a ligand, the extent of the competition between protons and the metal ion for the ligand must be considered. To simplify calculations in this situation, it has been found convenient to define a *conditional formation constant* K' by

$$K'_{ML} = \frac{[ML]}{[M]C_L} \tag{7-4}$$

where C_L is the analytical concentration of L in all forms other than ML. The term "conditional" indicates that this constant applies only under specified conditions. From Equation (7-4) and the relation

$$\alpha_{L^{2-}} = [L^{2-}]/C_L$$

K'_{ML} is related to K_{ML} by

$$K'_{ML} = \alpha_{L^2}K_{ML}$$

Since α_{L^2} varies with pH, K'_{ML} varies accordingly, and a series of values are obtained as the pH of the solution is changed.

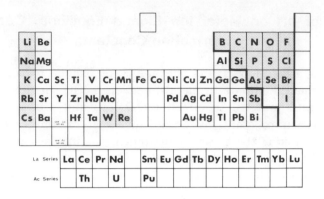

Figure 7-1. Elements determinable in some form by titration with EDTA. Shaded elements are determinable by indirect methods.

EXAMPLE 7-3

Calculate K' for the Mg^{2+} complex of succinic acid, H_2Su, at pH 4, 6, and 8. K_f for $MgSu = 110$.

$K'_{MgSu} = \alpha_{Su^{2-}} K_f = \alpha_{Su^{2-}} (110)$. At pH 4, $K' = (0.01)(110) = 1.1$; at pH 6, $K' = (0.71)(110) = 78$; at pH 8, $K' = (1.00)(110) = 110$.

In analytical titrations a large formation constant is desirable to ensure complete reaction between metal ion and ligand. A conditional formation constant of about 10^6 to 10^8 is generally adequate, though a larger value is often necessary when a visual indicator is employed.

7-3 MULTIDENTATE LIGANDS. ETHYLENEDIAMINETETRAACETIC ACID (EDTA)

Of the many multidentate ligands that have been investigated as analytical titrants over the past 35 years, the most important is ethylenediaminetetraacetic acid (EDTA), often written H_4Y. The tetrabasic anion of this acid, Y^{4-}, forms complexes with virtually all metal ions (Figure 7-1) and even to some extent with sodium ions. The metal atom is usually held in a 1:1 complex having four or five 5-membered rings (Figure 7-2). Such chelate ring formation contributes appreciably to the stability of the complexes.

The pK_a values (negative logarithms of the acid dissociation constants) for EDTA are 2.2, 2.7, 6.2, and 10.0. These values show the first two protons to be strongly, the third weakly, and the fourth very weakly acidic. The fraction of EDTA in each of the forms H_4Y, H_3Y^-, and so on can be calculated at any pH value by the method outlined in Section 5-2. The

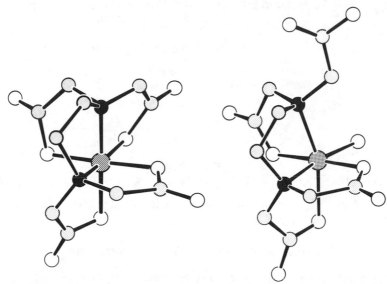

Figure 7-2. Molecular structures of cobalt and nickel complexes with EDTA. The nickel structure has an oxygen atom from a water molecule replacing one oxygen of EDTA in the sixth coordination position of the nickel. *Heavy shading*, metal; *solid*, nitrogen; *light shading*, carbon; *no shading*, oxygen. Hydrogen atoms not shown.

distribution[2]   of the various species as a function of pH is shown in Figure 7-3. The fraction $\alpha_{Y^{4-}}$ present as the tetravalent anion Y^{4-}, of particular importance in equilibrium calculations, can be calculated at various pH values by

---

[2]In solutions more acidic than about pH 2 the species H_5Y^+ and H_6Y^{2+} become prominent. (pK_a values are 0.96 for H_5Y^+ and 0.26 for H_6Y^{2+}.) Figure 7-3 is based on calculations in which, for simplicity, these species are neglected.

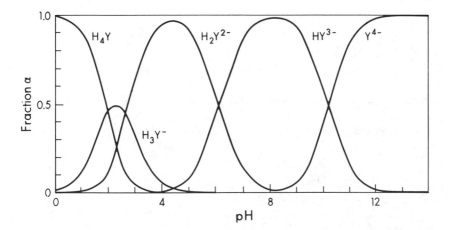

Figure 7-3. Fraction of EDTA present in various ionic forms.

Table 7-2. Fraction of EDTA as Y^{4-}, α_{Y4-}

pH	Fraction	pH	Fraction	pH	Fraction
2.0	3.7×10^{-14}	5.0	3.5×10^{-7}	9.0	5.2×10^{-2}
2.5	1.3×10^{-12}	6.0	2.2×10^{-5}	10.0	0.35
3.0	2.5×10^{-11}	7.0	4.8×10^{-4}	11.0	0.85
4.0	3.6×10^{-9}	8.0	5.4×10^{-3}	12.0	0.98

$$\alpha_{Y^{4-}} = \frac{K_1 K_2 K_3 K_4}{[H^+]^4 + K_1[H^+]^3 + K_1 K_2[H^+]^2 + K_1 K_2 K_3[H^+] + K_1 K_2 K_3 K_4} \quad (7\text{-}5)$$

where K_1, K_2, $\cdots$ are stepwise acid dissociation constants. Values for $\alpha_{Y^{4-}}$ at several pH values are given in Table 7-2.

Formation of Metal-EDTA Complexes

The formation constants of EDTA complexes of various metals, listed in Table 7-3, involve reactions of the type

$$M^{n+} + Y^{4-} \rightleftarrows MY^{n-4}$$

$$K_{MY} = \frac{[MY^{n-4}]}{[M^{n+}][Y^{4-}]} \quad (7\text{-}6)$$

If only the major species actually present in solution are considered, the formation of metal–EDTA complexes for divalent metals may be represented by the equations

$$M^{2+} + H_2Y^{2-} \rightleftarrows MY^{2-} + 2H^+ \qquad \text{pH 4 to 5}$$

$$M^{2+} + HY^{3-} \rightleftarrows MY^{2-} + H^+ \qquad \text{pH 7 to 9}$$

$$M^{2+} + Y^{4-} \rightleftarrows MY^{2-} \qquad \text{pH greater than 9}$$

At low pH values HMY^- or H_2MY may exist also.

Direct application of the values for formation constants in Table 7-3 to

Table 7-3. Formation Constants of EDTA Complexes at 20°C, Ionic Strength 0.1

Metal Ion	Log K_{MY}	Metal Ion	Log K_{MY}	Metal Ion	Log K_{MY}
Fe^{3+}	25.1	TiO^{2+}	17.3	Mn^{2+}	14.0
Th^{4+}	23.2	Zn^{2+}	16.5	Ca^{2+}	10.7
Hg^{2+}	21.8	Cd^{2+}	16.5	Mg^{2+}	8.7
Ti^{3+}	21.3	Co^{2+}	16.3	Sr^{2+}	8.6
Cu^{2+}	18.8	Al^{3+}	16.1	Ba^{2+}	7.8
Ni^{2+}	18.6	Al^{3+}	15.4	Ag^+	7.3
Pb^{2+}	18.0	Fe^{2+}	14.3	Na^+	1.7

estimate the completeness of a reaction may be somewhat misleading. Competing reactions, often encountered in practical situations, are not taken into account. For example, Y^{4-} will react with protons at most pH values, and metal ions may undergo hydrolysis or reaction with other ligands. We rarely need to know the concentration of every species in solution; what is generally of analytical interest is the completeness of the reaction between M and Y. With an approach similar to that described for ligand protonation, a quantity C_Y is defined as the total concentration of all forms of EDTA that are not coordinated to the metal, and a quantity C_M as the total concentration of all forms of the metal ion that have not reacted with EDTA. With these quantities the conditional formation constant K'_{MY} becomes

$$K'_{MY} = \frac{[MY]}{C_M C_Y} \tag{7-7}$$

where charges on ions are omitted for generality. For EDTA[3]

$$C_Y = [H_4Y] + [H_3Y^-] + [H_2Y^{2-}] + [HY^{3-}] + [Y^{4-}] \tag{7-8}$$

At constant pH the relative proportions of the various forms of uncomplexed EDTA are fixed, so that $[Y^{4-}]$ is a constant fraction of the uncomplexed EDTA (Table 7-2), or

$$[Y^{4-}] - \alpha_{Y^{4-}} C_Y \tag{7-9}$$

An *auxiliary complexing agent* is often added to a sample prior to EDTA titration. This is a ligand capable of forming complexes with the metal to be titrated that are too weak to interfere with quantitative reaction between the metal and EDTA, but strong enough to prevent the precipitation of metal hydroxides or basic salts. On addition of an auxiliary complexing agent, C_M is the total of the concentrations of the hydrated metal ion and the complexes of the metal with the auxiliary agent. Under usual analytical conditions of fixed pH and a concentration of auxiliary complexing agent higher than that of the metal ion, the proportions of the various species containing metal ions are fixed. Hence, combining the fraction α_M as hydrated metal ion, $[M]/C_M$, with Equations (7-6), (7-7), and (7-9), we obtain for EDTA

$$K'_{MY} = \frac{[MY]}{C_M C_Y} = K_{MY}\alpha_M\alpha_{Y^{4-}} \tag{7-10}$$

Equation (7-10) does not include provision for protonated metal–EDTA complexes of the type HMY or H_2MY, but does apply over a wide range of pH and K_{MY} values.

Because protonation and complexation side reactions impede the formation of metal–EDTA complexes, conditional formation constants are

---

[3]When the concentration of sodium salts in solution is high, Equation (7-8) should also include $[NaY^{3-}]$, a refinement not applied here.

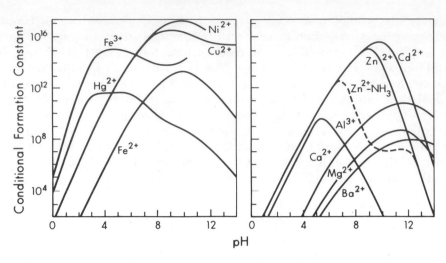

Figure 7-4. Conditional formation constants as a function of pH for metal–EDTA complexes. The dotted curve, Zn^{2+}-NH_3, represents zinc in the presence of $[NH_3]$ + $[NH_4^+]$ = 1 M. (Adapted from Ringbom.)

smaller than the formation constants in Table 7-3. Normally, the solution being titrated is buffered so that hydrogen ions displaced from H_2Y^{2-} or HY^{3-} upon formation of MY do not significantly change the pH, and hence $\alpha_{Y^{4-}}$ holds constant throughout the titration at some value less than unity. The value of α_M also depends on pH, since most metals forming EDTA complexes of analytical value form hydroxy complexes as well. In addition, these metals may form complexes with anions such as chloride, nitrate, or sulfate normally present in solution.

Figure 7-4 shows the relation between conditional formation constants and pH for a number of metal–EDTA complexes. The figure takes into account the effect of pH on $\alpha_{Y^{4-}}$ for EDTA and of the effect of hydroxy complex formation on α_M. In general, there is a pH region for each metal ion at which the conditional formation constant is at a maximum. In the case of zinc, for example, the conditional constant approaches the K_{MY} value 16.5 in the pH region 8 to 9, and falls off rapidly at higher or lower pH. The decrease at lower pH is due to competition of protons with zinc for Y^{4-}; that at higher pH is due to competition of hydroxyl ions with EDTA for zinc. The curves in Figure 7-4 represent the maximum values for the conditional constants when no auxiliary complexing agents are present. The concentration of an auxiliary agent when present is usually higher than that of the metal ion, and therefore the fraction α_M decreases as a function of pH and the nature and concentration of the auxiliary agent. When an ammine complex can be formed, the curves of Figure 7-4 for conditional formation constants are lowered further. The dotted curve labeled Zn^{2+}-NH_3, for example, indicates how the curve changes as a result of having in solution $[NH_3]+[NH_4^+]$ = 1 M. Many EDTA titrations are carried out in ammonia–ammonium chloride buffers. These buffers serve not only to adjust pH to a value where the conditional formation constant is large, but also to provide ammonia as an auxiliary complexing agent. Under these conditions the curve is calculated by use of equations similar to (7-3).

EXAMPLE 7-4

Calculate the conditional formation constant of 1×10^{-3} M NiY^{2-} in a buffer containing 0.05 M ammonium chloride and 0.028 M ammonia. Take values for log K_1, log K_2, ... log K_6 for the ammine complexes of nickel(II) to be 2.75, 2.20, 1.69, 1.15, 0.71, and –0.01.

From 7-3, $\alpha_{Ni^{2+}} = 1/(1 + 16 + 70 + 96 + 38 + 5 + 0.1) = 0.0044$. From the buffer ratio we calculate hydrogen ion concentration to be 1.0×10^{-9} M, and from Equation (7-5) or Table 7-2, $\alpha_Y = 0.052$. From Table 7-3 we take log $K_{NiY^{2-}}$ to be 18.6. Substitution of these values into Equation (7-10) gives $K'_{NiY} = 10^{18.6}$ $(0.0044)(0.052) = 9.1 \times 10^{14}$.

7-4 EDTA TITRATION CURVES

As in precipitation and acid-base titrations, diagrams of changes in concentration of the species titrated as a function of added titrant are helpful in estimating the species present at various points during the titration. In analogy to a pH titration curve, pM ($-\log[M]$) may be plotted against the volume of titrant. Under usual titration conditions, in which the concentration of metal ion is low compared with the concentrations of the buffer and the auxiliary complexing agents, the fractions $\alpha_{Y^{4-}}$ and α_M are essentially constant during the titration. The titration curve can then be calculated directly from the conditional formation constant, since it also remains constant.

EXAMPLE 7-5

A 10-mL sample of 0.001 M nickel(II) is titrated with EDTA. Calculate pNi after the following volumes of 0.001 M EDTA have been added under the conditions of Example 7-4: 0, 5.0, 9.0, 9.9, 9.99, 10.0, 10.01, 10.1, and 11.0 mL.

At start. From Example 7-4, $\alpha_{Ni^{2+}} = 0.004$, and $[Ni^{2+}] = \alpha_{Ni^{2+}}C_{Ni} = 4.4 \times 10^{-6}$. Therefore pNi $= 5.36$.

After 5.0 mL EDTA. Under the conditions of Example 7-4, K_{NiY} is 9.1×10^{14}. With so large a formation constant the reaction of nickel(II) with EDTA can be considered quantitative until near the equivalence point. Then mmol NiY can be assumed equal to mmol EDTA added; (0.001 mmol/mL)(5.0 mL) = 0.005 mmol. Mmol nickel(II) not complexed by EDTA = total mmol present – 0.005 = (10 mL)(0.001 mmol/mL) – 0.005 mmol = 0.005 mmol. C_{Ni} = 0.005 mmol/(10+5) mL = 3.3×10^{-4} M. From Equation (7-1), $[Ni^{2+}] = \alpha_{Ni^{2+}}C_{Ni} = (0.0044)(3.3 \times 10^{-4}) = 1.45 \times 10^{-6}$, and pNi = 5.84.

After 9.0 mL EDTA. As after 5.0 mL, total mmol nickel – (0.001 mmol/mL)(9.0 mL) = 0.01 mmol – 0.009 mmol = 0.001 mmol. C_{Ni} = 0.001 mmol/(10+9)mL = 5.3×10^{-5} M. $[Ni^{2+}] = (0.0044)(5.3 \times 10^{-5}) = 2.3 \times 10^{-7}$, and pNi = 6.64.

After 9.9 mL EDTA. C_{Ni} = [0.01 mmol – (0.001 mmol/mL) (9.9 mL)]/(10+9.9)mL = 5.0×10^{-6} M. $[Ni^{2+}]$ = $(0.0044)(5.3 \times 10^{-6})$ = 2.2×10^{-8}, and pNi = 7.66.

After 9.99 mL EDTA. C_{Ni} = [0.01 mmol – (0.001 mmol/mL)(9.99 mL)]/(10 + 9.99) mL = 5.0×10^{-7} M. $[Ni^{2+}]$ = $(0.0044)(5.0 \times 10^{-7})$ = 2.2×10^{-9}, and pNi = 8.66.

After 10.0 mL EDTA. This is the equivalence point, and the total EDTA present is equal to the total nickel(II). Most of the nickel(II) is bound as NiY^{2-}, and C_{Ni} is determined by slight dissociation of NiY^{2-}. At the equivalence point C_{Ni} = C_Y. On substitution into Equation (7-7), K'_{NiY} = $[NiY]/C_{Ni}C_Y$ = $[NiY]/C^2_{Ni}$. On rearrangement and insertion of appropriate values, C^2_{Ni} = $[NiY]/K'_{NiY}$ = $(0.01$ mmol/(10 + 10) mL)/(9.1×10^{14}) = 5.5×10^{-19}. Then C_{Ni} = 7.4×10^{-10}, $[Ni^{2+}]$ = $(0.0044)(7.4 \times 10^{-10})$ = 3.3×10^{-12}, and pNi = 11.5.

After 10.01 mL EDTA. Beyond the equivalence point EDTA is present in excess. The value C_Y is calculated first, followed by C_{Ni} from (7-10), and finally $[Ni^{2+}]$ and pNi as in the previous steps. Excess EDTA, C_Y,= $(0.01$ mL)(0.001 mmol/mL) in (10 + 10.01) mL, or 1×10^{-5} mmol/20.01 mL = 5.0×10^{-7} M. Substitution into (7-10) gives C_{Ni}= $[NiY]/K'_{NiY}C_Y$ = (0.010 mmol/20.01 mL)/(9.1×10^{14})(5.0 $\times 10^{-7}$) = 1.1×10^{-12}. $[Ni^{2+}]$ = $(0.0044)(1.1 \times 10^{-12})$ = 4.8×10^{-15}, and pNi = 14.3.

After 10.10 mL EDTA. C_Y = $(0.1$ mL)(0.001 mmol/mL)/(10 + 10.1) mL = 5.0×10^{-6} M; C_{Ni} = (0.010 mmol/20.1 mL)/(9.1×10^{14})(5.0 $\times 10^{-6}$) = 1.1×10^{-13}. $[Ni^{2+}]$ = $0.0044)(1.1 \times 10^{-13})$ = 4.8×10^{-15}, and pNi = 15.3.

After 11.0 mL EDTA. C_Y = (1.0 mL)/(0.001 mmol/mL)(10 + 11) mL = 4.8×10^{-5} M. C_{Ni} = (0.010 mmol/21 mL)/(9.1×10^{14})(4.8 $\times 10^{-5}$ = 1.1×10^{-14}. $[Ni^{2+}]$ = $(0.0044)(1.1 \times 10^{-14})$ = 4.8×10^{-17}, and pNi = 16.3.

Figure 7-5 shows curves for titrations of 10^{-3} M copper(II) at various pH values.

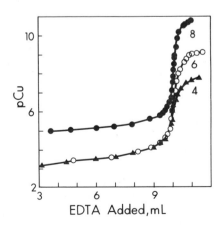

Figure 7-5. Titration curves of 10^{-3} M copper(II) with EDTA in a solution with $[NH_3]$ + $[NH_4^+]$= 0.1 M at various pH values as indicated. Data by J. Lipka and C. Johnson.

Before the end point the curves for copper at low pH values (4 to 6) coincide, there being no appreciable complexation between copper(II) and ammonia while the ammonia is tied up as NH_4^+. At higher pH values the pCu values increase in an amount determined by the stabilities of the various copper–ammonia complexes and the increasing concentration of free ammonia. Much above pH 10, where essentially all the buffer is present as ammonia, little additional pH effect would occur.

After the end point the positions of the curves depend not on the fraction α_M but only on K_{MY} and $\alpha_{Y^{4-}}$, because the excess EDTA overwhelms the effect of the more weakly complexing auxiliary ligand. Therefore, titration curves beyond the end point for a given metal depend only on the pH and not on the nature of the auxiliary complexing agent.

For the titration of a substance such as calcium the curves before the end point are essentially independent of pH. At low pH, α_Y and K'_{CaY} are so small, however, that no pM break occurs. At high pH, $\alpha_{Y^{4-}}$ approaches unity; consequently, the preferred range in pH for calcium titrations is 10 to 12. For titrations of iron(III) with EDTA, a low pH (3 to 4) prevents hydrolysis of iron(III); the low value is permissible here because the great stability of the complex FeY^- offsets increased protonation of the ligand.

7-5 INDICATORS

An important factor in the application of EDTA titration methods has been the development of indicators for metal ions that permit visual titrations to be carried out in dilute solutions. A metal ion indicator is usually a dyestuff that forms metal ion complexes of a color different from that of the uncomplexed indicator. The metal–indicator complex forms over some characteristic range in values of pM, exactly as an acid-base indicator forms a hydrogen ion complex over a characteristic range in values of pH. Since metal ion indicators are in general also hydrogen ion indicators, the acid-base equilibria, as well as the metal ion equilibria, must be considered for each indicator. A large number of indicators have been developed for various metal ions over a range of conditions. One of the first and most widely used was eriochrome black T.[4] Since it is easily oxidized [5] in solution, it has been largely replaced by the similar compound calmagite. Calmagite, [6] so named because of its excellent properties as an indicator for the titration of calcium and magnesium, has the structure

---

[4]Eriochrome black T is so named because its original use was as a color-fast black dye for wool. It was first proposed as an indicator for EDTA titrations by G. Schwartzenbach of Switzerland in 1948. The colors observed in solution during titration are red and blue, the black in the dye industry arising from interaction between the dye, chromium(III), and the wool fibers.

[5]Solid eriochrome black T is stable and is sometimes ground 1:100 with sodium chloride and added to titration solutions as a powder.

[6]Calmagite is 1-(1-hydroxy-4-methyl-2-phenylazo) -2-naphthol-4-sulfonic acid.

It is stable in aqueous solution, produces a sharper color change than eriochrome black T, and can be substituted for it without changes in procedure. For the sake of brevity we consider in detail only calmagite; the principles involved are similar for most other indicators.

Calmagite is a tribasic acid, which we designate H_3In. The first proton originating from the sulfonic acid group has a large dissociation constant and, since it is not involved in complexation, need not enter into the consideration of equilibrium. Thus the following acid-base behavior is significant:

$$H_2In^- \rightleftharpoons H^+ + HIn^{2-} \rightleftharpoons H^+ + In^{3-} \qquad (7\text{-}11)$$

$$\text{red} \qquad\qquad \text{blue} \qquad\qquad \text{red-orange}$$

In the pH range 9 to 11, in which the dye itself is blue, many metal ions form red 1:1 complexes as a result of chelation with the o,o–dihydroxyazo grouping. Calmagite forms intensely colored solutions and therefore is a sensitive detector for metals with which it reacts; for example, 10^{-6} to 10^{-7} M solutions of magnesium ion are a distinct red.

With magnesium ions the color-change reaction at pH values between 8 and 12 may be represented by

$$HIn^{2-} + Mg^{2+} \rightleftharpoons MgIn^- + H^+$$

$$\text{blue} \qquad\qquad\qquad \text{red} \qquad\qquad\qquad (7\text{-}12)$$

According to (7-11) and (7-12) the indicator color change is affected by the hydrogen ion concentration. From a practical viewpoint a *conditional indicator formation constant* K'_{MgIn}, which depends on the pH of the buffered solution, may be conveniently defined by

$$K'_{MgIn} = \frac{[MgIn^-]}{[Mg^{2+}]C_{In}} = K_{MgIn}\alpha_{In^{3-}} \qquad (7\text{-}13)$$

Here $[MgIn^-]$ is concentration of the magnesium–indicator complex; $[Mg^{2+}]$ is concentration of aquated magnesium ion, which is equal to $\alpha_{Mg^{2+}}C_{Mg}$; and C_{In} is concentration of indicator not complexed with magnesium ion, which, with neglect of $[H_3In]$, is $[H_2In^-] + [HIn^{2-}] + [In^{3-}]$. To relate K'_{MgIn} to the formation constant K_{MgIn}, we must know $[In^{3-}]$, that is, $\alpha_{In^{3-}}C_{In}$. Here $\alpha_{In^{3-}}$ is the fraction of the indicator in the completely ionized form:

$$\alpha_{In^{3-}} = \frac{K_2K_3}{[H^+]^2 + K_2[H^+] + K_2K_3} \qquad (7\text{-}14)$$

From (7-13)

$$\log K'_{MgIn} = pMg^{2+} + \log\left([MgIn^-]/C_{In}\right)$$

$$= \log K_{MgIn}\alpha_{In^{3-}}$$

where K_{MgIn} is the formation constant. The visual transition range occurs while [MgIn⁻] goes from about 0.1 to about 10 times the concentration of the uncomplexed indicator C_{In}. The range can be estimated from the K_2 and K_3 values for calmagite and the value 8.1 for log K_{MgIn}. Thus at pH 10, $\alpha_{In^{3-}}$ is calculated to be 3.9×10^{-3}, and log K'_{MgIn} is then 5.7. At this pH a visible color change of calmagite occurs approximately in the pM range 4.7 to 6.7. At pH values 8, 9, 10, 11, and 12, log K'_{MgIn} values are 3.34, 4.75, 5.69, 6.68, and 7.55.

The logarithm of the formation constant of calmagite with calcium is 6.1, a smaller value than with magnesium. The effect of this difference is that the conditional formation constant is not large enough at a pH of about 10 for calmagite to act as a sensitive indicator for calcium. Consequently, for a sharp end point in calcium titrations a small amount of magnesium must be present. The magnesium–EDTA complex has a smaller conditional formation constant (Figure 7-4) than the calcium, and therefore the magnesium-indicator–EDTA reaction and the color change do not occur until the calcium–EDTA reaction is complete. In the titration of solutions containing only calcium, magnesium can be added to either the titrant or the solution being titrated. When it is added to the solution being titrated, the amount added must be known; when added to the titrant before standardization, its presence is accounted for in the standardization step. The latter method is often employed when calcium carbonate is the primary standard.

Calmagite forms highly stable complexes with most transition metals, and therefore even traces of these metals may be sufficient to retain the indicator in the red MIn⁻ form. Often masking agents (Section 7-6) can be added to prevent such interference. For example, in the determination of calcium in limestone (Section 7-7) or in the determination of water hardness (Section 7-8) aluminum can be masked by addition of triethanolamine, and iron(II) and copper by addition of cyanide. Iron(III) must be reduced to iron(II) to accelerate the rate of reaction with cyanide to a useful level; this reduction is best accomplished with ascorbic acid.

An example of an indicator for EDTA titrations in acidic solutions, where calmagite cannot function, is xylenol orange. The structure is

In solutions of pH below 5.4, the indicator is yellow; above 7.4 it is red. Since complexes with xylenol orange of metals such as zinc are also red, a change in indicator color in zinc titrations can be seen only in solutions of pH 5.4 or below. Aqueous solutions of xylenol orange are stable for only a week or so; use of older solutions results in poor, drawn-out end points. Titrations using xylenol orange as indicator have been reported for aluminum(III), cadmium(II), cobalt(II), copper(II), mercury(II), and lead(II) in the pH range of 4 to 5. Some fourteen other metals can be titrated under defined conditions with this indicator. For example, zirconium(IV) can be titrated with EDTA in hot 1 M HCl. The solution must be kept hot to accelerate the otherwise slow reaction between zirconium(IV) and EDTA; the high acid concentration prevents interference from most metal ions.

Other Methods of Detection of End Point

Methods other than the use of visual indicators are available for the location of end points in complexation titrations. These include (1) photometric methods (Section 14-10), in which the intensity of the color of reactants or products is followed throughout the titration with a photometer and (2) potentiometric methods, in which the concentration of a metal ion species is followed by a selective electrode (Section 11-7) or through changes in potential of a mercury or silver indicator electrode (Section 11-6). These methods are usually chosen only when visual indicators are not available or when the titration is performed by an automatic titrating instrument.

7–6 APPLICATIONS OF EDTA TITRATIONS

Direct titrations with EDTA are commonly carried out by the use of disodium dihydrogen ethylenediaminetetraacetate, Na_2H_2Y, which is usually purchased as the dihydrate, $Na_2H_2Y \cdot 2H_2O$. The material is available in pure form, but when used as a primary standard it must be dried with care to prevent loss of part of the water of hydration at temperatures above 80°C. Even though EDTA forms strong complexes with many metals, much selectivity can be attained through control of pH and addition of reagents that block certain metals from reaction with the titrant. For instance, titration of a mixture of copper(II) and magnesium(II) with EDTA at pH 10 yields the sum of the two metals, whereas titration in a solution buffered at pH 4, where the magnesium–EDTA complex does not form (Figure 7-4), yields only copper(II). Most metals can be titrated in the pH range 9 to 10; auxiliary complexing agents are frequently added to prevent precipitation of the metal hydroxides. Another common addition is a buffer of relatively concentrated ammonia and ammonium chloride to provide a pH in this range and to supply ammonia as a complexing agent for those metal ions that form ammine complexes. Iron(III), bismuth, and thorium form complexes of exceptional stability with EDTA, and can be titrated in solutions of pH 2 or lower.

Considerable selectivity may also be achieved in complexation titrations by addition of masking agents. A *masking agent* is a reagent that decreases the concentration of a free metal ion or ligand to a level where it can no longer react with the titrant. For example, cyanide forms highly stable complexes at pH 10 with ions such as iron, copper, nickel, and cadmium, that prevent them from reacting in the titration of calcium magnesium, and bismuth with EDTA. Another example is fluoride, which allows titrations of metals such as zinc, calcium, and magnesium in the presence of aluminum through formation of stable AlF_4^-. Still another is triethanolamine [$N(CH_2 CH_2OH)_3$], which allows the titration of nickel by masking aluminum, iron, and manganese in ammoniacal solution. Masking agents are also used to prevent trace metals from blocking proper action of indicators and in analytical methods other than complexation titrations. In the determination of nickel in mixtures of iron and nickel through precipitation of nickel dimethylglyoxime, tartrate is added to prevent precipitation of iron hydroxide under the slightly alkaline conditions employed.

Back titration is the addition of a measured amount of a reagent in excess of that required for the reaction of interest, followed by titration of the excess with a standard solution of a second reagent. Thus a measured volume of standard EDTA solution might be added to a sample, and the uncomplexed EDTA titrated with a metal ion solution of known concentration. The metal ion in the titrant must form a weaker complex with EDTA than the metal being determined, or the titrant ion will displace EDTA from the original complex. Back titrations are useful for determining species such as chromium(III) and cobalt(III), which form complexes too slowly for direct titration, and for determining metals in precipitates (calcium in calcium oxalate, lead in lead sulfate, and so on).

A different type of back reaction is illustrated in Section 16-4, where the aluminum in a solution containing aluminum, magnesium, and zinc is determined by addition of a measured excess of EDTA and titration of the excess with standard zinc solution to a xylenol orange end point. Fluoride ion is then added to displace EDTA from the coordination sphere of the aluminum:

$$AlY^- + 4F^- \rightleftharpoons AlF_4^- + Y^{4-}$$

Since zinc and magnesium do not form fluoride complexes of appreciable stability, they do not react. The amount of EDTA released is equivalent to the amount of aluminum present. This EDTA is titrated with standard zinc solution to the same xylenol orange end point; the number of moles of zinc required is equivalent to the number of moles of aluminum in the solution.

Indirect determinations of several types can be carried out. Anions can be determined by precipitation as metal salts. The precipitate is filtered, dissolved in a measured excess of standard EDTA solution, and the remaining EDTA determined by titration with a standard metal ion solution. Alternatively, an excess of metal ion can be added, the precipitate removed by filtration, and the excess metal ion titrated in the filtrate. Among ions determinable in this way are sulfate as barium sulfate and phosphate as magnesium ammonium phosphate.

Although many other chelating agents have been investigated, few offer appreciable advantage over EDTA as a titrant. Nitrilotriacetic acid (NTA, $N(CH_2COOH)_3$) and diethylenetriamine ($H_2NCH_2CH_2NHCH_2CH_2NH_2$) are examples of inexpensive, moderately selective titrants that have limited use.

Primary Standards

EDTA is not often used as a primary standard, even though available in highly pure form. One reason is that the material must be dried at 80°C for several days to attain the precise composition of the dihydrate; another is that standardization against the metal being titrated reduces determinate errors in the procedure, especially in end-point selection. Many metals are available in high purity for use as primary standards — examples are copper, zinc, and iron. For calcium and magnesium, satisfactory standards include calcium carbonate, magnesium iodate tetrahydrate, and magnesium gluconate dihydrate.

Hardness of Water

Water supplies obtained from surface lakes or streams and from underground aquifers often contain minerals dissolved from the surroundings. The most common minerals are calcium and magnesium salts, principally bicarbonate and sulfate. Iron(II) also occurs in many underground waters. The presence of these salts significantly affects the efficiency of soaps and other detergents, and high levels of these ions produce what is known as hard water.

The hardness of water is customarily expressed in units of parts per million (ppm) of an equivalent amount of calcium carbonate. Since a liter of water weighs about 1 kg, 1 ppm corresponds to about 1 mg of calcium carbonate per liter. Calcium carbonate is used in this method of expressing hardness, even when the hardness is actually caused by calcium sulfate or magnesium salts. One reason for selecting calcium carbonate is that its formula weight, 100.09, is sufficiently close to 100 that parts per million of calcium carbonate can be taken to be equivalent to the molarity of the solution with respect to calcium multiplied by 10^5.

$$1 \text{ ppm} = 1 \text{ mg/L} = 0.001 \text{ g/L}$$

$$(0.001 \text{ g/L})/(100 \text{ g/mol}) = 1 \times 10^{-5} \ M$$

Thus, a water supply having a hardness of 100 ppm of calcium carbonate has a calcium concentration of 0.001 M. Water with a hardness below 100 is generally considered soft; water with a hardness above about 300 is considered hard. Waters with a hardness of 3000 to 4000 or more have been reported.

Calcium and magnesium in water can be determined readily by titration with EDTA at pH 10 using calmagite indicator; this method is universally employed by municipal water plants and many laboratories. Small amounts of certain other metals interfere by forming highly stable complexes with calmagite that block the indicator response. Copper and iron may be masked by addition of cyanide, and aluminum by triethanolamine.

7-7 DETERMINATION OF CALCIUM AND MAGNESIUM IN LIMESTONE WITH EDTA

Background

In this experiment a limestone sample is dissolved in hydrochloric acid, and the resulting calcium and magnesium salts titrated with EDTA with calmagite as indicator. The reactions are

(1) dissolution of limestone,

$$CaCO_3 + 2HCl \longrightarrow CaCl_2 + CO_2(g) + H_2O$$

$$MgCO_3 + 2HCl \longrightarrow MgCl_2 + CO_2(g) + H_2O$$

(2) titration of calcium and magnesium with EDTA at pH 10,

$$Ca^{2+} + HY^{3-} \rightleftharpoons CaY^{2-} + H^+$$

$$Mg^{2+} + HY^{3-} \rightleftharpoons MgY^{2-} + H^+$$

The calmagite indicator reaction is given by Equation (7-12).

Hydrochloric acid readily attacks pure calcium and magnesium carbonates. Care is needed to avoid loss of sample due to vigorous evolution of carbon dioxide. With natural limestones, other minerals (especially silicates and iron oxides) may slow dissolution of the last portions of carbonate, and warming may be necessary.

In the titration with EDTA the indicator calmagite works in the following way. When a small amount is added to a solution containing magnesium at pH 10, red $MgIn^-$ forms. As EDTA titrant is added, free magnesium ions in solution react with it. After all the free magnesium is complexed, the next portion of EDTA removes magnesium from the indicator complex, converting the indicator to the blue HIn^{2-} form. The pH must be near 10 because at pH values above 12 or below 8 the free indicator is red like the magnesium complex, and therefore no color change can be observed.

As discussed in Section 7-5, the calcium–calmagite complex is too weak to function as an indicator (log $K_{CaIn} = 3.7$ at pH 10). Calcium can, however, be titrated and an end point obtained when a trace of magnesium is present. In this experiment a small amount of magnesium is added to the titrant since the primary standard, pure calcium carbonate, contains none. The optimum pH range for titrations of calcium and magnesium with EDTA is about 9 to 11. In solutions more acidic than pH 9, formation of the metal–EDTA complex is incomplete, whereas in solutions above pH 12, magnesium hydroxide begins to precipitate.

A solution of approximate strength of the disodium salt of EDTA,

$Na_2H_2Y \cdot 2H_2O$, is prepared and standardized against primary standard calcium carbonate. In addition to having all the properties of a suitable primary standard, calcium carbonate has the added advantage that the element titrated, calcium, is a major component of limestone.

Procedure (time 4 to 6 h)

Preparation of EDTA Solution

To prepare an approximately 0.015 M EDTA solution, add 5 g of the disodium salt,[7] 20 mL of 1% $MgCl_2$ solution, and 3 mL of 6 M NH_3 to about 900 mL of water. In this experiment use only distilled water purified by passage through a column containing either a cation-exchange resin or a mixture of cation- and anion-exchange resins.

Weighing of Calcium Carbonate Standard and Sample

Weigh to the nearest 0.1 mg approximately 0.5 g of dry primary-standard $CaCO_3$ into a clean, dry 50-mL beaker. Transfer most of the $CaCO_3$ without loss to a dry funnel inserted in the top of a dry 100-mL volumetric flask. Transfer of the last particles is unnecessary. Do not use liquid in this transfer. Weigh the beaker and any remaining calcium carbonate to the nearest 0.1 mg, obtaining the weight of $CaCO_3$ in the flask by difference.

Using the same procedure as for the standard $CaCO_3$, weigh, to the nearest 0.1 mg, about 0.5 g of limestone unknown sample into a second dry 100-mL volumetric flask.

Preparation of Standard and Sample Solutions

For both standard and sample, tap the funnel to transfer most of the solid, and then rinse the last traces of solid through the funnels into the flasks with a milliliter or two of 12 M HCl, and then add 8 mL more of 12 M HCl to each flask by rapid dropwise addition. Warm if necessary to complete the dissolution.[8] A small residue of insoluble silicate minerals in the sample may be ignored. When dissolution is complete, remove the funnels, rinsing inside and out with demineralized water from a wash bottle as it is being removed. Fill each flask to the mark with demineralized water, stopper, and mix well.

Titration of Standards and Samples

Using a calibrated 10-mL pipet, measure aliquots of the standard calcium solution into each of three or four 200-mL conical flasks. Similarly, measure 10-mL aliquots of sample solution into three or four more conical flasks. Prepare a buffer solution of approximately pH 10 by dissolving 1 g of NH_4Cl in 20 mL of 6 M NH_3 and diluting to 100 mL.[9]

Titrate the calcium standards and samples in alternation. Immediately before

---

[7]The disodium salt of EDTA is the most commonly used form, being easier to prepare commercially and more soluble in water than the acid, H_4Y. Complete dissolution is somewhat slow, however, requiring 10 to 15 min. Occasional shaking speeds the process, as does gentle warming.

[8]Heating of volumetric flasks to moderate temperatures on a hot plate does not affect calibration, as the original volume is regained upon cooling to room temperature. For complete dissolution of limestone samples, heating for 30 to 60 min at 80 to 90°C is usually necessary.

[9]Use of volumetric flasks for preparation and storage of solutions of approximate concentration, such as buffer solutions, is not recommended.

each titration, measure 15 mL of NH_3–NH_4Cl buffer into a 10-mL graduated cylinder. Take a flask containing an aliquot to be titrated and the graduated cylinder containing the buffer to the laboratory instructor, who will add 50 mg of ascorbic acid to the flask, then the buffer, and finally 10 mL of a solution 0.4 M in NaCN (*Caution: Poison*) and 0.75 M in triethanolamine.[10] Swirl the flask to mix the contents after each addition. Add 4 to 5 drops of 0.1% calmagite indicator solution, and titrate with EDTA solution until the indicator changes from red to pure sky blue with no hint of red.[11]

Report the percentage of calcium in the sample.[12] Dispose of all solutions containing cyanide as directed by the laboratory supervisor.

Calculations

The concentration of EDTA solution is found by

$$M\ \text{EDTA} = \frac{(\text{g CaCO}_3)(V_{10}/100\ \text{mL})}{(\text{mol wt CaCO}_3)(\text{mL EDTA}/1000)}$$

where V_{10} is the calibrated volume of the 10-mL pipet. The percentage of calcium in the sample is then calculated by

$$\%\ \text{Ca} = \text{wt Ca}(100)/\text{wt sample}$$

$$\text{wt Ca} = (\text{mL EDTA}/1000)(M\ \text{EDTA})(\text{mol wt Ca})$$

$$\text{wt sample} = (\text{initial wt unknown})(V_{10}/100\ \text{mL})$$

7-8   DETERMINATION OF HARDNESS OF WATER WITH EDTA

Probably never in the history of chemistry has a method been adopted so quickly and universally as the EDTA titration for the hardness of water. In fairness, it should be added that never in history has there been a method so poor as the soap titration it displaced.

H. Diehl

Background

The determination of the hardness of water is discussed in Section 7-6. In this experiment a sample of hard water is titrated with EDTA, with calmagite as indicator. The procedure is essentially that for the determination of calcium and magnesium in limestone in Section 7-7.

---

[10]The cyanide solution may be dispensed by the laboratory instructor because of the hazard involved in its use. Other procedures may be advised by your instructor. Follow directions for handling and disposal carefully; contact with acid produces poisonous HCN gas.

[11]Most difficulties with indistinct end points arise from incorrect pH adjustment.

[12]The results could be reported in several ways, such as % $CaCO_3$, % $MgCO_3$, or % Mg. In accordance with common practice the approach employed here assumes that all the material titrated is calcium.

Procedure (time 3 to 4 h)

Prepare a solution of 0.015 M EDTA and a solution of calcium from primary-standard calcium carbonate as directed in Section 7-7. Using a calibrated 10-mL pipet, measure an aliquot of the standard calcium solution into each of three or four 200-mL conical flasks. Similarly, measure 100-mL samples of the water whose hardness is to be determined into three or four more conical flasks. Prepare a buffer solution of approximately pH 10 by dissolving 1 g of NH_4Cl in 20 mL of 6 M NH_3 and diluting to 100 mL.

Titrate the calcium standards and water samples in alternation. Add ascorbic acid, buffer, and 0.4 M NaCN–0.75 M triethanolamine as in the titration procedure for calcium in limestone. Then add 4 to 5 drops of 0.1% calmagite indicator solution, and titrate with EDTA solution until the indicator changes from red to pure sky blue with no hint of red.

If the hardness of the water under examination is unusually high or low, the volume of the samples should be adjusted to give buret readings in the range of 30 to 40 mL.

Calculations

Water hardness, expressed in parts per million or milligrams per liter of $CaCO_3$, can be calculated by

$$\text{hardness} = \frac{(\text{mL EDTA}/1000)(M \text{ EDTA})(\text{mol wt } CaCO_3)(1000)}{\text{vol sample in liters}}$$

PROBLEMS

7-1.   Define the following: (a) complex; (b) ligand; (c) coordination number; (d) unidentate; (e) multidentate; (f) chelate.

7-2.   Explain the difference between a labile and an inert complex.

7-3.   Write the chemical equations and stability-constant expressions corresponding to the stepwise formation of (a) $Ag(CN)_2^-$; (b) $Cu(NH_3)_3^{2+}$; (c) AlF_4^-.

7-4.   Write the chemical equations and the expressions for overall formation constants for the compounds in Problem 7-3.

7-5.   Explain how stepwise and overall formation constants are related.

7-6.   Explain how dissociation constants and stability constants are related.

7-7.   Explain how the conditional formation constant is related to the formation constant.

7-8.   Nickel ions form a series of complexes with the cyanide ion. In a solution containing nickel, what happens qualitatively to the value of $\alpha_{Ni^{2+}}$ as the cyanide concentratiion is increased?

7-9.   A solution is prepared by dissolving 0.023 mol of silver chloride in 1 L of 2.0 M ammonia. The free silver ion concentration is found to be 3.7 × 10^{-5} mg/L. Calculate the overall dissociation constant of the $Ag(NH_3)_2^+$ ion.

7-10. Copper ions form two complexes with ethylenediamine, $Cu(en)^{2+}$ and $Cu(en)_2^{2+}$, with formation constants of 5 × 10^{10} and 2 × 10^9. What fraction of free copper ion is present in a liter of 1.34 × 10^{-4} M $CuCl_2$ to which 0.01 mol of ethylenediamine has been added?

7-11.   If the K_1, K_2, K_3, and K_4 values for the ammine complexes of cadmium are 320, 91, 20, and 6.2, calculate the conditional formation constant for the reaction of EDTA with cadmium(II) at pH 9.5 with an ammonia concentration of 0.2 M.

7-12.   Define the term "conditional formation constant" and give an example.

7-13.   Calculate values of β_4 for the tetraammine complex of (a) copper(II), (b) cadmium(II). For copper(II), values of K_1, K_2, K_3, and K_4 are 1.1×10^4, 2.7×10^3, 6.3×10^2, and 3.0×10^1; for cadmium(II) they are 3.5×10^2, 1.0×10^2, 2.2×10^1, and 6.9.

7-14.   Calculate α_{Ag^+} for a solution in which the total concentration of silver(I) is 1×10^{-3} M and [CN$^-$] is 0.1 M. Calculate [Ag$^+$] for the system at equilibrium.
$K_1 = 4.6 \times 10^{15}$; $K_2 = 6.6 \times 10^4$.

7-15.   Calculate $\alpha_{Cu^{2+}}$ for a solution in which the total concentration of copper(II) is 1×10^{-5} M and [NH$_3$] = 0.01 M. Calculate [Cu^{2+}] for the system at equilibrium. See Problem 7-13 for K values.

7-16.   If β_1, β_2, β_3, and β_4 for the chloro complexes of silver are 2×10^3, 1.6×10^5, 2.5×10^6, and 1.3×10^6, what are the values of K_1, K_2, K_3, and K_4?

7-17.   Calculate [PO$_4{}^{3-}$] for a 0.05 M solution of phosphoric acid at (a) pH 5; (b) pH 8; (c) pH 12.

7-18.   Calculate [Ox^{2-}] for a 0.01 M solution of oxalic acid at (a) pH 8 and (b) pH 14.

7-19.   A common component of kidney stones is calcium oxalate, CaOx. With calcium assumed to be present in urine only as free Ca^{2+}, what is the highest calcium concentration that could exist at equilibrium in a urine sample of pH 5 containing a total oxalate concentration of 0.001 M?

7-20.   Solutions containing cyanide are often used for electroplating. If a 10.00-mL sample of such a solution is titrated with 0.03264 M silver nitrate and 41.64 mL of titrant is required to reach the end point, what is the concentration of cyanide in the solution? The reaction is

$$Ag^+ + 2CN^- \longrightarrow Ag(CN)_2{}^-$$

7-21.   Explain why careful pH control is *vital* for successful EDTA titrations. Explain why more metals can be successfully titrated with EDTA at pH 10 than at pH 4.

7-22.   From the K_a values for EDTA (H$_4$Y) identify the predominant species in solution at pH 5, 9, and 12.

7-23.   Calculate the conditional formation constants for the nickel complex of EDTA at pH values of 7, 9, 10, and 12.

7-24.   Explain what is meant by the term "auxiliary complexing agent" and give an example.

7-25.   A metal ion M forms a 1:1 complex with L ($K_{ML} = 6 \times 10^9$). For a solution 2×10^{-3} M in ML calculate [M] at equilibrium. What is the concentration of the metal ion if the solution is also 10^{-3} M in the ligand L?

7-26.   In a complexation titration what factors govern the magnitude of the change in metal ion concentration in the region of the equivalence point? How is this change affected by decreasing the concentration of the reactants?

7-27.   For the titration at pH 10 of 45.0 mL of 0.0100 M calcium chloride solution with 0.0100 M EDTA, calculate pCa values after the addition of 0, 10, 20, 30, 40, 44, 44.9, 45, 45.1, and 50 mL of EDTA. Plot pCa against volume of EDTA.

7-28.   For the titration of 30.0 mL of 0.0100 M cobalt(II) with 0.0100 M EDTA in a solution buffered at pH 5, calculate pCo values after addition of 0, 10, 20, 29, 29.9, 30, 30.1, 31, and 40 mL of EDTA. Plot a curve of pCo against volume of EDTA.

7-29. Repeat the calculations of Problem 7-28 for the titration of magnesium at pH 10. Plot pMg against volume of EDTA as in the preceding problem.

7-30. Explain how the indicator calmagite works in the titration of calcium with EDTA.

7-31. Explain why calmagite is not a suitable indicator for the titration of calcium when magnesium is not present.

7-32. Why is the pH of the sample solution adjusted to pH 10 in the EDTA titration of calcium? What color change is observed with calmagite in the vicinity of the equivalence point? What color change would be observed if the pH were 9; if the pH were 12?

7-33. If the conditional formation constant for the magnesium complex with calmagite at pH 10 is $10^{5.7}$, what is the true formation constant, that is, the constant under conditions where calmagite is completely in the In^{3-} form?

7-34. Define the term "masking agent" and give an example.

7-35. From Figure 7-4 indicate the pH range over which zinc can be titrated with standard EDTA in the presence of 0.1 M ammonia. Assume that the minimum conditional formation constant for a successful titration is 10^8. Repeat for cadmium, copper, iron, and aluminum in the absence of ammonia.

7-36. Explain how selectivity in an EDTA titration of a mixture of metal ions can sometimes be achieved by carrying out the titration at a different pH. Give an example.

7-37. Under what circumstances can a metal ion be best determined by a back titration? Give an example.

7-38. Both calcium and magnesium ions form stable complexes with EDTA. Why, then, is magnesium added to the EDTA titrant in the titration of calcium? Magnesium does not form complexes with ammonia. What happens to the value of $\alpha_{Mg^{2+}}$ as the concentration of an ammonia–ammonium chloride buffer is increased in a solution containing magnesium ion?

7-39. What would happen to $\alpha_{Cu^{2+}}$ under the same conditions as in the preceding problem?

7-40. In the determination of calcium and magnesium in a limestone, why must the solution be buffered at about pH 10? How does the buffer provide selectivity in the titration?

7-41. A 0.5234-g sample of pure, dry calcium carbonate was weighed and dissolved in hydrochloric acid. It was then diluted to 100.0 mL in a volumetric flask. A 10-mL aliquot of the solution required 38.64 mL of EDTA solution to reach the end point, according to the procedure of Section 7-7. What is the concentration of the EDTA solution?

7-42. Calculate the volume of 12 M hydrochloric acid required to just dissolve 0.5 g of calcium carbonate.

7-43. A limestone sample weighing 0.8574 g was dissolved in hydrochloric acid, and the solution diluted to 100.0 mL. A 10.00-mL aliquot of this solution required 36.20 mL of 0.02027 M EDTA for titration. What was the percentage of calcium carbonate in the sample? What was the percentage of calcium?

7-44. In most complexation titrations the adjustment of pH to a specific, appropriate value is critical. What volume of 6 M NH_3 should be added to 100 mL of a solution containing 8 mL of 12 M HCl to bring the pH to 10?

7-45. A 100.0-mL sample of river water required 4.61 mL of 0.01618 M EDTA for titration by the procedure of Section 7-8. What was the hardness of the water in milligrams of calcium carbonate per liter, in parts per million of calcium carbonate, and in milligrams of magnesium carbonate per liter?

REFERENCES

E. Bishop, Ed. *Indicators*, Pergamon, Oxford. 1972. Detailed description of how indicators function, plus data on the indicators in use today. Two chapters (120 pp) devoted to complexation indicators.

H. Diehl, *Quantitative Analysis*, Oakland Street Science Press, Ames, Iowa, 1970. Clear introduction to practical aspects of titrations with EDTA, especially of calcium and magnesium.

H. A. Laitinen and W. E. Harris, *Chemical Analysis*, 2nd ed., McGraw-Hill, N. Y., 1975. Advanced discussion of complexation reactions in analysis.

A. E. Martell and R. M. Smith, *Critical Stability Constants*, Plenum, N.Y., 1974–76. Tables of metal-complex equilibrium constants in four volumes.

D. D. Perrin, *Masking and Demasking of Chemical Reactions*, Wiley-Interscience, N.Y., 1970. Comprehensive treatment of the use of complexing reagents to allow selective titrations.

A. Ringbom, *Complexation in Analytical Chemistry*, Interscience, N.Y., 1963. General reference for metal titrations with EDTA.

G. Schwartzenbach and H. Flaschka, *Complexometric Titrations*, 2nd English ed., H. M. N. H. Irving, Translator, Methuen, London, 1969. A summary of conditions for the titration of most metals with EDTA, with detailed procedures.

8

GRAVIMETRIC ANALYSIS

As a beginning chemist, ... make a habit of weighing carefully, of pouring from one vessel to another without spilling and without missing the last drop, and of observing the small details which if overlooked often spoil several weeks of careful work.

J. J. Berzelius

In gravimetric analysis the critical final measurement is that of weight. Gravimetric analysis includes *precipitation* methods, in which the substance to be determined is precipitated as a slightly soluble compound either chemically or electrolytically, and *volatilization* methods (direct or indirect), in which a substance is removed from a sample by heating, reaction with an acid, or combustion. In this chapter we consider mainly precipitation methods.

Gravimetry is the oldest method of quantitative analysis. Somewhat more than a century ago gravimetric methods constituted all of analytical chemistry, and only with reluctance was the validity of titrimetric methods recognized. Today faster and simpler procedures continually displace sound but otherwise tedious procedures, so that gravimetric analysis occupies a less prominent place. Instruction in this topic is accordingly curtailed in keeping with its lesser importance.

In general, gravimetric methods are more time-consuming than titrimetry because more operations are required: precipitation, digestion, filtration, washing, and ignition, as well as weighing. These methods have the advantage of broad applicability (Figure 8-1) and may be applied in the same range, with about the same precision, as titrimetric methods. Coprecipitation errors, however, are more serious during weight measurements than during titrimetric precipitation reactions, owing to the error introduced by any nonvolatile contaminant. Despite the disadvantages, a gravimetric procedure may be preferred when only a few analyses are needed, or when the preparation and standardization of titrimetric reagents would be unduly time-consuming.

EXAMPLE 8–1

When a standard solution containing fluoride ion is required, titrimetric standardization is probably a poor choice, since few simple and accurate

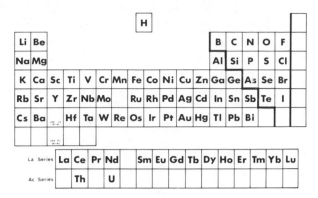

Figure 8-1. Elements determinable by gravimetric methods.

titrimetric methods are available. Gravimetric standardization by precipitation of lead chlorofluoride is probably the faster and more accurate method.

8-1 ATOMIC WEIGHTS: A HISTORICAL OVERVIEW

> Every substance must be assumed to be impure, every reaction must be assumed to be incomplete, every method of measurement must be assumed to contain some constant error until proof to the contrary can be obtained. As little as possible must be taken for granted.
>
> T. W. Richards (from Nobel Award address, 1914)

In the history of chemistry no other problem has undergone the continuous investigation accorded that of defining and measuring atomic weights. Work in this field has included some of the most fascinating chemistry and interesting personalities encountered in science. For about a century and a half virtually all the work on atomic weights involved gravimetric methods because of their inherent accuracy and wide applicability. In some ways a study of work on atomic weights provides a history of analytical chemistry.

The problem of atomic weights first revealed itself in 1803, when Dalton (1766–1844) proposed that continuous division of any substance would ultimately yield particles that were no longer divisible. For a half century after Dalton's suggestion, enormous confusion existed because, first, many doubted the existence of atoms. One reason for this lack of acceptance was probably that Dalton's defective experimental work discredited his ideas. A second source of confusion was that the chemists of the time were really determining combining, rather than atomic, weights, and they did not clearly recognize the difference between these two concepts for several decades. A third difficulty resulted from the use of different reference elements by different workers. Because atomic weights are simply a series of numbers indicating the relative weights of atoms, some one substance must be chosen as a point of reference. About 1810 Dalton suggested a value of 1 for hydrogen, in 1814 Wollaston suggested 10 for oxygen, in 1825 Thompson suggested 1 for oxygen, and in 1830 Berzelius suggested 100 for oxygen. In

1860 the proposal by J. S. Stas that the value of 16.0000 for oxygen be the standard won general acceptance, and the confusion finally diminished.

By 1840 about 100 determinations of atomic weights had been made. Although many others contributed, Berzelius (1779–1848) produced the most important work. He developed and refined methods of breaking down (analyzing) or building up (synthesizing) selected compounds so as to obtain combining ratios and thereby atomic weights. Table 8-1 lists some of his values, along with earlier values of Dalton and later ones of Stas and Richards.

From 1850 to 1890 about 500 determinations of atomic weights were carried out. This work required experimental technique of exceptional accuracy, and only a few investigators possessed the combination of intelligence, persistence, and dexterity necessary to achieve results worthy of general acceptance. One of these was Stas (1813–1891), who spent his life working in this field; his contributions were outstanding.

From 1890 to 1940 approximately 800 additional determinations were completed. This period virtually marked an end to determinations of atomic weights by chemical means.[1] The work of T. W. Richards (1868–1928) and his colleagues was monumental. Richards took up where Stas left off and, like Stas, based his work on silver. He found that even the most carefully purified potassium chlorate always contains some potassium chloride, that precipitated silver chloride contains traces of potassium chloride that cannot be removed by washing, and that metallic silver always contains some oxygen unless stringent precautions are taken. To avoid such problems, he devised new techniques for the purification and handling of chemicals. On discovering that water dissolves traces of silica from glass, he substituted platinum or quartz vessels. He developed the centrifuge, and also a bottling apparatus in which substances could be heated reproducibly in any gas or under vacuum and subsequently weighed without exposure to air.

Richards's first goal was to obtain the most accurate possible atomic weights of silver, chlorine, and nitrogen relative to oxygen. His measurements involved determination of the weight of silver chloride and of silver nitrate produced from a known weight of silver, followed by determination of the weight of silver chloride produced from a known weight of ammonium chloride. With the assumption of oxygen equal to 16.0000 and hydrogen equal to 1.0076 (determined by Morley in 1895), the atomic weights of silver, chlorine, and nitrogen could be calculated.

EXAMPLE 8-2

The following outline of a procedure used by Richards gives an idea of the meticulous work demanded.

Pure silver was obtained from silver nitrate that had been recrystallized 15 times and fused on lime in an atmosphere of hydrogen. The silver was dissolved in highly purified nitric acid, and silver chloride precipitated by the addition of highly purified sodium chloride. The reactions were carried out in red light to

---

[1] The last appears to have been that of fluorine by Scott and Ware in 1957. (See references at end of chapter.) The precise values reported today are obtained by mass-spectrometric measurements of isotope ratios.

minimize photodecomposition of the precipitate. The amount of silver chloride in the filtrate was measured with a nephelometer, a device invented by Richards. (This procedure marked the birth of instrumental analysis.) With every possible precaution the silver chloride precipitate was then filtered, washed, dried, and weighed.

Pure ammonium chloride was prepared from ammonium sulfate by treatment with permanganate to oxidize organic matter; ammonia was driven off by heating with pure calcium oxide and absorbed in purest hydrochloric acid. The solution was evaporated and the ammonium chloride sublimed. This was weighed and added to a solution of silver nitrate. The precipitate of silver chloride was then treated and weighed as before.

To determine the weight of silver nitrate obtained from pure silver, Richards dissolved silver in nitric acid that had been repeatedly distilled and diluted with highly purified water. The solution was evaporated in a gentle current of air. The silver nitrate was finally dried and weighed.

Richards was awarded the Nobel medal in 1914, primarily for determining the atomic-weight values of 107.881, 35.4574, and 14.0085 for silver, chlorine, and nitrogen. Once these weights were known, those of most of the other metallic elements could be determined, since the pure chlorides of the metals could be prepared and silver chloride precipitated. For example, with a divalent metal M, after the reaction

$$MCl_2 + 2Ag^+ \rightarrow 2AgCl(s) + M^{2+}$$

the silver chloride could be weighed. The weight obtained then allowed calculation of the atomic weight of M. The accuracy of all atomic weights determined in this way depends ultimately on the value used for silver.

In 1929 Giauque and Johnson discovered that oxygen atoms are not all identical, but vary in mass according to the number of neutrons in the nucleus. Thus isotopes of oxygen exist, that is, oxygen atoms of mass 17 and 18 as well as 16. Subsequently, the standard of oxygen equal to exactly 16 became unsatisfactory because the natural variations in the isotopic composition of oxygen were found to make its atomic weight uncertain by 1 part in 10 000. Since the atomic weight of other elements cannot be known more precisely than the standard to which they are compared, something had to be done. Part of the problem was that a standard had to be chosen that would be satisfactory to both chemists and physicists; also, any change should not be great enough to disrupt seriously the values already in use.

In 1959 and 1960 the International Union of Pure and Applied Chemistry (IUPAC) and the International Union of Pure and Applied Physics (IUPAP) agreed that the standard should no longer be natural oxygen equal to 16. Instead, the isotope carbon-12 was assigned a mass of exactly 12.0000, and all weights are now referred to this ultimate standard. Fortunately for chemists, since this agreement changed the former values by only about 1 part in 10 000, it was not severely disruptive.

A committee of IUPAC periodically evaluates the latest work on atomic weights and publishes an updated table of the most reliable values. Although we can never know true atomic weights, the values are becoming more and more precise, and revisions are becoming smaller. For some elements,

Table 8-1. Comparison of Some 1977 Atomic-Weight Values with Those Obtained by Early Workers

	Dalton (1810)	Berzelius (1836)	Stas (1890)	Richards and Coworkers (1923)	IUPAC (1977)
Hydrogen	1.1	1.007	1.002	---	1.0079
Carbon	11	12.23	11.99	12.000	12.011
Nitrogen	11	14.16	14.055	14.0080	14.0067
Sulfur	28	32.19	32.069	32.064	32.06
Chlorine	--	35.41	35.456	35.4561	35.453
Silver	114	108.13	107.93	107.876	107.868
Lead	206	207.11	206.905	207.19	207.2

especially sulfur and boron, natural variations in isotopic composition limit the precision of reported atomic weights.[2] Table 8-1 compares several values obtained by early workers with current recommended values.

8-2 PROPERTIES OF PRECIPITATES

In a sound gravimetric analytical method the precipitate must fulfill the following conditions: (1) Its solubility must not be large enough to give rise to significant error. (2) Its physical characteristics must be favorable. (3) It must be sufficiently pure and yield a form suitable for weighing. (4) Its stoichiometry should be known.

Solubility of the Precipitate

Solubility Product and the Common-Ion Effect

Under the conditions chosen for analysis the amount of sought-for substance remaining in the saturated filtrate and wash solution should be negligible. Since weighings are normally made to the nearest 0.1 mg, solubility losses should be less than 0.1 mg. In the usual procedure, then, conditions should be such that precipitation is 99.9% or more complete (quantitative).

Precipitation reactions are reversible:

$$MA(s) \underset{k_2}{\overset{k_1}{\rightleftarrows}} M^+ + A^-$$

That is, a solid is in equilibrium with its saturated solution. Continual interchange of ions goes on between the solid and the solution; the rates can be expressed as

---

[2]Most lithium in compounds obtainable in North America has an atomic weight considerably above that in the periodic table, owing to separation of the lithium-6 isotope. The natural abundance of lithium-6 is 7.5%, but some commercial material may contain much less. Atomic-weight changes of as much as 0.5% have been reported. This is sufficient to introduce significant error in analytical calculations based on the natural abundance.

$$\text{rate of solution} = k_1 \times \text{area of solid}$$

$$\text{rate of deposition of ions} = k_2 \times \text{area of solid} \times [M^+][A^-]$$

where k_1 and k_2 are proportionality constants. In a saturated solution the two rates are equal:

$$k_1 \times \text{area} = k_2 \times \text{area} \times [M^+][A^-]$$

By combination of k_1 and k_2 an overall constant, called the *solubility-product constant* K_{sp} can be written

$$[M^+][A^-] = k_1/k_2 = K_{sp} \tag{8-1}$$

The value of K_{sp} is a constant that depends on the temperature. The solubility-product constant is the most useful single piece of data in calculating how much precipitate will dissolve under various conditions. Gravimetric procedures always specify the addition of a slight excess of precipitating reagent. This slight excess normally reduces solubility by the so-called *common-ion effect*.

EXAMPLE 8-3

A common gravimetric procedure is the determination of sulfate by precipitation as barium sulfate with barium chloride. Suppose an excess were added to a sulfate solution to give a 0.001 M solution of barium chloride in 400 mL after the sulfate had reacted. How much barium sulfate would remain in solution?

$$[Ba^{2+}][SO_4^{2-}] = K_{sp} = 10^{-10} \qquad \text{(from table in Appendix)}$$

With $[Ba^{2+}] = 10^{-3}$, $[SO_4^{2-}] - 10^{-10}/0.001 = 10^{-7}$ M. Thus 400 mL of 10^{-7} M barium sulfate in solution amounts to $0.4 \times 10^{-7} \times 233 = 9 \times 10^{-6}$ g, a negligible quantity.

If just the stoichiometric amount of barium chloride had been added, that is, no excess of either sulfate or barium ions were present in the saturated solution, $[Ba^{2+}] = [SO_4^{2-}] =$ molar solubility $= \sqrt{10^{-10}} - 10^{-5}$ M. The barium sulfate dissolved in 400 mL would then be 0.0009 g, 100 times as much as in the previous calculation, and no longer negligible from an analytical point of view.

Calculations involving solubility products and the common-ion effect, as in the above example, are most reliable under the following conditions: (1) for dilute solutions, (2) with only a slight excess of reagent, (3) for highly insoluble precipitates, and (4) for precipitates having singly charged ions. The haphazard addition of an unwarranted excess of reagent to effect quantitative precipitation may have the opposite of the desired effect. For example, in the precipitation of silver ion with a solution containing chloride ion a slight excess reduces solubility losses in accordance with the common-ion effect. But a large excess causes the solubility of silver chloride to increase sharply through formation of chloro complexes such as $AgCl_2^-$ and $AgCl_3^{2-}$. Thus, in 0.001 M excess chloride the solubility of silver chloride is 1.1×10^{-7} M, which is close to the value calculated from the common-ion effect. In contrast, in 0.1 M excess chloride the solubility is about 4×10^{-6} M, hundreds of times larger than predicted from the common-ion effect and the solubility-product constant.

The Diverse-Ion Effect

Equation (8-1) is strictly applicable only to highly dilute solutions. In solutions less dilute the concentration terms should be replaced with activities (Section 3-1). Thus, strictly,

$$a_{M^+} \, a_{A^-} = K_{sp}$$

where a_{M^+} and a_{A^-} are activities of the ions M^+ and A^- in solution. Generally, activity $a = \gamma C$, where γ is the activity coefficient. By arbitrary agreement, in highly dilute aqueous solution γ is assigned a value of unity; a then becomes equal to C. Therefore (8-1) is the limiting case of the expression for the more general activity-product constant.

When solute concentrations are appreciable, the magnitude of the activity coefficient γ depends on the quantity and nature of all the ions in solution (Section 3-1). Serious error can result under these conditions from the use of solubility-product constants expressed in terms of concentration. The value of γ initially becomes smaller as the ionic strength of the solution increases.[3] The net effect of increasing concentrations of salts in solution is that ions become less effective in combining to form a precipitate, and therefore the solubility of the salt increases. This is known as the *diverse-ion effect*. Thus, if enough potassium nitrate or other salt is present in a solution to reduce the activity coefficient of Ag^+ and Cl^- ions to 0.5, the solubility of AgCl is increased by a factor of 2.

Effect of Other Chemical Equilibria

Ions in equilibrium with a precipitate may also be involved in other types of equilibria, all of which may increase the solubility of the precipitate. An example of an additional equilibrium is a complex-forming reaction with a foreign reagent. The more stable the complex ion, the more the solubility will be increased, the effect being most serious when the solubility-product constant is moderately large. Silver ion, for instance, forms relatively stable complexes with ammonia, so that silver chloride precipitation is completely prevented in a solution with a slight excess of ammonia. On the other hand, even though the solubility of silver iodide increases, precipitation remains quantitative because of the exceedingly small solubility-product constant. The quantitative effects of complex formation on equilibria are discussed in Chapter 7.

Complexation may also occur when an excess of a precipitating anion forms a soluble species; $AgCl_2^-$, for example, may form through reaction of AgCl with excess Cl^-. The extent to which solubility is increased can be

---

[3]The mean activity coefficient for 0.1 M KCl is 0.77, and that of 0.1 M $ZnSO_4$ is 0.15. The mean activity coefficient for KCl is the average of the K^+ and Cl^- ion values, and for $ZnSO_4$ the average of the Zn^{2+} and SO_4^{2-} ion values.

calculated from the solubility-product constant and the formation constant of the complex.

The solubility of many precipitates is strongly influenced by acids, bases, and hydrolytic effects. Normally insoluble metal hydroxides (iron(III) hydroxide, for example) dissolve at low pH values, as do normally insoluble salts of weak acids (CaC_2O_4, for example). They may also dissolve through reactions resulting in the formation of hydroxy complexes. Adjustment of pH can also often prevent precipitation of interferences.

EXAMPLE 8-4

Calcium oxalate, a component of kidney stones, becomes increasingly more soluble as pH decreases. What is its molar solubility at pH 9 (where species other than Ca^{2+} and $C_2O_4{}^{2-}$ are negligible) and at pH 3?

At pH 9:   solubility $= [Ca^{2+}] = [C_2O_4{}^{2-}]$

The molar solubility is then the square root of 2×10^{-9}, or 4.5×10^{-5} M.

At pH 3:

$$CaC_2O_4 \rightleftarrows Ca^{2+} + C_2O_4{}^{2-}$$

$$C_2O_4{}^{2-} + H^+ \rightleftarrows HC_2O_4{}^-$$

$$[Ca^{2+}][C_2O_4{}^{2-}] = K_{sp} = 2 \times 10^{-9}$$

$$\frac{[H^+][C_2O_4{}^{2-}]}{[HC_2O_4{}^-]} - K_2 = 6 \times 10^{-5}$$

$$[Ca^{2+}] = [C_2O_4{}^{2-}] + [HC_2O_4{}^-]$$

$$[C_2O_4{}^{2-}]/[HC_2O_4{}^-] = 6 \times 10^{-5}/10^{-3} = 0.06$$

$$[Ca^{2+}] = \text{solubility} = [C_2O_4{}^{2-}] + 16.7[C_2O_4{}^{2-}] = 17.7[C_2O_4{}^{2-}]$$

$$(\text{solubility})(\text{solubility}/17.7) = 2 \times 10^{-9}$$

$$\text{solubility} = 1.9 \times 10^{-4}~M$$

Effect of Other Factors

The solubility of most precipitates increases with temperature. For some the effect is moderate — $BaSO_4$ has a solubility of 2.8 mg/L at 25°C and 3.8 mg/L at 100°C. For others the effect is marked — silver chloride has a solubility of 1.7 mg/L at 25°C and 21 mg/L at 100°C.

The nature of the solvent also has substantial influence. Highly ionized inorganic salts normally are poorly soluble in organic solvents. Two salts useful for the gravimetric determination of potassium, $KClO_4$ and K_2PtCl_6, have solubilities that are too high to be analytically useful in water but sufficiently low in organic solvents.[4]

---

[4]The solubility of potassium perchlorate at 25°C is 20 g/L in water and 0.094 g/L in ethanol.

Time to achieve equilibrium may be another factor. *Supersaturation* is a condition in which a solution contains more dissolved solute than would be present at the equilibrium saturation value. For example, the quantitative precipitation of $MgNH_4PO_4$ requires several hours at a low temperature to overcome its tendency for supersaturation.

Particle size is another factor that influences solubility. This effect is not noticeable until particles of the order of 1 μm or less are involved. For such particles, solubility increases as size decreases.

In pure water some salts tend to form colloidal suspensions that pass through most filters. The particles in such suspensions may not coagulate or settle owing to surface charge caused by adsorbed ions. Being of like charge, the particles repel each other. Some precipitates when washed with water form colloidal suspensions; AgCl is one example. This process is called *peptization*. It can usually be avoided by adding an electrolyte to the wash water.

Physical Characteristics of Precipitates.
The Precipitation Process

Precipitation occurs in two stages — nucleation and crystal growth. *Nucleation* is the process of formation within a supersaturated solution of the smallest particles capable of growth to larger particles. *Crystal growth* consists in the continuing deposition of material, probably by adsorption, from solution onto nuclei and particles already formed.

The prime importance of nucleation lies in the fact that the average final particle size, that is, the physical character of a precipitate, is determined by the number of nuclei initially formed. If relatively few nuclei are formed, then the material to be precipitated must collect on these few. The resulting precipitate will consist of large particles with low surface area and easy filterability. The number of nuclei formed depends on the number of nucleation sites, their effectiveness in causing growth, and the extent of supersaturation.

In *homogeneous nucleation*, critical clusters of ions form spontaneously in a solution otherwise free of particles. This process probably does not occur in systems of analytical importance. In *heterogeneous nucleation*, critical clusters form with the aid of a second phase such as walls of the container or suspended particles. No reagent is absolutely pure, even one labeled "chemically pure" likely contains at least 0.01% insoluble material that can furnish a multitude of nucleation sites.[5] In analytical precipitation the number of nucleation sites is therefore countless and not amenable to experimental control. The effectiveness of nucleation sites varies. The most effective by far are those of the same composition as the material in supersaturated solution. Since in quantitative work seeding with such nucleating material cannot be used, the effectiveness of the sites is another

---

[5]R. B. Fischer (*Anal. Chim. Acta,* **1960,** *22,* 501) clearly demonstrated that a large number of nucleation sites are inevitably available from the container, traces of colloidal contaminants, and the air.

variable not amenable to experimental control. Consequently, the principal means by which an analyst can control nucleation, and hence the physical characteristics of the precipitate, is to control the extent of supersaturation.

The influence of supersaturation on nucleation is best understood in terms of the *supersaturation ratio* $(Q - S)/S$, where Q is the concentration of solute in the supersaturated solution at the instant before nucleation and S is the solubility at equilibrium. Other things being equal, the number of nuclei formed in the nucleation process increases in accordance with the supersaturation ratio. Realize that for any absolute value of supersaturation, the supersaturation ratio for various substances can differ markedly.

EXAMPLE 8-5

Calculate supersaturation ratios for $PbSO_4$ and PbS where $Q = 0.001\ M$. Take the solubility-product constants to be 2×10^{-8} and 1×10^{-28}.

For $PbSO_4$, S would be the square root of 2×10^{-8}, or $1.4 \times 10^{-4}\ M$, and the supersaturation ratio, $(0.001 - (1.4 \times 10^{-4}))/(1.4 \times 10^{-4}) = 6$. For PbS, S would be the square root of 1×10^{-28}, or 1×10^{-14}, and the supersaturation ratio, $(0.001 - (1 \times 10^{-14}))/(1 \times 10^{-14}) = 1 \times 10^{11}$. Thus for $PbSO_4$ the value is tiny, whereas for PbS it is enormous.

The solubility product for $PbSO_4$ is such that under normal conditions of precipitation the supersaturation ratio is low and few nuclei are formed. The material to be precipitated must then precipitate on these relatively few nuclei, and the individual particles form into crystals. For PbS, on the other hand, it is difficult to achieve conditions where the supersaturation ratio is not large; many nuclei form, and the precipitating ions must necessarily spread themselves among more particles. The resulting precipitate is extremely finely divided and amorphous.

For a single substance, say $BaSO_4$, the precipitation process can provide a variety of physical types of precipitates. Upon mixing of dilute solutions of barium chloride and sodium sulfate a coarsely crystalline precipitate will usually form. Upon rapid mixing of concentrated solutions a finely divided and even gelatinous precipitate can be obtained.

When two precipitates are formed under the same conditions of supersaturation, their physical characteristics are likely to be similar. The chemical properties of the ions or precipitate is of secondary importance. If the supersaturation ratio is kept low, few nuclei are formed, and coarsely crystalline precipitates can be produced; this low ratio is achieved by mixing the reagents slowly with efficient stirring, by using dilute solutions, and by raising the temperature or changing the pH to increase the equilibrium solubility. When few nuclei are present, supersaturation is released slowly, thereby allowing large particles to form. When many nuclei are present, supersaturation is rapidly released, causing many small particles to form.

Purity of a Precipitate. Coprecipitation

Clearly, a precipitate will be impure if two or more materials precipitate at the same time. For example, iron and aluminum hydroxide, having similar solubility-product constants, precipitate simultaneously when the pH is raised. For analytical work, conditions should be selected or separations performed so that only the single desired substance will precipitate. The purity of such single precipitates is determined by the extent of *coprecipitation*, the contamination of a precipitate by normally soluble materials. The subject of coprecipitation can be divided into two broad classes, adsorption and occlusion.

Adsorption

Adsorption is the attraction of solute particles to a surface by the action of electrostatic and van der Waals forces. For precipitates of relatively large surface area, adsorption is the main source of contamination. On the vast surface areas of flocculated colloids such as iron(III) hydrous oxide or aluminum(III) hydrous oxide, adsorption may account for a significant fraction of the total solid. In contrast, silver halide precipitates may also have large surface areas, but the extent of adsorption is relatively small. Thus surface area is only one contributing factor.

Adsorption occurs by several mechanisms; the most important is that in which adsorbed lattice ions interact with foreign ions as counterions. Thus if chloride is precipitated from solution by the addition of excess silver nitrate, the silver chloride precipitate will electrically attract some of the excess silver ions to localized sites of residual negative charge (primary adsorption). To maintain electrical neutrality, counterions of opposite charge (anions such as nitrate) are attracted to the resulting localized sites of residual positive charge and may either be absorbed or form a secondary diffuse (double) layer. On washing and ignition the adsorbed silver nitrate does not volatilize, and so causes error. On the other hand, if silver in solution is precipitated with excess sodium chloride solution, chloride ions will be electrically attracted to sites of residual positive charge, and counterions (cations such as sodium ion) attracted to the resulting sites of negative charge. In this case, if the silver chloride precipitate is washed with dilute nitric acid, the sodium ions may be replaced with hydrogen ions, and upon ignition the adsorbed HCl will volatilize. No error in weight of precipitate would result from this adsorption.

The following rules may be helpful in the prediction of the relative extent of counterion adsorption:

1. Other conditions being equal, the ion in highest concentration is adsorbed to the greatest extent.
2. Since adsorption on ionic precipitates is largely electrostatic in nature, other conditions being equal, a multicharged ion is adsorbed more strongly than a singly charged one.
3. Other conditions being equal, the foreign ion most nearly the same size as the lattice ion with the same charge is preferred. In this sense the lattice ion is ideal.
4. The *Paneth-Fajans-Hahn adsorption rule* states that an ion that forms a compound with a lattice ion with the lowest solubility is the one that will be adsorbed most strongly. Furthermore, such ions are adsorbed more strongly on the precipitate than on other surfaces. Hence crystal growth and not further nucleation is preferred once nucleation has occurred.

EXAMPLE 8-6

Suppose that salts of acetate and nitrate are in contact with a silver chloride precipitate. According to the Paneth-Fajans-Hahn rule, which will be adsorbed more strongly?

In this case acetate will be more strongly adsorbed than nitrate, since silver acetate is less soluble than silver nitrate (though both are normally soluble salts).

This rule explains why crystal growth is preferred once nucleation has occurred; ions of both silver and chloride in a supersaturated solution of silver chloride are adsorbed more strongly than other ions.

Because adsorption is a surface phenomenon, its effect can obviously be reduced by keeping the surface area small. Small particles have a larger ratio of surface to mass. Say you have a 1-cm cube, which has 6 cm^2 of surface. If you could divide it into cubes of 10^{-6} cm, you would increase the surface area to 6×10^6 cm^2. A surface factor such as adsorption is now a million times more important.

Contamination by adsorption causes error only if the adsorbed species is not volatilized during ignition. Once the nature of this type of coprecipitation is recognized, it can be minimized by (1) using dilute solutions so that the concentration of foreign ions is reduced; (2) reducing the surface area, that is, precipitating under conditions where few nuclei are formed and large crystals result; (3) carrying out a separation of ions that would be strongly adsorbed; (4) replacing strongly adsorbed with less strongly adsorbed ions; (5) lowering the oxidation number of a metal ion so that its charge is reduced; and (6) using hot solutions to reduce the amount of adsorption per unit area. In the most serious cases of adsorption the precipitate can be filtered, washed, redissolved, and a reprecipitation carried out in the presence of a now greatly reduced concentration of foreign ions.

Occlusion

Occlusion, the second class of coprecipitation, is internal contamination of a precipitate by normally soluble foreign substances. These substances may be adsorbed on the surface of a precipitate and left behind in lattice positions of the crystal as a result of crystal growth. Thus a nitrate ion adsorbed on the surface of barium sulfate may be surrounded and left in a sulfate position. Occlusion is, then, the result of adsorption on a former surface that is now no longer a surface. This type of coprecipitation is more significant than adsorption, since it affects all precipitates, not just those of large surface area. Barium sulfate, for instance, when precipitated from 10% sodium chloride solution contains about 1% sodium chloride in occluded form. The amount of adsorbed sodium chloride is negligible.

Because adsorption is the mechanism by which occlusion takes place, the extent of occlusion can be predicted by the same rules as for adsorption. One additional factor is important: the order in which reactants are mixed affects the extent and kind of occlusion. When barium sulfate is precipitated by adding barium chloride to sodium sulfate, an excess of sulfate ions will be present in solution during the formation and growth of small crystals. Adsorption of sulfate ions on the growing surface is strongly favored over

Figure 8-2. A photograph of two precipitates of potassium perchlorate formed by reaction of perchloric acid with potassium nitrate to illustrate the effect of order of mixing of reactants. In both, the same concentration of permanganate ion was present during precipitate formation. In one case perchloric acid was added to the potassium nitrate solution and in the other the order was reversed. From your knowledge of the mechanism of occlusion, identify which order of mixing was followed in the two cases. (See end of chapter for comment.)

that of other negative ions. This negative surface charge must be neutralized by the adsorption of positive counterions. Since the number of barium ions available during precipitate formation is small, foreign positive ions will be adsorbed and then occluded. Thus occlusion of foreign anions is minimized when crystal anions are in excess during the precipitation process. Similarly, occlusion of foreign cations is minimized when crystal cations are in excess during precipitate formation. Occlusion of either cations or anions, but not both simultaneously, can be reduced by the order chosen. A photograph of two precipitates to illustrate the effect of order of mixing on occlusion is shown in Figure 8-2.

Occlusion can be minimized by inhibition of adsorption through: (1) reducing the concentration of impurities, (2) reducing the charge, (3) raising the temperature, or (4) appropriate choice of order of mixing. Extreme cases may require a preliminary separation step.

An important way to reduce both adsorption and occlusion is through *aging*, or *digestion*. In this process the precipitate is allowed to remain in contact for a time with the solution from which it was precipitated (mother liquor). Since small particles are more soluble than larger ones, a solution in contact with a variety of particle sizes will become supersaturated with respect to the larger particles, saturated with respect to intermediate ones, and unsaturated with respect to the smallest. The small particles dissolve, liberating occluded material. This dissolved material recrystallizes on the larger particles, and since reprecipitation is slow, time is available for desorption before the crystal surface is covered by succeeding layers. Thus the precipitate particles grow larger and purer. Both because solubilities generally increase with temperature and viscosities decrease, the rate of aging is increased in hot solutions. Digestion of highly insoluble precipitates causes little change; the amount of material in solution is too small to be effective.

Mechanical entrapment, or *inclusion*, of portions of the solution inside the precipitate is not a serious source of contamination in most analytical precipitates. It can, however, cause contamination when relatively soluble materials are crystallized from concentrated solutions. Water in silver nitrate or in potassium perchlorate is an example.

In summary, contamination of a precipitate by coprecipitation is reduced when the supersaturation ratio is kept low. This is usually favored when the precipitating agent is added slowly, with stirring, to a hot, dilute solution of the sample. A still more effective way of controlling the supersaturation ratio is described in the next section.[6]

8-3 PRECIPITATION FROM HOMOGENEOUS SOLUTION

Quantitative precipitation requires low solubility of the precipitated compound. But it is difficult to ensure a low supersaturation ratio when precipitating a sparingly soluble compound; consequently, coprecipitation of contaminants may become serious. One way to reduce contamination of a precipitate is to control the supersaturation ratio during nucleation by adding the precipitating agent slowly and uniformly to the solution containing the sample, thereby avoiding a large local excess of reagent. An elegant technique of this type is *precipitation from homogeneous solution*. In this technique the precipitating reagent is generated homogeneously and slowly by chemical reaction in the sample solution. Nucleation occurs at an exceedingly low supersaturation ratio, and only crystal growth occurs subsequently.

One method involves formation of a metal hydroxide precipitate from an initially unsaturated homogeneous solution of a metal ion by gradually raising the pH to slowly decrease the solubility.[7] The experiment in Section 16-5 illustrates an application of this technique to the determination of aluminum, in which hydroxide is generated to precipitate aluminum hydroxide from a mixture of aluminum, zinc, and magnesium salts. Iron(III), aluminum, zinc, and magnesium all form insoluble hydroxides upon direct addition of base. Their solubilities, however, differ. Calculations based on solubility-product constants of the individual hydroxides show that iron(III) hydroxide begins to form from 0.1 M solutions of the metals at about pH 2.3, aluminum hydroxide at about pH 3.4, zinc hydroxide at about pH 5.5, and magnesium hydroxide at about pH 9.0. For the aluminum in 0.1 M solution to be precipitated quantitatively the aluminum ion concentration must be reduced to 10^{-4} M or less. To attain this lower concentration requires a pH of 4.8; at this pH the aluminum can be quantitatively precipitated before zinc or magnesium hydroxide begins to form. Iron(III) hydroxide, on the other hand, precipitates before aluminum hydroxide, but at a pH so close that selective precipitation is impossible. Iron, therefore, must be separated prior to precipitation of aluminum. (This separation is readily accomplished by extraction of the iron(III) into an immiscible solvent, as outlined in Section 16-5.)

---

[6]The problem of supersaturation ratio and nucleation is vital not only to chemists but also to others. Beekeepers know that the quality of honey in large part depends on nucleation and crystal size. Smooth honey consists of finely crystallized material, which forms when a large number of nuclei are present. The same is true for ice cream and some candies. To farmers, the nucleation of hail is a concern. In the bottling of soft drinks much attention is paid to controlling or preventing nucleation.

[7]Another method is generation of a reagent by a slow chemical reaction. For example hydrolysis of methyl oxalate produces oxalate ion, $C_2O_4^{2-}$. This ion can then react with a metal ion in solution to form an oxalate precipitate by precipitation from homogeneous solution.

The close control of pH required in this selective hydroxide precipitation is made possible by addition of a succinate–succinic acid buffer; the pH is then slowly increased to about 5 by adding urea to the solution and then heating it. Urea solutions slowly hydrolyze on warming to form ammonia and carbon dioxide:

$$(NH_2)_2CO + H_2O \rightarrow CO_2 + 2NH_3$$

The carbon dioxide, not being soluble in hot acid solutions, escapes. The ammonia remains, raising the pH by neutralizing part of the succinic acid. Because the pH increases gradually over several hours, the supersaturation ratio never becomes high, and the rate of nucleation remains low. These conditions are favorable for the growth of particles already in existence, resulting in a relatively dense, pure precipitate. Limiting the amount of urea added prevents the pH from rising to levels where the solubility products of zinc and magnesium hydroxide are exceeded.

8–4 STOICHIOMETRY

In gravimetric analysis several types of calculations may be encountered. In the simplest type the constituent sought is weighed in the same form in which the percentage by weight is to be expressed:

wt % = wt sought-for constituent $\times$ 100/(wt sample)

EXAMPLE 8–7

A 0.7777-g sample of limestone after ignition to 900°C weighed 0.4476 g. Calculate the percentage of carbon dioxide in the sample.

wt CO_2 lost from sample = 0.7777 – 0.4476 = 0.3301 g

$$\% \ CO_2 = \frac{0.3301}{0.7777} (100) = 42.45\%$$

More commonly, the percentage of the constituent sought is not directly equal to an observed weight (or difference in weights), and an appropriate conversion must be made.

EXAMPLE 8–8

A 0.5656-g sample containing a bromide salt is treated with excess silver nitrate. The silver bromide formed is filtered and weighed, yielding 0.7624 g of precipitate. Calculate the percentage of bromide in the sample.

The important reaction is $Ag^+ + Br^- \rightarrow AgBr(s)$. Thus, for each mole of bromide in the sample one mole of silver bromide is formed. Accordingly, from 0.7624 g of silver bromide there would be $(0.7624 \times 79.904/187.77)$ g of bromine. The percentage of bromine in the sample would be

$$\% \text{Br} = \frac{79.904 \text{ g Br/mol}}{187.77 \text{ g AgBr/mol}} \times \frac{0.7624 \text{ g AgBr}}{0.5656 \text{ g sample}} \, 100 = 57.36\%$$

The percentage of sodium bromide can be calculated in like manner:

$$\% \text{ NaBr} = \frac{(79.904 + 22.99) \text{ g NaBr/mol}}{187.77 \text{ g AgBr/mol}} \times \frac{0.7624 \text{ g AgBr}}{0.5656 \text{ g sample}} \times 100 = 73.86$$

The factor Br/AgBr or NaBr/AgBr in the foregoing example is called the *gravimetric,* or *chemical, factor,* defined as the ratio of the molecular weight (or its equivalent for a radical or ion) of the substance sought to that of the substance weighed. The same number of atoms for the constituent of interest must appear in both the numerator and the denominator of the ratio.

Through application of the appropriate gravimetric factor, results can be reported in a variety of ways. From the weight of $BaSO_4$ precipitate formed the results can be expressed as % S, % SO_2, % SO_3, % K_2SO_4, or % FeS_2; the appropriate factors for these cases are $S/BaSO_4$, $SO_2/BaSO_4$, $SO_3/BaSO_4$, $K_2SO_4/BaSO_4$, and $FeS_2/2BaSO_4$. Note that in the last factor the denominator has twice the molecular weight of $BaSO_4$.

Another type of calculation encountered in gravimetric work is that of estimating the amount of reagent to be added to a substance to ensure complete precipitation.

EXAMPLE 8-9

What weight of ammonium oxalate is required to precipitate the calcium from a 0.8-g sample of impure limestone?

Because one mole of calcium carbonate provides one mole of calcium, which in turn requires one mole of $(NH_4)_2C_2O_4 \cdot H_2O$, the maximum weight of $(NH_4)_2C_2O_4$ $\cdot H_2O$ (if we assume 100% $CaCO_3$) is

$$((NH_4)_2C_2O_4 \cdot H_2O/CaCO_3) \times 0.8 = (142/100) \times 0.8 = 1.14 \text{ g}$$

To ensure a slight excess, 1.2 g would be appropriate.

What volume of a 0.5 M solution of ammonium oxalate would be required?

$$\text{number of moles required} = 0.8/100$$

$$\text{vol 0.5 } M \text{ solution} = (0.8/100) \, (1000/0.5) = 16 \text{ mL}$$

Again, to ensure a slight excess, 18 to 20 mL would be appropriate.

8-5 TECHNIQUES OF FILTRATION, WASHING, AND IGNITION

After a test for complete precipitation by the addition of a small amount of precipitating reagent, the precipitate is separated from the solution phase,

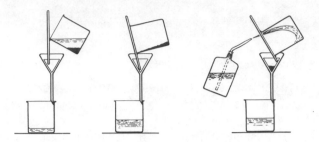

Figure 8-3. Operations of decantation and transfer of precipitate.

or mother liquor, by filtration and washed free of soluble salts. The filterability of a precipitate depends largely on how the precipitation was carried out.

Three steps are involved in the actual filtering operation — decantation, washing, and precipitate transfer. *Decantation* is the process of pouring the liquid above the precipitate through the filter. The solution is guided by a stirring rod, as shown in Figure 8-3. Next, the precipitate is washed in the beaker by adding, stirring, and decanting several portions of wash solution. Finally, the precipitate is transferred to the filter with a stream of wash liquid. A *policeman*, a flat rubber tip on the end of a stirring rod, is used to dislodge particles of precipitate adhering to the beaker.

Types of filters include paper, sintered glass, porous porcelain, and glass or asbestos fiber mats in Gooch crucibles. *Gooch crucibles* are made of porcelain and have a perforated bottom upon which a mat is laid. Glass fiber mats are available in ready-to-use form. They are more convenient than asbestos as long as the precipitate is being ignited at moderate temperatures, but the fibers fuse readily in the direct heat of a Bunsen burner. Sintered glass or porcelain crucibles with built-in porous bottoms can tolerate higher temperatures but are difficult to clean.

A precipitate can be washed with a relatively small volume of liquid used efficiently, that is, with a large number of small portions with complete drainage between additions. To ensure that washing is quantitative, test the wash liquid for completeness of removal of soluble impurities.

After the precipitate has been washed, the crucible and contents are dried and then heated to constant weight at an appropriate temperature. *Constant weight* is the weight to which a crucible returns reproducibly each time it is subjected to a controlled heating and cooling cycle. It is usually characterized by agreement of the weight within 0.2 mg after successive cycles. Only the final weight is used in calculations. If a burner or furnace is employed, this step is called *ignition*, usually at a temperature between 200 and 1000°C, determined by the nature of the precipitate. When a precipitate in a filtering crucible is ignited with a burner, the crucible is placed inside a larger one to protect the mat and precipitate from direct contact with the burner gases.

Although for most purposes filter paper is less convenient than a filtering crucible, for high-temperature ignitions it is less expensive than a sintered porcelain crucible. Also, it circumvents the problem of removal of ignited

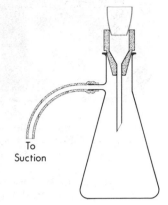

Figure 8-4. Cross section of suction filtration assembly with filter crucible.

To
Suction

material from the filter. The paper is burned off during ignition of the precipitate.

Procedure For Filtration Through a Filter Crucible

For filtrations other than through paper, a suction filtration assembly, as illustrated in Figure 8-4, is recommended. If suction is provided by an aspirator and the filtrate is to be saved, place a safety bottle between the aspirator and the flask to prevent accidental suckback of tap water into the flask. In any case, the crucible should be removed to break the vacuum before it is turned off.

Carry out the filtration operation in the following way. Place the filtering crucible, which has been brought to a known constant weight, in position over a clean suction flask, apply suction, and pour the supernatant liquid through the filter. Leave behind as much as possible of the precipitate so that it will not slow the flow through the filter. Use a stirring rod to guide the solution from the beaker tip to the crucible, to prevent splashing, and to prevent solution from running down the outside of the beaker when pouring is interrupted. Next transfer the precipitate to the crucible with a minimum of wash liquid. Use a policeman to loosen precipitate adhering to the beaker walls. After the precipitate has been quantitatively transferred, wash it with successive small portions of wash liquid to remove soluble salts. If the precipitate is sufficiently insoluble, use hot wash liquid to increase the solubility of impurities and speed the filtration. In some cases pure water is satisfactory; in others an electrolyte in the water is needed to prevent peptization. When necessary use a solvent mixture to reduce solubility losses. Keep each portion of wash liquid small, allowing it to drain completely before the next is added. If possible, test the completeness of washing to ensure that all the mother liquor has been removed.

Procedure for Filtration through Paper

Select ashless paper of the correct porosity and diameter and fold it in half, then in quarters (Figure 8-5).[8] Properly fold and fit the filter paper to the funnel

---

[8]Filter funnels are available with apical angles of either 58 or 60°. For a funnel of 60° the corners of the quarter-fold should not coincide, but be displaced about 3 mm in each direction. The second fold is made slightly off-center so as to form a cone that, upon opening of the larger quarter, forms an apical angle somewhat greater than 60°. For a 58° funnel the second fold is made off-center in only one direction; the apical angle will then be somewhat greater than 58°.

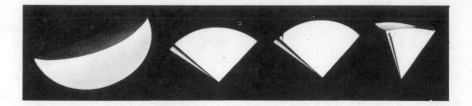

Figure 8-5. Method of folding filter paper. Fold along the diameter of the paper. Make a quarter-fold appropriate to the funnel to be used. Tear a small part of the corner for closer fit to the upper edge to the funnel.

so that it fits tightly to the funnel wall near the top and leaves a slight space farther down for better drainage. Seat the cone in the funnel, moisten it with water, and press the upper portion gently against the glass to make a seal. When a filter is properly fitted, the stem of the funnel will fill when water is poured into the funnel and remain full after the funnel has drained. Slight suction exerted on the underside of the filter by the weight of this liquid column markedly increases the filtration rate.

Procedure for Ignition of Filter Paper and Precipitate

After completing a quantitative washing and transfer of the precipitate, carefully lift the filter paper containing the precipitate. Fold the upper edges of the paper over, and then place the paper and contents in a crucible previously ignited to constant weight. Place the folded side of the paper downward so that it will not unfold while drying. Collect any small particles of precipitate adhering to the beaker by wiping with a small portion of moistened filter paper. Add it to the crucible. Place the crucible on a triangle over a burner as illustrated in Figure 8-6. Dry the paper first with a low flame to prevent spattering, and then slightly increase the heat to char the paper. Do not allow the paper to burst into flame, or precipitate may be lost. After smoking has ceased, increase the heat until all the carbon is removed. Carry out the final ignition at the maximum heat of the burner.

Figure 8-6. Position of crucible during drying and ignition of a precipitate.

8-6 APPLICATIONS: PRECIPITATION

The literature on the subject of gravimetry is extensive. In this and the following section only a brief indication is given of the variety of gravimetric methods. For a more thorough review of applications the reader is referred elsewhere. [9]

Precipitation processes in analytical chemistry involve mainly inorganic systems. They include the determination of halides by precipitation with excess silver nitrate. The determination of chloride by precipitation as silver chloride is described in Section 8-10. Since the method is subject to few inherent errors, high precision and accuracy are possible. A large number of multivalent metal ions can be determined by precipitation as hydrous oxides. For example, iron(III) can be treated with a slight excess of ammonia to precipitate iron(III) hydroxide (hydrous iron(III) oxide), which is extremely insoluble (solubility-product constant about 10^{-36}). Reprecipitation is ordinarily necessary to reduce coprecipitation. With the technique of precipitation from homogeneous solution a much purer precipitate can be obtained. Other common metal ions determined by a similar technique are aluminum, chromium, titanium, and zirconium. After filtration and washing, the precipitates are usually heated strongly to produce the metal oxide as the weighing form.

Sulfates, sulfides, sulfites, and sulfur compounds in general can be determined by conversion if necessary to the sulfate and precipitation with barium chloride. The barium sulfate produced is filtered, washed, and weighed. Coprecipitation can be a serious problem.

Silicate rocks are insoluble in most solvents; they are usually first fused with sodium carbonate. This procedure allows dissolution in acid owing to the increase in the proportion of sodium associated with the silica. The silica $(SiO_2))$ can then be determined by conversion to silicic acid, dehydration, filtration, ignition, and weighing. Since the silica is seldom pure, the weighed material is often treated with hydrofluoric acid to volatilize the silica as SiF_4, leaving behind the impurities. The difference in weight before and after HF treatment is equal to the weight of silica in the sample.

Systematic schemes for inorganic precipitation and separation have been developed. For example, sulfide separations for qualitative analysis involve precipitation equilibria of the type

$$M^{2+} + S^{2-} \rightleftarrows MS(s)$$

Those metal ions that form highly insoluble sulfides, such as copper, lead,

[9]I. M. Kolthoff, E. B. Sandell, E. J. Meehan, and S. Bruckenstein, *Quantitative Chemical Analysis*, 4th ed., McMillan, 1969, includes extensive discussion of many of the most common gravimetric methods. A. I. Vogel, *Textbook of Quantitative Inorganic Analysis*, 4th ed., Longman, New York, 1978, gives extensive coverage of gravimetric methods, including systematic procedures for most elements and other substances. *Standard Methods of Chemical Analysis*, 6th ed., N. H. Furman, Ed., Vol. I, Van Nostrand, Princeton, N.J., 1963, describes practical methods for the detection, separation, and determination of the elements in common natural and synthetic materials.

and cadmium, can be precipitated from strongly acidic solutions, others such as zinc from weakly acidic solutions, and still others such as manganese from alkaline solutions. Precipitation of sulfides from homogeneous solution by use of thioacetamide, which slowly hydrolyzes in solution to produce hydrogen sulfide, gives precipitates of superior physical properties and also reduces the extent of coprecipitation, often a serious problem.

Electrogravimetry, another general type of gravimetric analysis, involves electrodeposition of a substance, usually a metal, from solution onto an electrode, which is customarily weighed before and after deposition. This topic is treated in Section 12-1.

8-7 APPLICATIONS: VOLATILIZATION

A substantial number of substances can be separated by volatilization or distillation. The most common techniques involve the direct ignition of a sample, reaction with a reagent, or distillation.

Volatilization by Ignition

The analytical operation of *volatilization* consists in heating a sample in air or in an inert gas to remove a gaseous substance.   Probably the most widely applied determination by volatilization is that of water; many chemical and physical methods have been developed for its measurement, both direct and indirect. In the *indirect method* the sample is weighed, heated above the boiling point of water for about 2 h, cooled, and weighed again. The difference in weight is attributed to loss of water, and the results are calculated accordingly. For the method to be accurate, water must be the only volatile material in the sample, and no other reactions that would affect the weight, such as oxidation or hydrolysis, can occur.

EXAMPLE 8-10
Procedure for the Determination of Water in BaCl$_2$·2H$_2$O
    Weigh, to the nearest 0.1 mg, a weighing bottle and stopper that have been brought to constant weight by heating in an oven at 110°C. Add about 1 g of powdered hydrated barium chloride to the weighing bottle and reweigh. Place the unstoppered bottle and the stopper in the oven at 110°C for at least 1 h. Stopper the bottle and cool for at least half an hour. Weigh again to the nearest 0.1 mg. Repeat the heating, cooling, and weighing operations until constant weight is attained. Calculate the percentage of water according to (100)(loss in weight on heating/weight of original sample). For the pure dihydrate the result should be (2H$_2$O/BaCl$_2$·2H$_2$O) $\times$ 100, or 14.75%.

If reactions other than simply the evolution of water occur on heating, then a *direct* method specific for the volatilized compound should be employed. For instance, if hydrated magnesium chloride is heated to determine the water content, in addition to water vaporization, the following reaction occurs:

$$MgCl_2 + H_2O(g) \rightarrow MgO + 2HCl(g)$$

But, if the mixture of H_2O and HCl gases are passed over heated lead oxide, the reaction

$$PbO + 2HCl \rightarrow PbCl_2 + H_2O(g)$$

regenerates the water. If the gases are then carried in an air stream through a tube containing a water absorbent (such as a U-tube filled with magnesium perchlorate), the increase in the weight of the tube can be taken to be a direct measure of the water in the sample.

Direct ignition can be used to drive carbon dioxide from carbonates. An example is

$$CaCO_3 \rightarrow CaO + CO_2(g)$$

In the indirect determination of carbon dioxide in carbonate-containing materials, the measurement of either the loss in weight of the sample on heating or the increase in weight of a tube of alkaline absorbent for the evolved carbon dioxide gives a measure of the amount of carbonate present.

Volatilization Following a Chemical Reaction

Several of the many important applications of this method are mentioned briefly here. Some, but not all, involve gravimetry.

Carbon dioxide can be evolved from carbonate salts by the addition of excess acid:

$$CaCO_3 + H_2SO_4 \rightarrow CO_2(g) + H_2O + CaSO_4$$

The carbon dioxide can be quantitatively removed by warming the solution and passing a current of carbon dioxide–free air through it after passage through a drying agent. The carbon dioxide can be absorbed in a tube containing sodium hydroxide suspended on asbestos (Ascarite). The amount of carbonate present in the sample is then calculated from the gain in weight of the tube and contents upon carbon dioxide absorption.

Elemental analysis for *carbon and hydrogen* in organic compounds involves reaction of an organic sample at a high temperature with oxygen to form carbon dioxide and water. These gases can be absorbed by sodium hydroxide and magnesium perchlorate in preweighed tubes or measured with a gas chromatograph. Carbon in steel can be determined similarly, but higher temperatures are required.

The Kjeldahl method for *nitrogen* in organic samples involves the distillation of ammonia (Section 5-8).

Boron in borates can be separated through the distillation of methyl borate $B(OCH_3)_3$, which is formed when acidic borate solutions are heated with an excess of methanol.

Fluoride forms complexes with many metal ions, a characteristic that makes this element both difficult to determine and an interference in the determination of substances associated with it in a sample. It can be removed by distillation as H_2SiF_6 from sulfuric or perchloric acid solution at about 140°C.

Silica can be removed from silicates by reaction with excess hydrofluoric acid and heating:

$$SiO_2 + 4HF \rightarrow SiF_4(g) + 2H_2O(g)$$

Sulfides react with strong acids such as HCl to form H_2S, which can be absorbed and titrated with standard iodine solution (Section 10-4).

Arsenic can be distilled as the trichloride $AsCl_3$, as the pentabromide $AsBr_5$, or as arsine AsH_3, *antimony* can be distilled as $SbCl_3$, and *tin* as $SnCl_4$.

Mercury metal is relatively easily liberated from its compounds by heating with sodium carbonate in air and distilling the metal.

Salts of volatile strong acids can be determined by measurement of the residue following heating with sulfuric acid:

$$2NaCl + H_2SO_4 \rightarrow Na_2SO_4 + 2HCl(g)$$

Here sodium is determined by addition of excess H_2SO_4 and heating. The excess H_2SO_4 is volatilized along with the HCl, and the Na_2SO_4 is then weighed.

A frequently useful measurement is simply that of loss in weight on ignition. Ignition in air can remove *organic materials* as well as water. The determination of ash in coal is an example of this type of analysis.

8-8 ORGANIC REAGENTS

An enormous number of organic reagents have been employed for the precipitation and separation of metal ions. Under optimum conditions organic precipitants can provide exceptionally selective and sensitive methods for inorganic ions. In precipitation reactions, useful organic reagents react with the ion to be determined to produce compounds whose solubility is low enough to leave only a negligible amount in solution. The selectivity of the reaction can be enhanced by control of pH and of reagent concentration and by addition of masking reagents that prevent the precipitation of certain ions. Often a stepwise series of complexes form, with the metal ion displacing one or more hydrogen ions from each reagent ligand. When a neutral species ML_n is produced, its solubility is often low and precipitation occurs.

Three examples of organic precipitants are 8-hydroxyquinoline (oxine), ammonium   nitrosophenylhydroxylamine   (cupferron),   and   1-nitroso-2-naphthol; their structural formulas are

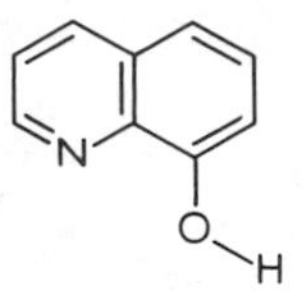

8-Hydroxyquinoline
(oxine)

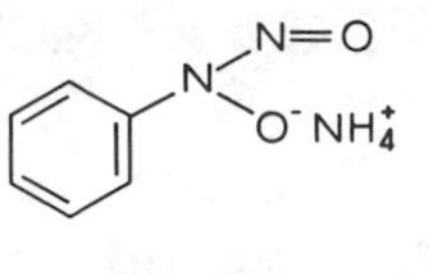

Ammonium
Nitrosophenyl-
hydroxylamine
(cupferron)

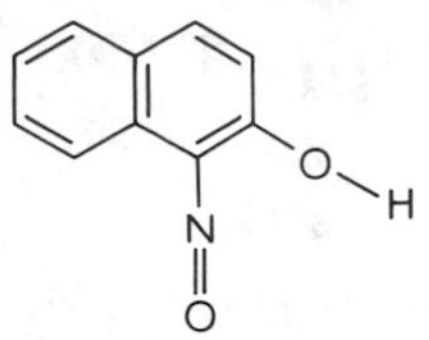

1-Nitroso-2-Naphthol

With all three a hydrogen ion (or other cation) is displaced from the ligand upon chelation with a metal ion to yield a neutral species insoluble in water:

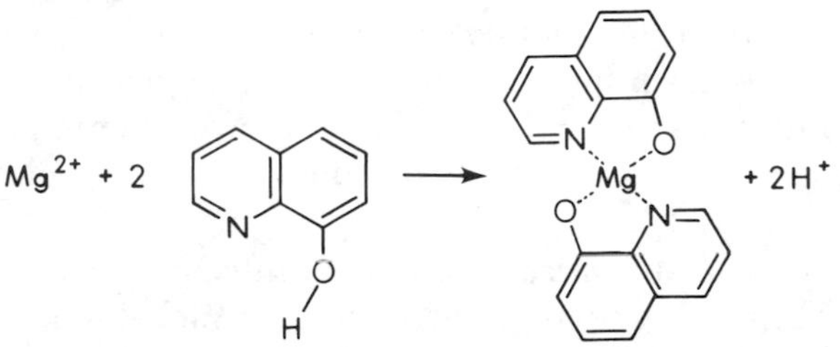

Thus 8-hydroxyquinoline (HOx) gives a precipitate with aluminum of the composition $AlOx_3$. With magnesium a precipitate of $MgOx_2 \cdot 2H_2O$ forms; the last two coordination positions around magnesium are occupied by water molecules because two ligands suffice to provide a complex of zero charge. Oxine reacts with many metal ions; some selectivity is possible through control of the pH.

Cupferron forms insoluble complexes with iron(III), vanadium(V), and titanium(IV) that serve to separate these metals from copper, lead, aluminum, nickel, and zinc. Precipitation is carried out in hydrochloric acid or sulfuric acid. The precipitates are rarely weighed directly, but instead after ignition to the metal oxides. An attractive and rapid method of separation prior to analysis is extraction of the complexes from aqueous acid into another solvent such as chloroform.

1-Nitroso-2-naphthol precipitates cobalt as $Co(C_{10}H_6ONO)_3$, a 3:1 complex, and is used principally to separate cobalt from nickel. The precipitate, which is always contaminated with products of decomposition, may be ignited to the oxide, Co_3O_4, and weighed; alternately it may be redissolved and the cobalt determined by oxidation-reduction titration or by electrodeposition.

The reagent dimethylglyoxime is highly specific for nickel; the nickel ion can displace a hydrogen ion from each of two dimethylglyoxime molecules to form highly insoluble nickel dimethylglyoxime. This reaction is discussed in the next section.

Organic reagents can also aid in separations by acting as extraction reagents (Section 16-2).

8-9   DETERMINATION OF NICKEL AS NICKEL DIMETHYLGLYOXIME [10]

Background

In the gravimetric determination of a substance the precipitating reagent should be as specific as possible for that substance. Although no known reagent is completely specific for a single element, dimethylglyoxime (DMG) is nearly so for nickel. In ammoniacal solution, nickel is precipitated quantitatively by DMG as a bright strawberry-red complex, $Ni(DMG)_2$. Palladium, though precipitated from acid solution as a yellow complex, $Pd(DMG)_2$, in the presence of ammonia is held in solution as a stable ammine complex. Of the other metals only copper in high concentrations and mixtures of cobalt(II) and iron(III) cause interference by precipitating along with nickel. Tartrate is added to complex any iron or aluminum present that otherwise would precipitate as the hydroxide when the solution is made alkaline.

The specificity of DMG arises from the unusual nature of the nickel chelate. Two DMG ligands are coordinated to each nickel ion, with the nickel and all four coordinating nitrogen atoms in the same plane:

Dimethylglyoxime (DMG) $Ni(DMG)_2$

In the crystals the nickel atoms are stacked one above another, with weak bonds between adjacent nickel ions. This unusual metal-metal bonding apparently provides sufficient stability in the crystal to give low solubility relative to other metal ions.

Nickel dimethylglyoxime is bulky when first precipitated; the precipitate also tends to creep up the walls of the vessel containing it. For these reasons, the amount of nickel precipitated should be limited to about 50 mg. The precipitate is stable and readily dried at 110 to 130°C; it begins to decompose at about 180°C.

Procedure (time 4½ to 6 h)

Prepare three Gooch crucibles for the filtration step by placing a borosilicate-glass filter pad in each. Place each crucible in a suction filtration assembly and wet the pad with water. Then pass a few drops of 0.1 M HNO_3

---

[10]We prefer this simpler name to the more accurate bis(dimethylglyoximato)nickel(II).

through it to remove acid- or water-soluble substances, followed by a few milliliters of water. Hold each crucible to the light to check for pinhole leaks. Dry the crucible plus pad in an oven at 110°C until constant weight ($\pm$ 0.2 mg) is reached.

Accurately weigh samples containing about 40 mg of nickel into 400-mL beakers; in a hood add 10 mL of 6 M HNO_3 to each, cover with a watch glass, and place on a hot plate to dissolve.[11] When dissolution is complete, dilute each sample to about 50 mL with water, and boil the solution gently for a short time to remove most of the nitrogen oxides. Allow the solution to cool, and add about 0.5 g of tartaric acid and 5 drops of bromocresol green indicator to each sample. Then add 6 M NH_3 slowly, with stirring, until the indicator just changes to an intermediate green; about 5 mL should be required.[12] Bring the solution back to yellow by dropwise addition of 6 M acetic acid.[13]

Add 75 mL of acetone to each beaker, and gently heat the solution to about 65°C.[14]   Slowly add 20 mL of a 2% solution of sodium dimethylglyoxime (NaDMG). Stir, and immediately add 3 M $NH_4C_2H_3O_2$ dropwise until the first permanent red precipitate of $Ni(DMG)_2$ forms.[15] Allow the precipitate to form and digest at 65°C for an hour. Do not cover.

Add 1 to 2 mL of NaDMG solution to the supernatant liquid to test for completeness of precipitation. When complete, add about 10 drops of a 10% solution of detergent, and cool to room temperature before filtering.[16]Transfer the precipitate, and wash it with small portions of a cold solution containing 1 mL of 6 M NH_3 and a few drops of liquid detergent in 500 mL of water. Dry the crucible plus precipitate to constant weight (again $\pm$ 0.2 mg) in an oven at 110°C.

Report the percentage of nickel in the sample. The formula weight of the nickel dimethylglyoxime precipitate, $Ni(DMG)_2$, is 288.93. In high-quality work the relative range for triplicate determinations should not exceed about 0.3%.

8-10 DETERMINATION OF CHLORIDE AS SILVER CHLORIDE

Background

This experiment provides practice in handling a colloidal precipitate and using a filter crucible. This type of crucible permits the precipitate to be dried at a low temperature; the temperature required to burn off

---

[11]If the samples contain nickel oxide substitute hydrochloric acid for nitric acid.

[12]If the indicator fades before a permanent color change is reached, add more indicator and continue.

[13]This is a convenient stopping point. Once the nickel has been precipitated, the precipitate should be filtered during the same laboratory period.

[14]The solutions must be warmed carefully, and not above 65°C, or bumping may cause loss of precipitate. A waterbath held at 60 to 65°C is useful in providing even, steady heat.

[15]Occasionally some precipitate of $Ni(DMG)_2$ will appear before the $NH_4C_2H_3O_2$ has been added, owing to the alkalinity of the solution. If the quantity of precipitate is small, the error is usually negligible.

[16]Should precipitation be incomplete, add about 10 mL more of DMG solution, stir, and allow to digest at 65°C for a half hour more. After again cooling to room temperature, decant the solution through the Gooch crucible with minimum disturbance of the precipitate. Retest the cooled filtrate for completeness of precipitation.

conventional filter paper causes appreciable decomposition of many precipitates, including silver chloride.

Chloride in an unknown sample is determined here by addition of a dilute solution of silver nitrate to a solution of the unknown to precipitate silver chloride:

$$Ag^+ + Cl^- \rightleftharpoons AgCl(s)$$

Precipitation is carried out from a solution slightly acidified with nitric acid. The precipitate is filtered in a Gooch crucible, washed, dried at a low temperature, and weighed as silver chloride.

Silver chloride is sensitive to light, reacting according to

$$2AgCl \rightarrow 2Ag(s) + Cl_2$$

When this reaction takes place before filtration, additional precipitate is formed, and the result is high:

$$3Cl_2 + 5Ag^+ + 3H_2O \rightarrow 6H^+ + 5AgCl(s) + ClO_3^-$$

After filtration, chlorine is lost and the result is low. Avoid exposing the precipitate to strong light, particularly direct sunlight; do not be concerned about normal laboratory illumination, which causes only a slight darkening.

Silver chloride does not undergo significant crystal growth during aging, but remains a colloidal precipitate. During aging the particles must be made to clump together so that they can be retained by a filter. This agglomeration is brought about by the presence of electrolyte, in this case silver nitrate and nitric acid. The wash water also must contain an electrolyte to prevent the silver chloride precipitate from dispersing (peptizing) during washing. A volatile electrolyte, nitric acid, prevents contamination of the precipitate.

Silver chloride has a significant temperature coefficient of solubility: 0.9 mg/L at 10°C, 1.7 mg/L at 25°C, 5.2 mg/L at 50°, and 21 mg/L at 100°C. For this reason it is filtered at or below room temperature.

With suitable precautions the determination of chloride by this procedure can be carried out with exceptional precision; the stoichiometry is exceedingly well behaved. Accordingly, this reaction is used extensively for halogen determinations in organic elemental analysis, and in inorganic analysis for oxyhalides as well as for halides. In fact, the superior properties of silver chloride as an analytical precipitate are responsible for its prominent historical position in the determination of atomic weights (Section 8-1).

Procedure (time 4½ to 5½ h)

Prepare three Gooch crucibles (for method see Section 8-9), and dry them in an oven at 110°C to constant weight. Use a desiccator during cooling if appropriate. Dry the unknown sample in a weighing bottle at 110°C for at least 1 h.[17] Accurately weigh 0.2-g samples of the unknown into 250-mL beakers. Dissolve in 125 mL of distilled water and add 2 mL of 6 M HNO₃. Calculate the

---
[17]Consult the laboratory instructor concerning drying of samples.

volume of 0.1 M AgNO$_3$ solution needed to precipitate the chloride in the unknown sample, assuming the sample to contain as much chloride as would pure NaCl. (See Example 8-9 for general approach.) Add slowly with stirring this volume plus an excess of about 5 mL. Heat the solution to 90 to 95°C to coagulate the precipitate. Allow the precipitate to settle, and test for completeness of precipitation by adding about 1 mL of 0.1 M AgNO$_3$ solution to the supernatant liquid. When precipitation is complete, cover the beaker with a watch glass, and allow the precipitate to digest for at least 1 h.[18]

Decant the supernatant liquid through the Gooch crucible with suction, and wash the precipitate in the beaker twice by adding portions of 0.01 M HNO$_3$, swirling, and decanting into the crucible. With the aid of a wash bottle with an upward pointing tip and containing 0.01 M HNO$_3$, quantitatively transfer the precipitate to the crucible; use a rubber policeman to loosen particles that may have adhered to the beaker walls. Continue washing the precipitate in the crucible until a test of the filtrate for silver ion is negative. Test by collecting a 5- to 10-mL portion in a test tube suspended with a thread in the filter flask and adding a few drops of about 0.1 M NaCl solution. When the test is negative, wash the precipitate with two or three small portions of water to remove HNO$_3$. Dry to constant weight at 110°C.[19] Report the percentage of chloride in the sample. In high-quality work the relative range for the results should not exceed 0.2%.

8–11   DETERMINATION OF SULFATE AS BARIUM SULFATE

Background

In this experiment the sulfate in a sample is determined by precipitation as barium sulfate. This salt is particularly susceptible to coprecipitation of foreign ions, and so care must be taken to keep the supersaturation ratio as low as possible during precipitation. This is accomplished by slow addition of a dilute solution of barium chloride with stirring to a hot hydrochloric acid solution of the sulfate salt. The barium sulfate is then digested, filtered, dried, and weighed. For high-quality results all operations must be carried out so that no more than a few tenths of a percent of the precipitate is lost.

Attention to several other details in addition to the supersaturation ratio is necessary to ensure quantitative results: (1) Hydrochloric acid added to the sulfate solution prevents the precipitation of barium salts of weak acids, such as carbonate or phosphate, and replaces adsorbed impurities. The acid volatilizes during the ignition step. (2) The presence of a slight excess of barium chloride lowers the amount of sulfate remaining in solution by the common-ion effect. (3) Digestion of the precipitate increases the average particle size and further reduces coprecipitation. In this operation a precipitate is allowed to stand in contact with the solution from which it was precipitated (mother liquor). For barium sulfate adequate digestion requires at least a half hour near the boiling point or several hours at room temperature.

---

[18]Prior to filtration no harm is done if the precipitate stands overnight or until the next laboratory period provided that the beaker is protected from light, dust, and HCl fumes.

[19]Upon completion of the experiment, solutions and precipitates containing silver may be collected for later recovery of the silver.

Procedure (time 4½ to 6 h)

Prepare a set of Gooch crucibles (see the procedure in Section 8-9 for the method of preparation). To dry the crucibles, set each inside a large conventional crucible positioned over a Bunsen burner on an iron ring and a clay triangle, and ignite for 10 to 15 min with the burner flame adjusted so that the bottom of the outer crucible is only faintly red.[20] Allow to cool to room temperature (~ ½ h) and weigh the Gooch crucible. Repeat the heating and weighing until each comes to constant weight. Each heating and cooling cycle takes about 1 h.

Weigh 0.5-g portions of a dry sulfate sample into clean, but not necessarily dry, 400-mL beakers and dissolve in 250 mL of distilled water. Add 3 mL of 6 M HCl. Heat the solution to nearly boiling. Add slowly an excess of 0.05 M $BaCl_2$ solution with efficient stirring (about 70 mL is required). Use a separate stirring rod without policeman for each sample. Allow the covered precipitate to digest near the boiling point for 30 min on a hot plate or over a low flame. Alternatively, allow to stand for at least several hours at room temperature. Test for completeness of precipitation by adding a few more drops of $BaCl_2$ solution to the liquid above the precipitate. If cloudiness appears, indicating that some sulfate is still in solution, add another 5 to 10 mL of $BaCl_2$ solution and digest as before.

When ready to filter, place a crucible in the filtration assembly and decant the clear, hot, supernatant liquid through the filter. With a policeman loosen the precipitate from the beaker walls and the stirring rod; then transfer the precipitate quantitatively to the crucible with the aid of a wash bottle having an upward-pointing delivery tube. Wash the precipitate and crucible walls with several small portions of water, allowing each portion to drain completely before adding the next. After washing, test a portion of the filtrate for complete removal of soluble salts by testing for chloride. To perform this test collect a 5- to 10-mL portion in a test tube suspended by a thread in the filter flask, and add 1 mL of 0.1 M HNO_3 and a few drops of $AgNO_3$ solution. Ignite the crucible to constant weight as before. It is unnecessary to wash the precipitate between weighings.

Calculate and report the percentage of sulfur in the sample.

Calculations

To obtain the weight of sulfur, multiply the weight of $BaSO_4$ precipitate by the gravimetric factor (at. wt S/mol wt $BaSO_4$). The relative range for the results should not exceed 0.4%.

PROBLEMS

8-1.   (a) The accepted values of atomic weights have changed greatly over the years. Why is this so? (b) Why was the standard of oxygen equal to 16 abandoned? What is the present standard? (c) How are the best modern values of atomic weights determined?

8-2.   What properties must a precipitate possess for it to be satisfactory for use in a gravimetric precipitation method?

8-3.   Define the solubility-product constant. What is its relation to solubility?

---

[20]Should a crucible be ignited too strongly, some of the glass fiber mat will melt, particularly near the edge. The maximum temperature should not exceed 500°C, where a porcelain crucible will just begin to glow red.

8-4.   Write the expressions for solubility-product constants for CaF_2, $Al(OH)_3$, $Ca_3(PO_4)_2$, $MgNH_4PO_4$, and Ag_2CrO_4.

8-5.   Calculate the solubility-product constant for each of the following compounds (solubility data provided): (a) TlCl, 0.323 g/100 mL; (b) $Pb(IO_3)_2$, 3.98×10^{-5} mol/L.

8-6.   Lead oxalate and calcium fluoride both have solubility-product constants of 3×10^{-11}. Which has the higher molar solubility and by how much?

8-7.   If the solubility-product constant of silver chloride is 1.78×10^{-10} and that of silver chromate 2.45×10^{-12}, which compound is more soluble in water and by how much?

8-8.   Calculate the molar solubilities for each of the following compounds (solubility-product constants provided): (a) CdS, 7.8×10^{-27}; (b) BaF_2, 1.7×10^{-6}; (c) BiI_3, 8.1×10^{-19}.

8-9.   Define (a) common-ion effect; (b) diverse-ion effect.

8-10.   If a small excess of a precipitating reagent is normally beneficial, why isn't a large excess always better?

8-11.   To 150 mL of a solution containing 0.50 g of Na_2SO_4 is added 50 mL of a solution containing 1.00 g of $BaCl_2$. Calculate the milligrams of sulfate ion remaining in solution after equilibrium has been attained.

8-12.   Define (a) adsorption, (b) occlusion, (c) digestion of a precipitate, (d) coprecipitation, (e) supersaturation.

8-13.   In the precipitation of barium as barium sulfate by addition of a solution of sodium sulfate, how can the occlusion of sodium ion be minimized?

8-14.   Explain the terms nucleation and nucleation site. What is the importance of the supersaturation ratio in nucleation?

8-15.   Calculate the supersaturation ratio at the instant of mixing equal volumes of 0.01 M solutions of barium chloride and sodium sulfate.

8-16.   If 50.0 mL of 0.100 M barium chloride solution is added to 100.0 mL of 0.0450 M sodium sulfate solution, which ion is most likely to be found in the primary adsorbed layer on the resulting precipitate? What is the concentration of Ba^{2+} in the solution at equilibrium; of SO_4^{2-}?

8-17.   How can changes in pH affect the solubility of a precipitate? Give an example.

8-18.   Describe what is meant by precipitation from homogeneous solution. Explain how urea may be used in this capacity.

8-19.   How can the supersaturation ratio be kept low during nucleation and crystal growth? Explain the terms homogeneous nucleation and heterogeneous nucleation. Which is the predominant process in analytical precipitations?

8-20.   Calculate the minimum pH at which a precipitate of magnesium hydroxide will first begin to form from a solution 0.015 M in magnesium ion.

8-21.   When a precipitate is digested, what is the effect of small particles being more soluble than large ones?

8-22.   List the criteria for selection of a wash liquid. Describe how a wash liquid should be used if solubility losses are to be minimized.

8-23.   What is the relative effectiveness of washing for the removal of ions that are (a) occluded, (b) adsorbed as a primary adsorbed layer, or (c) present as counterions on the surface of the precipitate?

8-24.   Define constant weight as used in analytical work. How is constant weight established in practice?

8-25.   A 0.1 M solution of NaCl has a mean activity coefficient of 0.77. Calculate the solubility of AgCl in 0.1 M NaCl using a solubility-product constant of 1×10^{-10}.

8-26.   The solubility of barium sulfate in water is 4.4 mg/L at 100°C. What volume of 100°C wash water could be used to wash a 0.5432-g barium sulfate precipitate before the solubility loss would amount to 0.10%? Assume the wash water to be saturated with barium sulfate upon leaving the filter.

8-27.   Explain why silver chloride precipitates as curds, and barium sulfate as fine crystals, even though the solubility-product constants of the two are about the same.

8-28.   A precipitate of AgCl weighing 0.5000 g is washed with 200 mL of 0.01 M HNO_3 at 25°C. (a) What percentage of the precipitate is lost through solubility? Assume the solubility of AgCl in 0.01 M HNO_3 to be the same as in pure water. (b) What percentage is lost if the temperature of the wash solution is 50°C? (c) How much wash solution can be used at 25°C without the loss exceeding 0.1%?

8-29.   What volume of 0.5 M ethyl oxalate, $(C_2H_5)_2C_2O_4$, would be required to be chemically equivalent to the calcium in a 0.5-g sample of limestone that is being analyzed by precipitation from homogeneous solution?

8-30.   Explain the term gravimetric factor. For improved precision of analysis is it preferable for this factor to be large or small? Calculate the gravimetric factors for the determination of (a) Cl as AgCl, (b) Ag as AgCl, (c) Ba as $BaSO_4$, and (d) Ni as $Ni(DMG)_2$ (Section 8-9).

8-31.   A 0.6443-g sample of a chlorinated hydrocarbon was burned to yield CO_2, H_2O, and HCl. The chloride was then precipitated as silver chloride. If the weight of silver chloride formed was 0.2432 g, what was the percentage of chloride in the sample?

8-32.   A 0.5555-g sample of fertilizer was analyzed for potassium by precipitation of the potassium with sodium tetraphenylborate, $NaB(C_6H_5)_4$, yielding 0.4623 g of potassium tetraphenylborate precipitate. Calculate the potassium content of the fertilizer as percentage of K_2O. What is the gravimetric factor for potassium in this reaction?

8-33.   Potassium can be determined by precipitation and weighing as potassium chloroplatinate, K_2PtCl_6. What is the gravimetric factor for the calculation of the results in terms of percentage of (a) K, (b) K_2O, (c) KCl, (d) $KCl \cdot MgCl_2 \cdot 6H_2O$?

8-34.   A 1.146-g sample of steel was burned in a stream of oxygen, and the carbon dioxide formed absorbed in a tube containing NaOH. If the increase in weight of the tube was 0.1001 g, what was the percentage of carbon in the steel?

8-35.   A 0.9732-g sample containing iron(III) ammonium sulfate yields upon analysis 0.1432 g of Fe_2O_3. What is the percentage of $Fe_2(SO_4)_3 \cdot (NH_4)_2SO_4 \cdot 6H_2O$ in the sample? What is the gravimetric factor?

8-36.   How many milliliters of 0.0500 M barium perchlorate are required to be chemically equivalent to the sulfate in 0.5217 g of Epsom salts, $MgSO_4 \cdot 7H_2O$? Write the reaction.

8-37.   A convenient method of standardizing solutions of fluoride salts is through precipitation of PbClF. If a 50.00-mL aliquot of a fluoride solution gives a precipitate of PbClF weighing 0.2644 g, what is the molar concentration of the solution?

8-38.   A 0.6123-g sample of a mixture containing only sodium chloride and potassium chloride was treated with an excess of sulfuric acid and evaporated to dryness. The residue weighed 0.7222 g. Write the equations for the chemical reactions. What is the percentage of sodium chloride in the original mixture?

8-39.†   A sample of highly purified sodium chloride weighing 4.617 12 g yielded 11.332 25 g of silver chloride. Taking the atomic weights of chloride and of silver to be 35.453 and 107.868, calculate the atomic weight of sodium from this experiment.

8-40. Why is it important to test for completeness of precipitation before and after filtration? What is the procedure?

8-41. Explain how water may be determined in a sample by (a) indirect and by (b) direct methods.

8–42.† A sample of wheat originally containing 11.6% moisture was dried and analyzed for nitrogen by the Kjeldahl procedure. The dry material was found to contain 1.463% nitrogen. What was the percentage of nitrogen in the undried ("as received") sample?

8-43. Why is dilute nitric acid rather than pure water used as a wash liquid for silver chloride precipitates?

8-44. A chemist requires 500 mL of 0.01 M nitric acid as wash liquid for a precipitate of silver chloride. How can he prepare the solution?

8-45. Phosphate can be quantitatively determined by precipitation as $MgNH_4PO_4$. What would be the effect on the solubility of this precipitate if (a) the pH of the solution at precipitation were changed from 5 to 10; (b) the concentration of ammonium chloride solution were changed from 0.1 to 1 M; (c) the temperature were lowered from 30°C to 20°C? Suggest a wash liquid that would minimize solubility losses.

8-46. List the advantages of the use of organic reagents in gravimetric analysis.

8-47. A 1.7462-g sample of a nickel alloy was analyzed by the procedure of Section 8-9. The weight of the empty crucible was 16.7176 g, and that of the crucible plus $Ni(DMG)_2$ precipitate, 17.7348 g. What was the percentage of nickel in the sample?

8-48. A 0.4767-g sample of an alloy containing only cobalt and nickel gave a $Ni(DMG)_2$ precipitate weighing 0.8102 g when analyzed by the procedure of Section 8-9. What was the percentage of cobalt in the sample?

8-49. In the analysis of the nickel in an alloy containing nickel, chromium, manganese, and iron a 0.1340-g sample yielded 0.5111 g of $Ni(DMG)_2$. What was the percentage of nickel in the alloy?

8-50. A 0.05 M barium chloride solution is needed in the determination of sulfate. What weight of $BaCl_2 \cdot 2H_2O$ is required to prepare 200 mL of this solution?

8-51. A 0.2312-g sample of an unknown was analyzed for chloride by precipitation and weighing of silver chloride. If 0.2030 g of AgCl was obtained, what was the percentage of chloride in the sample?

8-52. What volume of 0.10 M $AgNO_3$ is chemically equivalent to the chloride in 0.35 g of NaCl?

8-53. A 0.4892-g sample of impure ammonium sulfate yielded 0.8462 g of barium sulfate. What was the percentage of sulfur in the sample? What was the purity of the ammonium sulfate?

8-54. A 0.3054-g sample of the mineral chalcopyrite ($CuFeS_2$) yielded 0.6525 g of barium sulfate. What was the percentage of $CuFeS_2$ in the sample?

8–55.† A 0.5521-g sample of a sulfate salt contained 22.11% sulfur according to analysis by barium sulfate precipitation. Later it was found that the precipitate had been dried at only 110°C and contained 0.8% water at the time of weighing. What was the true percentage of sulfur in the sample?

8-56. What volume of 0.05 M $BaCl_2$ should be added to a fertilizer sample containing 0.30 g of $(NH_4)_2SO_4$ to precipitate the sulfate as $BaSO_4$ and provide a 10% excess of reagent? What would be the weight of $BaSO_4$ precipitate obtained?

REFERENCES

H. Diehl, *Applications of Dioximes to Analytical Chemistry*, G. F. Smith Chemical Company, Columbus, Ohio, 1940. Useful survey of the analytical scope of this interesting class of compounds.

R. B. Fischer, *Anal. Chem.* **1960**, *32*, 1127; **1961**, *33*, 1802. Studies on nucleation and precipitation from homogeneous solution.

J. F. Flagg, *Organic Reagents*, Interscience, New York, 1948; F. J. Welcher, *Organic Analytical Reagents*, Van Nostrand, New York, 1947. Systematic coverage of many organic reagents and their analytical applications.

N. N. Greenwood in *Treatise on Analytical Chemistry*, 2nd ed., I. M. Kolthoff and P. J. Elving, Eds., Wiley-Interscience, New York, 1978, Part I, Vol. 1, Chapter 8. Thorough discussion of atomic weights, including their reliability and variability.

J. L. Jones and L. C. Howick, *Talanta*, **1964**, *11*, 757. Description of the procedure for precipitation of Ni(DMG)$_2$ from acetone–water mixtures.

I. M. Kolthoff, E. B. Sandell, E. J. Meehan, and S. Bruckenstein, *Quantitative Chemical Analysis,*, 4th ed., Macmillan, New York, 1969, pp 198, 228, and 602.

H. A. Laitinen and W. E. Harris, *Chemical Analysis*, 2nd ed., McGraw-Hill, New York, 1974. More complete and rigorous background on solubility, the formation and physical characteristics of precipitates, and the purity of precipitates.

T. W. Richards, *Chem. Rev.* **1925**, *1*, 1. Clear description of the classic work of Richards and his coworkers.

A. F. Scott and M. Bettman, *Chem. Rev.* **1952**, *50*, 363. Comparison of the chemical and physical values of atomic weights for the monoisotopic elements.

A. F. Scott and W. R. Ware, *J. Amer. Chem. Soc.* **1957**, *79*, 4253. Description of the chemical determination of the atomic weight of fluorine by hydrolysis of perfluorobutyryl chloride and precipitation of the released chloride with silver nitrate.

F. Szabadvary, *History of Analytical Chemistry,* Pergamon, London, 1966, pp 139–144. Entertaining description of the early work of Berzelius on atomic weights.

W. Timmermans, *J. Chem. Educ.* **1938**, *15*, 353. Short biography of Jean Servais Stas of Belgium.

H. H. Willard and N. K. Tang, *J. Amer. Chem. Soc.* **1937** *59*, 1190; *Ind. Eng. Chem., Anal. Ed.*, **1937**, *9*, 357. First descriptions of urea as a reagent for precipitation from homogeneous solution.

Concluding comment on Figure 8-2: Both precipitates were formed in the presence of 0.001 *M* permanganate. Permanganate occlusion is minimized when crystal anions are present in solution in excess during crystal formation. Therefore the one of lighter color was formed by adding potassium nitrate solution to the perchloric acid solution. The darker one corresponds to more occlusion of permanganate and involved a reverse order of mixing. In view of the difference between these two precipitates, is it possible that the permanganate is inside the crystals through simple inclusion (mechanical entrapment) of solution?

9

OXIDATION-REDUCTION REACTIONS
AND EQUILIBRIA

The road to wisdom?—Well, it's plain and simple to express:

Err
and err
and err again
but less
and less
and less

Piet Hein

Oxidation-reduction, or *redox, reactions* are those in which electrons undergo interchange between reactants, usually in solution. They parallel acid-base reactions (Chapter 5), with transfer of electrons rather than protons. In Chapter 5 recall that for a proton-transfer reaction to occur two conjugate pairs must be involved — one to donate protons and the other to accept them. After reaction a new acid and a new base have been formed. Whether an acid-base transfer process can be utilized in analysis depends on the relative values of the ionization constants for the two conjugate pairs. The extent of reaction can be calculated from the ionization constants. In electron-transfer reactions the situation is analogous.

For a redox reaction to occur two redox conjugate pairs must be involved, one member of a pair to donate electrons and one member of the other pair to accept them. After reaction a new oxidant and a new reductant have been formed. For redox reactions a *redox conjugate pair* may be written

$$\text{oxidant} + n e \rightleftharpoons \text{reductant}$$

where oxidant is the electron acceptor and reductant the electron donor of the conjugate pair. Whether an electron-transfer process can be utilized in analysis depends in part on whether the equilibrium constant for the overall reaction is sufficiently favorable. The completeness of a reaction (equilibrium constant) can be calculated from the standard electrode potentials of the half-reactions involved (Section 9-4).

Recall from Section 3-2 that a reaction suitable for a quantitative titration must meet four basic requirements: be complete, rapid, of known stoichiometry, and have a detectable end point. Unlike acid-base reactions, most of which are rapid and of known stoichiometry, the majority of electron transfer reactions with favorable equilibrium constants are not suitable for titrimetry. Either they are slow or the stoichiometry is unsatisfactory owing to side reactions. It is fortunate, for example, that most of the possible reactions of oxygen with living organisms occur so slowly that equilibrium is never attained; if it were, human beings could not exist. Another complication is that calculated and actual equilibrium constants may differ considerably because solution conditions during a titration are seldom those under which standard electrode potentials are calculated or measured. In addition, electrode potentials can be no more than a guide, since an equilibrium constant specifies nothing about the rate of a reaction or its mechanism. A small value for a calculated equilibrium constant can certainly tell us that a reaction is *not* feasible; a favorable value can only indicate that the reaction *may be* feasible. Thus, although a knowledge of electrode potentials is of enormous assistance in understanding the chemical basis of redox analysis, additional information is often required before the utility of a reaction can be established.

9-1 DEFINITIONS

Oxidation is a process resulting in a loss of electrons. The species losing electrons undergoes an increase in oxidation number.[1] Examples of oxidation are Fe^{2+} to Fe^{3+} and S^{2-} to S^0.

Reduction is a process resulting in a gain of electrons. The species gaining electrons undergoes a decrease in oxidation number. Examples of reduction are Ag^+ to Ag and I to I^-. As in acid-base systems the strength of a species as an oxidant is reciprocally related to that of its conjugate as a reductant. Cerium(IV), for example, is a powerful oxidant; its conjugate, cerium(III), is a weak reductant.[2]

An *oxidizing agent* accepts electrons, causes oxidation of the donor substance, and is itself reduced. Similarly, a *reducing agent* donates electrons, causes reduction of the acceptor substance, and is itself oxidized. A *half-reaction* is an equation representing one redox conjugate pair. An example is

$$Fe^{3+} + e \rightleftarrows Fe^{2+}$$

A *redox reaction* must involve two redox conjugate pairs and therefore two half-reactions. The above definitions can be illustrated by the chemical reaction

---

[1]Bear in mind that oxidation numbers are charges assigned to atoms according to arbitrary rules. In covalent compounds especially, the oxidation number may have no relation to the actual electronic charge on the atom and may even be a fractional value.

[2]Throughout this chapter iron(III), cerium(IV), and other ions are often written Fe^{3+}, Ce^{4+}, and so on without implication as to the actual chemical entity in solution (see also Section 9-5).

$$6Fe^{2+} + Cr_2O_7^{2-} + 14H^+ \rightarrow 6Fe^{3+} + 2Cr^{3+} + 7H_2O \qquad (9\text{-}1)$$

The two half-reactions are

$$Fe^{3+} + e \rightleftarrows Fe^{2+}$$

$$Cr_2O_7^{2-} + 14H^+ + 6e \rightleftarrows 2Cr^{3+} + 7H_2O$$

In Equation (9-1), Fe^{2+} is the reducing agent and is oxidized. Its oxidation number increases by one. Dichromate, $Cr_2O_7^{2-}$, is the oxidizing agent and is reduced. The oxidation number of each chromium atom decreases from 6 to 3. Note that some substances involved in the overall reaction are neither oxidized nor reduced. In acid-base chemistry a substance such as bicarbonate may act as either an acid or a base depending on whether it donates or accepts a proton. Similarly, the oxidation number of an entity does not by itself tell whether the entity is an oxidant or a reductant. Thus the uranium in UO^{2+} is in an intermediate oxidation state and can act as an oxidant, being itself reduced to U^{3+}, or as a reductant, being itself oxidized to UO_2^{2+}. Other examples of species containing atoms in intermediate oxidation states include H_2O_2, I_2, Br_2, Cr^{3+}, VO^{2+}, Hg_2^{2+}, and MnO_2.

The *standard electrode potential* is a quantitative statement of the tendency in a redox conjugate pair for electrons to be accepted by the oxidant or given up by the reductant. Standard electrode potentials are discussed in Section 9-3; a table of values may be found in the Appendix.

9-2   OXIDATION–REDUCTION EQUATIONS AND REACTIONS

Balancing Redox Equations

For stoichiometric calculations in analysis the equation for the reaction must be known. For acid-base, precipitation, and complexation reactions, balancing the equations usually presents little difficulty. The same holds for many redox reactions; for the minority that may present a problem, the following systematic procedure is recommended:

1. Write a skeleton equation, with reactants on the left and products on the right. Postpone any decision about completeness of the reaction at equilibrium. Write, as far as possible, the various substances in the forms in which they predominate (water as H_2O, sodium chloride as Na^+ and Cl^-, and so on).

2. From your knowledge of chemical reactions, assisted as necessary by a table of standard electrode potentials, select the two redox conjugate pairs and identify the oxidant and reductant. Write the two redox half-reactions. One should be written as accepting electrons (reduction), and the other as donating electrons (oxidation).

3. Multiply each half-reaction by the smallest integer that will allow the total number of electrons donated by the reductant to equal the number accepted by the oxidant. *These numbers must be equal.*

4. Add the two half-reactions in the proportions established in the preceding step. Cancel substances to the extent that they appear in equal amounts on both sides.

5. Confirm the correctness of the equation by verifying that the net charge and the numbers of atoms of each type are the same on both sides.

EXAMPLE 9-1

In the determination of iron, tin(II) chloride is often used to reduce iron(III) chloride. Write the balanced equation for the reaction.

The skeleton equation is

$$Sn^{2+} + Fe^{3+} \rightleftarrows Sn^{4+} + Fe^{2+}$$

Tin(II) is the reductant and iron(III) the oxidant. The two half-reactions written in the appropriate direction are

$$Fe^{3+} + e \rightleftarrows Fe^{2+}$$

$$Sn^{2+} \rightleftarrows Sn^{4+} + 2e$$

To obtain the required number of two electrons on each side, we must multiply the iron half-reaction by 2:

$$2(Fe^{3+} + e \rightleftarrows Fe^{2+})$$

Adding the two half-reactions, we obtain

$$Sn^{2+} + 2Fe^{3+} + 2e \rightleftarrows Sn^{4+} + 2Fe^{2+} + 2e$$

$$Sn^{2+} + 2Fe^{3+} \rightleftarrows Sn^{4+} + 2Fe^{2+}$$

EXAMPLE 9-2

Iron(II) is titrated with a solution of dichromate in acid. Write the balanced equation for the reaction.

The skeleton equation is

$$Fe^{2+} + Cr_2O_7{}^{2-} \rightleftarrows Fe^{3+} + Cr^{3+}$$

The two half-reactions are

$$Fe^{2+} \rightleftarrows Fe^{3+} + e$$

$$Cr_2O_7{}^{2-} + 14H^+ + 6e \rightleftarrows 7H_2O + 2Cr^{3+}$$

A minimum of six electrons must be interchanged, and so

$$6Fe^{2+} \rightleftarrows 6Fe^{3+} + 6e$$

Adding, we obtain

$$6Fe^{2+} + Cr_2O_7{}^{2-} + 14H^+ \rightleftarrows 6Fe^{3+} + 7H_2O + 2Cr^{3+}$$

EXAMPLE 9-3

A permanganate solution is standardized by reaction with primary-standard sodium oxalate in acid solution. Write the balanced equation for the reaction.

The skeleton equation is

$$MnO_4^- + H_2C_2O_4 \rightarrow Mn^{2+} + CO_2(g)$$

The half-reactions are

$$MnO_4^- + 8H^+ + 5e \rightarrow 4H_2O + Mn^{2+}$$

$$H_2C_2O_4 \rightarrow 2CO_2(g) + 2H^+ + 2e$$

Since the least common multiple of 5 and 2 is 10, 10 is the minimum number of electrons exchanged. Multiplying the first half-reaction by 2, the second by 5, and adding, we have

$$2(MnO_4^- + 8H^+ + 5e \rightarrow 4H_2O + Mn^{2+})$$

$$5(H_2C_2O_4 \rightarrow 2CO_2(g) + 2H^+ + 2e)$$

$$2MnO_4^- + 5H_2C_2O_4 + 6H^+ \rightarrow 2Mn^{2+} + 10CO_2 + 8H_2O$$

In the addition, 10 H^+ were canceled.

In alkaline solutions hydroxyl rather than hydrogen ions are involved in some half-reactions, but the procedure for balancing is as outlined here.

Electron Affinity and Completeness of a Reaction

We next consider how to determine the degree of completeness of a redox reaction at equilibrium. In the three foregoing examples the reactions proceed quantitatively to the right, that is, the equilibrium constants for the equations as written are large. For other reactions, such as between iron(II) and iodine, the reaction progresses to some extent, but not quantitatively.

The extent of a reaction depends on how strongly the oxidants of the two redox conjugate pairs compete for electrons. Since the affinity of chlorine for electrons is far stronger than that of iron(III), the reaction between chlorine and iron(II) is quantitative. On the other hand, iron(III) has only moderately greater affinity for electrons than does iodine; hence the reaction between iodine and iron(II) is considerably less extensive. In general, a substance with a high affinity for electrons is referred to as a strong oxidizing agent; correspondingly it can be easily reduced. A substance with a strong tendency to give up electrons is referred to as a strong reducing agent and can be easily oxidized.

The relative electron affinity of the oxidant in a conjugate pair may be described in terms of the electrode potential of the pair. Consider a silver wire electrode in a solution of silver nitrate as shown in Figure 9-1. This is an example of a *half-cell*, a system consisting of an electrode and an electrolyte between which there can be an electrochemical half-reaction. At the interface of the silver wire and the solution an equilibrium can be established:

$$Ag^+ + e \rightleftarrows Ag(s)$$

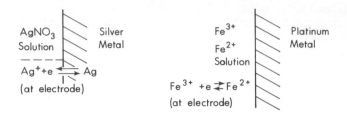

Figure 9-1. Examples of equilibria at single electrodes; metallic silver and silver nitrate in solution, and iron(III) and iron(II) nitrate in solution with an inert electrode.

That is, silver ions in solution tend to take on electrons from the electrode to form metallic silver, leaving the electrode with a more positive charge. In the opposite sense, atoms of silver tend to leave the electrode and enter the solution as silver ions. Free electrons, which cannot exist for any significant time in solution, are left behind to make the electrode more negative. Neither of these reactions actually takes place to a significant extent because the potential quickly builds to an equilibrium value. This value will be increasingly positive as the concentration of silver nitrate increases.

Similarly, a platinum (an inert metal) electrode can be immersed in a solution containing iron(III) and iron(II) salts to form a half-cell. The iron(III) ions will tend to remove electrons from the platinum, undergoing reduction to iron(II) and leaving the platinum with a positive charge. Meanwhile, iron(II) ions at the platinum surface tend to donate electrons to the platinum, undergoing oxidation to iron(III) and giving the platinum a negative charge. Again, equilibrium is quickly established without appreciable net reaction. The result of the two opposing tendencies is given by the net charge on the platinum electrode. This charge depends both on the nature of the redox couple and the relative concentrations of the species of the conjugate pair; thus the charge becomes more positive as the ratio of iron(III) to iron(II) in solution increases.

Which of two half-cells is more positive can be learned by combining them in an electrochemical cell as in Figure 9-2. The difference in potential between the two electrodes can then be measured. With similar concentrations of salts in the two half-cells of the figure, the difference in potential will be found to be small, the silver electrode being slightly more

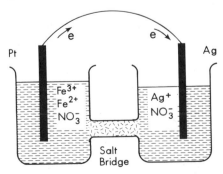

Figure 9-2. An electrochemical cell consisting of two half-cells connected in solution by a salt bridge and externally by a wire. The right half-cell is a silver electrode in a solution of silver nitrate, and the left a platinum electrode in a solution of iron(III) and iron(II) nitrate.

positive. That is, electrons will flow in the external wire toward the more positive (silver) electrode. At this electrode the reaction will be

$$Ag^+ + e \rightarrow Ag$$

and silver metal will deposit. With continued flow of electrons the silver ion concentration in solution decreases and in turn causes the potential to decrease.

Meanwhile in the other half-cell, electrons are furnished to the platinum electrode by the reaction

$$Fe^{2+} \rightarrow Fe^{3+} + e$$

As electron flow continues, the ratio of $[Fe^{2+}]$ to $[Fe^{3+}]$ decreases, and the potential at this electrode becomes more positive. Eventually, the potential of the silver electrode will decrease, and that of the platinum electrode increase, to the point where there is no difference between them. At this time the solution concentrations will be exactly the same as if they were obtained by mixing silver(I) and iron(II) in the appropriate concentrations and allowing the chemical reaction to come to equilibrium:

$$Ag^+ + Fe^{2+} \rightleftarrows Fe^{3+} + Ag$$

When reagents are mixed in solution, electron transfer takes place between ions rather than through an external wire as in an electrochemical cell. If in Figure 9-2 either $[Ag^+]$ or the ratio of $[Fe^{2+}]$ to $[Fe^{3+}]$ were lower, then the platinum electrode would be more positive, and electrons would flow in the opposite direction until equilibrium was achieved.

Knowledge of the potential of one electrode couple relative to that of another is essential to a practical understanding of redox reactions and electrochemical phenomena. Knowledge of the absolute potential of an electrode should be even more valuable. However, just as we cannot talk about elevation in absolute terms, but must express height relative to an arbitrary selected reference such as sea level, neither can we measure the absolute potential of either the silver or the platinum electrode in the above example. We need a reference point, or standard. For electrochemical work the standard hydrogen half-cell has been defined for this purpose.

9–3 STANDARDS OF POTENTIAL. THE NERNST EQUATION

The Primary Standard of Potential

By definition and agreement the primary standard of electrode potential is the standard hydrogen electrode. A value of 0.0000 V has been *assigned* to the potential that would be obtained at a platinum electrode for the equilibrium

$$2H^+ + 2e \rightleftarrows H_2(g)$$

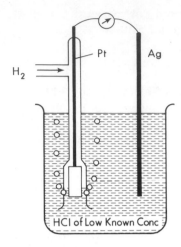

Figure 9-3. Electrochemical cell to determine the value of the potential of the silver–silver chloride electrode relative to a hydrogen electrode of low hydrogen ion concentration (not the standard hydrogen electrode).

under specified conditions. All other potentials are expressed relative to this zero value (analogous to elevation at sea level). From several lines of evidence the absolute potential for the hydrogen couple is probably not far from zero, the value arbitrarily agreed upon. The inert electrode surface in this half-cell is in contact with a solution of hydrogen ions and with hydrogen gas, both at an activity of unity.

The standard hydrogen electrode is an especially interesting primary standard in that it has never actually been used ! Any attempt to do so would meet with insurmountable difficulties.[3] This fact should not cause concern, since it in no way invalidates the data on standard electrode potentials. Potentials relative to this primary standard have been reliably established for hundreds of redox conjugate pairs, many of them to a precision of three or four significant figures, through the use of hydrogen electrodes of lower hydrogen ion activity. At low activities the relation between activity and concentration can be known to a high level of confidence. Also, secondary standards have been developed. Figure 9-3 shows a cell that provides a value for the useful secondary reference electrode, the silver–silver chloride reference couple. From the activity coefficient and the Nernst equation the potential corresponding to a hydrogen ion activity of unity can be calculated. The surface of the silver electrode is coated with solid silver chloride. The voltage of the cell is measured under conditions where *no current* is allowed to flow. The two half-reactions are

$$2H^+ + 2e \rightleftarrows H_2(g)$$

$$AgCl(s) + e \rightleftarrows Cl^- + Ag(s)$$

---

[3]One difficulty is that single ion activities cannot be determined; at best, a mean activity coefficient of the hydrogen cation and the anion associated with it is obtained. Even with reasonable assumptions the single ion activity coefficient of the hydrogen ion at an *activity of unity* cannot be established with adequate reliability. Also, for systems requiring a salt bridge, the junction potential at the interface between solutions of different compositions would present a formidable experimental barrier.

One advantage of this cell is that there is no interface between solutions of differing composition, and therefore no uncertainty is introduced into the measurements from junction potentials. The two most critical quantities in the cell are the activity of the hydrogen ion for the hydrogen electrode and the activity of the chloride ion for the silver–silver chloride electrode. These two can be known to a high level of reliability provided the concentration of hydrochloric acid is low enough (0.01 M or less) that activity coefficients are close to unity. With a hydrochloric acid concentration of 0.005 M the voltage of the cell is about 0.223 V, the silver metal being the more positive. In such cells potentials can be reliably measured under other than standard conditions. From these potentials the values that would be found under standard conditions can be obtained through the Nernst equation.

The Nernst Equation

It can be shown qualitatively (Figure 9-1) that for a half-cell of iron(III)–iron(II) solution in contact with a platinum electrode the potential increases with increasing concentration of iron(III) and decreases with increasing concentration of iron(II). Strictly speaking, the potential established at the electrode surface depends on the ratio of the activities of the two iron species in solution.

Theoretical and experimental studies (particularly by the German chemist Walther Nernst, about 1890) confirmed that the potential of an electrode in a half-cell for a reversible half-reaction is given by the relation known as the *Nernst equation*.[4] For a simple redox conjugate pair

$$ox + ne \rightleftharpoons red$$

the Nernst equation is

$$E = E^0 + \frac{RT}{nF} \ln \frac{a_{ox}}{a_{red}} = E^0 - \frac{RT}{nF} \ln \frac{a_{red}}{a_{ox}}$$

where E is the electrode potential, E^0 is the standard electrode potential, R is the gas constant of 8.314 joules/°C, T is absolute temperature (25°C = 298.15 K), n is the number of electrons in the redox half-reaction, and F is the faraday (96,487 C/equiv). Substituting the numerical values for RT/F in the expression along with $\log_e 10 = 2.303$ at 25°C, and choosing the form with the negative sign, we obtain

$$E = E^0 - (0.0592/n) \log (a_{red}/a_{ox})$$

At 29°C

$$E = E^0 - (0.0600/n) \log (a_{red}/a_{ox})$$

At 50°C

$$E = E^0 - (0.0641/n) \log (a_{red}/a_{ox})$$

---

[4]This is probably the most significant single equation in all of chemistry. It is essential for the understanding of the chemical basis of a large fraction of all analytical work.

In the material that follows, the expression at 25°C is rounded to 0.059 for simplicity. In exact calculations at this temperature, 0.0592 must necessarily be retained.

Assuming activity coefficients of unity, we obtain the Nernst equation in the commonly used form[5]

$$E = E^0 - \frac{0.059}{n} \log \frac{[\text{red}]}{[\text{ox}]} \tag{9-2}$$

By definition the *standard electrode potential* E^0 corresponds to the value of the potential E when the term log ([red]/[ox]) has a value of zero, that is, when the species in the log term are all at *unit activities* or the *ratio* of their activities is unity.

When the Nernst equation is written for a redox conjugate pair, any substances other than the oxidant or reductant that enter into the half-reaction must be included in the log term with their activities raised to the appropriate power. The activities of solids and of water as solvent are taken to be unity; for an ideal gas, unit activity is defined as a partial pressure of unity. We assume activity coefficients of unity in the remainder of this chapter unless stated otherwise.

EXAMPLE 9-4

Write the Nernst expression for each of the following half-reactions, taking the E^0 values from the table of standard electrode potentials in the Appendix.

a.
$$Ag^+ + e \rightleftarrows Ag$$

$$E = E^0_{Ag^+,Ag} - 0.059 \log (1/[Ag^+]) = 0.7994 - 0.059 \log (1/[Ag^+])$$

The activity of silver metal, the reduced form of the components of the half-reaction, is unity.

b.
$$Fe^{3+} + e \rightleftarrows Fe^{2+}$$

$$E = 0.771 - 0.059 \log \frac{[Fe^{2+}]}{[Fe^{3+}]}$$

c.
$$Cr_2O_7^{2-} + 14H^+ + 6e \rightleftarrows 2Cr^{3+} + 7H_2O$$

$$E = 1.33 - \frac{0.059}{6} \log \frac{[Cr^{3+}]^2}{[Cr_2O_7^{2-}][H^+]^{14}}$$

d.
$$2H^+ + 2e \rightleftarrows H_2(g)$$

$$E = 0.0000 - \frac{0.059}{2} \log \frac{P_{H_2}}{[H^+]^2}$$

---

[5]Another form is $a_{ox}/a_{red} = \exp [nF(E - E^0)/RT]$, which is convenient in thermodynamic applications.

At a hydrogen gas pressure of 1 atm

$$E = 0.0000 - \frac{0.059}{2} \log \frac{1}{[H^+]^2} = 0.0000 - 0.059 \log \frac{1}{[H^+]}$$

EXAMPLE 9-5
Calculate the potential of a silver metal electrode (such as in Figure 9-2) in 0.15 M silver nitrate solution.

$$E = 0.7994 - 0.059 \log (1/[Ag^+]) = 0.7994 - 0.059 \log (1/0.15) = 0.75 \text{ V}$$

EXAMPLE 9-6
Calculate the potential of a platinum electrode (such as in Figure 9-2) in a solution of 0.015 M iron(III) nitrate and 0.10 M iron(II) nitrate.

$$E = 0.771 - 0.059 \log ([Fe(II)]/[Fe(III)]) = 0.771 - 0.059 \log (0.10/0.015) = 0.72 \text{ V}$$

EXAMPLE 9-7
Calculate the potential of a platinum wire in a solution of 0.02 M Cr(III) and 0.05 M $Cr_2O_7^{2-}$ at pH 1.5.

$$E = 1.33 - \frac{0.059}{6} \log \frac{(0.02)^2}{0.05(3.2 \times 10^{-2})^{14}} = 1.14 \text{ V}$$

EXAMPLE 9-8
Calculate the potential of a cadmium electrode in 0.023 M cadmium chloride.

$$E = -0.403 - (0.059/2) \log (1/0.023) = -0.45 \text{ V}$$

EXAMPLE 9-9
Calculate the potential at 25°C of a silver–silver chloride electrode in which the hydrochloric acid concentration is 5.000 $\times$ 10^{-3} M. Make an exact calculation, assuming the activity coefficient of the chloride ion is 0.925 and the solubility product of silver chloride is 1.79 $\times$ 10^{-10}.

$$E = E^0_{Ag} - 0.0592 \log \frac{1}{a_{Ag^+}} = 0.7994 - 0.0592 \log \frac{1}{K_{sp}/a_{Cl^-}}$$

$$= 0.7994 - 0.0592 \log \frac{a_{Cl^-}}{K_{sp}} = 0.7994 - 0.0592 \log \frac{(0.925)(0.005)}{1.79 \times 10^{-10}} = 0.361 \text{ V}$$

Knowing the potentials of two half-cells, we can calculate the voltage of an electrochemical cell comprising the two.

EXAMPLE 9-10
Calculate the voltage of the electrochemical cell in Figure 9-2, using as initial concentrations the values in Examples 9-5 and 9-6.

From Examples 9-5 and 9-6 the silver electrode has a calculated potential of 0.75 V, and the platinum electrode in contact with the iron(III)–iron(II) solution has a potential of 0.72 V. Thus the potential difference in the two electrodes is 0.03 V, the silver electrode being slightly more positive.

When the cell in Figure 9-2 is short-circuited, electron flow occurs as shown. Eventually the potential of the silver electrode becomes equal to that of the platinum electrode. At this point each would have a potential somewhere between 0.72 and 0.75 V.

Secondary Standards of Potential

The electrochemical cell in Figure 9-3 may be used to establish the electrode potentials of silver–silver chloride reference half-cells. These half-cells have several characteristics that make them useful as secondary standard reference electrodes, the principal ones being their stability and convenience in day-to-day use. Mercury–mercury(I) chloride half-cells are also frequently used as reference electrodes; the assembly of these electrodes is described in Section 11-4. There are also many secondary standard hydrogen electrodes which depend on the hydrogen ion activity (which must be low).

The potential of each of the half-cells in Figure 9-3 can be established as follows. For the hydrogen electrode it can be calculated, with assumption of zero potential for the standard hydrogen electrode, from the values of the concentration and the activity coefficient of the hydrogen ion. The activity coefficient will be near unity in dilute solution; an estimate of its value can be calculated at low concentrations by the Debye-Hückel equation (Section 3-1) to the desired level of precision. For example, if the electrolyte in Figure 9-3 were 0.005 M hydrochloric acid, the activity coefficient of the hydrogen ion would be calculated to be close to 0.933. The potential of this hydrogen electrode at 25°C would be

$$E = 0.0000 - \frac{0.0592}{1} \log \frac{1}{a_{H^+}} = 0.0000 - \frac{0.0592}{1} \log \frac{1}{(0.933)(0.005)} = 0.1380 \text{ V}$$

In Example 9-10 the potential of a silver electrode in 0.005 M hydrochloric acid was found to be 0.361 V. The silver electrode, then, is more positive than the hydrogen electrode by roughly two tenths of a volt.

The potential of the hydrogen electrode can be calculated as indicated above, and the voltage of the cell carefully measured under conditions where no current flows so that the potential of neither electrode is disturbed by a net reaction occurring at the electrode. From the cell voltage and the potential of the hydrogen electrode the potential of the silver–silver chloride electrode may be determined, and its E^0 value has been calculated through the Nernst equation. The value of E^0 has been found to be 0.2223 V. Once the potential of the silver–silver chloride electrode has been established, it can then be used with other redox conjugate pairs to establish their E^0 values.

Several secondary standard mercury–mercury(I) chloride (calomel) reference half-cells have been developed. All contain saturated Hg_2Cl_2 and differ only in the potassium chloride concentration. For example, the *normal calomel half-cell* has a potassium chloride concentration of 1 M. The *saturated calomel electrode* (SCE) is saturated in potassium chloride (about 4 *M*). Calomel reference half-cells involve the half-reaction

$$Hg_2Cl_2(s) + 2e \rightleftarrows 2Cl^- + 2Hg$$

The saturated calomel electrode, the secondary standard most frequently employed for routine potential measurements, consists of mercury metal in contact with a solution saturated with both KCl and Hg_2Cl_2. The potential of this secondary standard, 0.244 V, is given by

$$E = E^0 - \frac{0.059}{2} \log \frac{1}{a_{Hg_2^{2+}}} = E^0 - \frac{0.059}{2} \log \frac{a^2_{Cl^-}}{K_{sp}} = E^0 - 0.059 \log \frac{a_{Cl^-}}{\sqrt{K_{sp}}}$$

where K_{sp} is the solubility-product constant for Hg_2Cl_2. Because of the high chloride ion concentration the cell potential is exceedingly stable and not significantly disturbed if a moderate current is passed through the electrode. The solubility of potassium chloride increases with temperature, and therefore the potential changes accordingly. In the measurement of potential, contact with the other half of an electrolytic cell is normally made by a solution of potassium chloride. Such a solution is called a *salt bridge*.

In most measurements of potential employed in chemical analyses, *differences* in potential between standards and samples are more important than potentials relative to the hydrogen electrode standard. Comparisons of solution potentials may be made readily as long as the potential of the reference electrode remains constant during the measurements. The stable potential of the saturated calomel electrode makes it admirably suited for comparison measurements.[6]

Sometimes electrode systems of constant potential other than the SCE or the silver–silver chloride electrode are used as reference couples. In such systems the potential must be constant but not necessarily measured or measurable for them to be analytically useful. An example is a platinum electrode in a solution containing both iodide and iodine in unknown concentrations.

Comment on the Meaning of "Potential"

Although it may seem obvious that an electrode with an excess of electrons has a negative charge and one with a deficiency of electrons a positive charge, a few decades ago there was intense disagreement about whether the sign of the potential of any given electrode should be positive or negative. After much argument consensus was finally reached by the International Union of Pure and Applied Chemistry (IUPAC) in Stockholm in 1953 that the term *electrode potential is invariant in sign*. It is used exclusively in the sense that the sign, positive or negative, should correspond to the electrostatic charge; that is, an excess of electrons means the charge is negative.[7] For convenience, we normally

---

[6]Unsaturated calomel reference half-cells are susceptible to a change in concentration through evaporation, but their potentials are less affected by temperature.

[7]The definitions of values for electromotive force (emf) for half-cells and cells are different from those for potential. Since we do not use the term emf in this book, we shall not be concerned with this distinction here. The references at the end of the chapter provide further information on this point.

use the term "potential" instead of "electrode potential," and it is interpreted as a sign-invariant quantity.

For a given half-cell, to calculate potential, we substitute the concentrations of reductant and oxidant directly into the Nernst equation (9-2). All E^0 values *with their accompanying signs* are taken from a table of standard electrode potentials (see Appendix). In this table E^0 for the silver ion–silver couple is given with a positive sign, $+0.80$ V, and E^0 for the zinc ion–zinc couple with a negative sign, -0.76 V.

Since the IUPAC agreement a quarter century ago the word potential is correctly used only with an invariant sign.

Potential in a Solution Containing Two or More Redox Conjugate Pairs

If a solution of tin(II) chloride is mixed with a solution of iron(III) chloride in hydrochloric acid, equilibrium is quickly reached:

$$2Fe^{3+} + Sn^{2+} \rightleftarrows 2Fe^{2+} + Sn^{4+}$$

The position of equilibrium is readily indicated by the potential that an inert electrode such as platinum acquires relative to a reference electrode when the two electrodes are immersed in a solution. Only a single potential can exist in any given solution at equilibrium. The value of this potential can be ascertained by substitution into the Nernst equation for either redox couple:

$$E = 0.771 - 0.059 \log \frac{[Fe^{2+}]}{[Fe^{3+}]}$$

$$E = 0.14 - (0.059/2) \log \frac{[Sn^{2+}]}{[Sn^{4+}]}$$

If iron(III) was initially present in excess, virtually no tin(II) will remain in solution at equilibrium. An appreciable concentration of both iron(III) and iron(II) will be present, however, and so it is convenient to choose the first of the two equations to obtain the equilibrium potential. Similarly, if an excess of tin(II) was initially present, the concentrations of both tin(II) and tin(IV) will be appreciable at equilibrium, and the second equation more convenient. But note that at all times for a given solution, E can have but a single value, no matter which of the two equations is selected. Generally, the concentrations of the species for one couple can be obtained more readily than for the other. As is seen in Section 9-4 the overall equilibrium constant for the reaction can be calculated from the E value at equilibrium.

A solution may contain three, four, or more reactants belonging to different redox conjugate pairs. In all cases, at equilibrium the potential of the solution and the concentrations of the species present must satisfy the Nernst equation for each of the redox couples.

9-4   OXIDATION-REDUCTION EQUILIBRIA

Recall that calculations of Examples 9-5 and 9-6 showed the silver electrode in the cell of Figure 9-2 to be 0.03 V more positive than the platinum electrode. Therefore, if the two half-cells were connected externally by a wire, electrons would flow toward the silver electrode, and silver ions would be reduced to the metal. At the other electrode an equal amount of iron(II) would be oxidized to iron(III). The overall cell reaction is a combination of the two half-reactions:

$$Ag^+ + Fe^{2+} \longrightarrow Ag + Fe^{3+}$$

In this cell electrons are interchanged by way of the external circuit rather than through direct interaction of the ions, as would occur if solutions containing Ag^+ and Fe^{2+} were mixed. The potential of the silver electrode would decrease from 0.75 V, and that of the platinum electrode increase from 0.72 V, until after a time they became identical. With no difference in the potential as a driving force, electrons would cease to flow, and no further net chemical reaction would occur.

The expression for the equilibrium constant for the above reaction is

$$K_{eq} = \frac{[Fe^{3+}]}{[Ag^+][Fe^{2+}]}$$

Since the activity of the silver metal is unity, it need not appear in the equilibrium-constant expression. At all times the potential of the silver electrode is given by the Nernst expression for the silver–silver ion half-reaction, and that of the platinum electrode by the iron(III)–iron(II) half-reaction:

$$E = E^0{}_{Ag+,Ag} - 0.059 \log\frac{1}{[Ag^+]}$$

$$E = E^0{}_{Fe^{3+},Fe^{2+}} - 0.059 \log\frac{[Fe^{2+}]}{[Fe^{3+}]}$$

Whatever the initial concentrations in the two half-cells, when the two electrode potentials become identical the current ceases to flow and equilibrium has then been attained. At equilibrium

$$0.799 - 0.059 \log (1/[Ag^+]) = 0.771 - 0.059 \log ([Fe^{2+}]/[Fe^{3+}])$$

where $[Ag^+]$, $[Fe^{2+}]$, and $[Fe^{3+}]$ are the concentration values in solution at equilibrium. From the above equality the two log terms can be collected on one side and the numerical values on the other. Then

$$\frac{0.799 - 0.771}{0.059} = \log \frac{1}{[Ag^+]} - \log \frac{[Fe^{2+}]}{[Fe^{3+}]}$$

Since subtraction of two log terms is equivalent to division of the quantities,

$$0.47 = \log \frac{[Fe^{3+}]}{[Ag^+][Fe^{2+}]}$$

Taking the antilogarithm of each side gives

$$2.97 = \frac{[Fe^{3+}]}{[Ag^+][Fe^{2+}]} = K_{eq} \tag{9-3}$$

This relation applies to the reactants in the two half-cells at equilibrium,[8] that is, when the potential difference is zero, independent of the initial concentrations. Note also that the only data required to calculate an equilibrium constant for the cell reaction are the E^0 values for the two half-reactions.

For a short-circuited cell composed of the two half-cells in Examples 9-5 and 9-6, it can be shown that the potential at equilibrium for each electrode would be 0.746 V and that the concentrations of Fe^{3+}, Ag^+, and Fe^{2+} would be 0.032, 0.082, and 0.132 M.

Other than pH measurements with a glass electrode only a small fraction of analyses use electrodes directly, and only a few of these entail appreciable reaction at an electrode surface. Coulometric analysis (Section 12-1) is the principal example, though other electrochemical methods are growing in importance. Most analytical applications of redox reactions involve reactions in solution. Even so, an understanding of these processes requires a knowledge of how to apply the Nernst equation. In the reaction of silver with iron(II), exactly the same equilibrium position would be attained, and much more quickly, by mixing the reactants in a single solution rather than separating them in two half-cells. Upon mixing, electron exchange would take place directly until equilibrium was reached. A platinum or silver electrode immersed in the solution could acquire but a single value of potential[9] which would be identical for both the redox conjugate pairs. An equilibrium-constant value can be calculated from any combination of half-reactions listed in a table of standard electrode potentials, even though the reaction may not be feasible analytically because of slow electron transfer or unsatisfactory stoichiometry. Nevertheless, knowledge of whether the equilibrium constant is favorable is necessary when an analytical method

---

[8]Another way of showing the relation between standard potentials for two redox couples and the equilibrium constant for their reaction is by

$$\frac{n(E_1{}^0 - E_2{}^0)}{0.059} = \log K_{eq}$$

where n is the number of electrons transferred from reductant to oxidant in the balanced equation.

[9]The potential could be measured by means of an appropriate cell, the other half-cell being a secondary standard reference electrode such as the SCE.

involving two redox species is being considered.

In calculating equilibrium constants, it is convenient to assume that the two reactants have been brought together in a single solution and that the system has attained equilibrium. The reaction is further assumed to be reversible, so that the forward and reverse rates are equal at equilibrium. Keep in mind that electrodes may or may not be present, and that *no net electrode reactions* are actually taking place. Recognize, however, that an electrode inserted into the solution could have but a single potential. Remember also that the sign of the potential is invariant and is given by the Nernst equation, for which the E^0 value also must retain the sign as it appears in the table of standard electrode potentials.

The procedure recommended for calculating an equilibrium constant from standard electrode potentials is as follows:[10]

1. Write the balanced equation for the proposed redox reaction.

2. Write the expression for the equilibrium constant.

3. Write the Nernst equation for each of the two redox conjugate pairs.

4. Set the two Nernst equations equal to each other. (Since the reactants are at equilibrium, the two potentials must be the same.)

5. Collect the numerical values on one side of the expression and the concentration (or activity) terms on the other. Collect the concentration terms so that they are are in the same form as the equilibrium constant.[11]

EXAMPLE 9-11

Calculate the equilibrium constant for the reaction

$$2Fe^{3+} + Sn^{2+} \rightleftarrows 2Fe^{2+} + Sn^{4+}$$

$$K_{eq} = \frac{[Fe^{2+}]^2[Sn^{4+}]}{[Fe^{3+}]^2[Sn^{2+}]}$$

$$E = E^0_{Fe^{3+},Fe^{2+}} - 0.059 \log \frac{[Fe^{2+}]}{[Fe^{3+}]}$$

$$E = E^0_{Sn^{4+},Sn^{2+}} - \frac{0.059}{2} \log \frac{[Sn^{2+}]}{[Sn^{4+}]}$$

In the equilibrium-constant expression, $[Fe^{2+}]$ and $[Fe^{3+}]$ are squared. Therefore the Nernst equation may be conveniently written

---

[10]Some prefer to calculate equilibrium constants using cell emf's in place of potentials. Exactly the same chemical conclusions are reached provided the calculations are carried out correctly. Disaster results if emf and potential values are mixed in the same set of calculations.

[11]A log term in the Nernst equation can be inverted with a change in sign preceding the log term, squared if the factor 0.059 is divided by 2, and so on. Thus

$$-\log x = \log (1/x)$$
$$\log x - \log y = \log (x/y)$$
$$\log x^n = n \log x.$$

$$E = 0.771 - \frac{0.059}{2} \log \frac{[Fe^{2+}]^2}{[Fe^{3+}]^2}$$

Setting this expression equal to that for the tin half-reaction gives

$$0.771 - \frac{0.059}{2} \log \frac{[Fe^{2+}]^2}{[Fe^{3+}]^2} - 0.14 - \frac{0.059}{2} \log \frac{[Sn^{2+}]}{[Sn^{4+}]}$$

Rearranging, we have

$$0.771 - 0.14 = \frac{0.059}{2} \log \frac{[Fe^{2+}]^2}{[Fe^{3+}]^2} - \frac{0.059}{2} \log \frac{[Sn^{2+}]}{[Sn^{4+}]}$$

Inverting the second log term and combining with the first, we obtain

$$0.771 - 0.14 = \frac{0.059}{2} \log \frac{[Fe^{2+}]^2[Sn^{4+}]}{[Fe^{3+}]^2[Sn^{2+}]} = \frac{0.059}{2} \log K_{eq}$$

$$2(0.771 - 0.14)/0.059 = \log K_{eq} = 21.4$$

$$K_{eq} = 2.5 \times 10^{21}$$

This large value of the equilibrium constant indicates that $[Sn^{4+}][Fe^{2+}]^2$ is large compared with $[Sn^{2+}][Fe^{3+}]^2$ and that at equilibrium the chemical reaction goes essentially to completion.

EXAMPLE 9-12

Calculate the equilibrium constant for the reaction

$$Sn^{2+} + 2V^{3+} \rightleftarrows 2V^{2+} + Sn^{4+}$$

$$K_{eq} = \frac{[V^{2+}]^2[Sn^{4+}]}{[V^{3+}]^2[Sn^{2+}]}$$

$$E = 0.14 - \frac{0.059}{2} \log \frac{[Sn^{2+}]}{[Sn^{4+}]}$$

$$E = -0.255 - 0.059 \log \frac{[V^{2+}]}{[V^{3+}]}$$

Note that for the vanadium couple E^0 is negative.

$$E = -0.255 - 0.059 \log \frac{[V^{2+}]}{[V^{3+}]} = 0.14 - \frac{0.059}{2} \log \frac{[Sn^{2+}]}{[Sn^{4+}]}$$

$$= -0.255 - \frac{0.059}{2} \log \frac{[V^{2+}]^2}{[V^{3+}]^2}$$

$$\frac{0.14 - (-0.255)}{0.03} = \log \frac{[Sn^{2+}]}{[Sn^{4+}]} - \log \frac{[V^{2+}]^2}{[V^{3+}]^2}$$

$$13.4 = -\log \frac{[Sn^{4+}][V^{2+}]^2}{[Sn^{2+}][V^{3+}]^2} = -\log K_{eq}$$

$$K_{eq} = 4 \times 10^{-14}$$

This exceedingly small value indicates that the reaction as written does not proceed to completion but, in fact, proceeds to a negligible extent.

EXAMPLE 9–13

Calculate the equilibrium constant for the reaction

$$Cr_2O_7^{2-} + 14H^+ + 6Fe^{2+} \rightleftharpoons 2Cr^{3+} + 6Fe^{3+} + 7H_2O$$

$$K_{eq} = \frac{[Cr^{3+}]^2[Fe^{3+}]^6}{[Cr_2O_7^{2-}][H^+]^{14}[Fe^{2+}]^6}$$

$$E = 1.33 - \frac{0.059}{6} \log \frac{[Cr^{3+}]^2}{[Cr_2O_7^{2-}] [H^+]^{14}}$$

$$E = 0.771 - 0.059 \log \frac{[Fe^{2+}]}{[Fe^{3+}]} = 0.771 - \frac{0.059}{6} \log \frac{[Fe^{2+}]^6}{[Fe^{3+}]^6}$$

$$1.33 - \frac{0.059}{6} \log \frac{[Cr^{3+}]^2}{[Cr_2O_7^{2-}][H^+]^{14}} = 0.771 - \frac{0.059}{6} \log \frac{[Fe^{2+}]^6}{[Fe^{3+}]^6}$$

$$\frac{1.33 - 0.771}{0.01} = \log \frac{[Cr^{3+}]^2[Fe^{3+}]^6}{[Cr_2O_7^{2-}][H^+]^{14}[Fe^{2+}]^6} = \log K_{eq}$$

$$K_{eq} = 7 \times 10^{56}$$

In titrations of iron(II) with dichromate, an important analytical method for iron in ores, the acid concentration is held constant. For purposes of calculation it is convenient to combine the hydrogen ion concentration with the equilibrium constant to obtain a *conditional equilibrium constant* (Section 7-2). For a fixed pH the conditional equilibrium constant becomes

$$K_{cond} = \frac{[Cr^{3+}]^2[Fe^{3+}]^6}{[Cr_2O_7^{2-}][Fe^{2+}]^6} = K_{eq}[H^+]^{14}$$

At pH 1.0 the conditional equilibrium constant is $(7 \times 10^{56})(0.1)^{14} = 7 \times 10^{42}$. At pH 0.6 the value would be $(7 \times 10^{56})(0.25)^{14} = 3 \times 10^{48}$.

9–5 PRECISION AND ACCURACY OF POTENTIALS.
FORMAL POTENTIALS

In practical measurements of potential the precision attainable is influenced by a variety of limitations. (1) Concentrations rather than activities are customarily used in calculations involving the Nernst equation. This practice may make a difference of 10 to 20% in the log term of the equation even for relatively dilute solutions. (2) One or more of the species involved in a half-reaction may form complexes with other species in solution. If the complexes are stable, large changes in potential, and therefore in oxidizing or reducing power, can be introduced. However, even weak complexes (for example, with chloride, nitrate, or sulfate) can make a difference of a factor of 10 to 100 or more in the activity of a reactant. (3) If a precipitate is formed as a result of a reaction, the solubility-product constant can be incorporated into the Nernst equation as is illustrated for silver chloride (Section 9-3). Uncertainty is introduced, since often the value of the solubility product may not be known within an accuracy of better than $\pm$ 10%. (4) Ordinarily in practical work the temperature need not be carefully specified or controlled, and room temperature is stated or simply assumed. (5) Experimental measurements of potential may have uncertainties arising from junction potentials and failure of an electrode to achieve complete equilibrium with the solution.

On the basis of these several sources of uncertainty, potentials precise to no better than a few tens of millivolts can be expected without special precautions and equipment. Moreover, a false sense of precision may be implied when the factor 0.0592 is used in the Nernst equation rather than simply 0.059 or 0.06.

Minor imprecision is of little practical consequence for most methods of analysis. Activity effects, however, may be sufficiently large to mislead when conclusions about equilibria are based on values of the standard electrode potential. Sounder conclusions can often be reached about equilibria by substituting for each activity term in the Nernst equation the product of activity coefficient and concentration:

$$E = E^0 - \frac{RT}{nF} \log \frac{a_{\text{red}}}{a_{\text{ox}}} = E^0 - \frac{RT}{nF} \log \frac{\gamma_{\text{red}}[\text{red}]}{\gamma_{\text{ox}}[\text{ox}]}$$

$$= E^0 - \frac{RT}{nF} \log \frac{\gamma_{\text{red}}}{\gamma_{\text{ox}}} - \frac{RT}{nF} \log \frac{[\text{red}]}{[\text{ox}]}$$

If the first two terms are combined for a particular set of conditions,

$$E = E^{0\prime} - \frac{RT}{nF} \log \frac{[\text{red}]}{[\text{ox}]}$$

Here $E^{0\prime}$ is the *formal potential*, defined as the value of E at unit concentrations of the reactants or at unit-concentration ratio. Because activity coefficients change, especially with changing ionic strengths, the value of $E^{0'}$ varies with the composition of the solution.

Complex formation may also affect the value of $E^{0'}$. For example, whereas the E^0 value for the iron(III)–iron(II) couple is 0.771 V, the formal potential is 0.70 V in 1 M HCl and 0.68 V in 0.1 M H_2SO_4. Iron(III), in other words, is a weaker oxidizing agent in these media than indicated by the E^0 value, owing to stabilization of iron(III) through formation of complexes with chloride or sulfate. The formal potential of the dichromate–chromium(III) couple is 1.00 V in 1 M HCl, 0.33 V lower than the E^0 value. A more realistic equilibrium constant for the reaction of dichromate and iron(II) in 1 M HCl solution is obtained with substitution of $E^{0'}$ values for the E^0 values in Example 9-13:

$$\log K_{eq} = \frac{1.00 - 0.70}{0.01} = 30$$

$$K_{eq} = 1 \times 10^{30}$$

This equilibrium constant is 26 orders of magnitude smaller than the one calculated from E^0 values. It is still enormous, however, and the reaction proceeds quantitatively to the right even under these conditions.

PROBLEMS

9-1.   Explain what is meant by a redox conjugate pair. How does it differ from an acid-base conjugate pair?

9-2.   Balance the following equations, adding water or H^+ as required:
$Bi + Fe^{3+} \rightleftharpoons BiO^+ + Fe^{2+}$
$Zn + Cr^{3+} \rightleftharpoons Zn^{2+} + Cr^{2+}$
$Br^- + Cr_2O_7^{2-} \rightleftharpoons Br_2 + Cr^{3+}$
$Ti^{3+} + BiO^+ \rightleftharpoons TiO^{2+} + Bi$
$HNO_2 + Br_2 \rightleftharpoons NO_3^- + Br^-$

9-3.   The preceding problem includes eight different half-reactions. Identify and arrange them in order, with the weakest oxidant at the top left and the strongest oxidant at the bottom left.

9-4.   Complete and balance the following equations:

$BrO_3^- + AsO_3^{3-}$ (in acid solution) $\longrightarrow$
$MnO_2 + Fe^{2+} + H^+ \longrightarrow$
$AsO_3^{3-} + MnO_4^- + H^+ \longrightarrow$
$FeSO_4 + NaVO_3 + H_2SO_4 \longrightarrow VOSO_4 +$
$Mn^{2+} + MnO_4^- + H_2O \longrightarrow MnO_2 +$
$Cr_2O_7^{2-} + I^- + H^+ \longrightarrow IO_3^- + I^- + H^+ +$

9-5.   Write the Nernst expressions for the following redox conjugate pairs: (a) iron(III)–iron(II); (b) mercury(I)–mercury(0); (c) mercury(0)–mercury(II); (d) ferrocyanide–ferricyanide; (e) permanganate–manganese(II); (f) sulfite–sulfate.

9-6.   Explain why the standard hydrogen electrode, though it is the primary standard of potential, has never actually been used directly.

9-7.   What is the relation between a half-cell and an electrochemical cell?

9-8.   What is a major advantage of the silver–silver chloride electrode in the practical determination of standard electrode potentials? How was the potential of this electrode related to that of the standard hydrogen electrode?

9-9.   Why is the saturated calomel electrode so widely used in the practical measurement of potentials? What are its disadvantages?

9-10.   Clearly distinguish between the terms electrode potential and standard electrode potential. Can the absolute value of an electrode potential be known?

9-11.   Calculate the value of the quantity $2.303\,RT/F$ at 0°C and at 100°C.

9-12.   For the following half-reactions write the Nernst equation for calculation of potential:
$$Zn^{2+} + 2e \rightleftarrows Zn$$
$$H_2SO_3 + 4H^+ + 4e \rightleftarrows S + 3H_2O$$
$$I_2 + 6H_2O \rightleftarrows 2IO_3^- + 12H^+ + 10e$$
$$Cd \rightleftarrows Cd^{2+} + 2e$$
$$Br^- + 2OH^- \rightleftarrows OBr^- + 2H_2O + 2e$$

9-13.   Calculate the potential an electrode would have if it were placed in a solution containing the following: (a) $0.12\,M\ I_2$ and $0.24\,M\ I^-$; (b) $0.5\,M\ H_3AsO_3$, $0.1\,M$ in H_3AsO_4 and $0.2\,M$ HCl; (c) $0.006\,M\ VO_2^+$ and $0.023\,M\ VO^{2+}$ at pH 1.5; (d) $0.04\,M\,Cr^{3+}$ and $0.001\,M\ Cr^{2+}$.

9-14.   Calculate the potential of a silver–silver chloride electrode in $0.1\,M$ NaCl at 25°C. Use the Ag^+–Ag couple and a value of 1×10^{-10} for the K_{sp} of AgCl.

9-15.   Calculate exactly the potential of a secondary standard hydrogen electrode at 25°C in $0.01\,M$ HCl. Assume an activity coefficient for H^+ of 0.914 and an activity of hydrogen of unity.

9-16.   Repeat the calculation of 9-15, but at 29°C.

9-17.   Repeat the calculation of 9-15, but with an activity coefficient of unity for H^+.

9-18.   Which solution would show the more positive electrode potential, $0.1\,M\ I_2$ and $0.2\,M\ I^-$ or $0.1\,M\ I_2$ and $0.01\,M\ I^-$; by how much?

9-19.   Which solution would show the more positive electrode potential, $0.01\,M\ H_3AsO_4$ and $0.02\,M\ H_3AsO_3$ at pH 2 or the same solution at pH 5; by how much?

9-20.   Calculate the potential at 29°C of a hydrogen electrode in which the electrolyte is $0.011\,M$ hydrochloric acid and the activity coefficient of the hydrogen ion is 0.912. What would be the potential at 0°C?

9-21.   Calculate the potential at 29°C of a silver–silver chloride electrode in which the hydrochloric acid concentration is $0.0110\,M$ and the activity coefficient of the chloride ion is 0.897. Use 1.79×10^{-10} for the solubility-product constant of silver chloride. What would be the value of the potential if the activity coefficients were assumed to be unity?

9-22.   A solution is prepared by mixing 40 mL of $0.03\,M$ iron(III) chloride with 20 mL of $0.03\,M$ tin(II) chloride. What would be the potential of a platinum electrode immersed in this solution? Would it be positive or negative relative to a saturated calomel electrode; by how much?

9-23.   If an electrochemical cell were prepared from two half-cells, one a cadmium metal rod in contact with $0.02\,M$ cadmium nitrate and the other a platinum rod in contact with $0.1\,M$ potassium iodide and $0.01\,M$ iodine, in which direction would electrons flow if the cell were short-circuited? What would be the potential of the more negative electrode?

9-24.   A 40-mL portion of 0.040 M tin(II) chloride is added to 40 mL of 0.030 M bromine solution. Write the equation for the reaction. Calculate the potential at equilibrium.

9-25.   Which mercury electrode will have the more positive potential, one in contact with a 1 M solution of potassium chloride and saturated with Hg_2Cl_2 or one saturated with respect to both potassium chloride and Hg_2Cl_2?

9-26.   A 40-mL portion of 1 M sulfuric acid solution, 0.03 M in iron(III) chloride and 0.01 M in iron(II) chloride, is mixed with 25 mL of 1 M sulfuric acid solution, 0.02 M in cerium(IV) sulfate and 0.02 M in cerium(III) sulfate. What is the potential at equilibrium?

9-27.   Excess zinc metal is added to a solution containing 0.1 M iron(III) chloride. Calculate the iron(III) concentration at equilibrium.

9-28.   Calculate the standard electrode potential for the reaction $AgI + e \rightleftharpoons Ag + I^-$. Use values of 1.0×10^{-16} for the solubility-product constant of AgI and 0.7794 V for the E^0 of the Ag^+–Ag half-reaction.

9-29.   Calculate the equilibrium constant for the reaction of the following: (a) Ce^{4+} with Fe^{2+} (b) $Fe(CN)_6^{-3}$ with iodide; (c) MnO_4^- with iron(II); (d) $Cr_2O_7^{2-}$ with iodide;(e) H_3AsO_4 with iodide;   (f) iron(III) with zinc metal to yield iron(II); (g) iron(II) with chromium(III).

9-30.   In the preceding problem, which of the equilibria would be suitable for analytical use on the basis of an equilibrium constant larger than 10^6?

9-31.   Calculate the equilibrium constant for the iron–dichromate reaction in 0.5 M H_2SO_4, using values of 0.68 and 1.15 for the formal electrode potentials of the iron and chromium couples.

9-32.   Calculate the equilibrium constant for the reaction of iron(II) chloride with iodine.

9-33.†   Mercaptans, RSH, are of considerable biological importance because of their reactivity with many metal ions. They form insoluble silver salts, RSAg. Calculate the solubility product of a silver mercaptide salt from the following data: the potential of a silver electrode in a 0.0500 M silver nitrate solution is +0.546 V compared with a reference electrode and +0.124 V compared with the same reference when in equilibrium with a 0.00110 M solution of the mercaptan saturated with its silver salt.

9-34.   Calculate the equilibrium constant for the reaction $Cr^{3+} + V^{2+} \rightleftharpoons Cr^{2+} + V^{3+}$.

9-35.   Calculate the equilibrium constant for $Cr^{3+} + V^{3+} + H_2O \rightleftharpoons VO^{2+} + 2H^+$.

9-36.†   A reversible organic redox half-reaction is represented in part as follows:
$$R + 2e + nH^+ \rightleftharpoons \text{products}$$
Suggest a method for determining experimentally (1) the number of hydrogen ions involved in the half-reaction; (b) the number of electrons.

9-37.   Most calculations involving the Nernst equation employ concentrations rather than activities. What is the justification for this practice?

9-38.   Define the term formal potential. Explain its use.

9-39.   Identify the following changes in oxidation number as either oxidation or reduction: (a) iron(III) to iron(II); (b) bismuth(III) to the metal; (c) bromide to bromine; (d) bromate to bromide; (e) oxygen to hydrogen peroxide; (f) H_2SO_3 to H_2SO_4; (g) oxygen to water.

9-40.   Give three examples of strong oxidizing agents, three of weak oxidizing agents, and three of strong reducing agents.

REFERENCES

G. Charlot and others, "Selected Constants," *Pure Appl. Chem.*, Supplement 2, IUPAC, Butterworth, London, 1971. Lists the values of standard electrode potentials for a large number of redox conjugate pairs. Also lists many formal potentials. Specifies the methods by which standard electrode potentials were obtained. Standard electrode potentials for most systems were measured from cells, but many were obtained from thermodynamic calculations involving free-energy data, from polarographic data, from potentiometric titrations, and occasionally from spectroscopic data.

H. A. Laitinen and W. E. Harris, *Chemical Analysis*, 2nd ed., McGraw-Hill, New York, 1974, Chapter 12. Gives additional background on the subject of electrode potentials. For those who prefer to use half-cell and cell emf values for Nernst calculations the IUPAC conventions are outlined. Discusses the effects of complex formation and of pH on electrode potential.

T. S. Licht and A. J. deBethune, *J. Chem. Educ.*, **1957**, *34*, 433. Gives the history of conventions for electrochemical signs.

10

OXIDATION-REDUCTION TITRATIONS

I believe that our souls are in our hands, for we do everything to the world with our hands.

Ray Bradbury

10–1 TITRATION CURVES

An *oxidation-reduction (redox) titration curve* consists of a plot of potential in a solution as a function of volume of added reagent. Such curves serve the same purpose as acid-base titration curves in that they show the changes occurring throughout a titration and are helpful in guiding the choice of end-point detection. With experience, from a titration curve the change in potential in the region of the equivalence point, and thereby the analytical utility of the system, can be seen at a glance. Redox titration curves may be plotted either from direct measurements of potential during a titration (Section 11-8) or from calculations using the Nernst equation (Section 9-3). Either standard electrode potentials or formal potentials may be used in these calculations.

Recall that for a redox reaction to occur, two redox half-reactions must be involved. The oxidant of one half-reaction must accept electrons from the reductant of the other. Consider the titration of a solution of iron(II) with a standard solution of cerium(IV). Cerium(IV) is one of the strongest analytical oxidants known, much stronger than iron(III). Therefore cerium(IV) will react quantitatively with iron(II). The two redox conjugate pairs are

$$Ce^{4+} + e \rightleftarrows Ce^{3+} \qquad\qquad E^0 = 1.61 \text{ V}$$

$$Fe^{3+} + e \rightleftarrows Fe^{2+} \qquad\qquad E^0 = 0.77 \text{ V}$$

For each of these half-reactions a Nernst equation can be written. The chemical reaction in the titration is a simple one-electron exchange:

$$Ce^{4+} + Fe^{2+} \rightleftarrows Ce^{3+} + Fe^{3+}$$

Although the titration is carried out in acidic solution, the calculated potentials are independent of pH.[1] The equilibrium-constant expression for the reaction is

$$K_{eq} = \frac{[Ce^{3+}][Fe^{3+}]}{[Ce^{4+}][Fe^{2+}]}$$

and its value, calculated as stipulated in Section 9-4, is 2×10^{14}. This large value confirms that equilibrium lies far to the right.

Calculations for the titration curve are of three types: one for potentials prior to the equivalence point, a second for potentials after the equivalence point, and a third for the potential at the equivalence point. The three types are illustrated in the following example.

EXAMPLE 10-1

Consider the titration of 30.00 mL of 0.05 M iron(II) sulfate with 0.05 M cerium(IV) sulfate in 1 M acid.

Initial Value of Potential

Initially, the solution presumably contains only iron(II) and no iron(III). The ratio $[Fe^{2+}]/[Fe^{3+}]$ would then be indefinitely large, and according to the Nernst equation the potential would be infinitely negative. But such a condition is not possible. In water the limiting negative potential is defined by the decomposition of water to form hydrogen. The presence of a few iron(III) ions produced by reaction of iron(II) with hydrogen ions, or with oxidizing impurities, will establish a potential at some low but unspecified value. We may conclude that the initial potential cannot be calculated.

Potential Before the Equivalence Point

At 1 mL of added titrant the 0.05 mmol (0.05 M × 1 mL) of added cerium(IV) reacts with 0.05 mmol of iron(II) to produce 0.05 mmol of iron(III) in 31 mL of solution. Since we started with 30 mL × 0.05 M, or 1.50, mmol of iron(II), we have 1.45 mmol remaining and the concentration of iron(II) is 1.45 mmol/31 mL, or 0.47 M. No excess cerium(IV) is present at equilibrium; its concentration is near zero. To calculate the potential of the solution, it is therefore prudent to apply the Nernst equation to the iron rather than the cerium half-reaction because both iron(II) and iron(III) are present in measurable amounts. Accordingly,

$$E = 0.77 - 0.059 \log \frac{[Fe^{2+}]}{[Fe^{3+}]} = 0.77 - 0.059 \log \frac{1.45/31}{0.05/31} = 0.68 \text{ V}$$

Note that in the log term the volume 31 mL appears in both numerator and denominator and thus is canceled. The potential depends on the *ratio* of iron(II) to iron(III), not on the dilution. All we need to consider in subsequent

---

[1] If no acid is present, both iron and cerium ions form complexes with the hydroxyl ion that cause the potential to change. These complexes become important at hydrogen ion concentrations below about 1 M. Complexation with anions such as chloride or sulfate also affect the formal potential.

calculations, then, is the ratio.

At 5 mL of added titrant the amount of iron(III) formed will be 5 mL $\times$ 0.05 M, or 0.25, mmol, and $1.5 - 0.25$, or 1.25, mmol of iron(II) will remain unreacted to give the ratio 1.25/0.25.

$$E = 0.77 - 0.059 \log (1.25/0.25) = 0.73 \text{ V}$$

At 15 mL the iron(II) present initially will be half-oxidized to iron(III). Therefore $[Fe^{2+}]$ will equal $[Fe^{3+}]$, and

$$E = 0.77 - 0.059 \log 1 = 0.77 \text{ V}$$

At this point the potential is equal to the standard electrode potential for the iron half-reaction.

As the titration proceeds, at 25 mL, $E = 0.81$ V, and at 29 mL, $E = 0.86$ V and continues to rise with added portions of titrant:

at 29.7 mL (99% completion), $E = 0.77 + 0.059 \log 100 = 0.89$ V
at 29.97 mL (99.9%), $E - 0.77 + 0.059 \log 10^3 = 0.95$ V
at 29.997 mL (99.99%), $E = 0.77 + 4(0.059) = 1.01$ V

Potential After the Equivalence Point

After the equivalence point the concentration of iron(II) is extremely small, approaching zero. Now, however, an excess of cerium(IV) is present, along with the cerium(III) produced by the reaction with iron(II). Since both cerium ions are present in significant amounts, the Nernst equation for the cerium half-reaction is the choice for calculation of potentials in this region. Again, only the ratio need be considered. For instance,

at 30.03 mL (100.1% completion), $E = 1.61 - 3(0.059) = 1.43$ V
at 30.3 mL (101%), $E - 1.61 - 2(0.059) - 1.49$ V
at 31 mL, $E = 1.61 - 0.059 \log (1.50/0.05) = 1.52$ V
at 35 mL, $E - 1.56$ V
at 50 mL, $E = 1.60$ V
at 60 mL (200%), $[Ce^{4+}] = [Ce^{3+}]$, and $E = 1.61$ V (the standard potential of the cerium titrant).

Potential at the Equivalence Point

To determine the potential at the equivalence point (30 mL in this case), we use the fact that at equilibrium the potential given by the Nernst equation for the iron(III)–iron(II) half-reaction is equal to that for the cerium(IV)–cerium(III) half-reaction. They must be equal because both couples are present in a single solution. Before and after the equivalence point we have found first one and then the other Nernst expression applicable. For the equivalence potential, both are needed:

$$E_{Fe} = E_{Ce}$$

$$E_{Fe} = 0.77 - 0.059 \log ([Fe^{2+}]/[Fe^{3+}])$$

$$E_{Ce} = 1.61 - 0.059 \log ([Ce^{3+}]/[Ce^{4+}])$$

Adding these two equations, we have

$$E_{Fe} + E_{Ce} = 2.38 - 0.059 \log ([Fe^{2+}]/[Fe^{3+}]) - 0.059 \log ([Ce^{3+}]/[Ce^{4+}])$$

$$2E = 2.38 - 0.059 \log \frac{[Fe^{2+}][Ce^{3+}]}{[Fe^{3+}][Ce^{4+}]}$$

By definition, at the equivalence point exactly equal quantities of Ce^{4+} and Fe^{2+} have been permitted to react, producing equal amounts of Ce^{3+} and Fe^{3+}. To the extent that the reaction is not complete, the number of unreacted Ce^{4+} and Fe^{2+} ions is the same. Therefore, substitution of $[Ce^{3+}] = [Fe^{3+}]$ and $[Ce^{4+}] = [Fe^{2+}]$ in the log term of the Nernst equation gives a log of unity, $2E = 2.38$ V, and $E = 1.19$ V. This is the equivalence potential.[2]

A curve of a similar titration is shown in Figure 10-1. This figure illustrates the gradual increase throughout the titration as the equivalence point is approached, the marked change in the region of the equivalence point, and the gradual change thereafter. Important to keep in mind is that for this example the changes in potential depend only on the ratio of the concentrations of the reactants.

The equivalence potential, 1.19 V, can be used to calculate the completeness of the reaction at the equivalence point. Since at all times

$$E = 0.77 - 0.059 \log ([Fe^{2+}]/[Fe^{3+}])$$

at the equivalence point

$$1.19 = 0.77 - 0.059 \log ([Fe^{2+}]/[Fe^{3+}])$$

$$\log ([Fe^{2+}]/[Fe^{3+}]) = (0.77 - 1.19)/0.059 = -7.1$$

$$([Fe^{2+}]/[Fe^{3+}]) = 8 \times 10^{-8}$$

At the equivalence point the iron(II) has been quantitatively converted to iron(III),

---

[2]The expression for the potential at the equivalence point may be more complex than for this example.

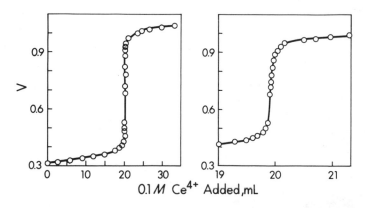

Figure 10-1. Titration of 20 mL of 0.1 *M* iron(II) with standard 0.1 *M* cerium(IV) solution in 0.1 *M* sulfuric acid. Shown is the entire titration curve along with an expanded scale of the equivalence-point region. Data by J. Crump and F. Williams.

or $[Fe^{3+}] = 30 \times 0.05/60 = 0.025 \ M$. Substituting this value for $[Fe^{3+}]$, we obtain

$$[Fe^{2+}] = 2.5 \times 10^{-9} \ M$$

At the equivalence point, therefore, only 1 part of iron(II) is left for every 10^7 parts of iron(III) produced in the reaction. Hence, the reaction is 99.999 99% complete (quantitative is 99.9% or better).

An oxidant as strong as cerium(IV) is clearly not required for iron(II) titrations, and so somewhat weaker oxidants might be considered. Figure 10-2 shows the titration curve obtained in a solution 1 M in hydrogen ions when vanadate $VO_2{}^+$ is substituted for cerium(IV) as the oxidant for iron(II). The reaction is

$$VO_2{}^+ + 2H^+ + Fe^{2+} \ \rightleftarrows \ VO^{2+} + H_2O + Fe^{3+}$$

Calculations for the vanadate–iron(II) system reveal that the iron(II) will be about 99% oxidized at the equivalence point. The equilibrium constant can be calculated to be about 6300. Since an equilibrium constant of about 10^6 is usually required, vanadate is just slightly too weak for this titration.

A solution of vanadate is suitable as a titrant only for reducing agents more powerful than iron(II). Chromium(II) might be such a reductant; in fact it is such a strong reductant that it can be oxidized quantitatively, not only with cerium(IV) and vanadate, but also with iron(III) and the far weaker oxidant vanadyl ion.

Curves can be calculated for any combination of half-reactions to show the changes in potential to be expected if the titrations were performed. Such calculations make clear that the main factor in determining the change in potential near the equivalence point is the difference in E^0 values. For those

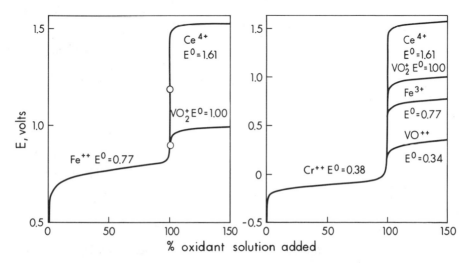

Figure 10-2. Redox titration curves of iron(II) with cerium(IV) and vanadate as oxidants, and of chromium(II) with cerium(IV), vanadate, iron(III), and vanadyl as oxidants.

combinations of oxidant and reductant that have large equilibrium constants, the general shape will be like those in Figures 10-1 and 10-2. The potential at half-reaction will usually (though not always) equal the E^0 value for the conjugate pair of the substance titrated, and the potential at 100% excess will usually equal the E^0 value for the conjugate pair of the titrant.

Curves for titration of an oxidant with a standard solution of reductant can also be calculated. During such titrations the potential becomes more negative rather than more positive.

For some systems the calculations can be more complex for a variety of reasons. Some of these are listed here.

1. Standard electrode potentials are often of less practical significance for titrations than are formal potentials (Section 9-5). Titration curves are more realistic if calculated with formal potentials. Even though the difference between $E^{0'}$ and E^0 values is frequently small, it may sometimes cause a major difference in the equilibrium constant (Section 9-5).

2. Hydrogen ions may be involved in one or both half-reactions. Solution potentials will then depend on the hydrogen ion concentration as well as on the ratio of reductant to oxidant. For instance, the strength of an oxidant such as dichromate decreases markedly as the pH rises.[3]

3. The same number of ions or molecules may not be involved in the two members of an oxidant-reductant conjugate pair. An example of this situation is $I_2 + 2e \rightleftarrows 2I^-$. In this instance the potential depends on dilution, in contrast with redox systems with equimolar stoichiometry. When a solution containing iron(III) and iron(II) is diluted 10 times, the ratio $[Fe^{2+}]/[Fe^{3+}]$ remains unchanged, as does the potential. But a solution containing iodine and iodide when diluted 10 times does undergo a change in the ratio $[I^-]^2/[I_2]$; the potential also changes (increases). For this example the potential at the midpoint in the titration is not the same as the standard potential for the iodine–iodide couple.

4. One or more of the oxidized or reduced forms in a reaction may be insoluble or form complex ions. In such cases the solubility products or stability constants must be incorporated into the calculation according to the Nernst equation.

5. The number of electrons lost per molecule or ion of the reductant may differ from the number gained per molecule or ion of the oxidant. The equivalence potential in such cases does not lie midway between the two E^0 values, but is skewed toward the E^0 of the half-reaction with the most electrons.

EXAMPLE 10-2

Calculate the potential at the equivalence point for the reaction

$$5Fe^{2+} + MnO_4^- + 8H^+ \rightleftarrows 5Fe^{3+} + Mn^{2+} + 4H_2O$$

---

[3]A classic example is the reaction of $Cr_2O_7^{2-}$ with H_2O_2. In acid the products are Cr(III) and O_2. In base, however, hydrogen peroxide will oxidize Cr(III) quantitatively to $Cr_2O_7^{2-}$.

$$E = 0.77 - 0.059 \log \left([Fe^{2+}]/[Fe^{3+}]\right)$$

$$= 1.51 - \frac{0.059}{5} \log \frac{[Mn^{2+}]}{[MnO_4^-][H^+]^8}$$

At the equivalence point, equilibrium for the reaction between iron(II) and permanganate lies far to the right, and the concentration of iron(III) is five times that of manganese(II) because the reaction between them was in the ratio 5:1. To the extent that reaction is not complete, five iron(II) ions remain for each permanganate ion. To calculate the equivalence potential, we combine the log terms for the iron and permanganate half-reactions so that the concentration terms cancel each other. Thus we must multiply the Nernst equation for permanganate at the equivalence point by 5 and add it to the one for iron:

$$5E_{eq} = (5 \times 1.51) - 5\,\frac{0.059}{5} \log \frac{[Mn^{2+}]}{[MnO_4^-]} - 5\,\frac{0.059}{5} \log \frac{1}{[H^+]^8}$$

$$E_{eq} = 0.77 - 0.059 \log \left([Fe^{2+}]/[Fe^{3+}]\right)$$

Adding these two expressions, we obtain

$$6E_{eq} = (5 \times 1.51) + 0.77 - 0.059 \log \frac{[Mn^{2+}][Fe^{2+}]}{[MnO_4^-][Fe^{3+}]} - 0.059 \log \frac{1}{[H^+]^8}$$

The first log term is zero, and hence

$$E_{eq} = \frac{(5 \times 1.51) + 0.77}{6} - \frac{0.059}{6} \log \frac{1}{[H^+]^8} = 1.39 - \frac{0.059}{6} \log \frac{1}{[H^+]^8}$$

The general case is

$$E_{eq} = \frac{n_1 E_1 + n_2 E_2}{n_1 + n_2}$$

Bearing in mind the sources of complication, detailed calculations for redox pairs of half-reactions indicate that for quantitative reaction the minimum difference in $E^{0\prime}$ values for two half-reactions is about 0.4 V. Consulting a table of E^0 or $E^{0\prime}$ values, we may conclude either that a reaction *will not be* complete at equilibrium or that it *may be*. Although to be of analytical utility a reaction must be rapid, the magnitude of an equilibrium constant says nothing about the time required to reach equilibrium. Questions about the stoichiometry of the reaction and means of detecting the equivalence point are best settled by laboratory observation.

As an example of the effect of pH on the position of equilibrium and the completeness of reaction, consider the reaction

$$Q + 2H^+ + 2Fe(CN)_6{}^{4-} \rightleftarrows H_2Q + 2Fe(CN)_6{}^{3-}$$

where Q stands for quinone and H_2Q for hydroquinone.[4]   The
equilibrium-constant expression is

$$K_{eq} = \frac{[H_2Q][Fe(CN)_6{}^{3-}]^2}{[Q][H^+]^2[Fe(CN)_6{}^{4-}]^2} = 2.2 \times 10^{11}$$

$$E = 0.70 - \frac{0.059}{2} \log \frac{[H_2Q]}{[Q][H^+]^2}$$

$$E = 0.36 - 0.059 \log \frac{[Fe(CN)_6{}^{4-}]}{[Fe(CN)_6{}^{3-}]}$$

With an equilibrium constant as large as 2.2×10^{11}, equilibrium would
be expected to lie quantitatively to the right. Note, however, that hydrogen
ions appear in the equilibrium-constant expression. Normally in a titration
the hydrogen ion concentration is fixed. By writing the equilibrium constant
in the alternate form

$$K_{cond} = \frac{[H_2Q][Fe(CN)_6{}^{3-}]^2}{[Q][Fe(CN)_6{}^{4-}]^2} = 2.2 \times 10^{11} [H^+]^2$$

we can clearly reveal the effect of pH on the position of equilibrium. In a
solution buffered at pH 0 ($[H^+] = 1$)

$$K_{cond} = 2.2 \times 10^{11} \times (1)^2 = 2.2 \times 10^{11}$$

At pH 5

$$K_{cond} = 2.2 \times 10^{11} \times (10^{-5})^2 = 22$$

At pH 10

$$K_{cond} = 2.2 \times 10^{11} \times (10^{-10})^2 = 2.2 \times 10^{-9}$$

These calculations demonstrate that in solutions of low pH the reaction is
quantitative to the right, at pH values in the region 5 to 6 it is incomplete in
both directions, and at pH 10 it is quantitative to the left. In this example
the reaction of quinone with ferrocyanide can be driven quantitatively to
either the right or the left simply by buffering the solution at appropriate pH
values.

---

[4]Quinone is one of the few organic compounds to undergo reversible reduction readily. The
standard potential is $+0.70$ V for the half-reaction

$$O{=}\langle\bigcirc\rangle{=}O + 2H^+ + 2e \rightleftarrows HO{-}\langle\bigcirc\rangle{-}OH$$

10-2 DETECTION OF THE EQUIVALENCE POINT. REDOX INDICATORS

There are three general methods by which the change in potential that occurs near the equivalence point in a redox reaction can be detected. The most usual is by potentiometry (Section 11-8). The second most common method is by a general redox indicator. The third is by specific or other indicators. The last two methods are discussed here.

General Indicators

Whereas an acid-base indicator responds to a marked change in pH, a redox indicator responds to a marked change in potential by being oxidized or reduced at potentials near the equivalence point. For example, an indicator for the iron(II)–cerium(IV) reaction must have a transition potential that allows it to be oxidized by cerium(IV) but not by iron(III). The indicator must have different colors in the oxidized and reduced forms, and at least one of the colors should be sufficiently intense that a minute amount of reagent can bring about a visible change. The half-reaction for the indicator conjugate pair should be reversible:

$$In_{ox} + ne \rightleftarrows In_{red}$$

where In_{ox} is the indicator in the oxidized form and In_{red} the indicator in the reduced form. For simplicity, in the following discussion charges on the ionic forms are omitted, and hydrogen ions are not included in the half-reaction. Also, the color intensities of the two forms of the indicator are assumed to be the same.

For a reversible indicator half-reaction, the Nernst equation can be written in the usual way:

$$E = E^0{}_{In} - (0.059/n) \log (a_{In_{red}}/a_{In_{ox}})$$

The standard electrode potential for the indicator $E^0{}_{In}$ will be the value of E when the log term is equal to zero; when aIn_{red} equals aIn_{ox}, E equals $E^0{}_{In}$. Since what the eye sees is the concentration of each of the colored forms, we must use concentrations rather than activities in the Nernst expression; therefore we deal with formal potentials rather than standard electrode potentials. Thus (Section 9-5)

$$E = E^{0\prime}{}_{In} - \frac{0.059}{n} \log \frac{[red]}{[ox]}$$

In a solution with a potential equal to $E^{0\prime}{}_{In}$, a redox indicator will exhibit an intermediate color because half the indicator is present in each of the colored forms. As with an acid-base indicator, the color perceived is determined by the relative amounts of the two colored forms present. In most cases an indicator 90% in the oxidized form will be visually indistinguishable from one 100% in that form. Similarly, an indicator 90% in the reduced form will be indistinguishable from one 100% in the reduced form. A gradation in color, however, will be perceived between these limits — from a ratio of about 10 to a ratio of about 0.1. Again as with acid-base indicators, this range is termed the *transition interval*. The potential corresponding to one extreme of the transition interval can be calculated when the ratio is 10 by

$$E = E^{0'}{}_{In} - (0.059/n) \log 10 = E^{0'}{}_{In} - (0.059/n) \text{ V}$$

When the ratio of the two colored forms is 0.1,

$$E = E^{0'}{}_{In} - (0.059/n) \log 0.1 = E^{0'}{}_{In} - (-0.059/n) = E^{0'}{}_{In} + 0.059/n \text{ V}$$

Thus the indicator range is $E^{0'}{}_{In} \pm (0.059/n)$ V, and a visible color change takes place typically over about 0.12 V in the neighborhood of the formal potential of the indicator when n is equal to 1.

When n is equal to 2, the transition range is about $E^{0'}{}_{In} \pm 0.03$ V.

Choice of an Indicator for Redox Titrations

As a first approximation the indicator chosen for a redox titration should have an $E^{0'}{}_{In}$ equal to the formal equivalence potential. Here too the situation parallels that in acid-base titrations, where the indicator chosen should have a pK_{In} equal to the pH at the equivalence point.

A number of satisfactory indicators have been developed; several of these behave reversibly, so that the Nernst equation can be applied. Table 10-1 lists redox indicators suitable over a range of potentials. Some involve hydrogen ions (for example, methylene blue and variamine blue) in the indicator half-reactions; for such indicators the transition range depends on the pH of the solution. Others, such as diphenylamine, undergo oxidation or

Table 10-1. Redox Indicators and their Characteristics

Indicator	Color Oxidized Form	Reduced Form	Usual Formal Potential*
$V(II)(1,10\text{- phenanthroline})_3{}^{2+}$	Colorless	Blue-violet	0.14
Phenosafranine	Red	Colorless	0.28
Methylene blue	Blue	Colorless	0.36
Variamine blue	Colorless	Blue	0.59
Diphenylamine sulfonic acid	Violet	Colorless	0.85
Erioglaucine	Red	Green	1.0
$Fe(II)(1,10\text{-phenanthroline})_3{}^{2+}$	Blue	Red	1.11
$Fe(II) (5\text{-}NO_2\text{-}1,10\text{-phenanthroline})_3{}^{2+}$	Blue	Violet	1.25

*The value of the formal potential $E^{0'}{}_{In}$ varies with the composition of the solution. It should be determined for the proposed titration system.

reduction in a complex way. For still others, electron transfer is irreversible, and the indicator may change color prematurely owing to temporary local excesses of titrant in the solution. The end point under these conditions is likely to be premature and less sharp.

Ferroin, $Fe(1,10\text{-phenanthroline})_3^{2+}$, is a highly stable, intensely red iron(II) complex that is reversibly oxidized to a blue iron(III) compound. The formal potential, $+1.11$ V, is in a range that makes it especially useful for titrations with cerium(IV).

Specific and Other Indicators

A *specific indicator* is useful only with a particular reagent. The most important example is starch solution, which forms a reversible blue complex with iodine. When iodine is titrated with a reducing agent, usually thiosulfate, starch must not be added until near the end point, or the end point will be diffuse. Experimental details on the use of starch as an indicator are given in Section 10-9.

Some redox titrants do not require an indicator. These include titrations of substances with solutions of potassium permanganate, which acts as its own indicator. Since permanganate is an intense purple-red and its reduction product, manganese(II), is essentially colorless, the first slight excess of permanganate in a titration denotes the end point. Potassium permanganate undoubtedly retains much of its popularity as a reagent despite several disadvantages because of this ability to serve as its own indicator. Cerium(IV) sulfate and iodine also form colorless reduction products, though the colors of the oxidized forms are less intense than the color of permanganate.

10-3 REAGENTS FOR PRELIMINARY OXIDATION OR REDUCTION

Prior to performing a redox titration an oxidation or reduction is frequently necessary to convert quantitatively the substance being determined to a known oxidation state. This preliminary adjustment of the oxidation state is generally accomplished by adding an excess of a reagent whose electrode potential is sufficiently high (or low) to effect the required oxidation (or reduction). The criteria for reagents for preliminary oxidation or reduction are that they (1) must react rapidly, (2) must react quantitatively, (3) must not introduce interference, and (4) must allow the excess to be removable or convertible to an unreactive form. Many reagents, though unsuitable as primary standards or standard solutions because of insolubility, instability, or nonstoichiometric reaction, may be useful for the preliminary adjustment of oxidation states. Such oxidants and reductants may be gases, insoluble solids, or substances in solution; occasionally, electrochemical methods are employed.

Oxidants for Preliminary Oxidations

The majority of preliminary oxidants are designed to convert other substances to their maximum oxidation number and so are classed as strong oxidants. Several are noted here, with brief comments.

Ozone, O_3 ($E \approx 2.1$ V), is a gaseous oxidant capable of oxidizing manganese(II) to permanganate. The excess is removed by bubbling with nitrogen.

Bromine, Br_2 ($E \approx 1.1$ V), and *chlorine*, Cl_2 ($E \approx 1.4$ V), can oxidize iodide to iodate. The excess may be removed by reaction with phenol, as discussed in Section 10-5 under potassium bromate.

Potassium peroxydisulfate, $K_2S_2O_8$ ($E \approx 2.0$ V), is a strong oxidant that requires silver ion as a catalyst. In fact, it is so strong that the excess is destroyed simply by boiling the solution, whereupon water is oxidized and oxygen liberated.

Sodium bismuthate, $NaBiO_3$ ($E \approx 1.6$ V), is a strong oxidant that can be used in nitric or sulfuric acid solution.

Potassium periodate, KIO_4 ($E \approx 1:7$ V), oxidizes manganese(II) to permanganate:

$$2Mn^{2+} + 5IO_4^- + 3H_2O \rightarrow 2MnO_4^- + 5IO_3^- + 6H^+$$

The iodate formed and the excess periodate may be precipitated as the mercury(II) salts and removed by filtration.

Perchloric acid, $HClO_4$ ($E \approx 2.0$ V), when hot and concentrated is a dramatically effective oxidant for the dissolution of refractory sulfide ores, the wet oxidation of organic matter, and the oxidation of chromium and vanadium. Upon cooling and dilution with water all oxidizing power is lost. *Caution*: If used incorrectly, it may be dangerously explosive!

Most oxidants are effective only in acidic solutions. *Hydrogen peroxide*, H_2O_2 ($E \approx 1.8$ V), is an efficient oxidant in basic solution. The excess is destroyed by boiling. *Solid sodium peroxide* is a vigorous oxidant in alkaline fusions.

Silver(II) oxide, AgO ($E \approx 2.0$ V), in acid solution is a powerful oxidant, silver(II) being reduced to silver(I). The excess is eliminated by warming the solution.

Reductants for Preliminary Reductions

Sulfur dioxide in acid solution is present as sulfite. The half-reaction is

$$SO_4^{2-} + 4H^+ + 2e \rightleftharpoons H_2SO_3 + H_2O \qquad E^0 = 0.17 \text{ V}$$

It reduces iron(III) quantitatively to iron(II). The excess in acid solution is removed by boiling.

Tin(II) chloride, $SnCl_2$ ($E \approx 0.1$ V), is the usual choice for reduction of iron(III) in hot hydrochloric acid solution. Only a slight excess is needed, the yellow color of iron(III) indicating when the correct amount has been added.

The slight excess can be destroyed by addition of an excess of mercury(II) chloride:

$$Sn^{2+} + 2HgCl_2 \longrightarrow Sn^{4+} + Hg_2Cl_2(s) + 2Cl^-$$

Chromium(II) chloride, $CrCl_2$ ($E \approx -0.4$ V), is one of the most powerful reductants available in aqueous solution. The excess is removed by bubbling with air, whereupon the chromium(II) is oxidized to chromium(III) by oxygen.

Hydrazine, N_2H_4 ($E \approx -0.2$ V), reduces arsenic(V) to arsenic(III), and reduces HgI_4^{2-} to the metal and is oxidized to nitrogen. The excess is destroyed by heating with sulfuric acid.

The *zinc reductor*, Zn ($E \approx -0.8$ V), employs zinc metal amalgamated with mercury to minimize evolution of hydrogen gas. The metal is placed in a column, and the acidified solution to be reduced is passed through it. This system is known as the *Jones reductor*. Zinc metal, not being selective, reduces many metals besides iron to low oxidation states where they can be determined by titration. After zinc reduction, vanadium can be titrated from oxidation state 2 to 4, chromium from 2 to 3, uranium from 4 to 6, titanium from 3 to 4, and molybdenum from 3 to 6.

The *silver reductor*, Ag ($E \approx 0.2$ V), is a selective reductor that employs metallic silver in conjunction with hydrochloric acid. The half-reaction is

$$AgCl(s) + e \rightleftarrows Ag(s) + Cl^- \qquad E^0 = 0.22 \text{ V}$$

A column of granular silver through which hydrochloric acid solutions of the species to be reduced are passed is known as a *Walden reductor*. Used as columns, metal reductors of this type are especially useful for selective reductions in mixtures of metal salts.

Liquid amalgams are alloys of metals such as sodium, lead, and bismuth with mercury. The reducing power of the amalgam is determined by the metals present. Sodium amalgam($E \approx -1.9$ V) is one of the most powerful. An advantage is that the oxidation product, sodium ion, normally causes less interference than does the ion of a heavy metal.

Removal of Organic Matter

In many samples the substance of interest is associated with organic matter that interferes with the measurement step. Removal of such material is often a challenging problem. Both dry and wet ashing techniques have been developed for this purpose. In dry ashing the organic matter is ordinarily removed by oxidation at a high temperature in contact with oxygen of the air. In wet ashing the organic matter is degraded through the use of iron(II)–hydrogen peroxide or of oxidizing acids at their boiling point, such as HNO_3, H_2SO_4, or $HClO_4$ (*with care!*).[5] Section 21-3 gives additional detail on these operations.

---

[5]See reference by Schilt at the end of the chapter for precautions in the use of perchloric acid.

10-4 TITRANTS AND PRIMARY STANDARDS

The most important oxidizing titrants in redox analysis are $K_2Cr_2O_7$, I_2, $KMnO_4$, $Ce(SO_4)_2$, KIO_3, and $KBrO_3$. The primary standards often used in conjunction with these oxidants are As_2O_3, KI, and $Na_2C_2O_4$.

The important reducing titrants are $Na_2S_2O_3$ (for titrations with I_2), $FeSO_4$, $CrSO_4$, and $Ti_2(SO_4)_3$. Ascorbic acid is an example of an organic reductant. The primary standards often used for the standardization of these reductants include $K_2Cr_2O_7$, KIO_3, and $KBrO_3$.

The subsections that follow discuss the range of applicability of several of these oxidants and reductants as titrants, along with special conditions for their use.

Potassium Dichromate

The overall half-reaction for potassium dichromate as an oxidant is

$$Cr_2O_7^{2-} + 14H^+ + 6e \rightleftharpoons 2Cr^{3+} + 7H_2O \qquad E^0 = 1.33 \text{ V}$$

The value of the standard electrode potential indicates that dichromate is a strong oxidant, though less strong than permanganate or cerium(IV). The half-reaction indicates that the potential is markedly affected by changes in hydrogen ion concentration. Thus, by control of pH the oxidizing power of potassium dichromate can be varied from strong to only moderately strong.

EXAMPLE 10-3

$$E = 1.33 - \frac{0.059}{6} \log \frac{[Cr^{3+}]^2}{[Cr_2O_7^{2-}][H^+]^{14}}$$

If the ratio $[Cr^{3+}]^2/[Cr_2O_7^{2-}]$ is unity, at pH 5

$$E = 1.33 - \frac{0.059}{6} \log \frac{1}{(10^{-5})^{14}} = 0.64 \text{ V}$$

At pH 0, $E = 1.33$ V.

In concentrated phosphoric acid solution, complexation of the chromium(III) species by phosphate anion increases the oxidizing strength even more than expected from the effect of pH alone.

Potassium dichromate is easily obtained in pure form and is stable; therefore the salt can be employed directly as a primary standard. Solutions are stable indefinitely and may even be boiled without decomposition. The electrode potential, though lower than that of cerium(IV), is in some instances an advantage; for example, chloride is not oxidized in the titration of iron(II) unless its concentration exceeds about 1 M. For this reason dichromate is the titrant of choice in almost all titrimetric determinations of iron in ores, steels, and slags where the samples are dissolved in hydrochloric acid. The reaction with iron proceeds rapidly and quantitatively even in dilute solutions. Details of the determination of iron with dichromate may be found in Section 10-7.

Potassium dichromate, for all its advantages, has some drawbacks. One is that the rate of reaction of dichromate with many reductants is often slow. Another is that reactions are sometimes complicated by the temporary existence of intermediates such as chromium(V) and chromium(IV). A third disadvantage is that in acid solution a mixture of species is present, including CrO_4^{2-}, $Cr_2O_7^{2-}$, and $HCrO_4^{-}$. A fourth is that the green of the chromium(III) formed during titrations requires an indicator of exceptionally intense color.

Other applications of dichromate include the determination of oxidants by the addition of a known excess of a standard iron(II) solution followed by titration of the excess. Other reductants can be determined by addition of an unmeasured excess of iron(III) followed by titration of the iron(II) formed.

An important biological application of dichromate is in the evaluation of the *chemical oxygen demand* of natural and waste water. This measurement provides an estimate of the oxidizable organic material available to aerobic organisms. The oxidizable material is proportional to the amount of dichromate consumed in a carefully specified and standardized procedure. Another important use is in the measurement of the ethanol content of the breath as a test for impaired driving. Dichromate oxidizes ethanol to acetic acid. In the Breathalyzer the extent of the reaction is measured by the decrease in intensity of the yellow dichromate solution compared with that of a standard solution; in the Alcolyser the green of chromium(III) is used in the measurement.

Iodine

Perhaps the most broadly useful half-reaction in redox analysis is that for the iodine-iodide couple:

$$I_2 + 2e \rightleftarrows 2I^- \qquad E^0 = 0.621 \text{ V}$$

An intermediate standard electrode potential permits the use of iodine as an oxidant for substances of low electrode potential and of iodide as a reductant for substances of high potential. Solutions of iodine may be used to titrate reduced materials directly, whereas oxidizing agents may be determined indirectly through oxidation of iodide to iodine. This is followed by titration of the iodine with a standard solution of sodium thiosulfate, $Na_2S_2O_3$:

$$I_2 + 2S_2O_3^{2-} \rightleftarrows 2I^- + S_4O_6^{2-}$$

The effective oxidizing power of iodine in practical titrations is slightly less than implied by the E^0 value of 0.621 V because it is tied up as the weak triiodide ion:

$$I_2 + I^- \rightleftarrows I_3^- \qquad K_{eq} = 710 = \frac{[I_3^-]}{[I_2][I^-]}$$

In virtually all analytical reactions involving iodine, iodide is present in excess. Consequently, the redox half-reaction is more realistically written

$$I_3^- + 2e \rightleftarrows 3I^- \qquad E^0 = 0.545 \text{ V}$$

According to this half-reaction the oxidizing power of iodine is about 0.08 V weaker when the formation of the triiodide complex is included.

Whether or not the I_3^- complex is considered, hydrogen ions are not involved in the iodine half-reaction, and therefore the potential of the couple would be expected to be independent of pH. This characteristic is found in practice at pH values lower than about 8; at higher values, however, species such as hypoiodite, IO^-, begin to form.

EXAMPLE 10-4

An analytically important reaction is that of arsenite with iodine:

$$H_3AsO_3 + H_2O + I_3^- \rightarrow H_3AsO_4 + 3I^- + 2H^+$$

$$K_{eq} = \frac{[H_3AsO_4][I^-]^3[H^+]^2}{[H_3AsO_3][I_3^-]}$$

Using E^0 values of 0.56 and 0.54 V for the arsenic and the iodine half-reactions, a calculation such as that of Section 9-4 yields an equilibrium constant of 0.22. A conditional constant can be written

$$K_{cond} = \frac{[H_3AsO_4][I^-]^3}{[H_3AsO_3][I_3^-]} = \frac{0.22}{[H^+]^2}$$

At pH 0, then, the conditional constant is 0.22, indicating that the reaction is not quantitative in either direction. If the iodine formed as the result of reaction were removed by, say, extraction or other chemical reaction, the iodide–arsenate reaction could be driven quantitatively to the left. At pH 5, however, the conditional constant would be 2.2×10^9. This large value indicates a reaction that is quantitative to the right.

Usually, the iodine half-reaction is little affected by complexing or precipitating reagents. Thus, the extent of reaction can often be effectively controlled by such auxiliary reagents.

EXAMPLE 10-5

Iron, an element widely dispersed in nature, may interfere with the determination of other materials. In particular, iron(III) may oxidize iodide according to

$$2Fe^{3+} + 3I^- \rightleftarrows 2Fe^{2+} + I_3^-$$

This reaction has a large equilibrium constant:

$$K_{eq} = \frac{[Fe^{2+}]^2[I_3^-]}{[Fe^{3+}]^2[I^-]^3} = 5 \times 10^7$$

If an anion such as fluoride or phosphate is added to a solution containing iron(III), the iron(III) ions are strongly complexed. For example,

$$Fe^{3+} + F^- \rightleftharpoons FeF^{2+} \qquad K_{eq} = \frac{[FeF^{2+}]}{[Fe^{3+}][F^-]}$$

Since the stability constant for the fluoride complex is large, the amount of iron(III) present at equilibrium with excess fluoride,

$$[Fe^{3+}] = [FeF^{2+}]/K_{eq}[F^-]$$

will be small. If a substantial excess of fluoride is added, the iron(III) ion concentration becomes so small that the reaction of iron(III) with iodide becomes negligible.

Iodine can be obtained in high purity through sublimation, and therefore it should be a suitable primary standard for the preparation of a standard solution. The volatility of iodine, however, makes direct preparation of standard solutions difficult. For this reason a solution of approximately the desired concentration is usually prepared and standardized against a primary standard such as arsenic trioxide (Section 10-10).

Sodium thiosulfate is not a primary standard; solutions of it are usually standardized iodimetrically against potassium dichromate or copper.

Standard iodine solutions are used to determine a wide range of substances, including arsenic, tin(II), and water (by the Karl Fischer method). It is widely applied in organic analysis — compounds such as ascorbic acid, mercaptans, and phenols can be titrated with standard iodine solutions, as can fats and vegetable oils containing carbon–carbon double bonds. Measurement of the number of double bonds, or the extent of unsaturation, is important in many foods.

EXAMPLE 10-6

Karl Fischer Determination of Water. An important application of iodine in analysis is in the determination of the free-water content of materials. In this method the sample is dissolved in a solvent such as methanol and titrated with a mixture of iodine, sulfur dioxide, and pyridine in methanol. The balanced equation may be written

$$I_2 + SO_2 + CH_3OH + 3C_5H_5N + H_2O \rightarrow 2C_5H_5NH^+I^- + C_5H_5NH^+SO_4CH_3^-$$

In this reaction iodine is reduced to iodide and the sulfur in sulfur dioxide is oxidized to sulfate; water is required for the reaction to proceed. Both pyridine and methanol are necessary to stabilize the reaction products. Certain methoxy ethers when substituted for methanol provide greater titrant stability; otherwise, the titer decreases a percent or so per day. No satisfactory replacement has been found for pyridine. The end point is indicated by the first excess of iodine that persists in solution for 30 s or so; this excess can be detected either visually or by constant-current potentiometry (Section 12-4). The titrant solution is usually standardized by titration of a known weight of pure water or of sodium tartrate dihydrate.

Commercial apparatus, in which the iodine is generated coulometrically (Section 12-1) and the end point determined by constant-current potentiometry, is available from several manufacturers. Liquid samples may be injected by syringe through a rubber septum into the titration cell to reduce contamination from atmospheric moisture. Overall precision of the method, either manually or with automatic instrumentation, is on the order of 1%.

Applications include the determination of water in organic compounds, both liquids and solids, as well as in inorganic substances. Compounds that are reduced by iodide or sulfur dioxide, or oxidized by iodine, will of course interfere.

Sodium Thiosulfate

Most strong oxidizing agents can be determined iodimetrically by addition of an excess of iodide and titration with thiosulfate of the iodine produced. The amount of iodine formed is equivalent to the amount of oxidant in the sample:

$$\text{oxidant} + 2I^- \rightarrow I_2$$

Precautions must be taken to avoid side reactions. For instance, iodine will slowly oxidize tetrathionate to sulfate, especially at high pH values. At low pH values thiosulfate may undergo disproportionation, and iodide may be oxidized by atmospheric oxygen:

$$O_2 + 4H^+ + 4I^- \rightarrow 2I_2 + 2H_2O$$

This reaction leads to what is called *oxygen error*. The side reactions may be avoided by carrying out the thiosulfate–iodine titration in the pH range 2 to 5. The problems created by the volatility of iodine can be reduced by titrating promptly, keeping the temperature low, and providing excess iodide to stabilize the iodine in solution as triiodide.

The reaction of starch with triiodide produces an intense blue that serves as a specific indicator. The triiodide ion appears to be just the right size to enter the helical structure of starch and thereby form a colored complex. The color of free iodine, red-brown in high and yellow in low concentrations, can also be used to indicate its presence, though it is not so easily seen as that of the complex with starch.

The use of thiosulfate as a standard reductant is limited almost exclusively to the titration of iodine. One reason for this limited use is that oxidizing agents stronger than iodine oxidize it to a mixture of higher oxidation states of sulfur. A second is that ions of transition metals (such as copper) decrease the stability of sodium thiosulfate solutions by catalytic air oxidation. A third is that acids make thiosulfate solutions unstable by promoting disproportionation to sulfite and elemental sulfur:

$$HS_2O_3^- \rightarrow HSO_3^- + S$$

Addition of a small amount of sodium carbonate to make the thiosulfate solution alkaline prevents this decomposition. Water for the preparation of standard thiosulfate solutions may be boiled to destroy sulfur bacteria, which find these solutions a favorable medium for growth. Alternatively, a bactericidal agent such as chloroform or a mercury(II) salt may be added. Other details of the preparation and use of standard solutions of thiosulfate are given in Section 10-9.

Any of several primary standards including iodine, potassium dichromate, potassium iodate, and electrolytic copper metal may be used to standardize thiosulfate solutions. Whenever possible, the primary standard should be the same substance that is to be determined. Pure copper metal, for instance, is the preferred primary standard when copper is to be determined iodimetrically.

Permanganate

Potassium permanganate has a venerable analytical history, having been utilized to titrate iron(II) as long ago as 1846. Although it has the advantage of high standard potential, 1.51 V, and can serve as its own indicator, permanganate also has several disadvantages. To serve as its own indicator, it must be intensely colored. In practice this means that the top (rather than the bottom) of the meniscus in a buret filled with permanganate must be read. This technique cannot provide the same degree of precision. A second disadvantage is that permanganate is not a primary standard. A third is that it slowly reacts with traces of organic matter always present in distilled water to produce manganese dioxide, which catalyzes further decomposition of the permanganate.

The reduction of permanganate to manganese(II) is not straightforward because several intermediate oxidation states (6, 4, and 3) can form. The half-reaction

$$MnO_4^- + 8H^+ + 5e \rightleftarrows Mn^{2+} + 4H_2O \qquad E^0 = 1.51 \text{ V}$$

indicates a dependence of potential on pH. In alkaline solution either MnO_2 or MnO_4^{2-} may form.

Complexing anions such as phosphate can stabilize the Mn(III) oxidation state. Thus a wide range in formal potentials is found with permanganate, varying with the pH and the complexing substances present.

Sodium oxalate is the most common primary standard for permanganate solutions. Owing to the complex mechanism of the reaction between permanganate and oxalate accurate results are obtainable only under controlled conditions. Although under optimum conditions positive and negative errors still occur, they effectively cancel each other. Divalent manganese and higher temperatures increase the rate of reaction, which otherwise is sluggish. At high temperatures, however, the decomposition of both permanganate and oxalate in solution introduces error . Also, side reactions such as formation of hydrogen peroxide may occur. Should hydrogen peroxide form, care must be taken that it remains in solution to react with more permanganate and is not lost by decomposition of peroxide to water and oxygen. Because decomposition of peroxide is accelerated by vigorous stirring, mixing during oxalate titrations must be gentle. The most reproducible results in standardization titrations are obtained by adding about 90% of the required amount of permanganate to an oxalate–sulfuric acid solution at room temperature. The mixture is allowed to stand until the permanganate color disappears; then the solution is heated to about 55°C, and the titration finished slowly.

10–5 OTHER OXIDANTS AND REDUCTANTS

Cerium(IV) sulfate is a strong oxidant ($E^0 = 1.61$ V) in solutions of low pH. Standard solutions of cerium(IV) do not oxidize chloride ion as does permanganate, and they are more stable. The half-reaction involves a simple, one-electron exchange, and thus no intermediate oxidation states are involved. Cerium(IV) can replace permanganate in most titrations and is superior on almost every count.

Solutions of cerium(IV) are usually prepared in sulfuric acid and then standardized. Although ammonium hexanitratocerate, $(NH_4)_2Ce(NO_3)_6$, is a satisfactory primary standard, other cerium(IV) salts may be used as starting materials including ammonium sulfatocerate, [$(NH_4)_2Ce(SO_4)_3 \cdot 2H_2O$], sulfatoceric acid ($H_2Ce(SO_4)_3$), and hydrated cerium(IV) oxide ($CeO_2 \cdot xH_2O$). All these substances must be dissolved in a considerable excess of strong acid to prevent hydrolysis of the cerium and formation of insoluble cerium(IV) hydrous oxides.

The oxidizing power of cerium(IV) is highly dependent on the nature of the anions in solution. This dependence is caused by the strong complexing ability of the cerium(IV) ion. The more stable the cerium(IV) complex with an anionic species, the lower is the potential of the cerium(IV)–(III) couple. Thus, the formal potential in 1 M perchloric acid is 1.71 V, in 1 M nitric acid 1.61 V, in 1 M sulfuric acid 1.44 V, and in 1 M hydrochloric acid 1.28 V.

Potassium iodate is a primary standard that in 3 to 9 M hydrochloric acid reacts with reductants according to the half-reaction

$$IO_3^- + 6H^+ + Cl^- + 4e \rightleftharpoons ICl + 3H_2O$$

Free iodine is often formed in the reduction process. As the titration is continued with iodate,

$$2I_2 + IO_3^- + 6H^+ + 5Cl^- \rightarrow 5ICl + 3H_2O$$

The equivalence point corresponds to the disappearance of the free iodine. Since starch does not function well at high concentrations of hydrochloric acid, the end point may be determined by following the color produced on extraction of the iodine into chloroform. This method of end-point detection requires that the solution be shaken after each addition of titrant. Potassium iodate is a convenient titrant particularly for arsenic(III), antimony(III), mercury(I), and thallium(I), which tend to react slowly with several of the other strong oxidants.

Potassium bromate is both a primary standard and a strong oxidant in acid solution, as evidenced by the electrode potential for the half-reaction:

$$2BrO_3^- + 12H^+ + 10e \rightleftarrows Br_2 + 6H_2O \qquad E^0 = 1.5 \text{ V}$$

The bromine is usually reduced further to bromide:

$$Br_2 + 2e \rightleftarrows 2Br^- \qquad E^0 = 1.08 \text{ V}$$

Thus potassium bromate can function as a brominating agent, acting as though it were a stable standard solution of bromine. For stoichiometric calculations involving bromate as titrant the following overall net reaction is applicable:

$$BrO_3^- + 6H^+ + 6e \rightleftarrows Br^- + 3H_2O$$

Methods of end-point detection are based on the first appearance of free bromine. The red-brown of bromine itself can be used, or indicators such as methyl red. Methyl red undergoes bromination after the equivalence point, followed by rapid conversion to a colorless product in an irreversible reaction.

Potassium bromate is important in the analysis of organic compounds. Since bromination reactions are frequently slow, a typical procedure is to add a measured excess of bromate and an excess of potassium bromide to the sample. The bromate is converted quantitatively to bromine, which reacts with the sample. On completion of the reaction, the excess bromine is then determined either iodimetrically or by titration with a standard solution of arsenic(III).

With phenol the net reaction is

$$C_6H_5OH + 3Br_2 \rightarrow C_6H_2OHBr_3 + 3HBr$$

Many other aromatic compounds react similarly, with one to three bromine atoms being substituted for hydrogen atoms on the molecule.

For unsaturated organic compounds, bromine adds to the double bond:

$$R_1CH{=}CHR_2 + Br_2 \rightarrow R_1CHBrCHBrR_2$$

Periodate is a strong oxidant in acid solution. The commonest application is in the Malaprade determination of hydroxyl groups on adjacent carbon atoms in an organic molecule (Section 10-10).

Several other oxidants have been developed as titrants: *hypochlorite* and *hypobromite*, useful in mildly alkaline solution; *vanadium(V)*, comparable to dichromate as an oxidant; *chloramine-T*, applicable in both acid and alkaline solution; *ferricyanide*, a primary standard and applicable in alkaline solution; *lead tetraacetate*, useful in organic analysis; and *iron(III)*, useful for the determination of strong reductants.

10-6 NOTE ON ORGANIC ANALYSIS

Although the same fundamental analytical techniques are applicable to both organic and inorganic compounds, inorganic examples are frequently chosen when illustrating the principles of analysis because they tend to be less sensitive to experimental variables. One difficulty associated with organic analysis by chemical reaction is that many compounds can undergo several reactions simultaneously; another is the number and variety of structures and combinations that possess similar chemical properties. Accurate, reproducible analyses usually require practice and exceptional technique.

The determination of organic compounds is in some respects an art as well as a science. In the past, brilliant scientific artists often outdistanced available fundamental knowledge to solve problems by intuition and extraordinary technique. The scientific basis for methods of organic analysis is constantly being explored and expanded, and existing methods are steadily being improved or replaced by more efficient ones.

Quantitative organic analysis may be conveniently divided into elemental and functional-group determinations. In *elemental analysis* a compound is broken down into simple measurable species by drastic treatment. An example is the determination of carbon and hydrogen in an organic compound by heating a few milligrams in a stream of oxygen until the carbon is converted to carbon dioxide and the hydrogen to water. These may then be measured instrumentally or the gas stream may be passed in sequence through a preweighed tube of magnesium perchlorate, which absorbs the water, and one of sodium hydroxide, which absorbs the carbon dioxide. From the increase in weight of the tubes the percentages of carbon and hydrogen in the sample are calculated.

In *functional-group analysis*, atoms or groups of atoms having specific reactivity are measured. Although any two atoms covalently bonded can be broadly considered a functional group, the term is generally applied more precisely to groups of atoms that undergo specific reactions, such as aldehydes($-CHO$), alkenes($-C=C-$), alcohols($-OH$), and nitro groups($-NO_2$). Both qualitative and quantitative determinations of functional groups are important; such information helps determine structure, reactivity, and purity. For example, compounds that have hydroxyl groups on adjacent carbon atoms such as glycerol or ethylene glycol may be oxidized selectively by specific reagents such as periodate, lead tetraacetate, and cerium(IV).

Owing to improved speed and convenience, applications of chemical methods of functional-group analysis have expanded steadily in recent years. Determinations of functional groups can be carried out by physical as well as chemical techniques. Physical methods, especially those entailing a variety of spectral measurements, have made possible many kinds of measurements never before envisioned. A modern chemist would be severely handicapped without access to the information provided by techniques such as spectroscopy, chromatography, and electrochemistry. Nevertheless, since physical methods require instruments, the details of operations are not so directly under the control of the analyst as in wet chemical methods. He must rely on an instrument, and as with the chemistry, unless he understands how it is intended to function, he will not be able to judge whether it is operating properly, what levels of precision and accuracy to expect, or how to prepare samples and standards to obtain optimum results.

10-7   DICHROMATE DETERMINATION OF IRON IN AN ORE

Background

In this experiment the iron in a sample of ore is determined by titration as iron(II) with a solution of potassium dichromate. The prereduction of iron(III) to iron(II) before titration is carried out with tin(II) chloride.

Iron, the second most abundant metal in the earth's crust, is the most important industrial element. Although many minerals contain iron, the important ore deposits consist of iron(III) oxide (hematite). For analysis, ore samples are first treated with concentrated hydrochloric acid. The iron oxides usually dissolve completely, leaving some insoluble residue, primarily silica. Occasional particles of the black mineral magnetite, Fe_3O_4, also will not dissolve. If an ore contains substances that could interfere in the analytical measurement (such as titanium), conditions must be adjusted so that their effects are minimized. This adjustment sometimes can be accomplished without physical separation. In this experiment tin(II) is used as a selective reductant. Titanium(IV), not being reduced by tin(II), does not interfere in the subsequent titration.

The reduction of iron(III) to iron(II) with tin(II) chloride is carried out in hot hydrochloric acid solution. The temperature should be 70 to 90°C to ensure rapid, complete reduction. In concentrated hydrochloric acid, iron(III) forms a series of chloride complexes; the disappearance of their yellow color indicates when reduction is complete. The sample is then cooled, and the slight excess of tin(II) chloride destroyed with mercury(II) chloride; this must be added rapidly, or it may be reduced to finely divided metallic mercury by local excesses of tin(II). Metallic mercury might then be oxidized by dichromate in the subsequent titration and thereby cause the results to be high.

Once the reduction of iron is complete, sulfuric and phosphoric acids are added. Sulfuric acid lowers the pH so that a high electrode potential of dichromate is maintained. Phosphoric acid serves two purposes. First, it forms a colorless phosphate complex, $FeHPO_4{}^+$, with iron(III) that prevents yellow iron(III) chloride complexes from interfering in the titration end point. Second, the formation of phosphate complexes decreases the electrode potential of the iron(III)–(II) couple, thereby shifting the iron(II)–dichromate equlilibrium to the right and ensuring more complete reaction. Also, the decreased potential of the iron couple causes the end point of the indicator, diphenylamine sulfonic acid, to be sharper and to coincide more closely with the equivalence point.

Diphenylamine sulfonic acid is colorless in its reduced form and violet in its oxidized form. The transition potential is 0.85 V. In the dichromate titration of iron the solution before the end point is green owing to the presence of chromium(III), and so the color change at the end point is from green to purple. Although the reaction between dichromate and diphenylamine sulfonic acid is slow in pure solution, iron(II) catalyzes the reaction, and the color transition is rapid and sharp. The indicator reaction is a two-electron oxidation to a vivid purple species that cannot be isolated, but slowly decomposes. This decomposition, which results in a gradual fading of the end-point color, causes no difficulty. The indicator blank is approximately 0.04 mL of dichromate titrant.

Even though the reaction between dichromate and iron(II) comprises several transient intermediate oxidation states of chromium, the overall process is rapid. The reaction mechanism is not completely understood. A glance at a table of standard electrode potentials confirms that the oxidation of iron(II) ($E^0 = 0.771$) by dichromate ($E^0 = 1.33$) should be quantitative. Chloride is not oxidized by dichromate as long as the chloride concentration does not exceed about 1 M.

In summary, the reactions for the dichromate determination of iron in an ore are

dissolution,

$$Fe_2O_3 + 6HCl \rightarrow 2FeCl_3 + 3H_2O$$

reduction,

$$2Fe^{3+} + Sn^{2+} \rightarrow 2Fe^{2+} + Sn^{4+}$$

removal of excess tin(II) after reduction,

$$Sn^{2+} + 2HgCl_2 \text{ (excess)} \rightarrow Sn^{4+} + Hg_2Cl_2(s) + 2Cl^-$$

and titration,

$$6Fe^{2+} + Cr_2O_7^{2-} + 14H^+ \rightarrow 6Fe^{3+} + 2Cr^{3+} + 7H_2O$$

Tin(II) chloride is ordinarily chosen for the prereduction step because the standard electrode potential of the tin(IV)–(II) couple ($E^0 = 0.14$) is not low enough to reduce the titanium(IV) often present in iron ores. Titanium is therefore not titrated by dichromate and consequently does not interfere.

Procedure (time 5 to 7 h)

Preparation of 0.02 M Dichromate Solution

Accurately weigh 2.6 g of reagent-grade $K_2Cr_2O_7$ in a weighing bottle. Transfer most of the salt into a 500-mL volumetric flask, and weigh the bottle plus residual salt. Add distilled water to the flask, allow the salt to dissolve, and then dilute to the mark and mix well.

Dissolution of Iron Standard

Weigh, to the nearest 0.1 mg, 0.2 to 0.25 g of electrolytic iron into 500-mL conical flasks.[6] Add 10 mL of 12 M HCl to each (in a fume hood), cover with a

---

[6]Iron wire also may serve as a standard, though it varies in quality. Since it is prone to rust, it should be polished with fine emery cloth and a clean towel before use.

small watch glass, and warm on a hot plate until dissolved. Add HCl, not exceeding 5 to 10 mL, as needed to replace that lost by evaporation.

Dissolution of Sample

Accurately weigh samples of iron ore into 500-mL conical flasks.[7] Add 10 mL of concentrated HCl to each in a fume hood, cover with a small watch glass, and warm on a hot plate until the iron oxide portion of the sample dissolves (about 1 h). When dissolved,[8] rinse the under side of the watch glass into the flask. Do not allow the samples to evaporate to dryness, or iron oxide may bake onto the glass and be difficult to redissolve. Add HCl if necessary to replace that lost by evaporation. If the sample must be stored after dissolution, cover the flask with a watch glass and place it in a fume hood.

Prereduction of Iron(III)

Prereduce the iron(III) to iron(II) as follows. Prepare a tin(II) chloride solution by dissolving about 5 g of $SnCl_2 \cdot 2H_2O$ in 35 mL of 6 M HCl. As this solution is readily air-oxidized, keep it covered, and discard it at the end of the laboratory period.

From this point carry each sample separately through the remainder of the procedure. Heat the sample to nearly boiling (fume hood). Remove from heat, and slowly add tin(II) chloride solution dropwise until the orange-yellow almost disappears. Rinse the sides of the flask with a small amount of water. Heat again to nearly boiling, and continue dropwise addition of tin(II) chloride solution until the yellow or yellow-green disappears.[9] Add 2 drops excess; any more may result in formation of metallic mercury instead of Hg_2Cl_2 in the next step. Should more than 2 drops of excess tin(II) chloride be added inadvertently, or should the completeness of iron(III) reduction be doubtful, add $KMnO_4$ solution dropwise until the yellow-green returns, and decolorize again by addition of tin(II) chloride with 2 drops excess. Cover the flask with a small watch glass, and immediately cool the solution to room temperature by holding the flask under a stream of cold water. Add 100 mL of freshly boiled and cooled water[10] and then 10 mL of saturated mercury(II) chloride solution (*Caution: Poison*). A white silky precipitate should form. If no precipitate is visible, or if a dark precipitate forms, discard the sample and take more care with a subsequent one.[11]

Allow the solution to stand 3 to 5 min.[12]

---

[7]The weight required for samples containing approximately 60% Fe_2O_3 is about 0.4 g. Consult the laboratory instructor for the range in composition of the samples provided.

[8]The iron oxide in the sample is dissolved when no dark specks of sample remain. A white or gray-white residue, which is silica, may be ignored. Dissolution may be hastened by gentle boiling with frequent swirling and by adding enough tin(II) chloride to reduce the iron(III) to iron(II). The occasional dark particles of magnetite that may be present can be ignored; these dissolve only slowly in HCl. A convenient test for magnetite is to hold a magnet to the outside wall of the flask.

[9]The color of the solution at this point will be determined by the nature and origin of the sample. In any case an abrupt change from strongly yellow to almost colorless within 1 to 2 drops of $SnCl_2$, will indicate complete reduction of the iron(III).

[10]Insertion of a stirring rod into the beaker or flask before the water is heated will reduce bumping during boiling.

[11]If no Hg_2Cl_2 precipitate forms, the amount of $SnCl_2$ added was insufficient, and the iron is probably not all reduced. A dark precipitate indicates the presence of finely divided mercury metal, the result of too·much $SnCl_2$.

[12]Quantitative reaction between tin(II) and mercury(II) requires several minutes. A long delay, however, may result in appreciable air oxidation of the iron(II).

Titration of Iron(II) with Dichromate

Add 10 mL of 3 M H_2SO_4, 20 mL of 6 M H_3PO_4, and 4 to 6 drops of sodium diphenylamine sulfonate indicator solution. Titrate with $K_2Cr_2O_7$ solution, swirling constantly and adding the titrant slowly near the end point.[13] The end point is the first permanent appearance of purple in the gray-green solution. Discard the completed titration solutions, which contain mercury, as directed by your instructor.

Calculate and report the percentage of iron in the sample.

Calculations

Calculate the molarity of the potassium dichromate solution from the weight of dichromate taken and also from the standardization against pure iron metal. The two values should agree within 2 to 3 parts in 1000. In the standardization against iron

$$\text{mol } K_2Cr_2O_7 \text{ required} = (1/6)(\text{mol Fe taken})$$

$$M\ K_2Cr_2O_7 = \frac{\text{mol } K_2Cr_2O_7}{\text{mL } K_2Cr_2O_7/(1000 \text{ mL/L})}$$

$$= \frac{\text{g Fe taken}}{(\text{g Fe/mol})(6)(\text{mL } K_2Cr_2O_7/(1000 \text{ mL/L}))}$$

Before reporting results for this determination, test your calculation procedure by working the appropriate problems at the end of the chapter.

10-8 PERMANGANIMETRIC DETERMINATION OF IRON IN AN ORE
Background

In this experiment a sample of iron ore is analyzed for iron by titration with a solution of potassium permanganate. The permanganate is prepared to an approximate concentration and then standardized against sodium oxalate.

The dissolution of iron ores and subsequent reduction of iron to the divalent oxidation state with tin(II) are discussed in the background given in Section 10-7. Two minor modifications are introduced here. One is that the reduced iron solution is diluted prior to titration so that oxidation of chloride by permanganate is minimized. The other is that manganese(II) sulfate is added along with phosphoric and sulfuric acids to accelerate the otherwise slow reaction of iron(II) with permanganate. A mixture of manganese(II) sulfate, phosphoric acid, and sulfuric acid is often called *Zimmermann-Reinhardt reagent.*

The reaction for the titration of iron with permanganate is

---

[13]To minimize air oxidation of iron(II), use only wash water that has been made oxygen-free by recent boiling and cooling, and keep the volume to a minimum. Rinse the flask walls only near the end point, using a medicine dropper.

$$5Fe^{2+} + MnO_4^- + 8H^+ \rightarrow 5Fe^{3+} + Mn^{2+} + 4H_2O$$

The primary reaction in the standardization of the permanganate titrant is

$$2MnO_4^- + 5H_2C_2O_4 + 6H^+ \rightarrow 2Mn^{2+} + 10CO_2 + 8H_2O$$

The overall reaction involves several transient intermediate oxidation states of manganese. The reaction mechanism, like that of the dichromate–chromium(III) system, is not wholly understood.

Procedure (time 5 to 7 h)

Preparation and Standardization of 0.02 M KMnO₄ Solution

Weigh about 3.0 g of $KMnO_4$ on a triple-beam balance, and dissolve it in about 1 L of distilled water. Store the solution in a glass bottle at least a day to allow the permanganate to oxidize any organic matter that may be present. Filter the aged solution through a large sintered-glass filter, using a suction-flask assembly.[14] Store in a glass bottle previously rinsed with a small portion of the filtered solution.

Dry 1.0 to 1.5 g of $Na_2C_2O_4$ for 1 to 2 h at 110 °C . Weigh, to the nearest 0.1 mg, samples of 0.22 to 0.28 g into 200-mL conical flasks. Dissolve each sample in 50 mL of water and add 30 mL of 3 M H_2SO_4. Calculate the approximate amount of $KMnO_4$ solution needed to react with a sample of sodium oxalate (35 to 45 mL). With stirring add about 90% of the calculated amount to the oxalate-sulfuric acid solution. Allow the solution to stand at room temperature until it turns colorless, a process that should take no more than 5 min. Warm the solution to 50 to 60°C. (At this temperature the flask will feel hot but not uncomfortable to hold.) Titrate slowly until the pink of excess permanganate persists.

Calculate the molarity of the permanganate.

Dissolution and Prereduction of Sample

Weigh, dissolve, and prereduce a set of iron ore samples as directed in Section 10-7, "Dissolution of Sample." Titrate each sample immediately after prereduction.

Titration of Iron(II) with Permanganate

Dilute the solution that has been reduced with tin(II) to 250 mL with freshly boiled and cooled water, add 25 mL of Zimmermann-Reinhardt solution, and titrate immediately with standard $KMnO_4$ to a faint pink.

Determine a blank by measuring 10 mL of 6 M HCl and 2 drops of $SnCl_2$ solution into a 500-mL flask. Add 10 mL of saturated $HgCl_2$ solution, dilute to 250 mL with oxygen-free water, add 25 mL of Zimmermann-Reinhardt solution, and titrate as for the sample.

Report the percentage of iron in the sample.

---

[14]The glass filter may be cleaned by rinsing, first with HCl and then several times with water.

Calculations

To calculate the molarity of the permanganate, the following may be used as a guide. At the equivalence point in oxalate titrations

$$\text{mol KMnO}_4 = \text{mol Na}_2\text{C}_2\text{O}_4(2/5)$$

The term (2/5) is derived from the balanced oxalate-permanganate chemical reaction, in which one mole of permanganate reacts with (5/2) moles of oxalate:

$$\text{mol Na}_2\text{C}_2\text{O}_4 = \frac{|\text{g Na}_2\text{C}_2\text{O}_4 \text{ taken}}{\text{g Na}_2\text{C}_2\text{O}_4/\text{mol}}$$

The overall expression for molarity, then, is

$$M \text{ KMnO}_4 = \frac{\text{mol KMnO}_4}{\text{mL KMnO}_4/(1000 \text{ mL/L})}$$

$$= \frac{(\text{g Na}_2\text{C}_2\text{O}_4)(2)}{(\text{g Na}_2\text{C}_2\text{O}_4/\text{mol})(5)(\text{mL KMnO}_4/(1000 \text{ mL/L}))}$$

At the equivalence point in iron(II) titrations, five moles of iron(II) is equivalent to one mole of permanganate.

10–9 IODIMETRIC DETERMINATION OF COPPER IN BRASS

Background

In this experiment the copper in a brass sample is determined by a method in which iodine is liberated. The sample is dissolved in nitric acid, and the solution boiled to remove most of the nitrogen oxides formed during oxidation of the metal. The residual nitrogen oxides, which must be entirely removed to prevent iodide oxidation, are eliminated by the addition of urea.

Iron, present in most brasses, may also cause oxidation of iodide. This interference is eliminated by the addition of fluoride, which forms a stable complex with iron(III). Ammonium acid fluoride also adjusts the pH to 3.5 to 4.5. Next an excess of potassium iodide is added, and the iodine formed titrated with sodium thiosulfate.

For this experiment copper metal is selected as the primary standard because it is the material being determined and is readily available in primary-standard quality as electrical wire.

The principal reactions involved in the analysis of copper are

dissolution of sample in dilute nitric acid,

$$3\text{Cu} + 8\text{HNO}_3 \rightarrow 3\text{Cu}^{2+} + 2\text{NO} + 4\text{H}_2\text{O} + 6\text{NO}_3^-$$

$$\text{Cu} + 4\text{HNO}_3 \rightarrow \text{Cu}^{2+} + 2\text{NO}_2 + 2\text{H}_2\text{O} + 2\text{NO}_3^-$$

removal of residual nitrogen oxides by addition of urea,

$$2NO + O_2 \rightleftarrows 2NO_2$$

$$2NO_2 + H_2O \rightarrow HNO_2 + HNO_3$$

$$2HNO_2 + O{=}C(NH_2)_2 \rightarrow 2N_2 + CO_2 + 3H_2O$$

neutralization of remaining nitric acid with sodium hydroxide, followed by pH adjustment and complexation of iron(III) with fluoride,

$$Fe^{3+} + F^- \rightleftarrows FeF^{2+}$$

addition of excess potassium iodide,

$$2Cu^{2+} + 4I^- \rightarrow Cu_2I_2(s) + I_2$$

titration of iodine with thiosulfate,

$$I_2 + 2S_2O_3{}^{2-} \rightarrow 2I^- + S_4O_6{}^{2-}$$

Precautions must be taken to avoid side reactions, as explained in the discussion of sodium thiosulfate in Section 10-4.

Procedure (time 4 to 5 h)

Preparation of Starch Indicator Solution

Mix 1 g of soluble starch with enough cool water to produce a thin paste.[15] (Dry starch does not dissolve readily in boiling water.) Add the suspension to 100 mL of boiling water, and boil the solution for about 2 min. Solutions of starch begin to decompose and give unsatisfactory end points after a day or two; preparation of a fresh solution each day is recommended.

Standardization of 0.1 M Na₂S₂O₃ Solution

Dissolve 25 g of $Na_2S_2O_3 \cdot 5H_2O$ and 0.1 g of Na_2CO_3 in a liter of distilled water. Add a drop of chloroform.

Weigh accurately 0.2-g samples of clean copper wire into 200-mL conical flasks.[16] Add 10 mL of 6 M HNO_3 to each in a fume hood, and cover with a watch glass. When dissolution is complete,[17] remove the cover and boil the solution in a fume hood for 2 to 3 min to remove most of the nitrogen oxides.[18]

---

[15]Stable commercial preparations, ready for use, are available from chemical suppliers. These work well and eliminate the need for frequent preparation of starch solutions.

[16]Trace impurities markedly increase the resistance of copper wire. For electrical use they are removed by electrolytic refining to a level well below a part per thousand, and therefore copper electrical wire is an excellent primary standard. The surface can be cleaned if necessary by polishing with clean toweling.

[17]Replace evaporation losses with water if necessary.

[18]Insufficient removal of nitrogen oxides in this step is a major source of error in this experiment.

Add 10 mL of water and 5 mL of 4% urea solution, and boil again for about 1 min. Cool.

When ready to titrate, add about 30 mL of water, and then add 2.5 M NaOH with swirling until a slight permanent precipitate of $Cu(OH)_2$ appears. This will require typically about 15 to 20 mL of NaOH, according to the amount of HNO_3 present. Swirl gently to avoid depositing $Cu(OH)_2$ precipitate on the upper walls of the flask. Add 1 to 2 g of ammonium acid fluoride, NH_4HF_2, and swirl until all precipitate is dissolved.[19]Tilt and roll the flask to dissolve particles of $Cu(OH)_2$ adhering to the walls. Cool, add 3 g of KI, and titrate immediately to near the end point with $Na_2S_2O_3$. When the solution becomes pale yellow or buff, add 5 mL of fresh starch solution and titrate to the first permanent disappearance of blue.[20]

Procedure for Sample

Weigh 0.2-g samples of brass into 200-mL conical flasks. Dissolve in HNO_3 and boil 2 to 3 min.[21] Add 10 mL of water and 5 mL of 4% urea solution, and boil again as directed for the standards. Cool. When ready to titrate, neutralize with sodium hydroxide and dilute to 50 mL with water. Add 1 g of ammonium acid fluoride and swirl until dissolved. Add 3 g of KI, and complete the determination as in the standardization. Remember to use alternation of standards and samples.

Calculate the molarity of the $Na_2S_2O_3$ solution; one mole of copper requires one mole of thiosulfate for titration. Calculate and report the percentage of copper in the sample.

10–10   ORGANIC FUNCTIONAL-GROUP ANALYSIS :
DETERMINATION   OF ETHYLENE GLYCOL BY PERIODATE
CLEAVAGE

Background

This experiment illustrates functional-group analysis by a chemical method. An organic compound, ethylene glycol, is determined by oxidative cleavage with potassium periodate. Ethylene glycol reacts with periodate (added as periodic acid) in a 1:1 mole ratio to yield formaldehyde and iodate:

$$\begin{array}{c} H_2COH \\ | \\ H_2COH \end{array} + IO_4^- \rightleftharpoons H_2O + \begin{array}{c} H_2C-O \\ | \quad\quad IO_3^- \\ H_2C-O \end{array} \longrightarrow 2H_2CO + IO_3^-$$

Under optimum conditions this reaction goes to completion, while side reactions are negligible ($< 0.1\%$).

The rate of glycol reaction, as with many organic reactions, is sluggish but increases with periodate concentration. Thus the reaction can be driven to completion in a reasonable time by an excess of periodate. When the total

---

[19]Ammonium acid fluoride serves as both a means of preventing iron interference and a means of adjusting the pH to the optimum range of 2 to 5.

[20]To avoid etching by HF, empty and rinse the flasks immediately after completing each titration.

[21]If the brass sample contains tin, a white precipitate of metastannic acid, $SnO_2 \cdot xH_2O$, will remain; it will not affect the titration and can be ignored.

amount of periodate added is known, and if the excess is treated in turn with a known excess of standard arsenite, titration of the remaining arsenite with standard iodine solution will give the amount of periodate left after all the glycol has reacted. The reactions are

$$IO_4^- + H_2AsO_3^- \text{ (known excess)} \longrightarrow IO_3^- + H_2AsO_4^-$$

$$H_2AsO_3^- + I_2 + H_2O \longrightarrow H_2AsO_4^- + 2I^- + 2H^+$$

The first part of the reaction of periodate with glycol to form a 5-membered ring is reversible. Equilibrium is attained rapidly because the rates of the forward and back reactions are faster than the irreversible reaction that follows. The equilibrium constant being only about 18 at pH 1 and about 1000 at pH 8, the reversible step is not of direct analytical use. Although the constant is larger at high pH values, even one on the order of 1000 would seldom be analytically useful, and the reaction at pH 1 becomes so only because the second part of the reaction is irreversible. As the complex breaks down, more of the intermediate is formed, until all the glycol is oxidized to formaldehyde.

The rate of decomposition of the 5-membered periodate-ring intermediate is the crucial factor in timing the experimental operations. At pH 1 and room temperature the time required for half the intermediate to decompose (the half-time) is about 1 min. At pH 8 it is several hours. Thus the kinetics of the reaction is such that, for the experiment to be completed within a reasonable time, a low pH is necessary even though the equilibrium constant for the initial reversible step of the reaction is unfavorable in acid solution. In 10 half-times (about 10 min at pH 1) the overall reaction is quantitative (99.9% complete). In the overall process, then, a slow, irreversible second step outweighs the effect of an unfavorable equilibrium constant for the first step.

Two side reactions may affect the analytical results: periodate may react further with formaldehyde, or it may decompose spontaneously. If the rate of a side reaction is 1000 times or more smaller than that of the desired reaction, it should not cause significant error in the analytical procedure. In this analysis enough time is allowed for the main reaction to go to completion (10 to 15 min). The reaction is next stopped, or *quenched*, by raising the pH with bicarbonate and then adding excess arsenite to react with the remaining periodate. Since the spontaneous decomposition of periodate (possibly photodecomposition) proceeds rapidly in bicarbonate solutions, arsenite must be added without delay once the pH has been raised. Because of this decomposition reaction the excess periodate cannot be titrated directly with arsenite, and so an indirect procedure must be used — addition of excess arsenite followed by titration with iodine.

This experiment also illustrates the method of back titration of excess standard reagent, a convenient technique for slow reactions. Several precise volumetric measurements are required; carelessness in any one will invalidate the results.

All experimental work other than the preparation of some of the solutions must be completed in a single session. The technique of alternation is recommended for the preparation and quenching steps as well as for the titrations. Because of the time required for the numerous volumetric measurements, it is suggested that this analysis be performed only in triplicate. Careful planning is essential.

Procedure (time 4 to 6 h)

Preparation of Standard 0.14 M Arsenite Solution from As₂O₃ (As₄O₆)

Weigh, to the nearest 0.1 mg, about 3.6 g of dry primary-standard As_2O_3 into a dry beaker, add a few drops of 2.5 M NaOH, and stir to a smooth paste. Then add 50 mL of 2.5 M NaOH and stir again.[22] When the As_2O_3 is dissolved, transfer the solution quantitatively to a 250-mL volumetric flask, and half-fill the flask with water. Add 15 mL of 6 M HCl, swirl, and then add 8 g of $NaHCO_3$.[23] Dilute to volume. (*Caution:* Arsenic compounds are poisonous.)

Preparation of 0.06 M Iodine Solution

Using a triple-beam balance, weigh into a glass vessel about 8.4 g of I_2 and 17 g of KI. Transfer into a 1-L storage bottle.[24] Add about 15 mL of water, and allow at least 30 min for the I_2 to dissolve, shaking occasionally.[25] When it is completely dissolved, dilute to 500 mL with water. Prepare starch indicator solution as directed in Section 10-9.

Preparation of Periodic Acid Solution

Using a triple-beam balance, weigh into a flask about 5.2 g of H_5IO_6. Dissolve it in 100 mL of water.[26] Prepare the solution on the same day it is to be used. Mix thoroughly, and keep covered to minimize evaporation during use.

---

[22] As_2O_3 is only slowly soluble in neutral or acidic solutions. The solid is not easily wetted; a thin paste, prepared initially with NaOH, speeds dissolution appreciably. The reaction is
$$As_2O_3 + 4OH^- \rightarrow 2HAsO_3^{2-} + H_2O$$
Upon addition of acid the predominate species becomes $H_2AsO_3^-$. Note that two moles of arsenite are formed from each mole of As_2O_3. The molecular unit for arsenic(III) oxide is As_4O_6; this formula is sometimes recommended in preference to As_2O_3. The latter, however, is widely used and causes no difficulty in analytical work where the first step is dissolution of the compound. Arsenic(III) in solution exists in several protonated forms, including $H_2AsO_3^-$, which we use here.

[23] $NaHCO_3$ is added to adjust the solution to pH 7 to 9. If too much or too little acid should be added, the amount of $NaHCO_3$ added here will not furnish the correct pH, and the stability of the arsenite solution will be impaired.

[24] I_2 is both volatile and corrosive. Use of the triple-beam or other rough balance, not an analytical balance, is essential. Always weigh I_2 in a beaker rather than on weighing paper, and clean up any spillage immediately.

[25] The solubility of I_2 in water, normally only about 0.3 g/L, is greatly increased if iodide is in solution to form triiodide ion, I_3^-. Although I_3^- is the major species in solution, we use I_2 in the equations for clarity. Even with iodide present dissolution is slow, and if the iodide concentration is reduced by dilution, the rate decreases still further. The intense color of iodine makes it difficult to tell when the last particles have dissolved.

[26] Periodic acid is a strong acid, and the anion is a strong oxidant. Therefore, periodic acid and periodates, like iodine, should not be handled near an analytical balance or brought in contact with reactive materials such as paper.

Analysis (time 2½ h)

Obtain a glycol sample. Dilute to volume in a 100-mL volumetric flask with distilled water, and mix.

1. Assemble nine clean 250-mL conical flasks. Label three as standards, three as blanks, and three as samples.

2. Into each of the flasks labeled "sample," pipet 10 mL of glycol sample solution.

3. Into each of the flasks labeled "blank," measure about 10 mL of distilled water.

4. Into each of the flasks labeled "standard," measure about 20 mL of distilled water.

5. At 2-min intervals, in alternation, pipet 10 mL of H_5IO_6 solution into the three sample flasks and the three blank flasks.

6. To one of the standard flasks add 3 g of $NaHCO_3$, swirl to partially dissolve the $NaHCO_3$, and immediately pipet 20 mL of arsenite solution. Repeat for the other two standards.

7. Ten to fifteen minutes after the beginning of Step 5 start the quenching step. Quench in alternation — first blank, first sample, second blank, second sample, and so on; that is, at 2-min intervals fill the 20-mL pipet with standard arsenite solution, then add 3 g of $NaHCO_3$ to each flask, swirl briefly, and follow immediately with the 20 mL of arsenite.[27]

8. Add 5 mL of starch indicator, and titrate[28] with I_2 solution to the first blue color that persists after swirling for 10 to 20 s.[29,30] Titrate in the following order: first standard, first blank, first sample, second standard, and so on.

Report the weight of ethylene glycol in the entire sample.

Calculations

The molarity of the I_2 solution is calculated from the standardization titrations and should be about 0.06 to 0.07 M. The weight of ethylene glycol in grams is calculated from the expression

$$\text{g glycol} = \frac{(V_{\text{sample}} - V_{\text{blank}})(M_{I_2})\,(\text{g glycol/mol})(100\text{ mL})}{(1000)(V_{10})}$$

where V_{sample} is the milliliters of I_2 solution required to titrate the sample, V_{blank} the milliliters of I_2 solution required to titrate the blank,[31] and V_{10} the

---

[27]The timing here is critical. Once the $NaHCO_3$ has been added, periodate begins to decompose rapidly. Add the arsenite solution as quickly as possible after the bicarbonate has been mixed with the flask contents. After the addition of arsenite, the solutions are relatively stable and, if necessary, can stand several hours before titration.

[28]Read the top of the meniscus.

[29]I_2 is somewhat volatile, even from aqueous solution. Errors from this source can be minimized by placing an inverted test tube on the buret and using alternation.

[30]The titration volumes should be in the following ranges: blank, 5 to 10 mL; standard, 40 to 45 mL; sample, 20 to 35 mL.

[31]Note that the blank in this analysis is not the same as an indicator blank in conventional direct titrations. It refers rather to a solution that contains all reagents except glycol and establishes the concentration of the H_5IO_6. V_{blank} should therefore not be subtracted from the volume of titrant used in the standardization.

calibrated volume of the pipet. In the calculations use values of Sample 1, Blank 1, and Standard 1 together; then Sample 2, Blank 2, and Standard 2; and so forth. If one of the titration results is rejected on the basis of the Q test (Section 1-5), substitute for it the value obtained nearest in time.

PROBLEMS

10-1.   The electrode potentials at equilibrium for each of the two half-reactions involved in an oxidation-reduction titration must always be identical. Explain by means of an example why this is so.

10-2.   Why is it important that the standard electrode potentials of the two half-reactions in a titration differ widely? When is it possible to determine two reductants in a solution with a standard solution of a single oxidant?

10-3.   Calculate the electrode potential of a solution that is 30% to the equivalence point in a titration of 0.1 M iron(II) with standard 0.1 M cerium(IV) in 1 M sulfuric acid.

10-4.   Which of the following pairs contains substances that would be expected to react with each other to an appreciable extent: (a) AsO_3^{3-} and $Fe(CN)_6^{4-}$; (b) AsO_3^{3-} and $Fe(CN)_6^{3-}$; (c) AsO_4^{3-} and $Fe(CN)_6^{4-}$; (d) AsO_4^{3-} and $Fe(CN)_6^{3-}$?

10-5.   Calculate the equilibrium constant for the reaction $Ce^{4+} + V^{2+} \rightleftarrows Ce^{3+} + V^{3+}$.

10-6.   Calculate the potential relative to the standard hydrogen electrode in the titration of 30 mL of 0.05 M V(II) after the addition of 0, 5, 15, 25, 29, 30, 31, and 40 mL of 0.05 M Ce(IV) solution. Plot the titration curve.

10-7.   Calculate the potential at the equivalence point in the preceding problem, and select an appropriate indicator for this titration. Also calculate the concentration of unreacted V(II) at the equivalence point.

10-8.   The reaction of cerium(IV) with iron(II) is commonly used as a straightforward example of a redox titration. Calculations for other pairs of half-reactions may be more complex for several reasons. List these reasons.

10-9.   What is the effect of the decrease in pH on the strength of permanganate as an oxidant; of iodate as an oxidant; of tin(II) as a reductant?

10-10.   Calculate the potential at the equivalence point in the titration of (a) iron(II) with bromine; (b) iron(III) with tin(II); (c) cerium(IV) with ferrocyanide. Select suitable indicators for these titrations.

10-11.†   Explain the theoretical feasibility of titrating a solution of H_3AsO_4 with a solution of iron(II) sulfate in a buffer of pH 3.

10-12.†   Calculate the equilibrium constant for the reaction $Sn^{4+} + 2I^- \rightleftarrows Sn^{2+} + I_2$.

10-13.†   If 20 mL of 0.1 M tin(II) chloride solution is added to 20 mL of 0.1 M iron(III) chloride solution, what will be the iron(III) concentration after reaction is complete?

10-14.   Calculate the change in potential from 0.1% before to 0.1% after the equivalence point in a titration of (a) iron(II) with cerium(IV), (b) iron(II) with iodine, (c) iron(II) with permanganate at a pH of 1.

10-15.   What is meant by a conditional equilibrium constant for a redox reaction? Give an example.

10-16.   Calculate the conditional equilibrium constant for the reaction
$$H_3AsO_3 + H_2O + I_2 \rightleftarrows H_3AsO_4 + 2H^+ + 2I^-$$
at (a) pH 2 and (b) pH 8.

10-17.   How does a redox indicator work? At what potential should an indicator change if the end point and the equivalence point are to coincide?

10-18.   How does starch function as an indicator in the iodimetric determination of copper?

10-19.   What is the transition interval for a redox indicator? How is a redox indicator chosen?

10-20.   Which oxidant used for preliminary oxidation has the most positive electrode potential? Which reductant used for preliminary reductions has the most negative electrode potential? How can selective prereductions be carried out?

10-21.   Why does the presence of phosphoric acid increase the strength of dichromate as an oxidant?

10-22.   What is chemical oxygen demand?

10-23.   What is the effect on the electrode potential of a solution containing iron(II) and iron(III) if a complexing agent is added that (a) complexes iron(II) selectively, (b) complexes iron(III) selectively, (c) complexes both iron(II) and (III) to the same extent?

10-24.   Why is sodium thiosulfate unsuitable as a primary standard? How are solutions of it standardized?

10-25.   Titanium(III) is oxidized to titanium(IV) by permanganate. What percentage of titanium in a 2.468-g sample would require 43.96 mL of 0.02317 M permanganate for titration?

10-26.   Butyl hydroperoxide (C_4H_9OOH) is determined by reaction with excess potassium iodide, followed by titration of the iodine formed with sodium thiosulfate:
$$C_4H_9OOH + 2I^- + 2H^+ \rightarrow C_4H_9OH + I_2 + H_2O$$
$$2S_2O_3^{2-} + I_2 \rightarrow S_4O_6^{2-} + 2I^-$$
A 0.6300-g sample containing butyl hydroperoxide required 36.40 mL of 0.1000 M sodium thiosulfate for titration. Calculate the percentage of butyl hydroperoxide in the sample.

10-27.   The molar solubility of iodine in water and dilute salt solutions is 1.34×10^{-3} M. Using an equilibrium constant of 710 for the equilibrium $I_2 + I^- \rightleftharpoons I_3^-$, calculate the total solubility of iodine ($[I_2]$ plus $[I_3^-]$) in 0.100 M potassium iodide.

10-28.† In a solution 0.01 M in KI, what concentration of I_2 is required to give a detectable blue with starch? Assume the color can be seen when the I_3^- concentration reaches 10^{-5} M.

10-29.   A solution of sodium thiosulfate was standardized by titration of the iodine formed when excess acid and potassium iodide were added to a 0.2500-g sample of pure potassium dichromate. If 41.02 mL of solution was required, what was the molarity of the thiosulfate?

10-30.   What is meant by oxygen error? Give an example.

10-31.   List several primary standards that can be used to standardize a solution of iodine.

10-32.   Is it possible to determine iron(III) iodimetrically? If so, write balanced equations for the reactions and suggest a means of end-point detection.

10-33.   What are the advantages of potassium bromate as a standard oxidant?

10-34.   Solutions of potassium iodate may be employed in place of the less stable reagent iodine. On addition of acid and excess iodide, iodine is formed quantitatively according to
$$IO_3^- + 5I^- + 6H^+ \rightleftharpoons 3I_2 + 3H_2O$$
If a 1.5447-g sample containing arsenic(III) requires 31.89 mL of 0.1121 M potassium iodate solution for titration of the arsenic(III) to arsenic(V), what is the percentage of arsenic in the sample?

10-35.   For the determination of cerium by reduction with standard ferrocyanide, 500 mL of 0.1 M ferrocyanide solution is required. How much solid $K_4FeCN_6 \cdot 3H_2O$ must be weighed?

10-36.   What is meant by elemental analysis; by functional-group analysis? How is functional-group analysis carried out?

10-37.   The tin from 0.3772 g of a sample of a metal alloy was dissolved in hydrochloric acid to give a solution of tin(II) chloride, which was titrated with 42.36 mL of 0.1007 M iron(III) chloride solution. Calculate the percentage of tin in the alloy.

10-38.   A 1.463-g sample containing octyl mercaptan ($C_8H_{17}SH$) was titrated with 26.24 mL of 0.03642 M iodine solution. The reaction is
$$2RSH + I_2 \rightarrow RSSR + 2HI$$
Calculate the percentage of octyl mercaptan in the sample.

10-39.   In the dichromate titration of iron, what is the purpose of adding (a) tin(II) chloride, (b) phosphoric acid, (c) mercury(II) chloride, and (d) sulfuric acid?

10-40.   A dichromate solution was prepared by weighing 2.488 g of potassium dichromate into a 500-mL volumetric flask, dissolving it, and diluting the solution to 500.0 mL with water. What was the molarity of the solution?

10-41.   A solution of potassium dichromate was standardized against iron metal by the procedure of Section 10-7. If 0.2317 g of iron required 38.72 mL, what was the molarity of the dichromate solution?

10-42.   A 0.5069-g sample of iron ore, analyzed by the procedure of Section 10-7, required 43.36 mL of 0.01711 M dichromate for titration. What was the percentage of iron in the ore?

10-43.   The iron in a 0.3412-g sample of iron ore, determined by the procedure of Section 10-8, required 38.62 mL of potassium permanganate solution. A 0.2611-g sample of sodium oxalate required 41.68 mL of the same titrant. What was the percentage of iron in the sample?

10-44.   A 0.5359-g sample of iron ore was dissolved, reduced, and titrated with 43.01 mL of a standard solution containing the equivalent of 2.054 g of pure potassium permanganate in 500.0 mL. What was the percentage of iron in the sample?

10-45.†   By mistake an analyst standardized a potassium permanganate solution with sodium oxalate, thinking it was oxalic acid dihydrate. On the basis of this standardization the iron content of an ore sample was reported to be 57.45% iron(III) oxide . What was the correct percentage?

10-46.   In the iodimetric determination of copper, why is urea added? Why do the nitrogen oxides need to be removed? Why is potassium iodide used in the determination? Give two reasons why ammonium hydrogen fluoride is added.

10-47.   A 0.4121-g sample of brass was dissolved, and after addition of excess iodide the copper present was titrated with 0.1073 M thiosulfate. If 38.40 mL of thiosulfate was required to the starch end point, what was the percentage of copper in the sample?

10-48.   A 1.620-g sample of copper ore, analyzed for copper by the method of Section 10-9, required 43.50 mL of 0.1080 M thiosulfate for titration. What was the copper content of the sample?

10-49.   In the periodate determination of glycol explain the term "quenching." Why must the arsenite solution be pipetted immediately after the addition of sodium bicarbonate?

10-50.   What is the purpose of the blank in the determination of ethylene glycol by oxidative cleavage? Is this a blank in the normally defined sense?

10-51. In the oxidation of ethylene glycol with periodate, why is a direct titration with periodate not carried out?

10-52. An arsenite solution was prepared by dissolving 2.473 g of arsenic trioxide and diluting the solution to 250.0 mL. A 25.00-mL portion required 40.17 mL of iodine solution. Calculate the molarity of the iodine solution.

10-53. Analysis of a sample of ethylene glycol by the procedure of Section 10-10 furnished the following data: blank titration, 11.76 mL; standard titration, 46.26 mL; sample titration, 32.69 mL. If the weight of As_2O_3 taken was 3.6019 g and the calibrated volume of the 10-mL pipet was 9.988 mL, what was the weight of glycol in the sample?

10-54. For the determination of ethylene glycol by the procedure of Section 10-10, 250 mL of standard 0.14 M Na_3AsO_3 solution is required. The storeroom provides primary-standard As_2O_3. How much of this solid will you need?

10-55. Clearly distinguish between the concepts of potential at equilibrium and at the equivalence point.

10-56. Write the equations for the reduction of iron(III) in (a) a Jones reductor, (b) a Walden reductor.

REFERENCES

E. Bishop, Ed., *Indicators*, Pergamon Press, New York, 1972, Chapter 8. A thorough background of the history of redox indicators and their mechanism of action and applicability.

N. D. Cheronis and T. S. Ma, *Organic Functional Group Analysis,* Wiley Interscience, New York, 1964.

F. E. Critchfield, *Organic Functional Group Analysis*, Pergamon Press, Elmsford, New York, 1962.

R. D. Guthrie in *Methods of Carbohydrate Chemistry,* Vol. I, *Analysis and Properties of Sugars,* R. L. Whistler and M. L. Wolfrom, Eds., Pergamon Press, Elmsford, N.Y., 1962, p 432. Detailed treatment of methods for cleaving hydroxyl groups on adjacent carbon atoms.

H. A. Laitinen and W. E. Harris, *Chemical Analysis*, 2nd ed., McGraw-Hill, New York, 1974. Chapter 15 describes experimental redox titration curves, the mechanism of action of some redox indicators, and catalyzed and induced reactions in redox chemistry. Chapter 16 describes reagents for preoxidation or prereduction. Chapters 17–20 discuss the characteristics and applicability of the important oxidants and reductants.

F. Margeuritte, *Compt. Rend.* **1846,** *22,* 587. First report of the use of permanganate for the determination of iron.

G. H. Schenk, *J. Chem. Educ.* **1962,** *39,* 32. Description of the procedure for the determination of glycol by periodate cleavage.

A. A. Schilt, *Perchloric Acid and Perchlorates*, G. F. Smith Chemical Co., Columbus, Ohio, 1979. Discussion of the analytical use of perchloric acid as an oxidant and of precautions necessary.

S. Siggia, *Quantitative Organic Analysis via Functional Groups*, 4th ed., Wiley, New York, 1978.

The references by Cheronis, Ma, Guthrie, and Siggia give extensive background to the important subject of organic analysis.

11

ELECTROANALYTICAL METHODS: POTENTIOMETRY AND SELECTIVE ELECTRODES

The Stockholm Commissions of the I.U.P.A.C. in 1953 distinguished between the ambivalent half-cell emf and the sign-invariant electrode potential. ... The fundamental properties of the electrode potential, notably that of sign invariance, have been ignored or misunderstood in many texts.

A. J. deBethune

Electroanalytical chemistry may be defined as the application of electrochemistry to analytical chemistry. Electroanalytical measurements encompass the four fundamental properties of potential E, current i, resistance R, and time, t. Many electroanalytical methods have been developed; the principal ones are listed in Table 11-1.

Electroanalytical methods are sensitive and selective, and, in forms such as electrochemical titrimetry, are highly precise and accurate. They are admirably suited to automatic, repetitive measurements, to closed-loop control, and to remote measurements.

A thorough understanding of the term potential is essential for this and the next chapter. Review is recommended of the definition (Section 9-1), the

Table 11-1. Principal Electroanalytical Methods

Measurement(s)	Method(s)
Potential (current = 0)	Potentiometry
Current (potential controlled)	Polarography, voltammetry
Resistance	Conductometry
Potential and volume (current = 0)	Potentiometric titration
Current and volume (potential controlled)	Amperometric titration
Resistance and volume	Conductometric titration
Potential and time (current $\neq$ 0)	Chronopotentiometry
Current and time	Stripping analysis, coulometry, electrodeposition

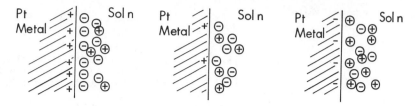

Figure 11-1. Schematic diagram of the interface between a platinum metal electrode and solutions in which platinum has gained, from left to right, a high positive potential (positive charge), a lower positive potential, and a negative potential (negative charge).

primary standard of potential (9-3), and the calculation of potential values by use of the Nernst equation (9-3). Briefly, the difference in potential between two points may be defined as the amount of electrical work needed to move a unit of positive charge from one point to another. When the charge is measured in coulombs and the electrical work in joules, the potential difference is measured in volts. A region of higher potential has a higher density of positive charge and is deficient in electrons.

11–1 POTENTIALS AT INTERFACES

An electrical potential develops at an *interface*, *boundary*, or *junction*, between phases, that is, between a solid and a liquid, between two liquids, between a liquid and a gas, and so on. A potential develops also when two solutions of different composition are in contact with each other. Figure 11-1 illustrates the development of potential between a platinum metal electrode (a chemically inert conductor), and a solution under three sets of conditions.

The measurement of potential is of interest because it is a convenient means of revealing the position of equilibrium in a system such as depicted in Figure 11-1. For example, a platinum electrode dipped into a solution 0.10 M in iron(III) and 0.01 M in iron(II) will almost instantaneously attain equilibrium with the solution. Recall from Section 10-2 that equilibrium in a solution is characterized by a stable composition and, for reversible systems, by the opposite reactions in the chemical equation proceeding at equal rates. In this example iron(III) reduction to iron(II) and iron(II) oxidation to iron(III) are both occurring at equal rates:

$$Fe^{3+} + e \rightarrow Fe^{2+} \qquad \text{and} \qquad Fe^{2+} \rightarrow Fe^{3+} + e$$

As a result no net reaction occurs either in solution or at the metal electrode; the electrode acquires a positive charge because the potential for this redox conjugate pair, given by $E = 0.771 - 0.059 \log (0.01/0.10)$, is 0.83 V more positive than the standard hydrogen electrode (Figure 11-2). In potentiometric analysis the purpose is to measure experimentally this

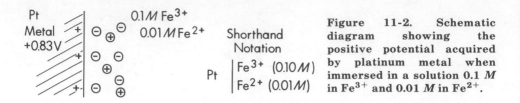

Figure 11-2. Schematic diagram showing the positive potential acquired by platinum metal when immersed in a solution 0.1 M in Fe^{3+} and 0.01 M in Fe^{2+}.

potential [1] by means of a pair of electrodes immersed in the solution and to determine how its value varies with changing composition of the solution. Potentials of analytical interest include those between a metal electrode and a solution and those generated at other interfaces.

The position of equilibrium as signified by the potential of an electrode in contact with a solution containing iron(III) and iron(II) ions is exactly the same whether the equilibrium is written as a reduction of iron(III) or as an oxidation of iron(II):

$$Fe^{3+} + e \rightleftarrows Fe^{2+}$$

$$Fe^{2+} \rightleftarrows Fe^{3+} + e$$

To depict electrochemical half-cells, as in Figure 9-4 and Figure 9-13, or combinations of half-cells, a shorthand convention may be employed to denote phases and phase boundaries. A single vertical line represents a boundary between an electrode phase and a solution phase, or between two different solution phases, for which a potential difference is to be taken into account. A double vertical line represents a boundary at which the potential difference is ignored or is considered to be reduced to zero by an appropriate salt bridge.

EXAMPLE 11–1

The half-cell in Figure 11-2 is represented by

$$Pt \,|\, Fe^{3+} \ (0.1 \ M), \ Fe^{2+} \ (0.01 \ M)$$

A half-cell involving the silver–silver ion couple is represented by $Ag \,|\, Ag^+$. A cell consisting of the two half-cells shown in Figure 9-2 is represented by

$$Ag \,|\, AgNO_3 \ (0.15 \ M) \,|\,|\, Fe^{3+} \ (0.015 \ M), \ Fe^{2+} \ (0.10 \ M) \,|\, Pt$$

For Figure 9-3 the representation is

---

[1]In connection with potential measurements, terms such as *electrode potential, reduction potential, oxidation potential, and oxidation-reduction potential* are seen. In all such terms the key word is *potential*, and each should be interpreted with identical sign, size, and meaning. When, on the other hand, terms such as half-cell emf, electrode emf, and so on are employed, the direction in which the half-reaction is written influences the sign. To minimize misunderstanding, we studiously avoid use of the term emf. Our concern is primarily with the analytical application of potentials and little with the use of electrochemical cells to generate power.

$$Pt \mid H_2(g)(1 \text{ atm}) \mid HCl \ (0.01 \ M), AgCl(s) \mid Ag$$

Since this cell contains no salt bridge, no double vertical lines are needed.

For a cell in which reaction occurs through application of an external voltage, the electrode at which oxidation occurs has the more positive potential and is called the *anode*; the electrode at which reduction occurs has the less positive potential and is called the *cathode*.

11–2 MEASUREMENT OF POTENTIAL

In the practical measurement of potential several aspects must be considered. (1) We are interested in the potentials of electrodes primarily in solutions at equilibrium. The presence of a half-reaction in a table of standard electrode potentials does not necessarily mean that an electrode is available whose potential will respond readily to changes in activities of the species. The dichromate–chromium(III) half-reaction is one for which useful potential measurements are difficult to obtain. (2) The potential of a single electrode, that is, the potential of a single half-cell, cannot be measured; a second electrode is necessary to complete the measurement system. (3) Where an interface is part of a cell, the potential across the interface must be included in the cell potential. (4) One half of a cell in potentiometric measurement systems usually consists of an *indicator electrode*, whose potential depends on the species and conditions in the solution of interest. The other half consists of a *reference electrode*, whose potential is independent of the conditions around it. The potential of a reference electrode has normally been established with reference to the standard hydrogen electrode. In some systems involving phase interfaces, two secondary reference electrodes are required to achieve an electrical contact that permits measurement — an example is the glass electrode system for the measurement of pH. (5) The operation of measuring a cell voltage should not affect the potential of the cell nor induce significant change in potential at the interface of interest. Moreover, the electrical current required by the measurement device must not be large enough to affect the cell equilibria significantly.

The Potentiometer

A basic instrument for measuring cell voltages without drawing current[2] is the *potentiometer* (Figure 11-3). In this device a stable current is passed

---

[2]The usual direct-current voltmeter requires current flow to operate. Such current would both disturb the conditions at the interface being observed and, owing to resistance across the cell, produce a voltage drop equal to the product of current and resistance (Ohm's law). A high-impedence voltmeter draws so little current that cell voltages are little affected, and it is now widely used in place of the potentiometer for most applications. We describe the potentiometer here because it provides a simple introduction to electrical circuits that is useful in later discussions.

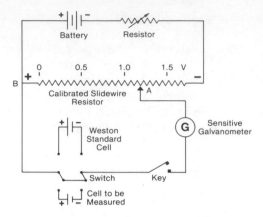

Figure 11-3. Schematic diagram of a potentiometer.

through a calibrated slidewire resistor that provides a known voltage drop along its length. The voltage drop is adjusted to a specific value through the use of a second resistor and a standard voltage, often from the Weston standard cell shown in Figure 11-4. With the Weston cell in the circuit and the slidewire set to read the voltage of the standard (1.0183 V at 25°C), the resistor is adjusted until no current flows through the galvanometer. Then the Weston cell is replaced in the circuit by the cell whose voltage is to be measured, and the slidewire contact is moved until the galvanometer again indicates no current flow. The position of the contact on the slidewire is then read directly as voltage. With a galvanometer of good quality as the detection device for null current, results precise to 1 mV are readily attained with cells whose resistance ranges from a few hundred to a few thousand ohms. When the resistance exceeds about a million ohms, the galvanometer must be replaced by a more sensitive current-indicating device containing electronic circuits to amplify the current that flows when the circuit is out of balance. In fact, voltmeters are now available that can directly measure voltages of cells with extremely high resistances (10^6 ohms and higher). The potentiometer has therefore been generally superseded by electronic systems, particularly in measurements of circuits containing selective electrodes. Details of the electronic circuits are outside the scope of this book.

Figure 11-4. The Weston cell.
$Cd(Hg) | CdSO_4(satd) | | Hg_2SO_4(satd) | Hg$

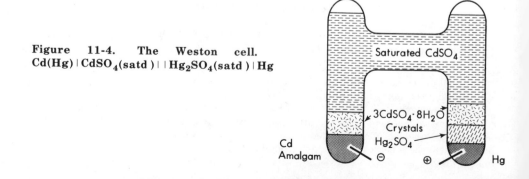

11–3 JUNCTION POTENTIALS

The internal connection between separate reference and indicator electrodes in the first two types of cells utilized in potentiometry (Section 11-5) is usually made by a junction that allows passage of ions (flow of current) but not physical mixing of the solutions. If the mobilities of the positive and negative ions diffusing across the boundary are unequal, a potential is created at the junction. The rate at which ions move across an interface depends on their mobilities. The hydrated proton, for example, is more mobile (350×10^{-5} cm/s) than the chloride ion (76.4×10^{-5} cm/s). Therefore, a liquid junction involving a solution of hydrochloric acid develops a potential because protons diffuse faster than chloride ions and give the region into which they diffuse a positive charge.

EXAMPLE 11–2

Figure 11-5 represents an interface between 0.01 M solutions of hydrochloric acid and potassium chloride. Hydrogen ions and potassium ions diffuse toward regions of lower concentration of their species; that is, the hydrogen ions diffuse into the potassium chloride solution and the potassium ions into the hydrochloric acid solution. The more highly mobile hydrogen ions move more rapidly into the potassium chloride solution than do the potassium ions into the hydrochloric acid solution. Hence, the potassium chloride side of the junction is positive.

Liquid-junction potentials are least when the ions in a solution of an electrolyte all have similar mobilities. The mobilities of the potassium ion (73.5×10^{-5} cm/s) and the nitrate ion (71.6×10^{-5} cm/s) are both close to the mobility of the chloride ion. Ordinarily, when it is desirable to hold a liquid-junction potential to a minimum, a salt bridge is used in which the potassium chloride concentration is several times higher than the electrolyte concentration in the solutions with which the bridge comes in contact. The junction potentials are then small, of the order of a few millivolts. By selection of appropriate mixtures of salts, junction potentials even smaller than those with potassium chloride can be attained.

The maximum possible junction potential is observed when either the cations or the anions of a system must carry the total charge across the

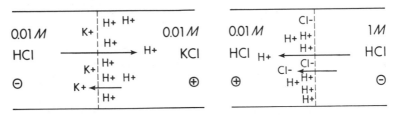

Figure 11-5. Diagram of the development of a junction potential at the interface between 0.01 M solutions of HCl and KCl, and between 0.01 and 1 M solutions of HCl. In the diagram at the left the higher mobility of hydrated protons relative to potassium ions gives the right side of the junction a more positive potential than the left. In the diagram at the right the more concentrated HCl solution is diffusing into the less concentrated one; here the potential on the left side of the junction is more positive than on the right.

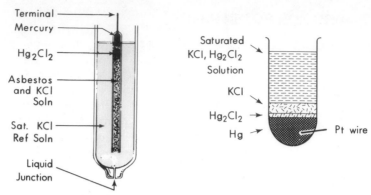

Figure 11-6. Diagram of a laboratory-prepared saturated calomel reference half-cell. Contact between the KCl solution and the test solution is made with a salt bridge; electrical contact between the voltage measuring device and the mercury is made through the platinum wire. A commercial version is shown alongside.

junction. When the oppositely charged ions are immobile, the junction potential E_j then becomes

$$E_j = 0.059 \log (a_2/a_1) \approx 0.059 \log (C_2/C_1)$$

where a_1 and a_2 refer to activities, and C_1 and C_2 to concentrations, of the mobile ion on opposite sides of the interface. The junction potential therefore changes by 2.303 RT/F (59 mV at 25°C or 60 mV at 29°C) for each tenfold change in a_1 relative to a_2 and is the same as that generated in a concentration cell. A glass electrode for measurement of pH is an example of a device containing a junction in which cations carry the entire charge. In general, for selective electrodes such as the glass electrode, ions of one charge type carry all the current. Thus in potential- measurement procedures employing cells containing selective electrodes, one junction is designed so that a single type of ion carries the maximum possible fraction of the charge across the junction. In procedures employing most other cells, such as platinum and saturated calomel electrodes, for potential measurement (Section 11-4) junction potentials are made as low as possible.

11–4 REFERENCE ELECTRODES

Practical measurements are taken with a cell that includes the system under measurement and one secondary-standard reference half-cell (two reference half-cells for measurement of a membrane potential). A reference half-cell or reference electrode should be convenient to prepare and use, and its potential should be reproducible and unaffected by the passage of small currents through the cell of which it is a part. Several secondary-standard half-cells are discussed in Section 9-3, including the hydrogen gas–hydrogen ion, silver–silver chloride, and calomel half-cells. The commonest of these is the saturated calomel electrode (SCE), the potential of which is 0.244 V more positive than the standard hydrogen electrode. The SCE is exceedingly easy to prepare. A simple form (Figure 11-6) can be assembled by adding to a tube mercury metal, a small amount of solid mercury(II) chloride, several grams of potassium chloride, and distilled water. To ensure a saturated solution, excess solid potassium chloride must be present at equilibrium.

The potential of the saturated calomel half-cell can be obtained by starting with the Nernst equation for the half-reaction:

$$Hg_2^{2+} + 2e \rightleftharpoons 2Hg(l) \qquad E^0 = 0.792 \text{ V}$$

$$E = 0.792 - \frac{0.059}{2} \log \frac{1}{a_{Hg_2^{2+}}} \qquad (11\text{-}1)$$

However, the solution is saturated with Hg_2Cl_2. In addition, the Hg_2^{2+} concentration is extremely low, owing to the low solubility of Hg_2Cl_2 and the high concentration (about 4 M) of chloride in saturated KCl. Appropriate substitution can be made for $a_{Hg_2^{2+}}$ in the Nernst equation from the solubility-product expression for Hg_2Cl_2:

$$a_{Hg_2^{2+}} \, a^2_{Cl^-} = K_{sp}$$

$$a_{Hg_2^{2+}} = K_{sp}/a^2_{Cl^-}$$

$$E = 0.792 + \log K_{sp} - (0.059/2) \log a^2_{Cl^-} = 0.268 - (0.059/2) \log a^2_{Cl^-}$$

$$= 0.244 \text{ V for saturated calomel at } 25°C$$

The value 0.268 includes the E^0 value for the mercury half-reaction and the solubility product for Hg_2Cl_2. The value 0.244 includes a small correction for the liquid-junction potential between the saturated KCl solution and a dilute aqueous solution of moderate acidity. Since at the high concentration of KCl in saturated solution the mean activity coefficient is substantially less than 1, the concentration of chloride ion and its activity can no longer be considered equal. This point should not cause concern; the important thing is that the electrode potential remains constant because at the high ionic concentration provided by saturated KCl the activity of chloride ion is essentially constant. This property is essential to a reference half-cell. Although the currents drawn upon measurements of potential are ordinarily slight, large currents can be accommodated by a saturated calomel half-cell without changing the chloride ion activity appreciably. With flow of current in one direction the reactions are

$$2Hg(l) \longrightarrow Hg_2^{2+} + 2e$$

$$Hg_2^{2+} + 2Cl^- \longrightarrow Hg_2Cl_2(s)$$

That is, a small additional amount of Hg_2Cl_2 forms, a small amount of potassium chloride dissolves, and the solution remains saturated, with the result that the potential is unchanged. With flow of current in the opposite direction, the reactions are the reverse of those above, or

$$Hg_2^{2+} + 2e \rightarrow 2Hg(l)$$

$$Hg_2Cl_2(s) \rightarrow 2Cl^- + Hg_2^{2+}$$

Here some Hg_2Cl_2 dissolves, and the additional chloride ions in solution cause a slight amount of additional KCl to precipitate. For reaction in either direction, the activity of chloride is so high and constant that the cell potential is not detectably disturbed. An extensive area of contact between Hg, Hg_2Cl_2, and aqueous KCl solution allows more current to flow without affecting the concentration of the species at the interface and thereby shifting the potential. In effect, it is the current density that must be kept low if shifts in potential with passage of current are to be minimized. An electrode that resists potential shifts with passage of current may be described as well *poised*.[3]

The principal shortcoming of the SCE as a reference is that the solubility of the potassium chloride changes substantially with temperature, and therefore the half-cell potential has a relatively large temperature coefficient. If 1 *M* KCl is substituted for saturated KCl in the calomel cell, the temperature coefficient is considerably smaller. Since temperature changes during analytical measurements of potential are not usually a problem, saturated half-cells are usually preferred.

For most potential measurements the SCE is connected by a salt bridge to a second half-cell of the system of interest:

$$Hg(l) \mid Hg_2Cl_2(satd) \; KCl \, (satd) \mid \mid solution \; of \; interest \mid indicator \; electrode$$

The voltage difference between the SCE and indicator electrode is measured with a potentiometer or voltmeter. From the overall voltage of the cell and the known potential of the SCE, 0.244 V, the potential of the indicator electrode can be calculated.

EXAMPLE 11–3

A cell containing an SCE reference electrode–indicator electrode pair has an overall voltage of 0.623 V, with the indicator electrode more positive than the calomel half-cell. What is the potential of the indicator electrode compared with that of the standard hydrogen electrode?

The SCE potential is 0.244 V. The indicator electrode is more positive by 0.623 V. Therefore the indicator electrode has a potential of +0.867 V compared with that of the standard hydrogen electrode, on the assumption that the liquid-junction potential is zero.

What is the potential of the indicator electrode if the voltage of the cell is 0.123 V, with the indicator electrode negative to the SCE reference half-cell?

The indicator electrode in this cell has a potential of +0.244 – 0.123 = +0.121 V compared with the standard hydrogen electrode (again on the assumption of zero liquid-junction potential).

---

[3]A solution is also said to be poised when it resists change in potential on addition of an oxidant or reductant. The situation is analogous to buffering in acid-base systems.

11–5 CELLS IN POTENTIOMETRY

Three distinct types of cells are used in direct potentiometry and in potentiometric titrations. The first contains an active metal indicator electrode, the second an inert metal indicator electrode, and the third two reference electrodes with a variable membrane potential between them.

In the first type one of the half-cells contains a metal electrode that undergoes reversible electron transfer with its ions in solution. The principal metals of this type are silver, mercury, copper, cadmium, and lead. Under suitable conditions the potentials of these metals accurately reflect the activities of the ions in solution. Most other metals, including iron, nickel, chromium, and tungsten, do not respond in a reversible way to ion activities, usually because they form oxide coatings or undergo other reactions at the metal–solution interface that yield nonreproducible potentials.

The silver electrode responds cleanly to changes in concentration of silver ions in solution. The silver–silver ion half-cell also can be used for the indirect measurement of numerous anions that form insoluble compounds with silver ions such as chloride, bromide, iodide, and sulfide. Section 9-3 illustrates the way to calculate the potential of this electrode in the presence of chloride by incorporating the chloride ion activity and the solubility product of silver chloride into the Nernst equation for the silver half-reaction.

Mercury metal is an excellent indicator electrode, not only for Hg_2^{2+}, but also for many other ions that form insoluble or slightly dissociated compounds with Hg_2^{2+}. As well, the mercury electrode responds in a predictable way to the concentration of metal ions that form complexes with EDTA (Section 7-3).

The second type of cell consists of a chemically inert indicator electrode such as platinum or graphite immersed in the test solution, a salt bridge, and a reference half-cell. The indicator electrode does not itself undergo oxidation but acts as an inert conductor that indicates the tendency of components of the solution to give up or take on electrons. Platinum is a useful indicator electrode for equilibria such as the iron(III)–(II), cerium(IV)–(III), and tin(IV)–(II) systems. The concentration (activity) ratio of the two oxidation states of a reversible couple determines the electrode potential for this half-cell, making it a useful indicator electrode for monitoring redox systems.

The third and most important type of cell consists of two reference half-cells separated by a junction material. An example would be

$$SCE \mid \mid test\ solution \mid junction\ material \mid SCE$$

The voltage measured is that between the two mercury metal electrodes in the reference half-cells. The potentials of the two saturated calomel half-cells are nominally the same and do not change significantly with time. The cell voltage, then, will be zero *except for the junction potential*, which is exactly what such cells are designed to measure. Imperative, of course, is that the potentials of the two half-cells remain constant while the junction potential is being measured.

11–6 DIRECT POTENTIOMETRY

Two basic techniques are practiced in analytical potentiometry: direct potentiometry and potentiometric titrations.

In *direct potentiometry*, potential is related directly to the amount of a solute present in a solution, usually by reading from a calibration graph of potential against concentration. Although it would be convenient if solute concentrations (more accurately, activities) could be obtained through direct substitution in the Nernst equation, this is impractical in analysis. In practice the voltage of a cell prepared from two half-cells is measured by a technique that involves negligible current flow. With knowledge of the electrode potential of the reference half-cell, the potential to be assigned to the remainder of the cell should in principle be obtainable by simple subtraction. However, the potential so obtained would normally include the junction potential in the overall cell voltage, and therefore this approach does not permit a direct calculation of ion activity.

A practical method of acquiring useful analytical data is to measure the voltages of two cells, identical except that for one of the measurements the half-cell containing the species under test is replaced with one containing a solution of that species at a known activity (standard solution). The potential of each such cell arises from three interfaces: one at the reference electrode, a second at the indicator electrode, and a third at the junction between the two liquids in the two half-cells. The potential at the reference-electrode interface and the potential generated at the liquid junction may generally be assumed to be the same in the two cells. This assumption is valid provided that the composition of the standard solution is close to that of the unknown. Two equations for the cell voltages can be written, from which the activity in the unknown can be computed. The actual values of the electrode potentials with the standard solution, the reference electrode, and the liquid junction are not required.

EXAMPLE 11–4

The silver ion concentration in a solution is to be found by measurement of voltage of a cell involving a silver electrode and a saturated calomel reference half-cell:

$$SCE \mid \mid AgNO_3 \text{ soln} \mid Ag$$

Suppose the cell reading is 0.400 V, the silver electrode having the more positive potential. If the unknown solution is replaced by a standard solution of 0.01 M AgNO$_3$, the voltage reads 0.422. Calculate the silver ion concentration in the unknown, assuming activity coefficients of unity.

The silver electrode responds to the silver ion activity in solution, becoming more positive with higher concentration (activity). The Nernst equation applied to the silver half-reaction is

$$E_{Ag} = E^0{}_{Ag} - 0.059 \log (1/[Ag^+])$$

For the cell containing the solution of unknown concentration,

$$E_{Ag} - E_{SCE} - E_j = 0.400 \text{ V.}$$

where E_{Ag} is the potential of the silver half-cell, E_{SCE} the potential of the saturated calomel reference half-cell, and E_j the potential of the liquid junction. Substituting the Nernst expression for E_{Ag} gives

$$E^0_{Ag} - 0.059 \log (1/[Ag^+]) = 0.400 + E_{SCE} + E_j$$

For the second cell

$$E^0_{Ag} - 0.059 \log (1/0.01) = 0.422 + E_{SCE} + E_j$$

Subtracting the first equation from the second, and canceling E^0_{Ag}, E_{SCE}, and E_j (assuming they remained unchanged during the two measurements), we obtain

$$0.059 \log (1/[Ag^+]) - 0.059 \log (1/0.01) = 0.422 - 0.400$$

$$\log \frac{1}{[Ag^+]} = \frac{0.022 + 0.12}{0.059}$$

$$[Ag^+] = 4.2 \times 10^{-3} \, M$$

A lower concentration than the standard is consistent with the slightly less positive potential of the unknown.

An improved technique is to use two calibration standard solutions, one lower and one higher in concentration than the solution to be determined. A further refinement is to prepare a calibration curve of cell voltage against log of concentration for a range of solutions of known concentration. The concentration of an unknown solution is then read from the graph. At low concentrations the slope of the line will probably be close to $0.059/n$, but at higher concentrations the slope decreases because the activity coefficient decreases with concentration.

EXAMPLE 11-5

A series of chloride-containing solutions ranging from $10^{-4}M$ to $1 \, M$ were measured in a cell comprising half-cells of saturated calomel and silver–silver chloride. The SCE was isolated from the test solution by a bridge of $1 \, M$ KNO_3 to prevent chloride contamination of the solution under test. The results are shown in Figure 11-7. As expected, the voltage becomes less positive with increasing chloride ion concentration.

Another technique involves the use of a *concentration cell*; one half consists of a reference electrode in the unknown solution, and the other half consists of a reference electrode of the same material in a solution of known concentration as near as possible to that of the unknown. On the assumption that the liquid-junction potential between two highly similar solutions is zero, the difference in potentials of the two electrodes then arises solely from the difference in the activities in the two solutions. The exceedingly high

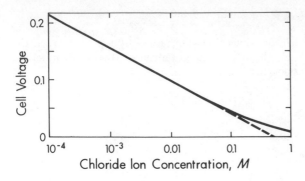

Figure 11-7. Calibration curve of cell voltage against sodium chloride concentration for a cell consisting of a saturated calomel reference half-cell and a silver-silver chloride half-cell. *Solid line, concentration; dashed line, activity.*

mobilities of hydrogen and hydroxyl ions can cause liquid-junction potentials to be appreciable, even when the concentrations of these ions are only moderate.

EXAMPLE 11–6

Calculate the concentration of silver ions in an unknown solution given that the voltage of the cell

$$Ag \,|\, 0.0100 \; M \; AgNO_3 \,|\,|\, AgNO_3 \; soln \,|\, Ag$$

is 0.008 V and the unknown solution is the more positive (and therefore more concentrated in silver ions).

The following equation applies:

$$E^0{}_{Ag} - 0.059 \log (1/0.01) + 0.008 = E^0{}_{Ag} - 0.059 \log (1/[Ag^+])$$

The $E^0{}_{Ag}$ values cancel each other, and on the assumption of activity coefficients of unity

$$(-0.12 + 0.008)/0.059 = - \log (1/[Ag^+])$$

$$[Ag^+] = 0.0136 \; M$$

If the activity coefficient of 0.01 M silver ion is 0.898 and that of the unknown assumed to be 0.889, what is the concentration of the unknown?

$$E^0{}_{Ag} - 0.059 \log [1/(0.01 \times 0.898)] + 0.008 = E^0{}_{Ag} - 0.059 \log [1/([Ag^+] \times 0.889)]$$

$$[Ag^+] = 0.0137 \; M$$

Even though the activity coefficients may be somewhat less than unity, the effect of the assumption of unity is small when the two solutions have similar concentrations.

To determine the composition of solutions, more measurements are made by direct potentiometry than by any other single technique. A comparison using a single standardizing solution is common practice. Nevertheless, this technique does not yield highly precise results. One reason for the imprecision is the difficulty of obtaining cell potentials with adequate accuracy and reproducibility. Also, the liquid-junction potential of two slightly different solutions can easily change by a millivolt or so instead of remaining constant as assumed. Thus an uncertainty of 2 mV for even careful measurements is not unreasonable. An uncertainty of $\pm$ 2 mV in the voltage in the preceding example would lead to a reported concentration that could be anywhere within 0.0126 to 0.0147 M, or an uncertainty of about 8%. Each millivolt of uncertainty in potential produces an uncertainty in concentration of about 4%. In general, careful measurements of the concentrations of solutions by direct potentiometry give results precise to about 5%. Moreover, at ionic strengths much above 0.1 the actual concentrations of ions increasingly deviate from activities.

11-7 SELECTIVE ELECTRODES

A measurement system using a *selective electrode*[4] consists of two reference half-cells, one on each side of a liquid or solid membrane that responds more or less selectively to a single species. One side of the membrane is in contact with a solution containing the ion to be determined, and the other side with a solution having a fixed activity of the ion:

ext. ref. half-cell | | test soln | membrane | ref. soln | | int. ref. half-cell

The reference half-cells make electrical contact with the two sides of the membrane where junction potentials are established. Two separate potentials are developed as a result of equilibrium between the solutions and the membrane surfaces, one on each side of the membrane. One junction involves contact of one side of the membrane with the reference, or internal, solution, at which the junction potential is assumed to be unchanging. The second junction involves contact of the other side of the membrane with the test solution. Since the two reference half-cells, as well as the reference side of the membrane, have unchanging potentials, changes in overall cell voltage arise from changes in junction potential at the test solution-membrane interface. A direct measurement with a selective electrode measures only the free forms of that species and not complexed, or bound, forms.

The Glass Electrode

The oldest and best known selective electrode is the glass electrode for measurement of hydrogen ion concentration. Probably more pH measurements are carried out each day than any other type of chemical measurement. Most of them are made potentiometrically with a glass indicator electrode. After fifty years of study its mode of operation is reasonably well understood, but even today improvements in glass

---

[4]The term *ion-selective* electrode is often used, but seems too narrow in light of modern developments in this field. Selective electrodes operating on the same basic principle are now available for substances as diverse as gases and enzymes. For purposes of illustration, much of the discussion here is concerned with ions.

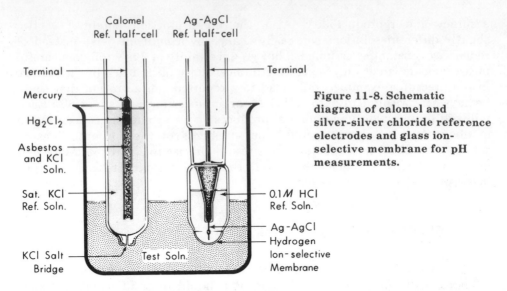

Figure 11-8. Schematic diagram of calomel and silver-silver chloride reference electrodes and glass ion-selective membrane for pH measurements.

membranes are largely empirical.

A typical pH electrode consists of a thin glass membrane, generally bulb-shaped, sealed onto the end of a glass tube. A constant pH is maintained on the inside of the bulb by filling it with a solution of often 0.1 M HCl. Contact between this solution and the voltmeter is through a reference electrode, often calomel or silver–silver chloride, called the *internal reference electrode*. For stability and convenience in handling, the solution of constant pH and the internal reference electrode are usually sealed into a closed unit (Figure 11-8).

The glass in the responsive tip of a glass electrode must have an appropriate composition. An early glass used for its pH sensitivity was composed [5] of 72% SiO_2, 22% Na_2O, and 6% CaO. An ion-selective glass membrane that has been soaked in water is viewed as consisting of three distinct regions — inner and outer surface layers of hydrated glass, about 0.1 μm thick, on either side of a dry glass layer about 50 μm thick. In the surface layers lithium or sodium ions at lattice sites in the glass have exchanged with hydrogen ions in solution. In the hydrated surface this reservoir of protons is in ion-exchange equilibrium with protons in solution. The glass membrane, then, may be viewed as follows:

test soln | hydrated layer | dry glass layer | hydrated layer | int. soln

The two junctions of interest are those between the test solution and the hydrated layer and between the internal solution and the other hydrated

---

[5]This composition, first recommended by D. A. MacInnes and M. Dole in 1930, was marketed for many years by the Corning Glass Works as Corning 015. It has the disadvantage of responding slightly to sodium ion, which leads to "sodium error," or "alkaline error," in solutions above pH 10. Modern electrodes are fabricated of glasses in which Na_2O is replaced with Li_2O, among other changes. These electrodes may be used over a pH range as great as 0 to 14 with little error from response to ions other than those of hydrogen.

layer. At these two junctions an equilibrium is quickly established at the surface, and potentials across the junctions result because protons are either taken up or lost from the surfaces, depending on the hydrogen ion activities in the solution and in the hydrated layers. The equilibrium is

$$H^+{}_{glass} \rightleftarrows H^+{}_{soln}$$

No methods are known for determining the absolute values of such potentials; but qualitatively speaking, if the hydrogen ion activity in solution is increased, protons transfer to the glass layer, and the potential of the glass surface becomes more positive.

It is important to recognize that no transfer of electrons is involved; that is, no redox process occurs in the familiar sense. A potential simply develops at the interface of interest as a result of an equilibrium established between the two phases. Quantitatively, the potential at the interface is given by

$$E = \text{constant} - 2.303\ RT/F \log (1/a_{H^+}) = \text{constant} - 0.059 \log (1/a_{H^+})$$

$$E = \text{constant} + 0.059 \log a_{H^+} = \text{constant} - 0.059\ \text{pH}$$

At the second, or inner, hydrated layer a similar potential is developed whose value is given by a similar expression. Because the activity of the hydrogen ion is held constant, the potential at the interface of the inner hydrated layer does not change. The net potential across the entire membrane is the difference in the potentials of the two hydrated layers:

$$E = [\text{constant} + 0.059 \log a_{H^+}\ (test)] - [\text{constant} + 0.059 \log a_{H^+}(ref.)]$$

$$= 0.059 \log \frac{a_{H^+}(test)}{a_{H^+}(ref.)}$$

From a practical point of view the two surfaces of a glass membrane do not behave exactly alike; they differ in response according to the history of the individual surfaces. These variations produce a small but observable potential difference across the membrane even when the solutions on each side are identical. This difference, called the *asymmetry potential*, is generally on the order of a few millivolts. Its magnitude varies from day to day and from one glass membrane to another. It is corrected for at the time an electrode is calibrated with a reference solution. The rate of change in asymmetry potential is sufficiently slow that rechecking during measurement is usually unnecessary.

Measurement of pH

Measurement of the pH of a solution entails immersion of the glass bulb, along with an external reference electrode, in the solution and observation of the potential developed between the two electrodes. The cell may be expressed by

SCE | | test soln | glass | 0.1 M HCl, AgCl (satd) | AgCl, Ag

The potential at each of the interfaces remains essentially constant when the test solution is changed except for the interface of the hydrated layer and the solution being measured. The overall voltage of the cell can then be written

$$E_{cell} = E_{Ag} - E_{SCE} - E_j(0.1\ M\ HCl) + E_j\ (soln) = E' + E_j(soln) \quad (11\text{-}2)$$

where E' includes the three potentials that are unaffected by the composition of the test solution: E_{Ag}, E_{SCE}, and $E_j(0.1\ M$ HCl). The value of E′ is influenced not only by the nature of the internal and external reference electrodes but also by the history of the membrane (asymmetry potential).

In summary, the potentiometric determination of pH involves measurement of a voltage between two reference electrodes. The internal reference electrode is retained by a glass membrane and is usually considered part of the indicator electrode. The outer surface of the membrane responds to changes in hydrogen ion activity of the test solution and produces the potential that is measured.

The resistance of glass used as electrode material is high, about 10^7 to 10^9 ohms. For this reason, and because the current drawn must be negligible to avoid affecting the measurement, the meter used to measure the potential must be highly sensitive. (A current of 10^{-10} A causes a voltage drop across 10^8 ohms of 10 mV, sufficient to produce an unacceptably large error in the measured potential.) Accordingly, the current read by the measurement circuit must be held to values low enough to ensure negligible error. Considerable amplification of these small currents is necessary before the signal can be read from the meter. Because the response of an amplifier of sufficient sensitivity normally varies somewhat with time, that is, drifts during operation, a means of correction is required. This correction for drift in the meter, as well as compensation for variations in the electrical characteristics of individual glass electrodes, is accomplished by calibrating the meter and electrodes in one or more solutions of known pH before measuring unknown solutions. The potential registered by a pH meter is affected also by the temperature of the solution, and so this effect, too, must be taken into account in accurate measurements. Therefore, a pH meter and an indicator electrode–reference electrode pair must be standardized before each set of measurements by placing the electrodes in a solution of known pH and adjusting the meter to read that pH value. In effect, a value for E′ in Equation (11-2) is determined by this procedure. After the first calibration in a series of measurements, most adjustments needed during restandardization compensate for drift in the electronics of the meter.

Since the factor 2.303 RT/F varies with temperature, most pH meters include a control that can be set to the temperature of the test solution.

When pH measurements involve a glass electrode, a finite amount of current must be drawn from the cell, even though the dry glass layer has an extremely high resistance. Present-day pH meters are designed to measure potentials in systems with a resistance of 10^{15} ohms or more. The current required by these meters approaches zero, and so the iR drop across the glass is negligible. As the dry glass layer is not permeable to hydrogen ions, the minimal conductivity that must exist results from slight shifts of cations (such as sodium or lithium) in the layer itself.

To be practicable, a glass electrode must be reasonably sturdy. Membrane thicknesses must compromise between mechanical strength and chemical resistance on the one hand and accurate and rapid pH response on the other. The membrane is susceptible to physical breakage and to chemical attack, especially from hydrofluoric acid and highly alkaline solutions. The normal life of a glass electrode of one to two years is shortened if the membrane is frequently allowed to dry.

For convenience of operation, commercial electrodes are available in which the glass and external reference electrodes are built into one unit. These so-called *combination electrodes* require less space, are easy to handle, and may be used to advantage with small volumes of sample solution.

From the Nernst equation the response of a glass electrode is predicted to be 0.0592 V per tenfold change in hydrogen ion activity at 25°C. Present commercial glass electrodes follow this relation closely over the pH range of about 1 to 13. Above or below these values, response tends to fall off and some error is introduced. The affinity of sites in the glass matrix for hydrogen ions over other cations is a major advantage of this electrode. Since other cations do not compete appreciably with hydrogen ions under most conditions, interference is rare.

Other Solid Membrane Electrodes

If the composition of the glass is changed, say to include some Al_2O_3, the potential of the hydrated layer may become less responsive to low levels of hydrogen ion and more responsive to other cations. For example, the potential of a specific glass surface may be affected by the equilibrium

$$Na^+ \text{ (glass)} \rightleftharpoons Na^+ \text{ (soln)}$$

Qualitatively, in this system the higher the sodium ion activity in the test solution, the more positive will be the hydrated surface layer of the glass. At sufficiently low hydrogen ion activity the potential is determined solely by the sodium ion equilibrium. For a cell of the type

$$\text{SCE} \mid \mid Na^+ \text{ test soln} \mid \text{glass} \mid 0.1 \ M \ \text{NaCl, AgCl (satd)} \mid \text{AgCl, Ag}$$

the potential is given by

$$E = \text{constant} + 0.059 \log a_{Na^+} = \text{constant} - 0.059 \ pNa$$

A glass containing about 5% Al_2O_3 responds at low hydrogen ion concentration to potassium ions in solution according to

$$K^+ \text{ (glass)} \rightleftharpoons K^+ \text{ (soln)}$$

The potential is given by

$$E = \text{constant} + 0.059 \log a_{K^+}$$

Because sodium and potassium ions are similar in chemical behavior, a high potassium ion concentration affects the potential of a sodium-sensitive glass in the same way as a high sodium concentration. Similarly, a high sodium ion concentration affects the potential of a potassium-sensitive glass. In both cases a weighting (correction) factor called the *selectivity constant* must be applied to obtain the net effect of both ions on the potential of the hydrated boundary layer. The selectivity constant indicates the relative response of an electrode to the two species. Incorporating this constant into the Nernstian expression for potential allows inclusion of the effect of this second ion:

$$E = \text{constant} + 0.059 \log (a_{Na^+} + a_{K^+} K_{Na^+,K^+})$$

where K_{Na^+,K^+} is the selectivity constant for a given electrode for sodium over potassium.[6] A typical value of this selectivity constant is 10^{-4}.

EXAMPLE 11–7

For $K_{Na^+,K^+} = 10^{-4}$, calculate the expected values of $(E - \text{constant})$ for several sodium ion activities at potassium ion activities of 1.0 and 0.1.

For $a_{Na^+} = 1.00 \times 10^{-3}$ and $a_{K^+} = 0.1$

$$E - \text{constant} = 0.059 \log (10^{-3} + 0.1 \times 10^{-4}) = -0.18 \text{ V}$$

a_{Na^+}	Value of $(E - \text{constant})$	
	$a_{K^+} = 0.1$	$a_{K^+} = 1.0$
0.1	−0.06	−0.06
0.01	−0.12	−0.12
10^{-3}	−0.18	−0.17
10^{-4}	−0.24	−0.22
10^{-5}	−0.28	−0.24
10^{-10}	−0.30	−0.24

The effect of potassium ion can be seen from this tabulation to be slight until the ratio of potassium to sodium ion becomes large.

For a glass lithium-sensitive electrode at high pH, interference from both sodium and potassium ions might need to be considered. In this event,

$$E = \text{constant} + 0.059 \log (a_{Li^+} + a_{Na^+} K_{Li^+,Na^+} + a_{K^+} K_{Li^+K^+})$$

---

[6]The general expression for an interference y^{z_y} in the measurement of x^{z_x} at 25°C is
$$E = \text{constant} + (0.059/z_x) \log [a_x + K_{x,y}(a_y)^{z_y/z_x}]$$
Here z_x and z_y are the charges on ions x and y. The value of the constant K is determined by several factors, chief of which are the relative mobilities of x and y in the membrane and the exchange equilibrium $y_{soln} + x_{membrane} \rightleftarrows y_{membrane} + x_{soln}$.

For present-day electrodes the value of the selectivity constant K_{Li^+,Na^+} is about 0.3, and that of K_{Li^+,K^+} about 10^{-3}.

The best hydrogen ion-selective glass electrodes have selectivity constants of about 10^{-15} for hydrogen ions over sodium ions. No other selective electrode has specificity even remotely approaching this level; the selectivity constants between metal ions for alkali metal-selective glass electrodes are many orders of magnitude larger. The glass pH electrode is responsive to changes in hydrogen ion concentration from about 1 M to strongly alkaline solutions, a range of about 14 orders of magnitude. For the alkali metals, on the other hand, it is satisfactorily responsive to changes in concentration from about 1 M to only about 10^{-5} M.

Junction membrane materials other than glass are possible for immobilization of either positively or negatively charged ions. Probably the most important practical example of such a material is a single crystal of lanthanum fluoride. It is an example of a *solid-state ion-selective electrode*. An interface with a fluoride-containing solution has the equilibrium

$$F^-(soln) \rightleftharpoons F^- \text{ (surface layer)}$$

For a cell containing two reference electrodes, separated by a crystal of LaF_3 as the membrane and with an internal solution with a fixed fluoride concentration, the potential is

$$E = \text{constant} - 0.059 \log a_{F^-} = \text{constant} + 0.059 \text{ pF}$$

Lanthanum fluoride has low solubility ($K_{sp} = 2 \times 10^{-22}$). Response to fluoride ion concentrations is satisfactory down to about 10^{-5} M (Figure 22-2). Hydroxide ion begins to interfere in the measurements at about 10^{-5} M, owing to formation of insoluble $La(OH)_3$:

$$E = \text{constant} - 0.059 \log (a_{F^-} + a_{OH^-}K_{F^-,OH^-})$$

Several other types of membranes for junction materials have been developed. If a mixture of Ag_2S and $AgCl$ is pressed into a disk and inserted between two solutions, the potential difference between two reference electrodes, one immersed in each solution, varies selectively with the silver ion activity or, through the solubility-product constant of $AgCl$, with the chloride ion activity. Similarly, by preparation of membranes containing silver bromide or silver iodide, selective response to bromide or iodide can be obtained. The silver iodide electrode also responds to cyanide ion activity at low iodide concentration. Silver sulfide alone is an excellent electrode for measurement of sulfide ion activities.

When silver sulfide is mixed with one of several other metal sulfides such as CuS, PbS, or CdS and pressed into a thin disk, the material functions as a membrane responsive to the second metal ion. Silver sulfide appears necessary as a matrix, or binder, for the other metal sulfides; additionally it aids in lowering the resistance of the membrane.

Liquid Membrane Electrodes

A *liquid membrane* consists of a porous inert support, such as a circle of filter paper, impregnated with a   liquid immiscible with water. The liquid contains a reagent capable of complexation with the ion for which selective response is desired. For instance, high-molecular-weight alkylamines and dinonylnaphthalene sulfonic acid form complexes with certain cations that are preferentially retained in the organic liquid rather than in the aqueous phase. An ion-exchange equilibrium is established between the ions of interest in the test solution and those in the organic liquid.

EXAMPLE 11-8

For the measurement of calcium ion concentrations the water-immiscible organic liquid di-*n*-octylphenylphosphonate containing the calcium complex of didecyl phosphoric acid may be used in a cell as follows:

$$SCE \mid \mid Ca^{2+} \text{ test soln} \mid \text{liq. membrane} \mid Ca^{2+} \text{ref. soln} \mid \text{ref. half-cell}$$
$$Ca^{2+}(\text{membrane}) \rightleftarrows Ca^{2+}(\text{soln})$$

and the potential is given by

$$E = \text{constant} + (0.059/2) \log a_{Ca^{2+}} = \text{constant} - (0.059/2) \, pCa$$

The selectivity constants are about 0.014 for strontium, 0.0016 for barium, 0.005 for magnesium, and 3×10^{-4} for sodium and potassium.

For liquid ion exchangers the selectivity constant is determined by several factors. First, a high selectivity constant requires that a stronger complex with the ion exchanger (the complexing agent) is formed by the ion of interest than by the interfering ions. Second, relative mobilities of the ion of interest and the interfering ion play a role, with high mobility of the ion of interest being favorable. A third factor is the equilibrium constant for the distribution of the ion between the aqueous and organic layers. Both the nature of the organic solvent and the complexing reagent influence this equilibrium. Hydrogen ions are virtually always at least a partial interference. For a calcium ion-selective electrode the useful pH range is from about 5 to 11.

A different type of liquid membrane consists of a neutral material that acts as a selective carrier for ions. One example is a membrane electrode containing valinomycin,[7] a molecule with a 36-membered ring, which provides the basis for a potassium-selective electrode. The selectivity constant K_{K^+,Na^+} of about 2×10^{-4} is some 100 times more favorable than that for the potassium-sensitive glass electrode.

---

[7]Valinomycin is one of a family of antibiotic drugs that appear to function by aiding transport of potassium through cell walls. The hydrophobic exterior of the molecule permits movement through the lipid-cell layers, while the potassium ion with its charge is safely buried within the valinomycin structure.

Other Membrane Electrodes

Electrodes have been developed for the measurement of several nonionic species. An example is the carbon dioxide electrode, which consists of a glass hydrogen ion-responsive electrode surrounded by a thin layer of a solution of sodium bicarbonate held by a gas-permeable membrane. The membrane permits carbon dioxide in the test sample to diffuse into the sodium bicarbonate solution, affecting the pH according to the equilibrium

$$CO_2 + H_2O \rightleftarrows H^+ + HCO_3^-$$

When carbon dioxide dissolves in the sodium bicarbonate solution the glass electrode responds to the resulting decrease in pH. Electrodes of this type have been developed for several other species of environmental and industrial importance, including ammonia, sulfur dioxide, hydrogen cyanide, hydrofluoric acid, and acetic acid.

Another class of membrane electrodes comprises the enzyme electrodes. A gel layer containing an enzyme such as urease is attached to an appropriate ion-selective electrode. Urease catalyzes the conversion of urea to ammonium ions. When urease is retained around a glass ammonium ion-selective electrode and the electrode placed in a solution of urea, the urea diffuses into the gel, where it is converted to ammonium ions and carbon dioxide. Some of the ammonium ions diffuse to the surface of the electrode, where their activity is measured. The potential developed at the glass membrane surface gives a measure of the concentration of the urea in the solution. The system must be calibrated with solutions of known concentrations of urea, and a calibration curve drawn of electrode response against urea concentration. In another design the gel layer can contain a fixed concentration of urea, and the electrode can be used to measure the concentration of the enzyme urease in solution.

Table 11-2 lists a selection of commercially available selective electrodes. Schematic diagrams of several membrane electrodes are shown in Figure 11-9.

The Internal Reference Solution

Keep in mind that in a selective-electrode system it is essential that the potentials of the two reference half-cells, the potential of the liquid junction between these half-cells, and the potential on one side of the junction material all remain constant. Provided these requirements are met, changes

Figure 11-9. Schematic diagrams of several membrane electrodes: from left, glass, LaF$_3$, enzyme, and liquid. From Laitinen and Harris, p 250. See references at end of chapter.

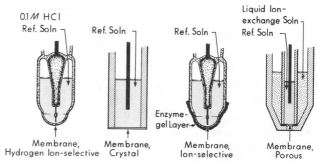

Table 11-2. Some Commercially Available Selective Electrodes*

Electrode	Type	Concentration Range, M	Interferences†
Ammonia (ammonium)	Gas-sensing	1 to 10^{-6}	Volatile amines
Bromide	Solid state	1 to 5×10^{-6}	S^{2-}, Br^-
Cadmium	Solid state	1 to 10^{-7}	Ag^+, Hg^{2+}, Cu^{2+}, Pb^{2+}, Fe^{3+}
Calcium	Liquid	1 to 10^{-5}	Na^+, Zn^{2+}, Pb^{2+}, Fe^{2+}, Cu^{2+}, Sr^{2+}, Mg^{2+}, Ba^{2+}, Ni^{2+}
Carbon dioxide	Gas-sensing	10^{-2} to 10^{-4}	Volatile weak acids
Chloride	Liquid	1 to 8×10^{-6}	ClO_4^-, I^-, NO_3^-, SO_4^{2-}, Br^-, OH^-, OAc^-, HCO_3^-, F^-
	Solid state	1 to 5×10^{-5}	Br^-, I^-, CN^-, S^{2-}
Chlorine	Solid state	3×10^{-4} to 10^{-7}	
Cupric	Solid state	Saturated to 10^{-8}	S^{2-}, Ag^+, Hg^{2+}, Cl^-, Br^-, Fe^{3+}, Cd^{2+}
Cyanide	Solid state	10^{-2} to 10^{-6}	S^{2-}, Br^-, Cl^-
Divalent cation (water hardness)	Liquid	1 to 6×10^{-6}	Na^+, Cu^{2+}, Zn^{2+}, Fe^{2+}, Ni^{2+}, Sr^{2+}, Ba^{2+}
Fluoride	Solid state	Saturated to 10^{-6}	OH^-
Fluoroborate	Liquid	Saturated to 3×10^{-6}	NO_3^-, Br^-, OAc^-, HCO_3^-, F^-, Cl^-, OH^-, SO_4^{2-}
Hydrogen sulfide	Gas-sensing	10^{-2} to 10^{-6}	
Iodide	Solid state	1 to 2×10^{-7}	S^{2-}
Lead	Solid state	1 to 10^{-7}	Ag^+, Hg^{2+}, Cu^{2+}, Cd^{2+}, Fe^{3+}
Nitrate	Liquid	1 to 6×10^{-6}	ClO_4^-, I^-, Br^-, HS^-, CN^-, HCO_3^-, Cl^-, OAc^-
Nitrogen oxide (nitrite)	Gas-sensing	10^{-2} to 5×10^{-7}	Volatile weak acids, CO_2
Perchlorate	Liquid	1 to 2×10^{-6}	I^-, NO_3^-, Br^-
pH	Glass	pH 0 to 14	Sodium (pH 14)
Potassium	Liquid	1 to 10^{-5}	Cs^+, NH_4^+, H^+, Tl^+, Ag^+, Na^+
Silver/sulfide	Solid state	Ag^+ or S^{2-}: 1 to 10^{-7}	Hg^{2+}
Sodium	Solid state	Saturated to 10^{-6}	H^+, Li^+, Cs^+, K^+
Thiocyanate	Solid state	1 to 5×10^{-6}	OH^-, Br^-, Cl^-, NH_3, $S_2O_3^{2-}$, CN^-, I^-, S^{2-}

*Orion Research Incorporated.
†The levels of interference range from extreme to slight.

in cell voltages can be related directly to changes in junction potential on the side of the junction material in contact with the test solution. For this reason, selective-electrode assemblies contain an *internal reference solution*. This solution must perform two functions. (1) It must contain a sufficient concentration of an ionic species to give a stable potential for the junction at the inner solution–membrane interface. (2) It must provide electrolyte for the inner reference half-cell.

In the usual pH glass-electrode assembly the inner reference half-cell consists of a silver–silver chloride electrode in contact with a solution of 0.1 M hydrochloric acid. The chloride ion activity is thereby relatively high and constant, and accordingly, the potential of the silver electrode is constant. Because the solution contains 0.1 M hydrogen ions, the high and constant hydrogen ion concentration ensures that the potential of the inner membrane surface is constant as well. This single electrolyte thus provides admirably the key ions needed to hold both potentials invariant.

Standard Solutions for Selective Electrodes

The mode of application of ion-selective electrodes in many analyses is to compare a potential obtained for a standard with that of the test solution. In a pH measurement a glass electrode–reference electrode pair is immersed in a solution of known pH (pH standard), and the potential of the readout device adjusted so that it reads the known pH. Then the sample solution is measured. With the assumption that no change occurs in either the asymmetry potential or in the liquid-junction potential between the saturated calomel reference half-cell and the standard or test solution, the pH of the test solution is given by

$$\text{pH} = \text{pH}_s + (E - E_s)(F/2.303\, RT)$$

where pH_s is the pH of the standard solution, E the voltage measured with the test solution, and E_s the voltage measured with the pH standard. This equation is the *operational definition of pH*.

Clearly, the accuracy of pH measurements with a glass pH electrode and a pH meter must rely on the accuracy of the pH standard solution and the care with which the standardization operations are performed. To obtain a pH value from this operational definition of pH, we need a standard of known pH. The values of pH_s for various standards have been made available by careful measurements of cells of the type

$$\text{Pt, } H_2(g) \mid \text{test soln, AgCl(satd)} \mid \text{Ag}$$

Because such cells have no liquid junctions, that source of uncertainty is eliminated. Typical test solutions include dilute solutions of KH phthalate and of buffers such as $H_2PO_4^-$–HPO_4^{2-}. Since activity coefficients of single ions cannot be known, reasonable assumptions about activity coefficients in the dilute solutions were selected so that as nearly as possible

$$\text{pH}_s = -\log a_{H^+} = -\log [H^+]\, \gamma_{H^+}$$

For pH values in the region 3 to 11 a series of primary pH standard solutions have been established.[8] The pH standards are specified to the nearest 0.001 pH unit and are, when meticulously prepared, accurate to 0.006 pH unit.

[8]Some of the most important buffers are KH tartrate (saturated at 25°C), 3.557; 0.05 M KH phthalate, 4.008; 0.025 M KHPO$_4$ and 0.025 M Na$_2$HPO$_4$, 6.865; 0.01 M Na$_2$B$_4$O$_7$·10 H$_2$O, 9.180; and 0.025 M NaHCO$_3$ and 0.025 M Na$_2$CO$_3$, 10.012. Two secondary pH standards have also been proposed: 0.05 M potassium tetraoxalate (KHC$_2$O$_4$·H$_2$C$_2$O$_4$·2H$_2$O), 1.679; and saturated Ca(OH)$_2$ at 25°C, 12.454.

A pH scale defined in terms of a single buffer is not the soundest, since the equation

$$pH = pH_s + (E - E_s)/0.059$$

is applied with the assumption that the liquid-junction potential is the same for the test and the pH standard solutions. The validity of this assumption improves as the pH value and the ionic strength of the standard approach those of the test solution. At pH values outside the range 3 to 11, the high mobility of the hydrated proton or the hydroxyl ions makes the liquid-junction values uncertain, and only secondary pH standards can be specified.

A specific set of pH_s values must be adopted for a specific temperature. In effect, for hydrogen ions there is a different standard state, and therefore a separate pH scale, for each temperature.

For cation- and anion-selective electrodes the problem of establishing single-ion activity coefficients arises just as it does for the activity of the hydrogen ion. On the basis of reasonable assumptions about activity coefficients, liquid-junction potentials, and asymmetry potentials, an operational definition of activity of a cation is

$$pM = pM_s + (E - E_s)n/0.059 \qquad (11\text{-}3)$$

where pM is the negative logarithm of the cation activity in a test solution and pM_s the value in a standard solution, E and E_s are the voltages for the test and standard solutions in the cell, and n is the charge on the ion. Similarly for anions

$$pA = pA_s - (E - E_s)n/0.059$$

Although reference standards for species other than the hydrogen ion are being worked out, they are not yet available with the precision and accuracy of the pH standards. For cations and anions generally, the assumption that values of liquid-junction potentials are the same for test and reference solutions is less likely to be valid than for pH measurements. For pH standards in the range 3 to 11 the contribution of the hydrogen or hydroxyl ions to liquid-junction potentials is generally negligible, whereas for other ion standards they normally make a significant contribution.

A source of uncertainty in measurements of ion activity arises from deviations of ion-selective electrodes from Nernstian response. For a tenfold change in concentration of a univalent ion at 29°C the change in membrane potential may be less than the 60 mV expected from the Nernst equation. Only a few of the glasses used in electrodes approach this theoretical response.

Involvement of an ion in an unsuspected chemical reaction such as ion-pairing, complexation, acid-base, or precipitation must be recognized as a possible source of misinterpretation. The sulfide-selective electrode, for example, responds only to the free S^{2-} ion, and in solutions of moderate or low pH, where the most prevalent species are HS^- and H_2S, it will always give values less than the total for sulfide.

In the *method of standard additions* small increments of the ion under test are added to a sample solution, and the potential is measured after each addition. The potentials are then plotted against concentration on semiantilog paper and extrapolated to zero added increments to obtain the total concentration in all forms of the ion under test. This approach permits low levels of some ions to be determined more precisely than by a direct single measurement. The method of standard additions can be applied to many analytical techniques. An example of the method, employing atomic absorption as the mode of measurement, is given in Section 15-8.

11-8   POTENTIOMETRIC TITRATIONS

Limitations of Direct Measurements

In potentiometric measurements involving real analytical systems, several sources of inaccuracy introduce uncertainties of a millivolt or more even with carefully taken data. Accordingly, relative uncertainties of 1 to 10% may be introduced even with painstaking work. Bear in mind that a single pH measurement normally has an accuracy and precision of a few hundredths of a pH unit.[9] Because an error of 0.01 pH unit will produce a relative uncertainty of over 2% in the hydrogen ion concentration, the concentration of an acid or base cannot be determined precisely by direct measurement with a pH meter.

Sometimes also the measurement of activity of a species does not yield exactly the kind of information sought. For example measurement of the pH of an acetic acid solution can give a measure of the hydrogen ion activity, but since acetic acid is only slightly dissociated, a pH measurement does not give directly the total concentration of available acid, which may be what is desired.

Scope of Potentiometric Titrations

The answer to the problem of imprecise data in direct potentiometry is frequently a *potentiometric titration*. Through the simple expedient of measuring potential as a function of added volume of a suitable titrant, highly precise results are possible. Exact values of potential are not important; the *change* in potential with volume of titrant is the significant matter here. If the changes are large with small additions of titrant near the equivalence point, precise data can be obtained with minimal attention to the precision of the potential measurement.

Potentiometric titrations can be carried out with all types of chemical reactions (redox, acid-base, complexation, and precipitation) and with all three types of indicator electrodes (active metal ion, inert, and membrane). For some titrations suitable indicators are not available, or indicators are unsuitable because the solution to be analyzed is colored or turbid. In these situations the titration can often be followed by monitoring the concentration

---

[9]Digital pH meters may provide readings to the nearest 0.001 pH unit. Although the last digit in such a readout may be warranted on the basis of the precision of the measurement, an accuracy to 0.001 pH unit is not obtainable.

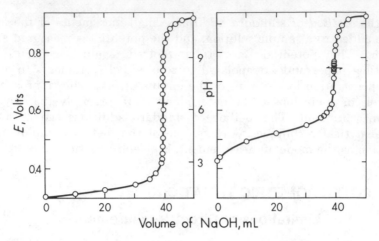

Volume of NaOH, mL

Figure 11-10. Experimental potentiometric acid-base titration curves of 40 mL of 0.1000 M HCl with 0.09924 M NaOH and 40 mL of 0.1000 M acetic acid with 0.09924 M NaOH using a hydrogen indicator electrode and a saturated calomel reference half-cell. Voltage readings for the cell are indicated on the left edge, and pH values on the right. Second-derivative end points are indicated on the curves. Data by W. E. Harris.

of a species potentiometrically. The pH during an acid-base titration can be followed with a pH meter and a reference electrode–pH electrode pair, or the potential during a redox titration monitored with a potentiometer or voltmeter and reference electrode–platinum electrode pair. A plot of pH or potential against volume of titrant added then makes possible precise location of the end point, often by inspection. Figure 11-10 shows experimental curves with a hydrogen indicator electrode.

Sometimes a section of the curve near the end point is expanded and plotted as rate of change of voltage against volume. Here the end point is located at the maximum of the curve. A plot of this type, known as a *first-derivative plot*, often gives a better estimate of an end point than a conventional plot.[10]

EXAMPLE 11–9

In a potentiometric titration the buret readings and the corresponding potentials were as shown in Table 11-3 along with the ratio $\Delta E/\Delta$vol. A plot of $\Delta E/\Delta$vol against volume gives a peak-shaped plot, the highest point of which is taken to be the end-point volume. A schematic representation of the titration curve and its first derivative is included in the table. See also Figure 11-15.

Occasionally a *second-derivative plot*, which passes through zero at the inflection point, is used instead of the first derivative to determine an end point still more precisely.

Theoretical titration curves for the precipitation of halides with standard

---

[10]In general for a series of buret volumes $v_1, v_2 \ldots v_n$ and corresponding meter readings $r_1, r_2, \ldots r_n$ the procedure is as follows. First tabulate the differences between successive meter readings, say $r_{20} - r_{19}$, and successive buret readings, say $v_{20} - v_{19}$. Then plot the quantity $(r_{20} - r_{19})/(v_{20} - v_{19})$ (usually written $\Delta r/\Delta v$) on the y axis against the average volume, $(v_{20} + v_{19})/2$, at that point.

Table 11-3. Example of Tabulation of Data for First-Derivative Curve

Buret Reading, mL	Potential, E	$\Delta E / \Delta$vol	Vol
10.10	1.45		
		0.05/0.02	10.11
10.12	1.50		
		0.20/0.02	10.13
10.14	1.70		
		0.10/0.04	10.16
10.18	1.80		

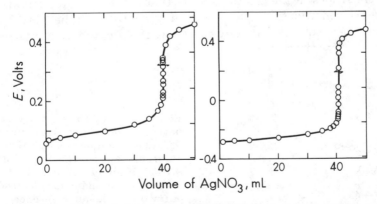

silver nitrate are derived in Section 4-1. Figure 11-11 shows experimental curves for titrations of chloride and iodide with silver nitrate, with a silver wire indicator electrode. Experimental points from the beginning to past the equivalence point are included. If the end point were all that were of interest, only points in the region of the equivalence point would need be recorded and plotted. The curves in the figure adhere closely to theoretical expectation.

Platinum is the preferred indicator electrode for most redox reactions because predictable titration curves are normally produced, provided one of the couples involved in the titration is reversible. Figure 11-12 illustrates an experimental titration curve of iron(II) with dichromate.

Potentiometric titrations of mixtures often yield two or more end points. On titration with standard base a mixture of two acids will yield two end points provided the acid dissociation constants differ significantly. Similarly, mixtures of reductants can be titrated with standard oxidant and several end points obtained provided the standard-potential values differ sufficiently. In precipitation reactions also, mixtures can be titrated and each component

Figure 11-11. Experimental potentiometric titration curves of chloride (40 mL of 0.09964 M KCl) and iodide (40 mL of 0.1023 M KI) with 0.1000 M silver nitrate using a silver indicating electrode and a saturated calomel reference half-cell. Second-derivative end points are indicated on the curves. (Note that for the iodide curve the voltage scale is twice as wide and that at the start of the titration the potential of the silver electrode is more negative than the saturated calomel electrode). Data by W. E. Harris.

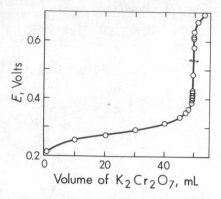

Figure 11-12. Experimental potentiometric titration curve of 50 mL of 0.09923 M Fe$_2$(NH$_4$)$_2$(SO$_4$)$_2$ in about 0.5 M phosphoric acid solution with 0.01667 M potassium dichromate. A platinum indicator electrode was used along with a saturated calomel reference half-cell. Second-derivative end point indicated. [Phosphoric acid reduces the observed voltage through complexation of iron(III).] Data by W. E. Harris.

determined provided the solubility products differ significantly. Figure 11-13 shows the curve obtained for the titration of a mixture of chloride and iodide with standard silver nitrate. The calculated curve is duplicated experimentally, and data for both chloride and iodide can be procured from the two end points.

Since the information of primary interest in potentiometric titrations is the *change* in potential, particularly near the equivalence point, it is unnecessary to know potentials relative to the standard hydrogen electrode. A convenient way to follow changes in potential is by a *differential potentiometric titration*. In this method the reference electrode in a reference electrode–indicator electrode pair is replaced by a second indicator electrode, which is isolated from the main bulk of the solution being titrated by enclosure in a small compartment. In a redox titration the second electrode could be a platinum electrode enclosed in a medicine dropper immersed in the solution. After a portion of titrant is added to the solution, the platinum electrode external to the medicine dropper will respond to the changed conditions in the solution, while that inside the dropper will reflect the previous conditions. A difference in potential will be observed. The dropper is then squeezed and released to replace the contents with the

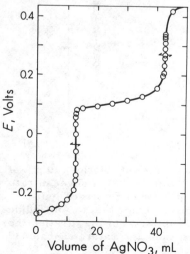

Figure 11-13. Experimental potentiometric titration curve for a 0.5111-g sample containing 20.70% chloride and 31.86% iodide with 0.1000 M silver nitrate, obtained with a silver metal indicator electrode and a saturated calomel reference half-cell. Second-derivative end points are indicated. Data by W. E. Harris.

external solution, titrant is added, and the difference in potential is again observed. At and near the equivalence point this difference is at a maximum, and a plot of volume of titrant against potential resembles a first-derivative curve (Section 11-9).

For many potentiometric titrations ion-selective electrodes can serve as indicator electrodes. For metal ion complexation reactions with EDTA, mercury metal can act as a generalized indicator electrode.

The concentration of many species can be monitored by a variety of methods other than measurement of pH or potential: measurement of current, light absorption, conductivity, temperature, or radioactivity — all can be put to analytical use. Several of these techniques are discussed in later chapters.

11-9 POTENTIOMETRIC TITRATION OF AN ACID MIXTURE

Background

In this experiment a mixture of two acids is titrated with standard base. Because the pH does not change abruptly enough at either equivalence point to permit accurate end points by visual indicators, a potentiometric titration with a pH meter and glass pH electrode–reference electrode pair is employed. A plot of pH against volume of base added yields a curve of the general shape of the one in Figure 5-4. From the curve the two end points, and thus the amount of each acid present, can be found.

A mixture of two acids can be analyzed with a precision of a few parts per thousand provided the dissociation constants differ by at least 10^3 to 10^4 and the weaker acid is not too weak. The molarity of the sodium hydroxide titrant is determined by titration of a standard acid. For highest precision the end points of the standardization and the acid being titrated are taken at the same pH; this procedure corrects for any error due to carbonate that may be present in the base. Because two end points at differing pH values are obtained in the titration of the mixture of acids, calculation of two different effective concentrations for the same sodium hydroxide titrant may be required (Figure 11-14).

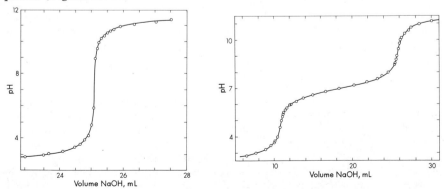

Figure 11-14. Plots of titration with 0.1 M NaOH of sulfamic acid and of a mixture of two acids having pK values of about 3 and 7. Data by D. Cox.

Procedure for Operation of a pH Meter [11]

1. Turn the operating switch to the standby (STDBY) position, and plug the instrument into an electrical outlet. Allow about 15 min for warmup. In the standby position, power is provided to the instrument, but the lead to the glass electrode is disconnected and the amplifier input is shorted. In this position the electrodes may be transferred from one solution to another, rinsed, or handled without damage to the instrument, while the amplifier remains ready for use. Keep the operation switch in the standby position between measurements.

2. To calibrate the meter, immerse the electrodes in a reference solution of known pH. Set the TEMP control to read the temperature of the solution. Turn the operating switch to pH, and adjust the CALIB control to set the meter needle to the pH of the reference solution.

3. Turn the operating switch to STDBY, and remove the electrodes from the reference solution. Rinse the electrodes, wipe them with a tissue, and immerse them in the solution to be measured. Turn the operating switch to "pH"; the instrument will now indicate the pH of the solution. The switch may be left in the pH position during the course of the titration as long as the electrodes are not removed from the solution.

4. When finished with the last titration, turn the operating switch to STDBY and unplug the instrument power cord. After rinsing the electrodes immerse them in a beaker of distilled water for storage.

Procedure for Acid Mixture

(time 5 to 7 h plus 1 to 2 h outside of laboratory for calculations)

Preparation of 0.1 M NaOH Solution

Place about 900 mL of water, freshly passed through a demineralizing column of ion-exchange resin, in a clean polyethylene bottle.[12,13]Add sufficient NaOH to make an approximately 0.1 M solution. Keep tightly covered to minimize absorption of CO_2.

Calibration of pH Meter

Before titrating each standard or sample, calibrate the scale of the pH meter as follows: Place 45 to 50 mL of a saturated solution of sodium hydrogen tartrate monohydrate (solubility at room temperature about 5 g/50 mL, pH 3.56) in a 250-mL beaker. Then follow the calibration steps in Item 2 in the *Procedure for Operation of a pH Meter*. Rinse the electrodes thoroughly with distilled water before immersing them in solution for measurement or calibration.

---

[11]Although the procedure outlined is for a Corning Model 5 pH Meter, it applies in general to most meters in teaching laboratories.

[12]The CO_2 dissolved in water is present predominately as H_2CO_3, HCO_3^-, and CO_2. Passage through a column containing an anion-exchange resin in the OH^- form effectively removes the CO_2 in all forms. Removal may be accomplished also by boiling or by bubbling nitrogen through the water.

[13]Because reagent-grade NaOH may contain considerable Na_2CO_3, it is frequently dispensed as a saturated solution (about 50% NaOH by weight), in which the solubility of Na_2CO_3 is low. In this experiment small amounts of Na_2CO_3 in the NaOH titrant are compensated for in the standardization operations. Nevertheless, care should be taken that CO_2 is not absorbed between the time of standardization and titration of the substances being determined.

Standardization of 0.1 M NaOH Solution

Weigh, to the nearest 0.1 mg, about 2.5 g of sulfamic acid, and transfer it to a 100-mL volumetric flask. Dissolve the acid in freshly demineralized distilled water, dilute to the mark, and mix.[14] Pipet 10-mL portions into 250-mL beakers, and add 40 to 50 mL of the demineralized water to each. Insert glass and reference electrodes into the solution so that the bulb of the glass electrode and the junction at the tip of the reference electrode are immersed in solution. Add a magnetic stirring bar, and position the electrodes and stirrer so that the bar does not strike the electrodes during operation.[15] Titrate with 0.1 M NaOH solution, recording the initial buret reading, and then add titrant until the pH meter indicates a change of about 0.2 pH unit. Record the buret and pH meter readings, and add titrant again until a further change of 0.2 pH unit is observed. Stir the solution thoroughly after each addition of base, and record the readings. Continue in this way until a pH of about 3 is reached. Between pH 3 and 8, record buret and pH values each 0.1 pH unit. Above pH 8, record values each 0.2 pH unit until about pH 11, when the titration may be terminated. For the second and subsequent titrations only data in the region around the equivalence point (between pH 3 and 10) are required, and so one can perform these titrations relatively quickly. Points outside the regions of the equivalence points are unnecessary. Alternate the sample and standard titrations.

Plot the titration data directly on a graph during the first titration to clarify the pH changes through the end-point region. Remember that points distant from the equivalence points are not useful in this determination.

Titration of an Unknown Mixture of Acids

Obtain a sample of an acid mixture. Quantitatively transfer the sample to a 100-mL volumetric flask, using a stream of freshly demineralized distilled water from a wash bottle to remove the last traces. Dilute the solution to volume with demineralized water, and mix. Pipet 10-mL portions of this solution into 250-mL beakers, add 40 to 50 mL of water to each, and titrate with 0.1 M NaOH. Record values each 0.2 pH unit up to pH 3, each 0.1 pH unit from pH 3 to 6 (region of first end point), each 0.2 pH unit from 6 to 8, each 0.1 pH unit from 8 to 10.5 (region of second end point) and finally each 0.2 pH unit until about pH 11. Again graph the first titration as it is completed, and perform subsequent titrations relatively rapidly, recording carefully the volume and pH values only near the two end points.

Plot the data according to the directions for calculations that follow. Selection of the end-point volumes is improved if the region around each equivalence point is plotted on an expanded scale. Buret corrections should be made to the volumes at the end points. The formula weight of H_2NSO_3H is 97.09.

Report the millimoles of each acid in the unknown mixture. From the titration curves, also estimate and report the dissociation constants of the two acids.

---

[14]Solutions of sulfamic acid slowly undergo hydrolysis on standing to give NH_4HSO_4; this does not affect the concentration of strong acid present, since the sulfamic acid is replaced by the equally strong HSO_4^- ion.

[15]A magnetic stirrer is convenient but not mandatory. Manual stirring with a glass rod after addition of each increment of titrant is completely satisfactory. The stirring rod is left in the titration vessel throughout the experiment; it can be used to remove single drops from the buret tip for addition to the solution.

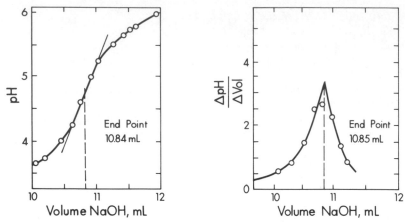

Figure 11-15.   First-derivative plots of curves shown in Figure 11-14. Data by D. Cox.

Calculations

Plot the data for each of the titrations, and draw a smooth curve through the points. The end points for the acid mixture are best located by determining the inflection points in the curves. This can be accomplished either by direct estimation from the curves or by first-derivative plots (Figures 11-14 and 11-15).

Sodium carbonate present in the NaOH titrant causes the molarity of the solution to vary as a function of the strength of the acid. For example, if one titrates a solution of NaOH containing Na_2CO_3 with a strong acid such as HCl, $CO_3{}^{2-}$ is converted to H_2CO_3 by the HCl, and the molarity of the base may be determined by the total NaOH plus Na_2CO_3 present in the solution. On the other hand, if one titrates a weak acid (pK 1×10^{-6}), Na_2CO_3 in the NaOH solution is converted only to $NaHCO_3$, and a different molarity is obtained. Thus molarity varies with pH. The best way to correct for carbonate error is to obtain two NaOH molarities from the standardization curves. The first is calculated from the volume of base required in the standardization, to produce a pH identical with the pH at the first end point in the titration of the unknown mixture. The second is calculated in like manner from the volume of base required in the standardization to produce a pH corresponding to the second end point. Using the NaOH molarity calculated at the lower pH value, calculate the millimoles of the stronger acid in the mixture. Similarly, using the NaOH molarity calculated at the higher pH, calculate the millimoles per gram of the weaker acid in the mixture.[16] Assume that both acids are monoprotic.

The acid dissociation constants for the two acids can be obtained from the midpoints of the titration plots. Determine the pK_a values for the two acids titrated, and record the values beside the graphs of the titrations. Calculate these values at the same time as those for composition.

---

[16]A common error in these calculations is to calculate the total millimoles of acid, then the millimoles of strong acid, and obtain the millimoles of weak acid by difference. The millimoles of weak acid must be calculated from the volume of base required from the first to the second end point.

11-10   POTENTIOMETRIC TITRATION OF CERIUM(IV) WITH POTASSIUM FERROCYANIDE

Background

In this experiment the concentration of a solution containing cerium(IV) in dilute sulfuric acid is determined by titration of an aliquot of standard ferrocyanide solution. The reaction is

$$Ce^{4+} + Fe(CN)_6{}^{4-} \rightleftarrows Fe(CN)_6{}^{3-} + Ce^{3+}$$

The potential at an inert platinum electrode in the solution increases according to the ratio of ferricyanide to ferrocyanide. After all the ferrocyanide has been oxidized, the potential changes rapidly to values readily computed from the ratio of cerium(III) to cerium(IV).

Cerium(IV), a powerful oxidant in a solution of strong acid (Section 10-4), is a useful analytical titrant. The electrode potential of the cerium(IV)–cerium(III) couple depends on the acid concentration of the solution, which controls the extent of hydrolysis of cerium(IV), and by the presence of complexing anions. For prevention of hydrolysis and subsequent precipitation of cerium(IV) hydroxide, cerium(IV) salts are first dissolved completely in concentrated acid and then diluted slowly with stirring. If concentrated acid is not used, the solution may become clear initially but precipitate cerium(IV) hydroxide later. A properly prepared solution of cerium(IV) in sulfuric acid is stable indefinitely and is thus a convenient titrant.

A vacuum-tube voltmeter (VTVM), a potentiometer, or a pH meter (a highly sensitive voltmeter) is suitable for measuring the voltage of a reference electrode–platinum indicator electrode pair. The reference electrode suggested for this experiment is a silver wire coated with silver chloride and immersed in a solution of sodium chloride. The potential of the electrode is determined by the silver(I) activity, which in turn is determined by the solubility product of silver chloride and the activity of the chloride ion. For an aqueous solution at 25°C, a chloride ion concentration of 0.1 M, and an assumption of an activity coefficient of 0.755, the Nernst equation may be written

$$E = E^0{}_{Ag^+,Ag} - 0.059 \log \frac{0.1 \times 0.755}{K_{sp}}$$

where K_{sp} is the solubility product for silver chloride, 1.8×10^{-10}, and 0.059 is the value of $2.303 \, RT/F$ at 25°C for a one-electron process. Substituting numerical values for E^0 and K_{sp} gives

$$E = 0.7994 - 0.059 \log (4.2 \times 10^8) = 0.289 \text{ V}$$

Electrical contact between the reference electrode and the titration

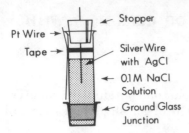

Figure 11-16. Assembly for silver–silver chloride reference electrode. The platinum wire taped to the assembly serves as a convenient indicator electrode.

solutions may be conveniently arranged through a ground-glass joint (Figure 11-16). Mixing of the solutions is generally undesirable. First, it results in loss of part of the solution being titrated, and second, both ferrocyanide and ferricyanide form insoluble precipitates with silver that affect the potential of the reference electrode. Despite the appreciable electrical resistance of the contact between the two solutions, accurate potential readings are possible because a VTVM requires only minute current.

One example of an application of the cerium(IV)–ferrocyanide reaction is in the indirect determination of reducing sugars. An excess of ferricyanide is added to a sugar sample in alkaline solution (pH 10), and the mixture allowed to stand for a length of time that depends on the sugar. Ferrocyanide is formed in an amount equivalent to the sugar present. After reaction the solution is acidified, and the ferrocyanide titrated with standard cerium(IV) solution.

Procedure (time 3 to 4 h)

Preparation of Standard Ferrocyanide Solution

Accurately weigh 1.7 g of $K_4Fe(CN)_6 \cdot 3H_2O$ in a beaker. Dissolve it in distilled water, and transfer the solution to a 100-mL volumetric flask. Add about 1 g of Na_2CO_3 and dilute to volume.[17]

Analytical Procedure

Obtain a sample containing cerium(IV) in solution. Quantitatively transfer it to a 250-mL volumetric flask, dilute to volume, and mix.

To prepare approximately 0.1 M NaCl for the reference electrode, weigh about 0.3 g to the nearest milligram, dissolve it in water, and dilute to volume in a 50-mL volumetric flask.

When ready to begin the titrations, obtain a VTVM[18] and reference-electrode assembly. Coat the silver wire to be used as a reference electrode with silver chloride by immersing it, along with a platinum wire, in 0.1 M HCl.[19] Connect the silver wire to the positive terminal and the platinum wire to the negative terminal of a 1.5-V battery, and allow current to flow for about 30 s. Disconnect the battery, rinse the assembly with distilled water, replace the cap, and partially fill the reference electrode with the previously prepared 0.1 M NaCl solution. Ensure that the ground-glass surfaces are completely wetted by the solution. Store with

---

[17]Sodium carbonate is added to stabilize the ferrocyanide, which decomposes slowly on standing in acid solution. The solution should be used within a few hours of preparation.

[18]The directions for this experiment apply in substance to most vacuum-tube voltmeters or pH meters equipped with a millivolt readout.

[19]If the assembly diagrammed in Figure 11-16 is used, simply remove the ground-glass cap and immerse the assembly in 0.1 M HCl. The platinum wire provides the second connection.

the end immersed in a beaker of distilled water, and rinse the outside with water before use.

Pipet 20 mL of the standard ferrocyanide solution into a 150-mL beaker, and add 25 mL of water. Turn on the meter to allow it to warm up for 10 to 15 min. With the leads shorted, zero the meter with the zero control. When ready to titrate, add about 5 mL of 3 M H_2SO_4 to the ferrocyanide solution, mix, and immerse the reference electrode in the solution. Only the platinum wire and the tip that includes the ground-glass junction of the reference electrode should come in contact with the solution being titrated. Connect the silver wire in the reference solution to the negative (reference electrode) terminal of the meter and the platinum wire to the positive (indicator electrode) terminal. Set the meter switch to (–) or (+) volts or millivolts, whichever causes the meter to read on scale.[20] The sign of the switch position is the sign of the potential of the platinum wire in the test solution relative to the reference electrode. Select a voltage-range setting of 1 or 2 V.

For the first sample titrate with cerium(IV) solution in increments of about 3 mL. It is worthwhile to plot the curve as the data are obtained. For succeeding samples, since only the values near the end point are important, the titrations may be carried rapidly to within 1 to 2 mL of the end point before the potentials are recorded. For these samples titrate dropwise near the end point. After each addition of titrant, stir the solution with the reference electrode, read the buret, and then read the meter. To aid anticipation of the end point, arrange the buret so that the titrant enters near the platinum wire.

Calculate and report the grams of cerium per milliliter of solution.

Calculations

Analytical Calculations

To determine the end point for each titration, plot the first derivative of the titration curve. Plot $\Delta E/\Delta$vol (volume) on the vertical axis against vol on the horizontal axis. Here ΔE is the difference between two successive potential readings, and Δvol the difference in the corresponding volume readings. On the horizontal axis vol is taken to be the average of the two values used to obtain Δvol. In Table 11-3 a plot of $\Delta E/\Delta$vol against vol exhibits a peak at 10.13 mL, which is the end point.

In calculation of the percentage of cerium(IV), the weight in grams of cerium(IV) per titration is obtained from the relation

$$\text{g Ce} = [\text{mL K}_4\text{Fe(CN)}_6][M \text{ K}_4\text{Fe(CN)}_6] \text{ (g Ce/mol)}/(1000 \text{ mL/L})$$

The weight of cerium per milliliter of solution is, then, the above value divided by the volume of the cerium required in the titration.

Supplementary Calculations

This experiment can provide additional useful information. From the titration data the formal potential of the ferricyanide–ferrocyanide couple and also the potential of the reference electrode can be computed. Both should be

---

[20]The (–) position may be required initially. After a small amount of titrant has been added, the potential will read zero. After this point, the (+) position should be used, because the solution potential has become positive with respect to the reference electrode. On vacuum-tube voltmeters the "DC Volts" position is the appropriate one.

reported relative to the standard hydrogen electrode. As indicated in Section 9-5, standard potentials are often not practicable because of unknown activity coefficients, complexation, and other effects, and in such cases formal potentials are frequently employed.

11–11 POTENTIOMETRIC TITRATION OF NTA IN A DETERGENT [21]

Background

The sodium salt of nitrilotriacetic acid, $N(CH_2COO^-Na^+)_3$ or NTA, is added to some detergents as a biodegradable alternative to phosphate for complexing calcium and magnesium. The amount added may be in the range 10 to 20%. In this experiment the weight percent of NTA in a household detergent is determined by potentiometric titration with $Cu(NO_3)_2$. Copper(II) forms a stable 1:1 complex with NTA and is thus a suitable titrant. The indicator electrode is a copper(II) selective electrode.

Procedure (time 3 to 4 h)

Preparation of Standard Copper Solution

Accurately weigh 0.3 g of pure copper metal into a small beaker. Add about 20 mL of 6 M HNO_3, and place it on a hot plate at about 60 to 80°C. Cover the beaker with a watch glass during dissolution to prevent loss of sample. When the copper has dissolved, heat to boiling and boil gently for 2 to 3 min to remove most of the nitrogen oxides. Cool to room temperature, and add 6 M NH_3 dropwise until the deep blue of $Cu(NH_3)_4^{2+}$ just remains. Transfer the solution quantitatively to a 500-mL volumetric flask. Add 50 mL of pH 10 NH_3–NH_4NO_3 buffer, dilute to volume, and mix. To prepare the buffer, add 25 mL of 6 M NH_3 and 2.2 g of NH_4NO_3 to 125 mL of water and mix thoroughly.

Analytical Procedure

Into a 150-mL beaker weigh a detergent sample of a size appropriate to its NTA content. (For detergents containing about 10% NTA, a sample of 2 g is sufficient.) Add 20 mL of water and warm slightly to dissolve the sample. Transfer quantitatively to a 100-mL volumetric flask. Add more water, swirling to complete the dissolution. Add two or three drops of t-butyl alcohol to suppress foaming, dilute to volume, and mix.

Pipet 20 mL of the sample solution into a 400-mL beaker. Add 20 mL of pH 10 buffer and 160 mL of water. Stir the solution gently with a magnetic stirrer and stirring bar. Perform a potentiometric titration, using a copper selective electrode, [22] and a saturated calomel or silver–silver chloride reference electrode. The first titration should be completed quickly and the data plotted to establish

---

[21]This experiment was suggested by H. Yeager, University of Calgary. For a time NTA has been banned in the U.S. because of indications that tumors in rats form when they are injected with NTA in large amounts. Recent evidence has not supported the earlier conclusions.

[22]Copper-selective electrodes are both expensive and fragile. Use care to neither buff the membrane surface nor contaminate it, which would impair the sensitivity. See Section 11-9 for instructions on operation of a pH meter; in this experiment it should be used on the millivolt scale. Since absolute values are not important, the meter does not have to be calibrated.

an approximate end point volume. Repeat the titration with additional samples. During these titrations record data through the equivalence-point region at 0.2-mL increments for about 10 increments. Wait about half a minute after each addition before recording the potentials.

Plot the data carefully in the region of the end point to obtain the end-point volume. A first-derivative plot may be made if desired (Section 11-10). Calculate the weight percent of NTA (mol wt 257.09) in the sample. NTA and copper react chemically in a 1:1 ratio.

PROBLEMS

11-1.   What are the three interfaces at which potentials of analytical interest are measured? How is the sign of potential defined?

11-2.   What is the function of a reference electrode? Give the shorthand representation of (a) a saturated calomel reference half cell; (b) a silver–silver chloride reference half cell.

11-3.   What is a potentiometer? What is the function of a Weston cell in the operation of a potentiometer? Which component ensures that negligible current is used in measurement of voltage with a potentiometer?

11-4.   Calculate the potential of a 2 M calomel reference half cell, using an activity coefficient of 0.55 for the chloride ion.

11-5.   What chemical reactions take place in a saturated calomel half cell when a small current is passed through it?

11-6.   What is direct potentiometry? What are the main sources of uncertainty and how are they minimized?

11-7.   If 0.2538 g of sodium chloride is dissolved in 50.0 mL of water to prepare a solution for a silver–silver chloride reference electrode, what is the potential of the electrode? Is it as well poised as the saturated calomel reference electrode?

11-8.   What errors arise if a cell voltage is measured with significant current flowing?

11-9.   What is a concentration cell? Give an example.

11-10.   What types of cells are used in direct potentiometry?

11-11.   What types of reactions can be monitored with inert electrodes?

11-12.   The potential of a silver electrode in contact with a 0.01 M solution of silver nitrate is +0.2043 V compared with a reference electrode. Assume activity coefficients of unity. (a) What potential would be expected if the solution were 0.002 M? (b) What would be the silver ion concentration of a solution if the potential were 0.2074 V? (c) What would be the silver ion concentration of a solution if the potential were 0.2084 V? (d) Measurement of a potential accurate to 1 mV requires care. What is the percentage difference in the concentrations calculated for the above two potentials? (e) If a precision in concentration of 1% is desired, how precisely would the potential have to be known? (f) If a precision in concentration of 1% is desired, how precisely would the potential have to be known for an ion such as Cu^{2+}?

11-13.   Define liquid-junction potential, and explain how it arises.

11-14.   What is the source of the junction potential arising from a 1 M HNO_3 solution in contact with a 1 M $NaNO_3$ solution? Which side of the junction would be more positive?

11-15.   When is it desirable to minimize junction potentials; to maximize them? How can they be minimized? How can they be maximized?

11-16.   The voltage of a cell with a saturated calomel reference and an indicator electrode pair is 0.721 V, with the indicator electrode the more positive. What is the potential of the indicator electrode relative to the standard hydrogen electrode?

11-17.   For the reaction $Cd(s) + 2AgCl \rightleftharpoons Cd^{2+} + 2Ag(s) + 2Cl^-$ (a) write the shorthand cell representation. (b) If the cell components are at unit activities, what are the potentials of each of the electrodes? (c) If the cell were short-circuited, in which direction would electrons flow?

11-18.   Figure 11-8 depicts a cell that can be used for the measurement of hydrogen-ion activity. (a) Identify the two reference electrodes. (b) Write the shorthand cell representation. (c) Give the purpose of two reference electrodes in a selective electrode system.

11-19.   Discuss how pH is measured with a glass electrode, including in your discussion all the components in the measurement system.

11-20.   In what way do selective electrodes differ from other electrodes with respect to the junction potential?

11-21.   Explain how an ion-selective electrode such as a glass electrode can be used to measure the activity of a species.

11-22.   What is meant by the selectivity constant of an ion-selective electrode? What determines the lower concentration limit of an ion that can be measured by an ion-selective electrode?

11-23.   Define the term asymmetry potential as applied to a pH electrode. How does this potential affect the use of a pH meter?

11-24.   Name three classes of materials used as membranes in selective electrodes. Give examples of materials used and the species each is designed to measure.

11-25.   Describe the components of a glass electrode. When it is used for the measurement of pH, is a redox reaction involved?

11-26.   Describe the components of a solid-state selective electrode.

11-27.   Describe the components of a liquid membrane selective electrode.

11-28.   Selective electrodes are useful for the measurement of substances other than ions. Explain how this is possible, using ammonia as an example.

11-29.   The internal reference solution in a selective electrode must perform two functions. What are they?

11-30.   What is the operational definition of pH?

11-31.   How is the pH scale defined? How may the problems of activity coefficient, junction potentials, and asymmetry potentials be minimized?

11-32.   How accurately can a pH value be measured in routine work; how accurately with special precautions?

11-33.   Suppose that the voltage of the cell in Figure 11-8 was 0.312 V, with a buffer of pH 8.00, and the glass electrode being more positive. In a test solution of unknown pH, the glass electrode became still more positive, and the cell voltage was 0.425. What was the pH of the test solution?

11-34.   In a solution containing Ce^{4+} and Ce^{3+} ions a platinum electrode has a potential of $+1.114$ V compared with an SCE. What is the ratio of the concentration of Ce^{3+} to Ce^{4+} ?

11-35.   When a cell with a fluoride ion-selective electrode was calibrated with a 0.010 M fluoride solution, a reading of 0.104 V. was obtained. A 3.2×10^{-4} M solution gave a reading of 0.194 V. A solution of unknown fluoride activity gave a reading of 0.152 V. What was the fluoride ion concentration of the unknown solution?

11-36. List the major advantages of potentiometric titrations over direct potentiometry.

11-37. A 25.0-mL portion of 0.03 M potassium bromide solution is diluted to 200 mL and titrated potentiometrically with standard 0.03 M silver nitrate using a silver indicating electrode and a saturated calomel reference half-cell. Assuming the solubility-product constant of silver bromide to be 4×10^{-13}, what is the value of $E_{Ag} - E_{SCE}$ at 1, 10, 20, 25, and 30 mL of added titrant? Sketch the general shape of the titration curve.

11-38. Referring to Section 3-1 as necessary, calculate the calcium activity in (a) a solution containing 3.3×10^{-4} M calcium chloride and (b) the same solution after enough sodium chloride has been added to make it 0.05 M in sodium chloride. How much would the potential of a calcium selective electrode be expected to change as the result of the addition of this quantity of sodium chloride? Would the potential shift be positive or negative?

11-39. In a cell like that in Figure 11-8, with a buffer of pH 4.006, a glass electrode is 0.2094 V more negative than the saturated calomel reference half-cell. Calculate the pH of a set of unknown solutions for which the voltages obtained were: -0.3011, -0.0710, -0.4829, and -0.6163 V relative to the SCE.

11-40. A 30-mL portion of 0.1 M solution of vanadyl ion is titrated potentiometrically with 0.1 M iron(II) solution according to $VO^{2+} + Fe^{2+} + 2H^+ \rightleftarrows V^{3+} + Fe^{3+} + H_2O$. Assume that the hydrogen ion activity is unity and the reaction attains equilibrium rapidly. (a) Calculate the potential of a platinum electrode in the solution relative to the standard hydrogen electrode after 10 mL of titrant has been added. What is the potential relative to the SCE? (b) Calculate the potential relative to the SCE at the equivalence point. (c) Sketch the approximate titration curve for this system. Name a suitable reference electrode and indicator electrode for this titration.

11-41. A potentiometric titration of 40 mL of 0.1 M Cr^{2+} with 0.1 M I_2 solution is carried out in acid solution. The reaction is $2Cr^{2+} + I_2 \rightleftarrows 2Cr^{3+} + 2I^-$. Calculate the potential of a platinum electrode relative to the SCE after the addition of (a) 10 mL of iodine solution and (b) 40 mL of iodine solution. On the basis of these two points, sketch the titration curve expected for this system. What is the expected equivalence potential?

11-42. In the presence of excess Cl^-, iridium(IV) occurs primarily in the form $IrCl_6^{2-}$. Stevina and Agasyan (*Zh. Anal. Khim.* **1965**, *20*, 351) advocated the determination of this species by potentiometric titration with Fe^{2+}. Given that for the half-reaction $IrCl_6^{2-} + 4e \rightarrow Ir + 6Cl^-$, $E^0 = +0.835$ V, (a) write the balanced overall equation for the reaction; (b) calculate the value of its equilibrium constant.

11-43. A mixture of potassium iodide and sodium chloride is titrated with silver nitrate in dilute nitric acid solution. In this titration, silver iodide precipitates first, giving an end point that can be detected by potentiometric means. Silver chloride precipitates next, giving a second end point. Between the first and second end points what ion is most likely to predominate in the primary adsorbed layer on the silver iodide precipitate?

11-44. In the potentiometric titration of Ce^{4+} described in Section 11-10, explain the role of the potassium ferrocyanide solution.

11-45. A known amount of strong acid is titrated with sodium hydroxide that contains a small amount of sodium carbonate. The molarity of sodium hydroxide is calculated twice with two volumes of sodium hydroxide. The first is the volume of sodium hydroxide titrant required to reach the pH 4 end point, and the second the volume required to reach the pH 8 end point. Explain the relation between the two calculated molarities of the sodium hydroxide.

11-46. A 1-g sample of an acid mixture took 22.43 and 38.37 mL of standard NaOH for titration to the first and second end points by the procedure of Section 11-9. If the molarities of the NaOH solution were 0.1058 and 0.1012 for titration to the first and second end points, how many equivalents of each acid were present in the sample?

11-47.   What additional information is needed to calculate the molecular weights of the two acids in the preceding problem?

11-48.   In the experiment of Section 11-9 suppose additional carbon dioxide were absorbed by the NaOH solution. How would the molarities of the solution as determined by the procedure used in the standardization of NaOH for the titration of the acid mixture be affected?

11-49.   A 5.220-g sample of an impure cerium(IV) salt was dissolved and diluted to volume in a 250-mL volumetric flask. The volume of cerium(IV) solution required to titrate a 19.95-mL aliquot of 0.0406 M ferrocyanide was 38.54 mL. Calculate the percentage of cerium(IV) in the sample.

11-50.   A cerium(IV) hydroxide sample was dissolved and diluted in a 250-mL volumetric flask. The data recorded were: weight cerium(IV) sample, 4.6741 g; volume standard ferrocyanide taken, 19.97 mL; weight potassium ferrocyanide trihydrate taken to prepare 100 mL of standard solution, 1.7284 g; volume cerium(IV) solution required to reach the end point, 41.20 mL. Calculate the percentage of cerium(IV) in the sample.

11-51.   Calculate the electrode potential of a solution 90% to the equivalence point in a titration of 0.2 M tin(II) with standard 0.1 M cerium(IV) in 1 M sulfuric acid.

11-52.   For a potentiometric titration of 50 mL of 0.1 M ferrocyanide with 0.1 M cerium(IV), calculate the potential of the indicator electrode against a standard hydrogen electrode (a) before any cerium(IV) is added, (b) at the midpoint, (c) at the equivalence point, and (d) 100% past the equivalence point. Plot the titration curve.

11-53.†   One mL of 0.1 M sodium benzoate was added to 50 mL of 0.1 M silver nitrate. The potential of a silver electrode in this solution measured against a reference electrode was found to be +0.300 V. After addition of 80 mL of 0.1 M sodium benzoate the potential was +0.164 V. Calculate the solubility product of silver benzoate.

11-54.   In the plot of the titration of iron(II) with dichromate shown in Figure 11-12, read from the graph the formal electrode potential of the iron(III)–iron(II) couple in the phosphoric acid solution (a) against the saturated calomel electrode and (b) against the standard hydrogen electrode.

REFERENCES

R. G. Bates, *Determination of pH*, 2nd ed., Wiley, New York, 1973. Discusses the definition of pH scales, including the operational definition of pH, liquid-junction potentials and activities, the establishment of pH standards, and the practical measurement of pH.

M. Dole, *J. Chem. Educ.* **1980**, *57*, 134. Describes the development of the glass electrode.

R. A. Durst, Ed., *Ion-Selective Electrodes*, Nat. Bur. Stand. (U.S.) Special Publication 314, 1969. Proceedings of a symposium.

H. A. Laitinen and W. E. Harris, *Chemical Analysis*, 2nd ed., McGraw-Hill, New York, 1974. Discusses the definition of pH scales and deals with electrode potentials, membrane potentials, ion-selective electrodes, and titration curves.

G. A. Rechnitz and N. C. Kenny, *Anal. Letters* **1970**, *3*, 509. Applies the copper selective electrode to the determination of NTA in detergents.

G. F. Smith, *Cerate Oxidimetry*, 2nd ed., G. F. Smith Chemical Co., Columbus, Ohio, 1964. Describes the use of cerium(IV) as an oxidant.

D. S. Tarbell and A. T. Tarbell, *J. Chem. Educ.* **1980**, *57*, 133. Traces the development of the pH meter from Helmholtz in 1881.

R. B. Whitmoyer, *Anal. Chem.* **1934**, *6*, 268. Ferricyanide is used to oxidize sugars such as glucose by addition of excess reagent and back titration with standard cerium(IV) sulfate.

12

OTHER ELECTROANALYTICAL METHODS

The electron is the only really pure chemical we have.

H. Diehl

The main topic of this chapter is the use of current in electroanalytical measurements. The measurement of resistance as an analytical method is also discussed briefly.

Measurement of electrical current (the number of electrons) per unit time passing through a solution is a highly versatile electrochemical procedure that can assist in making quantitative chemical measurements in a variety of ways. In *coulometry* (Section 12-1) current is used to carry out quantitatively a desired chemical reaction; the amount of electricity required is determined by the product of current and time at a current efficiency approaching 100%. In *electrodeposition* a substance is deposited on an electrode and the amount of material undergoing reaction is assessed by weighing the deposit. Either oxidation or reduction processes can be utilized. In several other electrochemical techniques the extent of reaction is held to a negligible level by means of microelectrodes, and the current flowing under these controlled conditions is measured as a function of the concentration of a solute under study. Two examples are polarography (Section 12-2) and amperometric titrations (Section 12-3).

12-1 QUANTITATIVE ELECTROLYSIS: COULOMETRY AND ELECTRODEPOSITION

In coulometric methods the quantity of electricity (in coulombs) required to carry a chemical reaction to completion is measured. At constant current the number of coulombs Q is the product of current i in amperes and time t in seconds, or $Q = it$. If the reaction of interest is 100% efficient, the passage of 96 487 C will cause the oxidation or reduction of one equivalent of a substance. The reaction can be carried out either directly at an electrode or indirectly by reaction with a second substance produced at an electrode.

For chemical reactions such as

$$Ag^+ + e \rightarrow Ag \text{ or } Fe^{3+} + e \rightarrow Fe^{2+}$$

the combination of one or more electrons with a higher oxidation state of a couple brings about reduction to a lower oxidation state. The extent of reaction can be determined by measurement of the amount of current passed in the process. The charge on the electron is 1.602×10^{-19} C. One mole of electrons is defined as the number equal to the number of carbon atoms in 12 g of carbon-12, or 6.022×10^{23} (Avogadro's number). The product of charge on an electron and number of electrons gives the number of coulombs required to bring about one mole of reaction for a one-electron process. This quantity, called the *faraday*, has been carefully determined[1] to be 96 487.

The weight of reactant in a quantitative electrochemical oxidation or reduction is related to the current passed by

$$wt = itM/nF$$

where M is the molecular weight of the reactant, F the faraday, and n the number of faradays in the reaction of one mole of the substance. High precision is possible provided (1) the amount of current is measured accurately and (2) the current is 100% efficient, that is, not consumed in side reactions. High sensitivity is possible because exceedingly small amounts of reagent can be generated.

In coulometry the amount of current may be measured under conditions of either constant current, where the potential may vary with time, or constant potential, where the current decreases with time.

Coulometric Titration (Constant-Current Coulometry)

In constant-current coulometry the current is maintained at a constant value. The reagent added or removed at the working electrode consists of electrons. As a result of chemical oxidation or reduction involving these electrons a chemical reagent is generated that reacts quantitatively with the substance under assay. Because the current is constant, the amount of reagent generated can be determined directly by measurement of time rather than volume. Coulometry is a convenient and precise way of producing a known amount of reagent. No standard solution or titrant-delivery system is required, and so automation of coulometric titrations is practicable. The end point is determined as in conventional titrations either by visual indicators or instrumentally.

A particular advantage of constant-current coulometry is that it permits the use of reagents that are either unstable in solution or difficult to prepare or store. The direct electrochemical generation of the reagent within the

---

[1]A special report on obtaining an accurate value for this extremely important constant gives some of the background behind the currently adopted value for the faraday. See H. Diehl, "High Precision Coulometry and the Value of the Faraday," *Anal. Chem.* **1979**, *51*, 318A.

sample solution makes preparation and standardization of titrant solutions unnecessary.

Because modern electronic instruments are capable of providing currents constant to 0.01% or better and of measuring the time of current passage to better than a part per million, the results of coulometric titrations can be highly accurate and precise. Any convenient method of end-point detection may be chosen. The titrant species is generated at a *working*, or *generator*, *electrode* in the solution[2] being analyzed. If products formed at the second, or *counter*, electrode interfere, it may be isolated from the test solution in a separate compartment, with electrical contact made through a fine glass frit or other junction that allows the current to be carried by ions in solution. Figure 12-1 illustrates schematically a coulometric system for the generation of a reagent.[3] By reversal of the applied potential either an oxidizing or reducing agent can be generated at the working electrode.

What if the substance being determined reacts directly at the working electrode instead of indirectly with the reagent produced at the electrode? Consider the constant-current coulometric titration of iron(III) with electrochemically generated titanium(III). At the beginning the current efficiency in terms of direct reduction of iron(III) may be close to 100% because of the relatively high concentration of iron(III). But, as the concentration falls near the equivalence point, there will be too few iron(III) ions in the vicinity of the cathode, even with efficient stirring, to carry the current being forced through the cell. At this stage constant current must be maintained by some other process. If no titanium(IV) were present, this process might be, say, reduction of hydrogen ions to form hydrogen, whereupon the current efficiency for iron(III) reduction might become significantly less than 100%. By the insertion of an appropriate intermediate such as titanium(IV), the current efficiency can be maintained at near the desired level. Titanium(IV) is reduced to titanium(III) much more readily than hydrogen ion is reduced to hydrogen. The titanium(III) produced is a

---

[2]Although procedures for the generation of a reagent external to the electrolytic cell have been worked out, they are seldom applied.

[3]Modern instruments incorporate more sophisticated arrangements and circuitry than illustrated here; however, the basic principles are unchanged.

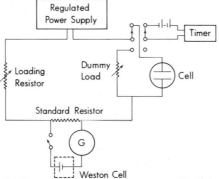

Figure 12-1. Schematic diagram of apparatus for constant-current coulometric titrations. From G. Marinenko and J. K. Taylor, *Anal. Chem.* 1967, 39, 1568; 1968, 40, 1645.

sufficiently strong reductant that it will reduce iron(III) readily to iron(II). Here, then, the titanium(III) ion acts as an intermediate reagent to ensure that each electron generated at the cathode reacts with either (a) an iron(III) ion directly or (b) a titanium(IV) ion to form titanium(III), which in turn reacts with the iron(III) in solution. In summary,

$$Fe^{3+} + e \rightarrow Fe^{2+} \qquad \text{(direct electrode reaction)}$$

$$Ti^{4+} + e \rightarrow Ti^{3+} \qquad \text{(production of the intermediate)}$$

$$Ti^{3+} + Fe^{3+} \rightarrow Ti^{4+} + Fe^{2+} \qquad \text{(chemical reaction)}$$

Thus, all electrons generated result in reduction of iron(III) to iron(II).

Applications

An example of a reagent produced with the generator electrode acting as an anode is the generation of bromine by electrolysis of a bromide-containing solution:

$$2Br^- \rightarrow Br_2 + 2e$$

The bromine produced can be used as a reagent to determine, say, the amount of an olefin. Bromine reacts with the double bonds in olefins (Section 12-6). Iodine, chlorine, cerium(IV), and iron(III) can be generated as reagents for the determination of reductants in solution. A major application of iodine so generated is in the Karl Fischer titration of trace water (Section 10-4).

In a solution containing copper(II) and substantial amounts of chloride ion the reaction

$$CuCl_4{}^{2-} + e \rightarrow CuCl_3{}^{2-} + Cl^-$$

produces the strong reductant $CuCl_3{}^{2-}$. Other reductants such as iron(II), chromium(II), and tin(II) can be similarly generated.

Coulometry is not limited to the generation of oxidizing or reducing agents. It can also be applied to acid-base reactions. Thus hydroxide ions can be produced at a cathode by electrolysis of water:

$$2H_2O + 2e \rightarrow H_2(g) + 2OH^-$$

Hydrogen ions can be produced at an anode by reversal of the cell current:

$$2H_2O \rightarrow 4H^+ + O_2(g) + 4e$$

For precipitation reactions, Ag^+ or $Hg_2{}^{2+}$ ions can be generated from silver or mercury metals for use as reagents. An example is the determination of chloride in serum or urine by coulometric titration with electrogenerated silver(I).

End points in coulometric titrations are usually obtained electrometrically. Potentiometric end points may be taken in the usual way with a reference half-cell, such as the SCE and an indicator electrode. The indicator electrode can be platinum for oxidation-reduction reactions, silver for argentimetric precipitation reactions, or a glass or other selective electrode for acid-base, complexation, or precipitation reactions. The potential of the indicator electrode responds as in a conventional potentiometric titration.

Other attractive end-point detection systems include amperometric methods (Section 12-3), constant-current potentiometry (Section 12-4), and photometric methods (Section 14-10).

EXAMPLE 12-1

It takes 9.729 min to titrate a sample of sodium carbonate coulometrically in an electrolytic cell with electrogenerated hydrogen ions. The generating current is 194.36 mA in a system incorporating platinum electrodes and an electrolyte of 1 M sodium sulfate. What is the weight of sodium carbonate in the sample? Assume that the end point in the titration is obtained when all the carbonate has been converted to carbonic acid.

$$\text{wt Na}_2\text{CO}_3 = \frac{(194.36 \text{ mA})(9.729 \text{ min})(60 \text{ s/min})(105.99 \text{ g Na}_2\text{CO}_3/\text{mol})}{(1000 \text{ mA/A})(2)(96\ 487)} = 0.06231 \text{ g}$$

Constant-Potential Coulometry

In constant-potential coulometry an electrolysis cell with three electrodes is used. One is a standard reference electrode such as the SCE through which no significant current is allowed to flow. A second is a working, or generating, electrode, which is maintained electronically at a preselected potential with respect to the reference electrode by a potential controller, called a *potentiostat.* Current for the chemical reaction of interest passes between the working electrode and a counter, or *auxiliary, electrode.* Electrolytic reactions occur at both the working and counter electrodes, one acting as cathode and the other as anode. The salt bridge from the reference half-cell is placed as close as possible to the working electrode so that the correct potential is maintained without the complication of a drop in potential caused by the electrical resistance of the solution. The current flowing between the working and counter electrodes is passed in series through a *coulometer,* a device for measuring the number of coulombs passed. Two kinds of coulometers are available, chemical and electronic. In a chemical coulometer a reaction such as deposition of silver, evolution of hydrogen and oxygen, formation of silver halide, or generation of hydroxide occurs in an amount exactly equivalent to the reaction at the working electrode. In an electronic coulometer the current is measured by, for example, continuous precise measurement of the voltage drop across a standard resistor. A voltage-to-frequency converter is then used to obtain the current-time product. Electronic coulometers are accurate and are easier to use than chemical coulometers.

Because in constant-potential coulometry the potential is selected so that only the species of interest undergoes electrochemical reaction, the concentration of that species decreases exponentially with time as the electrode reaction proceeds. The current decreases accordingly. With efficient stirring and sufficient electrode area a half-time of a few minutes can be achieved. For quantitative reaction (99.9% +) electrolysis for about 10 half-times is required. A typical reaction is complete in under an hour.

The working electrode at more negative potentials is often mercury metal, and at more positive potentials platinum or carbon. Either oxidation or reduction can be made to occur at the working electrode.

With controlled-potential coulometry it is possible to bring to completion one reaction and then, by an appropriate change in potential, to initiate another reaction. In this way mixtures can be analyzed in successive steps.

EXAMPLE 12-2

An acetic acid–sodium acetate solution containing a bromide salt is electrolyzed at a constant potential of $+0.16$ V against a saturated calomel reference half-cell by use of a silver anode and a platinum cathode. A silver electrode at $+0.16$ V against an SCE in a solution containing bromide produces the reaction

$$Ag + Br^- \longrightarrow AgBr(s) + e$$

At the platinum counter electrode the reaction is

$$2H^+ + 2e \longrightarrow H_2(g)$$

A chemical coulometer in which hydroxide ion is generated is operated separately in series with the electrolytic cell. The hydroxide ion generated is titrated with standard 0.01063 M hydrochloric acid after the coulometric reaction with bromide is complete. For one sample, 5.64 mL is required to reach the end point. How much bromide was in the sample?

$$\text{g AgBr} = 0.01063 \text{ mol/L} \times 79.90 \text{ g Br/mol} \times 5.64/1000 = 4.79 \times 10^{-3} \text{ g}$$

Electrodeposition

In electrodeposition the quantity of current passed is not measured. Electrolysis is allowed to proceed until the sought-for substance is quantitatively deposited on the electrode in pure form, at which time the increase in weight of the electrode is directly related to the amount of substance present. The method works best for metals such as copper, nickel, or silver. When a solution containing copper sulfate is electrolyzed between two platinum electrodes, the reaction at the cathode is

$$Cu^{2+} + 2e \longrightarrow Cu$$

and the reaction at the anode is

$$2H_2O \longrightarrow O_2 + 4H^+ + 4e$$

The copper metal plates out on the surface of the platinum cathode. If electrolysis is continued to quantitative depletion of the copper in solution, the amount originally present can be found directly by determining the increase in weight of the cathode. This is a classical method of analysis for copper or silver. Nickel can be similarly deposited from ammoniacal solutions. Lead can be deposited as the dioxide from nitric acid solution:

$$Pb^{2+} + 2H_2O \rightarrow PbO_2 + 4H^+ + 2e$$

The potential required to carry out an electrodeposition can be calculated from the expression

$$E_{app} = E_{anode} - E_{cathode} + iR + \omega$$

where E_{app} is the potential to be applied, E_{anode} and $E_{cathode}$ are potentials for the couples involved at the two electrodes, iR is potential drop caused by resistance of the cell solution between the two electrodes, and ω is the overpotential.

E_{anode} is determined by the oxidation reaction that occurs most readily in the solution. Correspondingly, $E_{cathode}$ is determined by the reduction reaction that occurs most readily. In water or an aqueous solution of an inert salt the reaction at the anode is oxidation of water to form oxygen, while the reaction at the cathode is reduction of water to form hydrogen. If transition metals are present, the potential at the cathode is often fixed by reduction of a species such as copper, silver, or cobalt.

The iR portion of the voltage required for electrolysis to occur is determined by the resistance of the solution, which depends on electrode size, spacing, and solution conductivity. Typical values in electrodeposition might be a current of 1 A and a resistance of 0.5 ohm. These values correspond to a voltage drop of 1 A $\times$ 0.5 ohm, or 0.5 V.

Overpotential arises when the passage of current is limited by kinetically controlled processes, such as rates of diffusion in solution or rates of electron transfer at the electrode. It represents a departure, as a result of flow of current, of the potential for a couple from the value calculable from the Nernst equation.

Concentration overpotential is the result of depletion or accumulation of species involved in a half-reaction at the surface of an electrode through passage of current. Thus the concentration at the electrode surface may differ from the bulk concentration of the solution. The overpotential increases as the current passed is increased, and decreases as the rate of stirring of the solution is increased.

Often one or more steps in the process of electron transfer at an electrode are slow; in this event the current may be determined by the rate of the reaction instead of equilibrium considerations. Application of a greater potential provides the additional energy required for the process to proceed more rapidly; this additional potential is called the *activation overpotential*. Values of overpotential vary with the nature of the electrode surface, the current density, the temperature, and the reaction taking place at the electrode.

Particularly important are the high overpotentials observed in the formation or reaction of oxygen and hydrogen at various electrode surfaces.[4] Large overpotentials may either help or hinder the analytical chemist. The high value associated with the formation of oxygen requires that a large additional potential be applied in many electrodepositions before the desired electron-transfer process will occur. This may permit other, less desirable, electrode processes. On the other hand, the high overpotential associated with hydrogen permits the reduction of many metal ions at potentials where hydrogen would otherwise interfere. Thus, in 0.1 M H^+, hydrogen ion reduction would be expected at

$$E = E^0_{H^+,H_2} - \frac{0.059}{2} \log \frac{P_{H_2}}{[H^+]^2} \approx \frac{0.0 - 0.059}{2} \log \frac{1}{(0.1)^2} = -0.06 \text{ V}$$

For the reduction of zinc(II) at a mercury electrode ($E^0_{Zn^{2+},Zn} = -0.76$ V) hydrogen ions would be expected to be reduced well before zinc ions. The same holds for all metals having potentials much below hydrogen in the table of standard potentials. But the overpotential of hydrogen on mercury is on the order of 1 V, sufficiently large that metals with potentials in the range 0 to –1 V against the standard hydrogen electrode can be electrolytically reduced at a mercury surface without interference from hydrogen.

EXAMPLE 12–3

The copper in a solution of 0.200 M copper sulfate is to be electrodeposited from 1 M nitric acid solution with two platinum electrodes (Section 12-7). Estimate the voltage required for electrolysis to proceed at a current of 1 A.

The potential of the cathode at the beginning of electrolysis can be calculated directly from the Nernst equation for the copper couple to be +0.32 V. After quantitative deposition the copper ion concentration must be at least 1000 times less than its original value. Suppose that at the end of the deposition it is 10^{-5} M; the potential is then given by

$$E = 0.34 - \frac{0.059}{2} \log \frac{1}{10^{-5}} = 0.19 \text{ V}$$

The potential at the anode is determined by the most readily oxidized species present, water:

$$2H_2O \rightarrow O_2 + 4H^+ + 4e$$

The Nernst equation for calculation of the potential at the anode for the formation of oxygen is

---

[4]Overpotentials are lower with a platinum electrode than with electrodes of other metals. For the discharge of hydrogen with a current density of 1 A/cm² on a smooth platinum electrode the overpotential is about 0.7 V, on a copper electrode about 1.3 V, and on platinized (coated with finely divided platinum) platinum only abut 0.05 V. For oxygen discharge with the same three electrodes and conditions the overpotentials are about 1.5, 0.8, and 0.8 V.

$$E_{anode} = E^0_{O_2,H_2O} - \frac{0.059}{4} \log \frac{1}{P_{O_2}[H^+]^4} = 1.23 - \frac{0.059}{4} \log \frac{1}{(0.2)(1)^4} = 1.22 \text{ V}$$

The partial pressure of oxygen may be taken initially to be 0.2, the level in the atmosphere. After O_2 is generated in solution for a time, this value will increase to near 1; if 1 is used, E_{anode} becomes 1.23 V instead of 1.22 V, a small difference.

The concentration overpotential at the cathode will depend mainly on the current density and the efficiency of stirring of the solution. If copper ions are not available at the cathode surface for discharge, then hydrogen will be discharged instead, with a lowering of the current efficiency. The activation overpotential for copper is negligible. At the anode the concentration overpotential is not significant, since the species undergoing oxidation is the solvent water. The activation overpotential at the anode is considerable; the discharge of oxygen on a smooth platinum surface requires about 1.5 V for a current of 1 A/cm^2.

Including the iR drop of 0.5 V, the required cell voltage for a current of 1 A/cm^2 of anode surface is then 1.23 – 0.19 + 1.5 + 0.5 = 3.04 V. The cell voltage must be somewhat higher than this if the resistance of the solution between the two electrodes is greater than 0.5 ohm.

As electrolysis proceeds, E_{anode} will change but little, since neither the pressure of oxygen nor the hydrogen ion concentration varies appreciably. However, as the concentration of copper ions decreases, the potential of the cathode will change.

If the voltage is held constant, the iR term will become smaller because the current flow must drop as copper ions are depleted at the cathode. The decrease in current will also decrease the overpotential. As electrolysis proceeds beyond the point where the copper has been quantitatively deposited, $E_{cathode}$ will continue to decrease to about –0.5 V. At this point reduction of hydrogen ions to hydrogen gas begins. This value corresponds to E for the hydrogen couple,

$$E = E^0_{H^+,H_2} - \frac{0.059}{2} \log \frac{P_{H_2}}{[H^+]^2} = 0.0 \text{ V}$$

minus the overpotential for hydrogen evolution at copper, 0.48 V at low current levels, or about –0.5 V. The potential of the cathode will therefore be controlled from here on primarily by the hydrogen ion concentration. The concentration of copper ion at the electrode surface is

$$-0.5 = 0.36 - \frac{0.059}{2} \log \frac{1}{[Cu^{2+}]}$$

or $[Cu^{2+}] = 10^{-29}$. Thus the copper is clearly quantitatively deposited before reduction of hydrogen ion begins.

EXAMPLE 12-4

What is the minimum time required for the electrodeposition of the copper from a 1-g sample containing 20% copper with a current of 1.4 A at 50% efficiency?

$$t = \frac{g \times n \times F}{i \times \text{mol wt}} = \frac{1 \text{ g} \times 0.2 \times 2 \times 96\ 487 \text{ A/s}}{1.4 \text{ A} \times 0.5 \times 63.546 \text{ g/mol}} = 868 \text{ s or } 14.5 \text{ min}$$

12-2 POLAROGRAPHY

In potentiometry and potentiometric titrations, potentials are measured under conditions designed to ensure that little or no current passes. Flow of current disturbs the values obtained. In coulometry and coulometric titrations, current is intentionally passed through an electrode until a reaction is brought to quantitative completion. In *voltammetry* current is measured as a function of an applied and changing voltage at a microelectrode, usually of platinum, gold, or carbon. When the electrode consists of mercury passing through a capillary as in Figure 12-2, the process is called *polarography*. In this case the electrode consists of a series of mercury drops that continually form and fall. In voltammetry, polarography, and amperometric titrations (Section 12-3) the current that flows as a result of the imposed potential is both small and limited in time; accordingly, no significant change occurs in the composition of the solution under study.

Typically, a voltage, which may vary with time, is imposed across an electrolytic cell consisting of a microelectrode as working electrode and a reference electrode of sufficient area that no concentration-dependent effects occur, and the current is then measured. If the applied potential is sufficiently large, an electrochemical reaction takes place. A recording of the current measured as a function of the imposed potentials in an unstirred solution produces a current-voltage curve called a *polarogram*, when a dropping mercury electrode acts as the working electrode.

Observing the current that will flow between a working and a reference electrode as a function of potential provides information as to the kind and concentration of species present in solution. Current flow at the working electrode can arise from transient or from long-term effects.

One kind of transient current is *charging,* or *nonfaradaic, current.* When a

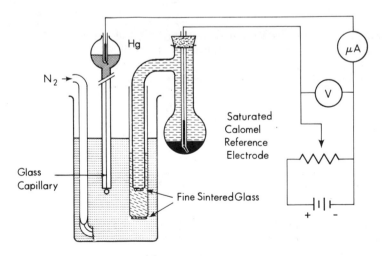

Figure 12-2. **Two-electrode polarographic cell   (25 × 100 mm) with a dropping mercury electrode (marine barometer tubing) and a saturated calomel reference half-cell in a 10-mL distillation flask.**

potential is applied to an electrode, a few electrons must flow to or from the electrode until it has acquired a charge corresponding to the applied potential, a process analogous to charging a capacitor. Most of the charge is acquired in a tiny fraction of a second. The current from this source may be large momentarily, but quickly becomes zero.

A second kind of transient current may occur when an electroactive species is present. For example, in a solution containing 0.001 M copper nitrate in 0.1 M potassium nitrate, copper ions in a thin film of solution next to the platinum surface will be reduced to copper metal when a sufficiently negative potential is applied. The thickness of the film depends on several factors, but is small under any set of conditions. The total number of copper(II) ions initially present in this thin layer of 10^{-3} M solution around the microelectrode will not be large, but during reduction a small measurable current is produced for a brief period (seconds or less). We will label this current $i_{initial}$; it is *faradaic* current.

For flow of current to continue in the long term, ions from the bulk of the solution must be transported to the region near the electrode surface. Movement of ions to an electrode can occur by any of the following three processes: (1) *Migration*. Under the influence of an electric field, positive ions move toward a negatively charged electrode and away from a positively charged one. (2) *Convection*. When a portion of the entire solution moves, material may be carried to an electrode surface by essentially mechanical means (stirring, for example). Convection can also arise from thermal or density gradients. (3) *Diffusion*. Molecules and ions in a solution are in continuous random motion, diffusing in every direction. Net diffusion proceeds from regions of higher to those of lower concentration, at a rate directly proportional to the *concentration gradient*, the concentration difference divided by the distance. The proportionality constant is called the *diffusion coefficient*. Concentration gradients of reacting species decrease with time.

The total current i_{total} at an electrode is, then, the sum of five processes:

$$i_{total} = i_{charging} + i_{initial} + i_{migration} + i_{convection} + i_{diffusion} \qquad (12\text{-}1)$$

Recall that the first two are transient, become quickly zero, and are of little interest for most polarographic studies. Here current from sources such as impurities or adsorption is neglected.

The migration current can either add to or subtract from the total current, depending on the charges of the ion and the electrode. That is, ions can be either attracted to or repelled from the electrode. When electrochemically active ions are attracted to the electrode, the current is enhanced. In practice a *supporting*, or *indifferent*, *electrolyte*, consisting of a relatively high concentration of nonelectroactive ions, is added to the solution to be analyzed. In our 10^{-3} M copper example the solution contains the salt potassium nitrate at a concentration 100 times that of the copper nitrate concentration. In such a solution both copper and potassium ions are attracted to a negatively charged electrode. Potassium ions are not reduced at any of the potentials in the example; they merely move to the electrode and form a layer near it in numbers just sufficient to neutralize the applied electric field relative to the bulk of the solution.[5] The layer is diffuse and a few ions in thickness. At sufficiently negative potentials, copper ions near the electrode are immediately reduced and therefore do not contribute to neutralization of the electric field. If the concentration of the supporting electrolyte is at least a hundred times that of the electroactive species, the migration current is negligible and need not be considered further. In the case under consideration, it can be assumed to be zero.

Convection currents in quiet solutions are not readily amenable to theoretical description.[6] If, however, the solution is not stirred mechanically or by density or thermal gradients, then convection and hence the current from this source is negligible. Density and thermal gradients in most solutions are negligible provided that measurements are carried out at constant temperature with small electrodes within a reasonable time (less than a minute).

Soon after a potential has been applied, under the conditions described, only the diffusion current continues to contribute to the overall current. Then

$$i_{total} = i_d$$

where i_d denotes diffusion current. As indicated, net movement from a diffusion process is proportional to the concentration gradient.

At a working electrode the concentration gradient will be at a maximum when the concentration at the surface of the electrode C_0 approaches zero.

In polarography the term diffusion current normally refers only to the limiting value, that is, the maximum possible diffusion current. This value is directly proportional to the concentration of the electroactive species in solution.

The dropping mercury electrode (DME) consists of a mercury reservoir attached to a small–bore capillary of such diameter that a new drop of mercury forms and falls every 3 to 6 s. Because the potential is reapplied with each drop, a decreasing concentration gradient with time is no longer a problem. The electrode has other attractive properties. For example, the surface area of each mercury drop is reproducible for a particular capillary; with constant renewal of the surface every few seconds, contamination of a drop by electrochemical adsorption or other reactions is removed when the drop falls. Another advantage is that it has a wide range of potential, especially in the negative region. The maximum negative potential (cathodic limit) is set by the reduction of hydrogen ions from water or acids. Since the standard potential for the hydrogen ion–hydrogen half-reaction is –0.244 V against the SCE, this means that in a solution with a hydrogen ion activity of unity the potential at which hydrogen ion reduction is initiated would be –0.244 V. At a pH of 10, hydrogen ions would be reduced at about –0.85 V

---

[5]Nitrate ions form a similar diffuse layer at a positively charged electrode.

[6]For certain systems, as with flow-through electrodes (tubular platinum, gold mesh) and rotating ring-disk electrodes, theoretical expressions have been derived that fit experimental data.

relative to the SCE. Fortunately, however, hydrogen ion reduction does not occur at a significant rate on a mercury surface until the potential is about 1 V more negative (see "overpotential" Section 12-1) than these values. Until higher potentials are applied, the passage of current is limited by a kinetically controlled electrode reaction rather than by diffusion. As a result the DME has a highly favorable cathodic potential limit, somewhat greater than –1 V relative to the SCE in acid solutions and about –2 V in alkaline solutions. Since the DME can be used at highly negative potentials without the complication of current from the discharge of hydrogen ions, this electrode is convenient for the study or analysis of many substances that could not otherwise be investigated.

Another property of mercury is that many metals are soluble in it, forming solutions called *amalgams*. The formation of an amalgam lowers the activity of the metal in the mercury drop, making it possible to reduce electrolytically the ions of highly active metals. At high pH values, for instance, sodium or potassium ions can be reduced from aqueous solution. In summary, the favorable cathodic potential range of mercury, along with other desirable properties, makes it the most useful of all electrode materials for electrolytic reductions in aqueous solutions.

The maximum positive potential of the mercury electrode (anodic limit) is normally established by the potential at which metallic mercury begins to be oxidized. $2Hg \rightarrow Hg_2^{2+} + 2e$. The E^0 for this half-reaction in aqueous solutions is $+0.79$ V ($+0.55$ relative to the SCE). Current from this oxidation of mercury becomes significant at about 0.4 V positive to the SCE. Therefore, the normal range in potential over which current-voltage curves can be applied analytically with the DME is about $+0.4$ to -1.8 V relative to the SCE. The anodic limit of mercury will be less positive than $+0.4$ V when the solution contains complexing anions, such as chloride ions.

Two complications arise with the DME as a result of the change in electrode size that takes place as each drop starts, grows, reaches maximum size, and then falls and a new one forms. One is the necessity of charging the continually forming new electrode surface. This process requires the movement of electrons in or out and therefore passage of a small current. For 10^{-3} M solutions of electroactive sample species the charging current constitutes about 1% of the diffusion current and can be neglected. This current decreases as the drop grows. For more dilute solutions the charging current is proportionately larger and must be taken into account.

The second complication is that just after each drop of mercury falls the electrode surface area is near zero, and the diffusion current then must also be near zero. As the drop grows in size, the area and therefore the diffusion current continually increase, both reaching a maximum at the instant just before the drop falls. Successive drops are normally reproducible, so that the change in current with time on a short time scale is highly reproducible even though the current is never steady. Averaging the current over time yields a value 6/7 of the maximum current.

The magnitude of the diffusion current at a DME can be derived by application of the laws of diffusion and the geometric characteristics of an expanding electrode. This derivation was made by Ilkovic in 1934. In its simplest form, the Ilkovic equation for faradaic diffusion current is

$$i_d = 708 \, nCD^{1/2}m^{2/3} \, t^{1/6}$$

where i_d is the maximum diffusion current in microamperes, n the number of electrons per mole of electrode reaction, C the concentration in millimoles per liter, D the diffusion coefficient of the electroactive species, m the mass of mercury flowing from the capillary in milligrams per second, and t the drop time in seconds. The quantity $m^{2/3}t^{1/6}$ is called the *capillary constant*.

For a given capillary and electroactive substance, all the quantities in the Ilkovic equation except concentration can be consolidated into a single constant. For a copper-containing solution we can write

$$i_d = k_{Cu} \, C$$

where k_{Cu} is the proportionality constant. That is, the diffusion current is directly proportional to the concentration.

Either reduction or oxidation can occur at a DME. A solution containing $K_3Fe(CN)_6$ will be reduced at dropping mercury electrode with a diffusion current equal to $k_{Fe(CN)_6^{3-}}C$.   Similarly, a second solution containing $K_4Fe(CN)_6$ may be oxidized with a diffusion current equal to $k_{Fe(CN)_6^{4-}}C$. A solution containing a mixture of the two will be either oxidized or reduced, depending on the potential.

Instrumentation

Modern polarographic instruments rely to a large extent on operational amplifiers to amplify current, maintain voltages, integrate currents or voltages, follow voltages, and differentiate signals. We do not consider these aspects of instrument design here, but emphasize the basic components that enable the concepts of polarography to be more readily introduced.

In its simplest form a polarograph is an instrument designed to apply a continually changing potential to a cell consisting of a DME and a saturated calomel reference half-cell. The current that results from reduction or oxidation of electroactive species in solution is usually plotted on a strip-chart recorder as a function of potential. Figure 12-2 depicts schematically a simple two-electrode polarograph.

A disadvantage of this cell arrangement is the voltage (iR) drop across the cell when the resistance of the solution is high. In potentiometric measurements the cell resistance may be a million ohms; with glass electrodes the resistance may be ten or a hundred million ohms. These high resistances are not of great concern in potentiometry because the measurements are made with zero or nearly zero current flow, so that iR drop is negligible. In polarography, however, both potential *and* current must be measured. Since

---

[7]The constant for the oxidant species k_{ox} is given by the relation $k_{ox} = i/(C_{ox} - C_{0ox}$, which includes a factor for the diffusion coefficient.

flow of appreciable current must take place, an iR drop must exist across the cell. Hence, the true potential of the DME cannot always be found by simply subtracting the potential of the SCE from the total voltage applied to the cell; often it must be corrected for iR drop. Resistance comes from several sources: the DME, the solution in the cell, the salt-bridge connection to the reference electrode, and the reference electrode itself. It is essential to be aware of the resistance of the salt bridge connecting the reference half-cell and the solution; reference electrodes commonly used with pH measurements have far too high a resistance for incorporation into two-electrode polarographs. Therefore both half-cells must be designed with sufficient cross-sectional area so that appreciable current can be drawn with negligible voltage drop. Normally a total resistance across a polarographic cell of more than a few hundred ohms produces a significant iR drop.

EXAMPLE 12-5

Calculate the iR drop across a polarographic cell having a current of 10 μA and a resistance of (a) 10 000 ohms; (b) 1000 ohms; (c) 100 ohms.

For 10 000 ohms the iR drop will be $10\,000 \times 10 \times 10^{-6} = 0.1$ V; for 1000 ohms, 10 mV; and for 100 ohms, 1 mV.

If the current were 3 μA, the iR drop for 1000 ohms would be 3 mV.

According to Example 12-5, if 0.5 V has been applied across the cell with the DME negative to the SCE, at a current of 10 μA the potential of the DME would be –0.4 V with a resistance of 10 000 ohms, –0.49 V with 1000 ohms, and –0.499 V with 100 ohms.

Occasionally, the resistance across the test solution cannot be reduced to a negligible level, particularly in nonaqueous solvents. When resistance is high, designs incorporating three electrodes are necessary. Here the potential of a DME is measured against a reference half-cell as before, but the current that flows through the mercury electrode on oxidation or reduction of an electroactive species is not made to flow through the reference half-cell. Instead, a third (auxiliary) electrode is inserted in the solution along with the DME to form an electrolytic cell through which the current flows. In this way the iR drop does not affect the potential measured between the reference and working electrodes. The potential difference between the auxiliary electrode and the DME now need not be known; the current flowing between them is measured as before. With three-electrode systems somewhat more complex instrumentation is required, but advances in electronics have made the apparatus both reliable and accurate. Most polarographs on the market today are equipped to operate with three electrodes.

Techniques of Polarography

The DME can be utilized for the analysis of both inorganic and organic compounds, either molecular or ionic, that undergo oxidation or reduction within the cathodic and anodic limits of the mercury electrode, the supporting electrolyte, and the solvent. Polarographic measurements are not confined to systems that undergo reversible redox reactions, though reversible systems do tend to yield sharper waves. Most organic reactions appear irreversible; exceptions include those involving the quinone class of compounds. [8]   Polarographic studies have been used for nonanalytical purposes such as characterizing the redox properties of many systems. For example, half-wave potentials may provide useful information on the rate and mechanism of a reduction process.

For organic substances an organic solvent may be necessary. In such cases high resistance may be a problem, since low solubility may not permit a high enough concentration of indifferent electrolyte, or ion-pairing may tie up current-carrying ions.

Quantitative polarographic measurements can be carried out in several ways. One is by an absolute method through direct application of the Ilkovic equation. If the capillary constant for the electrode, the value of n, and the diffusion coefficient for the electroactive entity are known, then the concentration of that species in the test solution can be calculated. This method is rarely used in real analyses.

In the comparison technique the current-voltage curve for a standard solution of the substance of interest is obtained, and from the diffusion current the ratio i_d/C is calculated. This quantity is normally about 4 μA for a millimolar solution of an electroactive substance with a one-electron redox reaction. The diffusion current for the test solution is then measured under as nearly the same conditions as possible. The concentration is calculated directly from the ratio of diffusion current to concentration. With this technique none of the capillary constant, n, or D have to be known. Since diffusion coefficients are temperature-dependent, the temperatures of the standard and sample solutions must be the same during the measurements.

A third method of measurement involves standard addition. Here the diffusion current for the test solution is measured. A known amount of the test substance is then added, and the diffusion current measured again. From the increase in the diffusion current the original concentration can be calculated. This technique is most precise when the amount added approximately doubles the original diffusion current.[9]

In general, the precision obtainable by polarographic means is in the range of a few percent in 10^{-2} to 10^{-4} M solutions. With care a precision of about 1% can be attained. The lower limit of concentration for analytical work is about 1×10^{-5} M.

In careful polarographic work a blank is run in which the magnitude of the charging current is determined and then subtracted from the diffusion-current values taken with the sample present (Figure 12-3).

---

[8]This apparent irreversibility occurs because polarographic measurements are relatively slow. By means of sufficiently rapid scanning with advanced equipment most organic reactions are reversible. The apparent irreversibility observed is generally caused by slow chemical reactions following electron transfer.

[9]A fourth method employs an internal standard. A second electroactive substance is added to the solution and a polarogram obtained. The capillary constant, the temperature, and the supporting electrolyte solution affect the diffusion current of both the added substance and the unknown to the same extent. The diffusion current of the unknown is related to that of the standard, as in the standard-addition method. This last method is restricted in that half-wave potential values must differ enough that the two polarographic waves are well separated.

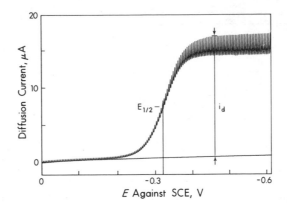

Figure 12-3. Polarogram of $10^{-3} M$ TlCl in 0.1 M KCl along with the charging current in 0.1 M KCl using a dropping mercury electrode. **Data by S. Pons.**

The solubility of oxygen gas in water in equilibrium with air gives the solution an oxygen concentration comparable to that of the sample species being studied. In practical polarography, particularly with a DME, the reduction of this dissolved oxygen in two steps over a potential range of 0.5 to –1.2 V, depending on pH, may cause interference. Removal is normally accomplished by bubbling oxygen-free nitrogen through the solution (Figure 12-2).

Platinum, gold, or carbon microelectrodes are often suited to voltammetric studies. Platinum electrodes are usually rotated at constant speed to enhance sensitivity and improve reproducibility.

Rotation markedly reduces the thickness of the diffusion layer; diffusion currents are normally about two orders of magnitude higher than for the DME. The anodic limit for the platinum electrode is the oxidation of water, about +1 V against an SCE:

$$2H_2O \rightarrow O_2 + 4H^+ + 4e$$

The cathodic limit is hydrogen reduction, which depends on the pH; at pH 7 it is about –0.8 V against an SCE.

One of the most common voltammetric determinations using either a platinum or a gold cathode is that of dissolved oxygen. The cathode is placed in a cell with a silver-anode reference half-cell and a solution of dilute potassium chloride. The electrodes and cell are separated from the solution under test by a thin gas-permeable membrane such as Teflon. A potential of about –0.8 V is applied to the platinum electrode,[10] which is sufficient to reduce the oxygen diffusing through the membrane to water. The current is proportional to the concentration of oxygen. Halogens and sulfur dioxide interfere, either through direct reduction at –0.8 V or by reaction with oxygen. Hydrogen sulfide acts as a poison on the electrode.

Carbon may be substituted for mercury or platinum as a voltammetric electrode; its anodic limit is about +1.3 V, and its cathodic limit about –1.3 V.

---

[10]This is often referred to as the Clark electrode, after its developer.

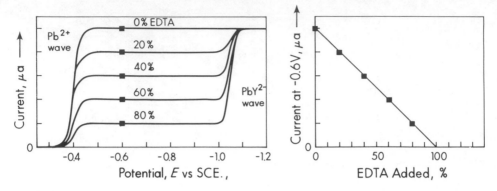

Figure 12-4. Current-voltage curves for Pb^{2+} and Pb^{2+}-PbY^{2-} mixtures, and a plot of the amperometric titration of Pb^{2+} with EDTA at –0.6 V against a saturated calomel electrode. The residual current has been subtracted from all readings. Diffusion coefficients of the two species are assumed to be the same.

12–3 AMPEROMETRIC TITRATIONS

The progress of a reaction can be followed during a titration by monitoring the diffusion current of a reactant or product. If diffusion current is measured as a function of amount of titrant, an *amperometric titration curve* is obtained. Since diffusion current is normally proportional to concentration, a plot of diffusion current against titrant volume consists of two straight lines that intersect at the end point. Amperometric titrations can be performed with many types of neutralization, precipitation, redox, and complexation reactions.

EXAMPLE 12–8

Consider an amperometric titration of a dilute solution of lead(II) with a standard solution of EDTA to form PbY^{2-}. The half-wave potential for lead(II) is about –0.4 V against the SCE, while that of PbY^{2-} is about –1.1 V at pH 4. If the potential of the indicator electrode (a DME or rotating platinum electrode) is set at a voltage somewhere in the diffusion-current region of lead(II), say –0.6 V, the current observed will be directly proportional to the Pb^{2+} concentration in solution. When EDTA is added, PbY^{2-} forms and the Pb^{2+} concentration decreases; accordingly, the diffusion current decreases. The diffusion current at –0.6 V continues to fall while standard EDTA solution is being added, until the end point is reached. After the end point the overall current, other than for the residual current, is zero. Figure 12-4 shows the current-voltage curves expected for lead(II) ions in solutions containing various mixtures of lead(II) and PbY^{2-}, along with the amperometric titration curve for the diffusion current at –0.6 V as a function of the percentage of EDTA added. The titration plot shows the current decreasing linearly with volume of titrant; the end point is obtained by extrapolation of this line to its intersection with the straight line representing the residual current at –0.6 V (or with the x axis, provided the residual current has been subtracted from all current readings).

A common type of amperometric titration is one in which the test substance is not electroactive, but the titrant is. Titration of EDTA with a

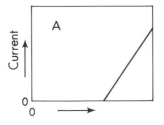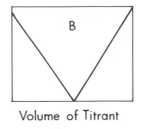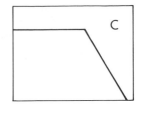

Figure 12-5.   Three types of amperometric titration curves: (*A*) titrant electroactive at potential used; (*B*) both titrant and substance titrated electroactive at potential used; (*C*) amperometric indicator.

standard solution of lead is an example. For this amperometric titration system the diffusion current at –0.6 V would remain at zero until the end point and increase linearly thereafter.

In amperometric titrations of electroactive sample species only a few readings of current are necessary prior to the end point; three or four are sufficient to establish the slope of the line so that it can be extrapolated to zero net current. In the reverse titration, as in Figure 12-5*A*, the situation is similar in that only a few readings of the residual current need be taken before the end point and three or four after to establish the slopes of the straight lines. Readings at or near the end point are unnecessary.

Precise end points are also possible when both the reagent and the test substance are electroactive at the potential used. Figure 12-5*B* illustrates this type of amperometric titration curve. The angle of intersection of the two straight lines may be sharp. Several other types have been developed. Figure 12-5*C* illustrates the type expected when neither sample nor titrant is electroactive and an amperometric indicator is employed. An indicator substance is chosen that is electroactive but does not react with the titrant until after the reaction of interest is complete.

Because the analyst can select the potential at which measurements are to be taken, he has considerable latitude in devising amperometric titration methods. For mixtures the correct choice of potential can provide a valuable measure of selectivity. The number and variety of useful amperometric titrations is large because only one species — reagent, test substance, or product — need be electroactive at the potential chosen. The electrode in these titrations does not respond to a particular species in solution except where conditions are arranged for it to do so. Many different ions or molecules give a diffusion current at some potential at either a platinum or mercury microelectrode. The fact that the method is not restricted to reversible reactions substantially extends its range of applicability.

For amperometric titrations the most attractive electrode, where applicable, is the rotating platinum electrode (RPE). It is used whenever possible at positive potentials, where the oxygen wave does not interfere. The electrode also performs the necessary stirring of the solution during a titration. Readings of diffusion current can thus be taken immediately after

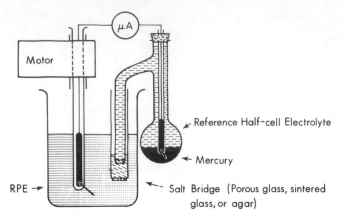

Figure 12-6. Apparatus for amperometric titration using a rotating platinum electrode and a reference half-cell with the correct potential in a 10-mL flask with a side arm. The RPE and reference half-cell are short-circuited through a sensitive low resistance current-reading device.

addition of a portion of titrant. Another advantage of this electrode is that it has a sensitivity about 100 times that of the DME, so that solutions of about $10^{-5}M$ can be titrated easily. Thus it is useful for the determination of traces with high precision.

One widely applied amperometric method using the RPE is the determination of traces of sulfhydryl-containing substances by titration with standard silver nitrate. Diffusion currents are read at -0.23 V against the SCE, a potential just positive enough to avoid interference from the oxygen wave. The titrations are carried out in ammoniacal solution to prevent chloride or traces of bromide from precipitating and interfering. The reaction is

$$Ag(NH_3)_2^+ + RSH \rightarrow RSAg(s) + NH_4^+ + NH_3$$

where RSH denotes a mercaptan. A large number of amperometric titrations can be carried out using halogen diffusion currents either before or after the end point of reactions involving iodine, bromine, bromate, or iodate as titrants. Here too, readings can be taken without interference from oxygen. Since many materials react with one or more of the halogens, such titrations are widely applicable.

For routine amperometric titrations with a rotating electrode it is convenient to employ a reference half-cell whose potential is the value that gives the desired potential for the indicator electrode. No external source of voltage is then needed, and the rotating electrode is merely short-circuited with the reference half-cell through a sensitive current-measuring instrument. The complete apparatus (Figure 12-6), is among the simplest of electrochemical measuring instruments.

In summary, amperometric titration with a rotating platinum electrode is a technique by which analyses for traces can be performed with high precision. Only in rare cases must corrections for dilution be applied. The procedure can be performed rapidly, typically within a couple of minutes.

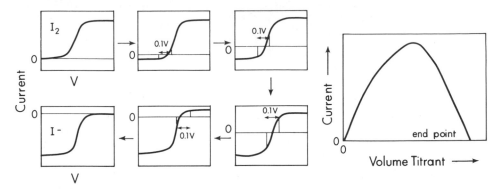

Figure 12-7. Schematic diagram of amperometric titration with two electrodes of iodine with standard sodium thiosulfate showing the current-voltage curves at six stages of the titration along with the overall titration curve.

Amperometric Titration with Two Indicator Electrodes

Suppose two small platinum electrodes are immersed in a solution and a difference in potential of about 0.1 V is applied between them. The more positive will act as an anode, and to the extent that oxidation can occur at this electrode, electroactive substances will be oxidized. The less positive will act as a cathode, and to the extent that reduction can occur at this electrode, electroactive substances will be reduced. Current cannot flow between the electrodes unless there is present a substance that can be oxidized at the anode and another that can be reduced at the cathode, with a potential difference of no more than 0.1 V. For example, if a pair of electrodes with a small potential difference is immersed in a solution containing both iodine and iodide, then at the more positive electrode iodide is oxidized to iodine, and at the more negative electrode iodine is reduced to iodide.

If a solution of iodine is titrated with a nonelectroactive reductant, initially only iodine is present in solution. Reduction of iodine would occur easily at the cathode, but not oxidation at the anode, because no species is present that will undergo oxidation at a potential only 0.1 V positive to iodine reduction. Since the amount of current flowing as a result of oxidation at the anode must equal that flowing as a result of reduction at the cathode, the total current must be zero (except for accidental traces of a reducible substance). As soon as a portion of the iodine has been reduced to iodide, both oxidation and reduction can occur and therefore a current will flow. The varying ratio of iodide to iodine during the titration will generate a current flow that will go through a maximum when the mixture is approximately 50:50. It will decrease thereafter, going to zero when the end point has been reached and all the iodine is reduced. At the end point no reaction occurs at the cathode. The complete titration curve is shown in Figure 12-7. Generally, only changes in current near the end point are of interest. In this example, the relation between current and volume of titrant just before the end point becomes nearly linear. With a nonelectroactive titrant the current becomes zero at the end point and remains so.

If the substance titrated, the titrant, and the product are all electroactive,

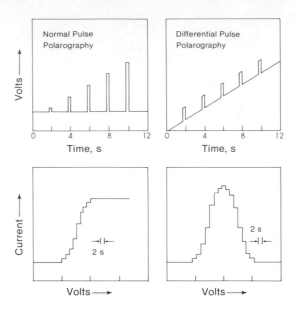

Figure 12-8. Schematic diagrams of the applied voltage as a function of time for normal pulse polarography and differential pulse polarography, along with typical current-voltage curves obtained by the two methods.

the current will fall to zero at the end point and rise again afterwards. The system of iron(II) with cerium(IV) is an example of such a titration. If neither the substance being titrated nor its product is electroactive at either cathode or anode, then the titration will exhibit zero current until the end point has been reached and rise thereafter owing to the electroactive titrant. The end points in all cases are sharply defined.

In this method the potentials of the electrodes adjust themselves by a flow of current until their difference equals the applied difference, in accordance with the current passing. The shape of the resulting titration curve depends on the applied voltage difference and on the particular electrochemical properties of the substance in solution.

12–4   OTHER ELECTROANALYTICAL TECHNIQUES

Modified Polarographic Systems

Several polarographic techniques have been developed that exploit modern electronics. In *pulse polarography* the potential is applied as a series of pulses of increasing magnitude. One pulse of less than a tenth of second is applied to each drop near the end of its lifetime. The current is measured in a short interval near the end of the pulse, when the charging current has fallen to a low level. This technique is more sensitive than ordinary polarography because the ratio of faradaic to charging current is more favorable. Detection limits are on the order of 10^{-6} M. Figure 12-8 indicates the general shape of the applied pulses and the current-voltage curves that

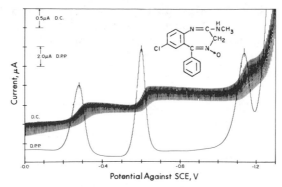

Figure 12-9. An example of a differential pulse polarogram of chlordiazepoxide, the active ingredient in the psychotherapeutic drug Librium, superimposed on a conventional direct-current polarogram. Sample is 100 μg in 3 mL of 0.2 M H$_2$SO$_4$ as supporting electrolyte. M. Hackman, M. A. Brooks, and J. A. F. deSilva, *Anal. Chem.* **1974,** *46* 1078.

result.

The technique of *differential pulse polarography* (DPP) uses ordinary polarography, in which the voltage is increased linearly with time, in combination with an additional small potential increase near the end of each drop life (Figure 12-8). Again, only a single pulse is applied per drop, and the difference measured between the current flowing just before the pulse is applied and just before the drop falls. In this type of measurement, charging current has even less effect because it has become small at the time of measurement (depending on the resistance of the solution). As a result, sensitivity is enhanced; detection limits of better than 10^{-7} M can often be achieved. Measurement of the difference between two potentials yields a waveform closely resembling the first derivative of the normal current voltage polarographic curve. In conventional polarography, when the half-wave potentials of two substances are close, the two waves may not be resolvable. Use of the first derivative of the current-voltage curve enhances resolution.

An example of the sensitivity of DPP is shown in Figure 12-9, where a conventional direct-current polarogram of the psychotherapeutic drug chlordiazepoxide, a tranquilizer, is compared with a differential pulse polarogram of the same solution. Both polarograms were scanned at the same rate of 5 mV/s with a 0.5-s mercury drop time. The differential pulse method is some 20 times more sensitive than conventional polarography, and peak heights and locations are more readily obtained.

In *alternating-current polarography* the voltage is increased slowly and linearly with time, but on top of this change in potential is imposed a sine-wave alternating voltage of small amplitude. Only the alternating current component of the current is measured. This gives the first derivative of the current-voltage curve. Both resolution and sensitivity are enhanced. Normally, several hundred pulses of 5 to 15 mV are applied per drop.

In *fast linear-sweep, (single-sweep* or *oscilloscopic) polarography* the desired potential range is swept during the lifetime of a single drop. A fast recorder, or oscilloscope, is required to record the current-voltage curve. The potential may be increased at rates of volts per second; a complete polarogram is obtained in the last half second of the drop life. The curves obtained are unlike those of conventional polarograms. The current arises partly from reduction of material in the small volume near the drop and partly from the slower process of diffusion.

In *cyclic-sweep voltammetry* the voltage is first increased and then decreased. This is primarily a tool for research studies of reaction rates and mechanisms and is little used in ordinary analytical measurements. Usually a species is first reduced and then oxidized in voltage scans at a hanging mercury drop or a microplatinum electrode. By varying the scan rate, information can be gained on the nature of chemical reactions occurring before or after an electrochemical oxidation or reduction. The method is also highly useful in studies of adsorption on electrode surfaces.

In *chronopotentiometry*, potential is monitored as a function of time under conditions where a constant current of a magnitude greater than the diffusion current is applied to a system. It is not a widely employed analytical technique.

Recall from Section 12-2 that the current flowing at an electrode is the sum of five processes. Suppose that a solution under test is not stirred, so that convective transfer of electroactive material to the surface of the electrode does not occur, and suppose also that electrical migration is eliminated by indifferent electrolyte. With the additional assumption that the charging current on the time scale normally used can be neglected (in practice it is usually corrected for by running a blank solution on another potentiostat), Equation (12-7) reduces to

$$i_{total} = i_{initial} + i_{diffusion}$$

In chronopotentiometry we are interested in the relation of potential to time during the period when the two current processes are equal to or less than the impressed current. When the impressed current, being larger than the diffusion current, has reduced the concentration of electroactive substance at the electrode surface to a certain level, current can no longer be sustained either by the concentration at the surface or by diffusion. For current to continue flowing at the impressed level, the potential must change so as to bring some other reaction into play. The constant current thus acts somewhat like a coulometric reagent that titrates the electroactive species in the small volume at the electrode surface. Usually the transition time of a system is reported. The *transition time* of a species may be defined as the time at which the maximum change in potential occurs; other conditions being equal, the square root of the transition time is proportional to concentration.

In summary, a number of newer methods based on classical polarography have been developed in recent years as a result of better and faster electronic means of applying and measuring currents and potentials. Basic to the intelligent use of them is a clear understanding of the processes that occur at a microelectrode when a voltage is applied, as in conventional direct-current polarography.

Stripping Analysis

Anodic stripping is a two-stage voltammetric method for the determination of traces of metal ions. Most commonly a microelectrode is employed in the form of a hanging mercury drop; occasionally a thin film of mercury on a small carbon or inert metal substrate is used.

In the first stage, metal ions in solution are electrolytically deposited in the drop of mercury to preconcentrate the metal. A sufficiently negative potential is applied to the mercury drop that the desired metal is reduced and forms an amalgam with the mercury. The solution is well stirred, and electrolysis allowed to proceed, typically for several minutes. The metal is thereby stripped from the solution and concentrated into a small volume of mercury.

In the second stage the potential of the mercury drop is made sufficiently positive that current flows in the opposite direction, oxidizing the metal back to its original ionic form. The current required to oxidize the metal is measured, and the amount of metal present calculated as in constant-potential coulometry. This technique is applicable to a substantial number of elements. It is highly sensitive; metals can be determined in volumes as small as a few milliliters at concentrations as low as 10^{-9} M.

Constant-Current Potentiomeric End Point-Detection

Potentiometric measurements are normally made with current flow approaching zero. Such measurements may take time when no reversible redox couple is present or when the solution is poorly poised,[11] that is, when one of the potential-determining substances is present at low concentration.

In *constant-current potentiometry* a small current, typically about 10 μA, is passed through an electrode pair and the potential difference during a titration is measured. In the region of the equivalence point the potential changes from a value determined typically by the species being titrated to a position where some other electrode reaction occurs. Thus the applied potential will increase until both cathodic and anodic reactions take place. The potential difference that is required to initiate reaction is measured during the titration. When both oxidized and reduced forms of a reversible couple are present, the 10 μA current will require but little difference in potential between the electrodes (Section 12-3). As the end point is approached, the concentration of one of the substances making up the redox conjugate pair falls to a low value. To provide the requisite current, the potential rises until some other reaction comes into play. Constant-current potentiometric titration with two indicator electrodes produces a titration curve with a maximum at the end point if both couples are reversible. In situations where the impressed current is small relative to diffusion currents,

---

[11]The poising capacity of a redox couple is analogous to the buffering capacity of an acid-base pair. If a solution contains only a low concentration of one of the components of a redox couple, the solution is poorly poised and equilibrium-potential values are attained only slowly.

the magnitude of the titration error is small.

This end-point system is most applicable when one or both couples are irreversible. Under these conditions the potential change does not go through a maximum at the end point but shifts from one value to another; it is often larger than in conventional potentiometry.

12-5 CONDUCTOMETRY

Unlike the preceding electrochemical methods conductometry is not concerned with either equilibria or reactions at electrodes involving specific half-reactions.

If two relatively large platinum electrodes are immersed in a solution of a nonelectroactive salt and a potential difference is applied between them, a small current will flow simply to charge the electrodes to the applied potential. After the electrodes have been charged, nothing more happens unless the voltage is sufficient to bring about an electrochemical change. If the direction of the applied potential is reversed well before the charge has been built up on the electrodes, then current (that is, charging current) will begin to flow in the opposite direction. Thus an applied alternating potential produces an alternating current totally associated with charging of the electrodes. Under the desired conditions this is nonfaradaic current. Provided the electrical capacitance between the two electrodes is relatively small, then the current that can flow depends on the resistance of the solution.

Since all ions in solution contribute to the conductivity, measurements of resistance give nonspecific information. *Conductance* may be defined as the reciprocal of resistance. The property of an ion that gives quantitative information about its relative contribution to the conductance is its mobility. The *mobility* of an ion is the rate at which it moves under the influence of an external force such as an electric field. In dilute solutions, when the concentration of an ion is doubled, its contribution to the conductance is doubled. When its mobility is also doubled, its contribution is similarly doubled. Accordingly, for a given ion or salt a linear relation exists between the measurement of conductivity and concentration.

Most applications of conductance involve aqueous solutions. Pure water is a poor conductor; being only slightly dissociated, the contribution of hydrogen ions and hydroxyl ions to the conductance is slight even though their mobilities are high, One application of conductance is to the determination of conducting impurities in purified water. In analysis the conductance method is typically used in the form of a titration for systems where the conductance differs significantly between components of the original solution and the products of reaction or of the titrant. Conductance measurements are also useful for assessing the degree of dissociation of acids, complexes, and so on.

Conductance changes linearly on addition of reagent, provided dilution is properly accounted for. Incomplete reactions result in curvature in plots of conductance against titrant volume in the region of the end point. For analytical applications, knowledge of absolute values of conductance is not required. Since the change in conductance is what enables extrapolation to obtain the end point, only relative values are important.

EXAMPLE 12-9

In a titration of hydrochloric acid with sodium hydroxide, the reaction may be written

$$H^+ + Cl^- + Na^+ + OH^- \rightarrow Na^+ + Cl^- + H_2O$$

Before the end point the solution composition is changing as hydrogen ions are being replaced with sodium ions. The hydrogen ion has a mobility[12] about seven times that of the sodium ion, causing the conductance to decrease sharply. After the equivalence point the conductivity increases sharply because both sodium ions and hydroxyl ions are being added in increasing amounts.

When a weak acid HA is titrated with sodium hydroxide, the reaction is

$$HA + Na^+ + OH^- \rightarrow Na^+ + A^- + H_2O$$

If the acid is so weak that ionization is negligible, the conductivity increases throughout the titration, but more sharply after the end point because the hydroxyl ion has a higher conductivity than does the anion of any weak acid.

In a precipitation reaction such as

$$Na^+ + Cl^- + Ag^+ + NO_3^- \rightarrow AgCl(s) + Na^+ + NO_3^-$$

the conductivity changes but slightly throughout the reaction and then increases after precipitation is complete.

12-6   CONSTANT-CURRENT COULOMETRIC TITRATION OF CYCLOHEXENE WITH ELECTROCHEMICALLY GENERATED BROMINE

Background

In this experiment the amount of cyclohexene in a methanol solution is determined by reaction with bromine. The reaction is The volatility of bromine makes preparation and storage of standard solutions difficult. These problems may be avoided by electrochemical generation of bromine through electrolysis of potassium bromide at a known constant rate in a mixture of acetic acid, methanol, and water.

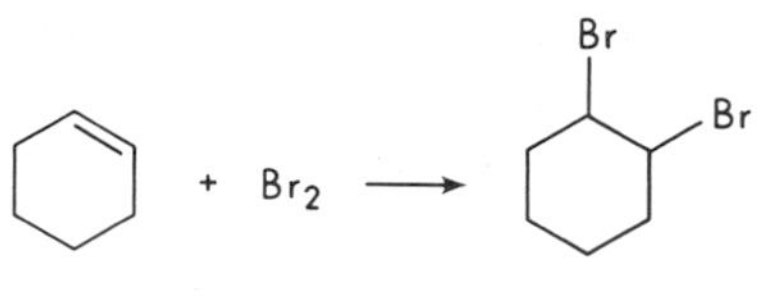

---

[12]At 25°C the mobility (in m/s $\times 10^7$) of the hydrogen ion is 349.82, of the sodium ion 50.11, and of the chloride ion 76.35. (R. P. Frankenthal in *Handbook of Analytical Chemistry*, L. Meites, Ed., McGraw-Hill, 1963, pp 5–30.)

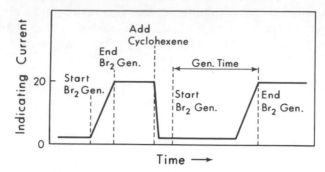

Figure 12-10. Schematic representation of change in current between indicator electrodes during blank and sample runs.

The equivalence point is located by measurement of the current flowing between two indicator electrodes across which a potential is applied. Recall from Section 12-4 that this method is based on the requirement that, for a current to flow between a pair of electrodes, one electroactive species must be capable of undergoing oxidation and another reduction at the applied potential. The magnitude of the current that flows depends on the nature of these electroactive species and on their concentrations. In this experiment a potential of 0.2 V is applied across the indicator electrodes. Before the equivalence point the possible reactions are the oxidation of bromide at one electrode and the reduction of hydrogen ion at the other. Because the potential required to produce bromine and hydrogen is considerably greater than 0.2 V, essentially no reaction takes place, and flow of current between the indicator electrodes is negligible. At the equivalence point the first excess of bromine in the solution enables the reaction at the cathode to be the reduction of bromine rather than hydrogen ions. Since a potential difference of 0.2 V between the two electrodes is more than sufficient to oxidize bromide at one and reduce bromine at the other, the equivalence point is marked by current flow between the indicator-electrode pair.

In practice it is convenient to generate bromine in the cell until a current of some suitable magnitude is reached in the indicator circuit, and then to add sample to the cell and generate bromine again until the same current is obtained. This method has two advantages: (1) the end point is more easily reproduced and (2) any impurities in the supporting electrolyte that react with bromine are corrected for. In effect, a blank has been run. The relation between the current in the indicator circuit and time is shown schematically in Figure 12-10.

Under the conditions of this experiment the addition of bromine to cyclohexene is slow unless a catalyst such as mercury(II) acetate is added. Alkenes (compounds containing two aliphatic carbons connected by a double bond) undergo bromination at different rates. Two that react slowly with bromine are acrylic acid ($CH_2=CHCOOH$) and acrylonitrile ($CH_2=CHCN$).

Procedure (time 3 to 4 h)

Obtain an unknown sample of cyclohexene. Place it in a dry 100-mL volumetric flask, dilute to volume with methanol, and mix well. Acquaint yourself with the experimental apparatus: identify the source of constant current, the amperometric end-point detection system, and the cell components.

Add electrolyte solution, 0.15 M KBr in 60:26:14 (volume percent) glacial acetic acid:methanol:water, to the cell until the generating electrodes are just covered. Add 0.1 g of mercury(II) acetate and a magnetic stirring bar. Adjust the stirring rate for maximum mixing without the bar striking the electrodes or splashing solution.

Set the voltage across the indicator-electrode pair to 0.2 V, and generate bromine at the anode until the current in the indicator circuit is 20 μA. Generate bromine in small increments as the 20-μA level is approached, waiting after each increment until the level of the indicator current stabilizes. Pipet 10 mL of sample into the cell, replacing the stopper immediately. Generate bromine until the indicator system shows 20 μA, using small increments near the end point as before. Record the time required.[13] Pipet a second sample into the same solution, and again generate bromine to the 20-μA end point. This operation may be repeated several times with the same electrolyte solution.

Calculate and report the milligrams of cyclohexene in the original 100-mL sample. This information can be obtained from the relation

$$\text{mg cyclohexane} = (\mu eq)\ \frac{0.08215}{2}\ (mg/\mu eq)\ \frac{100}{10}$$

The factor 2 is required in the denominator because two bromide ions must be oxidized for each molecule of cyclohexene. The microequivalents of bromine generated are related to the current in milliamperes and to the time in seconds by

$$\mu eq\ Br = (s)(mA)(10^3)/96\ 487$$

Operating Instructions for Coulometric Titration Apparatus [14]

1. Connect the cathode of the generator circuit (Figure 12-11) to the negative terminal and the anode to the positive terminal of the coulometer.

2. Turn on the power switch of the coulometer. (*Caution:* Do not touch the leads from the coulometer while the output (current generation) switch is on. Up to 300 V or more may be supplied to the leads at open circuit.)

3. Set the current control to a setting in the range of 5 to 10 mA. Set the microequivalent or time readout to zero.

---

[13]Some coulometers read directly in microequivalents of reagent generated rather than in time.

[14]The apparatus described here includes a Leeds and Northrup or equivalent coulometer, together with an amperometric end-point detection system consisting of a mercury cell as voltage source in series with a 1000-ohm potentiometer and a microammeter. Contact with the solution is made through two platinum wires.

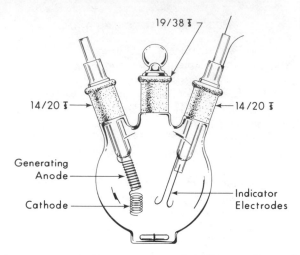

Figure 12-11. Arrangement of generator and indicator electrodes in a cell for constant-current generation of bromine.

4. When ready to generate bromine, turn on the output switch. Near the end point, turn the output switch off and on as needed to generate small increments of bromine, analogous to dropwise addition of conventional titrant near the end point.

5. After completing a titration, record the time or the microequivalents of bromine generated and reset the counter to zero. Another sample may then be added and the process repeated.

6. After completing a series of titrations, verify that the output and power switches are off. Disconnect the coulometer leads to the cell. Rinse the cell and electrodes thoroughly with water.

12–7 ELECTROLYTIC DETERMINATION OF COPPER IN BRASS

Background

Copper can be quantitatively electrodeposited by applying about 2 V across a cell containing a dilute solution of copper ions. The reactions are

$$Cu^{2+} + 2e \rightarrow Cu \qquad \text{at the cathode}$$

$$2H_2O \rightarrow O_2 + 4H^+ + 4e \qquad \text{at the anode}$$

A side reaction occurring at the cathode is

$$2H^+ + 2e \rightarrow H_2$$

Metals less active than copper must not be present. Iron(III), for instance, interferes with the quality of the deposit, which should be smooth and adherent. Spongy deposits, which result from the use of too high current densities or from improper acidity or absence of nitrate ion, often yield poor analytical results.

Procedure (time 3 to 4 h)

Weigh three 1-g samples of brass into 400-mL beakers. Add 25 mL of 6 M nitric acid and cover the beakers. Evaporate in a hood to a volume of about 5 mL, but do not allow to go to dryness. The digestion and evaporation steps require at least 1 h. Add 50 mL of water and 5 mL of concentrated sulfuric acid.[15] Neutralize the solution or filtrate with 6 M ammonia until a precipitate of copper hydroxide just persists. Add 10 mL of 6 M nitric acid and 1 g of tartaric acid. Transfer the solution quantitatively to a 500-mL electrolysis beaker. The volume at this point should be about 200 mL. Obtain three platinum cathodes and three anodes. Attach the electrodes to the electrolytic apparatus,[16] ensuring that the anode does not touch the cathode when it rotates. Place the beaker containing the copper solution on the apparatus at such a height that the cathode is almost, but not entirely, immersed. Start the motor to rotate the anode, and adjust the current to read 1.4 A. Continue the electrolysis until the blue color disappears (about 30 min), then add about 0.3 g of urea, and continue electrolysis for 5 min. Add enough water to raise the level of liquid and expose some fresh cathode surface for copper deposition. Continue deposition another 5 min. If further copper is deposited (indicated by changes on the fresh cathode surface), continue until no further deposit is seen on the cathode after the liquid level has been raised.

Holding the beaker with one hand, remove the support with the other. Lower the beaker slowly while washing the cathode gauze well with a stream from a wash bottle.[17] Turn the switch to "off" at the instant the circuit is broken. Remove the cathode, and wash it with water and then alcohol. Shake to remove excess alcohol, and allow to air-dry. Weigh it after a few minutes. Repeat the washing and drying to constant weight. To dissolve the copper, place the cathode in a beaker and add 6 M HNO$_3$ until the copper deposit is immersed. Allow to stand until the copper has dissolved and the electrode has taken on the color of platinum metal. Again wash with water and alcohol to constant weight. The difference is the weight of copper. The anode can be cleaned if necessary by dipping into dilute nitric acid containing some hydrogen peroxide.

Report the percentage of copper in the sample.

12-8   AMPEROMETRIC TITRATION OF ARSENIC(III) USING A ROTATING PLATINUM ELECTRODE

Background

Arsenic(III) may be titrated amperometrically with potassium bromate using a rotating platinum indicator electrode at a potential of +0.2 volts

---

[15]Should a precipitate of tin oxide or lead sulfate form, remove it by filtration and washing.

[16]Electrolytic analyzers can be purchased with from one to six positions. A direct current of up to five or more amperes is supplied to each position along with provision for stirring.

[17]Should a residue of HNO$_3$ be left on the cathode the copper would redissolve.

against the SCE. After the equivalence point the current increases linearly with concentration, owing to the formation of bromine from the interaction of excess bromate ion with bromide ion:

$$5Br^- + BrO_3^- + 6H^+ \rightarrow 3Br_2 + 3H_2O$$

By the use of a reference half-cell of the desired potential of $+0.2$ V, all that is necessary is to short-circuit the half-cell with the indicator electrode through a sensitive current-reading instrument.

A gravimetric titration procedure is also recommended to give better precision than dispensing titrant from a buret.

Procedure (time 4 to 5 h)

Prepare a phosphate reference half-cell by dissolving about 5 g of KNO_3, 4 g of KH_2PO_4, and 4 g of K_2HPO_4 in 50 mL of water. Add this solution, along with 2 to 3 mL of mercury and a gram or two of Hg_3PO_4, to a 10-mL side-arm flask like the one in Figure 12-6. Connect the reference half-cell to a salt bridge of appropriately low resistance as indicated in the figure. Short-circuit the half-cell by connecting the lead directly to that of the rotating platinum indicator electrode.

To prepare a standard solution of about 0.01 M potassium bromate, weigh (to the nearest 0.1 mg) the necessary quantity of pure, dry potassium bromate and dilute it by weight to about 100 g. Record the weight of the solution to the nearest 10 mg. Prepare a solution that is about 2 M in hydrochloric acid and 0.1 M in potassium bromide.

Transfer the unknown solution containing arsenic(III) to a 100-mL volumetric flask. Dilute to volume and mix well. Pipet 10 mL of the arsenic(III) solution into the titration vessel, and add 40 mL of the HCl–KBr solution. Fill a 5- or 10-mL syringe to be used for dispensing titrant with the standard bromate solution. Using a top-loading balance, weigh the syringe to the nearest milligram and record the weight. Add about 0.4 to 0.5 g of titrant solution to the cell. Retract the plunger slightly after each portion of titrant is added to minimize evaporation and loss of the drop that would otherwise remain on the delivery tip. Measure the current, and record the weight of the syringe. Add another portion of titrant, and continue the titration until four or five points have been recorded beyond the end point.

Plot weight of titrant against diffusion current. To determine the weight of titrant required to give the end point, draw straight lines through the points before and after the end point. Calculate the weight of arsenic(III) in the sample from the weight of titrant required and the concentration of the potassium bromate solution.

PROBLEMS

12-1.   In coulometric methods of analysis, what experimental quantities are measured or controlled?

12-2.   Explain the principles of constant-current coulometry as an analytical method. What are its advantages?

12-3.   Why is the counter electrode in a coulometric titration often isolated from the solution under measurement? How can it be isolated?

12-4.   Why is the addition of an intermediate in a constant-current coulometric titration generally desirable? Explain with an example.

12-5.   Write equations to illustrate the application of constant-current coulometric titrations to acid-base, redox, and precipitation reactions.

12-6.   Describe the major methods for the detection of end points in constant-current coulometric titrations.

12-7.   A 10.00-mL sample of arsenite was oxidized to arsenate by constant-current coulometric generation of iodine. If 2.463 min was required for complete oxidation at a current of 19.30 mA, what was the molar concentration of the arsenite?

12-8.   The chloride in a 25.12-mL sample of serum was determined coulometrically by electrogenerated silver. To reach the end point required 7.432 min at a current of 14.36 mA. Calculate the weight of chloride in the sample.

12-9.   To neutralize the n-butyric acid, $CH_3(CH_2)_2COOH$, in a 19.98-mL sample of a solution by electrogenerated hydroxyl ion took 4.236 min at a current of 14.36 mA. (a) What was the concentration of butyric acid in the solution? (b) How many moles were present; how many milligrams?

12-10.   In constant-potential coulometry three electrodes are required. Name and describe the function of each.

12-11.   (a) Explain what is meant by a chemical coulometer.
         (b) A chemical coulometer in which hydrogen ions are generated was placed in series with an electrolytic cell in which dichromate was coulometrically reduced with iron(II) as an intermediate. At the end point the hydrogen ions were titrated with 0.05123 M base, of which 6.44 mL was required. What weight of potassium dichromate was present in the sample?

12-12.   Explain what is meant by the term "overpotential". What factors influence the value of the activation overpotential?

12-13.   (a) Calculate the potential at the anode (relative to the standard hydrogen electrode) required for the formation of oxygen from a solution 0.2 M in hydrochloric acid.
         (b) Calculate the potential at the cathode required for the quantitative deposition of cadmium from a solution 0.02 M in cadmium chloride.

12-14.   Consider a solution of 0.003 M thallium chloride: (a) At what potential against the SCE would thallium(I) ions just begin to be reduced, with no interference from reduction of hydrogen ions? (b) At what potentials would the concentration of thallium(I) at the electrode surface be 10%, 1%, and 0.1% that of the bulk of the solution?

12-15.   What five main processes contribute to the current at an electrode? Explain how one of these contributions may be isolated in polarographic measurements.

12-16.   Define the term "charging current" in polarography. What effect does it have on the limit of sensitivity of polarographic measurements?

12-17.   By what three processes can solute in a solution be transported to the region near an electrode surface?

12-18.   Why is either a dropping mercury electrode or a rotating platinum electrode preferable to a stationary solid metal electrode in current-voltage measurements?

12-19.   List the functions of the supporting electrolyte in polarography.

12-20.   (a) In a current-voltage curve, why does the current rise to a plateau when an electroactive species is oxidized or reduced at the working electrode? (b) Why does temperature need to be controlled in polarographic work?

12-21. (a) What are the advantages of the dropping mercury electrode? (b) What reactions establish the maximum positive and negative potentials that can be used with it?

12-22.† Which of potassium iodide, bromide, or chloride has the widest useful range in potential as a supporting electrolyte in conjunction with a DME? Explain.

12-23. What is the Ilkovic equation? Define the various terms.

12-24. A commercial SCE cannot be used as the reference electrode in a 2-electrode polarograph. Why?

12-25.† What is the Clark electrode? How does it function? How does it differ from the gas membrane selective electrodes of Section 11-7?

12-26. Under what conditions does a two-electrode polarograph fail to provide valid data? How does a three-electrode design correct the deficiencies of the two-electrode system?

12-27. Name four methods of obtaining quantitative analytical results from polarographic measurements. What level of precision is normally expected in polarographic analysis?

12-28. (a) What is the purpose of a blank in polarography? (b) Why does dissolved oxygen generally have to be removed from the test solution before polarographic measurements are made? What is the usual procedure?

12-29. Describe several modified techniques designed to increase the sensitivity of conventional polarographic methods and to minimize the effects of the charging current.

12-30. How much time would be required to deposit 454 mg of copper metal by electrolysis from an acidic solution of copper sulfate if the current were 3.6 A and the current efficiency 55%?

12-31. The half-wave potential of cadmium(II) is about –0.6 V against the SCE. Sketch the expected amperometric titration curve at –0.8 V against the SCE for the reaction of cadmium with a strong complexing agent such as EDTA. Label the axes clearly.

12-32. The diffusion current for iron(III) can be measured at a potential equal to that of the SCE. Sketch the expected amperometric titration curve at 0 V for the titration of iron(III) with a standard solution of vanadium(II). Label the axes clearly.

12-33. The half-wave potentials of lead(II) and zinc(II) are –0.4 V and –1.0 V against the SCE. Sketch the expected amperometric titration curve at –0.7 V for a titration of a mixture of lead and zinc with a ligand that forms a stronger complex with zinc than with lead. Label the axes clearly.

12-34. A 5.47-g sample of a synthetic rubber latex was coagulated by addition to 75 mL of ethanol. The dodecyl mercaptan in the resulting solution was titrated amperometrically with 0.1010 M silver nitrate. The reaction is $Ag^+ + RSH \longrightarrow AgSR + H^+.$ The diffusion current for the silver ion at a rotating platinum electrode was recorded as a function of titrant volume. The following readings were obtained: 0 to 15.60 mL, no current; 15.85 mL, 1.73 μA; 15.91 mL, 3.22 μA; 15.97 mL, 4.81 μA. Calculate the percentage of dodecyl mercaptan (mol wt 202.4) in the rubber latex.

12-35. Calculate the capillary constant of a dropping mercury electrode for which the following data were obtained: time required to collect 10 drops, 31.8 s; weight of mercury collected in 5.00 min, 0.9784 g.

12-36. Compare the precision obtainable by direct polarographic measurement and by amperometric titration.

12-37. (a) Explain the reactions at the electrodes during a titration using two-electrode amperometric end-point detection. (b) How large is the current at the end point?

12-38. Explain how an end point is obtained in constant-current potentiometry.

12-39. Sketch the expected conductance titration curve for the titration of 20 mL of a 0.1 M solution of sodium hydroxide with 0.05 M hydrochloric acid. Label the axes clearly.

12-40. Sketch the expected conductance titration curve for the titration of a solution of sodium sulfate with barium chloride. Label the axes clearly. Indicate the reactions occurring at various stages in the titration.

12-41. (a) Why is potassium bromide added to the reaction mixture in the constant-current coulometric determination of cyclohexene as described in Section 12-6? (b) Why is mercury(II) acetate added?

12-42. Why is a phosphate reference half-cell chosen for the amperometric titration of arsenic(III) in Section 12-8?

12-43. A 5-mL portion of a 100-mL sample of a solution containing cyclohexene was analyzed by the procedure of Section 12-6. If a current of 6.43 mA was required for 483.6 s to reach the amperometric end point, how many milligrams of cyclohexene were present in the original sample?

12-44. Write equations for the reactions taking place at the anode and the cathode of the coulometer circuit during the bromination of cyclohexene (Section 12-6). Why is it unnecessary to isolate one of the generator electrodes in this reaction?

12-45. Write equations for the reactions taking place at each of the indicator electrodes used to locate the amperometric end point in the determination of cyclohexene by the method of Section 12-6 (a) before the end point and (b) after the end point.

12–46.† Devise a method for determining the equivalent weight of an unknown alkene. Could the molecular weight also be determined?

12-47. If an alkene that undergoes bromination relatively slowly were analyzed with the procedure and apparatus of Section 12-6, would the results likely be low or high? Why?

12-48.† Sketch the current-voltage curve to be expected at a rotating platinum indicator electrode for a 10^{-4} M solution of iodine. What would be a reasonable potential to use for an amperometric titration of this solution with sodium thiosulfate? (Sodium thiosulfate is not electroactive at the platinum indicator electrode.) Sketch the amperometric titration curve expected in the titration of 10^{-4} M iodine with a standard solution of sodium thiosulfate.

REFERENCES

A. M. Bond, *Modern Polarographic Methods in Analytical Chemistry*, M. Dekker, New York, 1980. Reviews modern pulse polarography and other polarographic techniques.

D. H. Evans, *J. Chem. Educ.* **1968**, *45*, 88. Discussion of the method of Section 12-6 for the determination of cyclohexene.

G. W. Ewing, *Instrumental Methods of Chemical Analysis*, 4th ed., McGraw-Hill, New York, 1975. Includes circuitry for potentiostats and electronic coulometers.

W. E. Harris, *J. Chem. Educ.* **1958**, *35*, 408. Gives the composition of reference half-cells (covering the range from +0.4 to –1.0 V against the SCE) that can be used in conjunction with amperometric titrations.

I. M. Kolthoff and J. J. Lingane, *Polarography*, 2nd ed., Interscience, New York, 1952. Extensive coverage of the theory, background, and applications of polarographic techniques.

D. Ilkovic, *Collection Czechoslav. Chem. Commun.* **1934**, *6*, 498. Solution of the problem of diffusion to the dropping electrode and derivation of the equation for the resulting current.

H. A. Laitinen and W. E. Harris, *Chemical Analysis*, 2nd Ed., McGraw-Hill, New York, 1975, Chapter 14. Gives background on electrolytic separations and determinations.

M. A. Lessler and G. P. Brierley in *Methods of Biochemical Analysis*, Vol 17, D. Glick, Ed., Interscience, New York, 1969, p 1. Calibrations and calculations involving the Clark oxygen electrode.

J. J. Lingane and L. A. Small, *Anal. Chem.* **1949**, *21*, 1119. Describes controlled-potential method for coulometric determination of small amounts of halides alone or in mixtures.

G. Marinenko and J. K. Taylor, *Anal. Chem.* **1967**, *39*, 1568; **1968**, *40*, 1645. Describes apparatus and conditions under which reagents can be generated with nearly 100% efficiency.

J. T. Stock, *Amperometric Titrations*, Interscience, New York, 1965. Background and applications of amperometric methods of analysis are described in detail.

13

REACTION-RATE METHODS IN ANALYSIS. RADIOCHEMISTRY

Nature, it seems, is the popular name for milliards and milliards and milliards of particles playing their infinite game of billiards and billiards and billiards.

Piet Hein

Chemical reactions proceed at a wide range of rates, from the almost instantaneous combination of an aquated proton with a base to the immeasurably slow reaction of molecular oxygen with carbon at room temperature. To have analytical value, a chemical reaction must reach equilibrium rapidly enough to be practical. *Reaction-rate*, or *kinetic*, *methods* can exploit information obtained from the rate of chemical reaction as well as the position of equilibrium. Rate measurements are valuable when a reaction is inconveniently slow to reach equilibrium, has an unfavorable equilibrium constant, or is accompanied by side reactions.

Three important factors that affect rates are concentration of the reactants, temperature, and catalysts. To understand how reaction rates can be applied to quantitative determinations we consider examples of methods of accelerating slow reactions, followed by examples of systems in which rates are followed to determine concentration.

13-1 THE CONCEPT OF RATE OF REACTION

Consider a process in which a reagent R undergoes a reaction to give a product P:

$$R \rightarrow P$$

The rate at which this reaction proceeds is customarily expressed in terms of the change in concentration of the reactant (or product) with time:

$$\text{average rate} = \frac{\text{change in concn of R}}{\text{elapsed time}} = \frac{\text{change in concn of P}}{\text{elapsed time}}$$

The rate can be formulated, with the use of the symbol Δ to denote change, according to

$$\text{average rate} = \frac{\Delta P}{\Delta t} = \frac{[P]_2 - [P]_1}{t_2 - t_1} = -\frac{\Delta R}{\Delta t} = -\frac{[R]_2 - [R]_1}{t_2 - t_1}$$

Conventional in writing rate equations is to choose sign values so that the rate of reaction is a positive quantity. Thus, because R is decreasing with time, $[R]_2 - [R]_1$ is negative, and a negative sign before expressions involving R results in a positive value for the rate.

EXAMPLE 13–1

The concentration of a reactant R decreases from 0.100 M to 0.072 M in 5.5 min. What is the average rate of reaction over this period?

$$\text{Average rate} = -\frac{[R]_2 - [R]_1}{t_2 - t_1} = -\frac{(0.072) - (0.100)}{5.5} = 5.1 \times 10^{-3} \ M/min$$

When the rate of reverse reaction (formation of R from P) is negligible, either because the reaction is irreversible or the concentration of P is extremely small, the rate of disappearance of R or appearance of P is given by

$$\text{rate} = k[R]$$

where k is a proportionality constant called a *first-order rate constant*. This equation shows the reaction rate to be linearly dependent on [R]; if the concentration of R is doubled, the rate is doubled. As R undergoes reaction to form P, the rate will decrease as [R] decreases, and a plot of [R] against time will be a smooth curve like the one in Figure 13-1. The rate at any given point on the curve could be expressed in the same way as the average rate, $-\Delta R/\Delta t$, but with extremely small Δ values. The limiting value of $-\Delta R/\Delta t$ is represented by $-dR/dt$, where d signifies an infinitesimal value of Δ.[1] The rate of reaction of R at a particular concentration is the negative of the derivative at that concentration:

$$\text{reaction rate} = -dR/dt$$

An equation relating reaction rate and concentration is called a *rate law*. The rate law for any reaction must be established by experimental measurement; it cannot be predicted from the equation for the reaction. A *first-order rate law* is one in which the rate of a reaction depends on the concentration of only a single species to the first power. The dissociation of N_2O_4 to two NO_2 molecules is an example; another is the decay of a radioactive species such as carbon-14 to nitrogen-14 and a beta particle. The overall, or total, order of a

---

[1] In the terminology of differential calculus, $-dR/dt$ is the derivative of the concentration of R with respect to time.

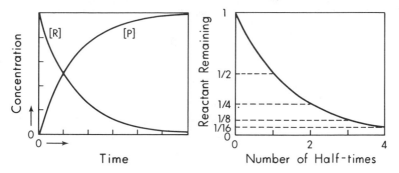

Figure 13-1. **Variation with time of the concentrations of reactant R and product P and relation between half-time and fraction of reactant remaining.** From H. A. Laitinen and W. E. Harris. *Chemical Analysis*, McGraw-Hill, New York.

reaction is generally determined by summing the exponents of the species involved in the rate law.[2]

EXAMPLE 13-2

For the reaction $2Br^- + 2H^+ + H_2O_2 \rightarrow Br_2 + 2H_2O$ the rate law was determined experimentally to be $d[Br_2]/dt = k[H_2O_2][H^+][Br^-]$

What is the order of the reaction with respect to each of the reactants? What is the overall order?

The reaction is first order in H_2O_2, H^+, and Br^-. Overall the reaction is third order. Note that the order in Br^- and H^+ found by experiment is first order even though both have coefficients of 2 in the balanced equation. In another reaction one or the other might be found to be second order, or perhaps 1/2 or 3/2 order.

For analytical purposes it is often important to be able to calculate the extent of a reaction by measurement of the change in concentration of a species after some specified time. These changes can be calculated from the integrated form of the differential equation. For a reactant R the integrated form of the first-order rate equation is

$$kt = 2.303 \log ([R]_0/[R]) \tag{13-1}$$

where $[R]_0$ is the initial concentration of R at time $t=0$ and $[R]$ the concentration at a specified time t. Often the rate of a reaction is reported in terms of its rate constant k; occasionally it is reported as the *half-life* for the process, that is, the time it takes for half the sample present at any given time to react. To elucidate the relation between the half-life $t_{1/2}$ and the rate constant k for a first order reaction, let $[R] = 0.5 [R]_0$ and $t = t_{1/2}$ in (13-1):

$$kt_{1/2} = 2.303 \log ([R]_0/0.5 [R]_0) = 2.303 \log 2 = 0.693$$

---

[2]Many rate laws involve the sum of several terms, each of which may have its own rate constant and order. Such complicated systems are not considered here.

$$t_{1/2} = 0.693/k \tag{13-2}$$

Thus if either $t_{1/2}$ or k is known, the other can be calculated.

EXAMPLE 13–3

Iodine-131, sometimes used to treat thyroid tumors, has a half-life of 8.1 days. What is the rate constant for the decay reaction of iodine-131?

Rearranging (13-2) and substituting, we obtain $k = 0.693/(8.1 \text{ days} \times 24 \times 60) = 5.9 \times 10^{-5} \text{ min}^{-1}$. Note that the rate constant is expressed here in minutes; other units such as seconds are often reported.[3]

When a first-order rate constant is 10 s^{-1} or greater, the reaction is quantitative in less than a second; when it is 10^{-3} s^{-1}, the reaction requires 100 min for completion.

EXAMPLE 13–4

For a first-order reaction with a rate constant of 10 s^{-1}, what is the time required for the reaction to be 99.9% complete?

The time can be calculated from (13-1) according to

$$10t = 2.303 \log ([R]_0/([R]_0(1 - 0.999))) = 2.303 \log (1/0.001) = 2.303 (3) = 6.9$$

$$t = 6.9/10 \text{ s}^{-1} = 0.69 \text{ s}$$

With a rate constant of 0.10 s^{-1} the time required would be $6.9/0.10 \text{ s}^{-1} = 69 \text{ s}$. Although this would be too slow for a direct titration, addition of excess reagent followed by titration of the excess with a more rapidly reacting substance (a back titration) might prove a viable method.

Many chemical reactions are second or higher order. An example of a second-order reaction is

$$A + R \rightarrow P$$

where the rate is given by

$$\text{rate} = k[A][R]$$

In this case the reaction is first order in both A and R, since the concentration of each is raised to the first power in the rate equation; the overall reaction is second order. A second-order rate constant has the dimensions of the inverse of the product of time and molarity; an example is $s^{-1} M^{-1}$. But keep in mind that the order of a reaction bears no relation to the coefficients in the balanced equation. Because the rates of second-and higher-order reactions generally depend on the concentrations of two or more

---

[3]The SI system specifies that all rate constants be expressed in units of seconds. On occasion this can lead to numbers so large as to be difficult to comprehend (the half-life of uranium-238 is $1.42 \times 10^{17} \text{ s}$).

species, the corresponding rate equations are seldom used in analysis. Instead, the concentrations of all species except the one to be determined are fixed at a level sufficiently high that they remain essentially unchanged. The resulting reaction is called *pseudo first order*, the term "pseudo" denoting that the overall rate is artificially first order, being determined only by the concentration of the species under measurement.

13-2   ENZYME-CATALYZED REACTIONS

Catalysts are substances that increase the rate of reaction but are not required for the stoichiometry of the reaction. They act by lowering the energy required (the activation energy) to permit a reaction to take place.

Enzymes are proteins that function as catalysts for biochemical reactions. They are critical in many biochemical processes, and so the determination of the concentrations of particular enzymes in the body is an important type of clinical diagnostic test. Furthermore, because enzymes are highly specific as to *substrates*, the compounds whose reactions they catalyze, the substrates themselves are often determinable by addition of an appropriate enzyme and measurement of the reaction rate. The general expression for an enzyme-catalyzed reaction in its simplest form can be written

$$E + S \underset{k_2}{\overset{k_1}{\rightleftarrows}} ES \overset{k_3}{\rightarrow} E + P$$

where E is the enzyme, S the substrate, ES an enzyme–substrate complex, and P the reaction product. The equilibrium concentration of the complex depends on the magnitudes of the rate constants k_1, k_2, and k_3.[4] By addition of enzyme to a substrate in an amount sufficiently small that the substrate concentration can be considered constant, the rate of reaction is proportional to enzyme concentration. A plot of reaction rate against enzyme concentration, then, is linear during this period, and from it the enzyme concentration can be determined. When the concentration becomes too high, the rate begins to fall off, and a straight-line relation is no longer observed (Figure 13-2).

Similarly, when a fixed excess of enzyme is added to a substrate, the rate of reaction is linearly proportional to the substrate concentration at low concentrations, and the amount of substrate present can be determined (Figure 13-2). Here, too, at high concentrations the rate becomes independent of concentration and cannot be used analytically.

---

[4]The *Michaelis constant* K_M defined by $(k_2 + k_3)/k_1$, is an equilibrium-like (pseudo equilibrium) constant for the steady-state formation of ES.

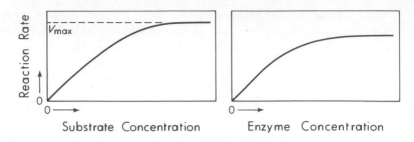

Figure 13-2. Relation between rate of an enzymatic reaction and concentration of substrate or enzyme. V_{max} is the maximum rate attained by the system. From H. A. Laitinen and W. E. Harris, *Chemical Analysis*, McGraw-Hill, N.Y.

13–3   APPLICATIONS OF RATE METHODS

Single-Component Analysis

Reactions of an order higher than first order are difficult to relate straightforwardly to the concentration of a sought-for substance; consequently they are seldom used in chemical analysis. Fortunately, concentrations of all reactants except one can often be made sufficiently high (or constant in the case of a catalyst) that the rate is controlled by the sought-for substance. This type of reaction is an example of one that is pseudo first order.

Concentrations can be derived from rate data by various methods. Most of them involve measurement of a change in concentration of a substance with time. Precise results can be assured only by employing methods capable of following variations in concentration with time and by carefully controlling such parameters as temperature of the solution. Nevertheless, rate methods have applications in several important areas. In general, either differential or integral measurements can be made to determine the amount of a sought-for substance R present in a sample.

Initial-Rate Methods

The *initial-rate method* (sometimes called the *differential method*) involves instantaneous measurement of the change in concentration of some species with time, $-dR/dt$ or dP/dt, immediately after the reactants are mixed. From the initial rate and the rate constant, measured under the same conditions, $[R]_0$ can be calculated. If the change in concentration is followed instrumentally, say electrochemically or spectrophotometrically, the slope of the initial straight line can be obtained electronically and the concentration calculated by a microprocessor or minicomputer. The response time of such electronic systems can be extremely fast, and reactions with half-lives as short as a millisecond can be employed analytically with a precision of 1 to 2% in the results.

Initial-rate methods have several advantages. (1) Complications from slower side reactions are usually slight. (2) Reverse reactions have little effect because of the small amount of product present during the measurement.

(3) Reactant concentrations are nearly constant and, being at an initial high level, give a higher rate and correspondingly higher precision than can be achieved later. (4) They may enable measurement of reactions with formation constants too small for conventional equilibrium methods.

EXAMPLE 13-5

The enzyme lactate dehydrogenase (LDH) catalyzes the reaction between lactate and nicotinamide adenine dinucleotide (NAD^+) to form pyruvate and the reduced form of NAD^+, NADH. Certain forms of LDH occur preferentially in heart muscle, and appear in the blood stream if the heart is damaged. Thus measurement of their concentrations in the blood stream is an indicator of possible heart attack.

For a system such as this, an initial rate method works well. Since NADH has an absorption maximum at 340 nm, where NAD^+ absorbs radiation but little, this system provides a convenient direct way of following the reaction. Care must be exercised, however, to ensure that standards are run under the same conditions of pH, ionic strength, and so on, since these factors greatly influence rate data. Also, the presence of substances that inhibit (slow down) or activate (speed up) the effects of catalysts must be watched for because they can have a profound influence on rates. Thus, traces of some metals may increase the activity of an enzyme manyfold, while other metals may inhibit its activity almost completely. An example is the enzyme isocitric dehydrogenase (ICD), which catalyzes the conversion of isocitric acid (ICA) to α-ketoglutarate (α-KG) according to

$$ICA + NAD^+ \longrightarrow \alpha\text{-KG} + NADH$$

Manganese at 10^{-5} M levels strongly activates ICD; addition of EDTA removes the metal from the enzyme and kills its activity almost completely. Certain other metals, including Hg^{2+} and Pb^{2+}, block the enzyme even at trace concentrations and inhibit its activity.

Integral Rate Methods

Integral rate methods involve measurement of changes in concentration over finite time intervals under conditions where the ratio $\Delta C/\Delta t$ represents the reaction rate. The measurements are made either over a preselected interval (fixed-time method) or over a preselected concentration change (variable-time method).

In the *fixed-time method* the concentration of substance I (which could be reactant or product) is measured at two preselected times, t_1 and t_2. The rate of change in concentration over the time interval $t_2 - t_1$, or Δt, is given by

$$\Delta I/\Delta t = k'[R] \tag{13-3}$$

where k' is the pseudo-first-order rate constant, that is, the rate constant of one reactant when the concentrations of all others are constant. Since Δt is constant, it can be combined with the rate constant to give

$$\Delta I = k''[R]_0$$

This equation tells us that the initial concentration of the sought-for substance $[R]_0$ is directly proportional to the change in $[I]$ over the preselected time interval. The advantage of this type of measurement is that the fraction of reaction occurring does not depend on the initial concentration, whether or not the two time values are on the linear portion of the curve in Figure 13-2.

A reaction occurring at a moderate rate can often be initiated by addition of a reagent, allowed to proceed for a measured period of time, and then quenched by altering conditions through cooling, changing the pH, adding a catalyst inhibitor, or other means. The products are then analyzed to obtain the necessary information.

In the *variable-time method* the time interval Δt is measured at two preselected concentrations of I. Here $\Delta[I]$ is fixed, and Δt is the measured variable. Incorporation of $\Delta[I]$ into the constant of Equation (13-3) gives

$$1/\Delta t = k'''[R]_0$$

where the initial concentration of R is now proportional to $1/\Delta t$. This method, in contrast to the fixed-time method, requires that the reaction be linear with time over the interval Δt, or significant error may be introduced. Standards should therefore be chosen whose concentrations are near those of the samples.

The variable-time method is most suited to measurements of the concentration of catalysts such as enzymes, where $[I]$ represents a constant fraction of a reaction; under these conditions the catalyst is being regenerated.

EXAMPLE 13–6

The substrate glucose can be determined with high specificity in biological fluids by measuring the extent of the reaction with oxygen:

$$C_6H_{12}O_6(\text{glucose}) + O_2 + H_2O \longrightarrow C_6H_{12}O_7(\text{gluconic acid}) + H_2O_2 \quad (13\text{-}4)$$

This reaction is catalyzed by the enzyme glucose oxidase; since the catalyzed reaction requires over an hour to go to completion, a measurement may be made on samples and standards at some fixed time, say 15 min, after addition of the enzyme. A calibration curve for a set of standards is prepared, and the unknown concentration read from the curve. The measurement is usually by spectrophotometer. Neither reactants nor products in Equation (13-4) absorb radiation in the visible region, but the hydrogen peroxide formed can be coupled with o-toluidine to produce a colored product.

Kinetic Analysis of Mixtures

If we have two compounds, similar except for different reaction rates, we may be able to determine each in a mixture by rate analysis without prior separation. Consider the first-order reactions between sample constituents A and B and a reagent R:

$$A + R \longrightarrow \text{products} \qquad \text{and} \qquad B + R \longrightarrow \text{products}$$

If R is present in a large, and therefore constant, concentration relative to A and B, the reactions are pseudo first order with respect to both A and B. Letting k_A be the pseudo-first-order rate constant for the reaction of A, the rate of disappearance of A is then

$$- \frac{d[A]}{dt} = k_A[A]$$

In integrated form the equation becomes

$$2.303 \frac{\log [A]_t}{\log [A]_0} = -kt$$

where $[A]_t$ and $[A]_0$ are the concentrations at time t and at the initial time. Similar equations can be written for the reaction of B.

If the ratio k_A/k_B is 463 or greater, the faster reaction (A in this case) can be shown to be 99% complete before the slower reaction is 1% complete. Thus at some instant A can be determined in the presence of B from the combined rate with a precision of 1%. If the rates are closer together, more complex methods may be resorted to; details are available in advanced textbooks. See the references at the end of the chapter.

EXAMPLE 13-7

Mixtures of two closely related alcohols can be analyzed for each component by following the rate of reaction with acetic anhydride in pyridine as solvent. Examples include mixtures of 1- and 2-propanol, 1,2-propanediol and 1,3-butanediol, and 3-amino-1-propanol and 1-amino-2-propanol. The reaction is

$$ROH + (CH_3CO)_2O \rightarrow CH_3COOR + CH_3COOH$$

The course of the reaction can be followed by withdrawal of portions at timed intervals and titrating the acetic acid formed with sodium hydroxide. The reaction rate depends on the alcohol — from a few minutes to several hours may be required for collection of adequate data.

In summary, knowledge of the factors affecting reaction rates can guide the adjustment of conditions or the selection of a catalyst to make a system analytically useful. The determination of concentrations by direct measurement of rates of reaction provides an additional important tool for the estimation of the components in difficult-to-separate mixtures.

13-4 RADIOCHEMISTRY

Although most chemistry deals with bonding electrons, the nuclei of atoms also provide valuable information about the kinds and concentrations of atoms in a sample. Certain naturally occurring nuclei undergo spontaneous disintegration, or decay, to yield products that give data on the kind and number of nuclei originally present. Such unstable nuclei are said to be *radioactive*. Other, normally stable nuclei can absorb particles such as neutrons and so be converted to radioactive nuclei.

The processes by which nuclei decompose include alpha decay, beta decay, and fission. In *alpha decay* a particle is ejected from an unstable nucleus; the nuclear charge is thereby decreased by two and the nuclear mass by four. The ejected particle, having the mass and charge of a helium nucleus, is called an alpha particle. In *beta decay* an electron is ejected; the nuclear charge is increased by one, but the mass is left nominally unchanged.

An example of alpha and beta decay is the decay of $^{238}_{90}U$ and its radioactive daughter, $^{234}_{92}Th$:

$$^{238}_{90}U \rightarrow \ ^{234}_{92}Th + ^{4}_{2}He \qquad\qquad t_{1/2} = 4.51 \times 10^9 \text{ years}$$

$$^{234}_{90}Th \rightarrow \ ^{234}_{91}Pa + \beta \qquad\qquad t_{1/2} = 24.1 \text{ days}$$

In *fission* an atom splits into two fragments of comparable mass, plus usually one or more neutrons. (Alpha decay is not considered fission.) Spontaneous fission occurs only in the heaviest elements, primarily uranium-235, plutonium-239, and uranium-233. Uranium-238 does not undergo spontaneous fission. However, fission can be induced in uranium-238 and many other heavy isotopes by bombardment with neutrons, protons, or alpha particles. Although neither fission nor alpha decay is of major importance for analytical measurements, beta decay provides a valuable handle for both qualitative and quantitative investigations.[5]

The process of nuclear decay is often accompanied by the emission of electromagnetic radiation of extremely high energy and short wavelength, called *gamma radiation*. Gamma rays may be considered beams of photons, as can all electromagnetic radiation, but of extremely short wavelength and high energy. Whereas a photon of visible light has an energy of about 3 eV (electron volts), a typical gamma photon energy may be 10^6 eV (1 MeV). The emission of photons from the nucleus of an atom does not affect either the charge or mass number of the atom, but does reveal the existence of quantized levels of nuclear energy, analogous to those of electrons in atoms. The energies of the emitted photons are a valuable source of analytical information. Gamma radiation can be measured in several ways; by use of a detector capable of measuring radiation as a function of energy (an energy-dispersive spectrometer), extensive data about the number and kind of nuclei present can be obtained. Thus, radioactivity is characterized by two measurable properties, the rate of decay and the energy of the ejected particle.

Several methods of measuring the number and energy of ejected particles are available. One method of measuring the level of beta emission is with a *Geiger counter*. This device typically consists of a tube containing two

---

[5]The radiation produced on bombardment of a surface with alpha particles is diagnostic of the elemental composition of the surface, and can be used for both qualitative and quantitative analysis. Early unmanned lunar landings used alpha-backscattering experiments to establish the surface composition.

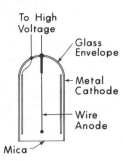

Figure 13-3. Schematic diagram of a Geiger tube.

electrodes and filled with a gaseous mixture of argon and an organic compound such as ethanol (Figure 13-3). One end is covered with a thin mica window to allow low-energy beta radiation to pass into the tube. From 1000 to 1500 V is applied across the two electrodes. An entering beta particle ionizes several molecules of the gas inside the tube; electrons and positively charged gas ions are produced. The generated electrons are drawn to the positive electrode with high velocity and attain sufficient energy to cause further ionization of gas molecules along the way. The resulting shower of electrons cascades to the positive electrode, creating a pulse of current through the circuit. This pulse trips a counting device and is recorded. After the pulse the effective voltage drops, ionization ceases, and the tube again is ready to count. The interval from the time a beta particle enters the tube until the system is ready for another count is called the *dead time*, which lasts about 200 μs. A beta particle entering during this time is not counted.

The Geiger counter remains important because of its simplicity, ruggedness, and dependability. The major disadvantage is that it does not measure the energy of the entering particles. Other methods of measurement of radiation should be mentioned. In a *scintillation counter* the radiation is caused to strike a compound (such as an anthracene crystal) that will emit visible light on impact. The intensity of the light produced is measured by a photomultiplier tube. A *proportional counter* is similar to a Geiger counter, but has practically no dead time, is more stable, and can give a measure of the energy of the particles entering it. However, it requires considerably more extensive and sophisticated electronics. *Photographic film* is blackened by radioactivity, and so the extent of blackening can be exploited as a measure of the amount of radioactivity. Though not important in quantitative analysis, it is used in "film badges" for personnel exposure control and in special applications such as the recording of rare events in certain studies of cosmic rays.

The limit of detectability of radioactivity is so low (since single atoms can be counted) that radioactivity methods provide analytical chemists with one of their most sensitive tools. Because most natural nuclei do not spontaneously decay, they are often made radioactive by bombardment with neutrons. By measurement of the beta and gamma radiation emitted by the artificially produced radioactivity over a period of time, information as to the kind and number of atoms initially present can be gathered by neutron-activation analysis, discussed in Section 13-5.

The time at which a particular radioactive atom will decay is completely unpredictable. On the other hand, a sufficiently large number of atoms of a radioactive isotope will behave statistically in an exactly predictable manner, and a definite fraction will undergo radioactive decay in a given time. The rate of decay by this random process is given in terms of the *half-life*, the time required for half the nuclei of a radioactive isotope to decay, and is related to the rate constant for the process according to Equation (13-2). Analytical applications of radioactivity depend upon the predictability of the rate of nuclear decay when large numbers of atoms are considered. The key relation is that the number of disintegrations of a large number of nuclei, measured per unit time, is directly proportional to the total number of atoms in a given sample of radioactive isotope.

Half-lives of radioactive elements vary from less than 10^{-6} s to more than 10^9 years. Rates at all levels can be useful analytically, although those with either extremely long or short half-lives entail considerably more effort and expense than do those in the range of a few seconds to a few years. Section 13-6 describes an experiment in which picogram quantities of thorium-234 are isolated by carrier precipitation on iron(III) hydroxide and then measured by counting the beta radiation produced.

13-5 APPLICATIONS OF RADIOCHEMISTRY TO ANALYSIS

In one application of radiochemistry to analysis, radioactive elements initially present in the sample are measured. In a second, more important application, radioactivity is introduced into the sample in some way. As examples of the latter, we consider here two methods of this type, neutron activation and isotope dilution.

Neutron-Activation Analysis

In neutron-activation analysis a sample is bombarded with a stream of thermal (low velocity, therefore low kinetic energy) neutrons. A fraction of the incoming neutrons are absorbed, or captured, by nuclei of the sample, forming a new nucleide with the same atomic number as the parent but with an increase of one in atomic mass. The product nucleus possesses excess energy in an amount equal to the binding energy produced on absorption of the neutron plus the kinetic energy of the neutron. It can release this excess energy by (1) emitting gamma radiation or (2) emitting one or more neutrons, protons, beta particles, or alpha particles, or (3) both processes. The resulting radiation or particle emission is measured and related to the elemental composition of the sample. The neutron source is usually a nuclear reactor, in which neutrons are produced as a product of fission of an element such as $^{235}_{92}U$. When an atom of uranium-235 captures a neutron, it undergoes fission to produce two lighter nuclei plus an average of two or three more neutrons. These neutrons can strike other uranium-235 nuclei,   forming more neutrons in a chain reaction. Thus a large number of neutrons (high flux) can be produced in a small volume.

The steps in performing an analysis by neutron activation are as follows. First the sample is placed in a small container, often of polyethylene or aluminum, and exposed to a high flux of thermal neutrons for a time. The sample size and time of exposure vary with the concentration of the sought-for element in the sample and its radiochemical properties; typically a 0.1- to 1-g sample might be exposed for a few minutes to a few hours. Next, the sample is moved to a counting station, typically a proportional or scintillation counter, allowed to stand for a measured time to permit species of shorter half-lives to decay, and counted. The count rate at the energy of the sought-for species is proportional to the amount of that species present in the sample. Because levels of neutron flux vary with time, sample position, and sample composition, standards as similar in composition to the sample as possible are run, and the composition of the unknown calculated from the standard.

Neutron-activation analysis is most applicable to the determination of trace elements. The accuracy and precision of the method are on the order of 2 to 3%. Some approximate detection limits are Fe and S, 1 mg/g; Si, O, and N, 0.1 mg/g; Ca and Cr, 0.01 mg/g; Mg, K, Ni, and Sn, 1 μg/g; Cl, Zn, and Cd, 0.1 μg/g; Na, Al, Cu, As, I, Hg, and U, 0.01 μg/g; and V, Au, and Mn, 1 pg/g. Several types of interferences may occur. One type is interference from nuclear reactions that produce the same product $[^{27}_{13}\text{Al (n,}\gamma)^{28}_{13}\text{Al and}^{28}_{14}\text{Si (n,p)}^{28}_{13}\text{Al}]$,[6]. Another is interference from unresolved gamma radiation from different reactions of similar energies.

Important advantages of neutron activation are (1) high sensitivity for many elements, (2) simultaneous multielement capability, (3) nondestructive examination of the sample (significant in forensic chemistry and moon-rock analyses, for instance), and (4) minimal sample handling and pretreatment. The major disadvantages are (1) a neutron generator (usually a nuclear reactor [7]) must be available for sample irradiation and (2) high concentrations of certain elements cause high background radiation that can interfere in the measurement of trace elements. Often this problem can be alleviated by allowing the irradiated sample to stand for a day or two until the short-lived isotopes decay to a level where they no longer interfere.

Activation analysis can sometimes solve the question of whether bullets involved in different crimes came from the same manufacturer. Antimony, silver, aluminum, copper, cadmium, tin, manganese, and cobalt have all been identified in fragments of lead bullets in varying concentrations that can differentiate among manufacturers.

---

[6]These equations for nuclear reactions are written in a widely used shorthand form. In the first equation aluminum-27 absorbs a neutron to form aluminum-28 and gamma radiation as products; in the second equation silicon-28 absorbs a neutron, yielding as products aluminum-28 and a proton.

[7]The SLOWPOKE reactor, manufactured by Atomic Energy of Canada Ltd., is a convenient source of neutrons for activation analysis. Requiring less than a kilogram of uranium-235 as fuel, and using ordinary water as moderator and radiation shield, the SLOWPOKE costs less than many present-day large laboratory instruments such as NMR or mass spectrometers.

Isotope-Dilution Analysis

In *isotope-dilution analysis* a radioactive form of the compound of interest is added to the sample, and then the compound is isolated in pure form from the sample. Although the isolation procedure need not be quantitative, the separated portion of the compound must be pure. From the weight of the isolated compound, and knowledge of the weight and radioactivity of the radioactive form added initially, the quantity of compound in the original sample can be calculated from the relation

$$\% \ x = (\frac{\text{wt of } x \text{ initially present}}{\text{wt of radioactive } x \text{ added}} \times \frac{\text{specific activity of added } x}{\text{specific activity of isolated } x} - 1)(100)$$

Here *specific activity* refers to radioactivity per unit weight, such as counts per minute per gram. This method is particularly useful in biological systems where quantitative isolation of a component of interest may be difficult or impossible.

13–6 CARRIER PRECIPITATION: SEPARATION OF THORIUM-234 FROM URANIUM-238 AND MEASUREMENT BY RADIOACTIVITY

Background

Most of the experiments in this book involve chemical reactions with favorable equilibrium constants. Conditions for quantitative reaction are often not feasible, however, or conventional methods of calculating chemical equilibria cannot be applied. In such systems useful analytical results may still be obtained if a standard and sample are handled under as nearly identical conditions as possible, and all operations are carried out in the same way on both. Results are then calculated by comparison of the sample with the standard. When reactions are nonquantitative, the extent of reaction is much more sensitive to slight variations in procedure than when reactions are driven to completion. The determination of thorium-234 is an example of nonquantitative recovery of a substance to be measured.

Radioactivity and Thorium-234

This experiment illustrates some of the characteristics of radiochemical analysis, which requires knowledge of both chemical properties and nuclear behavior. The first 103 elements include 272 stable isotopes, 58 naturally occurring radioactive isotopes, and about 1200 artifically produced radioactive isotopes. For tracer work, radioactive isotopes of about 60 elements can be purchased.

Most of the problems met in radiochemical analysis are those of any analytical procedure. Separations may incorporate operations such as precipitation, ion exchange, liquid-liquid extraction, and distillation. After separation the amount of radioactive species present is determined by counting the radioactivity in an instrument such as a Geiger counter.

Uranium-238, the principal isotope of naturally occurring uranium[8] (99.3% uranium-238, 0.7% uranium-235), is radioactive and decays to thorium-234:

$$^{238}_{92}U \longrightarrow ^{234}_{90}Th + ^{4}_{2}He$$

A helium nucleus, called an *alpha particle*, is produced in the process. Uranium-238 has a half-life of 4.51×10^9 years. Thorium-234, also radioactive, then decays to protactinium-234:

$$^{234}_{90}Th \longrightarrow \beta^- + ^{234}_{91}Pa$$

Here a beta particle is produced in addition to protactinium-234. The half-life of thorium-234 is 24.1 days. Protactinium, in turn, has a half-life of 1.17 min. It decays to uranium-234:

$$^{234}_{91}Pa \longrightarrow \beta^- + ^{234}_{92}U$$

Although uranium-234 also is radioactive, its half-life of 2.47×10^5 years makes further decay extremely slow. It decays to thorium-230, which has a half-life of 8.0×10^4 years.

Uranium-238 is normally in secular equilibrium with thorium-234. *Secular equilibrium* is the limiting case of radioactive equilibrium in which the half-life of a *parent* (here uranium-238) is many times greater than that of a *daughter* (here thorium-234). Under such conditions the radioactivity of the parent shows no appreciable change during a period equal to many half-lives of the daughter. After a period of at least 10 half-lives of the daughter, the relation that applies is

$$N_1 t_2 = N_2 t_1 \qquad (13\text{-}5)$$

where N_1 is the number of atoms of the parent element having a half-life t_1 and N_2 is the number of atoms of the daughter having a half-life t_2. Secular equilibrium of thorium-234 with uranium-238 is established within 0.1% in about 8 months. In turn, secular equilibrium of protactinium-234 with thorium-234 is established in about 12 min.

Separation of Thorium-234

The quantity of thorium in the sample of this experiment is so minute that conventional methods of handling the material are impossible. Instead, a small amount of iron(III) is added to the sample solution, and the pH increased so that a gelatinous precipitate of iron(III) hydroxide forms. This

---

[8]Some commercially available uranium has different isotopic composition because uranium-235 has been removed for nuclear applications.

precipitate has an enormous surface area that tends to adsorb ions. In alkaline solution thorium(IV) is adsorbed extensively, even though the solubility-product constant for thorium hydroxide may not have been exceeded.

Because adsorption of thorium-234 by iron(III) hydroxide precipitate is not quantitative, experimental conditions must be carefully controlled and reproduced. Carrying a standard solution of thorium-234 through the same process as the sample compensates for incompleteness of scavenging, radioactive decay, and counting geometry. The simplest way to prepare a standard solution of thorium-234 is to determine by conventional chemical means the concentration of uranium in a solution of a uranium salt in secular equilibrium with thorium-234. The concentration of thorium in this solution can then be computed from the concentration of uranium and the half-lives of thorium-234 and uranium-238. Since this standard solution of thorium-234 contains uranium, for reproducible separation conditions the sample also must contain uranium. The continuous production of thorium-234 must be taken into account. Accordingly, the time of separation is recorded, and the decay is accounted for through a calculation involving the thorium-234 half-life. Since the half-life is 24 days, the time of separation need not be noted to closer than the nearest hour or two. In the separation, upon addition of ammonium carbonate uranium first precipitates as yellow ammonium diuranate, $(NH_4)_2U_2O_7$. With an excess of carbonate a soluble complex, probably $UO_2(CO_3)_3^{4-}$, forms and the uranium precipitate redissolves, leaving the iron hydroxide precipitate.

Measurement of Thorium-234 Radioactivity

This analysis might be performed by measuring the absolute disintegration rate of the thorium-234. Then no standard would be needed. But absolute counting requires prodigious effort; ordinarily, a comparison of the activities of a sample with a standard is far easier. The observed count rate differs from the absolute rate for the following reasons: (1) Some devices for measuring radioactivity can count two disintegrations separately if they occur at or nearly at the same time. A Geiger counter, however, counts such a dual event as one, because after each count there is a short interval during which the instrument does not respond. Corrections for simultaneous decay so as to give a more accurate count rate can be made by application of statistics and the laws of probability. (2) Activity from sources other than the sample may be counted. Cosmic rays and natural and artificial radioactivity in the surroundings contribute to the count rate. This background count must be measured separately and subtracted from the total. (3) The positioning of a sample under the counter tube is critical. Thorium-234 emits beta rays randomly in all directions. With a sample in the normal position beneath the end window of a Geiger tube less than 50% of the radiation is intercepted. (4) Elastic collisions of beta particles with other molecules can result in a change of direction. These changes are classified as backscattering, forescattering, and sidescattering. (5) The mass of a sample itself absorbs some of the beta particles. This self-absorption is reduced by keeping the mass of the sample small.

Overall losses in count rate owing to losses from precipitation, geometry, scattering, and self-absorption can be calculated with the knowledge that 750 betas are emitted per minute per milligram of uranium from thorium-234 in secular equilibrium with aged uranium. Actually, beta particles from thorium-234 are of such low energy that they cannot penetrate the end window of a Geiger tube. But, since thorium-234 decays to the short-lived species protactinium-234 (half-life 1.18 min), which emits betas of high energy, a count of these particles is acceptable; 750 of these also are emitted per minute per milligram of uranium.

In the following procedure, quantitative results depend on proper use of a standard. For reliable analyses, preparation and counting conditions for both samples and standards should be reproduced meticulously. Because radioactive decay is a random process, the indeterminate aspects of the data can be treated in a statistically ideal way.

Reliable results can be expected only if the separation of thorium-234 is carried out simultaneously in the standards and sample. Plan your operations before coming to the laboratory so that the precipitation and filtration steps can be completed in one period.

Procedure (time 3 to 5 h)

Separation

A standard thorium-234 solution may be provided in the form of an aged uranium solution of specified concentration.

Obtain a thorium-234 unknown sample, and quantitatively transfer it to a 100-mL volumetric flask, using 1 mL of 6 M HNO_3 during the washing stage to more completely transfer the thorium-234 from the sample tube.[9] Dilute to volume. Pipet 10-mL portions into four 50-mL beakers. Pipet 10-mL portions of the standard thorium solution directly into another set of four 50-mL beakers. Add 3 drops of 0.03 M $FeCl_3$ solution to each portion. Warm the solutions on a hot plate to about 60°C. Add 0.5 M $(NH_4)_2CO_3$ dropwise, several drops at a time, to precipitate $Fe(OH)_3$ and $(NH_4)_2U_2O_7$. Continue the dropwise addition, swirling the solution until the yellow precipitate of $(NH_4)_2U_2O_7$ dissolves, and then add about 1 mL excess. Carry all samples through each stage at the same time. Do not use stirring rods.[10]   At the completion of this stage, only a small amount of red-brown $Fe(OH)_3$ should remain. Leave the beakers and contents at 60°C for 4 to 5 min to coagulate the precipitate.

Tilt the beakers at an angle of about 30° so that the precipitate settles to the lip side, and allow them to cool for 15 to 30 min. Record the hour and date.[11]   Do not stir the solutions. Obtain a filter assembly (Figure 13-4), and clean the frit by

---

[9]The adsorption of thorium on the walls of the sample tube is decreased by the use of a dilute HNO_3 solution. Quantitative transfer requires rinsing of the tube.

[10]The additional surface provided by stirring rods should be avoided; swirling the beaker and contents provides sufficient mixing.

[11]Because thorium-234 is both continually decaying and continually being formed from the parent uranium, the time of separation of thorium from uranium must be recorded. Noting the time of separation makes it possible to allow for changes in concentration of the thorium due to decay.

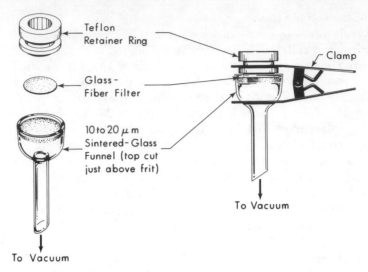

Figure 13-4. Sintered-glass filter assembly for collection of radioactive precipitates.

passing several milliliters of 6 M HCl through it.[12]   Rinse with several portions of distilled water.

When ready to filter, center a glass-fiber filter mat on the 10- to 20-μm sintered-glass support. Clamp the retainer ring in the center of the filter mat and apply vacuum. Decant the liquid from the first beaker into the retainer ring. Try not to disturb the precipitate until as much as possible of the supernatant liquid has been filtered, and then transfer it in one motion. Rinse the beaker with a few 1-mL portions of 0.1 M $(NH_4)_2CO_3$, using a plastic wash bottle with a delivery tube that points upward. Transfer each rinse portion to the filter. Allow air to be drawn through the precipitate for a minute or two after the last rinsing. Transfer the filter mat to a counting planchet with the aid of a spatula, without touching the $Fe(OH)_3$. The precipitate should be face up and the mat should lie flat. Allow the precipitate to air-dry. When the filtration operations have been completed, remove residual $(NH_4)_2CO_3$ from the funnel by drawing a milliliter or two of 6 M HCl through it. Finally, wash the funnel with water.

Counting

Consult your laboratory instructor for information on operation of the Geiger counter.

Obtain an initial background count (blank) by placing an empty planchet plus filter mat in the topmost position beneath the Geiger tube. (*Caution!* The end of the tube is fragile.) Count for 3 min. Then count in order for 3 min each: first standard, first sample, second standard, second sample, third standard, through the set.[13]   Repeat the count for each and then count the blank again for 3 min. [14]

---

[12]This treatment with HCl dissolves any $(NH_4)_2CO_3$ that may be plugging the filter from previous filtrations. Repeat the treatment between runs if a slow filtration rate indicates plugging.

[13]Reproducibility of sample placement is crucial. Position each filter mat in the planchet so that the precipitate is centered under the tube of the Geiger counter.

[14]If time is available, a third count of the standards and samples may be performed to improve precision.

The amount of thorium-234 present in a sample is directly proportional to the net count rate. No correction for decay is needed at this stage, since the half-life is 24.1 days and standards and samples were separated and counted at the same time.

Calculate and report the picograms of thorium in the sample. Also report the date and time of the separation so that allowance for decay in the sample can be made in the results reported.

Calculations

Weight of Thorium-234 in the Sample

First calculate the weight of thorium in the standard in picograms per 10 mL ($1 \text{ g} = 10^{12}$ picograms) as follows: For secular equilibrium

$$N_{Th} = t_{Th}N_U/t_U$$

where N_{Th} and N_U are the number of atoms of thorium-234 and uranium-238 present and t_{Th} and t_U are the half-lives of the two isotopes (24.1 days and 4.51×10^9 years).

Since the number of atoms present is proportional to the number of moles, the concentration of thorium-234, M_{Th}, in moles per liter is given by

$$M_{Th} = \frac{24.10 \text{ days}}{(4.51 \times 10^9 \text{ years})(365.2 \text{ days/year}) \, M_U}$$

Here M_U, the moles of uranium-238 per liter of standard, may be expressed by

$$M_U = \frac{(\text{g U/mL std})(0.993 \text{ g}^{238}\text{U/g U total})(1000 \text{ mL/L})}{\text{g}^{238}\text{U/mol}} = \frac{(\text{g U/mL})(0.993)(1000)}{238.0}$$

The 0.993 is included because only 99.3% of natural uranium occurs as uranium-238. Therefore, the moles of thorium-234 per liter of standard is

$$M_{Th} = \frac{(1.463 \times 10^{-8})(\text{g}^{238}\text{U/mL})(0.993)}{238.0}$$

$$= (6.105 \times 10^{-11})(\text{g U/mL})$$

and the weight of thorium-234 in the standard in picograms per liter is

$$\text{wt Th}_{std} = (6.105 \times 10^{-11})(\text{g U/mL})(234.0 \text{ g Th/mole})(10^{12} \text{ pg/g})$$

$$= (1.429 \times 10^4)(\text{g U/mL})$$

Next, calculate the weight of thorium-234 in the sample. The relation between the weights of standard and sample and their corresponding counts being linear,

$$\frac{\text{wt Th}_{smp}}{\text{wt Th}_{std}} = \frac{C_{smp}}{C_{std}}$$

where C_{smp} and C_{std} are the net counts of sample and standard. Since the same volume, V_{10}, of both sample and standard was taken, the total weight in the initial 100 mL, or 0.1 L, in picograms is

$$\text{wt Th}_{smp} = \frac{(C_{smp})(0.1)(1.429 \times 10^4)(\text{g U/mL})}{C_{std}} = (1.429 \times 10^3)\frac{C_{smp}}{C_{std}}(\text{g U/mL})$$

Statistical Calculations

Calculate the standard error of the average counts per minute (cpm) and the 95% confidence limits that would be expected on the basis of the indeterminate errors associated with only the radioactivity-measurement part of the experiment. Also calculate the overall efficiency of the process, including losses due to separation, counting geometry, and radioactive decay. Assume that the beta emission rate from protactinium-234 in secular equilibrium with 1 mg of uranium is 750/min. Calculate the overall efficiency from the relation

$$\% \text{ overall efficiency} = \frac{(\text{cpm for std})(100)}{(\text{mg U})(750 \text{ cpm/mg U})}$$

where mg U is the milligrams of uranium per 10-mL aliquot. Calculate also the efficiency of separation and counting (corrected efficiency) by correcting the factor 750 to account for radioactive decay occurring between the time of separation and the time of counting. This correction is determined from the relation

$$\log N_0 = \log N_t + 0.301 (t/t_{1/2})$$

where N_0 is the corrected count rate, N_t is the net cpm for the standard, t is the elapsed time in days between filtration and counting, and $T_{1/2}$ is 24.1 days. The overall and the corrected efficiency would be the same only if counting immediately followed filtration.

$$\% \text{ corrected efficiency} = \frac{(N_0)(100)}{(\text{mg U})(750)}$$

To calculate the standard error of the average counts per minute and the 95% confidence limits for the calculated weight of thorium, using as a basis only the indeterminate errors associated with counting the radioactivity, apply the following (see also Example 13-8): (a) The standard deviation s of a single radiochemical measurement is equal to the square root of the total number of counts observed. (b) The average value of n measurements is more reliable than a single measurement by a factor of the square root of n:

$$s = (\text{average count}/n)^{0.5}$$

(c) The count rate (counts per unit time) and its standard deviation are obtained by dividing the total count and its standard deviation by the time of counting. (d) The standard deviation of the sum or difference of two numbers is equal to the square root of the sum of the squares of the individual standard deviations. (e) The relative standard deviation of the product or the quotient of two numbers is equal to the square root of the sum of the squares of the individual relative deviations.

EXAMPLE 13–8

Calculation of results for a set of radiochemical measurements. (Each count is made for 3 min. The letters in parentheses refer to the corresponding statement preceding the example.)

	Blank Count	Sample Number	Sample Count	Standard Number	Standard Count
	255	1	10 261	1	7 963
		2	10 536	2	lost
		3	10 176	3	8 013
		4	10 350	4	8 140
		1	10 183	1	7 985
		2	10 518	2	-----
		3	10 301	3	8 110
	297	4	10 444	4	8 119
Total	552		82 769		48 330
Average count	276 ± 12 (b)		10 346 ± 36 (b)		8 055 ± 37 (b)
Average cpm	92 ± 4 (c)		3 449 ± 12 (c)		2 685 ± 12 (c)
Net cpm	--		3 357 ± 13 (d)		2 593 ± 13 (d)

The standard deviation for net cpm is the square root of $(12)^2 + (4)^2 = 13$(d). The relative standard deviation in the ratio (net count rate for sample)/(net count rate for standard) equals the relative standard deviation in the weight of thorium. The relative standard deviation is equal to the square root of $(14/3357)^2 + (13/2593)^2$ or 0.0063 (e). The absolute standard deviation $= \pm (0.0068)$ (wt of thorium). The 95% confidence limits are estimated to be two times the absolute standard deviation.

PROBLEMS

13-1.   If the concentration of the product of a reaction increases from zero to 0.0023 M after 47.2 min, what is the average rate of reaction?

13-2.   For the reaction $2I^- + H_2O_2 + 2H^+ \rightarrow 2H_2O + I_2$ write the expression for the expected rate of appearance of iodine in the solution. What is the reaction order for each reactant?

13-3.   For the reaction $OH^- + C_2H_5I \rightarrow C_2H_5OH + I^-$ write the expression for the expected rate of hydrolysis. What is the order of the reaction with respect to pH?

13-4.   If the half-life for the decomposition of the periodate glycol complex (Section 10-10) at pH 1 and 25°C is 1.0 min, what is the first-order rate constant?

13-5.   For the reaction $2H_2O_2 \rightarrow 2H_2O + O_2$ the first-order rate constant for the decomposition of hydrogen peroxide at a particular temperature is 1.7×10^{-4} s^{-1}. How much time is required for the decomposition to be 99.9% complete?

13-6.   For the reaction $C_6H_5N_2Cl \rightarrow C_6H_5Cl + N_2$ the rate constant at 50°C is 1.4×10^{-3} s^{-1}. How long is 10 half-lives for this reaction?

13-7.   Often the rate of a proposed reaction is too slow to be used directly in analysis. How might such a reaction be made analytically useful?

13-8.   What are the advantages of initial-rate methods of kinetic analysis?

13-9.   Explain what is meant by (a) fixed-time and (b) variable-time integral methods of kinetic analysis.

13-10.   Explain how the following techniques work: (a) neutron-activation analysis; (b) isotope-dilution analysis.

13-11.   What are the differences and similarities between a Geiger counter and a scintillation counter for measuring radioactivity?

13-12.   A radioisotope decays at a rate such that ¼ the original activity remains after 68 min. Calculate the decay constant and the half-life of the isotope.

13-13.   Suppose that you wish to carry out some experiments with fluorine-18 (half-life 1.87 h). By scheduled airline the fastest delivery time from the nuclear reactor to your laboratory is 12.5 h. How much fluorine-18 would you have to order for 1 mg to arrive at your laboratory?

13-14.†   How fast would you have to work if you wished to separate protactinium-234 from a sample of thorium-234 and measure its activity? Assume you could obtain a satisfactory count rate if 1% of the initial protactinium-234 remained at the end of a 1-min counting period. The half-life of protactinium-234 is 1.18 mm.

13-15.   Two thorium-234 samples were counted for 10 min each, giving rates of 1124 and 2050 counts per minute. The background, counted for 10 min, gave 440 cpm (the counter was contaminated). Calculate the ratio of the activities in the two samples and the standard deviation of this ratio.

13-16.   The radioactive isotope cesium-137 decays by the loss of a beta particle to form the radioactive isotope barium-137. The half-life for the process is 30.23 years. Barium-137 is unstable and emits a gamma ray to generate a stable form of barium. The half-life for this process is 2.55 min. Write the radioactivity-decay equation for the cesium-137 isotope. If a solution contains 1.00×10^{-5} M of cesium-137 per liter, calculate the molar concentration of the radioactive barium-137 in the solution.

13-17.   In the thorium-234 determination of Section 13-6 the uranium sample used to prepare the thorium standard solution must be aged at least 8 months after any chemical processing before use as a standard. Explain why.

13-18.   What fraction of a sample of tritium (half-life 12.3 years) would remain after 10 years?

13-19.   A sample of wood found in an excavation site had a carbon-14 disintegration rate of 9.5 cpm/g. If carbon-14 (half-life 5760 years) at equilibrium with the atmosphere has a specific disintegration rate of 15.3 cpm/g, how old is the wood?

13-20.   Why is a standard solution carried through the same procedure with the sample in the radiochemical determination of thorium (Section 13-6)?

13-21.   (a) Why would a student who ran two replicates of standard and sample in the thorium-234 determination (Section 13-6) probably have poorer results than one who ran four replicates? (b) Would repeated counting of the beta activity make up for fewer replicates? Why?

13-22.   A solution containing thorium-234 was analyzed by the procedure of the experiment in Section 13-6, which yielded the following data: average counts for 3 min for standard and sample were 9651 and 11 423; average count for 3 min for the blank was 291; the standard thorium-234 solution contained 2.147 mg/mL of uranium. How much thorium-234 was present in the entire sample solution?

13-23. Calculate the weight of protactinium-234 (half life = 1.18 min) in secular equilibrium with 2.00 g of uranium.

13-24. Calculate the percentage of uranium-234 in natural uranium from the half-lives and the percentage of uranium-238 present.

13-25. In the experiment in Section 13-6 the radiation count is the beta emission from the reaction $^{234}_{91}Pa \rightarrow \beta^- + ^{234}_{92}U$ (half-life = 1.18 min)
Why is this activity detectable even after several weeks?

13-26. If a pure sample of protactinium-234 were isolated, what fraction of it would remain after 10 min?

13-27. A phosphate fertilizer containing phosphorus-32 as a radioactive tracer has a count rate of 800 cpm/g. How long will it take for the count rate to become 100 cpm/g? The half-life of phosphorus-32 is 14.5 days.

13-28. Artifacts found in a prehistoric Indian campsite in Saskatchewan had a carbon-14 disintegration rate of 5.24 cpm/g. If carbon-14 in equilibrium with the atmosphere has a specific disintegration rate of 15.3 cpm/g, how old are the artifacts? The half-life of carbon-14 is 5760 years.

13-29. What properties must parent and daughter radioactive elements possess if secular equilibrium is to be attained in a sample?

13-30. In the experiment in Section 13-6, how does one avoid the potential interference from precipitation of uranium(VI) in the precipitation of thorium-234 on iron(III) hydroxide?

13-31. What is the purpose of the blank count taken on the Geiger counter before and after the counting of the samples and standards in the thorium-234 determination of Section 13-6?

13-32. List the methods by which a radioactive nucleus can decay.

13-33. A piece of charred bone found in the Arctic had a carbon-14 disintegration rate of 14.1 cpm/g. Carbon-14 in equilibrium with the atmosphere has a specific disintegration rate of 15.3 cpm/g. What is the age of the bone? The half-life of carbon-14 is 5760 years.

13-34. Complete the following equations for radioactive decay:
(a) $^{235}U \rightarrow {}^{94}Y + 2n +$ (b) $^{210}Bi \rightarrow {}^{4}He +$ (c) $^{38}Cl \rightarrow {}^{38}Ar +$ (d) $^{30}P \rightarrow {}^{30}Si +$

13-35. Cobalt-60 decays by emission of a beta particle. How long would it take for a sample of cobalt-60 to decay to 0.1% of its original activity? Its half-life is 5.2 years.

13-36. Strontium-90 has a half-life of 28 years. What fraction would remain after 6 years? The half-life of tritium is 12.3 years.

REFERENCES

H. B. Mark, Jr. and G. A. Rechnitz, *Kinetics in Analytical Chemistry*, Interscience, New York, 1968. A general treatment of the applications of kinetic methods to analysis.

H. L. Pardue, *Clin. Chem.* **1977**, *23*, 2189. A classification and critical survey of kinetic methods in clinical analysis.

G. Friedlander, J. W. Kennedy, and J. M. Miller, *Nuclear and Radiochemistry*, 2nd ed., Wiley, New York, 1964.

S. Sethi and R. S. Rai, *Z. Anal. Chem.* **1970**, *252*, 5. Describes separation of thorium-234 by adsorption on insoluble iodates.

14

INTRODUCTION TO SPECTRAL METHODS: ULTRAVIOLET AND VISIBLE SPECTROPHOTOMETRY

I passed a light of 32 candle power perpendicularly through two sheets of glass after which I found its intensity to have been halved to 16 candle power ... when I passed it through ten sheets it was still as intense as that from one candle... Since all the layers absorb like portions, it is obvious that the light will always be diminished in geometric progression.

P. Bougouer, 1729,
quoted by D. Malinin and J. H. Yoe,
J. Chem. Ed. 1961, *38*, 129.

Broadly defined, *spectral methods* of making chemical measurements involve the encoding of analytical information by the interaction of electromagnetic radiation with a chemical species in a sample. A measurement can be visualized as a series of operations as illustrated in Figure 14-1. The signal is decoded and the output sent to a readout device such as a recorder or meter. The parameters that carry the encoded information are the frequency (which carries qualitative information) and the radiant power (which carries quantitative information). This chapter deals primarily with the quantitative aspects.

In general, spectral measurements may be classified as either (1) emission, in which the sample upon excitation emits radiation whose intensity is proportional to the concentration of the species present, or (2) absorption, in which light from an independent source is passed through a sample and the amount absorbed which is proportional to the concentration of the interesting species is measured. Experimentally, the operations prior to the measurement step may closely resemble those of titrimetric analysis.

Figure 14-1. Schematic illustration of the series of operations required for a spectral method of analysis.

Samples and standards are prepared, separations performed if necessary, pH and other conditions adjusted, and perhaps aliquots taken or dilutions made. Finally, a measurement of radiant energy must be completed.

Spectral methods of analysis are of immense value to all branches of chemistry; probably one third of all measurements made in chemical laboratories entail the use of information encoded in radiant energy. They are particularly important for the examination of traces of material in samples, for studies when only a small sample is available, and for qualitative analysis. Except for a few specialized techniques such as photometric titrations, spectral methods are generally less precise than, for example, titrimetric methods. In trace analysis, however, precision on the order of a percent normally is acceptable. Literature on the subject is extensive.

14-1 CHARACTERISTICS OF ELECTROMAGNETIC RADIATION

Electromagnetic radiation, or *light*,[1] has a dual nature. It may be regarded sometimes as a wave propagating through space and at other times as being composed of particles or photons. A *photon* is a discrete bundle of energy. When large numbers of photons are involved, the concept of the wave nature of light is convenient; when a detailed examination of the interaction of radiation with individual atoms is undertaken, the concept of photons as discrete bundles of energy is essential.

Wave Properties of Radiation

Electromagnetic radiation may be viewed as being composed of (1) an oscillating electric field that travels through space, including vacuum, and (2) an associated oscillating magnetic field. The magnetic field is at right angles to the electrical, and both are perpendicular to the direction of motion of the light. The electric-field component interacts with electrons in matter, and only the electrical part of the wave motion is usually depicted, as in Figure 14-2.

Several wave parameters should be defined. *Frequency* is the number of sine-wave oscillations that occur per second. It is not affected by the medium through which the wave is passing. Frequency is expressed in cycles per second; the unit is the *hertz* (1 Hz = 1 cycle/s).

Wavelength is the distance that light travels in one complete cycle (Figure 14-2). Unlike frequency, wavelength depends on the medium through which the light passes and is at its maximum in vacuum. The product of frequency ν and wavelength λ is the *velocity of light c* in vacuum:

$$\text{frequency} \times \text{wavelength} = \nu\lambda = c = 2.99792 \times 10^{10} \text{ cm/s}$$

---

[1]Electromagnetic radiation covers a wide range of frequencies, from gamma radiation through the radio-frequency region, whereas the term light is usually thought of as the visible region of the electromagnetic spectrum (and perhaps also the regions adjoining the visible). In this book we frequently use the term light in the broad sense for convenience.

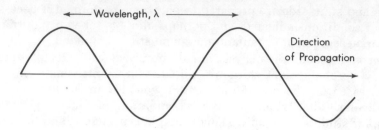

Figure 14-2. The oscillating electric-field component of light.

$$\approx 3 \times 10^{10} \text{ cm/s}$$

The velocity of light is not a constant but, like the wavelength, varies according to the medium. The *refractive index* of a medium is the ratio of the velocity of light in vacuum to the velocity in that medium. Thus for water this ratio is about 1.334, varying somewhat with the temperature of the water and the wavelength of the light. For a typical glass the ratio is about 1.53, and for air, 1.0003, depending slightly on temperature, pressure, and wavelength. [2]   The dependence of refractive index on wavelength is highly important in the design and use of optical instruments.

Wavenumber is defined as the number of sine-wave oscillations per centimeter. It is the reciprocal of wavelength when that parameter is expressed in centimeters. Like wavelength, wavenumber depends on the medium.

As to the particulate nature of light, the *energy E of a photon* can be characterized as proportional to the frequency ν; the proportionality constant h is called *Planck's constant*, expressed in joule-seconds. Thus

$$E = h\nu$$

where $h = 6.626 \times 10^{-34}$ J-s. This direct relation between energy and frequency is convenient. The energy of a photon is occasionally stated in terms of wavelength, but the inverse relation between energy and wavelength is not so convenient and is complicated by the fact that wavelength varies with the medium.

Although frequency can be considered a more fundamental property of light than wavelength or wavenumber, both of which vary with the medium, frequencies unfortunately are nearly always inconveniently large numbers. For example, the frequency of light at the blue end of the visible region of the spectrum (300 nm) is about 10^{15} cycle/s, or 1 kiloteraHz. Therefore chemists usually work with wavelength or wavenumber.

It is sometimes more convenient to speak of energy than frequency. Thus, for a frequency of 10^{15} Hz the energy would be

---

[2]Mirages in deserts and on the ocean are partly the result of differences in the refractive index of air layers of differing temperature.

$$(10^{15}/\text{s}) \times 6.626 \times 10^{-34} \text{ J-s} = 6.626 \times 10^{-19} \text{ J}$$

One electron volt of energy is equal to 1.6×10^{-19} J, or 1.24 μm. Hence, light with a frequency of 10^{15} Hz, or 300 nm wavelength, corresponds to $6.626/1.6 = 4.1$ eV. This is an amount of energy comparable to that in many chemical bonds.

The Electromagnetic Spectrum

A *spectrum* is the distribution of wavelengths, or frequencies, of radiant energy present in a system.

The range of energies covered by the electromagnetic spectrum is exceedingly wide, ranging from multimillions of electron volts for cosmic rays to nano (10^{-9}) electron volts for radio frequencies and energies of interest in nuclear magnetic resonance (NMR). The range is so wide that a unit suitable in one region becomes unsuitable in others. Hence, several different units by which radiation may be specified become reasonable for certain regions of the spectrum. Table 14-1 summarizes the regions of the spectrum and their characteristics.

As this table indicates, the electromagnetic spectrum is huge, covering about 20 orders of magnitude in both frequency and wavelength. It can be broken into two main subdivisions. The first, called the *electronic region*,

Table 14-1. Regions of the Electromagnetic Spectrum

Region	Frequency Range, Hz	Wavelength in Air	Type of Interaction
Cosmic or gamma rays	$>10^{20}$ (energy meV)	$<10^{-12}$ m	Nuclear
X-ray	10^{20} to 10^{16}	10^{-3} to 10 nm	Inner-shell electron transitions
Far ultraviolet	10^{16} to 10^{15}	10 to 200 nm	Electron transitions
Ultraviolet	10^{15} to 7.5×10^{14}	200 to 400 nm	Electron transitions
Visible*	7.5×10^{14} to 4.0×10^{14}	400 to 750 nm	Outer-electron transition
Near infrared	4.0×10^{14} to 1.2×10^{14}	0.75 to 2.5 μm	Vibrational transitions
Infrared	1.2×10^{14} to 10^{11}	2.5 to 1000 μm	Vibrational or rotational transitions
Microwave	10^{11} to 10^{8}	0.1 to 100 cm	Rotational transitions
Radio frequency	10^{8} to 10^{5}	1 to 1000 m	Nuclear-spin transitions
Acoustic	20 000 to 30	15 to 10^{6} km	Molecular motion
ac	60 (power)	5000 km	Electrical

*Human vision is confined to about 400 to 750 nm. There is evidence that other species may possess a wider visual range. Bees, for example, appear to be able to see in the ultraviolet region; some flowers that are white to the human eye absorb light in the ultraviolet and are therefore visible to them.

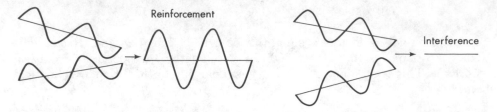

Figure 14-3. Schematic diagram to illustrate reinforcement and interference of two beams of light.

ranges from ac currents up to the border of infrared radiation. In this region the radiation can be directly propagated through wires or tubes, and the frequencies can be observed by direct methods. The electronic region includes microwave-oven frequencies and NMR. The second main subdivision is the *optical region*, from infrared through X-ray. The narrow region of visible light, 400 to 750 nm, though less than an order of magnitude, is the most extensively exploited for chemical measurements. Almost all regions, however, have been developed for particular applications. Instrumentation tends to be the simplest in the visible region. Electromagnetic radiation can interact with matter to give rise to different types of phenomena: *emission, absorption, luminescence, scattering, reflection,* and *refraction,* but not all of these have been applied to or developed for all regions of the spectrum.[3] In the X-ray region the interaction involves the innermost electrons, those in the K or L atomic shells. In the far-ultraviolet region the photons have sufficient energy to affect the midshell electrons. In the ultraviolet and visible it is the valence electrons, that is, the ones involved in the formation of chemical bonds, that are affected. In the infrared, molecular vibrations are involved. In the far infrared where the energies are still lower, both molecular rotations and low-energy vibrations are involved, and in the microwave region, only molecular rotations and nuclear-spin properties.

If two beams of light of the same wavelength and in phase are brought together, the amplitude of the sine wave increases and they reinforce each other. If the two beams are out of phase, they can partially or completely cancel each other. Figure 14-3 is a schematic diagram to illustrate these concepts.

14-2 ABSORPTION AND EMISSION OF PHOTONS

Interaction with Atoms

An understanding of the interaction of radiation with matter can be built upon the knowledge of the structure of atoms and molecules. Consider

[3]Significantly, two of the interactions have been designated to be the fundamental standards of frequency and length. For cesium the frequency of one energy transition has been assigned the value 9.192 631 77 × 10⁹ Hz. For krypton-86 a spectral line in the visible region has a value of 605.780 210 5 nm, thus 1 650 763.73 wavelengths in vacuum is equal to 1 m. Currently, definition of the meter in terms of the velocity of light is being discussed, and the present standard may be superseded.

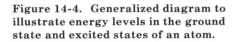

Figure 14-4. Generalized diagram to
illustrate energy levels in the ground
state and excited states of an atom.

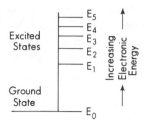

neutral sodium atoms in a dilute sodium vapor. Under ordinary conditions
nearly all the sodium atoms will be in the ground state, that is, with two
electrons in each of the $1s$ and $2s$ levels, six in the $2p$ level, and one in the $3s$
level. The first 10 electrons need not concern us for the moment. The
electron of interest in chemical reactions is the valence electron in the $3s$
level. This electron will also normally be in its ground state. If the sodium
vapor is irradiated with white light (radiation consisting of all visible
wavelengths), the atoms may absorb photons at 589.00 or 589.59 nm. These
photons are of precisely the energy required to cause the $3s$ electron to be
transferred to a higher energy level, in this case one of the two closely spaced
$3p$ levels. Such atoms are said to be excited. The electron in the excited atom
tends strongly to return to its normal $3s$ state. In the process of return it
must either emit a photon of the same energy as was absorbed or lose the
energy through thermal collision.

One of the basic principles of quantum theory is that an atom or
molecule can accept only certain amounts of energy, that is, the electron can
only occupy certain orbitals. A $3s$ ground-state electron in a sodium atom
must absorb exactly the right amount of energy if it is to move into the $3p$
orbital. As it goes from one energy state to another, the right amount of
energy must always be absorbed or emitted in one event; no more or less can
be transferred. Should the energy of the photon irradiating the sodium vapor
be too small, the photon is not absorbed. However, should the photon energy
be too large, the photon may be almost instantaneously absorbed, sufficient
energy from the photon being retained to bring about excitation, and the
remainder being emitted as a photon of lower energy. If the energy is larger
than that necessary to cause ionization, the additional energy may be
imparted to the liberated electron as kinetic energy.

In a sodium atom many excited states are possible. Energy from outside
can raise an electron from the $3s$ to the $4p$, $5p$, or other energy level, or from
the $3p$ to the $4p$ level and so on. The return to the ground state may not
necessarily occur in a single event. Thus a transition from the $4p$ to the $3s$
level may take place in two steps: $4p$ to $3p$ and then $3p$ to $3s$. When the
source of radiation is highly energetic, the absorbed and emitted radiation
may contain thousands of discrete frequencies, with transitions involving
inner as well as outer electrons. Even so, not all conceivable transitions can
occur; some are forbidden by the rules of quantum theory. The transitions
can be visualized in a generalized way as shown in Figure 14-4.

Atoms do not usually remain in an excited state for long, but return to
the ground state. The energy lost will exactly equal the difference between

the energy of an electron in its excited state and that in its state after transition. The electronic energy E is proportional to the frequency of the emitted radiation and is equal to $h\nu$. For example, if an electron were to undergo transition from an excited state E_3 to a state of lower energy E_2, the frequency of the radiation emitted would be

$$\nu_{(3\text{-}2)} = (E_3 - E_2)/h \tag{14-1}$$

The radiant energy is emitted in a random direction, that is, emission is generally unrelated to the initial direction of movement of the photon. This randomness is important in analytical work.

Because only sodium atoms contain 11 protons and 12 neutrons, the electronic energy associated with the E_3 state of sodium is unique to those atoms. The E_2 state is similarly unique, since no other atom has exactly the same energy of excitation. Furthermore, the difference in energies, $E_3 - E_2$, will be unique to the sodium atom and characteristic of it. For the several excited states several electronic transitions are possible, each one being uniquely characteristic. The particular distribution of energy levels of an atom gives rise to its atomic spectrum. Emission spectra are obtained when the transition is from a higher to a lower level, and absorption spectra for the reverse process. Although the intensities are not necessarily the same, for dilute atomic vapors the energy transfers have the same values, and so emission and absorption spectra have the same set of wavelengths.

To excite electrons in an atom requires energies in the region of one or more electron-volts. This minimum normally corresponds to light in the visible or UV regions of the spectrum and is sufficiently energetic to cause transition of a bonding electron from its ground state to an excited state. Transition of inner electrons requires larger amounts of energy and correspondingly higher frequencies. If an atom loses an electron, the energy levels of the resulting ion are shifted, and a different absorption and emission spectrum results.

Interaction with Molecules

Electronic transitions are associated with the electrons in atoms or molecules. In addition, molecules such as CH_3Cl may absorb energy and be converted thereby to higher *vibrational* energy levels. Such absorptions are quantized just as are electronic transitions, but smaller amounts of energy are required to bring them about. Also, absorption and emission of photon energy may be involved in the conversion to quantized energy of rotation of a molecule. These *rotational* energy differences are even smaller than the vibrational ones.

The differences between vibrational and rotational energy levels are much less than between electronic transitions. Therefore, an absorption line corresponding to a single electronic energy change in a molecule will be spread into a band of many lines because of the large number of closely spaced vibrational and rotational levels that are possible.

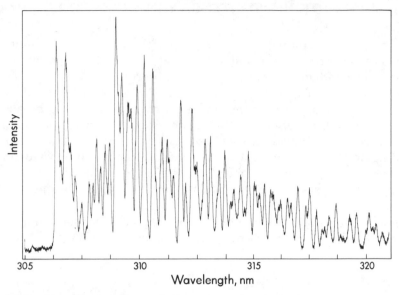

Figure 14-5. A portion of the emission spectrum of the OH radical in a plasma, showing one electronic transition at 306 nm and associated vibrational transitions. Spectrum by G. Horlick.

$$h\nu \; = \text{electronic transition} \pm \text{vibrational transition} \pm \text{rotational transition}$$

Vibrational and rotational absorptions and emissions represent only small differences in energy levels, and therefore all lines corresponding to these transitions are near electronic transition lines. Figure 14-5 is a spectrum of an electronic transition with the associated vibrational fine structure arising from vibrational transitions. Figure 14-6 is an example of a rotational spectrum.

When a molecular gas is compressed, the many fine lines in the molecular band spectrum begin to coalesce because of broadening arising from intermolecular interactions. In a liquid a large number of additional interactions are introduced, and only broad bands of absorption are observed. Band absorption is characteristic of an electronic transition when the atoms or molecules are close together.

Figure 14-6. Spectrum of carbon dioxide showing numerous Q-branch rotational transitions in the region of 4 μm. Spectrum by J. Hoyle.

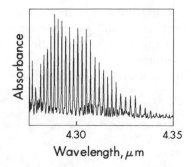

14-3 ABSORPTION AND INTENSITY INFORMATION

Part of the information of interest for quantitative measurements encoded in electromagnetic radiation relates to radiant power. Consider, therefore, how radiant power changes with absorption.

When a beam of light of all frequencies (white light) is passed through a sample, the radiant power of the emergent beam is decreased by absorption of portions of the radiation resulting from interaction with the sample. This is the encoded information. The reduction in radiant power will differ according to the frequency. For the measurements to have quantitative meaning it is important to control the effects of reflection from surfaces, scattering from suspended particulate matter, and absorption by the container or the solvent.

First consider radiation of a particular frequency incident on a sample. If part of the radiation is absorbed and the remainder transmitted,

$$P_0 = P_a + P_t$$

where P_0 is the radiant power of the incident radiation, P_a that of the absorbed radiation, and P_t that of the transmitted radiation.

The Beer-Lambert Relation

In making quantitative measurements it is important to know how P_0 and P_t are related to amounts or concentrations of materials being analyzed. The relation among these quantities for monochromatic radiation is expressed by the Beer-Lambert law:[4]

$$\log (P_0/P_t) = abC = A = \log (100/\% \text{ transmitted}) \tag{14-2}$$

where a is a proportionality constant called the *absorptivity*, b is the path length of the sample in the radiation path, and C is the concentration of the absorbing species. According to this law, P_t decreases exponentially with concentration or with path length. The term $\log (P_0/P_t)$ is given the symbol A and the name *absorbance*. Thus (14-2) can be symbolized most simply by

$$A = abC \tag{14-3}$$

In experimental measurements of concentration the path length b is held constant, and so absorbance is proportional to concentration. Beer's law may be regarded as a simple statement of the fact that reduction in intensity of radiation is proportional to the number of absorbing entities in the path of a radiant beam. Successive equal increments of sample in the beam absorb

---

[4]Beer discovered the relation between absorbance and concentration in 1852. Lambert is generally credited with establishing the relation between absorbance and path length in 1760, although Bougouer reported the basic idea through experiments with a candle and several sheets of glass in 1729. We use the term Beer's Law throughout to denote the relation in Equation (14-2).

equal fractions of the radiant energy available to each increment. Although Beer's law is primarily applied to analytical measurements in the ultraviolet, visible, and infrared regions of the spectrum, it also is applicable in all frequency ranges and for all materials that absorb radiant energy.

Since the value of the transmitted radiation can vary from zero to P_0 the absorbance can vary from zero to infinity. (In practice, absorbance values higher than 2 or lower than about 0.2 are avoided for the reason that precision deteriorates with values at either extreme.)

The proportionality constant a is characteristic of a substance for a given wavelength. Where the sample is a solution, solute-solvent interaction also plays a role. Thus absorptivity is a property of the substance, whereas absorbance varies with both concentration of the substance and dimensions of the container. If b is given in centimeters and C in moles per liter, the constant a is called the *molar absorptivity* ϵ with units of liters per mole-centimeter.[5]   Molar absorptivity is directly useful where comparisons are to be made of the absorption of several substances of known molecular weight.

When the molecular weight of the absorbing material is unknown, as upon the isolation of a new compound, it is appropriate to express concentration as grams per liter. In this event the constant a is called the *specific absorptivity*.

Equation (14-3) is somewhat oversimplified in that it contains no terms to indicate that absorptivity maydepend on temperature or solvent. Shifts in temperature-dependent equilibria will cause changes also, especially in saturated solutions. For some regions of the spectrum, temperature has a profound effect (NMR, for example) but for most measurements in the visible and ultraviolet regions its effects are minor, particularly if direct comparison is being made with a standard at the same temperature.

Generally, the effect of changing solvents cannot be predicted. Changes in the interaction of solutes with solvents and in the molar concentration (due to changes in volume) are usually minor. Solvent selection is often restricted by solubilities. The choice is critical in certain regions of the spectrum in that the solvent should be transparent to radiation at the frequencies of interest. In the ultraviolet, for instance, aromatic solvents are completely unsatisfactory below 300 nm. Chloroform is limited to about 245 nm, and water itself is no longer transparent below 190 nm.

Limitations to Beer's Law

According to Equation (14-3) a plot of absorbance A against concentration C, with all other factors constant, should result in a straight line passing through the origin. Such straight lines are indeed usually obtained over at least a limited concentration range. But on occasion the line may curve upward or, more commonly, downward at higher concentrations.

---

[5]Several other terms can be encountered. *Optical density* and *extinction* are earlier names for absorbance. *Extinction coefficient* and *absorbancy index* have the same meaning as absorptivity. *Molar extinction coefficient* is the same as molar absorptivity. The *transmittance* T is the ratio (not the log) of P_t/P_0. Thus $A = \log(1/T)$.

For useful quantitative measurements it is desirable, though not essential, that the line be straight. From a calibration curve prepared by measurements of a series of standards, the concentration of a sample can be read with little error.

Beer's law tends to hold well for incident radiation from a narrow band of wavelengths. When the band is broad, however, the absorptivity a becomes an average of the values over the band, and deviations may result. The value of a at the peak maximum will also be less for a broad wavelength region than for the optimum narrow region.

The limitations in linearity between A and C at fixed b may be considered under three headings: real, chemical, and instrumental.

Real Limitations

Beer's law is applicable only to dilute solutions. When the concentration of a solute exceeds about 0.01 M, absorbing species become so crowded that solute-solute interactions begin to affect the interaction of the species with the incident radiation. Nonlinearity then results.

Molar absorptivity is also influenced by the refractive index of the solution. Significant changes in refractive index resulting from changes in solute concentration produce a nonlinear Beer's-law plot. In dilute solutions such effects are negligible.

On occasion fluorescence (Section 14-8) from a sample can produce spurious signals. *Matrix effects* result from components of the sample that either decrease or enhance the measurement. Such matrix effects, if serious, may make it difficult to match the standards and samples. An example of a general matrix effect is the effect on absorption values obtained by atomic absorbance when high levels of electrolyte are present in the sample. The increase in viscosity of a solution when electrolyte is present results in a decreased rate of delivery of solution to the atomic-absorption flame and thereby a decreased signal. This is not a real deviation from Beer's law but the result is the same.

Chemical Limitations

If a light-absorbing substance, say $Cr_2O_7^{2-}$ ion, is in chemical equilibrium with other substances with different light-absorbing properties, say $HCrO_4^-$ ion, and the position of equilibrium depends on concentration,

$$Cr_2O_7^{2-} + H_2O \rightleftarrows 2HCrO_4^-$$

nonlinearity will be observed. The absorbance due to chromate is proportional to its concentration, and that due to dichromate proportional to its concentration. However, the equilibrium between these two species depends on the pH and the chromate concentration in the solution. Even at a fixed pH and with more dilute solutions, a higher fraction of the total may be in the form of chromate, and there is an apparent (though not real) deviation from Beer's law. If the solution is buffered at a high or a low pH, such deviations can be made small.

This sort of apparent deviation from Beer's law can be predicted from a knowledge of the chemistry of the systems and of the molar absorptivities as a function of wavelength. Frequently the steps needed, such as buffering, to overcome apparent deviations also can be predicted.

In infrared measurements, hydrogen bonding in alcohols can affect the relation between concentration and absorbance. Phenol, for example, gives two infrared absorptions, one corresponding to free OH and the other to hydrogen-bonded OH groups. The relative intensities of these two absorptions depend on concentration.

Instrumental Limitations

Strictly, Beer's law applies only to an absorbing system in which the incident radiation is of a sufficiently narrow band of wavelengths that it can be termed *monochromatic*. When the incident radiation is more broad (*polychromatic* radiation) and absorptivity changes with frequency, the absorbance at two wavelengths is given by the equations

$$A' = a'bC = \log (P_0'/P_t')$$

$$A'' = a''bC = \log (P_0''/P_t'')$$

If both wavelengths are observed simultaneously, then the observed absorbance is

$$A_{obs} = \log [(P_0' + P_0'')/(P_t' + P_t'')]$$

The log term is not proportional to concentration unless a' is equal to a''. With increasing difference between a' and a'' the relation between absorbance and concentration becomes more and more nonlinear, with absorbance being less than proportional to concentration.

In a practical sense, since a band of frequencies must be used, it is desirable to select for measurements a region of wavelengths in which absorptivity changes but little with frequency.

Another possible instrumental limitation involves stray light. *Stray light* is any radiation reaching the detection and readout system of an instrument that is not of the wavelength setting of the monochromator. It may or may not have passed through the sample. Stray light usually leads to negative deviations from Beer's law. At one extreme the level can be constant, in which case the observed absorbance is

$$A_{obs} = \log [(P_0 + P_s)/(P_t + P_s)]$$

where P_s is the intensity of the stray light. Since the log term in this expression is a smaller quantity than $\log (P_0/P_t)$, the observed absorbance is low. Another possibility is that the stray light is some fraction of P_0. In this circumstance the observed absorbance is

$$A_{obs} = \log [(P_0 + \Delta P_0)/(P_t + \Delta P_0)]$$

Again, a low value is obtained, and the results are more seriously in error when the absorbance and the intensity of stray light is high.

EXAMPLE 14–1

If the fraction of stray light in a spectrophotometer is 0.1 of P_0 and the true absorbances of three samples are 0.3, 1, and 3, the observed absorbance values are 0.26, 0.74, and 1.04, giving errors of 13, 26, and 65%. If the fraction of stray light were 0.01, the observed absorbance values would be 0.30, 0.96, and 1.96, giving errors of 0.1, 4, and 35%. Thus the effects of stray light are, as would be expected, more serious at high absorbance values.

Other physical and instrumental limitations include nonreproducible reflections from surfaces of cells, mirrors, and lenses, differences or changes in cell length, and scattering of light from suspended particles (turbidity) in the solution. At low absorbance values the limitations are primarily concerned with turbidity of the solution (light-scattering effects), impurities that absorb radiation slightly, the effects of cell position, and the cell dimensions. At high absorbance values the instrumental errors become large because of increased effects of dark current and stray light, and limitations in the stability of the radiation source. These topics are discussed in Section 14-5.

14–4 PRECISION OF MEASUREMENTS

Even for systems that obey Beer's law, precise measurements cannot be made at concentrations that are either too low or too high. The limits are set by the smallest change in radiation intensity that can be detected. Qualitatively, at low concentrations the difference between P_0 and P_t is small and the precision will be poor. At high concentrations little radiant energy is transmitted through the sample, and the difference between zero and P_t is small relative to P_0 (which still must be measured).

Ideally, the smallest detectable difference in absorbance, ΔA, should be the smallest possible fraction of the actual absorbance A. To determine the absorbance for which the ratio $\Delta A/A$ would be expected to be at a minimum, it is necessary to differentiate Beer's law twice and to set the second differential equal to zero. When these mathematical steps are performed,[6] it turns out that if the uncertainty in transmittance is proportional to the square root of the transmittance then the optimum value of transmittance is 10% (absorbance 1.0). For constant absolute uncertainty in transmittance, the optimum value of transmittance is 36.8% (absorbance 0.434). With either assumption the curves of relative error against absorbance exhibit broad

---

[6]For constant absolute uncertainty in transmittance dT, the expression for the relative error is obtained from differentiation of Beer's law as follows (since absorbance is proportional to concentration):

$$dC/C = dA/A = d \log T/\log T = 0.434 \, dT/T \log T$$

If the uncertainty in transmittance is proportional to the square root of transmittance, then dT can be replaced by $k\sqrt{T}$, where k is the proportionality constant. The above expression then becomes $dC/C = 0.434 \, k/(\sqrt{T} \log T)$.

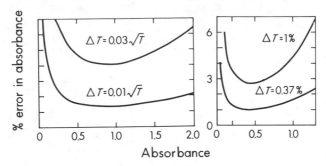

Figure 14-7. Relative error in absorbance as a function of absorbance for an uncertainty in transmittance proportional to its square root or for constant absolute uncertainty in transmittance.

minima, and both types of curves are observed experimentally.

Figure 14-7 shows curves for two levels of error expected on the basis of the foregoing two assumptions. On the assumption that the error in reading transmittance is proportional to its square root, the optimum is broad and ranges from about 0.4 to 1.6, with the minimum at 1.0. The error increases rapidly outside the optimum range for low absorbance values and more slowly for high values. On the assumption that the error in transmittance is constant, the relative error in concentration is near the minimum over a narrower range of absorbance from about 0.2 to 0.8, with the optimum at 0.43. The error increases rapidly in both directions outside the optimum range. In either case it should be recognized that a relative error of 1% in reading the transmittance results in a relative error at the optimum in absorbance that is greater than 1%.

Neither of the assumptions is likely to apply in detail to any real instrument, and error curves of the type shown in Figure 14-7 would be expected to vary with the instrument. The optimum value of absorbance for best precision depends on the particular aspects of noise that limit the precision. If the readout device has limited resolution, the error curve will exhibit an optimum at a relatively low value of absorbance. With the improved resolution provided by a device such as one with a digital display, the optimum value of absorbance is higher. The precision becomes poor at low absorbances because the variability in the readings is often proportional to the signal, owing to variability in positioning of the sample cell. At high absorbances the variability may be nearly independent of the signal but dependent on noise in the dark current or in the efficiency of the instrument in eliminating stray light.

Figure 14-8 shows experimental error curves for two instruments, the Spectronic 20 (designed for routine measurements) and the Cary 118 (designed for research applications). Much of the variability in the readings for the Spectronic 20 arises at low absorbance values from cell positioning and at high absorbance values from limited resolution of the readout device. The Cary 118 yields a broader error curve, with minimum error at higher absorbance values.

Whenever feasible, concentrations of measured solutions should be

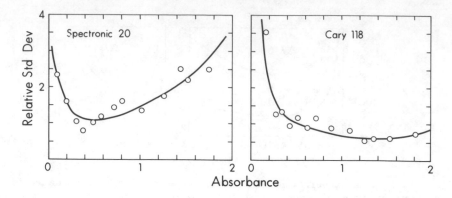

Figure 14-8. Experimental curves of relative error in absorbance as a function of absorbance for two commercial spectrophotometers. Data tabulated by S. May.

chosen so that they are read within the optimum range for the instrument. If the concentrations are too high, the remedy is simply quantitative dilution to within the optimum range. In some instances the choice of another wavelength of lower absorptivity is an effective alternative.

Beer's-law plots are convenient for evaluating the analytical utility of a system and as calibration curves. If the plot of absorbance as a function of concentration is linear, recognize that those points near the beginning and end of the line have significantly greater relative errors than those in the middle. Assignment of equal precision to all data points is the normal simplifying assumption when the position of the straight line is being deduced. The best line is obtained in most cases from a graphical plot or by a least-squares calculation (Section 22-4).

14-5 INSTRUMENT COMPONENTS FOR THE ULTRAVIOLET AND VISIBLE REGIONS

Most radiant-energy measurements are taken in the ultraviolet and visible regions of the spectrum. The radiation in these regions is sufficiently energetic to bring about electronic transitions of outer, or valence, electrons. For a sample in the gaseous state the absorption spectrum will consist of a series of narrow, well defined lines or bands. For a sample in solution, broad regions of absorption occur that may not be highly selective but, on the other hand, can be superior for quantitative measurements.

Instruments for the measurement of absorbance of solutions typically have five components: (1) a stable source of radiant energy, (2) a device to isolate a restricted wavelength region, (3) a transparent container for the sample, (4) a transducer to convert the transmitted radiant energy to an electrical signal, and (5) a signal-recording device or readout.

Two basic types of measurement are ordinarily made. The first is for qualitative information from a plot of absorbance as a function of wavelength, called a *spectrum*. The second is to obtain quantitative

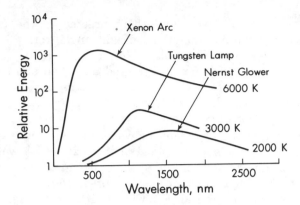

Figure 14-9. Radiant energy from several sources as a function of wavelength.

information by recording absorbance at a selected wavelength as a function of various known concentrations to provide a working analytical curve. In most analytical work these are followed by measurement of the absorbance of a sample to determine the concentration of the substance of interest.

Sources of Radiant Energy

The most common source of visible radiation is the tungsten-filament lamp. Other sources include xenon lamps in the visible region and hydrogen or deuterium lamps in the ultraviolet. The energy emitted from any incandescent source is controlled by its temperature. A plot of radiant power emitted by these sources as a function of wavelength yields a curve with a maximum (Figure 14-9), and the overall energy output is proportional to the fourth power of the absolute temperature. For a source such as tungsten, as the temperature increases, the maximum increases and moves to shorter wavelengths. At room temperature this maximum is seen at about 10 μm, a relatively long wavelength, whereas at the usual tungsten-filament temperatures of around 3000 K it occurs around 1200 nm. Thus most of the energy from a tungsten filament is in the infrared.[7]

Isolation of Wavelength: Monochromators

For either measurement of wavelength or isolation of a band of wavelengths, a device is needed to resolve radiation into its component spectra. A *monochromator* is such a device; it consists normally of a system of slits, lenses, and a dispersing element such as a prism or a grating.

Prisms

A prism monochromator exploits the fact that the refractive index of a medium depends on, or is a function of, the wavelength. Light of a range of wavelengths enters the system through an entrance slit, which is the focal point of a lens. The light from the entrance slit is converted to a parallel

---

[7]Sunlight, which originates from the solar surface at about 6000 K, reaches a maximum at about 500 nm.

beam by a lens or curved mirror and then directed to the prism. When radiation strikes the prism at other than a right angle, refraction occurs on both entering and leaving. The extent of refraction depends on the wavelength, longer wavelengths being refracted less than shorter ones. The dispersed radiation is then focused by a second lens on a slightly curved surface containing an exit slit. Rotation of the prism causes any desired wavelengths to fall upon the exit slit.

A prism gives nonlinear dispersion of wavelengths, that is, wavelengths are distributed nonlinearly along the plane of the exit slit. Wavelengths near the red (longer) wavelength end of the spectrum tend to be more crowded, while those near the blue end are more spread out. The material of construction of the prism is critical. For wavelengths in the region from 350 to 800 nm, glass is the material of choice. It is superior to quartz because of the greater change in refractive index with wavelength. Since glass absorbs too much of the radiation below about 350 nm to be useful, quartz is substituted in the range 200 to 350 nm.

Gratings

A grating consists of a series of parallel, closely spaced (1000 to 20 000 or more/cm) grooves ruled on a transparent plate or a mirrored surface.[8] A beam of polychromatic radiation, or white light, can be dispersed into a spectrum of its component wavelengths by passage through a *transmission grating* or reflection from a *reflection grating*. Each ruling on a grating acts like a miniature slit that functions as a small independent source of light. When light emerges from a slit, it radiates forward in all directions. For a multitude of such equally spaced light sources, most of the wavelengths are destroyed by interference. At certain angles, however, certain wavelengths are reinforced because the difference in path length from adjacent rulings is equal to the wavelength or some multiple of it. This reinforcement occurs at angles where the difference in path lengths is equal to the wavelength. Reinforcement will also occur at multiples of the difference in pathlength. The relation between the wavelength of light reflected and the angle at which it is reflected from a grating surface can be expressed for the simplest designs by the grating equation,

$$n\lambda = d \sin \theta \tag{14-4}$$

where n is an integer called the *order* of the interference, λ the wavelength, d the spacing of the grating (the distance between centers of adjacent grooves), and θ the angle at which reinforcement is observed. With white light composed of a multitude of wavelengths, reinforcement occurs at different

---

[8]Producing an accurately ruled grating is difficult and expensive. The gratings in most commercial instruments are replicas, prepared by coating a grating with plastic, then stripping the coating and making the surface reflective by alumizing it. A recent method of preparing accurate gratings is by coating a substrate with a photosensitive layer, which is exposed to an interference pattern formed by two monochromatic laser beams. Photographic development of the layer produces parallel grooves that are coated with a reflective layer. These *holographic* gratings, as they are called, produce less stray light than conventional gratings.

angles for different wavelengths. Longer wavelengths reflect at higher angles: with a particular grating light of 200-nm wavelength might be observed at 11.5°, light of 300 nm at 17.4°, of 400 nm at 23.5°, of 500 nm at 30°, and of 600 nm at 36.8°. If this dispersed radiation is focused on a plane surface containing an exit slit, a spectrum consisting of a series of images of the light from the entrance slit is observed. Any wavelength desired can be selected by rotating the grating. Dispersion is greatest when the spacing is of the order of the dimensions of the wavelengths to be isolated. Accordingly, gratings utilized for infrared radiation have coarser rulings than do those for visible radiation.

Light of the same wavelength can be observed at several different angles that conform to the several different orders [values of n in (14-4)]. In most instances only the first-order beam is exploited. Orders can overlap; thus radiation of 600 nm in the first order appears at the exit slit at the same angle as 300 nm in the second order or 200 nm in the third order. With more lines per unit distance on the grating the angle of interference is larger and resolution is higher. There is a limit, however; should the wavelength be greater than the spacing, the grating acts like a mirror. Another limitation is that, though close spacing is desirable for highest resolution of close wavelengths, it curtails the range of wavelengths that can be covered. In infrared instruments, for example, with a wide range of wavelengths to be covered, two or three gratings are often needed to achieve the sought-for results.

A grating, unlike a prism, has linear dispersion, that is, a given grating produces radiation of nearly constant bandwidth regardless of the wavelength.

The most important characteristic of a grating is its resolving power. *Resolving power* $\lambda/\Delta\lambda$ is the ability of a grating to separate two closely spaced frequencies. Here λ is the average wavelength of the closest pair of lines that are resolvable and $\Delta\lambda$ is their wavelength difference. Resolving power can be estimated by multiplying the order n by the number of lines on the grating. Thus, the resolving power of a 50-mm grating with 1000 lines per millimeter for a first-order spectrum is 50 000; for 500-nm light this would be 500/50 000, or 0.01 nm. In a practical sense, resolution is limited by the size of the entrance slit rather than by the theoretical resolving power of the grating.

Slits

The slits of a monochromator play an important role in determining its quality. They should possess sharp edges which lie parallel in the same plane. Slits can be either fixed or adjustable by a micrometer. The entrance slit allows radiation to enter the monochromator. This radiation is rendered parallel by a lens or mirror and directed to the prism or grating. In one widely used design the same mirror also focuses the light diffracted from the grating onto the focal plane containing the exit slit (Figure 14-10). Spectral lines are formed as images of the entrance slit; lines of varying frequency can be brought to focus on the exit slit by rotation of the prism or grating. The entrance and exit slits should be of a size such that the image of the entrance

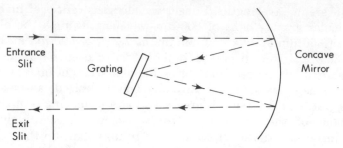

Figure 14-10. One arrangement of a grating monochromator.

slit will just fill the exit opening when the monochromator setting corresponds to the appropriate wavelength. Movement in either direction should give decreased intensity, with zero intensity being reached at a displacement of one slit width.

The spectral region falling on the sample can be narrowed by decreasing the width of the slits. Although a minimum slit width is desirable when spectral detail is sought, the reduced intensity of the radiation makes precise quantitative measurements more difficult (Figure 14-11). Less expensive spectrophotometers have fixed slits.

To take a measurement, one of two basic approaches is applied. The first is to set an exit slit identical with the entrance slit so as to pass a single spectral line or a narrow spectral region. Different wavelengths can be observed by rotation of the grating. The second approach is to place a photographic plate in the focal plane and photographically record the spectrum to give simultaneous measurements of a range of wavelengths.[9]

Filters

Filters are simple devices for restricting wavelengths. They are normally described in terms of the wavelength of maximum transmission and

---

[9]Recently developed photodiode arrays with large-scale integrated circuits instantly provide electronically the same kind of information as a photographic film.

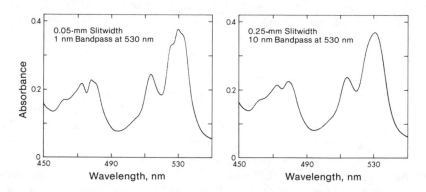

Figure 14-11. Illustration of effective, or spectral, bandwidth of a wavelength-restricting device on the spectra obtained. Data by B. Kratochvil.

the effective bandwidth. *Effective*, or *spectral, bandwidth* is the wavelength region within which the radiant power is one half or more of the maximum.

Absorption filters restrict radiation by absorbing certain parts of the spectrum. The most common type is made of colored glass and has an effective bandwidth of 50 to 250 nm. Filters with narrow band-passes absorb radiation significantly even at the wavelength of maximum transmission.

Interference filters permit the passage of a narrow band through the use of destructive interference to remove the unwanted frequencies. The spectral bandwidth may be as narrow as 10 nm. This filter is constructed of two partly transparent, partly reflective surfaces separated by an extremely thin transparent layer. A portion of the white light passing through the first surface is reflected back and forth between the surfaces. The spacing is set so that if the reflected portion from the second surface is of the desired wavelength it will be in phase with, and thus reinforce, incoming light of the same wavelength. Other wavelengths, being out of phase, will suffer destructive interference and not be transmitted. Interference filters are excellent devices for wavelength selection when only a few wavelengths are needed in an instrument. When a wider choice of wavelengths is necessary, a monochromator is recommended.

An instrument incorporating a filter is called a *colorimeter*, or sometimes a *filter photometer*.

Interferometers

Another device for obtaining spectral information is the *Michelson interferometer*. It is useful for the technique of *time-domain* or *Fourier-transform, spectroscopy*, particularly in the infrared region. The spectral information obtained must undergo a mathematical operation by computer called Fourier transformation before it can be interpreted in the usual manner. The principal advantages are increased sensitivity and the measurement of all frequencies simultaneously. The signal-to-noise ratio of a device of this kind is superior to monochromators using slits and single wavelengths.

Sample Containers

Sample containers, or *cuvettes*, range from inexpensive test tubes to costly quartz cells. For routine work in the visible region the container is typically a test tube, marked so that it can be positioned in the light beam in the same way each time to reduce variations from striations or scratches in the glass. When several solutions are to be compared, use of a number of matched cuvettes can avert the necessity of frequent emptying and refilling of a single container (Section 14-12).

For more precise work and for measurements in the ultraviolet region, cells with flat quartz windows are used. Such cells are often fabricated so that the solution path length can be known to 0.1%. Path lengths of 1, 2, 5, and 10 cm are common. For ultraviolet wavelengths quartz-cell windows are needed, since ordinary glass absorbs all but the longest wavelengths of ultraviolet.

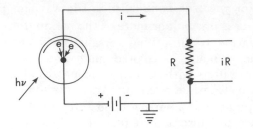

Figure 14-12. Schematic design of a circuit for a simple phototube detector.

Detectors

Once information encoded on a sample has been obtained in the form of variations in radiant power with wavelength or concentration, it must be converted to some measurable electrical, mechanical, or chemical parameter. This conversion device, called a *detector*, should respond to frequencies over a broad range, be sensitive to low levels of light, respond rapidly, have a low noise level, and produce an electrical signal that can be amplified. It is also desirable that the signal produced be directly proportional to the radiant power of light impinging on the detector.

Photoemissive Tubes

In a simple *phototube*, light impinges on a photocathode in an arrangement schematically shown in Figure 14-12. The electrodes are sealed in an evacuated glass envelope. Incident photons impart sufficient energy to electrons on the photoemissive surface on the cathode to liberate them from the surface, that is, at the photoemissive surface–vacuum interface. A small anode, held at a positive potential, collects the released electrons, which return to the cathode through an external electrode circuit that measures the current.

The photoemissive surface of the cathode usually consists of an alkali metal such as cerium or an alkali metal oxide. Not all coatings are equally responsive to all frequencies of light. When radiation in the ultraviolet region is to be measured, the tube window is of quartz so that short wavelengths can reach the photoemissive surface. Response limits may, therefore, be governed by the window material rather than the surface.

The most widely used detector in the UV and visible regions is the *photomultiplier tube*. It utilizes the same type of photosensitive cathode as a simple phototube. The emitted electrons are not collected directly at an anode, however, but focused with an electrostatic field onto the first of a series of electrodes, called *dynodes*, each at a successively more positive potential. At each dynode stage, every electron that impinges on the surface typically releases four or five electrons to yield a gain of four or five. A photomultiplier tube may have from 7 to 20 stages and an overall amplification of 10^5 to 10^8. The current from the final stage is fed into an amplifier circuit, the output voltage of which is a measure of the small initial current. This voltage may be read from a meter, traced on a recorder, or compared with a reference voltage.

EXAMPLE 14-2

A typical signal impinging on a photomultiplier tube in a spectrophotometer is a million photons per second. If the amplification factor of the tube is 10^6, what is the output current?

10^6 anode pulses per second $\times$ 10^6 electrons per pulse $\times$ 1.6×10^{-19} C per electron = 1.6×10^{-7} C/s (amperes). Thus the output signal from a photomultiplier tube is a small current measurable by conventional techniques.

At operating voltages photomultiplier tubes must be protected from even momentary exposure to intense sources of light because they are easily damaged by the photon overload.

A small output signal called *dark current* is continually produced by a photomultiplier tube, even without illumination. The main cause of dark current is thermal emission of electrons from dynode surfaces. The emission of such electrons, called *thermionic emission*, can be reduced by cooling the photomultiplier tube.

Shot Noise

For low levels of illumination the preferred technique is usually photon-counting in which the number of anode pulses per unit time is measured. Photons are emitted from most chemical sources as a random process, that is, photons arrive at random times at the photocathode. Thus, variations in the rate of arrival of photons are inherent in the signal. For such a random process the standard deviation of the rate is equal to the square root of the number of photons. For 10 000 photons/s the standard deviation would be 100, and the relative standard deviation 100/10 000 = 0.01, or 1%. For 100/s the standard deviation would be 10%. Thus the relative standard deviation decreases as the counting rate increases. *Shot noise* is defined as the variability in a signal arising from the randomness associated with the photons constituting the signal.

Photoconductive Detectors

Photoconductive detectors are made from a plate of metal on which a thin layer of a semiconductor has been deposited, such as selenium on an iron base. A thin transparent layer of silver is deposited on the semiconductor to act as a collector electrode. Photons striking the selenium cause excitation of electrons and so increase the electrical conductivity of the material. Electrons in the selenium move toward the silver–selenium interface and are picked up by the silver. Thus a voltage is generated across the selenium without the need for an external power supply. When the external resistance is low a current flows between the iron and silver that is nearly proportional to the light intensity. Even though the requirement for low resistance makes amplification of the response more difficult, the device is attractive because it is inexpensive and durable. Photoconductive detectors are the answer when light intensities are high and when ruggedness is important.

A disadvantage of photoconductive detectors is that they show fatigue effects; at high constant levels of incident radiation the current slowly decreases with time. They also have appreciable temperature coefficients and

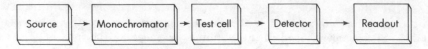

Figure 14-13. Components of an ultraviolet-visible spectrophotometer.

consequently must be warmed up before use. The response curves of photoconductive detectors are broader than for photoemissive tubes, especially in the green and yellow regions of the spectrum. Response times are sufficiently slow that modulation (Section 14-6) is inappropriate.

14–6 SPECTROPHOTOMETERS

Figure 14-13 shows the basic components for a *spectrophotometer*, an instrument for measuring the spectral distribution of the radiant energy absorbed by a sample. Radiation from a stable source, usually a tungsten lamp for the visible region, passes through a monochromator (prism or grating), which allows only a narrow range of wavelengths to strike the sample at one time. Rotation of the prism or grating permits selection of the desired wavelength band. Because the monochromator enhances selectivity (reduction of spectral interference) as well as sensitivity, it is the most vital part of the instrument. The radiation then passes through the sample, and the unabsorbed portion falls upon a photodetector, which produces an electrical signal proportional to the radiation intensity.

Figures 14-13 and 14-14 illustrate a single-beam, single-cell arrangement. To make a measurement with this arrangement, three operations must be performed either directly or indirectly. First, the detector readout scale must be set for infinite absorbance (0% transmission for the dark current), next it must be set for zero absorbance (100% transmission, P_0), and finally, a reading must be obtained for the intensity of transmitted light for the sample P_t. To measure P_0 requires a blank in the light path; to measure P_t requires

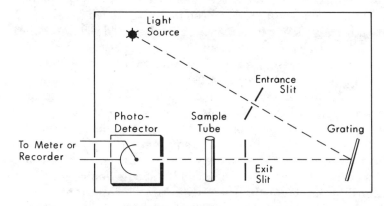

Figure 14-14. Schematic diagram of a grating spectrophotometer.

the sample in the path. Thus an interchange of sample and blank, or reference, solutions is needed. For a single-beam arrangement both the source and the detection system must be highly stable. In practice single-beam arrangements are not common; several "double" arrangements are available — double photocell, double beam, double monochromator, or double wavelength.

Double-Photocell Design

One way stability can be improved is by incorporation of a second photocell to monitor and compensate for variations in the source. An example of this type of arrangement is the Bausch and Lomb Spectronic 20. Newer models of this instrument, though essentially of single-beam design, have a second photocell and associated circuitry solely for that purpose.

Spectrophotometer designs of this type have virtually completely replaced filter photometers for general laboratory work in the range 340 to 620 nm. The dispersion device typically is a reflection grating of 600 lines per millimeter that is rotated for wavelength selection. Fixed entrance and exit slits isolate a band about 20 nm in width. Controls are provided for setting zero and 100% transmission and for selecting the wavelength. The output from both the reference and measuring phototubes is determined, and the difference is amplified for readout. With regulated lamp voltage and a double-phototube arrangement, fluctuations in lamp intensity, formerly a serious problem, cause little difficulty.

Double-Beam Design

In the *double-beam in time* arrangement, after light from a source is passed through the monochromator in the usual way, the radiation emerging from the exit slit is switched alternately between two pathways, one containing a reference and the other the sample, by means of a device such as a rotating mirror. The two beams are focused on the same detector, which receives a signal alternately from the two channels. The magnitude of the alternating-current signal obtained depends on the difference in intensity of the two beams. When the rate of alternation is many times per second, the effects of source variations are greatly reduced. The reference and sample sides of the instrument are identical except that the reference cell contains the sample matrix only, without the component to be measured. Factors other than sample absorption that reduce the radiation intensity, such as reflection from cell surfaces and light-absorbing impurities, are thereby cancelled.

A beam chopper such as a rotating mirror does not send half the radiation to each channel; there is an interval during which the beam is sent to neither. This interval may be used by the instrument to monitor the level of the dark current. Chopping the radiation into a succession of pulses has several advantages. (1) With a fast-response detector the signals from the two channels can be sorted out electronically. (2) Alternation permits electronic elimination of noise and stray radiation which is not chopped. (3) The amplifier can be tuned to respond only to signals of the chopped frequency. The Cary Model 118 spectrophotometer is an example of this type of arrangement.

In the *double-beam in space* arrangement, two separate light paths are created by a beam splitter and mirrors, and the two beams are sent through sample and reference cells. Two separate matched detectors are used. The ratio of the intensities is measured, and the light beam may be chopped before it is split. The Cary Model 15 is an example of this type of arrangement.

Double-Monochromator Design

Single-stage monochromators suffer from the problem that it is difficult to reduce stray radiation to less than about 0.1% of the total radiation. Recording double-beam spectrophotometers of better quality have two monochromators, either two prisms or a prism and a grating in series with an intervening slit. The two dispersing devices must track each other closely; the slit between the two acts as an exit slit for the one and the entrance slit for the other. This design drastically reduces stray light.

Dual-Wavelength Design

Some instruments have been designed so that two beams of light of different wavelength can be made to pass through the same cell in alternation. Two monochromators are needed, and one or two light sources. Instruments of this design can be used in several different ways. (1) Two components of different spectral properties in the cell can be measured as a function of time. This modification is useful in kinetic studies and process control. (2) One wavelength can be fixed at a part of the spectrum where absorption is negligible and used like a dual-beam instrument to minimize the effects of variation in lamp intensity, but now with the same cell. (3) Both wavelengths can be changed with time to obtain a spectrum. If the two wavelengths are offset by several nanometers, the recording displays the difference in absorbance for the two wavelengths and yields a derivative-like curve. Such curves are useful when the spectrum of the substance to be measured overlaps that of an interference.

Overall Spectrophotometer Design

For the practical operation of a spectrophotometer several controls are needed: (1) a wavelength control, normally a mechanical linkage to rotate the prism or grating, which causes the desired wavelength to fall on the sample container; (2) an adjustment device for the slit widths (not always included); (3) a zero adjustment or dark-current control to set the readout to 0% transmission when the detector is not illuminated; (4) a sensitivity control to adjust the readout to 100% transmission (zero absorbance) when a blank, or reference, solution is in the light path; (5) other output controls as necessary; (6) for an instrument with the capability of recording a spectrum of absorbance as a function of wavelength, recorder controls along with controls

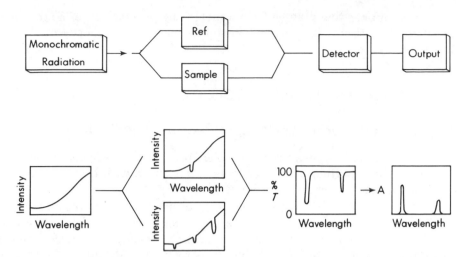

Figure 14-15. Schematic diagram of a double-beam spectrophotometer, illustrating output as a function of wavelength at the various stages.

for selection of scanning speed, wavelength, and absorbance ranges. Most practical spectrophotometers employ one of the double-beam designs. The arrangement indicated schematically in Figure 14-15 summarizes the overall principles of spectral measurement.

This discussion has considered only systems in which the monochromator precedes rather than follows the sample; the reverse is sometimes employed. The intensity of a visible-source output is never constant with wavelength, usually increasing in intensity at longer wavelengths until a maximum is reached in the infrared region (Figure 14-9). Part of the radiation from the source is absorbed by the reference or solvent. During passage of the light through the sample solution, additional absorption occurs. Not indicated in the diagram is the fact that detector response varies with wavelength. In the detector the ratio of the intensities of the transmitted light from the sample and the reference are compared at various wavelengths, and a curve is produced as shown. Note the relative heights of the peaks at the various stages.

There are three ways to obtain the ratio of the signal intensities when the signals have changing absolute levels and differences as in this example. (1) The first is through direct calculation by an attached digital or an analog computer that evaluates the intensities of the reference beam and the sample beam. (2) The second involves electronic null with a double-beam instrument. The signal from the reference side is used to control or adjust either the slit width or the detector gain so that the reference output has a constant value of 100% transmittance. The resolution and the precision of the measurements may vary according to the wavelength region. (3) The third way involves optical null, in which the slit is programmed to hold the reference intensity constant as a function of wavelength. This method requires a source with an intensity that is reproducible from day to day.

The output reading may be from a meter, digital readout, or a strip-chart recorder, sometimes as percent transmittance but more usually in absorbance.

14–7 QUANTITATIVE METHODOLOGY

Criteria for an Analytical Spectroscopic Method

For quantitative measurements the procedure basically consists in comparing the absorbance at a particular wavelength with that of a standard solution or a series of standard solutions.[10]

Applications are limited to species that either absorb appreciable light or can be made to react with or produce absorbing species. When a reaction is completed the color-developing reagent and the reaction product should be stable long enough to obtain reliable measurements. Although few chemical reactions are entirely specific, many are highly selective. Selectivity can be improved in many instances through the use of masking agents, control of pH, or adjustment of oxidation number. In difficult cases a separation may be necessary.

The wavelength chosen for measurements should be one at which unwanted substances do not absorb the radiant energy. In practice this may not always be possible, and compromises may be required. For example, it may be necessary to work on the shoulder of an absorption peak if an interference absorbs radiation at the peak maximum.

Quantitative Analysis of Mixtures

Spectrophotometry offers the substantial advantage that two or more light-absorbing constituents can often be determined in a sample without the need for a chemical separation. Interferences are less likely with a narrow band of incident radiation. Even when several substances in a mixture absorb radiation appreciably at the wavelengths of maximum absorption of each other, reliable measurements can be obtained by setting up and solving simultaneous equations. This method depends on the fact that absorbances are directly additive.

Consider the simplest case of a mixture of two light-absorbing substances (Figure 14-16). Absorbance is measured at two selected wavelengths for the mixture and for known solutions of each substance alone. Two equations are then set up:

$$A_1 = a'_{\lambda_1} bC' + a''_{\lambda_1} bC'' \tag{14-5}$$

$$A_2 = a'_{\lambda_2} bC' + a''_{\lambda_2} bC'' \tag{14-6}$$

---

[10]In the visible region such a comparison can be made visually with what is called the *standard-series technique*. With visual techniques the precision is low. Section 14-11 illustrates this approach.

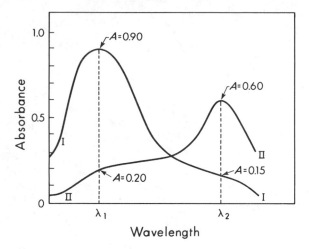

Figure 14-16. Spectra of two compounds and the wavelengths at which absorbance can be measured for simultaneous analysis.

where C' is the concentration of the first substance, C'' that of the second, and λ_1 and λ_2 are the first and second wavelengths. The path length b is the same for all measurements and so cancels out. The molar absorptivities a' and a'' are obtained from measurements on separate standard solutions of the two substances at the two wavelengths. From the measured values of A_1 and A_2 the two unknowns C' and C'' are calculated.

For the highest precision in the analysis of such a mixture the wavelengths should be chosen with care. Least interference, highest sensitivity, and best precision are attained if the wavelengths are chosen where absorbances are maximal and overlap of the two spectra is minimal. Therefore, the spectra of the two substances separately must be known. A practical application of this method is illustrated in Section 14-14.

In principle this kind of analysis can be extended to include three substances by measurements at three wavelengths and solving three equations in three unknowns, or four substances at four wavelengths, and so on. When the number of substances is large, narrow absorption bands favor minimal mutual overlap. Such bands are more likely in the infrared than in the ultraviolet or visible regions. This approach to a four-component mixture in the infrared is applied in Section 15-9.

14-8 FLUORESCENCE AND PHOSPHORESCENCE

Recall from Section 14-1 that after a molecule has absorbed a photon corresponding to some electronic transition the excited molecule generally returns to the ground state with the reemission of a photon. In some cases direct return to the ground state is not the most probable process. Instead, return may occur in steps involving photons of lower energy than that absorbed. This process of emission of lower-energy photons is called

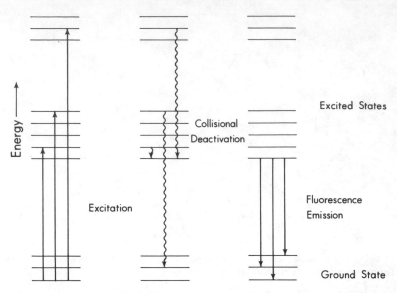

Figure 14-17. Schematic diagram to illustrate the processes of excitation and emission in fluorescence.

fluorescence. The time scale is about 10^{-15} s for absorption of a photon, followed by emission in less than 10^{-6} s. Fluorescence can be employed from the X-ray region through the infrared.

Figure 14-17 illustrates the processes of excitation and emission in fluorescence. When a molecule or atom absorbs energy by a process of excitation an electron is raised from the ground state to an electronically excited state. In the excited state it may be in any one of many excited vibrational levels. The excited molecule can lose energy by any of several different pathways. The one of importance to fluorescence involves rapid transfer of the energy of vibrational excitation to other molecules, which places the molecule in the vibrational ground state of the electronic excited state. A molecule in the second level of electronic excitation may have an energy equal to a vibrationally excited first level. Then the most likely process to occur is *internal conversion,* in which the molecule switches to a vibrationally excited state. The vibrational energy can now be rapidly lost until the molecule is in the vibrational ground state of the first excited electronic energy level. Both the internal conversion and relaxation processes are rapid. If the molecule subsequently returns to the electronic ground state by emission of a photon, the photon will be of lower energy than the initially absorbed photon, and the radiation produced will appear at a longer wavelength.

Another process by which excitation energy is released in an analytically useful way is *phosphorescence.*[11] Here the emitted radiation does not appear until 10^{-6} to 100 s after photon absorption. The reason for this delay

[11]Several other pathways are also possible. *Resonance fluorescence* involves the direct reemission of a photon of energy equal to that absorbed. Nonradiative processes in which energy is transferred through collision with other molecules also occur.

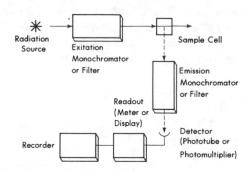

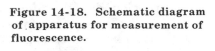

Figure 14-18. Schematic diagram
of apparatus for measurement of
fluorescence.

may be explained as follows. Electrons in a molecule normally have their
spins paired.[12]   Upon excitation of an electron to a higher energy state its
spin is still paired with another electron. After it relaxes to the first excited
state the electron may change spins and become unpaired. This electron is
said to be in a vibrationally excited level of the triplet state T_1. Vibrational
energy is rapidly lost, and the molecule is left in the lowest vibrational level
of T_1.   The lifetime of the triplet state is much longer than that of the
singlet states; if return to the ground state by photon emission takes place,
the radiation may appear only after a long time (10^{-6} to 100 s). This process
is the basis of the phenomenon of phosphorescence.

Instrumentation

Two variables may be adjusted to provide selectivity when fluorescence is
applied to useful measurements: (1) selection of the optimum wavelength for
the absorption of radiation by a sample and (2) selection of the optimum
wavelength for the fluorescent emission of photons. The fluorescent radiation
is emitted in all directions, independent of the exciting shorter wavelength.

The components of an instrument for measuring fluorescence, a
fluorimeter, are similar to those of a spectrophotometer but in a configuration
(Figure 14-18) that follows logically from the nature of the process of
fluorescence. First, the excitation source must produce intense emission in the
ultraviolet or short-wavelength visible region; a zenon or a mercury-discharge
lamp is often employed. Light from the excitation source is passed through a
monochromator or filter to provide the desired wavelength for excitation. The
excited sample becomes the source for a second monochromator, usually
located at right angles to the first, that isolates the fluorescent radiation of
longer wavelength. The second monochromator should of course exclude
radiation of the exciting frequency (one reason for the right-angle
configuration). Also, the first monochromator should not pass radiation of the
fluorescent frequency, or scattering of the exciting radiation by the cell and
contents may result in spurious signals.

The radiant power of the incident beam must be stable, but in contrast
with normal spectrophotometry its actual value is not measured. The radiant
power of the fluorescence signal is measured with respect to an arbitrary

[12]Paired electrons may be said to be in a *singlet* state. An unpaired electron is said to be in a
triplet state.

standard. The standard is placed in the sample tube, and the output reading adjusted to some convenient value. Then the standard is replaced by the sample and a reading taken. Zero fluorescent emission is adjusted with the sample tube filled with solvent only.

When the exciting wavelength is in the ultraviolet region, then the sample holder must be of quartz to transmit such wavelengths. To compensate for a turbid or highly absorbing solution, an angle of less than 90° is occasionally used. In such systems the fluorescence in the layer of sample next to the incident beam of the lamp is the principal source of measurable energy.

The observed intensity of fluorescence is proportional to the number of molecules that absorb the exciting radiation. Therefore, for this part of the process a Beer's-law relation is valid. The observed radiant power depends on the *quantum efficiency* of the process of fluorescence, the fraction of excited molecules that fluoresce. It also depends on a geometry factor, since radiation from only one direction is actually measured. Although the exciting radiation penetrates the sample solution, absorption by the sample prevents its uniform distribution along the path length.

In summary, the intensity of fluorescence is proportional to two main factors: the radiant power of the exciting radiation and the concentration of the fluorescing species.[13]   Fortunately, the radiant power of the exciting radiation need not be measured and therefore can be made extremely high. The detection limit is set. by the noise level of the detector circuit. High-energy radiation sometimes creates a problem with chemical photodecomposition at high source intensities. Nevertheless, fluorescence techniques may be used at concentrations three or four orders of magnitude lower than ordinary absorption methods because sample emission is measured directly, rather than as the difference between two signals. A working curve involves a plot of radiant power as a function of concentration; it is often, though not always, linear at low concentrations.

Fluorescence techniques tend to be relatively specific, two stages of selectivity being available. Phenanthrene, for example, can be determined in the presence of anthracene by use of an excitation wavelength of 265 nm, which causes both to fluoresce. The phenanthrene fluorescence can be measured at 350 nm. Since anthracene fluoresces at 400 nm, it does not interfere.

Phosphorescence provides an additional variable, time, that can increase specificity. Thus a phosphorescent substance can be measured in the presence of a fluorescent one by irradiating the sample with a burst of light, waiting a microsecond or so until the fluorescence has decayed, and then recording the phosphorescence signal. Response is enhanced at low temperatures, and so samples are often frozen in liquid nitrogen. Phosphorescence measurements are most often made on biological samples.

---

[13]For low absorbance values (less than about 0.01) the radiant power of the fluorescence F is given by $F = kP_0 abC$, where k is a constant related to the fluorescence efficiency and depends on the solute, solvent, temperature, other solutes, and the fraction emitted that reaches the detector; P_0 is the radiant power at the excitation frequency; a is the molar absorptivity at the exciting frequency; b is the pathlength for the exciting radiation; and C is the concentration of the fluorescing species.

14-9 APPLICATIONS OF ULTRAVIOLET AND VISIBLE METHODS

The applications of ultraviolet and visible measurements (200 to 800 nm) cover a broad range. They are applied to materials that absorb radiation directly and to those that can be made to undergo a reaction involving a light-absorbing reactant or product. A major fraction of all practical analytical measurements involve spectral data. The methods are normally rapid as well as sensitive and can usually be applied to small amounts of either inorganic or organic material, often at a trace level. Extensive treatises are available on the subject, and reviews of recent developments continue to appear.

The Nature of Absorbing Species

With organic compounds bands of absorption in the ultraviolet or visible regions arise from groups of atoms called *chromophores*, which contain electron configurations capable of being excited by low or moderate energies. Unsaturated organic compounds (those which contain one or more double bonds) absorb radiation because the double bond is a chromophore. Compounds with a single double bond (ethylene, $CH_2=CH_2$, is the simplest example) absorb radiation strongly at about 190 nm and have molar absorptivities of 5 000 to 10 000. Such absorptivities are typical of many organic molecules with chromophore groups. Compounds with conjugated double bonds, that is, two double bonds separated by a single bond between adjacent carbon atoms (butadiene, $CH_2=CH-CH=CH_2$, is the simplest example), also absorb strongly but at longer wavelengths of about 220 nm. Benzene (C_6H_6, the simplest aromatic compound) with three conjugated double bonds has three regions of absorption at about 180, 200, and 255 nm. The larger the number of double bonds in conjugation, the longer is the wavelength at which absorption occurs. Aromatic spectra tend to show many well defined peaks from superimposed vibrational transitions.

Other chromophores include the carbonyl, carboxyl, nitro, and azo groupings. When the chromophores are conjugated, electronic transitions occur more readily and absorption appears at longer wavelengths. Thus a molecule with only a single ketone chromophore absorbs radiation at about 280 nm, but if the ketone group is conjugated with a double bond in an unsaturated ketone, absorption takes place at about 320 nm.

With inorganic compounds three main types of absorption occur in the ultraviolet and visible regions.

1. Certain oxidation states of the eight metals (titanium through copper) in the first transition series, and also of several elements lying below them in the second and third transition series, absorb radiation in the visible region. These elements have partly filled d orbitals, and absorptions arise from transitions involving the electrons in these orbitals. Solvent and ligand interactions play a substantial role in determining both the shape and intensity of the relatively broad absorption bands.

2. The lanthanide and actinide elements have partly filled f orbitals, and the ions of many of these elements exhibit spectra from transitions involving the $4f$ (lanthanides) or $5f$ (actinides) electrons. The f electrons are located in inner orbitals and are largely shielded from interaction with the solvent or ligands in solution. The absorption bands are consequently much sharper than would otherwise be the case, and the spectra consist of sharp, well defined peaks.

3. In charge-transfer complexes one species acts as an electron donor and another as an electron acceptor. Transition of an electron from an orbital of the donor species to an orbital of the acceptor can result from absorption of radiation. The iron(III)–thiocyanate complex, mentioned in Section 4-2 as an indicator for thiocyanate titrations, is an example of a charge-transfer complex. In this complex the iron acts as an electron acceptor. Other well known examples are the tris(*ortho*-phenanthroline) iron(II) and triiodide ion complexes. Charge-transfer complexes have high molar absorptivities, of the order of 10^4.

Preliminary operations for quantitative analysis by measurement of light absorption in the visible and ultraviolet regions generally resemble those in titrimetric analysis. Samples and standards are prepared, and usually aliquots are taken prior to the final measurement step. The samples may be solids, liquids, or gases; a variety of cells for containing liquids and gases during spectral measurements is available.

Analysis

Measurements of ultraviolet and visible radiation are widely utilized for the determination of both organic and inorganic substances. The molar absorptivities are usually such that materials can be measured at the part-per-thousand level or less with a precision of a few percent. Concentrations in the range 10^{-4} to 10^{-5} M are often determinable. Interferences can frequently be avoided through judicious choice of wavelength for the absorption measurements. Further selectivity can many times be obtained chemically by adjustment of the sample conditions. In unfavorable cases measurements at more than one wavelength may make it possible to take into account or correct for an interference.

To aid in selecting the optimum wavelength to be used in analytical measurements, a plot of absorbance as a function of wavelength should be obtained. Sensitivity will be most favorable when a wavelength corresponding to an absorbance maximum for the sought-for substance is chosen. Wavelengths at which absorbance is changing rapidly should normally be avoided, since slight shifts in the monochromator setting during a series of measurements can introduce significant error. The relation between absorbance and concentration for the system should be determined to ascertain the range over which Beer's law applies. A plot of this kind is called a *standard*, or *calibration, curve*. Although small changes in solvent composition, temperature, or ionic strength may have only minor effects on the molar absorptivity, the standard curve should nevertheless be determined as nearly as possible under the same conditions as are the measurements on the sample. Should a Beer's-law plot not be linear, the calibration curve should be defined with a substantial number of points.

When the precision of a spectral measurement is limited by the precision with which the readout device can be read, and noise is not evident on the readout device, then a scale-expansion technique (differential spectrophotometry) can usefully be applied to improve the precision of results. Such techniques, by covering the range of interest in more detail, may give improved precision with dilute solutions. When sources of noise or fluctuations in the source are clearly present, the slight improvement that scale-expansion techniques can provide is not worthwhile.[14]

Fluorescence

Because many compounds that absorb radiation do not fluoresce the number of substances determinable by fluorescence methods is not so large as by absorption methods. The technique is nevertheless applied to a large number of organic and inorganic substances. Recall from Section 14-8 that fluorescence analysis can be applied to highly dilute solutions (typically 10^{-6} M), and is highly selective. An example is the determination of aromatic polynuclear hydrocarbons in the atmosphere. Benzo[a]pyrene, a compound of five fused benzene rings, has been known as a carcinogen since 1934. Traces in the air can be determined by trapping in a solvent of sulfuric acid, where a cation that absorbs radiation at 520 nm is formed. When the solution is exposed to radiation of 520-nm wavelength, a fluorescent peak at 545 nm appears and can be used for quantitative measurements. Although a number of similar compounds absorb radiation at or near 520 nm, few fluoresce at 545 nm. Thus the method is both highly sensitive and specific.

Determination of the Nature of a Complex

Spectral measurements can be used to determine the composition in solution of a complex formed between two entities. One such method is *Job's method of continuous variations*. In this procedure a series of mixtures are prepared in which two constituents are present at varying concentrations, but their sum is held constant. The absorbances of the mixtures are measured. The difference is then determined between the measured absorbance of each mixture and the absorbance the solution would have from the two substances in the absence of a reaction. This difference is plotted as a function of the mole fraction of the solutes. Normally, a maximum appears in the curve at the mole fraction corresponding to the complex that forms. Thus a maximum at a mole fraction of 0.5 would indicate a 1:1 complex, at 0.33 a 1:2 complex, at 0.25 a 1:3 complex, and so on.

EXAMPLE 14–3

The reaction of iron(III) with sulfosalicylic acid in acid solutions gives a violet complex that is a sensitive indicator for the presence of iron. The nature of the complex can be determined by a Job's plot as follows.

---

[14]See references by J. D. Ingle at end of chapter.

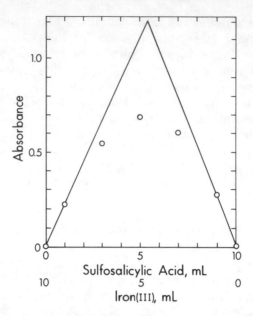

Figure 14-19. Continuous-variations plot for the 1:1 complex formed between iron(III) and sulfosalicylic acid. Data by C. Johnson.

Prepare 100 mL of 0.01 M solutions of sulfosalicylic acid and iron(III) sulfate, both in 0.1 M perchloric acid. To five 50-mL volumetric flasks transfer 1, 3, 5, 7, and 9 mL of the sulfosalicylic acid, and then 9, 7, 5, 3, and 1 mL of the iron(III) sulfate solution. Dilute to volume with 0.1 M perchloric acid. The sum of the concentrations of the sulfosalicylic acid and the iron(III) sulfate in each flask is a constant as required by the method. Allow the reaction to proceed to equilibrium (at least 30 min).

Determine the spectrum over the range 350 to 750 nm of the solution containing 5 mL of sulfosalicylic acid and 5 mL of iron(III) sulfate. Determine the absorbance for each of the solutions at the wavelength of maximum absorbance. Determine the difference in absorbance for the complexed and uncomplexed iron(III) solutions at the five concentrations. Plot the data as a function of mole fraction of iron(III) (Figure 14-19). Determine the formula of the complex from the position of the maximum.

14–10 PHOTOMETRIC TITRATIONS

Spectral measurements typically have a precision (Figure 14-8) of one to several percent. Therefore, concentrations of solutes cannot normally be determined to a precision of a part per thousand by direct measurement of absorbance. But by the simple expedient of measuring absorbance as a function of amount of a suitable titrant in a *photometric titration*, precision equivalent to that of conventional titrimetry can be achieved. Absolute values of absorbance are not required; only the change in absorbance with amount of titrant need be recorded. If the changes in absorbance are large for small

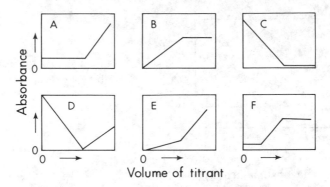

Figure 14-20. Several schematic photometric titration curves. (A) Titrant absorbs radiation at wavelength of measurement; (B) product absorbs, titrant does not; (C) substance titrated absorbs, titrant does not; (D) substance titrated and titrant absorb, product does not; (E) substance does not absorb, product and titrant absorb, but the titrant more strongly; (F) a mixture of two substances, neither of which absorb, but the second forms an absorbing product.

increments of titrant, highly precise end points can be obtained with only moderate precision of the absorbance measurement.

In a photometric titration the substance titrated, the titrant, or the product must absorb radiation at the wavelength used. Under these conditions, since absorbance is linearly proportional to concentration, a plot of absorbance against volume of titrant yields two straight lines that intersect at the end point. If the equilibrium constant for the reaction is not highly favorable, curvature will be noted in the region of the end point. In such cases the straight-line portions far from the end-point region are extrapolated. Thus, readings near the end point are not essential. Several types of photometric titration curves are shown schematically in Figure 14-20.

The photometric titration technique has several advantages over conventional methods of end-point detection. First of all, readings at the end point are unnecessary. Second, only the change in absorbance is important. Thus absorbing but unreactive substances at the wavelength used need not interfere. The absorbance of such substances must not, of course, be high enough to make absorbance measurements imprecise. Third, only one of substance, reagent, or product need absorb radiation. Thus photometric titrations can be extended to materials that are not themselves light-absorbing. Often it is acceptable to choose a wavelength that does not correspond to an absorption maximum, or where the colors before and after the end point do not greatly differ.

It must be kept in mind that, even though Beer's law may be obeyed, two straight lines are obtained only when dilution by titrant is properly accounted for. During titration the volume of solution continually increases. Accordingly, plots of absorbance against volume bend toward the volume axis, and the end-point intersection would be in error if the dilution effect were not taken into account. Correction for dilution is accomplished by

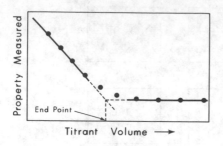

Figure 14-21. Selection of an end point by extrapolation of linear portions of a photometric titration curve.

multiplying the absorbance by the factor $(V + v)/V$, where V is the initial volume of sample solution and v the volume of titrant added. Through the use of a microburet, or of a gravimetric buret where weight rather than volume of titrant is measured, the volume of titrant can be small and the dilution effect may well be negligible. The dilution effect also may be negligible even with substantial volumes of titrant provided the absorbance changes strongly with a small amount of reagent so that only data near the equivalence point need be acquired.

For a photometric titration to be practical, the reaction must fulfill the usual requirements of speed and of known stoichiometry. The reaction need not have a particularly large equilibrium constant, however. A reaction that has a small constant will be incomplete in the region of the equivalence point. Excess titrant may drive it to completion. Hence points somewhat distant from the equivalence point can aid in accurate end-point determinations. For such reactions, the end point is obtained by extrapolation of the linear portions of a curve such as shown in Figure 14-21.

14–11 AMMONIA BY THE NESSLER STANDARD–SERIES METHOD

> The best education is to be found in gaining the utmost information from the simplest apparatus.
>
> A. N. Whitehead

Background

The determination of a trace of ammonia described here employs a comparison method of measurement called the *standard-series* technique. It is the first of several experiments designed to illustrate the principles and techniques involved in the measurement of radiation absorption as an analytical tool. The standard-series technique is one of the oldest, simplest, and most reliable of all methods of color measurement. For ammonia it is fast and extremely sensitive. The precision, however, is poor; 1 part in 20 is about as much as can be expected. The light source is ordinary laboratory illumination, and the detector is the eye. The technique is therefore inexpensive and straightforward.

Precision in Measurement

In laboratory work the precision needed for each operation must always be recognized. For this experiment each measurement need be made to only about a few parts per hundred. To apply corrections to pipet volumes, for instance, is unnecessary, and the ammonium chloride for the standard solution of ammonia need be weighed only to the nearest milligram.

Reagent Impurities in Trace Analysis

A serious difficulty in trace analysis is the presence of interfering impurities at the trace level. In the determination of ammonia at the level of parts per million, even small amounts of ammonia or ammonium salts in the distilled water or sodium hydroxide can lead to high results. This type of problem is especially serious when trace amounts of common substances are being determined. The most satisfactory procedure to reduce the determinant error is to purify and use only reagents in which the concentration of the substance of interest is too low to interfere significantly.

The Nessler Method for Ammonia

The method for the determination of ammonia discussed here is a modification of the original procedure developed by Nessler in 1856. The color-forming reagent, Nessler's reagent, is prepared by first dissolving potassium iodide and then adding enough mercury(II) chloride to the solution to just initiate the precipitation of mercury(II) iodide:

$$HgCl_2 + 4I^- \;\rightleftharpoons\; HgI_4^{2-} + 2Cl^-$$

The solution is then made strongly alkaline with potassium hydroxide or sodium hydroxide. The determination of ammonia is based on the reaction between this alkaline mercury(II) iodide solution and an alkaline solution containing ammonia or ammonium salts:

$$2HgI_4^{2-} + NH_3 + OH^- \;\rightleftharpoons\; NH_2Hg_2I_3 + 5I^- + H_2O$$

The insoluble $NH_2Hg_2I_3$ appears as an intense yellow colloid that is the basis for the measurement. Since the system is not a true solution, the relation between absorbance and concentration is not linear; under these conditions the standard-series technique is the most efficient for the measurements. The color comparison is made in long, thin, flat-bottomed tubes called *Nessler tubes*. The final comparison with a set of standards closely spaced in concentration can be made readily in a special rack for holding the tubes.

Because the sensitivity of the Nessler technique is so high, even distilled water is far too impure for the preparation of standards and reagents. The preparation of ammonia-free water, formerly onerous, is now accomplished simply by passing distilled water through a bed of strong acid cation-exchange resin in the hydrogen form.

Determination of traces of ammonia is important in the field of water pollution. The ammonia content of water supplies is a reliable index of suitability for general use. The technique can also be used for measurement of traces of ammonium salts or ammonia in gases. Yet another application is to measure traces of protein nitrogen by analysis of the distillate from an alkaline-permanganate distillation of the protein-containing material. Traces of nitrates or nitrites can be measured by reduction with suitable alloys, followed by distillation and measurement of the ammonia formed.

Procedure (time 3 to 4 h)

Preparation of Standard Ammonium Chloride Solution

To obtain ammonia-free water, pass distilled water through a cation-exchange column.[15] Use this purified water throughout. Test a sample of the water with Nessler's reagent before use by adding 10 mL of 2.5 M NaOH[16] to a Nessler tube. Dilute to the mark with a portion of the water, mix, and add about 4 mL of Nessler's reagent as directed under Procedure for Analysis.[17]

Dissolve 0.250 g of NH_4Cl in water, and dilute to 50 mL in a volumetric flask. Pipet 10 mL of this solution into a 50-mL volumetric flask and dilute to volume. Similarly, pipet 10 mL of the diluted solution into a 50-mL volumetric flask, and dilute to volume with water. In the same way, pipet 10 mL of the doubly diluted solution into a 50-mL volumetric flask, and dilute to volume with water. The resulting standard solution should contain about 0.04 mg of NH_4Cl/mL.

Procedure for Analysis

Obtain an unknown sample, and quantitatively transfer it to a 250-mL volumetric flask. Dilute to volume. Also obtain five Nessler tubes, a rack to hold them for viewing, and a 2-mL Mohr pipet.

First comparison. To a Nessler tube half-filled with ammonia-free water, add 1 mL of the sample and 10 mL of 2.5 M NaOH solution (NH_3-free), and dilute to volume with water.

Similarly, to four Nessler tubes half-filled with water, add 0, 0.3, 1.0, and 3.0 mL of standard. Add 10 mL of 2.5 M NaOH to each tube, dilute to volume, stopper, and mix by inverting each tube about five times. With a medicine dropper add approximately 4 mL of Nessler's reagent to each tube. The high density of the reagent makes mixing unnecessary. Allow the tubes to stand for 10 min to complete color development, and then visually compare the intensity of the color in the sample tube with that of the four standards. For easy comparison, place the standard tubes in alternate spaces in a Nessler rack; the sample tube then may be placed between the standards. The rack is used with the reflecting glass sloping away from the observer and toward a source of light, preferably diffuse daylight.

---

[15]A drop (no more) of 3 M HCl added to each liter of distilled water before passage through the column ensures that any NH_3 present is converted to $NH_4{}^+$ and held on the resin. If a mixed-bed resin is used, the HCl is unnecessary.

[16]The NaOH solution should be prepared from solid NaOH and ammonia-free water. Stock solutions of NaOH solution provided in the laboratory frequently are not sufficiently free of ammonia for this experiment.

[17]A yellow color may be due to contaminants in either the water or the equipment. Glassware must be rinsed thoroughly after cleaning because most synthetic detergents give a strong yellow color or form a precipitate with Nessler's reagent.

Second comparison. Prepare a set of standards as follows. To four Nessler tubes, add 1.0, 1.3, 1.7, and 2.0 mL of standard. To each tube add 10 mL of 2.5 M NaOH, dilute to volume, mix, and add Nessler's reagent as before.

Prepare sample as follows. To the fifth tube add 10 mL of 2.5 M NaOH and a predetermined volume of sample selected in the following way. If the first comparison showed the ammonia concentration of the sample to be equivalent to (a) less than 0.3 mL of standard, use 4 to 7 mL of sample in the second comparison; (b) between 0.3 and 1.0 mL of standard, use 2 to 3 mL of sample in the second comparison; (c) between 1.0 and 3.0 mL of sample, but nearer 1.0 than 3.0, use 1 mL of sample in the second comparison; (d) between 1.0 and 3.0 mL of standard, but nearer 3.0 than 1.0 mL, dilute 20 mL of sample to 50 mL in a volumetric flask, and use 1 mL in the second comparison; (e) more than 3.0 mL of standard, dilute 10 mL of sample to 50 mL in a volumetric flask. In the last case repeat the first comparison, and on the basis of the results proceed to the second comparison. Remember to allow for this dilution in the calculations.

Dilute the sample to volume in the Nessler tube, mix, and add Nessler's reagent as before. Visually compare the intensity of color with that of the four standards.

Third comparison. Prepare a third set of four standards. The lowest standard concentration should be that concentration whose color intensity is just below that of the sample in the second comparison. The highest standard concentration should be that concentration whose color intensity is just above that of the sample in the second comparison. Use the volume of sample that was used in the second comparison. Add NaOH and Nessler's reagent to the sample and standard as before. Small differences in color intensity can be seen more readily with natural than with artificial light. The best sensitivity is observed when the final matching is with about 1 mL of standard solution.

Record the volume of standard whose color intensity most closely matches that of the sample solution.

Report the results as micrograms of ammonia per milliliter of solution in the 250-mL flask containing the sample. Neither averages nor medians are applicable in this experiment.

Calculations

Calculate the ammonia concentration in the sample from the experimental procedure as follows.

When the color intensities in the two Nessler tubes match, as in the final color comparison or balance,

$$\mu g \text{ ammonia in sample} = \mu g \text{ in standard}$$

The standard contains an initial weight in grams of NH_4Cl in 50 mL that has undergone four dilutions (original plus three). The original 50 mL contains (mol wt NH_3 × g NH_4Cl)/(mol wt NH_4Cl), or (mol wt NH_3 × g NH_4Cl × 10^6)/(mol wt NH_4Cl × 50)$\mu g/mL$. As a result of the dilutions the micrograms of NH_3 in the final comparison tube are given by

$$\mu g \text{ } NH_3 = (NH_3/NH_4Cl) \times g \text{ } NH_4Cl \times (10^6/50)$$

$$\times (10/50) \times (10/50) \times (10/50) \times V_{std}$$

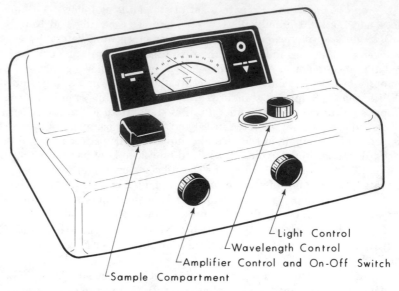

Light Control
Wavelength Control
Amplifier Control and On-Off Switch
Sample Compartment

Figure 14-22. **External view of Bausch and Lomb Spectronic 20 spectrophotometer.**

where V_{std} represents the milliliters of standard present in the final color comparison. This quantity also gives the micrograms of ammonia present in the final comparison tube that contains V_{smp} mL of the sample solution. Since V_{smp} mL of the sample was used, for 1 mL of sample in the 250-mL volumetric flask the weight of NH_3 in micrograms is

$$\mu g\ NH_3/mL = (1/V_{smp}) \times (NH_3/NH_4Cl) \times g\ NH_4Cl$$

$$\times\ (10^6/50) \times (10/50) \times (10/50) \times (10/50) \times V_{std}$$

$$= (V_{std}/V_{smp}) \times (NH_3/NH_4Cl) \times g\ NH_4Cl \times 160$$

14–12   THE SPECTRONIC 20 AND SAMPLE-TUBE MATCHING

Operating Instructions for Bausch and Lomb Spectronic 20

1. Turn on the instrument by rotating the amplifier control clockwise. (See Figure 14-22 for location of controls.)

2. Set the wavelength control to the desired wavelength. With the amplifier control, set the meter needle to zero on the percent-transmittance scale. The setting corresponds to infinity on the absorbance (optical density on older instruments) scale.

3. Dry the outside of the matched sample tubes with a lintless towel or tissue. Insert a tube containing a blank into the sample compartment. Position the tube in the instrument with the aid of the index mark. Close the cover of the compartment.

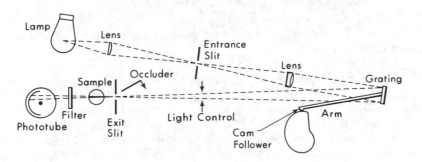

Figure 14-23. Diagram of the optical path of Bausch and Lomb Spectronic 20 spectrophotometer. (The second phototube for monitoring changes in intensity of the light source not included.)

4. Rotate the light control until the meter reads 100 on the percent-transmittance scale (zero on the absorbance scale). Figure 14-23 shows the optical path of the spectrophotometer.

5. Remove the blank tube and recheck the zero reading. Replace the blank tube with one containing a standard or sample. Read the absorbance directly and record it. The cover of the sample compartment should be closed for all readings. Variations of ± 1% in the readings are normal.

6. If readings are to be taken at another wavelength, remove the sample tube and insert the blank tube. Turn the light control counterclockwise before changing the wavelength setting; otherwise, increased sensitivity of the photodetector at the new wavelength may result in a signal of sufficient magnitude to damage the meter. Set the new wavelength.

7. Repeat Steps 4, 5, and 6 until readings at all desired wavelengths have been taken. Readjust the light control whenever the wavelength setting is changed. The zero percent-transmittance setting (dark current) also should be checked periodically and readjusted as needed with the amplifier control.

Readings taken at 10- to 20-nm intervals are sufficient to outline an absorption curve except at absorption peaks or shoulders, where additional points may be needed to characterize the curve more completely. Near the ends of the spectral range of the instrument (below about 350 and above 650 nm) a 100% transmittance reading may be impossible to obtain with the blank.

Sample Tubes (Cuvettes) in Spectrophotometry

Several different types of sample containers are used in spectrophotometry. Less expensive instruments are designed to use test tubes for liquid samples. To ensure that the solution path length is the same for sample and standard solutions, a matched set of tubes is needed; the use of a single tube for all measurements, although possible, is inconvenient. Tubes are matched by placing a solution of intermediate absorbance in each and comparing absorbance readings. One tube is picked arbitrarily as a reference, and others are selected that give the same reading within 1%.

Procedure (time 1 h)

Selection of Matched Tubes for Bausch and Lomb Spectronic 20 Spectrophotometer

Tubes should be matched in a separate operation before any spectrophotometric experiments are begun. Follow the procedure outlined below.

Obtain a supply of 13- by 100-mm test tubes that are clean, dry, and free of scratches. Half-fill each tube with a solution prepared by dissolving about 2 g of $CoCl_2 \cdot 6H_2O$ in 100 mL of 0.3 M HCl.[18] Set the wavelength to 510 nm on the spectrophotometer. (See operating instructions for Bausch and Lomb Spectronic 20 at the beginning of this section.) With the amplifier control, set the instrument to read zero. Place a vertical index mark near the top of one of the tubes, and insert the tube into the sample compartment. Adjust the light control so that the meter reads 90% transmittance. Using this tube to check periodically the 90% reading, insert the other tubes and record their transmittance. If the percent transmittance is within 1% of the initial tube, place an index mark so that the tube can be inserted in the same position every time. If is not within 1%, rotate the tube to see whether it can be brought into range. In future measurements insert each tube in the same position relative to its index mark. Choose a set of seven tubes that have less than 1% variation in reading. (Select nine if the experiments in Sections 14-14 or 16-3 are to be carried out.) Retain these tubes for subsequent photometric work, and return the remainder. To compensate for variations between instruments, use the same one for both tube matching and experimental work.

14-13   SPECTROPHOTOMETRIC DETERMINATION OF TRACE IRON AS BIPYRIDINE COMPLEX

Background

In the determination of a trace of ammonia by the Nessler procedure the color intensity is estimated by eye. Since the human eye responds to energy only in the spectral region from about 400 to 750 nm, any substance that absorbs light within this range interferes with the procedure. Compounds that absorb radiant energy outside this range are not a problem.

High sensitivity is obtained in the Nessler method by use of a long column of liquid. Sensitivity may also be increased by use of a narrow band of radiation as a light source. For instance, for a material that absorbs only yellow light, a light source emitting only yellow light will give best results. Incident radiation from other regions of the spectrum does not enhance, but reduces, the sensitivity. Restricting the wavelength band incident on a sample has the additional advantage of increasing selectivity; species that absorb radiation outside the incident band cannot interfere. Greater precision (up to about 1 part per 100) can be gained by substituting a photosensitive detector for the eye and a more reproducible light source for

---

[18]This solution is recommended because it is stable, has a broad absorption band at about the center of the visible region, and transmits about 50% in a 1-cm cell.

daylight. An additional advantage of reduction in the number of spectral interferences in an analytical method is the lessened need for separations.

Bipyridine for the Analysis of Trace Iron

Bipyridine (bipy) forms an intensely red complex with iron(II) that may be exploited to determine iron concentrations in the range of parts per million. The reaction is

$$3\text{bipy} + \text{Fe}^{2+} \rightleftharpoons \text{Fe(bipy)}_3{}^{2+}$$

The reagent and complex have the structures

The molar absorptivity (a in Beer's law) of the complex is 8650 L/mole-cm at 522 nm, the wavelength of maximum absorption. The complex forms rapidly, is stable over the pH range 3 to 9, and may be used to determine iron(II) concentrations in the range of 0.5 to 8 ppm. Iron(III), if present, must be reduced to iron(II) to produce the colored species. A suitable reagent for this purpose is hydroxylamine hydrochloride, $NH_2OH \cdot HCl$.

The concentration of iron in the sample could be calculated from Beer's law if the molar absorptivity and solution thickness were known. It is preferable, however, to prepare one or more standards and compare absorbance readings of the sample and standard solutions, since this technique minimizes the effects of instrument and solution variation.

Although spectrophotometric methods are normally accurate to about 1%, higher accuracy and precision can be obtained with more sophisticated instruments. In most cases an accuracy of 1% at the level of milligrams per liter is sufficient. The standard in this experiment, $FeSO_4 \cdot (NH_4)_2SO_4 \cdot 6H_2O$, though not a primary standard, has a purity greater than 99%, which is adequate.

Procedure (time 3 to 4 h)

Preparation of Standard Iron Solution

Weigh enough $FeSO_4 \cdot (NH_4)_2SO_4 \cdot 6H_2O$ to prepare 250 mL of a solution 0.002 M in iron. Transfer the salt to a 250-mL volumetric flask, dissolve it in water, add 8 mL of 3 M H_2SO_4, dilute to volume with distilled water, and mix. Pipet 10 mL of this solution into a 100-mL volumetric flask, add 4 mL of 3 M H_2SO_4, and dilute to volume.[19]

Measurement of Absorption Spectrum

To determine the absorption spectrum of the iron–bipyridine complex, first measure the absorbance of one of the standard solutions at 20-nm intervals over the visible spectral region. (The procedure is given in Section 14-12.) Plot absorbance (vertical axis) against wavelength (horizontal axis) to ascertain the

---

[19]The iron concentration of this standard solution should be known to within about 0.5%.

spectrum. Place the plot in your laboratory notebook. From this graph select the wavelength to be used for measurement of the unknown solutions.

Analytical Procedure

The procedure outlined here is for a liquid sample of, say, ground or stream water. Transfer the entire sample to a 100-mL volumetric flask, add 4 mL of 3 M H_2SO_4, mix, dilute to volume, and mix again. Pipet a 10-mL portion into each of three 50-mL volumetric flasks. Into another set of three flasks, pipet 10-mL portions of the standard iron solution. For a blank add 0.4 mL (about 10 drops) of 3 M H_2SO_4 to a seventh 50-mL volumetric flask.[20]

Add to each of the seven flasks, in order, 1 mL of 10% hydroxylamine hydrochloride solution, 10 mL of 0.1% bipyridine solution, and 4 mL of 10% sodium acetate solution,[21] mixing after each reagent is added. Dilute to volume. Using matched tubes, read the absorbance of each solution several times at the wavelength of maximum absorbance. Set the absorbance reading to zero with the blank solution between each group of six sample readings.

Report the total weight in milligrams of iron in the original sample.

Calculations

A comparison method may be used only if a system follows Beer's law. For the iron(II)–bipyridine system, in the concentration range 0.5 to 8 ppm, the absorbance is proportional to the amount of iron present, and

$$\frac{A_{std}}{A_{smp}} = \frac{[Fe]_{std}}{[Fe]_{smp}} = \frac{(mg\ Fe/100\ mL)_{std}}{(mg\ Fe/100\ mL)_{smp}}$$

where A_{std} is the average absorbance reading of the standard and A_{smp} is that of the sample. The concentration of iron in the final standard solution, then, is given by

$$(g\ std)\ \frac{at.\ wt\ Fe}{mol\ wt\ FeSO_4(NH_4)_2SO_4 \cdot 6H_2O}\ \frac{V_{10}}{250}\ (1000\ mg/g) = mg\ Fe\ in\ 100\ mL$$

Substitution of this value and the average absorbance readings for standard and sample in the preceding equation gives the total weight (in milligrams) of iron in 100 mL of sample.

EXAMPLE 14–4

By use of the foregoing procedure, average absorbances for a standard and a sample were 0.337 and 0.428. Assuming that the standard was prepared from 0.413 g of $FeSO_4 \cdot (NH_4)_2SO_4 \cdot 6H_2O$, how many milligrams of iron did the sample contain?

First calculate the milligrams of iron in the standard solution:

---

[20]A 100-mL volumetric flask may be used if convenient; in that case use 0.8 mL of 3 M H_2SO_4.

[21]The sodium acetate plus sulfuric acid gives an acetic acid–sodium acetate buffer in the pH region of about 4.5 to 5.

$$\text{mg Fe} = \frac{\text{at. wt Fe}}{\text{mol wt FeSO}_4\cdot(\text{NH}_4)_2\text{SO}_4\cdot6\text{H}_2\text{O}} \ (0.413 \text{ g})(1000 \text{ mg/g}) = 58.8 \text{ mg Fe/250 mL}$$

$$(58.8 \text{ mg}) \ 10 \text{ mL/250 mL} = 2.35 \text{ mg Fe in 100 mL}$$

$$\frac{A_{std}}{A_{smp}} = \frac{(\text{mg Fe/100 mL})_{std}}{(\text{mg Fe/100 mL})_{smp}}$$

$$(\text{mg Fe/100 mL})_{smp} = (0.428/0.337) \ 2.35$$

$$= 2.99 \text{ mg/100 mL}$$

14-14 SIMULTANEOUS SPECTROPHOTOMETRIC DETERMINATION OF COBALT AND NICKEL AS EDTA COMPLEXES

Background

Spectrophotometric analysis of complex mixtures without prior separation is often possible (Section 14-7). This experiment is designed to illustrate the application of the principles of spectrophotometry to a sample containing two absorbing substances. In mixtures of cobalt and nickel many complexes of the two metals have broad absorption spectra that overlap appreciably. Nevertheless, both can be determined by measuring the absorbances at two selected wavelengths provided the molar absorptivity of each absorbing species at the two wavelengths is measured. Both cobalt and nickel react with EDTA at pH values above 4 (Table 7–3). The resulting complexes are so stable that almost no other common ions interfere with development of the color, and the relation between absorbance and concentration of each complex is linear. Although the complexes (and therefore the intensity of the colors) are stable, the rate of reaction of EDTA especially with cobalt is slow. Solutions consequently have to be warmed long enough after addition of EDTA to ensure complete reaction.

Since the molar absorptivities of these complexes in the visible region are small (less than 100), the procedure described here is applicable only to relatively high concentrations of these metals.

Procedure (time 4 to 6 h)

Preparation of Standard Solutions of Cobalt and Nickel

Accurately weigh into a 100-mL volumetric flask enough powdered cobalt metal to prepare 100 mL of approximately 0.05 M solution. Add 15 mL of dilute HNO_3, and place on a hot plate until dissolution is complete. Do not stopper the flask! Neutralize the acid present by the slow addition of 2.5 M NaOH until a permanent precipitate of cobalt hydroxide remains after swirling. Next add dropwise the minimum amount of acetic acid necessary to just redissolve the precipitate. [22] Dilute to volume. Similarly prepare 100 mL of 0.05 M nickel solution. [23,24]

Analytical Procedure

If the sample is a solid, weigh accurately an amount corresponding to 0.1 to 0.2 g each of cobalt and nickel, and transfer to a 100-mL volumetric flask. The dissolution procedure depends on the composition of the sample. If it is a mixture of oxides, add 8 mL of concentrated HCl, and dissolve on a hot plate. If the sample is a liquid omit this step.

Add 2.5 M NaOH to the appearance of a permanent precipitate as directed above. Add dropwise the minimum amount of acetic acid necessary to clear the solution. Swirl the solution well after each addition to allow time for the precipitate to dissolve, dilute to volume, and mix.

Pipet 20-mL aliquots from the 100-mL volumetric flask containing the sample into two 50-mL volumetric flasks. Pipet 20-mL aliquots of cobalt standard and of nickel standard into four more 50-mL volumetric flasks (two for cobalt, two for nickel). Prepare 100-mL of approximately pH 9.2 buffer, 1 M in NH_4Cl and 1 M in NH_3. Add 15 mL of this buffer to each volumetric flask, 1.9 g of the disodium salt of EDTA (0.005 mol) to the samples, and 0.8 g (0.002 mol) to each of the cobalt and nickel standards. The EDTA is most readily added with the aid of a dry powder funnel. Remove the stoppers, and warm the flasks on a hot plate overnight to complete formation of the complex. Complexation is indicated by a marked color change. Note the pH of each solution at this point; it should be in the range 9 to 10. If below 9, add another 5 mL of buffer and return the solution to the hot plate for another hour before proceeding. Cool to room temperature and dilute to volume.[25] Prepare a blank containing 3 mL of buffer and 0.4 g of disodium EDTA in 25 mL of solution .[26]

Determine the absorbance at various wavelengths in the visible region for one cobalt and one nickel standard solution, taking readings at 20-nm intervals from 350 to 650 nm. (Refer to Section 14-12 for the procedure.) Plot the cobalt and nickel spectra on a graph, absorbance on the vertical axis and wavelength on the horizontal axis. From the curves choose two wavelengths for analysis. Then measure the absorbance of each of the standard and sample solutions at the two wavelengths selected. Take at least five sets of readings of the solutions at each wavelength. After each set place the blank solution in the sample compartment, and readjust the absorbance to read zero.

Report the percentages of cobalt and nickel in the sample. If the samples were liquid, report the percentages of cobalt and nickel assuming a sample of 1 g was taken.

---

[22]Sometimes a small amount of precipitate will remain. This is probably a fatty acid (less than 0.1%) added in the preparation of the metal powder to aid in nucleation of the metal particles. Filtration is usually not required.

[23]If small amounts of insoluble residue remain after the dissolution of nickel powder, obtain another sample of metallic nickel. The residue is probably NiO, which does not dissolve readily in HNO_3.

[24]If insufficient NaOH is added, the acid remaining may exceed the capacity of the ammonia buffer to be added later.

[25]Some white solid may appear upon cooling. This is probably excess EDTA, which should cause no difficulty. Allow any solid to settle before removing samples for spectral measurements.

[26]The blank need not be prepared in a volumetric flask.

Calculations

Set up two equations in two unknowns analogous to (14-5) and (14-6). The molar absorptivities are calculated from the absorbance readings of the standards. By inclusion of the weights of cobalt and nickel taken, the absorbance readings of the standards, the final volumes of solution, and a single value for b, the equations can be rearranged to

$$\% \text{ Co} = \frac{(A_1 A_{Ni_2} - A_2 A_{Ni_1})\ (\text{g Co})(100)}{(A_{Ni_2}A_{Co_1} - A_{Ni_1}A_{Co_2})(\text{g sample})}$$

$$\% \text{ Ni} = \frac{(A_1 A_{Co_2} - A_2 A_{Co_1})\ (\text{g Ni})(100)}{(A_{Co_2}A_{Ni_1} - A_{Co_1}A_{Ni_2})(\text{g sample})}$$

where the symbol A_{Ni_2} is the absorbance of the standard nickel solution at the second wavelength, and so on.

14–15 PHOTOMETRIC TITRATION OF IRON(III) WITH EDTA

Background

Salicylic Acid as a Photometric Indicator for Iron(III)

In this experiment, salicylic acid acts as a photometric indicator for the titration of iron(III) with EDTA. The carboxylic acid portion of the molecule is a moderately strong acid ($pK_1 = 2.97$), while the hydroxyl proton is very weak ($pK_2 = 13.4$). Several transition-metal ions form chelate complexes with the salicylate anion:

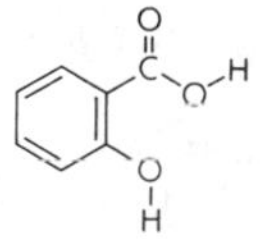

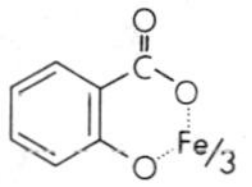

For iron(III) the final complex has three salicylate anions (Sal^{2-}) attached to each iron(III) ion. The log of the overall equilibrium constant for the formation of the complex is 34.5. In acid solution, competition for the phenolic (ROH) oxygen by hydrogen ion shifts the equilibrium to the point where EDTA can compete successfully with salicylate for iron(III), even though the log of the formation constant for the iron(III)–EDTA complex is only 25.1. The reaction for the titration may be written

$$FeSal_3{}^{3-} + H_3Y^- + 3H^+ \rightleftarrows FeY^- + 3H_2Sal$$

The iron–salicylate complex is blue-violet, and the iron–EDTA complex is faint yellow. Control of the pH during the titration is essential. If the pH is too low, extensive protonation of salicylate will occur, and formation of the iron–salicylate complex will be repressed; if the pH is too high, iron(III) hydroxide will precipitate. In this experiment the pH is adjusted first with hydrochloric acid and then maintained at about 2.2 with a buffer of chloroacetic acid, $CH_2ClCOOH$ (pK = 2.3), and its sodium salt during the titration.

The end point is located by measurement of light absorption as a function of added reagent. Because of the high molar absorptivity at the absorption maximum, and the large formation constant for the iron(III)–salicylate complex, absorbance readings at that wavelength remain high until near the end point. Just before the end point the absorbance readings fall rapidly; subsequently they level out at a low value. The EDTA is standardized against pure calcium carbonate by a method similar to that in Section 7-7.

Procedure

Preparation of EDTA Solution

Prepare an approximately 0.015 M EDTA solution by adding 4 g of the disodium salt of EDTA ($Na_2H_2Y \cdot 2H_2O$) and 3 mL of 6 M NH_3 to about 750 mL of water.

Weighing of Calcium Carbonate Standard and Iron Sample

On a triple-beam balance weigh approximately 0.5 g of dry primary-standard $CaCO_3$ into a clean, dry 50-mL beaker. Take the beaker and contents to an analytical balance, and weigh to the nearest 0.1 mg. Handling the beaker with rubber finger cots or a strip of paper, transfer most of the $CaCO_3$ without loss to a dry funnel inserted in the top of a dry 100-mL volumetric flask. Transfer of the last particles is unnecessary. Do not use liquid in this transfer. Weigh the beaker and any remaining calcium carbonate to the nearest 0.1 mg, obtaining the weight of $CaCO_3$ in the flask by difference.

Using the same procedure, weigh, to the nearest 0.1 mg, about 0.6 g of iron unknown sample into a second dry 100-mL volumetric flask.

Preparation of Standard and Sample Solutions

For both standard and sample, rinse the last traces of solid through the funnels into the flasks with a milliliter or two of 12 M HCl, and then add 8 mL more of 12 M HCl to each flask by rapid dropwise addition. Warm if necessary to complete the dissolution. A small residue of insoluble silicate minerals in the sample may be ignored. When dissolution is complete, remove the funnel, rinsing inside and out with demineralized water from a wash bottle as it is being removed. Fill the flask containing the standard to the mark with demineralized water, stopper, and mix well. Neutralize the sample solution with 6 M NaOH solution until $Fe(OH)_3$ just begins to form as a permanent precipitate. Carefully clear the solution by adding a drop or two of 6 M HCl, warming the solution if necessary. Cool and dilute to volume.

Titration of Standards

Pipet 10-mL aliquots of the standard calcium solution and 10-mL portions of 0.02% $MgCl_2$ solution into 200-mL conical flasks. Immediately prior to each titration add about 6 mL of 6 M NH_3 solution and 4 or 5 drops of calmagite indicator solution. Titrate with EDTA solution until the indicator changes from red to pure sky blue with no hint of red. Run a blank titration on a solution containing 1 mL of 6 M HCl and an accurately pipetted 10-mL portion of the MgCl$_2$ solution. Add NH$_3$ and indicator and then titrate as before. Subtract the volume required for the blank titration from the volume required for the sample to obtain the net volume of EDTA solution.

Selection of Wavelength for Titration

Select a suitable wavelength for the titration from spectra of typical solutions before and after the end point. Determine the spectra as follows. Pipet a 10-mL aliquot of the sample into a 50-mL volumetric flask, dilute to volume, and mix. Pipet a 10-mL aliquot of this solution into a 250-mL volumetric flask.[27] Add 2 mL of 6% salicylic acid[28] in methanol and 1.0 mL of chloroacetate buffer. Dilute to volume. Determine the approximate absorption spectrum for this solution with a spectrophotometer. (Operating instructions for the Bausch and Lomb Spectronic 20 are given in Section 14-12.) Now add EDTA solution to a portion of the iron(III)–salicylate solution until the violet color disappears, and then 1 to 2 mL more. Determine the absorption spectrum for this solution. Plot the two absorption spectra on a single graph. The optimum wavelength for the photometric titration is the one at which the difference between the absorbances is largest.

Titration of Sample

Pipet 10-mL aliquots of the original iron solution into 250-mL beakers, dilute to 50 mL with distilled water, and add 1 mL of 6% salicylic acid in methanol and 5 mL of chloroacetate buffer. Add standard EDTA solution from a buret until the color begins to fade. Remove a portion of the solution from the beaker and measure the absorbance, which should be in the range 0.5 to 0.8. Return the solution to the beaker. Add 0.5-mL portions of EDTA, reading the absorbance after each addition until three or four readings have been taken before, and three or four beyond, the end point. Plot absorbance against volume of EDTA added. Determine the end point from the intersection of the linear portions.

Report the percentage of iron in the sample.

PROBLEMS

14-1.   How does the velocity of light depend on refractive index? Calculate the velocity of light in air, given that the refractive index of air is 1.0003.

14-2.   (a) What is the wavelength in micrometers of electromagnetic radiation of frequency 6.122×10^{13} Hz? What is the wavenumber? (b) Calculate the frequency in Hz for light of 355-nm wavelength.

14-3.   Explain what is meant by reinforcement and by interference of beams of light.

14-4.   Clearly distinguish between the terms (a) absorbance and absorptivity and (b) specific and molar absorptivity.

14-5.   What types of limitations can give rise to a failure of absorbance and concentration to be linearly related? Show how at a fixed pH the fraction of chromate can change in the dichromate–chromate equilibrium.

---

[27]This serial dilution gives a final concentration equal to 0.008 of the original.

[28]A substantial excess of salicylic acid over iron ensures that all the iron will be in the form of Fe(Sal)$_3^{3-}$. If desired, 5-sulfosalicylic acid can be substituted for salicylic acid; being water soluble, the sulfonic acid derivative eliminates the need for methanol as solvent.

14-6.   When does stray light become especially significant in absorbance measurements?

14-7.   A 0.00351 M solution of a compound in water transmits 45.0% of the light at 530 nm in a cell of pathlength 2.50 cm. (a) What is the molar absorptivity of the compound? (b) What is the concentration of a solution of this compound that would transmit 74.8% under the same conditions?

14-8.   A solution of 0.0040 M copper dithizone transmits 66% of the light at a particular wavelength in a cell 1.5 cm thick. (a) What is the absorbance? What is the molar absorptivity? (b) What is the concentration of a copper dithizone solution that transmits 30% under the same conditions?

14-9.   A 2.0 $\times$ 10^{-4} M solution of a new compound has a transmittance of 48.3% in a cell that has a path length of 2.0 cm. What is the concentration of a solution having a transmittance of 28.2% under the same conditions? What is the molar absorptivity of the compound?

14-10.   A colored substance S, molecular weight 150, was found to have an absorption peak at 405 nm. A solution containing 3.03 mg/L had an absorbance of 0.842 when examined in a 2.50-cm cell. What weight of S is contained in 100 mL of a solution that has an absorbance of 0.768 at 405 nm when measured in a 1.04-cm cell?

14-11.   The radiant power of a monochromatic light beam is decreased to 18% of its original value on passing through 3.00 cm of a 2.14 $\times$ 10^{-4} M solution of picric acid at a given wavelength. What is the molar absorptivity of picric acid at this wavelength?

14–12.†   An acid HA has a dissociation constant of 1 $\times$ 10^{-4}. The molar absorptivities of the undissociated species are 30 and 0 at 500 and 650 nm, and those of the anion are 0 and 200 at 500 and 650 nm. Calculate and plot the relation between the absorbance of a solution of HA in a 1-cm cell at 650 nm and a concentration over the range 0.1 to 0.001 M in (a) pure water, (b) a solution buffered at pH 2, and (c) a solution buffered at pH 4. Carry out a similar set of calculations and plots for the absorbance at 500 nm.

14-13.   The base form of a weak base (ionization constant 1 $\times$ 10^{-4}) has an absorption maximum at 520 nm; the conjugate acid form does not absorb radiation at this wavelength. A Beer's-law plot was prepared by adding varying known amounts of base to a series of volumetric flasks, followed by 10 mL of pH-10 buffer and dilution to volume with distilled water. The absorbance of each solution was measured at 520 nm. Sketch the general shape expected for the Beer's-law plot from these measurements. Explain your conclusions.

14-14.   (a) Explain what is meant in a practical sense by monochromatic light. (b) List four types of devices for limiting the wavelength of light passing through a sample for spectrophotometric measurements. Describe how each type works.

14-15.   (a) Compare and contrast prisms and gratings as devices to restrict wavelength. (b) What are the advantages of monochromators over filters for isolating bands of radiation?

14-16.   Estimate the uncertainty in an absorbance reading taken from Figure 14-8 for readings of 1.6, 1.0, 0.4, and 0.1, taken with the Spectronic 20 spectrophotometer.

14-17.   In spectrophotometers what are the advantages of doubling (a) the monochromators; (b) the detectors?

14-18.   Cadmium forms a chelate complex with oxine that is soluble in chloroform and has a molar absorptivity of 3200. A 1.256-g sample of a rock was dissolved, the cadmium complex extracted quantitatively into chloroform and diluted to 100 mL, and the percent transmittance measured. A value of 31.6% was obtained in a 2-cm cell. What was the percentage of cadmium in the sample?

14-19.   A compound used in ball-point-pen ink has an absorption peak at 623 nm. A solution containing 1.05 mg/L has an absorbance of 0.521 when measured in a 1-cm cell. What is the concentration of the dye in a solution that has an absorbance of 0.321 when measured in a 2-cm cell?

14-20.   Explain briefly why spectrophotometric analyses are generally, but not always, carried out at wavelengths of maximum absorbance of the substance being determined.

14-21.†   Chromium and manganese in steel may be determined by dissolution and then oxidation to chromate and permanganate, followed by spectrophotometric measurement at 450 and 525 nm. A solution containing 12.0 μg of manganese per milliliter gave a percent transmittance reading of 76.1 at 450 nm and of 32.0 at 525 nm; another solution containing 201 μg of chromium per milliliter gave readings of 40.0 and 72.0% at 450 and 525 nm. What was the chromium and manganese content of a solution that, under the same conditions, gave transmittance readings of 36.1 and 45.0 % at 450 and 525 nm?

14-22.   What are the advantages of a calibration curve over comparison with a single standard in a spectrophotometric analysis?

14-23.   When recording a spectrum with a single-beam spectrophotometer, why is it desirable to readjust the readout to 100% T with a blank solution in the cell every time the wavelength is changed?

14-24.   The molar absorptivity of a colored uranium complex is 14 300 at 453 nm. (a) What is the absorbance expected for a solution containing 0.4 ppm (mg/L) of uranium if all the uranium is converted to the complex and measured in a 10-cm cell? (b) What is the percent transmittance of this solution?

14-25.   Sketch the optical path of an instrument used for making (a) spectral measurements, such as a single-beam spectrophotometer, and (b) fluorescence measurements.

14-26.   Solutions of varying amounts of aluminum were prepared. 8-Quinolinol was added, and the complex extracted with chloroform. The chloroform extracts were diluted to 50 mL, and the fluorescence was measured. For aluminum concentrations (μg/50 mL) of 2, 4, 6, 8, 10, 12, 14, 16, and 18 the fluorescence readings were 10, 19, 28, 37, 45, 53, 60, 66, and 71. Plot the fluorescence reading against the aluminum concentration. Does the calibration curve follow the expected linear relation? Over what concentration range is the approximation valid? Why is the calibration curve for fluorescence analysis often linear over only a small range of concentrations?

14-27.   Why is fluorescence often more selective than UV or visible spectrophotometry for determining a specific compound in a mixture of several compounds of similar chemical structure?

14-28.   In fluorescence spectroscopy two monochromators or filters are usually employed. How are the wavelengths to be used in analysis selected?

14-29.   Sketch the shape of the photometric titration curve expected for the titration of the iron(III)–salicylate complex with EDTA according to Section 14-15. On the same graph sketch the shape expected if the molar absorptivity were 10 times less.

14-30.   Sketch the photometric titration curves expected for the reaction A + B $\rightleftharpoons$ C + D if B were the titrant and, at the wavelength used, radiation were absorbed by (a) A only; (b) C only; (c) A and C equally.

14-31.   What is the advantage of a photometric titration over direct spectrophotometry for quantitative work?

14-32.   In the photometric titration of iron a chloroacetate buffer is required. If 2 g of NaOH is added to 24 g of monochloroacetic acid, $CH_2ClCOOH$, how many moles of the chloroacetate salt are formed? How many moles of chloroacetic acid remain? What is the ratio of the two?

14-33. Why is white light rather than monochromatic light acceptable in the Nessler method for the determination of ammonia? Why are long glass tubes used for the measurement rather than a small cell as in a spectrophotometer?

14-34. A 5.0-mL sample treated according to the Nessler procedure for ammonia gave a color intensity that matched the tube containing 1.2 mL of standard ammonium chloride solution (0.263 g $NH_4Cl/50$ mL). What was the NH_3 concentration in the sample in micrograms per milliliter?

14-35. A 100-mL sample of Lake Ontario water in a Nessler tube gave the same color intensity as 1.3 mL of a standard solution prepared from 0.261 g of ammonium chloride and diluted according to the Nessler procedure for the determination of ammonia. How many micrograms of ammonia did the sample contain? How many micrograms were present per milliliter?

14-36. A sample of ammonium chloride weighing 0.4194 g was dissolved and diluted to 500 mL. A 25.00-mL aliquot was diluted to 1 L, and a 25.00-mL aliquot of that solution diluted to 250 mL. What was the NH_3 concentration of the final solution in micrograms per milliliter?

14-37. A bottle of reagent-grade potassium chloride was tested for ammonia by the Nessler procedure. A solution containing 10.2 g exhibited the same color in a Nessler tube as 1.4 mL of a standard solution prepared from 0.286 g of ammonium chloride and diluted by the procedure of Section 14-11. What was the ammonium ion concentration in the potassium chloride in micrograms per gram?

14-38.† If the solution used for the tube-matching procedure absorbs little light, or if the solution absorbs most of the incident light, tubes may appear to be matched when they are not. Why?

14-39. The iron in a sample of stream water was determined by the procedure of Section 14-13. The data recorded were: weight of iron(II) ammonium sulfate, 0.3151 g; absorbance reading of standard, 0.452; absorbance reading of sample, 0.513. What was the concentration of iron in the stream water?

14-40. Why is hydroxylamine hydrochloride added during the spectrophotometric determination of trace iron?

14-41. In the cobalt–nickel determination, why are solutions heated overnight after addition of EDTA and before measurement?

14-42. A 1.596-g sample was analyzed for cobalt and nickel by the procedure of Section 14-14. The weight of pure cobalt metal taken was 0.3247 g and of pure nickel metal, 0.3014 g. The average absorbances recorded for the sample, cobalt standard, and nickel standard were 0.621, 0.538, and 0.097 at wavelength λ_1 and 0.644, 0.118, and 0.496 at wavelength λ_2. What were the percentages of cobalt and nickel in the sample?

14–43.† A mixture of two similar alkaloids, W (mol wt 270) and Q (mol wt 245) was separated from a mushroom. A 24-mg/L sample of W gave absorbance readings of 0.850 at 290 nm and 0.320 at 370 nm in a 5-mm cell. A 41-mg/L sample of Q gave readings of 0.440 and 0.730 under the same conditions; the mixture gave readings of 0.625 and 0.690. Calculate the molar concentration of W and Q in the mixture.

14–44.† Explain how a mixture of two substances might be determined simultaneously through use of a dual-wavelength spectrophotometer.

14–45.† Cobalt and nickel can be determined simultaneously by spectrophotometry as 8-hydroxyquinoline complexes. The analytical wavelengths are 365 nm and 700 nm. The absorbance of a standard solution containing 1.00×10^{-4} M cobalt as its

8-hydroxyquinoline complex was 0.353 at 365 nm and 0.0429 at 700 nm when measured in a 1-cm cell. The absorbance of a standard solution containing 1.00×10^{-4} M nickel as its 8-hydroxyquinoline complex was 0.323 at 365 nm and 0.000 at 700 nm. What were the concentrations of cobalt and nickel in a solution whose absorbance was 0.724 at 365 nm and 0.0710 at 700 nm?

14–46.† Calculate the equilibrium constant for the reaction

$$FeSal_3^{3-} + H_3Y^- + 3H^+ \rightleftharpoons FeY^- + 3H_2Sal$$

(*Hint*: Write the equilibrium expressions for the acid dissociation of EDTA and salicylic acid and for the formation of complexes of iron with EDTA and salicylate. Combine them to give the expression for the equilibrium constant).

REFERENCES

D. F. Boltz, Ed., *Colorimetric Determination of Non-metals*, Wiley, New York, 1958. Extensive discussion of applications in the visible region of the spectrum.

G. W. Ewing, *Instrumental Methods of Chemical Analysis*, 4th ed., McGraw-Hill, 1975, Chapters 3–12. Background on optical methods of analysis, including instrumentation.

J. B. Headridge, *Photometric Titrations*, Pergamon Press, Elmsford, New York, 1961.

J. B. Ingle, *Anal. Chim. Acta* **1977**, *88*, 131. Step-by-step procedure for the analysis of the precision characteristics of spectrophotometers. For two common instruments the author concludes that the precision of measurement is limited by reproducibility of cell positioning and by noise that is independent of the photon signal. Precision is not limited by the fundamental signal shot noise. Additional papers on this topic by Ingle and others include *Anal. Chem.* **1974**, *46*, 2161; **1972**, *44*, 1375; and **1975**, *47*, 1231. For an early paper, see M. T. Gridgeman, *Anal. Chem.* **1952**, *24*, 445.

C. K. Mann, T. J. Vickers, and W. M. Gulick, Harper and Row, New York, 1974, Chapter 16. Includes a discussion of precision in photometric measurements.

M. L. Moss and M. G. Mellon, *Anal. Chem.* **1942**, *14*, 862. First paper on determination of iron with bipyridine.

J. Nessler, *Chem. Ztg.* **1856** 27, (N.F.I.), 529. Procedure for the determination of a trace of ammonia described for the first time.

E. B. Sandell, *Colorimetric Metal Analysis*, 3rd ed., Wiley (Interscience), New York, 1959. Thorough background on inorganic colorimetric trace analysis, with selected and tested procedures for the determination of traces of metals.

F. D. Snell and C. D. Snell, *Colorimetric Methods of Analysis*, Vols. 1–4, Van Nostrand, Princeton, New Jersey, 3rd ed., 1948–1954. Numerous procedures of applications to inorganic and organic systems.

P. B. Sweetser and C. E. Bricker, *Anal. Chem.* **1953** 25, 253. Describes the photometric titration of iron, copper, and nickel with EDTA.

W. C. Vosbergh and G. R. Cooper, *J. Am. Chem. Soc.* **1941**, *63*, 437. Describes a modification of Job's method of continuous variations applied to the determination of a complex.

15

OTHER SPECTRAL METHODS OF ANALYSIS

The ability to measure is one of man's great capabilities.

F. D. Rossini

15-1 INFRARED SPECTROPHOTOMETRY

Much of what has been said about the spectrophotometry of the ultraviolet and visible regions of the spectrum can be transposed directly to the infrared, since the basic principles are the same. There are, however, several significant differences at the practical level. One obvious difference is in the nature of the process under observation. In the infrared region, vibrational, rather than electronic, transitions are observed. Lattice vibrations of ionic crystals are broad and are not important for analytical applications. Therefore we limit discussion to compounds with covalent bonds, especially vibrations of organic substances. Not all covalently bonded molecules produce infrared-absorption spectra; a change in dipole moment is also required. Molecules such as H_2, N_2, or O_2, for example, do not undergo a change in dipole moment and do not give an infrared spectrum.

Electronic absorption spectra tend to involve but few transitions; frequently only a single one is important. Especially in solutions, electronic transitions are broadened by numerous vibrational, rotational, and solvent interactions that results in broad spectral bands. In the infrared region, where vibrational spectra are observed, the transitions are not usually broadened but remain relatively narrow. Since an infrared spectrum can comprise a large number of narrow absorption bands, it is likely to be highly specific. A methyl functional group, for example, possesses CH-stretching, CH_3-bending, hydrogen-hydrogen scissoring, and twisting vibrational modes, in addition to a carbon-carbon stretching vibration if the carbon in the methyl group is attached to another carbon. Each of these vibrational transitions is characteristic of a methyl group.

The frequency of a vibrational absorption band increases with increasing bond energy and with decreasing masses of the atoms. Vibrational energy

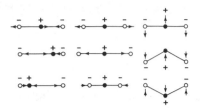

Figure 15-1. Modes of vibration in a molecule of carbon dioxide.

absorption does not entail displacement of the center of mass of a molecule, which would merely mean translational movement of the whole molecule. Each vibrational absorption is accompanied by several rotational levels of only slightly different energies; thus an absorption band is observed rather than a narrow absorption line. In infrared measurements a nonpolar solvent is usually selected to reduce interference from solvent absorption.

Figure 15-1 illustrates several of the vibrational modes that can occur in a molecule of carbon dioxide — symmetric stretching, scissoring, and asymmetric stretching. Figure 15-2 is an experimental infrared spectrum of carbon dioxide. For more complex molecules some of the other vibrational modes of energy transfer are described as twisting, wagging, rocking, and so on.

Infrared spectra are ordinarily plotted as transmittance against wavelength or wavenumber (Section 14-1).

Apparatus

The instrumental components usually employed in the various regions of the infrared spectrum differ from those in the ultraviolet and visible, although the basic designs are similar. Since the material for windows, lenses, prisms, cells, and so forth must transmit radiation in the infrared region, optical glass, which does not transmit well beyond 2.5 μm, is ruled out. Front-surface concave mirrors are usually substituted for transmission lenses. Materials that transmit in the infrared include sodium chloride, potassium bromide, and calcium fluoride; all can be used as prisms, cells, and lenses. Being water-soluble to varying extents, they must be protected from

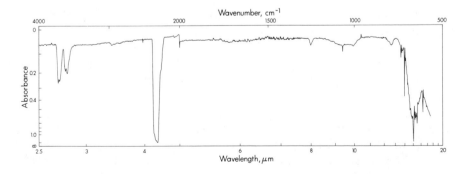

Figure 15-2. Infrared spectrum of carbon dioxide. Spectrum by C. Johnson.

moisture. For this reason, and also because water absorbs radiation strongly in much of the infrared region, water is rarely chosen as solvent. The most widely employed solvents include carbon tetrachloride and carbon disulfide.

Sources of Radiation

The most common source of infrared radiation is the *Nernst glower,* an electrically heated rod or hollow tube of sintered oxides of cerium, zirconium, thorium, or yttrium. For the near infrared, an ordinary tungsten lamp such as is used in the visible region can be the source; for the far infrared (beyond about 30 μm) a heated rod of silicon carbide called a *globar* is used.

Detectors

Detection systems are substantially different from those required in the ultraviolet and visible. The energies involved are substantially lower and consequently incapable of generating free electrons that can be measured in a photomultiplier tube. Accordingly, thermal detectors are necessary. The *thermocouple* is a thermal detector that produces a measurable response as the result of the heating effects of incident photons. Although it is widely used, it is limited by low sensitivity and slow response time. To minimize heat losses through conduction, the thermocouple is mounted in an evacuated housing with a small infrared-transparent window. Only changes in temperature are measured; absolute values are unimportant. The actual site of absorption is usually a small piece of blackened gold foil, welded to the junction of two wires of dissimilar metals that produce a large thermoelectric voltage. The apparatus is usually designed so that the signal from the detector is chopped several times per second; the resulting alternating voltage is less subject to interferences and more readily amplified than a dc signal.

For the near-infrared region, where photon energy is greater, a lead sulfide photoconductor detector is preferable. In this cell the photons excite electrons to a state that increases the conductivity of the lead sulfide. Lead sulfide detectors function satisfactorily at room temperature; most detectors of this type must be cooled strongly with liquid nitrogen or helium to operate effectively.

Monochromators

Many infrared instruments are of double-beam, single-wavelength design. Usually two wavelength-restricting devices, either a filter plus a grating or two gratings, are employed to encompass the wide frequency range. Figure 15-3 is a sketch of an infrared spectrophotometer, and Figure 15-4 is an example of an infrared spectrum recorded on an instrument of this kind. Typically, one grating is designed to cover the range from 2 to 5 μm, and a second monochromator is brought into play by an interchange mechanism to cover 5 to 16 μm. For still longer wavelengths additional gratings with as few as two or three lines per millimeter are necessary. Because much of the energy from infrared sources lies in the short-wavelength infrared region, the radiation tends to be strongly scattered. For example, at 1.5 μm there may be 100 times as much energy as at 10 μm. With measurements at 10μm the

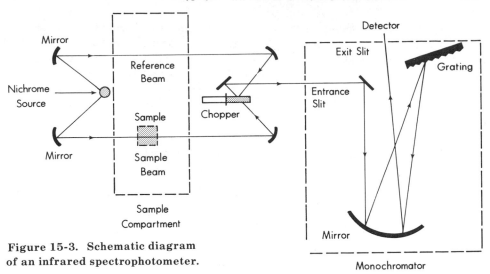

Figure 15-3. Schematic diagram
of an infrared spectrophotometer.

scattered radiation from the 1.5-μm region can interfere seriously. This background interference is a major problem in the design of infrared instruments. It can be reduced by placing filters in the optical path.

In the far-infrared region, background thermal radiation from the sample and instrument components can be a significant problem. If the source radiation is chopped before being passed through the sample and monochromator, the alternating-current component can readily be separated electronically from the background direct current and measured.

Sample Cells

Infrared spectra are often run on solids and pure liquids as well as on solutions. To accommodate this range of sample forms, a variety of sample-handling techniques have been developed. Only a few of the important ones are mentioned here. Solutions are usually placed in a cell of variable pathlength, typically 0.1 to 1 mm, with infrared-transparent windows of NaCl. The pathlength is controlled by the thickness of spacers placed between the windows before assembly. For pure liquid samples a drop may be squeezed between two NaCl plates to provide a sample only a few

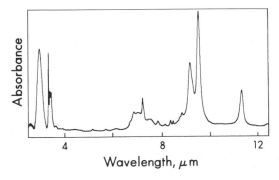

Figure 15-4. Infrared spectrum of ethanol. Spectrum by J. Hoyle.

micrometers thick. Solid samples can be run as solids provided the particle size is such that light-scattering is avoided. The particles must be suspended in a medium of about the same refractive index as the solid so that reflection at the surface of the particles is reduced. One method is to grind the sample with a long-chain hydrocarbon (Nujol) to form a *mull*, which is then placed between two NaCl plates for spectral measurement. Another method is to grind a small portion of the sample with a quantity of KBr and compress a portion of the mixture into a thin transparent wafer by application of high pressure. The disk is then placed in the spectrometer for study.

Applications

Infrared spectra of molecules with covalent bonds can provide much information about molecular structure that is useful for the identification and characterization of compounds. Through the examination of a large number of spectra of known materials, numerous correlations of specific absorption peaks with particular functional groups have been accumulated. Several generalizations can be made.

Most applications of infrared absorption are for qualitative purposes in the identification of substances or the determination of structures. Infrared spectra can often reveal both the specific functional groups present and their general location in a molecule. Thus they are an extremely useful aid to identification. The spectra of many thousands of compounds have been catalogued. This information can be used to match spectra peak by peak. The specificity of absorption patterns in the "fingerprint" region (7 to 15 μm) is especially valuable.

The infrared region from 0.7 to 4 μm includes the stretching vibrations of bonds between hydrogen and heavier atoms and thus is useful in the identification of functional groups that contain hydrogen. The region from 4 to 6.5 μm includes the vibrations of double and triple bonds. Beyond 6.5 μm distortions and bending vibrations occur, including CH–bending. In the far infrared (beyond about 25 μm) the absorptions correspond to vibrational modes with heavy atoms or groups of atoms, as well as rotational modes.

Tables and charts of infrared wavelengths and functional-group correlations have been prepared. When using such tables keep in mind that the spectrum of a compound may be affected by the conditions under which it is observed. These variations make it necessary to be cautious about assigning structures to unknown materials. Some variations are due to changes in the chemical background, whereas others may be due simply to instrumental variations such as in slit widths or scanning rates. Hydrogen bonding can have measurable effects on spectra; for example, the frequency of $C=O$ stretching is reduced when a solute containing this group forms hydrogen bonds with a hydroxylic solvent. Another example is formation of a dimer of a carboxylic acid. The frequency of the $C=O$ stretching mode in carboxylic acid monomers is in the range 5.6 to 5.71 μm, while that in dimers is in the range 5.81 to 5.85 μm. Carboxylic esters absorb radiation at about 5.76 μm.

Because of the large number of vibrational interactions possible in most molecules, two compounds will be highly unlikely to exhibit identical infrared spectra. Thus, comparison of a spectrum of an unknown compound with a spectrum in a compilation of spectra can provide valuable evidence for the identification of an unknown. For positive identification a sample of the pure material can then be obtained and its spectrum recorded on the same instrument under the same conditions as for the unknown. Identical spectra then provide almost certain identification.

From a quantitative point of view, spectroscopy is not so readily applicable to the infrared as to the UV-visible region. Although Beer's law still applies, instrumental deviations are more likely. To reduce solvent absorption, a combination of high solute concentrations and thin cells is often resorted to. Typical sample pathlengths of 10 to 1 mm or less are used in the near infrared. With high solute concentrations deviations from Beer's law are more likely because of intermolecular interactions. Furthermore, the narrow absorption peaks of the infrared can cause difficulty in quantitative measurements. If the peak width is of the same order as the slit width of the instrument, nonlinear concentration-absorption plots are likely. Also, because a large number of absorption peaks are present in the spectra of most molecules, overlap of adjacent peaks is likely to create interferende. For quantitative measurements the techniques are similar to those applied in the visible region of the spectrum. Owing to the greater significance of deviations from Beer's law at higher concentrations, calibration curves of absorbance against concentration are necessary, the measurements being obtained with care.

The potassium bromide disk technique is often employed for quantitative measurements. The absorbance of the sample is compared with standards prepared from known amounts of the substance in potassium bromide. To improve precision an internal standard such as potassium thiocyanate may be ground with the sample and potassium bromide. The thiocyanate absorption at 4.70 μm is measured and the ratio of this value to that of the sample at its peak is plotted against the sample concentration. Inherent in the mixing of solids are substantial sampling errors (Section 21-2) that are rarely recognized.

Quantitative organic analysis of volatile compounds by way of infrared has become less important with the advent of gas chromatography. The union of gas-chromatographic and infrared techniques is attractive. Sample components can be separated and measured by gas chromatography, and their identity confirmed by infrared analysis. More and more, a major role of infrared spectroscopy lies in the qualitative identification of substances and the establishment of structures.

At longer wavelength regions absorptions caused by rotational transitions in covalent molecules give rise to microwave spectra. The most useful region is from 8 to 40 GHz (1 GHz = 10^9 Hz) and admirably complements the infrared for the identification of small molecules. For large molecules the number of interactions becomes so great that the spectrum appears as a mass of lines that are too weak to be of value. Accordingly, quantitative applications are few.

15-2 NUCLEAR MAGNETIC RESONANCE (NMR)

Deep inside the submicroscopic atom is the even tinier nucleus. Many atomic nuclei behave like spinning micromagnets; when placed in an external magnetic field these micromagnets attempt to line up either parallel or antiparallel to the field. This behavior is analogous to that of a gyroscope in the earth's gravitational field. If a radio-frequency field of exactly the right energy is applied to these nuclei, the energy can be absorbed by the nuclear micromagnets and cause them to go from one orientation to another. Nuclear resonance was first observed in 1946.[1] Although all nuclei have both charge and mass, only about half the isotopes of the elements have a spinning nuclear charge that generates a small magnetic field.

The most widely utilized isotopes are those of hydrogen, fluorine-19, nitrogen-15, phosphorus-31, and carbon-13. By far the most information has been gathered with hydrogen nuclei (protons), owing to the sensitivity of the proton NMR signal. Since hydrogen is present in virtually all organic compounds, it is a useful general probe. Carbon-13 measurements are also of value in understanding structural details of organic molecules.

The energy required to produce an NMR signal corresponds to a wavelength of the order of 1 cm and a frequency of 10^8 Hz (100 MHz), about 100 000 times lower than the vibrational energies involved in the infrared region. The detector must be sensitive to these low frequencies, which are in the radio-frequency region.

From the magnetic moment of the spinning nucleus it can be calculated that in a magnetic field of 14 092 gauss a proton will absorb or emit energy of 60-MHz frequency; this is called the *Larmor frequency* for the particular field and nucleus. All hydrogen nuclei in a magnetic field of 14 092 gauss absorb energy at about this frequency. The slight variations in absorption frequency that occur are due to small differences in the actual magnetic fields around the different nuclei because of small secondary fields generated by the electrons surrounding them. These small locally induced fields partly overcome the effect of the much larger external field imposed on the nuclei. Thus they respond to magnetic fields slightly less than would be expected from the magnitude of the applied field. The induced field depends on the electron density around the nuclei; the higher the density, the larger is the field. Because its magnitude depends on the chemical environment, the shift in absorption frequency due to the induced field is called the *chemical shift*. Chemical shifts are often reported as relative values for convenience.

The phenomenon of chemical shifts in nuclear resonance was discovered in 1949 and by 1956 chemists were becoming the principal beneficiaries of NMR. It is so significant to chemistry because the chemical shift of the nucleus of an atom, say a hydrogen atom, depends on the particular molecule in which the hydrogen atom is present. NMR has revolutionized the identification and

---

[1]Two teams of physicists — Bloch, Hansen, and Packard at Stanford University and Purcell, Torrey, and Pound at Harvard University — shared the Nobel Prize in physics for this discovery. The discovery of the chemical shift, made by Knight in 1949, made possible the extraction of chemical information from NMR measurements.

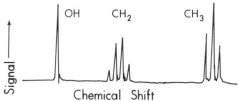

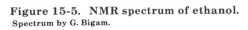

Figure 15-5. NMR spectrum of ethanol.
Spectrum by G. Bigam.

characterization of molecules.

Chemical shifts are minute; in 10^8 Hz proton chemical shifts may cover a range of 1000 Hz. It is often desirable to measure chemical shifts of the order of 0.1 Hz, which in 10^8 means a measurement precise to 1 part in 10^9. This imposes severe demands on the instruments.

An NMR spectrum supplies information on (1) chemical shifts, which reflect the chemical environment, (2) spin-spin splitting, and (3) peak areas. *Spin-spin* splitting arises from the effects of nearby protons on the energy levels of the protons being observed. Thus, in a molecule containing the ethyl group, CH_3CH_2-, the methyl protons appear as three closely spaced absorption peaks rather than as a single peak, owing to splitting by the CH_2 protons.

The *peak areas*, or the absorbance, indicate the number of protons of a given type or provide the basis for quantitative measurements in mixtures. Qualitative NMR analysis uses all three types of information and is enormously powerful for the identification of structures of organic compounds.

As a simple example of a proton NMR spectrum, consider that for ethanol (CH_3CH_2OH). Ethanol contains hydrogen atoms in three chemically different environments, each of which absorbs radio-frequency radiation at slightly different frequencies. The extent of absorption when plotted as a function of frequency gives information about the immediate environment of that particular hydrogen nucleus. The proton NMR spectrum of ethanol is shown in Figure 15-5.

The basic instrument components for NMR measurements are (1) a magnet to provide a uniform magnetic field, (2) a radio-frequency transmitter and receiver, (3) a sample holder, and (4) the necessary associated electronics for data acquisition and treatment. In modern instruments, highly sophisticated electronics is required to control the radio-frequency transmission, detect the frequencies absorbed, amplify the signals, and present the information to the operator. The data-acquisition systems may include a minicomputer to collect and process the radio-frequency signals to extract the maximum chemical information. In Figure 15-6 schematic diagrams of a spinning nucleus and an NMR instrument are shown.

To measure frequency differences of 1 part in 10^9 requires that the magnetic field be both constant and homogeneous. Improved electronic stability, and thereby improved precision, is gained if the frequency is not varied but held constant and the scan performed by varying the overall

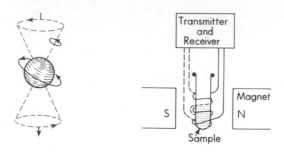

Figure 15-6. Schematic diagrams of spinning micromagnet in a nucleus and an NMR instrument.

magnetic field slightly through the use of *sweep coils*. As the magnetic field is changed, a spectrum is obtained. Since the total magnetic field is not known exactly, the chemical shifts are measured relative to some reference. An example of a reference is $(CH_3)_4Si$. The electron density around the hydrogen nuclei of $(CH_3)_4Si$ is greater than around the hydrogen nuclei in most organic compounds, thus providing an isolated line.

The magnet is the heart of the instrument. Homogeneity of the magnetic field is essential; it is improved by spinning the sample to average out inhomogeneities. For observation of slight differences in radio-frequency absorption, the magnetic field should be both large and uniform. Resolution and sensitivity improve as the field is intensified. With conventional electromagnets the largest fields that can be developed are about 23 000 gauss, an immense value. The generation of even stronger magnetic fields has recently become practical through the development of superconducting solenoids. Particular alloys of niobium and titanium at liquid helium temperatures lose all resistance to the flow of electricity even in a large magnetic field. A large current can then flow with no generation of heat. Thus with superconductivity larger currents and magnetic fields are possible. Magnetic fields now are attainable that are about four times the maximum with conventional electromagnets (and about a million times as strong as the earth's magnetic field). These larger magnetic fields have increased the spread in the frequencies, thereby improving resolution, and at the same time give higher sensitivity with small amounts of material. Combined with stronger fields, the enormously greater sophistication and reliability of present-day electronics and computers are permitting the study of more complex molecules as well as smaller amounts of material.

Many NMR instruments now incorporate the *Fourier transform* (FT) mode of operation, in which the sample is irradiated with a range of frequencies simultaneously at high intensity for a brief period, say 50 μs. The radiation is then turned off, and as the sample returns to equilibrium, a signal is generated as a function of time. The advantage of FT is that an entire spectrum may be obtained in approximately a second instead of several minutes. The results of advances in instrumentation that have taken place in two decades is illustrated by the spectra in Figure 15-7.

About 1% of naturally occurring carbon is carbon-13. For this reason, and

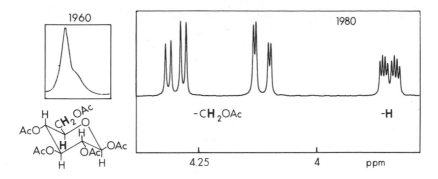

Figure 15-7. A comparison of a portion of two spectra of glucose pentaacetate, one taken with early NMR instrumentation at 60 MHz and the other with modern instrumentation at 400 MHz. The absorptions result from the three hydrogen atoms associated with the grouping $HCCH_2OAc$. From R. U. Lemieux, R. K. Kullnig, H. J. Bernstein, and W. G. Schneider, *J. Am. Chem. Soc.* 1958, *80*, 609. The 1980 spectrum by T. Nakashima.

because the sensitivity of the carbon-13 nucleus is less than that of the proton, the overall sensitivity of carbon-13 measurements is about 0.1% of the sensitivity of proton NMR. The development of FT techniques has enabled the recording of carbon-13 spectra of high quality. The information from carbon-13 NMR spectra complements that from proton NMR. For example, decane contains numerous CH_2 and CH_3 protons that give a proton spectrum of little value for structural studies. With carbon-13 NMR each carbon in the decane chain produces distinct, well separated peaks.

Although small molecules such as ethanol give simple spectra, more complex molecules may give complex spectra reflecting hundreds of absorptions. In such complex molecules the NMR information obtained from atoms other than hydrogen, such as carbon-13, nitrogen-14, and fluorine-19, can be crucial in the interpretation of their structures.

15–3 X-RAY FLUORESCENCE ANALYSIS

When a large amount of energy is transmitted to an atom, such as by collision with an energetic electron, an electron can be completely removed from even an inner shell of the atom. An electron from an outer shell may then fall into the vacant position. The energy involved in this change is large and is emitted as exceedingly short wavelength radiation (0.01 to 10 nm) called *X-rays*. A beam of X-rays produced in this way can be utilized as a source of radiation to cause excitation of atoms in a sample. During absorption of the X-rays an electron from an inner shell may be removed from the atom. The excited atom may return to the ground state by a series of transitions through intermediate energy levels, each of which results in photon emission in the X-ray region. Since these emissions are of less energy and therefore of longer wavelength than the absorbed photons, the process is similar to fluorescence in the UV-visible region.

The emitted radiation may be analyzed by (1) *wavelength dispersion* with a crystal of, say, ammonium dihydrogen phosphate or lithium fluoride as the dispersing element or by (2) *energy dispersion*, in which the spectrum of photon energies are all allowed to fall on a detector simultaneously, and the energies of the various photons are analyzed by a multichannel pulse-height analyzer. The preferred detector for both wavelength- and energy-dispersive instruments is a lithium-drifted silicon or germanium semiconductor. This detector is extremely sensitive, but must be kept thermostated in liquid nitrogen (–196°C) to decrease background noise and maintain efficiency. It can measure the energy of single photons. Energy-dispersive instruments are becoming increasingly popular for many regions of the spectrum because of the simplicity and sensitivity of the design.

X-ray fluorescence analysis has the advantage that it is nondestructive. It can be used for the analysis of works of art, valuable coins, and forensic materials. Several elements can be determined in a few minutes, on only a tiny amount of material. The accuracy and precision are of the same order as other spectrometric techniques. The major disadvantages are that elements lighter than sodium cannot be determined readily, that lower concentrations are not so readily determined as by other spectral methods, and that the instruments are relatively costly. It also should be kept in mind that the technique deals primarily with the surface of the sample, and that the composition of the outermost layer of a material may differ appreciably from that of the internal layers.

This field of spectroscopy is extremely important for elemental analysis. Related fields include X-ray photoelectron spectroscopy (ESCA), ultraviolet photoelectron spectroscopy, electron-impact spectroscopy, and auger-electron spectroscopy.

15–4 RAMAN SPECTROSCOPY

If a molecule is exposed to radiant energy insufficient to bring about an electronic transition but with far more than enough energy to cause vibrational transitions, scattering of the radiation may occur that can be analytically useful as *Raman spectroscopy*. Suppose a sample of carbon tetrachloride is irradiated with radiation of 488 nm, and the radiation intensity of the sample is observed at right angles to the incident beam as a function of frequency. An intense line will be seen at 488 nm, with radiation appearing at several other distinct wavelengths in that vicinity. If the spectrum observed is plotted as radiant power against *change* in frequency (rather than wavelength or frequency), for carbon tetrachloride there will be lines at 218, 314, and 459 wavenumbers greater than that corresponding to 488 nm; these are called *anti-Stokes lines*. Also seen will be three lines that are 218, 314, and 459 wavenumbers less than corresponding to 488 nm; these are called *Stokes lines*.

Although light of 488 nm is insufficient to bring about electronic excitation in the molecule, it is 100 times more than enough to bring about vibrational changes. A small fraction of the molecules will therefore interact with the generously available supply of energy in the incident photons. To interact with a photon, a molecule obtains a small fraction of the energy of the photon and undergoes vibrational excitation. The photon, still possessing a major fraction of its original energy, now proceeds in a random direction. Its frequency is only slightly different from the original, since its energy loss has been slight. This slight difference is measured; the lines corresponding to the difference are Stokes lines. Thus Stokes lines result from a process in which a molecule absorbs some of the energy from the incident photons. A somewhat less common kind of interaction is one in which a portion of the vibrational energy of a molecule is transferred to a passing photon. The molecule goes from a higher to a lower vibrational energy level, having added a small amount of energy to the already energetic photon and then sending it on its way in a random direction. This interaction gives rise to the less intense anti-Stokes lines. In terms of wavenumber, which is proportional to frequency, the relation is

$$\text{wavenumber of Raman line} = \text{wavenumber of incident radiation}$$
$$\pm \; \Delta \; \text{wavenumber}$$

For radiation of 488 nm the wavenumber is 20 492 cm^{-1}.

The wavenumbers of observed lines in normal Raman spectroscopy do not depend on the frequency of the incident radiation. Accordingly, the Raman spectrum observed with 632.8-nm incident radiation, plotted as change in wavenumbers, is the same as with 488-nm incident radiation, though the intensities differ.

For Raman lines to be sharp, the incident light must be monochromatic. The recent development of high-intensity laser sources of monochromatic light has largely solved this formerly difficult problem in Raman instrumentation. An argon laser gives a line at 488 nm, and is the most common source, provided that fluorescence is not induced in the sample. A helium-neon laser gives a line at 632.8 nm. Although this less energetic radiation is not so likely to induce fluorescence, the probability of interaction with molecules also decreases, and photomultiplier tubes are less sensitive in this region of the spectrum.

Like infrared spectra Raman spectra provide infrared information about the vibrational interactions in molecules. The particular excitations, however, are often not the same as those with the much gentler photons used in infrared work. Moreover, certain molecules that do not give rise to infrared spectra will yield Raman spectra. For instance, the symmetrical stretch vibration of carbon dioxide is IR-inactive but Raman-active.

Raman spectroscopy is carried out in the visible region of the spectrum, where the instrumental problems are different from those in the infrared region. Glass cells and optics can be employed, the detector can be a photomultiplier tube, and water and a wide variety of other liquids make satisfactory solvents. Excitation sources, however, have been a problem. Powdered samples can be examined directly without the need for potassium bromide pellets. Smaller amounts of sample are required, since a laser beam is small in cross section. Raman spectroscopy is presently utilized primarily for qualitative information and only rarely for quantitative information.

15–5 ARC AND SPARK EMISSION SPECTROSCOPY

The electrons of atoms heated to high temperatures absorb sufficient thermal energy to be excited to higher levels of electronic energy. On returning to their ground state they emit photons. Depending on the energy transitions involved, the resulting radiation may appear in either the visible or the ultraviolet region. Analytical information is encoded in both the frequency and intensity of the emitted radiation.

Atoms can be thermally excited in several ways. One involves the use of a dc arc. A sample of a metal can be made into an electrode, or a nonmetallic material placed in a small hollow in the end of a carbon rod. Solutions can be evaporated to leave a residue of solute on the end of the rod. Two electrodes are then placed in holders a few millimeters apart. A current of about 10 A at 50 V is passed between the electrodes to produce an electric arc. Electrons emitted from the cathode bombard the anode and raise the temperature to several thousand degrees. At these temperatures positive ions boil off and migrate toward the cathode. A cathodic layer is produced that contains a high density of ions and atoms. The radiation emitted from this region of the arc corresponds to emission from neutral atoms and ions of the element. This emission then passes through a prism or grating, and the spectrum is typically captured on a photographic plate. An emission spectrum may consist of thousands of narrow lines of varying intensity, each corresponding to some energy difference between two levels in an atom or ion.

To isolate the multitude of resulting spectral lines requires prisms or gratings, slits, and lenses of extraordinarily high quality. The resolving power of present-day optics is such that lines differing by a fraction of a nanometer can be distinguished.

Provided the excitation conditions are properly controlled, the energy emitted is proportional to the number of atoms excited. The energy can best be measured by photographing the spectrum, since a photographic emulsion integrates the amount of light falling on it. It also detects radiation simultaneously over the entire spectral range. The density of blackening of the film is proportional to the log of the exposure over a considerable range. Film quality and development time and temperature are important variables.

Another type of detector system consists of direct measurement of incident radiation by photomultiplier tubes placed behind exit slits in the monochromator at positions corresponding to specific emission lines. One tube is required for each line measured. Quantitative results are greatly improved by incorporation of an internal standard. In this technique the line density of the sample is compared with the line density of an element of known concentration. Internal standards reduce errors caused by matrix effects from other species in the sample. For reliability the ratio of the emission radiant power of the standard and the sample should not vary with the conditions of excitation. For quantitative work, fractional distillation may cause problems. Similarly, self-absorption may occur, that is, absorption of radiation of the wavelength of the emission spectral line by the vapor of the element can cause complications.

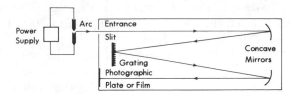

Figure 15-8. Schematic diagram of a spectrograph.

Instrumentation in spectroscopy can be highly sophisticated. Figure 15-8 is a schematic diagram of a emission spectrograph incorporating a diffraction grating for dispersion of the spectrum and a photographic plate as the detector. Figure 15-9 is an example of two emission spectra. Detection limits are especially favorable, and up to 70 elements can be detected simultaneously. The complexity of the line spectrum obtained from some mixtures can be both an advantage and a disadvantage.

Two other methods of exciting a sample between electrodes are often used. One is an ac arc with 60-cycle current, where an arc between the electrodes is formed and broken 120 times per second. The second is a dc spark, where a high-voltage spark is produced by discharging a 30 000-V condenser across the two electrodes. With this technique the apparent temperature in the discharge is approximately 10 000°C.

15–6 FLAME EMISSION SPECTROSCOPY

A flame is a much gentler source of excitation than an arc or spark because at the lower temperatures of flames only easily produced electron transitions occur. Consequently the spectra exhibit many fewer lines. The measuring apparatus can therefore be simplified, and the optical components of absorption spectrophotometers can be used. The sample is introduced, or aspirated, into the flame as fine droplets of solution produced in a chamber called a *nebulizer*. In the flame the solvent evaporates, the solute is dissociated into atoms, and the atoms are excited to higher energy levels. The emission on return to the ground state is then observed.

The source in a flame photometer consists of a flame carefully regulated as to size and temperature by a controlled flow of fuel and oxidant gases. Methane, propane, hydrogen, and acetylene are common fuels; air, oxygen, and nitrous oxide are employed as oxidants. The fuel-oxidant combination is

Figure 15-9. Example of high-voltage spark emission spectra of stainless steel (top) and brass in the region from about 300 to 330 mμ. Spectra by J. Seto and G. Horlick.

selected on the basis of the temperature desired. For example, the temperature of a flame of air and acetylene is around 2000°C, of acetylene and oxygen around 3200°C, of hydrogen and air around 2000°C, and of propane and air around 1900°C. The temperature varies with the position in the flame.

Since the sample must be introduced into the flame at a constant rate, the design of the nebulizer for sample introduction is important. The nebulizer draws sample solution into a chamber where it is broken into fine droplets and mixed with a portion of the oxidant gas. The resulting mist is introduced into the flame, where the solvent evaporates, leaving fine solid particles of metal salts or oxides. These particles vaporize and dissociate to produce thermally excited atoms, which then emit photons upon return to the ground state.

At about 2000°C only a few elements, principally the alkali and alkaline earth metals, are sufficiently energized that emission can be observed. At about 3000°C somewhat more than 30 elements emit enough radiation to be analytically useful. Because only a few lines are observed for each element, the spectra are simple. The most prominent lines reflect transitions from one of the two lowest excited states to the ground state.

For accurate quantitative measurements the emission intensity must be proportional to the amount of the element in solution. The intensity of the flame emission depends on several factors other than the total concentration of sample present in the burner. In fact, virtually any variation in the flame may influence the radiant power. The emitted radiant power is proportional to the number of atoms in the flame that are thermally excited. The fraction of excited atoms increases with increasing flame temperature. Experimental variables that affect the fraction include (1) the rate of aspiration of sample solution, which depends on the oxidant flow rate and the solution viscosity, (2) the fraction of the aspirated solution that reaches the flame, which depends on viscosity and surface tension, (3) the fraction of droplets reaching the flame that is evaporated, which depends on flame temperature, gas flow rate, and droplet sizes, and (4) the fraction of solid particles of salt that is converted to atoms, which depends on flame temperature, gas flow rate, and particle size. It is therefore essential that solution viscosity for samples and standards be the same, that flow rates of oxidant and fuel be controlled so as to fix the flame temperature and aspiration rate, and that the nebulizer performance be steady.

In making quantitative measurements a blank solution is used to measure the emission from the flame background (such as from OH bands) and from impurities in the sample solution. A working curve is then prepared from measurements on a series of standards, the net radiant power being plotted as a function of concentration.

There is an optimum flow rate of the sample solution to the nebulizer for each element. At high flow rates, flame energy is lost because a large volume of solvent must be vaporized; hence the temperature falls and the intensity of the emission decreases. The flow rate at maximum intensity depends on the excitation and ionization potentials of the atom. For instance, solutions containing lithium salts reach a maximum emission intensity at lower flow rates than those containing sodium or potassium. Potassium in particular is more effectively excited at lower temperatures. Anions can affect the emission intensity greatly. An example is perchlorate, which enhances the emission of all metals. Phosphate and sulfate, on the other hand, often depress the emission. Some cations interfere through enhancement of the emission intensity of other elements.

Flame spectroscopy finds its widest application in the routine determination of sodium, potassium, and calcium because of the sensitivity and freedom from interference. Some instruments are single channel and require careful flame control for reproducible results. Others are dual channel, in which an internal standard such as a lithium salt is added to standards and samples, and a lithium line monitored with a second detector at the same time as the lines for sodium or potassium. Any changes in flame conditions are likely to affect both emissions similarly.

15–7 ATOMIC ABSORPTION

Knowledge of the phenomenon of atomic absorption (AA) has a long history. The Fraunhofer lines in the sun's spectrum caused by absorption of radiation by cooler atoms in the outer reaches were observed more than a century and a half ago. Only in recent years, however, has the analytical instrumentation been developed to exploit this phenomenon.

As the name implies, the analytical information is encoded by an absorption process. The flame is located between a radiation source of optimum wavelength and a detector in a manner analogous to a spectrophotometric cell. The sample solution is first drawn into the burner-nebulizer by a stream of air or other oxidant such as nitrous oxide, and then mixed with fuel. Most of the sample solution passes into the nebulizer chamber as relatively large droplets that settle and then pass down the drain tube. The remaining solution is carried by the air-fuel mixture as a mist to the burner head, where the solvent evaporates and the solute is dissociated into atoms by the heat of the flame. The number of atoms reaching this point in the operation constitutes only a small fraction of the total. For high sensitivity, as many as possible of these atoms present in the flame should absorb radiation from the source. The ideal source for this purpose should be of high intensity at the wavelength needed for the element being determined, and of low intensity at all other wavelengths. The nearest approach to this ideal is a lamp whose cathode contains the element being determined. On being heated, atoms of this element emit energy at the wavelengths most likely to be absorbed by that same element in the flame. In this way, selectivity as well as sensitivity is provided because other elements in the sample generally do not absorb radiation close enough to the chosen wavelength to interfere in the measurement. Background interference is reduced by introduction of a monochromator containing a grating (or sometimes a quartz prism) between the sample and the detector.

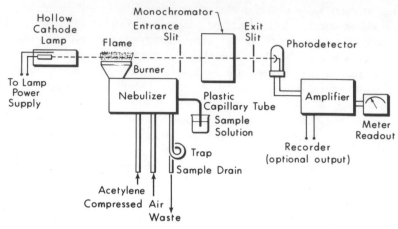

Figure 15-10. Schematic diagram of an atomic-absorption spectrophotometer.

When a beam of light from a source of radiation is arranged so that it passes through the sample in the form of an atomic vapor, part of the radiation is absorbed and part transmitted. The amount absorbed is proportional to the concentration of the atoms of interest in the high-temperature vapor. Atomic absorption is the inverse of emission spectroscopy. In all emission techniques the sample is excited, and the emitted radiation measured. In atomic absorption the element of interest is not excited in the flame, but merely converted to atoms in the ground state. In this state the atoms are capable of absorbing radiation of the same frequency that would be emitted were the element excited. Dissociation is almost always brought about by burning the sample in a flame, but sometimes by heating in a small furnace. A minute proportion of the atoms will be excited; in a flame at 3 000°C only one zinc atom in about 10^9 is in a level other than the ground state. In high-temperature flames 1 to 2% of sodium atoms may be excited. Since the proportion of ground-state atoms does not change appreciably, high precision is achievable.

The apparatus for atomic-absorption analysis normally consists of a source lamp, chopper, flame, monochromator, and detector, as diagrammed in Figure 15-10. The optical path of an instrument of this kind is shown in Figure 15-11. The apparatus is arranged so that the detector reads alternately a signal for first the lamp plus flame and then the flame alone. The signal can be related to the amount of light absorbed by the elements in the flame. In this instrument the flame-gas solution replaces the conventional solution of ordinary spectrophotometric analysis. Chopping the radiation from the hollow cathode lamp is necessary to get rid of emission from excited atoms in the flame, which would otherwise produce a background that varied with concentration. Either mechanical (rotating shutter) or electrical (pulsed hollow-cathode output) chopping can be employed.

The source lamp in an atomic-absorption instrument must supply radiation of sufficient intensity to excite an appreciable number of the ground states of the atoms of interest to higher levels. It also should, if possible, avoid

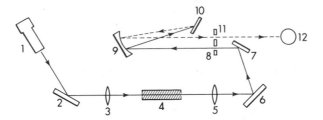

Figure 15-11. Diagram of optical path in the Perkin-Elmer Model 290B atomic-absorption spectrophotometer: (*1*) hollow-cathode lamp; (*2*) mirror; (*3*) lens; (*4*) flame; (*5*) lens; (*6*) mirror; (*7*) mirror; (*8*) slit; (*9*) parabolic mirror; (*10*) grating; (*11*) slit; (*12*) photodetector. Components *8* through *11* constitute the monochromator.

excitation of atoms of other elements in the same wavelength region. Both these objectives are admirably achieved by use of a hollow-cathode lamp. These lamps are low-pressure, low-temperature, cold cathode tubes that produce emission lines with spectral bandwidths of the order of 10^{-5} nm. The spectral bandwidth of an absorption line in the flame is of the order of 10^{-3} nm and thus broad compared with the source emission line. The cathode contains the same element as that which is to be determined in the sample. When a current flows between the cathode and anode, positive ions of the metal are knocked from the cathode surface. These ions collide with neon or argon atoms contained in the lamp, leaving the metal atoms in an excited state, from which they can then emit their characteristic radiation as they return to the ground state. Such lamps can furnish an exceedingly intense source at precisely the wavelength that can be absorbed by atoms in their ground state. Spectral interferences are therefore minor. Nevertheless, a monochromator is included to isolate the particular line whose intensity is reduced, just as a monochromator isolates a wavelength region in conventional spectrophotometry. The monochromator in an AA system rejects other lines and reduces flame background emission. The process is outlined in Figure 15-12.

Atomic-absorption spectroscopy is particularly suited to the measurement of small amounts of elements, usually metals, in a sample. The method is unusually simple, sensitive, and selective.

The usual procedure for dissociating an element from its initial matrix is to dissolve the sample and then aspirate the solution into a flame. For increased sensitivity a small portion of sample is sometimes placed in a

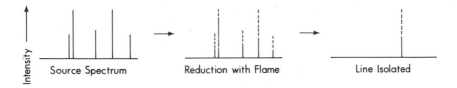

Figure 15-12. Schematic diagram of an atomic-absorption measurement.

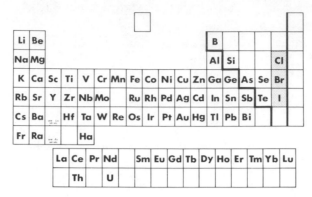

Figure 15-13. Elements determinable by atomic absorption. Shaded areas indicate elements determinable by indirect methods.

graphite tube, warmed to evaporate any solvent, and then heated strongly and rapidly in an electric furnace to vaporize the sample. The vapor is carried into the optical path of an AA instrument, and the absorbance of the flame measured. Solutions more concentrated than 10^{-3} to 10^{-5} M should be diluted to this range. Only a few milliliters are required.

A large number of elements can be determined by atomic-absorption techiques (Figure 15-13). Many applications to analysis of trace metals in a variety of organic, inorganic, and biological systems have been developed and are replacing slower, more tedious techniques. Atomic absorption has been found exceptionally satisfactory for the determination of magnesium in cast iron in the 0.002 to 0.1% concentration range and of silver, zinc, copper, and lead in cadmium metal in the 0.004 to 0.4% range. Beer's law describes the absorption quantitatively, provided that the flow rate of gases, viscosity and surface tension of the solutions, and the nebulizer performance all remain reproducible and constant.

Although atomic absorption has developed rapidly as a convenient analytical method, it must be applied with care. Instrument drift due to changes in lamp intensity and monochromator settings with time may affect results. Also, because of the extreme sensitivity of the method, trace contamination from reagents, water, and surroundings during sample preparation and analysis may create problems. Essential is frequent rechecking of instrument operation with standard solutions and care in the preparation and handling of both standards and samples.

As with flame emission, nonspectral sources of interference may become important. Vaporization interferences such as that of phosphate in the determination of calcium can be reduced by the addition of another element such as lanthanum. The lanthanum then ties up the phosphate so that the vaporization of calcium is not impeded. Atoms such as sodium or potassium can cause ionization interference, which results in enhanced emission of other easily ionized metals. This interference is reduced by addition of an alkali metal that is not being determined to samples and standards.

In summary, atomic-absorption and flame-emission methods are more selective than titrimetric methods. It is possible to quantitatively analyze for elements in the presence of many others without the need for separation. A disadvantage is that these methods provide only elemental analysis and no information on molecules, since molecular structure is destroyed in the process.

An important new excitation source for emission spectroscopy is the inductively coupled plasma (ICP) discharge. Sample solution is nebulized in a stream of inert gas, usually argon, and passed through a quartz tube surrounded by a coil through which a radio frequency is passed. The high frequency dissociates the argon into a stable plasma (a conductive gaseous mixture of ions and electrons). The temperature of the plasma is high (up to 10 000°C). With the high temperature and inert atmosphere of an ICP, chemical interferences are rare. This characteristic, along with the stability of the source, makes it useful for multielement quantitative analysis. Sensitivities are generally better than for dc arc or spark sources.

15-8 ATOMIC–ABSORPTION DETERMINATION OF TRACE COPPER IN NICKEL METAL

In this experiment, trace copper in a sample is determined by the method of multiple standard additions, in which varying amounts of standard are added to a series of solutions containing a constant amount of sample. A plot of absorbance against the amount of standard can then be used to determine the amount of copper in the sample.

Procedure (time 2 to 3 h)

Preparation of Solutions

Sample Solution. Weigh, to the nearest 0.1 mg, about 0.4 g of the sample into a 100-mL volumetric flask. Add 10 mL of 6 M HNO_3, and warm gently on a hot plate until dissolved. Cool, dilute to volume with water that has been passed through a cation or mixed-bed ion-exchange column to remove ionic impurities (particularly copper), and mix.

Standard Copper Solution. To prepare a concentrated copper solution, dissolve 0.12 to 0.14 g of pure copper metal, weighed to the nearest 0.1 mg, in a 100-mL volumetric flask with 10 mL of 6 M HNO_3.[2] Dilute to volume with demineralized water and mix. Prepare a standard copper solution 100 times more dilute by serial dilutions of 10-mL portions to 100 mL.

Standard-Addition Solutions. With a buret, measure, to the nearest 0.01 mL, 0, 2, 5, 10, 15, and 20 mL of the standard copper solution into six 50-mL volumetric flasks. Pipet 10 mL of the sample solution into each flask. Dilute the contents of each flask to volume with demineralized water, and mix.

Take the six filled flasks, a 50-mL beaker, and a wash bottle of demineralized water to an atomic-absorption spectrophotometer.

---

[2]Do not use a standard exceeding 0.15 g, or the absorbance readings will be excessively high and introduce error.

Measurement of Absorption

Adjust the fuel (usually acetylene) and air settings for the burner to the proper values. After the flow of fuel and air is well established (about 5 s), light the burner. *Caution*: If the mixture is not allowed to establish itself in the burner before ignition, a small explosion may occur inside the burner chamber. Never leave the flame unattended.

Consult the laboratory instructor concerning the wavelength setting, and optimize it as directed.

Aspirate water from a 50-mL beaker (blank). Set the absorbance reading to zero.

Replace the blank with the most concentrated solution. Turn the control dial so that the instrument reads a specified value at the upper end of the absorbance scale (usually 100).[3]

Again aspirate the blank and concentrated solutions to ensure that the readings are reproducible.

Aspirate the other solutions and record their readings.[4] Take each reading 5 to 10 s after beginning aspiration. Repeat each aspiration and reading several times. Recheck the settings of the blank and concentrated solution after every group of readings, and reset if required.

Aspirate water for a few minutes before shutdown to clean the capillary and chamber.

Report the percentage of copper in the sample. Report both the graphical and least-squares values, along with the standard deviation of the result calculated by the least-squares method. Indicate which value is to be graded.

Calculations

Using an $8\frac{1}{2}$- by 11-in. sheet of accurately ruled graph paper, construct a plot of instrument readings (which are proportional to absorbance) against volume of copper solution. Draw the best straight line through the points. The plot will not go through zero on the vertical axis because of the copper from the sample present in each solution. The simplest way to measure the copper concentration of the sample is to extend the plot until it intersects the horizontal axis. If the horizontal axis is continued to the left of zero, the concentration of the sample can be obtained from the point of intersection. Figure 15-14 illustrates this method of plotting.

In calculating the percentage of copper in the original sample, recall that the sample was dissolved in a volume of 100 mL.

Calculation of Copper in a Sample by the Least-Squares Method

Measurements that are linearly related to concentration and for which the slope or an axis intercept of the straight line is to be determined can be analyzed by either least-squares calculation or by graphical methods. The mathematical method usually takes longer; it is generally more precise but not necessarily more accurate. It can provide a value not only for the intercept but also for its standard

---

[3]An unexpectedly low reading for a sample may be caused by a fouled burner slot, an obstruction in the sample capillary, or an incorrect wavelength setting. Notify the instructor if one of these problems is suspected.

[4]The values may be read directly from a meter or a digital readout or may be recorded on a strip-chart recorder.

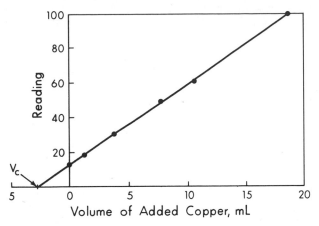

Figure 15-14. Example of plot of data obtained by method of multiple standard additions. The intercept is proportional to the copper contributed by the sample.

deviation.[5]   This experiment affords an opportunity to compare these two methods of data evaluation. It is recommended that the graphical procedure be completed first, so that the drawing of the line will not be prejudiced by prior knowledge of the least-squares results.

The equation for the volume of standard copper solution corresponding to the sample (the horizontal-axis intercept) is given by

$$V_c = \frac{(\Sigma A)(\Sigma V^2) - [\Sigma(AV)]\,(\Sigma V)}{(n)[\Sigma(AV)] - (\Sigma A)(\Sigma V)} \tag{15-1}$$

where V_c is the horizontal-axis intercept, ΣA the sum of the readings, ΣV^2 the sum of squares of the volumes of standard copper used, $\Sigma(AV)$ the sum of products of corresponding pairs of meter reading–copper volume sets, ΣV the sum of volumes of standard copper solution, and n the number of solutions measured. An example of a tabulation is given in Table 15-1; the calculation follows:

---

[5]Further information on fitting straight-line relations is provided in Section 22-4.

Table 15-1. Tabulation of Data for Least-Squares Calculation

Volume of Standard Copper Solution, V	V^2	Average Reading, A	AV
0	0	13.0	0
2.03	4.12	21.5	43.64
4.99	24.90	35.0	174.65
9.99	99.80	56.5	564.44
15.01	225.30	77.0	1155.77
20.00	400.00	100.0	2000.00
ΣV 52.02	ΣV^2 754.12	ΣA 303.0	$\Sigma(AV)$ 3938.50

$$V_c = \frac{(303.0)(754.12) - (3938.5)(52.02)}{(6)(3938.5) - (303.0)(52.02)} = 3.001 = 3.00 \text{ mL}$$

$$\text{Concn of Cu std} = \frac{0.100 \text{ g Cu} \times V_{10} \text{ mL} \times V_{10} \text{ mL}}{100 \text{ mL} \times 100 \text{ mL} \times 100 \text{ mL}}$$

$$= 1.00 \times 10^{-5} \text{ g/mL (for a standard of 0.100 g)}$$

$$\text{g Cu from sample} = (3.00 \text{ mL})(1.00 \times 10^{-5} \text{ g/mL}) = 3.00 \times 10^{-5}$$

Thus 10 mL of the sample solution contained 3.00×10^{-5} g of Cu. The original 100 mL, then, contained 3.00×10^{-4} g of Cu.

$$\%\text{Cu in sample} = (0.000300 \text{ g}/1.0000 \text{ g})100 = 0.030\% \text{ (1.000-g sample)}$$

Calculation of Standard Deviation for Least-Squares Result
The error in the value of V_c determined above can be estimated as follows.

1.   Calculate the slope m of the least-squares line:

$$m = \frac{\Sigma A}{\Sigma V + nV_c} \qquad \text{(15-2)}$$

2.   Calculate the values of A for each value of V:

$$A_{\text{calc}} = m(V + V_c) \qquad \text{(15-3)}$$

3.   Find values of d for each point:

$$d = A - A_{\text{calc}} \qquad \text{(15-4)}$$

4.   Find $\Sigma d^2/(n - 2)$, the variance of the readings, for the data.

5.   Calculate the variance s^2 of V_c:

$$s^2 = \frac{\Sigma d^2}{n - 2} \frac{\Sigma(V + V_c)^2}{(\Sigma A_{\text{calc}})^2} \Bigg/ \left(\frac{(\Sigma(V + V_c))^2}{n\Sigma(V_c)^2} - 1 \right) \qquad \text{(15-5)}$$

Calculate the standard deviation s of V_c by taking the square root of the variance s^2. Applying the above steps to the data in Table 15-1, s is estimated as follows:

$$m = 303.0/(52.02 + (6 \times 3.00)) = 303.0/70.02 = 4.327$$

The data for steps 2 to 4 are summarized in Table 15-2. Then

$$\Sigma d^2/(n - 2) = 1.432/4 = 0.3581$$

$$\text{variance} = (0.3581)\frac{1120.24}{91\,791} \frac{4902.80}{6 \times 1120.24} - 1 = \frac{(0.3581)(1120.24)}{(91\,791)(1.370\,94 - 1)} = 0.01178$$

$$\text{Standard deviation in } V_c = \sqrt{0.01178} = 0.109 = 0.11$$

Table 15-2. Summary of Data for Calculation of Standard Deviation

V	$V + V_c$	A_{calc}	d	d^2	$(V + V_c)^2$
0	3.00	12.98	0.02	0.0004	9.00
2.03	5.03	21.76	0.26	0.0676	25.30
4.99	7.99	34.57	0.43	0.1849	63.84
9.99	12.99	56.21	0.29	0.0841	168.74
15.01	18.01	77.93	0.93	0.8649	324.36
20.00	23.00	99.52	0.48	0.2304	529.00
	70.02	302.97		1.4323	1120.24

The value for V_c, then, is 3.00 ± 0.11 mL. With the assumption of negligible error in the weights of the copper standard and the sample, the error in the percentage of copper reported is proportional to the error in the calculation of V_c. Therefore, the answer and its uncertainty would be 0.030 ± 0.001 % Cu.

15-9 INFRARED SPECTROPHOTOMETRIC DETERMINATION OF FOUR SUBSTITUTED BENZENES
Background

In Section 14-14 a two-component mixture is resolved by making measurements at two wavelengths and setting up and solving two equations. A chemical separation prior to measurement is thereby made unnecessary. With sufficient data, spectrophotometric techniques of this type can be used to analyze mixtures of several components. Thus, one could in principle analyze three-, six-, or ten-component mixtures by making measurements at three, six, or ten wavelengths and solving three, six, or ten simultaneous equations. As the number of components increases, however, the quality of the instruments becomes more critical, and the arithmetic more onerous. In this experiment the tedium of solving four simultaneous equations is relieved through the use of a computer. Also, the application of infrared spectrophotometry to the resolution of a complex mixture is illustrated.

For this experiment sets of spectra run on a high-resolution infrared spectrophotometer are supplied. The samples comprise spectra of mixtures of p-dichlorobenzene, p-dibromobenzene, p-bromomethylbenzene, and p-xylene that have been dissolved in a decane–bromoform solvent. The infrared spectra of these substituted benzenes are nearly identical in the wavelength region 12.2 to 12.7 μm (Figure 15-15). In this region infrared absorption by these compounds is due to symmetric out-of-plane deformation vibrations of adjacent hydrogen atoms on the ring. The vibrations are at right angles to the plane of the ring and give rise to intense absorption when the hydrogen atoms are vibrating in phase.

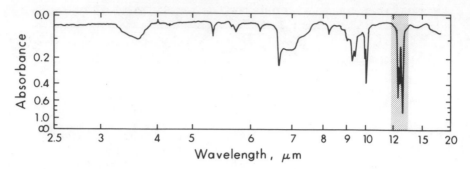

Figure 15-15. Infrared spectrum of a mixture of *p*-xylene, *p*-dichlorobenzene, and *p*-dibromobenzene in the ratio 3:1:2. The spectra are the same except for the shaded region. (Note that the plot of absorbance against wavelength involves a logarithmic scale and is inverted relative to normal practice in the ultraviolet and visible regions.) Spectrum by R. Swindlehurst.

Measurement of Absorbances from Recorded Spectra

In addition to the mathematical problems associated with quantitative multicomponent analysis, several points associated with infrared spectra should be considered. A problem in the extraction of reliable absorbance information from recorder tracings is deciding on the most accurate positions for the base lines and the absorbances at the absorption maxima. The position of the base line in a recorded spectrum is generally subject to error because of noise. It may also drift with wavelength. Another effect of noise is that the absorbance at the absorption maximum often does not coincide with maximum excursion of the recorder pen. With noise of short period the maximum absorption is likely to be less than the maximum excursion; with noise of longer period it may be either greater or less. When absorbances are being read from a recorded spectrum, the maximum of each peak should be drawn as though noise were absent. Peak maxima may be distorted also by interferences. In the sample spectrum in Figure 15-16 the maximum for xylene may be distorted by the small absorption peak of *p*-bromomethylbenzene at about 12.7 μm.

Procedure (time 2 to 5 h)

Obtain prerecorded infrared spectra of a sample mixture and of the four pure (standard) components. Bearing in mind the influence of noise on maxima and base lines, read the absorbance and base line of each standard at the four wavelengths corresponding to each of the four peak maxima. Thus, in Figure 15-16 four absorbance readings would be recorded from the spectrum of Standard 1, *p*-dichlorobenzene, at wavelengths of 12.24, 12.38, 12.50, and 12.64 μm. Readings of the base line would also be taken at the same four wavelengths, and each subtracted from the corresponding spectral reading. This procedure would result in four absorbance readings at four wavelengths for Standard 1. The procedure would be repeated at the same four wavelengths for each of the other three standards and then for the unknown mixture. Sixteen absorbance values, each the difference between the peak and base-line readings, for the standard curves and four for the mixture must be obtained.

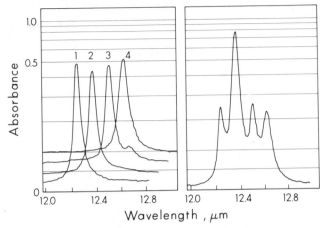

Figure 15-16. Infrared spectra of each of four standards (*left*) and of a mixture of the four (*right*). Wavelength region corresponds to shaded portion of Figure 15-15. The composition in weight percent of the standards is (*1*) *p*-dichlorobenzene, 0.300; (*2*) *p*-dibromobenzene, 0.472; (*3*) *p*-bromomethylbenzene, 0.388; and (*4*) *p*-xylene, 0.354. Spectra by R. Swindlehurst.

Calculate the concentrations of the four components by solving four equations in four unknowns, either manually or by the use of a small computer program.[6] Set up the equations as outlined for the two-component mixture in Sections 15-7 and 14-14.

PROBLEMS

15-1.　What are the differences and similarities between the instrumentation used for the ultraviolet and infrared regions of the spectrum?

15-2.　Why are infrared spectra especially useful for qualitative identification? Why are they applied less widely than spectra from the visible region for quantitative analysis?

15-3.　Why may there be more noise in infrared than in ultraviolet or visible spectra?

15-4.　What modes of vibrations are to be expected in a molecule of CO_2; of C_2H_5OH?

15-5.　How may samples be prepared for measurement in the infrared region? What special precautions are necessary?

15-6.　What is the wavelength and wavenumber of the infrared radiation that can be absorbed by a molecule with the following normal-mode vibrational frequencies: (a) 7.236×10^{13} Hz, (b) 2.549×10^{14} Hz, (c) 4.78×10^{13} Hz, and (d) 1.096×10^{14} Hz?

15-7.　A transition between two vibrational energy levels of a molecule produces an infrared absorption peak centered at 6.43 μm. (a) Calculate the wavenumber of this band and the frequency of the vibrations. (b) Calculate the energy separation between the two vibrational levels.

15-8.　For an absorbance reading of 0.005, calculate the minimum liquid concentration determinable (in milligrams per milliliter) in 0.025-mm cells for (a) chloroform, (b) methylene chloride, and (c) benzene. The infrared specific absorptivities in

[6]A minicomputer or microprocessor, if available, can be readily programmed to execute this calculation.

milligrams per milliliter are 900 for $CHCl_3$ at 1216 cm^{-1}; 1320 for CH_2Cl_2 at 1259 cm^{-1}; and 4900 for C_6H_6 at 1348 cm^{-1}.

15-9. For the preceding problem calculate the cell thickness required to give a transmittance of 0.600 for a 0.1% solution of each liquid at the wavelength of analysis. Assume solution densities of unity.

15-10. Define the terms chemical shift, Larmor frequency, and spin-spin splitting.

15-11. What are the three types of information available from an NMR spectrum?

15-12. (a) Would *n*-propyl and isopropyl alcohol be expected to have the same proton NMR spectrum? Explain. (b) Would the carbon-13 NMR spectrum of the two be the same? Explain.

15-13. Describe the basic components of the instruments used to obtain an NMR spectrum. What are the advantages of using Fourier transform NMR?

15-14. (a) Which electrons of the sample atom are involved in analytical X-ray measurements? (b) Which radiochemically stable element would be expected to have the most energetic X-rays?

15-15. Explain what is meant by X-ray fluorescence. List the principal advantages and disadvantages of this technique.

15-16. For the analysis of samples of Portland cement a series of standards were prepared containing 100, 75, 50, 25, 10, and 0 $\mu g/mL$ of Na_2O, along with a second set containing the same weights of K_2O. Each standard solution also contained 6300 $\mu g/mL$ of calcium (expressed as CaO) to compensate for the influence of calcium on the flame-emission readings. The emission intensities for sodium (measured at 590 nm) were 100, 87, 69, 46, 22, and 3%, and those for potassium (measured at 768 nm) were 100, 80, 58, 33, 15, and 0%. Three different 1.000-g samples of Portland cement were weighed, dissolved in acid, and diluted to 100 mL. For the three samples the emission readings at 590 and 768 nm were 28 and 69, 58 and 51, and 42 and 63. Calculate the percentage of sodium and potassium in the samples of Portland cement, expressed as Na_2O and K_2O.

15-17. What are the major characteristics and advantages of Raman spectra?

15-18. Explain what is meant by Stokes and anti-Stokes lines.

15-19. Briefly describe the various methods whereby sample excitation can be brought about in arc and spark spectroscopy. What electronic process occuring in the sample atoms produces analytically useful information?

15-20. Describe the basic components of instruments used for obtaining emission spectra.

15-21. (a) In flame photometry, how is the temperature of a flame controlled? (b) What are the advantages in the use of this less energetic excitation source relative to several other types of spectroscopy?

15-22. (a) What processes occur when a sample is introduced into a flame? (b) What experimental factors govern the intensity of the radiant power?

15-23. Draw a schematic diagram showing the essential components of an atomic-absorption spectrophotometer.

15-24. Briefly describe the operation of a hollow-cathode lamp. What are its principal advantages?

15-25. Why is the sample placed ahead of the monochromator in atomic-absorption spectroscopy? What type of relation between absorbance and concentration might be anticipated if it were located after the monochromator?

15-26. When a flame serves as the atomization system in atomic-absorption spectroscopy, what interferences affect the results? How can spectral interferences be minimized?

15-27. Outline the applications, advantages, and disadvantages of atomic-absorption spectroscopy.

15-28. Describe the differences and similarities of atomic-absorption and flame emission spectroscopy.

15-29. Why are atomic-absorption measurements less sensitive to flame temperature than flame emission photometric methods?

15-30. How can the sensitivity of atomic-absorption measurements be increased?

15-31. A chemist has available instruments for atomic-absorption spectroscopy and for flame emission spectroscopy. Which method would be preferred for (a) trace analysis of transition elements; (b) alkali metals. Explain your choice.

15-32. What is the method of multiple standard additions? Why is it valuable in atomic-absorption spectroscopy?

15-33. Briefly describe how the final solutions are prepared for the determination of copper by atomic absorption using the method of multiple standard additions.

15-34. Why do atomic emission and absorption spectra contain narrow lines instead of broad bands?

15-35. In the determination of trace copper in nickel metal by atomic absorption, explain why a satisfactory calibration plot cannot be prepared from dilute solutions of pure copper salts.

15-36. A 10.00-mL portion of filtered tap water was diluted to 50.00 mL with pure water and analyzed by atomic absorption for copper. The analysis showed a reading 1.73 times that for a standard copper solution that was prepared by dissolving 0.3142 g of copper wire in nitric acid, diluting the solution to 100 mL, and then repeating a 10- to 250-mL dilution three times. Calculate the concentration of copper in the tap water in milligrams per milliliter.

15-37. In the analysis of copper in an ore by atomic absorption, a sample solution is prepared to contain 1.23 g of ore per 100 mL, and the standard solution is prepared to contain 2.00×10^{-3} g of copper per 100 mL. Into one flask 10.00 mL of sample solution is pipetted, and into a second, 10 mL of sample solution and 10 mL of standard solution. Both flasks are diluted to 100 mL. An atomic-absorption spectrophotometer gave readings of 0.421 and 0.863 for the two solutions. What is the percentage of copper in the ore?

15-38. Compare atomic absorption on the basis of sensitivity, selectivity, and precision with (a) titrimetric, (b) radiochemical, (c) colorimetric, and (d) flame emission methods.

15-39. A 0.9125-g sample of powdered nickel was analyzed for copper by the procedure of Section 15-8. The weight of copper used for the standard was 0.2214 g. The calibrated volume of the pipet was 9.980 mL. The readings for 0, 2.01, 4.96, 10.02, 14.98, and 20.13 mL of the standard copper solution were 40.7, 46.2, 55.5, 71.0, 86.2, and 100.0. Calculate the percentage of copper in the sample (a) by the graphical method; (b) by the least-squares method.

15-40. In the infrared computer experiment (Section 15-9) explain why the absorbance of each standard must be measured at four wavelengths.

15-41. Contrast the least-squares and graphical methods of calculating results for copper by atomic absorption according to the directions in Section 15-8.

15-42. A 0.8550-g sample of powdered nickel was analyzed for copper by atomic absorption (Section 15-8). The sample was dissolved in nitric acid and then brought to volume in a 250-mL volumetric flask. The standard solution was prepared by dissolving 0.1743 g of pure copper metal with nitric acid and diluting it to volume in a 100-mL volumetric flask, followed by two 10- to 100-mL dilutions. A 10-mL portion of the sample solution was pipetted into each of six 50-mL volumetric flasks, followed by 0, 2.00, 5.03, 10.01, 15.04, and 20.02 mL of the standard solution into the six flasks. The contents were then diluted to volume, and the solutions measured by atomic absorption. Average readings of 40.0, 46.1, 55.2, 71.3, 86.0 and 100.0 were obtained for the six solutions. Calculate the percentage of copper in the sample by (a) the graphical method and (b) the least-squares method.

15-43.   A 1.65-g sample of wild duck feathers was ashed. The ash was dissolved in hydrochloric acid, diluted to 25 mL, and analyzed for lead by atomic absorption. The analysis showed a reading 1.32 times as great as that of a lead solution prepared by dissolving 0.336 g of lead, diluting the solution to 100 mL, and then repeating a 10- to 250-mL dilution three times. What is the percentage of lead in the feathers?

REFERENCES

Analytical Chemistry, Fundamental Reviews. Reviews of recent developments appear every even year. The April 1980 issue includes reviews of over 5000 original articles appearing in the previous two years. The reviews appear under the following headings: Electron Spectroscopy; Ultraviolet and X-Ray Excitation; Electron-Spin Resonance; Emission Spectroscopy; Flame Emission, Atomic Absorption and Atomic Fluorescence; Molecular Fluorescence, Phosphorescence, and Chemiluminescence Spectrometry; Infrared Spectrometry; Mossbauer Spectrometry; Nuclear Magnetic Resonance Spectrometry; Raman Spectrometry; Ultraviolet and Light Absorption Spectrometry; and X-Ray Spectrometry.

H. H. Bauer, G. D. Christian, and J. E. O'Reilly, Eds., *Instrumental Analysis*, Allyn and Bacon, Boston, 1978, Chapters 6–15. Covers the various fields of analytical spectroscopy. Each chapter written by experts in the individual area.

E. P. Bertin, *Introduction to X-ray Spectrometric Analysis*, Plenum, New York, 1978.

G. W. Ewing, *Instrumental Methods of Chemical Analysis*, 4th ed., McGraw-Hill, New York, 1975. Provides extensive discussion of the applications of the various methods of spectral analysis.

D. A. Skoog and D. M. West, *Principles of Instrumental Analysis*, 2nd ed., Saunders/HRW, Philadelphia, 1980, Chapters 4–15. Treats the subject of spectroscopy.

16

SEPARATIONS IN ANALYSIS: SINGLE-STAGE AND OTHER TECHNIQUES

... separation methods form the basis of chemistry, and the definition of a pure chemical substance ultimately depends on separative operations.

Arne Tiselius

16-1 BACKGROUND AND CLASSIFICATION

To this point the theory and the experiments themselves have emphasized techniques and measurements that do not require extensive prior treatment of the analytical sample other than dissolution and adjustment of reaction conditions. In the procedures, interferences have been coped with either by selective methods of analysis or by the addition of reagents that block interference through complexation, change in oxidation state, change in pH, and so on. Ideally, measurement techniques would be specific enough that only the substance of interest in the sample would be measured, and no other parts of the sample would interfere. Many samples of practical interest, however, contain other substances that cannot be prevented from interfering in the available methods of measurement. The separation of these substances from the material of interest then becomes mandatory.

Many different properties such as size, density, vapor pressure, solubility, and rate of chemical reaction can serve as the basis for separation. In most separation techniques a chemical or physical change is brought about in a single-phase (gas, liquid, or solid) system that leads to the formation of a second phase. Alternatively, the components of a sample can be distributed between two phases that comprise mainly substances that are not part of the sample. The desired sample component is preferentially concentrated in one of the phases, and the two phases are then physically or mechanically isolated. Separation processes can be classified in several ways: on the basis of the phases involved, on whether a separation can be achieved in a single stage, or on which of several specific processes is used. Broadly, the separation processes of analytical significance are *precipitation, volatilization,*

Table 16-1. Classification of Equilibrium Separation Processes Used in Analysis

| Phases Involved | Typical Processes | |
	Single Stage	Multistage
Gas and solid	Condensation, sublimation, volatilization, sorption	Chromatography
Gas and liquid	Condensation, distillation, evaporation, sorption	Chromatography, distillation
Liquid and solid	Precipitation, dissolution, electrodeposition	Chromatography
Liquid and liquid	Distillation, extraction, electrodeposition	Chromatography, extraction

extraction, and *chromatography*. Such *two-phase* separations can be described in terms of equilibrium properties such as solubility, vapor pressure, or distribution ratio. After distribution equilibrium of sample components between the two phases is established, the phases are separated in either a *single stage* or a *multistage* operation with repeated equilibrations and phase separations. A classification of these processes is summarized in Table 16-1.

Not included in this table are those separation processes of little analytical utility. Thus separations involving two solid phases are omitted, even though such processes as sieving, flotation, and magnetic separations are highly significant in other contexts. Omitted also are diffusion processes involving constituents of a gaseous phase with differing molecular weights, despite their importance in some nonanalytical systems.

Some separations can be achieved within a single phase. *Single-phase* separations are based on nonequilibrium properties of the sample components such as diffusion coefficient. In a single phase, for example, large ions, proteins, and colloidal particles can be separated under the influence of an electric-field gradient. The process, called *electrophoresis*, is especially important in biologically related subjects. Highly complex mixtures can be resolved by this process. Owing to limitations of space this important subject is not developed here. A somewhat similar method, used particularly for protein-containing and other high-molecular-weight substances, involves application of centrifugal force in an ultracentrifuge.

In analytical separations the object is to reduce the concentration of the interferences to a level of negligible influence without significant loss of the material of interest. The extent of a separation is given by the *separation factor*, defined as the ratio of the recovery of an unwanted substance to the recovery of the desired substance.[1]   Quantitative separation is normally considered to be achieved with a separation factor of 0.001 and a *recovery*

---

[1]We use here the definition of separation factor given by Sandell (*Anal. Chem.* **1968**, *40*, 834). The term separation factor is defined by some as the ratio of distribution ratios; such a definition does not directly indicate the extent of a separation.

factor (the fraction of the substance present in the sample that is isolated) of 0.999. For analysis of traces, the separation factor may need to be smaller; the recovery factor may also be smaller.

In this chapter nonchromatographic techniques are discussed. These involve mainly single-stage separations, which are feasible when the separation factor is sufficiently favorable. In Chapter 8 one kind of separation, precipitation, is examined as a single-stage operation for gravimetric analysis. This chapter includes consideration of the completeness of the precipitation reaction, the factors affecting solubility, purity, and formation of precipitates, and the separation of phases by filtration. Traces of substances can sometimes be separated and recovered through the technique of *carrier precipitation.* For example, a trace of thorium so small as to be irrecoverable by conventional precipitation and filtration can be recovered by a carrier precipitate of iron hydroxide (Section 13-6). This technique is employed most often in radiochemical analysis. Precipitation from homogeneous solution is a selective fractional-precipitation method for determining one substance in the presence of closely similar interferences.

If the separation factor is small, one or two operations may be sufficient to provide the separation required. When the technique involves physical isolation of the two phases from each other before a second equilibration is carried out, it is called a *batch* technique. If the separation factor is near unity, the single-stage technique may have to be repeated on a sample a number of times or continuous multiple operations may be required before the desired degree of separation is achieved.

Fractional distillation may be applied to a sample of a mixture of volatile substances having appreciable vapor pressures. On being heated the several constituents will be distilled at different times and can be collected separately.

In this and the succeeding chromatography chapters the experiments chosen to illustrate separations are also designed to broaden experience with a range of measurement techniques.

16–2 LIQUID-LIQUID EXTRACTION

In *liquid-liquid (solvent) extraction* a solution containing the sample is brought into contact with an immiscible solvent. The first phase is usually water, and the second phase an organic solvent, poorly soluble in water. Equilibrium of solute between the two phases is established by shaking or otherwise mixing the two. Separation of traces as well as large amounts of substances can be quickly and simply effected with high efficiency in a test tube or a separatory funnel (Figure 16-1).

Liquid-liquid separations have been applied to the extraction of organic substances for many decades; application to inorganic substances is more recent.[2]   Hydrated ionic materials generally are not highly soluble in

---

[2]The need for materials of high purity in connection with nuclear reactors initiated extensive studies on methods for the large-scale removal of trace impurities from metals. An example is purification of uranium as $UO_2(NO_3)_2$ by liquid-liquid extraction.

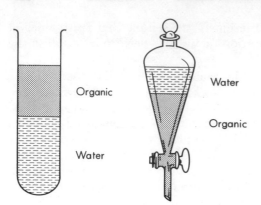

Figure 16-1. Examples of laboratory apparatus for separations by liquid-liquid extraction with the organic phase lighter (test tube) or heavier (separatory funnel) than water.

organic solvents, in contrast with neutral organic substances. Hydrated metal ions can be dehydrated as well as converted to an uncharged state by interaction with an appropriate organic reagent; both steps are important for increased solubility in nonpolar organic phases. Neutralization of ionic charge can also be achieved by direct reaction of the metal ion with negatively charged ligands. If a metal ion is converted to an uncharged species, not only is its solubility in most organic solvents enhanced, but its solubility in water is reduced.

An example of the formation of a readily extractable species is the reaction of iron(III) with hydrochloric acid. Iron(III) forms a series of chloride complexes:

$$Fe^{3+} + Cl^- \rightleftarrows FeCl^{2+}$$

$$FeCl^{2+} + Cl^- \rightleftarrows FeCl_2^+$$

$$FeCl_2^+ + Cl^- \rightleftarrows FeCl_3$$

$$FeCl_3 + Cl^- \rightleftarrows FeCl_4^-$$

$$FeCl_4^- + Cl^- \rightleftarrows FeCl_5^{2-}$$

$$FeCl_4^- + H^+ \rightleftarrows HFeCl_4$$

The uncharged acid $HFeCl_4$ is readily extracted[3] from an aqueous hydrochloric acid solution by a dipolar organic solvent such as methyl isobutyl ketone (MIBK) or diethyl ether. The $HFeCl_4$–MIBK system is applied to the separation of iron(III) from other metals in Sections 16-4 and 16-5.

More commonly, uncharged species are formed by displacement of protons from the coordination sites on ligands by the hydrated metal ion. An example is the complex formed between magnesium and 8-hydroxyquinoline.

---

[3]Methyl isobutyl ketone(MIBK), $CH_3C(O)CH_2CH(CH_3)_2$, extracts iron as effectively as does the long-used solvent ethyl ether, $CH_3CH_2OCH_2CH_3$, but has a more favorable distribution ratio and is less flammable.

The larger and the more hydrophobic the neutral complex, the more its solubility in water is decreased and its solubility in the organic solvent increased.

Uncharged species are also formed by *ion pairing*, in which a large ion of low charge is extracted along with an ion of opposite charge into an organic phase. As an example, cobalt(III) thiocyanate can be extracted quantitatively into chloroform as an ion pair with the tetraphenylarsonium ion $(C_6H_5)_4As^+$:

$$Co^{3+} + 4SCN^- \rightleftarrows Co(SCN)_4^-$$

$$(C_6H_5)_4As^+Cl^- + Co(SCN)_4^- \rightleftarrows [(C_6H_5)_4As^+Co(SCN)_4^-] + Cl^-$$

Liquid-liquid extractions are carried out by batchwise techniques (either single or multistage), by countercurrent methods, and by continuous procedures. The completeness of a separation can be forecast from the distribution equilibrium. The equilibrium constant of most practical value in liquid-liquid extraction is the *distribution ratio D*, defined as the ratio at equilibrium of the sum of concentrations of all forms of a substance in the nonaqueous phase to the sum of concentrations in all forms in the aqueous phase:

$$D = C_{org}/C_w$$

where C_{org} is the analytical concentration in the organic solvent and C_w the analytical concentration in the aqueous phase.

For the particular case of iron(III) and the chloride ion, the distribution ratio is

$$D = \frac{[HFeCl_4]_{org}}{[Fe^{3+}] + [FeCl^{2+}] + [FeCl_2^+] + [FeCl_3] + [HFeCl_4] + [FeCl_4^-] + [FeCl_5^{2-}]}$$

The distribution ratio varies according to the experimental conditions. For the iron(III)–chloride system a change in pH or chloride ion concentration drastically changes the proportion of the total amount of substance in the neutral extractable form. The distribution ratio may therefore change over many orders of magnitude when experimental conditions are changed.[4] Figure 16-2 illustrates the change in distribution ratio for iron(III) as a function of hydrochloric acid concentration. The distribution of iron between water and MIBK goes through a broad maximum at a hydrochloric acid concentration of about 6 *M*. The decrease at higher acid concentrations has been attributed to an increased solubility of MIBK in the aqueous acid, which causes the solvent properties of the two phases to become similar.

---

[4]A quantity that is insensitive to changes in experimental conditions is the *distribution constant* K_D, defined as the ratio of the concentrations of a *particular solute species* in the two phases. Thus the distribution constant for $HFeCl_4$ between MIBK and water has a large value according to $K_D = [HFeCl_4]_{org}/[HFeCl_4]_w$. On the other hand, the value of K_D in this system for an ionic species such as $FeCl_2^+$ is minute. The distribution constant is a more fundamental but less practical quantity than the distribution ratio.

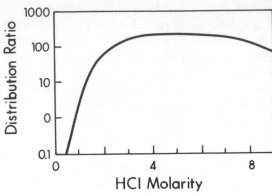

Figure 16-2. Distribution of iron(III) between MIBK and water as a function of hydrochloric acid concentration in the water.

When an uncharged complex forms between a metal ion and an organic chelating reagent, changes in pH can have profound effects on the distribution ratio. For example, dithizone (HDz) behaves below pH 12 as a monobasic acid with an ionization constant of about 3×10^{-5}. It reacts with metal ions such as lead(II) according to

$$Pb^{2+} + 2HDz \rightleftharpoons Pb(Dz)_2 + 2H^+$$

For this system the distribution ratio $(C_{Pb})_{org}/(C_{Pb})_w$ depends heavily on the concentration of the reactants and on pH. At low pH values the proportion of lead present as the neutral complex is small, and the distribution ratio is correspondingly low. Lead becomes more easily extracted as the pH rises, and so it is normally extracted from a solution of nearly neutral pH. Figure 16-3 illustrates the effect of pH on the extraction of several metal 8-hydroxyquinolates into chloroform. It is clear from this figure that solely by changing the pH nickel can be separated from manganese, for example.

Distribution ratios can also be greatly altered by adding a second complexing (masking) agent. For example, addition of EDTA to a lead–dithizone solution will greatly reduce the distribution ratio through

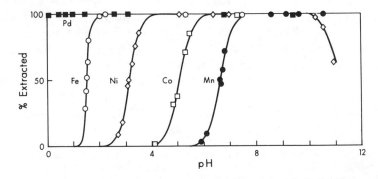

Figure 16-3. Effect of pH on the extraction of palladium(II), iron(III), nickel(II), cobalt(II), and manganese(II) from water into 0.01 M 8-hydroxyquinoline in chloroform. From J. Stary, *The Solvent Extraction of Metal Chelates*, Macmillan, N.Y., 1964. By permission.

formation of the water-soluble, exceedingly stable lead–EDTA complex.

Selectivity of extraction among metal ions can be achieved by control of pH or reagent concentration or by use of masking agents to enhance differences between metal ions. The choice of solvents and the rate of extraction provide additional variables to control a separation. The organic solvent chosen for an extraction should not be appreciably soluble in water. It should favor a high distribution ratio for the desired material and a low ratio for the interfering substances. Extraction is facilitated if the solvent has low viscosity so that it mixes easily with the water layer. The density of the solvent should be significantly either more or less than that of water so that the two phases separate quickly after mixing. If the density of the organic solvent is greater than unity, and therefore the bottom layer is to be removed, a separatory funnel can be used for the extraction. If the density is less than unity, a test tube is used and the upper solvent is drawn off with a pipet (Figure 16-1).

The extent of a separation can be calculated directly from the distribution ratio and the volumes of the two phases. In general, the fraction remaining unextracted in the water layer f_w is

$$f_w = \frac{V_w}{V_w + DV_{org}} \tag{16-1}$$

where V_w is the volume of the water layer and V_{org} the volume of the organic solvent layer. Examination of this equation reveals that the larger the value of the distribution ratio, the larger is the volume of organic solvent, and the smaller the volume of water, the larger is the extent of extraction. For the special case where $V_w = V_{org}$, $f_w = 1/(1 + D)$. Figure 16-4 shows how the fraction f_w changes with the distribution ratio for this special case.

If after equilibration with the aqueous layer the organic layer is removed and replaced by a fresh portion of solvent, then the fraction remaining in the water layer after this second extraction f_2 is

Figure 16-4. Fraction of solute in the aqueous phase as a function of the distribution ratio where $V_{org} = V_w$.

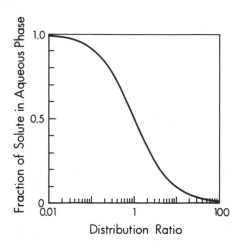

$$f_2 = \frac{f_1 V_w}{V_w + D V_{org}} = \left[\frac{V_w}{V_w + D V_{org}} \right]^2$$

After a third such extraction

$$f_3 = \left[\frac{V_w}{V_w + D V_{org}} \right]^3$$

and after n extractions

$$f_n = \left[\frac{V_w}{V_w + D V_{org}} \right]^n \tag{16-2}$$

Equation (16-2) indicates that the efficiency of an extraction improves (1) nearly in proportion to the volume of the organic extractant, (2) nearly in inverse proportion to the volume of the aqueous phase, and (3) exponentially as the number of extractions. Therefore, several smaller portions of organic solvent provide more efficient extraction than a single large portion. A substance can be extracted quantitatively with a relatively small volume of organic solvent provided the value of D is not too small. Complete separation of the phases after equilibration is essential for the calculation of f_n values from experimental results.

EXAMPLE 16-1

What fraction of iron(III) remains unextracted from 20 mL of 6 M hydrochloric acid after extraction with 10, 20, or 50 mL of MIBK? Assume a distribution ratio of 100.

With 10 mL of MIBK the fraction unextracted $f_w = 20/(20 + (10 \times 100)) = 0.02$, indicating 98% extraction. With 20 mL, $f_w = 0.01$. With 50 mL, $f_w = 0.004$, a nearly quantitative extraction (99.6%).

EXAMPLE 16-2

Repeat the calculations of Example 16-1 using a distribution ratio of 10, such as occurs in 1.4 M hydrochloric acid.

With 10 mL of MIBK, $f_w = 20/(20 + 10 \times 10) = 0.17$. With 20 mL, $f_w = 0.09$. With 50 mL, $f_w = 0.04$.

In this example, in no instance is the extraction quantitative.

EXAMPLE 16-3

Suppose iron(III) is to be quantitatively extracted from 20 mL of a hydrochloric acid solution under conditions where the distribution ratio is 10. Could a higher percentage of the iron be removed if four successive extractions were made with (50/4)-mL portions instead of a single 50-mL portion?

$$f_4 = \left[\frac{20}{20 + 10(50/4)} \right]^4 = 0.00036$$

Thus the use of 50 mL in four portions gives an extraction that is better than 99.9%. Even with only three (50/4)-mL portions the extraction is 99.7% complete.

Suppose 10 extractions of 2 mL each were carried out. In this case $f_{10} = 0.001$, a quantitative extraction with a total volume of only 20 mL. Clearly the extracting solvent is used more efficiently in small portions.

After an extraction has been completed, the organic solvent containing the substance of interest may be washed with two or three small portions of the aqueous phase (4 M HCl in the preceding example) to remove small amounts of impurities.

One of several courses may be chosen for determination of the separated substance. The most direct is measurement of the amount of solute in the organic phase. But sometimes this is not feasible, and the solute must be returned to a fresh aqueous phase by one of several techniques. One technique, *back extraction*, involves shifting the distribution ratio, usually by changing the pH, so that the solute will preferentially return to a fresh portion of pure initial phase. Two or three such extractions are usually required to ensure completeness of the transfer. Another technique useful when the solvent is volatile and the solute is not is to add water and evaporate the solvent.

When separation of two substances that are both extractable to varying degrees is desired, batch solvent extraction is inefficient.

EXAMPLE 16–4

A sample contains two substances A and B with the greatly different distribution ratios of 0.1 and 10. If equal volumes of aqueous and organic phase are used in a batch extraction process, what is the best separation that can be achieved?

After the first extraction $f_w = 0.91$ for A and 0.091 for B, with 9% of A and 91% of B recovered in the organic phase. The separation factor is then $9/91 = 0.10$. This is the smallest and thus most favorable separation factor that can be achieved.

What is the separation factor when B has just been quantitatively recovered?

After the third extraction f_w is 0.75 for A and 0.0008 for B. This represents the minimum number of extractions for quantitative recovery of B. The separation factor at this stage is $(1 - 0.75)/(1 - 0.0008)$, or 0.25, much poorer than the separation factor after the first extraction.

Countercurrent Liquid-liquid Extraction

Example 16-4 illustrates the serious shortcoming of batch extractions for selective separations.[5] By a simple change in procedure, however, separations can be made highly selective.

In *countercurrent liquid extraction* the sample is placed in the first of a series of tubes containing an aqueous phase. An organic solvent is added to this tube, and after equilibration with the aqueous phase it is transferred to the next. If the volumes of the two phases are equal, the fraction transferred is $D/(1 + D)$, and the fraction of solute retained by the first tube is $1/(1 +$

---

[5]It can be calculated that for quantitative extraction of one component into an organic phase and at the same time quantitative retention of a second component in the aqueous phase in a single extraction, a minimum ratio of distribution ratios of about 10 000 is necessary. Such a difference can rarely be achieved.

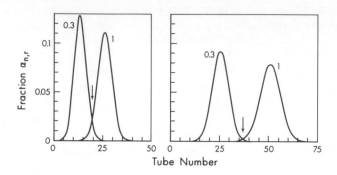

Figure 16-5. The distribution of two solutes in a series of tubes after 50 and after 100 equilibrations and transfers. The distribution ratio for one solute is 1.0 and for the other is 0.33. Adapted from W. E. Harris, *Rev. Anal. Chem.* 1971, *1*, 7.

D). Fresh solvent is added to the first tube. After equilibration in both the first and second tubes the organic solvent is transferred from the second to the third tube and from the first to the second, and again a fresh portion of solvent is added to the first. At this stage $[1/(1 + D)]^2$ remains in the first tube, $[D/(1 + D)]^2$ is transferred to the third tube, and it can be shown that $2D/(1 + D)^2$ remains in the second. This process is continued through as many stages as desired.

EXAMPLE 16–5

What is the distribution of 100 units of a solute in a series of countercurrent extraction tubes after four equilibrations and transfers if the distribution ratio is exactly 2 and the volumes of the two phases are equal?

After one equilibration and transfer the first tube contains $100 \times 1/(1+2) = 33.3$ units, and the second contains $100 \times 2/(1+2) = 66.7$ units. After two equilibrations and transfers the first tube contains $(1/3)^2 \times 100$. The second tube has had some solute transferred out and some transferred in; the amount transferred out is $(2/3) \times 66.7 = 44.4$ units, and that transferred in is $(2/3) \times 33.3 = 22.2$ units. Thus the second tube still contains 44.4 units, and the third tube contains 44.4 units. After three equilibrations and transfers the first four tubes contain 3.7, 22.2, 44.4, and 29.6 units. After four equilibrations and transfers the amounts are 1.2, 9.9, 29.6, 39.5, and 19.8 units in the first five tubes.

A second extractable solute will also be transferred from tube to tube in amounts corresponding to its distribution ratio. If the ratios differ, this second substance will eventually be present largely in one group of tubes, while the first will be present in another group. This separation will occur even when the difference in their distribution ratios is small because the process can be continued through as many equilibration stages as are required to separate the two substances.

Figure 16-5 illustrates this distribution for two solutes with distribution ratios of 1.0 and 0.33 after 50 and after 100 equilibrations and transfers. The amount of overlap clearly is less as the number of equilibrations is increased. The broadening is less than proportional to the number of equilibrations, whereas the separation is proportional to the number (see Problem 16-25).

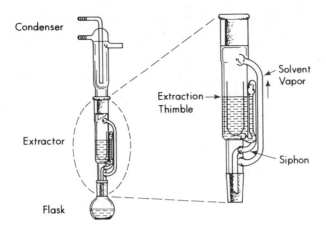

Figure 16-6. Soxhlet extractor.

Countercurrent separations are most easily carried out when the difference in distribution ratios of the materials separated is considerable and when the distribution ratio of one of two solutes is near unity. A large difference in distribution ratios by itself does not ensure an efficient separation, however. Two solutes with distribution ratios of 1000 and 2000 (or 0.003 and 0.006) cannot be easily separated even though their distribution ratio values differ by a factor of 2. On the other hand, distribution ratios of 1 and 2 represent a highly favorable case, since neither solute is strongly held by either the aqueous or the organic phase.

In liquid-liquid extraction procedures the objectives are to recover components quantitatively and in high purity. Although these objectives can never be attained absolutely, they can often be approached closely. Sometimes a compromise between purity and quantitative recovery is necessary.

Other Liquid-liquid Extraction Methods

Continuous Extraction

When the distribution ratio is low or when equilibration is achieved slowly, *continuous extraction* may be appropriate. The *Soxhlet extractor* (Figure 16-6) is an example of apparatus designed for this purpose. The solvent is recycled by evaporation and condensation, and the condensed solvent is used repeatedly to extract the sought-for component from a sample in a thimble, a deep narrow cup constructed of heavy filter paper. When the solvent level in the sample compartment reaches a certain height, a built-in syphon empties the solution into the receiving flask, where the solvent is recycled.

Countercurrent Cascade (Counter Double-current) Extraction

In *countercurrent cascade extraction* the organic phase is transferred from one equilibration stage to the next as in ordinary countercurrent extraction, and at the same time the aqueous phase is transferred from one stage to the next in the opposite direction. Thus both phases are transferred, but in *opposite directions*. This technique for separations is. the most powerful of all batch methods. More complete separations with quantitative recovery can be achieved than by the simpler countercurrent technique. Countercurrent cascade techniques are of considerable importance in industrial separations processes. One application is in multistage distillations.

Applications

The variety of analytical applications is large; only a few examples are mentioned here.

Simple molecules such as fatty acids or hydrocarbons can be separated by liquid-liquid extraction. The standard method for the determination of fats in cereals, for instance, entails a continuous procedure with ethyl ether in a Soxhlet extractor. Simple inorganic molecules can also be extracted. Iodine, bromine, mercury(II) chloride, arsenic(III) choride, and molybdenum(V) thiocyanate all can be transferred from water to solvents such as chloroform.

Other extractable materials may be partly dissociated in the water phase, the dissociated species being in equilibrium with the neutral molecule. Dithizone in water has the equilibrium

$$H\,Dz \;\rightleftarrows\; H^+ + \;\; Dz^-$$

in accordance with an ionization constant of 3×10^{-5}. The neutral molecule H ·Dz is highly soluble in an organic solvent. The distribution ratio for this molecule between chloroform and water is 2×10^5. At a pH of, say, 9, appreciable amounts of dithizone will be present in both phases — in the organic phase as neutral H Dz and in the water phase primarily as the Dz^- ion. Similar equilibria occur with other extraction reagents such as 8-hydroxyquinoline, cupferron, and dimethylglyoxime as well as with neutral metal chelate complexes.

Following are some examples of particular metal ion–reagent combinations used for extraction.

Beryllium can be extracted from neutral aqueous solution as its acetylacetone complex into chloroform. EDTA can serve as a masking agent for metal ions, particularly aluminum. Excess reagent extracted by the solvent can be back extracted into water by raising the pH with sodium hydroxide.

Copper can be extracted from slightly alkaline solutions with diethyldithiocarbamate, and from slightly acidic solutions with 2,9-dimethyl-1,10-phenanthroline.

A procedure for the extraction of *iron* as the chloro complex from aqueous HCl into MIBK is given in Section 16-4. Iron can also be extracted as iron(II) 8-hydroxyquinolate into chloroform.

Lead can be extracted with dithizone from neutral or slightly alkaline solution into chloroform. *Silver, mercury,* and *copper* can also be extracted with this reagent from aqueous solutions of about pH 1.

Nickel can be extracted from water into chloroform as the

dimethylglyoxime complex (Section 16-3); few other metal ions interfere.

Uranium (as the UO_2^{2+} ion) can be extracted with 8-hydroxyquinoline as the complexing agent. Interfering elements such as iron and aluminum can be masked by EDTA. Uranium can also be extracted as $UO_2(NO_3)_2·2TBP$, where TBP denotes tri-*n*-butyl phosphate.

16-3   DETERMINATION OF TRACE NICKEL BY LIQUID-LIQUID EXTRACTION FROM COPPER AND SPECTROPHOTOMETRIC MEASUREMENT

Background

In this experiment a small amount of nickel is separated from a large quantity of copper by extraction of nickel dimethylglyoxime, $Ni(DMG)_2$, from water into chloroform. Although a single extraction does not quantitatively separate all the nickel, high-quality results can be obtained by carrying standards through the same procedure and preparing a calibration curve. Beer's law is followed by $Ni(DMG)_2$ in chloroform over a wide range in concentration, and consequently a graph of absorbance against concentration should be linear. The absorbances of all solutions should be read relative to a blank prepared by carrying a portion of water through the extraction and the sample treatment. This method compensates for nickel impurities in the reagents.

Thiosulfate is added prior to extraction to form $Cu(S_2O_3)_2^{2-}$, an anionic complex that is not extracted into chloroform. Tartrate is added to form a complex with any iron(III) present. Hydroxylamine hydrochloride is added to prevent oxidation of $Ni(DMG)_2$ to a nickel(IV) complex with dimethylglyoxime, which has a different absorption spectrum.

Procedure (time 4 to 6 h)

Preparation of Solutions

Accurately weigh into a 100-mL volumetric flask enough nickel powder to prepare 100 mL of approximately 0.04 *M* solution. Add 15 mL of 6 *M* HNO_3, and heat the mixture on a hot plate until dissolution is complete. Do not stopper the flask! Neutralize with 2.5 *M* NaOH to the first appearance of a permanent precipitate of nickel hydroxide. Add 6 *M* acetic acid dropwise, shaking after each addition, until the precipitate just dissolves.[6] Dilute to volume and mix. Pipet a 10-mL aliquot into a 250-mL volumetric flask, and dilute to volume to give a standard solution containing about 100 ppm of nickel.

If the nickel sample is an alloy, weigh, to the nearest 0.1 mg, a sample containing from 1 to 4 mg of nickel. Dissolve the sample, neutralize with NaOH and acetic acid, and then proceed in the same manner as for the standard nickel powder. [7]

If the nickel sample is a solution of a nickel salt, transfer it quantitatively to a

---

[6]This operation adjusts the pH so that the capacity of the buffer to be added later will not be exceeded.

[7]If the sample contains more than 5% nickel, a 10- to 100-mL dilution of the initial solution may be necessary to provide the appropriate concentration.

100-mL volumetric flask, and neutralize with NaOH and acetic acid as directed for the standard nickel powder. Dilute the resulting solution to volume, and mix. The sample solution should contain 1 to 4 mg of nickel per 100 mL.

To prepare a buffer of pH 6.5, dilute 8.7 mL of 6 M acetic acid to 100 mL, and then add 10 mL of this solution to 40 mL of water containing 15 g of sodium acetate.

Analytical Procedure

Following the procedure outlined in Section 14-12, select nine matched 13- by 100-mm test tubes for the spectrophotometer.

Obtain nine 25- by 150-mm test tubes. Into five of the tubes deliver 0.5, 1.0, 2.0, 3.0, and 3.5 mL of standard nickel solution from a buret, recording the exact volume of each. Measure distilled water into the five tubes with a 10-mL Mohr pipet to give a total volume in each of 10 $\pm$ 0.1 mL;[8] into a sixth tube pipet 10.0 mL of distilled water to act as a blank. Into each of the remaining three tubes pipet 10 mL of the unknown sample solution using a delivery pipet.

To each test tube add in succession (with vigorous swirling after each addition) 0.5 g of sodium tartrate, 5.0 mL of pH 6.5 buffer, 2.5 g of sodium thiosulfate, 1.0 mL of 10% hydroxylamine hydrochloride in water, and 2.0 mL of 1% dimethylglyoxime in ethanol.[9] With a delivery pipet, measure 10 mL of chloroform into each test tube.[10] Carefully add a magnetic stirring bar designed for stirring in test tubes to one of the solutions, place the tube on a magnetic stirrer, [11] and adjust the stirrer motor so that the two phases mix thoroughly for about 20 s. Try to perform the same amount of mixing in each test tube. Allow the phases to separate. With a 10-mL Mohr pipet transfer 5 to 6 mL of the chloroform layer to a dry 10-cm qualitative filter paper of medium porosity, and collect the filtrate [12] in a spectrophotometer sample tube. To minimize evaporation[13] cover the solution in the tube with a layer of about 5 mm of water.

Measure the absorbance of each of the solutions at 420 nm, using the extracted blank solution to set the instrument to zero absorbance. Record several measurements for each solution.

---

[8]The Mohr pipet is calibrated in fractions of a milliliter. Remember to deliver only the volume between the 0- and 10-mL marks. Do not dispense the volume in the tip below the 10-mL mark. The volume of solution in each tube should be the same within 0.1 to 0.2 mL to give a uniform final volume for extractions.

[9]The order of addition is important. Throughout the procedure alternate standards and samples. A convenient stopping point, if desired, is after addition of the buffer. Thiosulfate and subsequent reagents should be added the same day the extraction and measurement steps are performed. About 1½ h is required to complete the experimental work after the addition of thiosulfate.

[10]Evaporation of chloroform, of concern later, is not a problem here because it is protected by the lighter layer of water.

[11]A wooden block on the stirrer motor with a hole the diameter of a test tube in its center is an aid in proper positioning of the test tubes.

[12]Chloroform, being immiscible in water, will not pass through a water-wetted filter paper so do not moisten the filter paper before filtration.

[13]Since in the measurement step it is the *concentration* of nickel–DMG in the chloroform layer that is recorded, loss of a few drops of solution is not critical. A change in concentration due to evaporation of solvent will cause error, however, and so the tube should be kept covered after filtration.

Calculations

Plot the absorbance readings of the standard nickel solutions (vertical axis) against the volume of standard nickel solution added to each test tube (horizontal axis). Draw the best straight line through the points; the line should pass through the origin. Read from the line the volume of standard nickel solution corresponding to the absorbance values of each of the three replicate unknown sample solutions.

Calculate the milligrams of nickel present per milliliter of the standard solution; from this value calculate and report the total milligrams of nickel in the sample.

16-4   DETERMINATION OF ALUMINUM BY LIQUID-LIQUID EXTRACTION FROM IRON AND TITRATION WITH EDTA

Background

A general discussion of metal–EDTA complexes is given in Section 7-4. A source of difficulty in the determination of aluminum with EDTA is that formation of the complex takes place only slowly, especially if the aluminum is present in the form of hydroxide complexes. Conditions must therefore be carefully controlled if titration of this element with EDTA is to be successful.

The formation constant for $Al^{3+} + Y^{4-} \rightleftarrows AlY^-$ ($K_f = 10^{16.1}$) does not indicate the extent of formation of AlY^- in water because hydrogen ions from water compete with EDTA for sites on the EDTA anion and hydroxide ions from water compete with EDTA for sites in the coordination sphere of aluminum. Similarly, zinc forms a hydroxide complex, $Zn(OH)^+$, of appreciable stability ($K_f = 10^{4.3}$) that should be taken into account; if the pH becomes too high, $Zn(OH)_2$ ($K_{sp} = 5 \times 10^{-18}$) may precipitate. Recall (Section 7-3) that for EDTA the species Y^{4-} constitutes a significant fraction of the total amount of EDTA present only when the pH is 10 or greater. Below pH 10, protonated forms of EDTA predominate. The pH of the solution, then, largely determines the fraction of the total aluminum or zinc in a sample that is present in a complex with EDTA.

Recall that chemists, to facilitate calculation of the extent of complex formation at various pH values, introduced the concept of *conditional* formation constants (Section 7-2). These are formation constants that apply under specified conditions, usually of pH. The expression for the constant for the zinc–EDTA system is

$$K_{cond} = \frac{[ZnY^{2-}]}{C_{Zn}C_{EDTA}}$$

where C_{Zn} is the molar concentration of zinc present in all forms other than ZnY^{2-} and where C_{EDTA} is the molar concentration of EDTA present in all forms other than ZnY^{2-}. This equation is useful in determining whether a reaction under the conditions used has a sufficiently large equilibrium constant to be quantitative, that is, whether the ratio of C_{Zn} to $[ZnY^{2-}]$ is

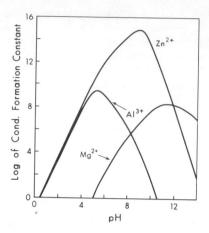

Figure 16-7. Plots of the logarithm of the conditional formation constants for aluminum(III), zinc(II), and magnesium(II) with EDTA as a function of pH.

0.001 or less at the equivalence point.

Plots of pH against the log of the conditional formation constants for the zinc–EDTA and aluminum–EDTA systems are shown in Figure 16-7. Note that for zinc the conditional constant reaches a maximum at about pH 9. At lower values hydrogen ions from water compete successfully with zinc for sites on the EDTA anion, whereas at higher values complex formation between zinc(II) and hydroxide ion becomes predominant. With aluminum, complexation by hydroxide is much stronger (for $AlOH^{2+}$ $K_f = 10^{8.8}$), and so the conditional constant for aluminum with EDTA begins to decrease at lower pH values.

The minimum conditional formation constant necessary for quantitative reaction at the part-per-thousand level is about 10^8 (Section 7-6). Figure 16-7 indicates that if we wish to utilize the aluminum–EDTA complex for the quantitative determination of aluminum, we must operate within a rather narrow range around pH 5. Zinc, on the other hand, can be titrated over the much wider range of pH 4 to 12.

Differences in conditional formation constants sometimes can be used directly. For instance, zinc can be titrated at pH 10 or 11 without interference from aluminum because at these high values the conditional constant for zinc is still large, whereas that for aluminum is sufficiently small that essentially no complex is formed.

In this experiment a solution containing aluminum, iron, zinc, and magnesium is analyzed for aluminum. First, the iron is removed by solvent extraction from aqueous hydrochloric acid into methyl isobutyl ketone. Next, the remaining mixture of aluminum, zinc, and magnesium is treated with EDTA in excess of the amount required to complex the aluminum and zinc present. The pH is adjusted to 5 with hexamethylenetetraamine, at which value the magnesium complex of EDTA does not form appreciably. The excess EDTA is then titrated with zinc chloride solution. At this point all the EDTA is present as a complex with either zinc or aluminum. An excess of fluoride ion is added, and the solution boiled a minute or two. Since fluoride forms a stable complex with aluminum (β_4 for $AlF_4^- = 10^{19}$) but not with zinc, it will displace EDTA from the coordination sphere of aluminum. The amount of EDTA released is equivalent to the amount of aluminum in the sample. This EDTA is titrated with standard zinc solution; the number of moles of zinc required in the titration is equivalent to the number of moles of aluminum in the sample. The use of fluoride to displace aluminum in this way is an example of a type of *masking* reaction (Section 7-6).

Hexamethylenetetraamine[14] is chosen to adjust the pH for two reasons. (1) It is a sufficiently weak base that the presence of a local excess during its addition will not cause hydrolysis of aluminum. Hydrolysis occurring when a stronger base such as ammonia is used results in formation of polynuclear complexes containing Al—O—Al bonds that are slow to revert back to mononuclear forms even in strongly acidic solution. The presence of these species blocks complexation of aluminum by EDTA. Consequently, a strong base should not be used to adjust the pH of an acidic solution containing aluminum prior to EDTA titration. (2) Hexamethylenetetraamine, with a pK_b of 8.9 (pK_a of $BH^+ = 5.1$), provides effective buffering at pH 5, the optimum value for formation of the aluminum–EDTA complex. The indicator selected for this titration, xylenol orange, is described in Section 7-5.

Procedure (time 5 to 9 h)

Preliminary Operations

Preparation of 0.025 M ZnCl$_2$. Weigh, to the nearest 0.1 mg, approximately 0.82 g of pure granulated zinc metal into a 100-mL beaker. Slowly add 5 mL of 6 *M* HCl, covering the beaker with a watch glass to avoid loss of solution. When dissolution is complete, rinse the liquid on the watch glass into the beaker with 10 to 15 mL of water, transfer the contents of the beaker quantitatively to a 500-mL volumetric flask, fill to the mark, and mix.[15]

Preparation of 0.05 M EDTA. Dissolve about 9.3 g of the disodium salt of EDTA, $Na_2H_2Y \cdot 2H_2O$, in 500 mL of water. This solution does not need to be standardized.

---

[14]Hexamethylenetetraamine is an unusual polycyclic cage molecule containing four nitrogen atoms, each connected to the other three by a methylene —CH$_2$— bridge. The compound is inexpensive, being readily produced by heating a mixture of formaldehyde and ammonia. It has been exploited for a variety of purposes, ranging from an intermediate in the production of Bakelite plastics and RDX high explosives to a urinary antiseptic.

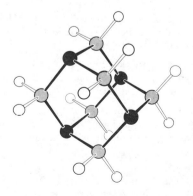

[15]If desired the standard ZnCl$_2$ solution remaining on completion of this experiment can be saved for standardization of the EDTA solution used in the determination of zinc and nickel by ion exchange (Section 19-8). Since the concentration is somewhat lower than that suggested in the zinc-nickel experiment, a 20-mL aliquot may be advised.

Preparation of Sample. Obtain a liquid sample, dilute to volume in a 100-mL volumetric flask, and mix. Pipet 10-mL aliquots of this solution into each of three 25-mm test tubes, and add 12 mL of 12 *M* HCl to each.

Extraction of Iron(III)

Add a few milliliters of 6 *M* HCl to about 150 mL of MIBK in a 250-mL conical flask and shake to saturate the MIBK with the aqueous phase. Allow the phases to separate.[16] When removing MIBK portions for extractions,[17] ensure that no aqueous phase is taken along with the MIBK. Carefully add a magnetic stirring bar designed for stirring in test tubes to one of the sample solutions, and then add about 10 mL of the HCl-saturated MIBK.[18] Place the tube on a magnetic stirrer[11] in a hood, and adjust the motor so that the two phases thoroughly mix for 15 to 20 s. Shut off the stirrer motor and allow the phases to separate.[19] Withdraw the MIBK layer with a disposable pipet, and transfer it to a 25-mm test tube. Repeat the extraction with fresh portions of MIBK until all the yellow iron(III) chloride has been removed from the water layer.[20] Combine all the MIBK extracts from a single aliquot in one tube. After the last extraction, wash the upper surfaces of the tube with a small portion of MIBK and add this portion to the collected MIBK extracts.

Scrub aluminum from the MIBK extracts by adding about 5 mL of 6 *M* HCl and mixing as before. Carefully remove and discard the MIBK layer. Take care that none of the aqueous layer is removed and discarded along with the MIBK, or aluminum will be lost. Wash the HCl fraction with two additional 2- to 3-mL portions of MIBK, combine the aqueous layers from the extraction and scrubbing operations in a 100-mL volumetric flask, dilute to volume, and mix.

Repeat the extraction process on the second 10-mL portion of the original sample, and place it in a second 100-mL volumetric flask. Finally, repeat the process on the third portion to provide a third extracted aliquot.

Minimize exposure to MIBK vapor; discard all used MIBK solutions in the containers provided.

Titration of Aluminum

Pipet 10-mL portions of the extracted solution from one of the volumetric flasks into 200-mL conical flasks. Add approximately 25 mL of 0.05 *M* EDTA

---

[16]HCl has a small but finite solubility in MIBK. This step ensures that aqueous phase from the sample solution is not dissolved in the MIBK. Also, colored impurities in the MIBK may be removed by this pretreatment.

[17] Do not waste this solution. Here, as in all experimental work, take from the reagent stock only the amount specified. Not only are many chemicals expensive, but problems of disposal and pollution are aggravated when excessive amounts are taken, only to be wasted.

[18]At this point one sample should be carried entirely through the extraction operations to provide practice in the technique. High concentrations of MIBK vapors are irritating to mucous membranes; use this solvent with adequate ventilation.

[19]After equilibration has been established in the first extraction, the color of the initially yellow water layer will decrease in intensity, and that of the initially colorless MIBK layer will become yellow or slightly orange. The mixing should be continued until no further change is seen in the colors of either phase. After the second extraction the aqueous layer may turn a faint green or brown due to impurities. This will not affect the extraction and may be ignored.

[20]The MIBK layer for the third extraction should be colorless or nearly so. If not, perform another extraction. A color that persists in the aqueous layer after several extractions may be due to impurities in the MIBK that were not completely removed in the initial saturation step.

solution to each flask and swirl to mix; then add 10 mL of 3 M hexamethylenetetraamine solution, and boil for 1 to 2 min to ensure complete formation of the aluminum–EDTA complex. Remove from heat, add 4 or 5 drops of 0.2% xylenol orange indicator solution, and titrate the hot solution with 0.025 M Zn^{2+} to a definite color change. Record the buret reading. Add 25 mL of a saturated solution of sodium fluoride, swirl to mix, and boil again for 1 to 2 min. Remove from heat, add 2 or 3 more drops of xylenol orange indicator, and immediately titrate again.[21]

Repeat this procedure on sets of aliquots taken from the volumetric flasks containing the second and third extractions. Calculate and report the weight in grams of aluminum in the entire sample.

Calculations

Calculate the weight of aluminum present in each titrated aliquot from the molarity of the standard zinc chloride solution and the volume of zinc solution required for titration from the first to the second xylenol orange   end point for each aliquot. Then calculate the total weight in grams of aluminum in the original sample.

16–5 DETERMINATION OF ALUMINUM BY LIQUID-LIQUID EXTRACTION FROM IRON AND PRECIPITATION FROM HOMOGENEOUS SOLUTION

In this experiment the analysis for aluminum in the same mixture of salts as in the procedure of Section 16-4 is performed; this time aluminum is separated from zinc and magnesium by precipitation from homogeneous solution. The analysis step involves filtration of the precipitated aluminum hydroxide, followed by ignition and weighing as Al_2O_3. The iron is separated from the mixture by liquid-liquid extraction as in Section 16-4. The principles of liquid-liquid extraction of metal complexes and of precipitation from homogeneous solution are discussed in Sections 16-2 and 8-3, and the experimental techniques of gravimetric analysis in Section 8-5.

Procedure (time 7 h)

Preliminary Operations

Obtain a sample, dilute to volume in a 100-mL volumetric flask, and mix. Pipet 10-mL aliquots of this solution into each of three 25-mm test tubes, and add 12 mL of 12 M HCl to each.

Extraction of Iron(III)

Extract the iron(III) from each of the aliquots by the procedure given in Section 16-4, with one modification. When the extractions are complete, combine the aqueous layers from the extraction and scrubbing operations for each aliquot in each of three 250-mL beakers rather than in 100-mL volumetric flasks.

---

[21]The volume of zinc chloride solution required for calculation of the aluminum present is that required to go from the end point of the first titration to that of the second.

Precipitation of Aluminum

Add to the solution in each 250-mL beaker a few grains of $NaHSO_3$[22] and a few drops of thymol blue indicator. Add 6 M NH_3 dropwise with constant stirring until the indicator turns yellow, and then 5 to 10 drops more.[23] Add 3 g of succinic acid, 5 g of NH_4Cl, and 3 g of urea.[24] Stir until most of the solids dissolve (succinic acid dissolves slowly). Remove the stirring rod, cover with a watch glass, and leave overnight on a hot plate set at low heat.

Filtration and Ignition of Aluminum Precipitate

Filter the solutions through medium-porosity paper (Section 8-5 and Figure 8-5). Test for completeness of precipitation by adding a few drops of NH_3 to the clear filtrate. Prepare about 500 mL of 1% succinic acid wash solution made neutral to methyl red with NH_3, and wash the precipitate five to ten times. Finally, wash the filter paper with distilled water until a test for chloride in the filtrate is negative.

A small amount of precipitate will adhere to the beaker walls. Loosen as much of it as possible with a policeman and transfer it to the filter paper; add about 1 mL of dilute HCl to the beaker to dissolve the remainder.[25] Add 10 mL of water, then a drop of methyl red, and neutralize with NH_3. Transfer this precipitate to the filter paper with the 1% succinate wash solution.

Place the aluminum precipitate and filter paper in a crucible ignited previously to constant weight ($\pm$ 0.2 mg) (Section 8-5). Position the crucible at an angle of about 45° (Figure 8-6) and about 2 cm above a cool flame about 1 cm high. Heat gently at first, and then more strongly after the paper has charred. Finally, heat at full burner flame for 20 to 30 min. Cool for 30 min, weigh, and reheat to constant weight.[26]

Report the weight in grams of aluminum in the entire sample.

---

[22]Sodium hydrogen sulfite, $NaHSO_3$, is added to reduce oxidizing impurities in the system, which may otherwise destroy the thymol blue indicator.

[23]The pH at the point of indicator change is about 2.5 to 3.0; the additional 5 to 10 drops of NH_3 raises it to 6 or 7. If insufficient NH_3 is added here, the amount generated by urea hydrolysis will not raise the pH enough to precipitate all the aluminum. If a precipitate appears, it will usually dissolve when succinic acid is added in the next step. If dissolution is not complete after the addition of succinic acid and mixing, add 6 M HCl dropwise with stirring until no solid remains, and then continue.

[24]The amount of urea added should be sufficient to raise the pH upon hydrolysis to about 5. If no precipitate appears after heating, reneutralize with NH_3, add another 5 g of urea, and heat again for 24 h.

[25]Although most of the precipitate will be granular and coarse, a small amount will form a thin, adherent film on the beaker walls that is difficult to remove with a policeman. This can be dissolved with a little hydrochloric acid and reprecipitated with ammonia. Since the amount of precipitate is small and interfering ions are no longer present, direct reprecipitation is acceptable.

[26]Use a small flame during the early stages of ignition; do not rush the process. If the precipitate and filter paper are ignited too rapidly at a high temperature, some elemental carbon may form that is hard to remove. Also, portions of charred paper with adherent particles of precipitate may be carried out of the crucible. The final precipitate should be white. *Caution!* Red-hot crucibles are apt to break if handled with cold tongs.

PROBLEMS

16-1.   What are the four main separation processes of analytical significance?

16-2.   Give examples of single-stage and multistage separations.

16-3.   Define and explain what is meant by (a) separation factor; (b) recovery factor.

16-4.   Define distribution ratio. Why is it a more practical quantity than the distribution constant to use in the description of separation by liquid-liquid extraction? What is the relation between distribution ratio and distribution constant?

16-5.   Assume that the neutral conjugate-base species of hydroxylamine is soluble in both water and chloroform and that the conjugate acid is soluble only in water. (a) Define the distribution constant for the neutral conjugate-base species between chloroform and an aqueous phase. (b) Define the distribution ratio for hydroxylamine between chloroform and an aqueous phase. (c) Sketch the distribution ratio for hydroxylamine as a function of pH of the aqueous phase.

16-6.   Why does the distribution ratio of iron(III) chloride between MIBK and water go through a maximum that depends on the hydrochloric acid concentration?

16-7.   List several ways in which selectivity can be achieved in a separation.

16-8.   (a) What solute properties favor extraction from water into an organic solvent? (b) List four ways in which the affinity of a species for a nonaqueous solvent may be increased so that appreciable extraction from water is made possible.

16-9.†   Since carboxylic acids dissociate in aqueous solution, it is more convenient to treat their extraction quantitatively in terms of distribution ratio rather than distribution constant. Derive an equation that gives the distribution ratio of a carboxylic acid in terms of the distribution constant, the ionization constant, and the hydrogen ion concentration. Assume that the acid does not dissociate to a measurable extent in the organic phase.

16-10.   How does the efficiency of an extraction depend on (a) the volume of the extractant, (b) the volume of the aqueous phase, and (c) the number of extractions?

16-11.   In making a separation, how can a small volume of organic extractant be used most efficiently? Give an example of a procedure for such a separation.

16-12.   The distribution ratio of a compound between toluene and an aqueous phase is 3.51. Calculate the fraction f_w of the compound remaining in the aqueous phase after 100 mL of solution has been extracted with three 50-mL portions of toluene.

16-13.   A chemist wishes to remove an alcohol from an aqueous solution by extraction into diethyl ether so that spectrophotometric measurements can be made at a wavelength at which the alcohol absorbs radiation. With 100 mL of the aqueous solution of the alcohol and 100 mL of ether, and a distribution ratio for the alcohol in this solvent pair of 4.0, (a) what fraction of the alcohol remains in the aqueous phase if a single extraction is performed with 100 mL of ether; (b) what fraction remains if the ether is divided into five 20-mL portions and a multiple extraction is performed?

16-14.   A 50-mL portion of a solution buffered at pH 6 and containing 0.001 mole of zinc(II) was extracted with 100 mL of diethyl ether that was 0.05 M in 8-hydroxyquinoline. After separation of the two phases it was determined by atomic-absorption spectroscopy that 90% of the zinc had been extracted. Calculate the distribution ratio for the zinc–8-hydroxyquinoline complex in this system.

16-15.   The compound benzophenone has a solubility in hexane of 0.50 mg/mL and a solubility in water of 0.10 mg/mL. What is the distribution ratio for benzophenone distributed between 50 mL of hexane and 100 mL of water?

16-16.   The distribution ratio for the extraction of the strontium–8-hydroxyquinoline complex from an aqueous phase of pH 11 into chloroform is 10. If the volumes of the aqueous and chloroform layers are equal, how many extractions are needed to remove 99.99% of the strontium from the aqueous portion?

16-17.   Codeine is distributed between 10.0 mL of diethyl ether and 50.0 mL of 10^{-3} M aqueous sodium hydroxide, giving 13.0 mg of codeine in the ether phase. If the distribution ratio between these two solvents is 45, what weight of codeine remains in the aqueous phase?

16-18.   The deposits of phosphate rock that occur in nature contain a large fraction of the total uranium reserves. In the manufacture of phosphate fertilizers from these deposits, the uranium dissolves in the phosphoric acid produced from the rock. The uranium in the phosphoric acid (about 0.1 g/L) can be extracted and recovered by liquid-liquid extraction with octaphenyl phosphoric acid, H_2PO_3OR, in a kerosene-type solvent. If the distribution ratio is 8.4, what ratio of organic solvent to phosphoric acid must be taken to extract 99% of the uranium from the process acid?

16-19. p-Nitroaniline, $NO_2C_6H_4NH_2$, has a pK_b of 7.2. The distribution constant for the neutral species between chloroform and water is 50. Sketch the relation between the distribution ratio and pH. From the definition of distribution ratio show that after one equilibration the fraction of the neutral compound in the aqueous phase is equal to $(1/D)(V_{org}/V_w) + 1$.

16-20.   The distribution ratio between toluene and water for a compound X, D_x, is 90 and for compound Y, D_y is 0.02. (a) What fraction of X and Y remain in each phase after one equilibration with 50.0 mL of each solvent? (b) If the aqueous phase is extracted with a second 50.0-mL portion of toluene, what fraction of X and of Y remain in the aqueous phase? (c) What fraction remains in the combined toluene extracts after three such extractions?

16-21.   Explain why batch extraction techniques are inefficient for the selective separation of two substances with similar properties.

16-22.   How does countercurrent liquid-liquid extraction differ from batch extraction? Why is it more efficient?

16-23.   How does countercurrent cascade separation differ from countercurrent extraction?

16-24.   What is meant by back extraction? How can it be brought about?

16-25.   Carefully measure the peak separation and the peak base widths in Figure 16-5. Will the separation of the peaks be doubled when the number of equilibrations is doubled? How much will the peak widths have increased?

16-26.   Under a given set of conditions, X and Y have distribution ratios D_X and D_Y of 2.0 and 0.2 between toluene and water. They are to be separated by a countercurrent extraction procedure using 20-mL portions of toluene and water. (a) What fraction of X is present in each of the phases after 1, 2, 3, and 4 equilibrations; what fraction of Y? (b) If the initial ratio of X to Y was unity, what is the ratio after 1, 2, 3, and 4 equilibrations?

16-27.   (a) Calculate the fraction of a solute that would be extracted from 100 mL of an aqueous phase into 50 mL of an immiscible organic solvent if the distribution ratio of the solute is 5. (b) Calculate the fraction that would remain in the aqueous phase after a total of five extractions with fresh 50-mL portions of the organic solvent.

16-28.   A 0.1548-g sample of an alloy was dissolved and analyzed for nickel by the extraction procedure of Section 16-3. The weight of nickel taken for the standard was 0.2407 g, and 1.02, 2.07, and 3.46 mL of standard solution gave absorbance readings of 0.140, 0.288, and 0.476. If the absorbance of the sample was 0.350, what was the percentage of nickel in the alloy?

16-29. In the determination of nickel in Section 16-3 what are the advantages of a calibration curve over comparison with a single standard? Which method would be the choice for samples covering a range of concentrations? Which would be the choice for quality control of a single product of unvarying composition?

16-30. In the determination of a trace of nickel by extraction of a DMG complex, what is the purpose of adding sodium thiosulfate; hydroxylamine hydrochloride; sodium tartrate?

16-31. In the determination of nickel by extraction of its DMG complex followed by spectrophotometric measurement (Section 16-3), what effect would each of the following have on the result? (a) Some of the chloroform is spilled while transferring it to the extraction tube. (b) Some is spilled while transferring it into the measurement cuvette. (c) The pH of the aqueous phase is 1 instead of 6.5 during the extraction. (d) The filter paper is moistened before filtering the extract.

16-32. In a sample originally containing aluminum, zinc, iron, and magnesium, the aluminum is determined by EDTA titration according to the procedure of Section 16-4. What is the purpose of the addition of sodium fluoride; MIBK; hexamethylenetetraamine?

16-33. In the liquid-liquid extraction of iron(III) chloride in Section 16-4, why is the aqueous solution held at 6 M, in hydrochloric acid?

16-34. In the determination of aluminum by precipitation from homogeneous solution, explain the role of urea, including chemical equations as necessary.

16-35. (a) For a distribution ratio of 30, how many extractions are needed to remove at least 99.9% of the iron(III) from a 10-mL aqueous aliquot into MIBK by the procedure of Section 16-4? Assume that 10 mL of MIBK is used for each extraction. (b) How many extractions would be necessary if 3 mL of MIBK were used per extraction?

16-36. Write the chemical equations for each of the steps in the determination of aluminum by the procedure of Section 16-4.

16-37. A sample of aluminum analyzed by the titration procedure of Section 16-4 required 9.73 mL of zinc nitrate solution to titrate the EDTA released from aluminum through the addition of fluoride. If the molarity of the zinc solution was 0.0500 M, what was the total weight of aluminum in the initial sample?

16-38. A 10.014-mL aliquot taken from a 100-mL sample containing an aluminum salt yielded 0.3412 g of aluminum oxide (from the procedure of Section 16-5). What was the total weight of aluminum in the sample?

16-39. Calculate the pH at which copper(II) hydroxide would just begin to form (a) from a 0.1 M solution of copper(II) nitrate and (b) from a 0.001 M solution.

16-40. A rock sample weighing 1.8320 g gave a precipitate of iron(III) oxide and aluminum oxide weighing 0.5420 g. The mixed oxide was dissolved, the iron(III) removed by extraction with MIBK, and the aluminum titrated with 24.82 mL of 0.02012 M EDTA. What was the percentage of iron in the sample?

16-41. A sample of 2.1460 g containing salts of magnesium, aluminum, iron, and zinc was dissolved, diluted to volume in a 100-mL volumetric flask, and analyzed for aluminum by the procedure of Section 16-4. A 10-mL aliquot of the solution was transferred to a 25-mm test tube; 12 mL of 12 M HCl was added followed by 10 mL of MIBK. The two phases were equilibrated, allowed to settle, and then the MIBK phase was removed. The yellow of the aqueous layer indicated that some iron still remained in that phase. The extraction procedure was repeated twice, after which the aqueous phase was water-white. The MIBK extracts were combined and scrubbed with 2 mL of 6 M HCl. That 2-mL portion was then combined with the other aqueous phase and diluted to volume in a 100-mL volumetric flask. A 10-mL aliquot of this solution was transferred to a 200-mL conical flask, along with 25.00 mL of 0.0500 M EDTA and 10 mL of 3 M hexamethylenetetraamine, and the solution heated to boiling. Indicator was added, and the solution titrated to the xylenol orange end point with 0.02500 M zinc solution, 8.71 mL being required. A 25-mL portion of sodium fluoride solution was then added, and the solution again boiled and titrated to the xylenol orange end point with the zinc solution, 10.10 mL being required this time. What was the total weight of aluminum in the sample?

REFERENCES

A. K. De, S. M. Khopkar, and R. A. Chalmers, *Solvent Extraction of Metals*, Van Nostrand-Reinhold, London, 1970. A survey of liquid-liquid extraction methods for application to the determination of metals. Contains many references to the original literature.

L. Gordon, M. L. Salutsky, and H. H. Willard, *Precipitation from Homogeneous Solutions*, Wiley, New York, 1959.

W. F. Hillebrand, G. E. F. Lundell, H. A. Bright, and J. I. Hoffman, *Applied Inorganic Analysis*, 2nd ed., Wiley, New York, 1953. Contains detailed procedures for gravimetric analyses of naturally occurring materials.

I. M. Kolthoff, E. B. Sandell, E. J. Meehan, and S. Bruckenstein, *Quantitative Inorganic Analysis*, 4th ed., MacMillan, New York, 1969, Chapter 12. Gives a description of organic precipitants.

J. Kragten, *Analyst* **1974**, *99*, 43. Describes the use of hexamethylenetetraamine for pH adjustment prior to determination of aluminum by addition of excess EDTA and back titration.

H. A. Laitinen and W. E. Harris, *Chemical Analysis*, 2nd ed., Wiley, New York, 1974, Chapters 22 and 23. Gives more complete and rigorous background on liquid-liquid extraction.

G. H. Morrison and H. Frieser, *Solvent Extraction in Analytical Chemistry*, Wiley, New York, 1957. A valuable introduction to the use of liquid-liquid extraction in separations.

W. Nielsch, *Z. Anal. Chem.* **1956**, *150*, 114. A study of the determination of small amounts of nickel in copper salts by the method of Section 16-3. (In German.)

R. Pribil and V. Veseley, *Talanta* **1962**, *9*, 23. A study of the determination of aluminum by complexation. Includes a critical investigation of the use of fluoride to determine aluminum by displacement of a chelating agent.

J. Stary, *The Solvent Extraction of Metal Chelates*, MacMillan, New York, 1964. Gives both theory and practice of extraction. Many extraction curves are shown.

H. F. Walton, "Separations by Distillation and Evaporation" in *Standard Methods of Chemical Analysis*, F. J. Welcher, Ed., 6th ed., Vol. 2A, 1963.

H. H. Willard and N. K. Tang, *J. Amer. Chem. Soc.* **1937**, *59*, 1190; *Anal. Chem.* **1937**, *9*, 357. Early papers on the technique of precipitation from homogeneous solution.

17

INTRODUCTION TO CHROMATOGRAPHY

True knowledge can only be acquired piecemeal, by the patient interrogation of nature.

Sir Edmund Whittaker

17-1 BASIC CONCEPTS

About four million natural or synthetic compounds are currently known. Of these about 70 000, mostly organic, are produced commercially. Although these hundreds of thousands of compounds exhibit a countless variety of properties, in such numbers many of them must necessarily closely resemble each other. *Chromatography* is a technique for carrying out separations in which a mobile phase passes over an immobilized liquid or solid. Such separations have been carried out for many decades; the earliest were forerunners of modern paper chromatography.[1] Chromatographic techniques have proven to be the most effective means of coping with the exceedingly difficult separations required for the analysis of complex materials such as those encountered in biological systems.

In Chapter 16, separations are described as usually requiring two phases. The phases are mechanically separated, and the one containing the material of interest is usually retained. The theory of separation processes, therefore, relates primarily to heterogeneous equilibria and the rates at which they are attained. Chromatography is the outstanding *physical* technique for carrying out separations; these separations do not rely on chemical changes such as precipitation, complexation, or oxidation-reduction. In the chromatographic process the two phases are a *mobile phase*, which may be gas or liquid, and a *stationary phase*, which may be liquid or solid. The four possible combinations of phases lead to the four main types of chromatography — gas-liquid, gas-solid, liquid-liquid, and liquid-solid, where the first-named

---

[1]Observations extend possibly back to Pliny 2000 years ago. At the turn of the century D. T. Day used adsorption chromatography; shortly afterward M. Tswett, investigating plant pigments, gave chromatography its name. Developments then stagnated for about 30 years. During World War II ion-exchange techniques were extensively developed. Enormous impetus was given to the field by the development of gas chromatography in 1952 and more recently by high-pressure liquid chromatography.

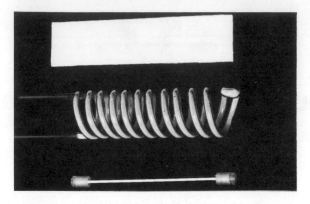

Figure 17-1. Examples of chromatographic beds: thin-layer, gas-liquid, and high-pressure liquid. See Figure 18-1 for another type.

phase is mobile and the second-named is stationary. The stationary phase is in the form of a bed. If a liquid, it is dispersed on an inert support consisting of a porous packing material of large surface area or the inner walls of a tube or capillary; if a solid, it is simply an inert material of large surface area. The mobile phase is caused to filter or percolate through the bed of stationary phase. For a gaseous mobile phase it must be enclosed in a column; for a liquid mobile phase the bed can be either a thin layer on a support or enclosed in a column. Three examples of chromatographic columns are shown in Figure 17-1.

As an introduction to the chromatographic process recall from Section 16-2 that the distribution of a substance between two immiscible phases is represented by the distribution ratio D, the ratio of the concentrations of the substance in the two phases. If a single substance is added to one phase s in contact with a second m and the system allowed to come to equilibrium, the amount of the substance in the first phase is given by $C_s V_s$, where C is concentration and V is volume of phase s. Similarly, the amount in the second phase is $C_m V_m$, and the ratio of the amounts in the two phases is

$$k = \frac{C_s V_s}{C_m V_m} = D \frac{V_s}{V_m} = \frac{\text{amt s}}{\text{amt m}} = \frac{n_s}{n_m}$$

where the ratio[2] k is the *partition ratio, mass distribution ratio*, or *capacity factor*, and n is the number of moles. The fraction of of the total solute in phase s is $k/(1+k)$, and the fraction in phase m is $1/(1 + k)$.

Now consider a column packed with particles of a stationary phase s (liquid or solid) through which a mobile phase m (gas or liquid) can percolate. Suppose that a small amount of a substance is placed on the leading end of the column and the mobile phase is caused to move along the column at a rate sufficiently slow that solute equilibrium is established between the two phases

---

[2]In liquid-liquid extraction the distribution ratio is the ratio of the concentrations involved. In chromatography concentrations are seldom measured or measurable. Therefore the partition ratio is defined in terms of the total amounts in the stationary and mobile phases.

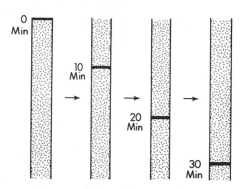

Figure 17-2. Idealized diagram illustrating location of a substance on a chromatographic column at various times. Velocity of mobile phase 6 cm/min; column length 40 cm; partition ratio 4.

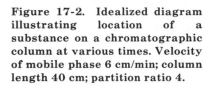

at all times. For the moment consider only ideal circumstances, in which all processes are ignored that can give rise to band broadening, such as diffusion or irregularities of flow rate across the bed. The substance then moves along the column at a rate determined by its partition ratio, since it can move only when in the mobile phase. The rate of travel of solute is given by the product of the rate of travel of mobile phase and the fraction of the solute in the mobile phase.

EXAMPLE 17-1

A small amount of a substance is placed on a column 40 cm long, and mobile phase passed through the column at a rate of 6.0 cm/min. On the assumption of ideal behavior, what is the location of the solute on the column after 0, 10, 20, and 30 min if the partition ratio is 4.0? How long would be required for the solute to appear at the outlet of the column?

The fraction in the mobile phase is $1/(1+4) = 0.2$. The rate of travel of the solute along the column is then $6 \times 0.2 = 1.2$ cm/min. At the beginning (0 min) the solute would be at the top of the column; after 10 min it would have travelled $1.2 \times 10 = 12$ cm; after 20 min, 24 cm; and after 30 min, 36 cm. The solute would be eluted from the column in $40/1.2 = 33$ min.

Figure 17-2 schematically illustrates this example.

The *retention time* is the time required for the maximum concentration of solute to appear at the end of the column. (In Example 17-1 the retention time is 33 min.) It can be readily obtained experimentally by monitoring the amount of solute appearing in the mobile phase as it is eluted from the column. The *retention volume* V_R is the volume[3] of mobile phase required to elute a sample component to its maximum concentration.

If a solute is neither dissolved in nor absorbed by the stationary phase, the partition ratio is zero, and the fraction in the mobile phase is then unity. The solute under these conditions travels along the column at the same rate as the mobile phase. The retention volume in this case is equal to the volume of mobile phase present in the column at any given time and is called the *dead-space volume, void volume,* or *hold-up volume* V_m. A *chromatogram* is a plot of concentration of solute in the mobile phase emerging from the column (the eluate) against the volume of eluate (Figure 17-3). The relation between

---

[3]If the flow rate in Example 17-1 were 2 mL/min, the retention volume would be 33.3×2, or 66.6 mL.

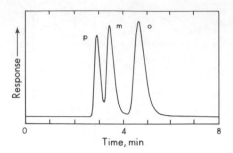

Figure 17-3. A typical chromatogram showing concentration of solutes in the eluate as a function of volume of eluate. *o-*, *m-*, and *p*-Nitroaniline were separated by high-pressure liquid chromatography using a 20-cm column of Amberlite XAD-2 with 90% methanol–10% water as the mobile phase. Chromatogram by F. Cantwell and D. Brown.

the retention volume V_R and the partition ratio k is

$$V_R = V_m + kV_m = V_m(1 + k)$$

$$k = (V - V_m)/V_m \qquad\qquad (17\text{-}1)$$

With the retention volume and the dead-space volume known, this equation permits calculation of the partition ratio for a substance under particular column conditions. The value of the retention volume varies according to the partition ratio of the solute; V_m is, of course, independent of solute and depends only on the column design. If a mixture of two substances is placed at the inlet of a column and mobile phase is passed through it, the components will move through the column independently at rates controlled by their partition ratios and the linear velocity of the mobile phase. Separation occurs if the partition ratios are sufficiently different, and the components of the mixture can then be eluted one after another from the column.

The extent of separation relates directly to the extent of the differences in the partition ratios. Ideally, the extent to which two components move apart from one another increases in proportion to the distance traversed, whereas the extent of spreading increases only in proportion to the square root of the distance. Thus they separate more rapidly than they spread. Recall Figure 16-5, which shows that separation is more complete with a larger number of equilibration stages. The same principle operates in chromatography.

When the amount of stationary phase is known or measurable, the *specific retention volume* V_g, the net retention volume per gram of stationary phase, can be calculated as $V_g = V_R'/g$.

For two solutes a quantity called the *relative retention* α is

$$\alpha = \frac{V_2 - V_m}{V_1 - V_m} = \frac{V_2'}{V_1'} = \frac{k_2}{k_1} \qquad\qquad (17\text{-}2)$$

where V_2 and V_1 refer to the retention volumes of the two solutes (subscript R omitted for simplicity).

In an idealized description, several factors are ignored that are present in real systems. They include eddy diffusion, molecular diffusion, and resistance to mass transfer. All these factors result in spreading of the solute along the

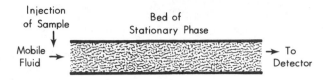

Figure 17-4. Schematic diagram of the chromatographic process.

axis of the bed (Section 17-3).

The chromatographic process is illustrated schematically in Figure 17-4. A small volume of sample is introduced or injected into the mobile phase at the leading end of the column, and each component of the sample is carried along the column, largely independently of the others, toward the outlet. This is, in essence, *elution*.[4]   Usually the material eluted from the column is sent to a detection or measuring system; in other cases the separated materials are examined while still on the bed or in the column.

The chromatographic process is similar in many respects to the process of countercurrent liquid-liquid extraction (Section 16-2). The various components in a sample move through a series of separation stages essentially independently of each other at a rate depending on their individual distribution ratios. With time they become distributed among more and more stages, that is, there is increased spreading as they are transferred through the series. Two main aspects of the process are of practical significance: (1) the average *rate of movement* of sample components along a column and (2) the extent of *spreading* of the components while traversing the column.

The average rate of movement of a component depends directly on the flow rate of the mobile phase and on the distribution equilibrium of the component between the moving and stationary phases, that is, on the partition ratio k, which determines the fraction f_m in the mobile phase. A particular molecule can move significantly toward the outlet of a column only during the times it is in the mobile phase. It is repeatedly transferred back and forth between the two phases; the average rate of movement depends on the relative amounts of time in the two phases. In the limit, with a k of zero, a component stays entirely in the mobile phase, and its average rate of movement is the same as that of the mobile phase. At the other extreme, components with extremely large k values are strongly retained by the stationary phase and move only slowly from the inlet end of the column.

The extent of spreading of a component can be observed either from the point of view of distribution within a column, as in thin-layer chromatography, or elution from a column, as in gas or high-pressure liquid chromatography. The distribution of a solute within a column is commonly referred to as a *band*, or *zone*, of solute; its location is described in terms of *band position* and *bandwidth*. Thus a band may be defined as a record of the concentration of a substance as a function of distance on a column.

More commonly, elution from a column is considered, and here the

---

[4]Other chromatographic processes include *frontal* and *displacement* techniques. Although of value industrially, they are of little analytical significance and are not considered here.

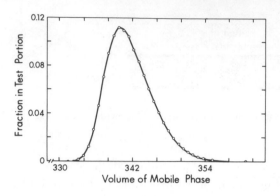

Figure 17-5. Chromatogram of the elution of sodium ions from a cation-exchange resin (Dowex 50) with 0.7 M HCl. From R. H. Betts, W. E. Harris, and M. D. Stevenson, *Can. J. Chem.* 1956, *34*, 65. With permission.

observed distribution is referred to as a *peak*, which may be defined as a record of concentration of a substance in the eluate as a function of its volume. As solutes move along a column bands quickly assume nearly Gaussian shapes, as in Figure 16-5; the eluted peaks are therefore also Gaussian. Asymmetric peaks are usually the result of a small number of separation stages, or of two or more mechanisms operating to retain substances. These may result in either *leading*, or more commonly *tailing*, of peaks. Tailing is illustrated in Figure 17-5.

For chromatographic bands, such as in thin-layer chromatography, the retention parameter of importance is the R_f value, which is the ratio of the distance from the band maximum to the distance from the solvent front, both measured from the origin.

17-2 BAND BROADENING

For most effective separation, samples must be injected onto the inlet of a column as a sharp band. As the mobile phase carries the sample components through the column, the band for each substance progressively spreads, or broadens. Band broadening results mainly from several independent nonequilibrium effects. Modern theories of band broadening were first developed from a systematic study of band-broadening effects in gas chromatography and later extended to other types of chromatography.

The separation of solutes cannot succeed unless they have different partition ratios. The bands for the individual solutes will then separate from each other more rapidly than they broaden (Figure 16-5). In some cases sample spreading may occur on injection or in the detection system; in efficient systems and with careful technique these effects should be insignificant. Total broadening can be calculated statistically by adding the effects of the several independent contributions through summing variances. *Variance* is the square of the standard deviation; here the standard deviation is that of a Gaussian curve.[5] Individual variances from several sources can be added to obtain the overall variance. For n independent contributions to band

---

[5]For a Gaussian peak the base width is defined as 4σ, and the standard deviation is measured as one half the peak width at 0.607 of the maximum height.

Figure 17-6. A Gaussian chromatographic peak showing retention volume, peak width, dead-space volume, and standard deviation.

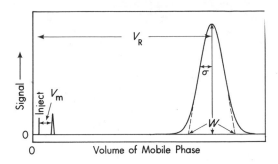

broadening the overall variance σ^2 is

$$\sigma^2 = \sigma_1{}^2 + \sigma_2{}^2 + \sigma_3{}^2 + \cdots + \sigma_n{}^2$$

EXAMPLE 17-2

The standard deviations of three independent sources of band broadening in a chromatographic process have values of 0.14, 0.071, and 0.014 mL. What is the overall standard deviation?

$\sigma^2 = (0.14)^2 + (0.071)^2 + (0.014)^2 = 0.0248$, and σ is the square root of 0.0248, or 0.16 mL.

The most important factor, then, is the one contributing the term 0.14 mL; it contributes $(0.14)^2/0.0248 = 0.79$ of the total. The second factor contributes $(0.071)^2/0.0248 = 0.20$, and the third, 0.008. Thus the third factor can be neglected; even though its standard deviation is 10% that of the largest factor, its contribution to the overall standard deviation is less than 1%.

To express the overall broadening that occurs in a chromatographic column, the most important unifying concept is that of the theoretical plate — a concept borrowed from distillation theory. The *number of theoretical plates n* in a column is defined by

$$n = 16 \, (V_R/W)^2 = L/h$$

where V_R is retention volume, W is peak width (measured as in Figure 17-6), L is column length, and h the *height equivalent to a theoretical plate*. A *theoretical plate* may be regarded as the length of column that produces as much change in composition as would a single stage in a distillation column or a single equilibration in countercurrent extraction. Recognize that a chromatographic column is not separated into distinct stages, and a theoretical plate does not correspond to a discrete unit — the concept is simply a mathematical convenience. The quantities used to calculate the number of theoretical plates n in a column that corresponds to a particular Gaussian chromatographic peak are shown in Figure 17-6. The plate height h is equivalent to σ^2/L, the variance per unit length.

EXAMPLE 17-3

Measure V_R, V_m and W on Figure 17-6. Assuming a column length of 10 cm, calculate the number of theoretical plates n, the plate height h, and the partition

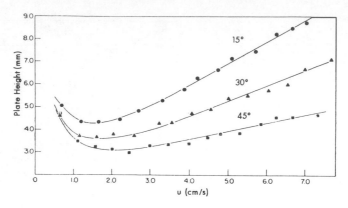

Figure 17-7. Effects of temperature and flow velocity on plate height in gas chromatography for a 1-m column containing 39% Carbowax 400 using samples of 2,3-dimethylbutane. From J. J. Duffield and L. B. Rogers, *Anal. Chem.* **1960**, *32*, 340.

ratio k.

The volumes, all given in arbitrary units of chart length, are 4.73, 0.37, and 1.38.

$$n = 16(4.73/1.38)^2 = 188$$

$$h = 10/188 = 0.053 \text{ cm}$$

$$k = (4.73 - 0.37)/0.37 = 11.8$$

Other conditions being equal, the smaller the plate height, the less the spreading of the solute on a column, and the more effectively can two solutes of similar partition ratios be separated.

The number of theoretical plates and the plate height for a column have been found experimentally to vary with the conditions under which the column is operated. The temperature and the flow velocity of the mobile phase are variables of particular significance. Figure 17-7 shows typical minima in plate height–velocity curves. Factors that contribute in varying ways to overall plate height are examined in the following subsections.

Effects of Flow Pattern (Eddy Diffusion)

The flow pattern affects the plate height because, in a column filled with a packing, not all solute molecules travel the same path from column inlet to outlet. Part of the effect is schematically shown in Figure 17-8. This diagram illustrates that solute molecules in different flow channels will not all travel the same distance in traversing a segment of column. Some will lag behind and others will be carried forward, spreading the solute in the direction of flow as it moves through the column. Spreading also results from the differences in flow velocities between and within flow channels. Individual solute molecules can diffuse between channels and move at varying rates. A single molecule is random in its motion — sometimes virtually stopped in its

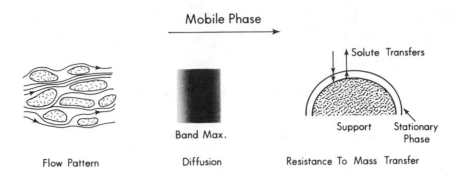

Figure 17-8. Schematic illustration of effects that contribute to band broadening.

forward movement, at other times moving at the mobile-phase velocity, and at still other times moving even more rapidly than the mobile phase. The plate-height contribution to the overall band spreading from flow-pattern effects depends on the uniformity of the packing particles, the dimensions of the particles, channels, and column, the flow-rate differentials among channels, and the diffusion coefficient of the solute in the mobile phase D_m. As would be expected, there is no dependence on partition ratio. In early chromatographic literature the flow-pattern effect was expressed as a small factor independent of flow rate u; more extended and careful study, however, has revealed a certain amount of dependence on flow rate.[6] At low flow velocities diffusional processes largely overcome the differences arising from the various flow-pattern effects; at high velocities diffusion plays a minor role, and the factor eventually approaches a constant value. For simplicity we consider it a constant.

Molecular Diffusion

At a given velocity of flow of mobile phase, solute molecules on the average spend the same amount of time being carried by the mobile phase from column inlet to outlet. Depending on the partition ratio, they also spend more or less of their time in the stationary phase. At the band maximum the concentration of solute is higher than at other locations in the column. During the period required for separation and elution there is opportunity for diffusion in both the forward and reverse directions. Since net diffusion is from regions of high to regions of low concentration, material moves away from the band maximum (Figure 17-6). The extent of this movement increases with the time spent in the mobile phase, which means high flow velocities are desirable. High diffusion coefficients increase the extent of

---

[6]The flow-pattern contribution h_e to plate height is now given by an expression such as
$$h_e = Ku/(D_m + \alpha d_p u)$$
where K and α are constants depending on the packing characteristics, D_m is the diffusion coefficient in the mobile phase, d_p is the average particle diameter of the packing, and u is flow velocity. At low velocities the term becomes nearly proportional to flow velocity, and at high velocities it approaches the constant value $K/\alpha d_p$ as a limit.

movement, and therefore spreading due to molecular diffusion is higher in gaseous than in liquid mobile phases, and it is of much more importance in the mobile than in the stationary phase. Overall, band spreading from molecular-diffusion effects is inversely proportional to flow velocity.

Resistance to Mass Transfer

Consider a substance with a large partition ratio. Upon injection most of the solute will be held by the first portion of the stationary phase at the column inlet — in the first "plate." When mobile phase is passed through the column, it carries forward a tiny fraction of the solute to fresh stationary phase, where it is retained. At the same time the solute–stationary phase mixture at the column inlet is exposed to fresh mobile phase, which takes up another tiny fraction of solute. It too is transferred forward to enter the adjacent portion of stationary phase. By a continuation of such processes virtually all the solute must be transferred back and forth between the stationary and mobile phases numerous times as the band proceeds along the column. If there are several hundred equilibration stages, or plates, then there must be several hundred transfers to and from the stationary phase. Transfers require a finite time, distribution equilibrium not being instantaneous; this requirement is referred to as *resistance to mass transfer*. With continuous flow of the mobile phase, distribution equilibrium will never be totally attained, but it will be approached more closely the lower the flow rate. The rate of the approach to equilibrium depends strongly on diffusion rates and processes in the mobile and stationary phases. Since within a given phase net diffusion is toward a region of lower concentration, if a solute is more concentrated on the surface layer of the stationary phase, it will diffuse into the lower layers. This process will occur as the solute enters the stationary phase, and the reverse process as the solute leaves the stationary phase (Figure 17-8). Because diffusion takes time, the solute lags behind on both entrance into and exit from the stationary phase.

In the mobile phase an analogous phenomenon occurs. If the layer next to the stationary phase is more concentrated in solute than are more remote layers, there will be net diffusion into the less concentrated regions. Here also, the solute needs time to enter and leave the mobile phase; it is carried forward in the first instance and lags behind in the second. Transfer of material between the mobile and stationary phases occurs when there is a difference in activity of a solute[7] in the two phases.

Since the driving force favoring the establishment of equilibrium and the transfer of material from one phase to another is one of diffusion, equilibrium will obviously be more nearly approached when the diffusion coefficients in the two phases are large and the diffusion distances are short. Diffusion distances are minimized when the interfacial area between the mobile and stationary phases is at a maximum. For this reason the stationary phase is

---

[7]Recall that activity is the product of concentration and activity coefficient (Section 3-1). Since the activity coefficients are different in different phases, at equilibrium the concentrations in the two phases are different. The ratio of the two activity coefficients for a particular entity in two phases is equal to the distribution constant (Section 16-2).

spread as a thin layer on a support of large surface area, and the ratio of mobile to stationary phase is as small as possible.

Solutes have large diffusion coefficients in gaseous mobile phases, normally a few tenths of a square centimeter per second.[8] In liquid mobile phases diffusion coefficients are about four or five orders of magnitude smaller, typically 10^{-5} or 10^{-6} cm^2/s. Resistance to mass transfer is accordingly many orders of magnitude higher in liquid than in gaseous mobile phases. Typical liquid or ion-exchange stationary phases have even smaller diffusion coefficients, in the range 10^{-7} to 10^{-10} cm^2/s.

The number of transfers a solute will undergo between the stationary and mobile phases on its passage through a column depends on the partition ratio for the solute. If the partition ratio is large, many such transfers both to and from the mobile phase are required with consequent increased opportunity for band broadening. If the partition ratio is small, fewer transfers are required and the broadening is less. Whether the partition ratio is large or small, the plate-height contribution is proportional to the mobile-phase flow rate u.

Overall Band Spreading

The overall bandwidth of a solute that has traveled some distance along a chromatographic column is the sum of the effects of flow pattern, molecular diffusion, and resistance to mass transfer. The overall general equation (the van Deemter equation) for plate height in its simplest form is

$$h = A + B/u + Cu \qquad (17\text{-}3)$$

where A, B, and C are constants. In this equation the term containing B (diffusion term) is inversely proportional and the term containing C (mass-transfer term) is directly proportional to mobile-phase flow velocity. It follows that the optimum flow velocity for minimum plate height must be some intermediate value (Figure 17-7). Similarly, diffusion coefficients appear in both numerator and denominator through the constants B and C, again leading to an optimum of some intermediate value. In addition to these two factors, the optimum flow velocity depends on the values of the packing parameters and the partition ratio; these factors affect the constants A, B, and C (Example 17-4). It is important to recognize that band broadening does not affect the retention volume because the average retardation is unaffected.

EXAMPLE 17-4

The detailed equation for plate height in gas chromatography is

$$h = \frac{K_u}{D_m + \alpha d_p u} + \frac{2\gamma_m D_m}{u} + \frac{2\gamma_s D_s k}{u} \quad \frac{q\,k}{(1 + k)^2} \frac{d_s^2 u}{D_s} + \frac{\omega_m d_p^2 u}{D_m}$$

---

[8]Some examples of diffusion coefficients at 25°C and 1 atm are helium in hydrogen. 1.69 cm^2/s; methane in hydrogen, 0.732 cm^2/s; benzene in hydrogen, 0.371 cm^2/s; carbon tetrachloride in hydrogen, 0.349 cm^2/s; and benzene in helium, 0.384 cm^2/s.

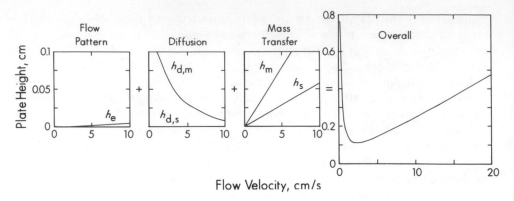

Figure 17-9. Variation in overall plate height and the five terms contributing to plate height as a function of flow velocity for a particular solute and column in gas chromatography.

where K and α are constants that depend on the packing characteristics, D_m is diffusion coefficient in the mobile phase, d_p is average packing-particle diameter, γ_m and γ_s are obstructive factors that depend on the packing, D_s is diffusion coefficient in the stationary phase, k is partition ratio, d_s is thickness of the film of stationary phase, q is a configuration factor, and ω_m is a mass-transfer broadening factor. Calculate the relation between mobile-phase velocity and the plate-height contribution for each of the terms in the overall plate-height equation, using the following values for the constants: $K = 2 \times 10^{-4}$ cm^2, $\alpha = 0.2$, $\gamma_m = \gamma_s = 0.5$, $\omega_m = 2$, $d_p = 0.02$ cm, $q = 2/3$, $D_m = 0.3$ cm^2/s, $D_s = 10^{-6}$ cm^2/s, $d_s = 2 \times 10^{-4}$ cm, and $k = 1.0$.

The answer is given in the following table and in Figure 17-9 as a van Deemter plot.

Plate-height Contribution, cm

Velocity cm/s	Flow Pattern	Molecular diffusion Mobile	Molecular diffusion Stationary	Resistance to Mass Transfer Mobile	Resistance to Mass Transfer Stationary	Overall Plate Height
0.2	0.0001	1.50	5×10^{-6}	0.001	0.0005	1.506
0.5	0.0003	0.600	2×10^{-6}	0.003	0.001	0.604
1.0	0.0007	0.300	1×10^{-6}	0.007	0.003	0.311
2.0	0.001	0.150	5×10^{-7}	0.013	0.005	0.169
5.0	0.003	0.060	2×10^{-7}	0.033	0.013	0.109
10.0	0.006	0.030	1×10^{-7}	0.067	0.027	0.130
20.0	0.011	0.015	5×10^{-8}	0.133	0.053	0.212
50.0	0.020	0.006	2×10^{-8}	0.333	0.133	0.492
100.0	0.029	0.003	1×10^{-8}	0.667	0.267	0.966

Suppose calculations like those in the foregoing example were made, but with different values of diffusion distances, diffusion coefficients, and the various constants. The van Deemter curves would be similar, but the flow rate and plate height at the optimum would be different. Then suppose you have the drastically different condition of a liquid instead of a gas mobile phase. Even though the general shape of the curve would be similar, the optimum flow rate would now be expected to be about five orders of magnitude smaller, because diffusion coefficients are about five orders of magnitude smaller. In fact, the optimum velocity is so low that it is virtually never employed in liquid chromatography.

In general, the most effective separations are obtained through the use of long columns containing small particles of uniform size coated with a thin film of stationary phase. Each of these conditions favors large plate number and small plate height.

Other Factors

Variation in the temperature of operation can materially affect the efficiency of a chromatographic separation in both gas and liquid chromatography. Diffusion coefficients are markedly increased by increasing the temperature, and consequently the several terms in the equation for plate height can be affected. In gas chromatography particularly, the partition ratio and consequently the retention volume also can be drastically changed by varying the temperature. In addition, a change in temperature affects the viscosity of the mobile phase, thereby altering flow rates unless independently controlled. No simple overall relation applies between temperature and the quality of a chromatographic separation.

An assumption basic to most chromatographic theory is that each component of the sample behaves independently of the others. The partition ratio is assumed to have the same value at the band maximum as elsewhere, including the head or tail. In reality, interaction between solute molecules may cause the partition ratio to be somewhat smaller than expected, giving more rapid elution and slightly smaller retention volumes.

The activity of the surface of solid stationary phases is likely to vary from point to point. These nonuniformities tend to cause asymmetry in peaks such as tailing.

Noncolumn factors, too, may affect the results of the chromatographic process. These include the methods of placement of the sample on the column and of measurement of the material leaving the column. Ideally the sample should be placed at the inlet of the column with negligible band width — in the limit a sample injection volume of zero. In practice, for negligible band broadening the maximum sample volume should not exceed about $V_R/(2\sqrt{n})$, where V_R is retention volume and n the number of theoretical plates. Whether band broadening for a particular column is negligible, then, depends on the particular solute.

EXAMPLE 17-5

A column has a plate number of about 2000 for two solutes whose retention volumes are 30 and 300 mL. What is the maximum sample injection volume for each solute if band broadening from this source is to be minimal?

For the first solute the maximum sample injection volume is $30/2\sqrt{2000} = 0.34$ mL. For the second solute it is $300/2\sqrt{2000} = 3.35$ mL.

In a sample containing both solutes the smaller volume, 0.34 mL, would be indicated — a more stringent requirement than if the injection volume for the second were the limiting factor. (Note that in gas chromatography the volume refers to that of the vaporized sample.)

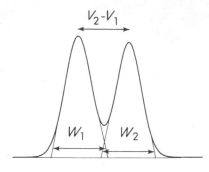

Figure 17-10. Two chromatographic peaks to illustrate the quantities in the calculation of resolution.

The variance arising from the detection system has the same limitation as for injection. The maximum effective volume of the detector should not exceed $V_R/(2\sqrt{n})$ and thus the effects of large detector dead volume are more significant when retention volumes are small.

17–3 THE SEPARATION OF MIXTURES: RESOLUTION

The best measure of the degree to which a chromatographic column achieves a given separation is *resolution R*, defined as the difference in retention volumes of two peaks divided by their average peak width:

$$R = \frac{V_2 - V_1}{0.5(W_2 + W_1)} \tag{17-4}$$

where V_2 and V_1 are retention volumes, and W_2 and W_1 are peak widths. Figure 17-10 illustrates the quantities involved. Better resolution is achieved when two peaks are farther apart and narrower.

For *closely spaced* peaks, resolution can also be expressed by the relation

$$R \approx \frac{\sqrt{n}}{4} \frac{k}{1+k} \frac{(k_2/k_1) - 1}{k_2/k_1} \tag{17-5}$$

where n is the average number of plates for the two peaks and k is the average of the individual partition ratios k_1 and k_2.[9] This equation makes clear that, although resolution improves as the square root of the number of theoretical plates in a column, plate number is not the only parameter of importance. The ratio k_2/k_1 should be large for best resolution. It should be pointed out that the partition-ratio factor $k/(1 + k)$ is sensitive to column dead space, decreasing strongly with an increase in dead-space volume.

EXAMPLE 17–6
Calculate the value of $k/(1 + k)$ for two columns having dead spaces of 1 mL/g and 10 mL/g of stationary phase, and for a solute having a specific retention volume of 1 mL/g of stationary phase. Retention volumes in the two cases are then 2 mL and 11 mL and the partition ratios are 1 and 0.1. From Equation 17-5 the factor $k/(1 + k)$ for the first column is $1/(1 + 1) = 0.5$; for the second column $0.1/1.1 = 0.091$. Thus

---

[9]The quantity k_2/k_1 is equal to V_2'/V_1', the ratio of the net retention volumes for the two substances, and is therefore also equal to the relative retention.

Figure 17-11. The relation between partition ratio and number of theoretical plates required for a resolution of unity, for two solutes having a k_2/k_1 ratio (relative retention) of 1:1. From S. H. Tang and W. E. Harris, *Anal. Chem.* 1973, *45*, 1977.

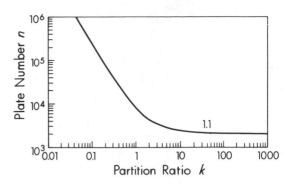

the factor $k/(1 + k)$ (and resolution) is at least five times smaller when the dead space is ten times larger.

In this example, to offset the adverse effect of larger dead space in the second column, the number of theoretical plates would have to be increased by a factor of $(0.05/0.091)^2$, or 30 times. If the first column had 1000 theoretical plates and achieved a satisfactory level of resolution, one with 30 000 plates would be needed to achieve the same level of resolution in the second. Alternatively resolution could be regained by using conditions under which the partition ratio were larger.

For best resolution, therefore, column dead space must be minimized; otherwise, columns with much larger numbers of theoretical plates are required to achieve the same resolution. Figure 17-11 depicts the relation between plate number and partition ratio for a resolution of unity and k_2/k_1 of 1:1. This figure makes clear that for optimum separation the partition ratio should not be too small. At the same time, keep in mind that large partition ratios require large retention volumes and hence long elution times. In practice, the time needed to achieve a given separation is an important consideration. It can be calculated that the best resolution can be achieved in the shortest time if partition ratios are in the range of about 1 to 10.

A solid support only partly covered with the thin film of stationary phase will show variable affinity for solute molecules. The varying thickness of the film on irregular particles also leads to variability within the column. The thickness of the liquid stationary phase is important in terms of sample capacity. For a lightly loaded column a sample of even moderate size will flood the column, and separation will be less efficient. Moreover, enough sample must be injected to be adequate for the operation of the detection system. Sample capacity is therefore another important consideration.

PROBLEMS

17-1.   (a) How does chromatography differ from other physical methods of separation?
(b) Name the four principal types of chromatography. What phases are involved in each type?

17-2.   Define the term partition ratio. How is it related to the distribution ratio D of solvent extraction (Section 16-2)?

17-3.   Why is the partition ratio defined in terms of total amount of solute rather than in terms of concentration?

17-4.   For a solute with a partition ratio of 0.1, what percentage would be in the mobile phase in a chromatographic column?

17-5.   A solute upon equilibration between a stationary and a mobile phase has a partition ratio of 84. If the ratio of the stationary- to the mobile-phase volume is 0.19, what is the distribution ratio?

17–6.†   A solute has a distribution ratio D of 5.00 between a stationary and a mobile phase. The volume of the stationary phase V_s is 2.00 mL and that of the mobile phase V_m is 3.00 mL. If the linear velocity of the mobile phase is 1.00 cm/s, what is the linear velocity of the solute? What will be the retention volume for the solute?

17-7.   What is meant by band, peak, tailing, and chromatogram?

17-8.   Define the terms retention volume, retention time, dead-space volume, net retention volume, specific retention volume, theoretical plate, and number of theoretical plates.

17-9.   What is the relation between retention volume and partition ratio? A solute with a partition ratio of 9.7 is passed through a column containing 2.2 g of stationary phase and a dead-space volume of 2.4 mL. What is the retention volume; the net retention volume; the specific retention volume?

17-10.   Distinguish between a chromatographic peak and a chromatographic band.

17-11.   Explain why the partition ratios of solutes must be different for a successful separation. Need they be highly different?

17-12.   Name the principal dynamic effects in chromatography that lead to band spreading. For each of these effects specify the experimental conditions that reduce their magnitude.

17-13.   What is the van Deemter equation? Explain its significance.

17-14.   In gas chromatography the plate height depends on the flow rate. As the flow rate is decreased from a high value, the plate height decreases, passes through a minimum, and then increases as the flow rate continues to be decreased. What process in the column is responsible for the increase in plate height as the flow rate decreases? Why does this contribution to peak spreading depend on flow rate?

17-15.   Explain what is meant by resistance to mass transfer. What is its significance in chromatography? Under what conditions is it more significant in gas chromatography than in liquid chromatography?

17-16.   What is the significance of flow-pattern effects and molecular-diffusion effects in chromatography?

17-17.   Compare the terms separation factor (Section 16-1) and resolution.

17-18.   Why is molecular diffusion more important in gas chromatography than in liquid chromatography?

17-19.   (a) Measure the retention volumes of cyclohexane and benzene in Figure 18-9. Also measure the dead-space volume. (b) Calculate the partition ratios for these two substances and their relative retention.

17-20.   The standard deviation of four independent sources of band broadening are 1.4, 2.6, 0.7, and 0.2 mm. What is the overall band width expressed in the same units?

17-21.   (a) How does the diffusion coefficient qualitatively affect the magnitude of eddy diffusion, molecular diffusion, and resistance to mass transfer? (b) Should diffusion coefficients be large or small for optimum chromatographic separations? Explain.

17-22. How does the partition ratio qualitatively affect the magnitude of eddy diffusion, molecular diffusion, and resistance to mass transfer?

17-23. How does flow velocity affect plate height? Explain why the optimum flow velocity can be expected to be many orders of magnitude lower in liquid chromatography than in gas chromatography.

17-24. Why should the sample injection volume be made as small as possible? What constitutes negligible sample injection volume?

17-25. If the retention volume of a solute is 44 mL and a column has 1600 theoretical plates, what is the maximum sample injection volume?

17-26. A chromatographic peak has a retention time of 60 s. (a) If the column has 1000 theoretical plates, what will be the width of the band in units of seconds? (b) If the column is 50 cm long, what is the plate height?

17-27. Calculate the number of theoretical plates for the peak in Figure 17-6.

17-28. A gas chromatogram of a tranquilizer drug showed a peak at 3.5 min with a width corresponding to 0.9 min. What was the number of theoretical plates in the 1.5-m column? What was the height equivalent to a theoretical plate?

17-29. The retention volumes of two organic impurities extracted and concentrated from a water sample were 102 and 112 mL. If the column had 310 theoretical plates, what is W for each peak? What is the resolution of the two peaks?

17-30. The effectiveness of a chromatographic separation depends on the location of the peak maxima upon elution (retention volume) and on the width of the peaks. List four factors that affect these two parameters, and indicate the direction and extent of each effect.

17-31. Define resolution.

17-32. A column is used to separate two closely spaced substances under conditions where it exhibits 1800 theoretical plates. The average partition ratio is 3.6 and the relative retention is 1.4. Calculate the resolution of the two substances.

17-33. Other conditions being equal, how is the resolution of two closely spaced peaks affected by (a) doubling the theoretical plate number; (b) changing the average partition ratio from 10 to 5; (c) changing the relative retention from 1.1 to 1.2?

17-34. A mixture of butane and pentane was injected into a gas-chromatographic column with the following results. Air peak maximum appeared after 0.70 min with peak base width equivalent to 0.20 min; butane maximum appeared after 3.5 min with base width equivalent to 0.80 min; pentane maximum appeared after 4.8 min with base width equivalent to 0.90 min. (a) What is the partition ratio for butane on this column? (b) What is the resolution of butane and pentane?

17-35. (a) After making the appropriate measurements on the two bands shown in Figure 16-5, calculate the resolution to be expected if a mixture of these two substances were eluted from a chromatographic column. (b) How much improvement in resolution would be expected for a quadrupling of the number of equilibration stages?

REFERENCES

L. S. Ettre and A. Zlatkis, 75 *Years of Chromatography*, Elsevier, New York, 1979. Review of background events leading to early work in chromatography.

H. A. Laitinen and W. E. Harris, *Chemical Analysis*, 2nd ed., McGraw-Hill, New York, 1974, Chapter 24. Deals with the theoretical foundations of chromatography.

18

GAS CHROMATOGRAPHY

...though Martin and Synge (1941) suggested the use of gas-liquid partition chromatograms for analytical purposes, no work has been reported along these lines. This paper ... describes the application of the gas-liquid partition chromatogram to the separation of volatile fatty acids.

A. T. James and A. J. P. Martin, 1952. from the first paper describing gas chromatography.

Gas chromatography is a technique for carrying out the separation and measurement of mixtures of materials that can be volatilized. It is a separation method in which the mobile phase is a gas; it includes both *gas-liquid chromatography*, in which the stationary phase is a nonvolatile liquid, and *gas-solid chromatography*, in which the stationary phase is a solid of large surface area. The materials examined may be gases, liquids, or solids that have appreciable vapor pressures at temperatures up to a few hundred degrees. Introduced in 1952, gas chromatography was adopted probably more quickly and widely than any other technique in the history of analytical chemistry. The rapid adoption was the result of a combination of circumstances. Previous methods for the analysis of hydrocarbons were inaccurate as well as slow, and gas chromatography could separate large numbers of closely related substances quickly, required but a miniscule sample, and could not only achieve separations but also (in combination with any of several detection devices) furnish quantitative data. Figure 18-1 is a schematic diagram of a gas chromatograph.

18–1 FACTORS AFFECTING SEPARATION

The Carrier Gas

The mobile phase in gas chromatography is the *carrier gas*, an inert material whose sole purpose is to transport sample components from an injection point at the head of the column through the column to a detector. Because the partition ratio is virtually independent of the carrier gas, separation effectiveness will be influenced but little by the choice of gas.

As mobile phases, gases in general have far lower viscosities than liquids.

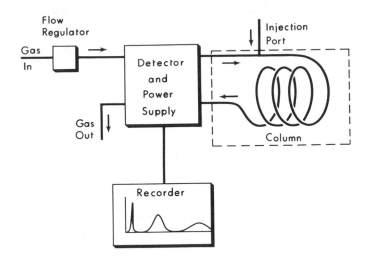

Figure 18-1. Schematic drawing of a gas chromatograph.

Consequently, the pressure drop over the length of the column can be modest for a relatively high flow rate, even with a tightly packed column. Recall from Section 17-2 that diffusion coefficients in gases are typically several orders of magnitude higher than in liquids. A gaseous mobile phase, then, offers far lower resistance to mass transfer than does a liquid, and provides an optimum flow velocity several orders of magnitude higher. Separations are thus possible in shorter times. A further general advantage of gases as mobile phases arises from their lower mass. This means that sample components are associated with less extraneous material, making them easier to monitor and measure in the effluent. The fact that the carrier gases do not react with solutes may sometimes be their principal disadvantage. Since partition ratios cannot be affected by a change in gas, they must be controlled by optimal choice of stationary phase to achieve the best separation.

Helium, though relatively expensive, is the usual choice for carrier gas because it is inert and has higher thermal conductivity than any sample component except hydrogen. With helium as carrier gas in conjunction with a thermal-conductivity detector a small amount of a sample component can yield a large signal. Diffusion coefficients in helium are high, with the results that the effect of resistance to mass transfer is minimized, flow rates can be higher without loss of theoretical plates, and analysis times can be short. *Hydrogen* is more chemically reactive than helium and may explode with air if mishandled. Nevertheless, it has the advantages of high thermal conductivity, high diffusion coefficients for solutes, and most important, a viscosity less than half that of other carrier gases. Flow rates of several milliliters per minute are typical in columns of about 5-mm internal diameter. *Nitrogen,* though essentially inert and inexpensive, has an unfavorable thermal conductivity and yields low diffusion coefficients for solutes.

When the mobile fluid is a liquid, the amount emerging from the column outlet can be stated directly and unambiguously. When it is a gas, however, both temperature and pressure must be specified when reporting the amount of emerging carrier gas. Since a pressure drop across the length of the column is necessary to cause forward movement of the gaseous phase, the pressure at the inlet must be higher than at the outlet. The total volume of mobile phase at various points along the column is then less than that measured at the column outlet. For this reason, in gas chromatography the retention volume must be multiplied by an appropriate correction factor[1] called the *pressure-gradient correction factor.*

Usually, the volume of carrier gas is measured at room temperature, and the column operated at a higher temperature. A temperature correction T_c/T should then be applied, where T_c is the column temperature and T the absolute room temperature, both in Kelvin.

The Stationary Phase

Liquid Stationary Phases

In gas-liquid chromatography the composition of the liquid stationary phase is the principal variable available for improving separations in difficult mixtures. Careful selection of the liquid can drastically affect partition ratios and, more important in difficult separations, the *relative* values of partition ratios as well. The liquid should not react chemically with sample components and should have low vapor pressure at the maximum temperature of use. Much effort has gone into the search for liquid phases that are chemically and thermally stable and have low vapor pressures at high temperatures.[2]

Liquid stationary phases have been classified qualitatively into four groups. (1) *Paraffinic* phases have few or no reactive functional groups and typically are high-molecular-weight hydrocarbons of low volatility (squalane). (2) *Nonpolar* phases (other than paraffinic) have a large fraction of the molecule consisting of nonpolar groups (silicone oil). (3) *Polar* phases have a large fraction of the molecule occupied by polar groups (polyethylene glycol).

---

[1]The pressure-gradient correction factor j is

$$j = \frac{3\,[(P_i/P_o)^2 - 1]}{2\,[(P_i/P_o)^3 - 1]}$$

where P_i and P_o are inlet and outlet pressures. When V_R is multiplied by j the quantity obtained is the same as would be observed if the mobile phase were not compressible. Say the inlet pressure is 964 mm of mercury and the outlet pressure is 742 mm; the value of j is then

$$j = \frac{3[(964/742)^2 - 1]}{2[(964/742)^3 - 1]} = 0.865$$

Similarly, for $P_i/P_o = 1.1, j = 0.95; 1.2, j = 0.91; 1.3, j = 0.86;$ and $1.4, j = 0.83.$ For a value of j of 0.865 and a volume of gas measured at the column outlet to the peak maximum of 47.6 mL, the retention volume would be 47.6 × 0.865, or 41.2 mL.

[2]The approximate maximum temperature of use for several important liquid stationary phases are dinonyl phthalate, 100°C; squalane, 150°C; apiezon grease, 200°C; and silicone gum rubber SE-30, 300 to 350°C. Molten inorganic materials can tolerate even higher temperatures. For example, lithium chloride has been used at about 500°C. The temperatures at which some liquids used as stationary phases have a vapor pressure of 10^{-6} g/mL are as follows: squalane, 160°C; dinonyl phthalate, 145°C; tricresylphosphate, 163°C; glycerol, 101°C. S. J. Hawkes and E. F. Mooney, *Anal. Chem.* **1964**, *36*, 1473.

(4) *Specific* phases interact with certain classes of compounds; for example, stationary phases that incorporate silver nitrate substantially increase the retention volumes of unsaturated compounds through interaction of silver ions with carbon-carbon double bonds.

To achieve optimum separations with a liquid stationary phase there are several important considerations. The first is the manner in which the film of liquid is spread on the solid support. A support only partly covered with stationary phase will show variable affinity for solute molecules. Varying thickness of the film on irregular particles also leads to variability within the column. A second consideration is the viscosity. The higher the viscosity, the larger is the resistance to mass transfer in the stationary phase and the less efficient the separation. A third important experimental variable is the amount of liquid phase loaded onto the solid support. High loadings (20 to 30 g of liquid per 100 g of solid support) lead to thick films of liquid, high resistance to mass transfer in the liquid, and a loss of theoretical plates. On the other hand, low loadings (2 g or less of liquid per 100 g of support) result in excessive column dead space, an unfavorable $k/(1 + k)$ factor, and less resolution than might be expected from the improved plate number. A loading of 5 to 10% is a typical compromise. The loading is also important in terms of sample capacity. A sample of even moderate size will flood a lightly loaded column and so inhibit efficient separation. On the other hand, the size of sample must be adequate for operation of the detection system. A fourth consideration is the inert support for the liquid phase. The support may not be entirely covered with the liquid film, particularly at low loading. Retention of solute may then be the result partly of dissolution in the liquid and partly adsorption by the solid. Sometimes the surface of a support is chemically treated (silanized) to minimize adsorption. Examples of commercial solid supports are Chromosorb W, glass beads, and Fluoropak, which typically have surface areas of the order of a square meter per gram. Uniform size and shape of the support particles is essential for small plate heights; the usual mesh size is in the range of 60 to 80 (about 200 μm).

Solid Stationary Phases

Solid stationary phases such as silica gel, alumina, charcoal, or molecular sieves are ordinarily stable to at least 500°C. Their partition ratios depend on the extent of physical adsorption of the solute molecules. Only a monomolecular layer is expected by this process; therefore large surface areas are necessary. All surfaces have certain sites that tend to adsorb solute molecules more strongly than others, this tendency normally causing tailing of peaks.

The Column

A critical aspect of gas-liquid chromatography is preparation of the column. The solid support must possess a uniform structure of high surface area. The support, along with the stationary phase it carries must be packed carefully into a long tube so that channeling or plugging is minimal.

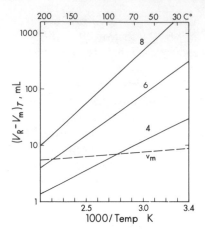

Figure 18-2. Specific retention volume as a function of temperature for butane, hexane, and octane with apiezon L as stationary phase. From H. W. Habgood and W. E. Harris, *Anal. Chem.* 1960, *32*, 450.

Temperature

In gas chromatography temperature is the most important of the variables under the control of the experimentalist; it has a critical effect on the partition ratios and consequently on the retention volumes of solutes. Figure 18-2 shows the relation between net retention volume and temperature for three hydrocarbons. Clearly, retention volume can be altered by orders of magnitude by simply altering the temperature. As a rough guide, an increase of 30°C approximately halves the retention volume. Changes in temperature also markedly affect diffusion coefficients in both gaseous and liquid phases. Gaseous diffusion coefficients increase by about the 1.8 power of the absolute temperature, and those in many liquids often by an even larger factor. Thus with an increase in temperature, band spreading due to molecular diffusion increases, and the effects of resistance to mass transfer are reduced. At the same time the viscosities of gases are increased, so that higher inlet pressures are required to maintain a given flow rate. Hence the pressure-gradient correction factor also is affected.

It should be realized that at any single temperature no flow rate can be selected that will give the maximum number of theoretical plates for every solute in a mixture.

Sample Injection

Sample injection is an arrangement by which a solid, liquid, or gaseous sample is transmitted as a short pulse into the carrier-gas stream before it enters the column. The sample should be vaporized and carried to the leading end of the column in negligible time. It is usually injected through a silicone rubber septum with a microsyringe and needle. The silicone rubber must be soft enough to be readily penetrated by the needle and not leak after the needle is withdrawn. It must also be able to withstand the maximum temperature of column operation without significant deterioration. Figure 18-3 is a photograph of a syringe and two different silicone rubber septa.

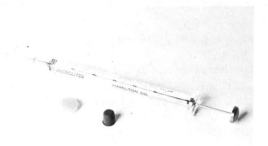

Figure 18-3.   Microsyringe for sample injection in gas chromatography and silicone rubber septa.

18-2   DETECTION AND IDENTIFICATION

Detectors

A *detector* monitors the composition of the carrier-gas stream as it leaves the column. A large number of detectors have been developed for gas chromatography. The most widely used at the present time are the *thermal-conductivity, flame-ionization*, and *electron-capture detectors*. For special cases, rapid-scan mass spectrometers and infrared spectrometers are also used. We discuss here the thermal-conductivity detector in more detail than the others because of its widespread usage.

The thermal-conductivity cell played a key role in the rapid adoption and development of gas chromatography. Simple, sensitive, and stable, this detector has contributed in a major way to the explosive growth of this method of analysis. The device (Figure 18-4) and associated equipment responds to the small amounts and low concentrations of sample components in the stream of carrier gas leaving the column. A significant advantage is that

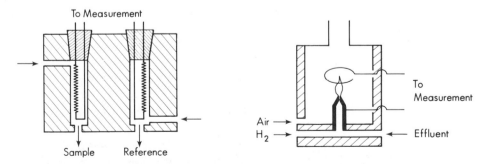

Figure 18-4.   Schematic diagrams of thermal-conductivity and flame-ionization detectors. In the thermal-conductivity cell the carrier gas enters the chamber of the reference side of the detector by diffusion; for more rapid response the gas flows through the chamber on the sample side.

it provides a recorder response proportional to concentration of a substance in the effluent from a column.

In this detector a fixed current is passed through pairs of inert wire filaments having a high temperature coefficient of resistance. One wire is located in the carrier gas alone as it enters the column; the other is located in the carrier gas as it leaves the column, with or without sample components.[3] Gases of differing thermal conductivity cool the two filaments to differing extents. The difference in resistance between the filaments is measured and recorded as a function of time or volume of carrier gas. The thermal effect is at a maximum when the difference in thermal conductivity of the carrier gas and the sample components is large. The high thermal conductivities of hydrogen and helium — 7.1 and 5.5 times that of air, and some 10 times that of a typical sample constituent such as benzene — give them a considerable advantage in terms of detector sensitivity. The carrier gas cools the heated filament strongly on the reference side at all times, but on the sample side to the same extent only when pure carrier gas is emerging from the column. When the carrier gas contains a sample component, the filament is cooled less and the resistance increases. The change in resistance is measured and the signal sent to a recorder. To make the signal-to-noise ratio as high as possible, matched filaments are used, and the difference in their response is measured. High sensitivity is possible; response to samples of a fraction of a milligram is ample provided the retention volume is not too large. The detection limit is about 5 μg per milliliter of sample gas.

The thermal-conductivity detector is nonspecific in that it responds to any solute that alters the thermal conductivity; its response to a wide variety of solutes is approximately but not exactly the same. In quantitative measurements response factors must therefore be determined and applied.

Several types of highly sensitive ionization detectors have been developed. The flame-ionization detector (Figure 18-4) is a widely used example of this type. If the carrier gas is hydrogen or mixed with hydrogen and then burned in air or oxygen, few ions are produced and the conductivity of the burner gas is low. If an organic compound is eluted, it is thermally decomposed in the flame, yielding a large number of ions. These can be collected at an electrode, and the ion current between the collector and the jet for the gas can be measured. The small currents of around 10^{-10} A require electronic amplification of the signal, which is approximately proportional to the number of carbon atoms injected into the flame. In contrast with the thermal-conductivity detector, this detector shows little response to molecules such as carbon dioxide, water, nitrogen, and oxygen, making it especially useful for trace analysis of atmospheric samples. The limit of detection, about 1 ng per milliliter of sample gas, is more than a thousand times better than that of the thermal-conductivity detector.

The electron-capture detector ionizes the carrier gas with a beta-emitting radioactive source. The electrons produced on ionization are collected at a

---

[3]Alternatively, the gas on the reference side of the detector may be supplied as a separate gas stream. Since the composition of the gas on the reference side is unchanging, it need not actually pass over the filament as it does on the sample side.

positive electrode and measured. Solutes that react with (capture) these electrons cause a reduction in the number that reach the electrode. The difference is measured and fed to a recorder. The limit of detection is extremely high, about 1 pg per milliliter of sample gas. This detector is also selective. Its sensitivity to halogen-containing compounds makes it applicable to the detection of traces of pesticides such as dieldrin, DDT, and lindane.

Among other detectors one of the first was a microtitrimeter for the titration with base of eluted fatty acids. As another example, extremely high sensitivity is obtainable when the sample contains radioactive substances that can be measured with a detector sensitive to radioactivity.

Identification of Sample Components

Powerful assistance in identification has been obtained by examining the column effluent by the use of a mass, infrared, or ultraviolet spectrometer. The combination of a mass spectrometer and a gas chromatograph is especially effective. A fraction of the column effluent can be sent directly to the inlet of the mass spectrometer, and with even tiny gas-chromatographic peaks enough solute is obtained to yield a mass spectrum of the substance. When the mass spectrum and the retention characteristics both agree with those of a known material, identification is virtually certain.

In addition to the use of large instruments such as mass spectrometers to identify substances eluted from columns, comparison of the elution characteristics of an unknown mixture with those of known compounds for well characterized samples is common. Provided that operating conditions such as temperature and flow rate are carefully reproduced, the retention volume for a given substance also will be reproducible and can serve as a means of identification. The identity of the components of a mixture can usually be established with greater certainty by comparison with known materials and by the use of two or more different kinds of liquid stationary phases. In some instances this can be time-consuming; also, it is not always practicable to have a sufficient variety of known compounds on hand. In these instances chromatographic data can be plotted in various ways. One useful fact is that for a homologous series of compounds (such as C_nH_{2n+2}) a plot of the log of the net retention volume against number of carbon atoms in the molecule yields a straight line.[4] Hence the retention-volume characteristics of many homologs can be established by examining only a few in the series. Figure 18-5 is an example of this type of plot. The lower homologs tend to deviate somewhat from the linear relation.

A systematic identification technique utilizing relative retentions involves the *Kovats retention index*. In this system the normal alkanes are assigned retention-index values of 100 times the carbon number; *n*-hexane is 600, *n*-octane 800, and so on. These are the reference points in the retention-index system (Example 18-1). Be aware that the particular retention volumes shown in Figure 18-5 are applicable only to the column on which they were measured. If the column or the experimental conditions are changed the retention volumes will change, giving a different linear relation.

---

[4]To save time semilog graph paper can be used directly.

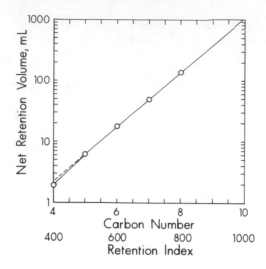

Figure 18-5. **Relation between net retention volume and carbon number for** *n*-**alkanes.** From data of A. G. Polgar, J. J. Holsst, and S. Groennings, *Anal. Chem.* **1962,** *34,* 1226.

EXAMPLE 18–1

With the column and conditions of Figure 18-5, a substance is eluted between *n*-hexane and *n*-heptane with a retention volume of 34.6 mL. The dead-space volume is 4.6 mL. From the figure, estimate the retention index for this substance.

The retention index is 653. A somewhat more precise value can be calculated from the equation for the straight line.

18–3 QUANTITATIVE ANALYSIS

The output signal from a chromatographic detector as a function of time is normally fed to a data collector to provide a permanent record. This is often a recorder having as output a strip of paper on which the output signal is traced as a function of some property such as time or volume of carrier gas. Plots obtained in gas chromatography consist of a series of peaks. The signal from a detector such as thermal-conductivity or flame-ionization is linearly proportional to the amount of a given solute in the carrier gas. Thus the total amount of a particular solute eluted from a column is directly proportional to the peak area. Since the proportionality constant varies for different solutes, it must be determined empirically for each substance through the use of known amounts.

Quantitative information about a peak can be obtained in a variety of ways. Several years ago manual methods of peak evaluation were applied widely, but the trend is for peak-area integrations to be carried out by automatic digital methods, especially for routine measurements. Of the manual methods, height, height-width, and related techniques are the most common; occasionally planimetry is employed for poorly resolved peaks.

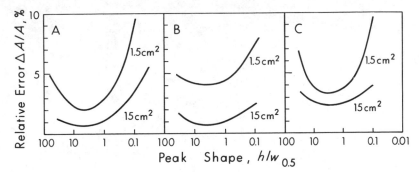

Figure 18-6. Relative error in measurement of area as a function of peak shape for peaks of 1.5- and 15-cm^2 area by (A) height-width, (B) planimeter, and (C) cut-and-weigh techniques.   See first reference at end of chapter.

Manual Integration Techniques

Height-Width

The area of a Gaussian peak can be found by measuring its height and width and then multiplying the product of these measurements by a proportionality factor. The value of the factor depends on the fraction of the height at which the width is measured.

The precision and accuracy of the measurement of peak heights and widths rely heavily on technique. Four major operations, and consequently four sources of indeterminate experimental error, are entailed: placement of the base line, measurement of the height from the base line, positioning of the measuring instrument at a predetermined intermediate height, and measurement of the width at that height. If all four errors are combined statistically for peaks of constant area but of varying height-to-width ratio, the relative error in peak area reaches a minimum at a height-to-width ratio of about 3 (Figure 18-6). For flat peaks the width measurement is subject to large relative error. As expected, for both sharp and flat peaks the relative error is larger than for nearly optimum shapes. Also, for peaks of a given shape the relative error in height-width measurements decreases with increasing area as the square root of the area.

Although the width can be measured at any fraction of the height, it is slightly advantageous to measure the width of broad peaks at half height and that of sharp peaks near the base line. When the areas of a full range of smooth Gaussian peaks are to be measured, the best single practical value to choose is measurement of peak width at one quarter of the peak height. The peak area is then given by 0.753 × height × width.

Height Alone

The evaluation of peaks by height measurement alone requires only placement of the base line and measurement of the height. The precision of this method is therefore inherently superior to measurement of peak area — better by an order of magnitude for some narrow peaks. Unfortunately, peak heights are sensitive to small instrumental and operational variations, and so

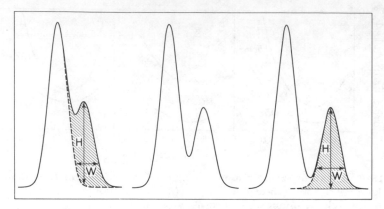

Figure 18-7. Pairs of peaks with resolution values of 0.5, 0.75, and 1.0.

determinate rather than indeterminate errors limit the precision. This technique is of most value in carefully controlled routine measurements. Since measurement of peak areas is much less sensitive to determinate variations, it is preferred in nonroutine work.

Perimeter Methods

The perimeter methods of planimetry and of cutting and weighing are appropriate for peaks of irregular shape as well as for Gaussian peaks.

When a planimeter is used, there is again an optimum peak shape for minimum error; the relative error, however, does not increase so drastically when peaks become flatter or sharper as it does in the height-width method (Figure 18-6). For peaks of a given shape the relative error decreases with increasing size according to the three-quarter power of the area. The planimeter technique is therefore best suited to peaks of large area.

The cut-and-weigh method is highly vulnerable to variations in paper density. It is most appropriately applied to peaks of small area.

Overlapping Peaks

In the retrieval of quantitative information from chart recordings, overlapping peaks frequently present a problem. For a peak appearing as a shoulder on a larger peak, the outline of the larger peak should be sketched in at a position where it would be expected to fall if the smaller peak were absent (Figure 18-7). The sketched-in outline then becomes a steeply changing base line for the smaller peak. In determination of the area of the smaller peak, perimeter techniques are preferable to height-width measurements because of the distortion of both height and width. With improved resolution, as in Figure 18-7, the outline of the overlapping peak can be completed in the normal way. For intermediate types either method may be tried.

Applications

Many thousands of compounds have appreciable vapor pressures, especially when heated to 200 or 300°C, and therefore can be separated and measured by gas-chromatographic techniques. Compilations of retention characteristics of large numbers of materials on a variety of stationary phases have been prepared. Gas chromatography is exceedingly widely used in research, industry, agriculture, and medicine for the determination in complex mixtures of gases, fatty acids, alcohols, aldehydes, amines, esters, ethers, hydrocarbons and their derivatives, ketones, phenols, and even metal complexes that can be volatilized. Even polymers may be examined by the technique of *pyrolysis gas chromatography*, in which substances of high boiling point are decomposed by heat and their gaseous degradation products separated. Through the use of large columns and preparative gas chromatography, samples of highly pure materials can be prepared. Monitoring of industrial processes with automated gas chromatographs is also an important application.

18-4 TEMPERATURE PROGRAMMING

For a sample containing a pair of solutes it is possible to select, usually by trial and error, a temperature that yields optimum separation on a given column by *isothermal chromatography* (chromatography in which the column temperature is constant). However, for the complex mixtures often encountered in real samples, several difficult-to-separate pairs of substances may be present, and a single set of operating conditions cannot provide optimum resolution for all of them. For a given set of conditions the most volatile compounds in a mixture may be eluted, yielding a group of poorly resolved peaks near the beginning, whereas the least volatile require lengthy times to be eluted, yielding broad, low, unmeasurable peaks. Nearly optimum conditions for such separations can be achieved by *programmed-temperature chromatography*. In this technique the temperature of the column is increased at a controlled rate during the separation and analysis. The column is operated at a low temperature at the start, where the more volatile components are eluted with reasonable values of the partition ratio. The less volatile remain in the stationary phase near the inlet of the column. As the temperature is raised, the column successively reaches temperatures at which individual components begin to be eluted at an appreciable rate. Each solute tends therefore to be eluted at a temperature optimum for separation from other solutes. The peaks obtained are generally well resolved, neither excessively narrow nor excessively broad, and are readily measured (Figure 18-8).

In programmed-temperature chromatography it is inappropriate to speak of retention volume. Instead, the term *retention temperature* is used, the temperature of the column at the solute peak maximum. The retention temperature can be calculated from isothermal retention volumes, along with heating and flow rates. In programmed-temperature chromatography the theoretical plate number cannot be calculated by the usual means. Instead it is calculated according to

$$n = 16(V_{T_R}/W)^2$$

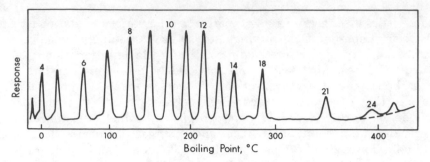

Figure 18-8. Programmed-temperature chromatogram of a wide-boiling-range mixture of *n*-alkanes. The numbers correspond to the carbon numbers of the alkanes. From F. T. Eggertsen, S. Groennings, and J. J. Holst, *Anal. Chem.* **1960** *32*, 904.Reprinted by permission of the American Chemical Society.

where n and W refer to plate number and peak width as before, and V_{T_R} is the retention volume a solute would have if it were injected and eluted with the column held at the retention temperature. Experimentally, plate numbers obtained by isothermal chromatography and by programmed-temperature chromatography are comparable.

Temperature programming offers a significant advantage in that the limit of sample volume is $V_{T_0}/(2\sqrt{n})$, where V_{T_0} is the isothermal retention volume at the initial temperature. Thus by selecting a low initial temperature, it is possible to take advantage of low-temperature sample injection without incurring the disadvantage of the large retention volumes that would be obtained in isothermal chromatography.

18-5   SEPARATION AND ANALYSIS OF A MIXTURE OF BENZENE AND CYCLOHEXANE

Background

In this experiment a mixture of cyclohexane and benzene is separated and analyzed by gas chromatography. The column is of copper tubing (1 m × 5 mm) packed with DC 200 (a silicone oil) and 5% Bentone (a clay) on Chromosorb W (a diatomaceous earth support) and operated at 75°C. The close boiling points of benzene and cyclohexane, 80.1 and 80.7°C, make separation by distillation difficult. Although with a thermal–conductivity detector peak areas are proportional to the amount of a given constituent, they are not the same for like quantities of *different* substances. The composition of a mixture must therefore be calculated from measurements of peak-area ratios of the components of a known mixture and of the sample. This experiment illustrates the technique of using a known mixture to determine the response ratio of two compounds when the amount of sample is not, or cannot be, measured precisely. In addition, it provides experience in both the handling of moderately volatile substances and the use of a strip-chart recorder for the collection of data.

Procedure (time 2 h)

Preparation of Standard

To prepare an accurately known mixture of cyclohexane and benzene, weigh about 3 g of each into a weighing bottle. (*Caution*: benzene is toxic.)[5] Use a weighing technique that takes into account evaporation of the first component added. [6]   Obtain the sample and a syringe.[7]

Injection of Solution into the Gas Chromatograph

Immersing the tip of the syringe in the sample, withdraw more than is actually needed for an injection (2 to 10 μL). Inverting the syringe, push in the plunger to free the sample of air bubbles and bring the volume indicator to the appropriate sample setting. With the syringe still inverted, wipe the needle with a tissue, and retract the plunger to include a sample of air of approximately three to five times the volume of liquid (if capacity of syringe permits). When ready to inject the sample, insert the needle through the injection port and immediately push the plunger all the way down. Withdraw the needle. The insertion, injection, and withdrawal should be done in one smooth, continuous operation. Evaporate remaining traces of sample from the syringe by moving the plunger back and forth a few times. Time on the instrument can be saved by practicing the technique for injection on a spare injection port before attempting a run.

In normal chromatographic practice one sample is injected and the entire chromatogram obtained before the next is injected. Mark the injection points and mixtures injected on the chart to avoid confusion during interpretation[8] (Figure 18-9).

Obtain three or four satisfactory chromatograms of the standard and of the sample solutions in alternation.

Measurements

Determine peak areas by height-width measurements, as explained in Section 18-3. Measure the width at quarter height. Calculate and report the weight percent of cyclohexane in the sample. Cyclohexane has a shorter retention time than benzene.

---

[5]Benzene can be absorbed by both inhalation and through the skin. Although the principal danger is as a chronic poison through long-term exposure, it should be handled with care. R. Smith in *J. Chem. Ed.* **1980**, *57*, A85, discusses the toxicity of benzene.

[6]One method is to weigh about 3 g of benzene to the nearest 0.2 mg into a clean, dry weighing bottle. Remove the cover for about the length of time required to add a portion of cyclohexane (a few seconds); re-cover and reweigh. Record this weight, and quickly add about 3 g of cyclohexane (previously weighed approximately into a test tube or flask) to the bottle; re-cover and reweigh again. The total weight of benzene lost by evaporation is then twice the amount measured during the first removal of the cover.

[7]A syringe capable of accommodating enough air to give a discernible recorder signal is required. One of at least 2.5-μL capacity with a 12-mm needle is suitable.

[8]If an injection should be defective (or thought to be so), wait for the length of time required to elute the benzene before injecting another sample. Since some sample usually enters the column even in poor injections, immediate injection of another sample may result in overlapping peaks. Do not inject a second sample even if no air peak is observed. The amount of air injected is somewhat independent of the amount of sample, and therefore the absence of an air peak is not a reliable indicator of absence of sample.

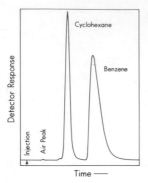

Figure 18-9. Chromatogram illustrating the separation of a mixture of benzene and cyclohexane. Data by D. Cunningham.

Calculations

Calculate the composition of the sample in weight percent from the standard and sample peak-area ratios and the known composition of the standard. In this calculation, allowance must be made for the fact that all substances do not cause the same size of detector signal per unit weight of material. Assuming linearity between detector signal and amount of material, apply the following formula to calculate weight percent of cyclohexane in the mixture:

$$P_{cy} = \frac{100 \, g_{cy}}{g_{cy} + (r_s/r_x) \, g_{bz}}$$

where P_{cy} denotes weight percent cyclohexane in the unknown, g_{cy} the weight of cyclohexane in grams, g_{bz} the weight of benzene in the standard, r_s the ratio of the peak area of the cyclohexane to the area of the benzene in the standard, and r_x the ratio of the peak areas in the unknown mixture.

Several parameters of the separation system can be obtained from the chromatograms. The following calculations are suggested:

1. Calculate the number of theoretical plates for a cyclohexane and a benzene peak. The number is given by $16(t_R/W)^2$, where t_R is the retention time and W the peak width at the base line in the same units of time[9]   (Figure 18-9).

2. Calculate the resolution R of the two peaks by dividing the difference in the retention volumes of two peaks by the average peak width $(V_2 - V_1)/0.5(W_2 + W_1)$. Retention volumes and peak widths can both be expressed in units of time for this calculation.

3. Calculate the partition ratio for each component — the ratio of net retention volume to dead-space volume $(V_R - V_m)/V_m$.

PROBLEMS

18-1.   In gas chromatography the effectiveness of a separation depends but little on the choice of the carrier gas, and yet the choice is important. What considerations govern this choice?

---

[9]With a recorder chart operating at constant speed the time is directly proportional to distance along the chart. Therefore, t_R and W can be most conveniently measured in centimeters on the chart paper.

18-2. When the particular choice of carrier gas has little effect on the extent of a gas-chromatographic separation, how are selective separations achieved?

18-3. What considerations govern the choice of stationary phase in gas chromatography?

18-4. In gas chromatography why is temperature such an important experimental variable?

18-5. Calculate the pressure-gradient correction factor for an inlet pressure of 1.62 atm and an outlet pressure of 1 atm.

18-6. Explain how a thermal-conductivity detector operates. What characteristics make it so valuable?

18-7. Why is hydrogen or helium the preferred carrier gas with a thermal-conductivity detector? Why is nitrogen rarely used as a carrier gas?

18-8. Explain how a flame-ionization detector operates. Compare its applicability with that of the thermal-conductivity detector.

18-9. Explain how an electron-capture detector operates. For what applications is it especially useful?

18-10. Briefly describe the various techniques by which a substance eluted from a chromatographic column can be identified.

18-11. Define the Kovats retention index. With the same column and conditions as in Figure 18-5 an alkane is eluted with a retention volume of 123 mL. If the dead-space volume is 5.8 mL, estimate from the figure the retention index for the substance.

18-12.† On a column with a dead-space volume of 7.6 mL, heptane has a retention volume of 38.9 mL, octane of 86.4 mL, and that of an unidentified substance 55.6 mL. Either consult a textbook on gas chromatography or develop the appropriate equation for the straight-line relation on a log scale, and then calculate the precise value of the retention index for the substance.

18-13. In the quantitative calculation of the composition of a sample, why must a proportionality constant be determined for each substance?

18-14. What techniques are used to measure peak areas? What are their relative merits?

18-15. What are the critical measurements in the determination of peak areas by height-width measurements?

18-16. Compare the advantages and disadvantages of determining the amount of a substance by height-width, height alone, and perimeter methods.

18-17. The relative standard deviation in the peak area measured by the height-width technique for a 30-cm^2 peak having a height-to-width ratio of 10 is 0.6%. (a) What error would be expected for a peak with an area of 1.2 cm^2 and the same height-to-width ratio? If a planimeter were used, the 30-cm^2 peak would be measured more precisely (relative standard deviation 0.5%). (b) What error would be expected for the 1.2-cm^2 peak by this method?

18-18. What would be the effect on peak shape and the precision of measurement for sharp peaks and flat peaks (a) if the recorder chart speed were decreased; (b) if the recorder sensitivity were increased?

18-19. A compound is not eluted from a gas-chromatographic column in a reasonable time. Independent measurements indicate that its partition ratio is about 100 000. What changes in experimental conditions could be recommended to decrease the retention time?

18-20. Define programmed-temperature chromatography. What are its advantages?

18-21. Define retention temperature.

18-22.   How is plate number defined in programmed-temperature chromatography?

18-23.   A gas chromatogram was run on a mixture of ketones. The peak for methyl ethyl ketone appeared after 5.2 min, that for MIBK after 6.8 min, and that for air (not retained by the stationary phase) after 1.6 min. The carrier-gas flow rate was 30 mL/min. (a) What was the partition ratio for methyl ethyl ketone and for MIBK? (b) What additional information is needed to calculate the resolution of the two ketones? (c) How could column conditions be modified to give smaller retention volumes?

18-24.   List the three factors that contribute in a major way to the broadening of peaks eluted from a gas-chromatographic column.

18-25.   What purpose could an internal standard serve in gas chromatography?

18-26.   A gas chromatogram provides both qualitative and quantitative information on a sample. What characteristics of the chromatogram provide this information?

18-27.   In the gas-chromatographic determination of cyclohexane in Section 18-5, what technique is used to compensate for the fact that the volume of injected standard or sample is not accurately known?

18-28.   A sample containing a mixture of 1-chlorobutane and 1-chloropentane was separated and determined by the procedure of Section 18-5. The standard mixture was prepared to contain 2.643 g of 1-chlorobutane and 3.006 g of 1-chloropentane. For a standard run the first peak had an area of 21.41 cm^2 and the second an area of 27.41 cm^2. For a sample run the first peak had an area of 18.36 cm^2 and the second an area of 17.96 cm^2. What was the percentage of chlorobutane in the sample?

18-29.   In gas chromatography, capillary or lightly loaded columns can provide a large number of theoretical plates. They have the drawback of a large dead space, which may nullify this advantage. Calculate the resolution (Section 17-3) expected for two substances having specific retention volumes of 2.0 and 2.1 mL/g of stationary phase for two columns, one of which has 2000 theoretical plates and a column dead space of 1 mL/g of stationary phase and a second having 20 000 theoretical plates and a dead space of 100 mL/g of stationary phase. Which column provides the more effective separation?

18-30.   On a particular column at temperatures of 30, 50, and 70°C the dead-space volumes are 8.0, 7.5, and 7.0 mL, hexane has retention volumes of 229, 110, and 64 mL, and butane 25.9, 20.1, and 15.5 mL. (a) Plot the log of the net retention volume as a function of the inverse of the absolute temperature for these two compounds. By extrapolation estimate the retention volumes of hexane and of butane at 100°C. (b) Plot the data as in Figure 18-5 at 70°C for the two compounds, and estimate the retention volume of heptane, octane, and nonane at 70°C.

REFERENCES

D. L. Ball, W. E. Harris, and H. W. Habgood, *Anal. Chem.* **1968**, *40,* 129; **1968**, *40*, 1113; *J. Gas Chromatogr.* **1967**, *5*, 613. A study of the precision with which peak measurements can be made.

L. S. Ettre and A. Zlatkis, *75 Years of Chromatography*, Elsevier, New York, 1979. Background events leading to early work in chromatography are reviewed.

W. E. Harris and H. W. Habgood, *Programmed Temperature Gas Chromatography*, Wiley, New York, 1966. Theoretical and practical foundations for the subject are described.

H. A. Laitinen and W. E. Harris, *Chemical Analysis*, 2nd ed., McGraw-Hill, New York, 1974. Chapters 24 and 25 deal with chromatography.

D. A. Leathard and B. C. Shurlock, *Identification Techniques in Gas Chromatography*, Wiley, New York, 1970.

A. B. Littlewood, *Gas Chromatography*, 2nd ed., Academic Press, New York, 1970.

19

LIQUID CHROMATOGRAPHY

The adsorption phenomena observed during the filtration through the powders are very interesting. The solution flowing down through the lower part of the tube is at first completely colourless, but then becomes yellow (carotine), until at the top of the inulin column a green band is formed, and beneath this a yellow layer slowly develops. If pure ligroin is filtered through the column, both stripes begin to expand and move.

M. Tswett

Gas and liquid chromatography complement each other admirably. Gas chromatography is largely limited to materials or their derivatives that have significant vapor pressures and are stable at the temperatures necessary for separation. This limitation can hardly be construed as severe, since it includes about half of all known compounds. Nevertheless, another about equally large group of materials do not have significant vapor pressures. For the separation of these, liquid-chromatographic techniques often are the answer. Recall that in gas chromatography the carrier gas acts mainly as a transport medium, and so the stationary phase plays a major active role in determination of the partition ratio. In *liquid chromatography* the mobile phase may also play a prominent role.

Since in liquid chromatography both the mobile and stationary phases can be chosen to meet particular needs, a wide range of separation techniques is available. One classification includes ion exchange, adsorption on solid surfaces, liquid phases chemically or physically attached to the support material, and permeation. Classification can also be made on the basis of the conditions employed, as in high-pressure chromatography, thin-layer chromatography, and so on. With recent improvements in the areas of stationary phases, detection, and pumps, liquid chromatography has become one of the most powerful tools of the analytical chemist.

In liquid chromatography the diffusion coefficients of sample constituents in liquid mobile phases are several orders of magnitude smaller than in gas chromatography. Consequently, from a practical point of view flow velocities at or below the minimum in the van Deemter curve of plate height against velocity are not observed. Figure 19-1 is a plate height–velocity curve for a column used in liquid chromatography. The curve levels off at higher flow velocities instead of continuing to rise as would be expected from the straightforward operation of the effect of resistance to mass transfer.

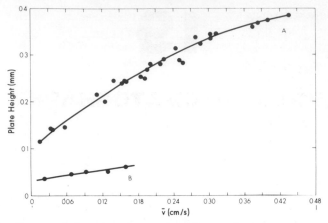

Figure 19-1. Curve of plate height against flow velocity of mobile phase. Column 15- ×0.41-cm Amberlite XAD-2 (a porous synthetic adsorbent solid), particle size 15.6 to 44 μm. From R. G. Baum, R. Saetre, and F. Cantwell, *Anal. Chem.* 1980,*52* 15.

19-1 ION–EXCHANGE CHROMATOGRAPHY

In *ion-exchange chromatography* the mobile phase is usually an aqueous solution of electrolyte, and the stationary phase a solid with sites capable of exchanging ions. Various clays were early recognized as being able to retain or exchange ions such as potassium or calcium. A series of sodium aluminosilicate minerals termed *zeolites* have been the basis of water-softener systems that exchange sodium ions for calcium or magnesium ions.

Natural ion exchangers do not have the predictable properties and convenience of the synthetic ones. Synthetic ion-exchange resins can be described as inert, insoluble polymers containing charged sites that hold ions by electrostatic attraction. Most commercial ion exchangers are organic polymers; a common type is a copolymer of styrene and divinylbenzene. The degree of cross-linking between polystyrene chains is determined by the amount of divinylbenzene.[1] High cross-linking produces a more rigid resin that is less inclined to shrink or swell upon change in composition of the solution with which it is in contact.

[1]Dow Chemical Co. and Rohm & Haas Co. are two major producers of ion-exchange materials. Dow designates the percentage of divinylbenzene in each resin by an x and a number immediately after the identifying number of the resin (8% is typical).

Table 19-1. Some Commercial Ion-Exchange Resins

Type	Exchange Site	Rohm & Haas	Dow
Strong acid	$-SO_3^-$	Amberlite IR-120	Dowex 50
Weak acid	$-COO^-$	Amberlite IR-50	--
Strong base	$-CH_2N^+(CH_3)_3$	Amberlite IRA-400	Dowex 1
Strong base	$-CH_2N^+(CH_3)_2(C_2H_5OH)$	Amberlite IRA-410	Dowex 2
Weak base	Polyamine	Amberlite IR-45	Dowex 3

Ion-exchange sites on resins may consist of strong or weak acidic or basic groups. Common ones are $-SO_3^-$, $-COO^-$, $-CH_2N^+R_3$, and $-CH_2N^+R_2H$, where R is an organic group such as $-CH_3$ or $-C_2H_5OH$. In cross-linked polystyrene resins these groups are usually attached to benzene rings. Several typical resins are listed in Table 19-1.

A strong-acid resin with, say, $-SO_3^-$ exchange sites is conveniently represented by R^-. If the positive ions that provide electrical neutrality are all protons, the ion-exchange resin in the acid form is represented by RH. An example of an ion-exchange equilibrium between sodium ions and a resin in the acid form is

$$Na^+ + RH \rightleftharpoons RNa + H^+$$

The distribution ratio for this equilibrium is $[RNa]/[Na^+]$, where $[RNa]$ is the concentration of sodium ions in the resin. This ratio varies with the composition of the mobile phase, as can be seen from the equilibrium constant for the reaction:

$$K_{eq} = \frac{a_{H^+}\, a_{RNa}}{a_{Na^+}\, a_{RH}} = \frac{[H^+][RNa]}{[Na^+][RH]} \frac{\gamma_{H^+}\gamma_{RNa}}{\gamma_{Na^+}\gamma_{RH}}$$

where a denotes activity and γ denotes activity coefficient. The distribution ratio then is

$$D = \frac{[RNa]}{[Na^+]} = K_{eq}\frac{[RH]}{[H^+]}\frac{\gamma_{Na^+}\gamma_{RH}}{\gamma_{H^+}\gamma_{RNa}}$$

Thus the distribution of sodium ions between the ion-exchange resin and the solution depends strongly on the composition of the mobile phase as well as on the value of the equilibrium constant. Increasing the acid concentration has a marked effect, as expected from the mass-action law, but activity-coefficient variations with ionic strength of the solution are also important. Changing the *co-ion*, the anion accompanying the sodium ion in solution, from one univalent anion, say chloride, to another such as nitrate has little effect on the distribution ratio under nearly all solution conditions. For divalent or multivalent ions, however, the nature of the co-ion or other solutes may have notable effects through hydrolysis or complex formation. For example, zinc ions form both chloro and hydroxyl complexes. Here it is convenient to write

$$[Zn^{2+}] = \alpha_{Zn^{2+}} C_{Zn^{2+}}$$

where $\alpha_{Zn^{2+}}$ is the fraction of the total analytical concentration $C_{Zn^{2+}}$ present as the simple hydrated ion. The ion-exchange equilibrium involves the displacement of two protons:

$$2RH + Zn^{2+} \rightleftarrows R_2Zn + 2H^+$$

The distribution ratio for this equilibrium is

$$D = \frac{[R_2Zn]}{C_{Zn}} = K_{eq} \frac{[RH]^2 \, \alpha_{Zn^{2+}} \, \gamma_{Zn^{2+}} \, \gamma^2_{RH}}{[H^+]^2 \, \gamma^2_{H^+} \, \gamma_{R_2Zn}}$$

The distribution ratio is now greatly affected not only by the acid concentration but also by the nature of the acid, the presence of complexation agents, and the more significant activity-coefficient factor of the divalent zinc ion.

For metals generally, the factor α_M, which is important in the separation of one ion from another, depends strongly on the extent of hydrolysis and on the kind and concentration of any complexation agents that are present. Thus the separation is strongly influenced by the composition of the mobile phase, including that of the co-ion.

To indicate the relative preference of an ion-exchange resin for one ion over another, say Na^+ over H^+, the term *selectivity coefficient* is used . For the Na^+–H^+ pair it is defined by the ratio

$$\frac{[H^+][RNa]}{[Na^+][RH]} = K_{eq} \frac{\gamma_{Na}\gamma_{RH}}{\gamma_{H^+}\gamma_{RNa}}$$

where the selectivity coefficient is normally written to be larger than unity.

The selectivity of a resin for various ions is governed primarily by the charge on the ion and its hydrated radius. The higher the charge and the smaller the hydrated radius, the higher is the charge density, and the higher the resin affinity for an ion. In general, resin affinity increases with increasing cross-linking of the resin, with increasing charge on an ion, and with increasing atomic number for ions in the same group in the periodic table. Either specific bonding effects or nonspherical ion shapes may, however, throw individual species out of sequence. Table 19-2 lists the selectivities of two resins. Although the range in relative ion affinities is only one or two orders of magnitude, repeated equilibrations are possible through the use of a resin packed in a column, and thus exchange of ions on the resin for ions in solution can be made quantitative.

The *exchange capacity* of a resin is defined as the number of milliequivalents of protons or hydroxyl ions per gram of dry resin that can undergo reaction with ions in solution. For typical strong-acid

Table 19-2. Affinity of Resins for Selected Ions*

Ion	Hydrated Radius, nm	Relative Affinity
Ag^+	0.34	8.5
Cs^+	0.33	3.2
Rb^+	0.33	3.2
K^+	0.33	2.9
NH_4^+	0.33	--
Na^+	0.36	2.0
H^+	0.28	1.3
Li^+	0.38	1.0
Ba^{2+}	0.40	11.5
Pb^{2+}	0.40	9.9
Sr^{2+}	0.41	6.5
Ca^{2+}	0.41	5.2
Mg^{2+}	0.43	3.3
ClO_4^-	0.34	10.0
NO_3^-	0.34	3.0
$C_2H_3O_2^-$	---	18
I^-	0.33	18
Br^-	0.33	3.5
Cl^-	0.33	1.0
F^-	0.35	0.1
OH^-	---	0.5

*For sulfonic acid (cation exchanger) and quaternary ammonium (anion exchanger) resins with 8% cross-linking. Values are relative to $Li^+ = 1.00$ or $Cl^- = 1.00$.

cation-exchange resins this capacity is of the order of 2 meq/g, and for strong-base resins about half that.

Ion-exchange columns are simple to prepare and use; an example, illustrated in Figure 19-2, consists of a glass column with a plug of glass wool in the bottom to retain the bed of resin and a stopcock or pinch clamp to control the flow of mobile phase. For columns such as this, in which the flow is maintained by gravity, the particle size of the resin must not be too small, or the flow will be slow and separation times will be unacceptably long. Another factor determining the time required for separation is the rate at which the exchange of ions takes place. This rate is generally governed by the resistance to mass transfer in and out of the ion-exchange beads and in and out of the film of mobile phase surrounding them. Since the diffusion coefficients involved are tiny compared with those in gas chromatography, the optimum flow rate is much lower.

To increase the rate of exchange, it is important to decrease the distance that ions must traverse to reach exchange sites in the resin particles. This decrease may be achieved by placing ion-exchange sites only on the surface of small, solid resin beads. Such resins, called *pellicular* ion exchangers, have exceptional separating ability but only limited exchange capacity, and so can be used only with samples of a few micrograms. Because the small size of the beads (on the order of tens of micrometers in diameter) creates high

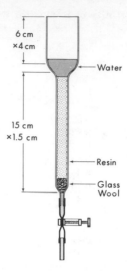

Figure 19-2. Ion-exchange column for analytical use.

resistance to flow of mobile phase, gravity is usually insufficient and pumps become necessary.

In contrast to gas chromatography, in ion-exchange chromatography there are few broadly applicable detection systems for monitoring the effluent from a column. The principal cause of this difficulty is that the solutes are necessarily associated with a relatively large mass of mobile phase. A common practice is to collect small successive portions of the eluate and analyze them for the substance of interest. Alternatively, small-volume flow-through cells in a spectrophotometer or refractive-index meter may be used to detect eluted peaks on a continuous-flow basis.

Applications

Although ion-exchange resins can be used in several different ways, most procedures involve passage of a solution through a column of resin. The resin may be used to convert one substance completely into another, or it may be used to separate two ions in a mixture by a true chromatographic process based on a slightly different affinity of the resin for each of the ions.

In one application a salt is completely converted to an equivalent amount of acid by passage through a sulfonic acid resin in the hydrogen form. The reaction is of the type

$$M^+ + A^- + RH \rightleftharpoons RM + H^+ + A^-$$

The acid HA formed can be titrated with standard base provided the ionization constant is larger than about 10^{-5}. Similarly, by use of a strong-base anion-exchange resin a salt can be converted to an equivalent amount of base, and the base, if sufficiently strong, titrated with standard acid. By such methods the total salt content of a sample can be determined after the acid or base has been completely eluted from the column. Such reactions can also be applied in the direct preparation of a standard solution of an acid by weighing a known amount of a primary standard such as potassium chloride and passing it through a column of sulfonic acid resin in the hydrogen form to produce an equivalent amount of hydrochloric acid. A somewhat similar type of application involves the conversion of sulfates or phosphates to chlorides by use of the resin in the chloride form. In subsequent measurement steps, chloride is often less likely to cause interference in the measurement of metal cations such as iron(III) and calcium.

A further application involves removal of interfering ions prior to analysis. In the determination of phosphate in phosphate rock, for example, ions such as calcium, magnesium, and iron can cause serious problems in the subsequent analysis. By passage of the dissolved phosphate sample through a strong-acid cation resin in the hydrogen ion form these ions can be replaced by hydrogen ions.

Still another important application of ion exchange involves the separation of metal ions. For this purpose either cation- or anion-exchange resins may be employed. Thus for chemically similar metal ions separation may be improved by addition of complexation reagents to the mobile phase. Selectivity coefficients can be varied by taking advantage of differences in formation constants and, with some substances such as the lanthanides, through differences in the rate of formation of complexes with ligands such as EDTA. For example, with anion-exchange resins, varying concentrations of hydrochloric acid may be added to form anionic chloro complexes of the metal ions, which are then separated. Chloro complexes form with almost all transition-metal ions lying in the central part of the periodic table. Separations become especially simple if one metal ion forms an anionic complex while the other remains as the cation. Then all that is required of the resin is separation of ions of opposite charge — an easy operation (Section 19-8).

Ion exchange is also suited to analyses for traces of ionic substances (Figure 19-3). If a large volume of solution is passed through a bed of cation- or anion-exchange resin under conditions where the partition ratio for the ion under study is large, traces of the ion will be retained strongly at the head of the column. Following this concentration step the substance can be eluted in a small volume by changing conditions to lower the partition ratio.

Ion exchange provides an economical and convenient way to remove ionic impurities from water for use in the laboratory. If water containing a salt such as sodium chloride is passed through a cation-exchange resin bed in the acid form, the sodium ions are replaced by hydrogen ions to yield hydrochloric acid. If this solution is then passed through an anion-exchange resin in the basic form, the chloride ions are replaced by hydroxide ions and the result is the formation of water. The ionic impurities can be similarly replaced by the use of a mixed bed containing both cationic-and anionic-exchange resins.

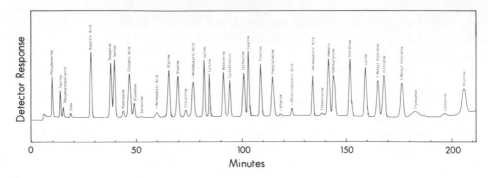

Figure 19-3. An ion-exchange chromatogram illustrating the separation of a mixture containing 20 mmol of each of 40 biologically important compounds. The eluent is lithium citrate buffer; five different pH values were utilized during elution from a single column of 6-mm diameter.

On a commercial or household scale, ion exchange is employed to soften water by exchanging undesirable calcium, magnesium, and iron ions for sodium. A strong-acid cation resin in the sodium form is placed in a column, and hard water passed through it. Since the partition ratios for divalent ions such as calcium and magnesium are greater than for sodium, they replace sodium ions on the resin. When the resin is almost completely converted to the calcium–magnesium form, it is regenerated by passage of a concentrated solution of sodium chloride. By mass action the high concentration of sodium ions in solution displaces calcium and magnesium ions from the resin, reconverting the resin to the sodium form for reuse. Ion-exchange resins may be cycled in this way almost indefinitely.

In a recently developed technique called *ion chromatography* the problem of detecting the eluted species in the presence of a large background of the electrolyte used for elution is solved by passing the sample through two columns. A small quantity of sample is injected onto a column of pellicular ion-exchange resin, and the ions of interest separated by elution with a salt, strong acid, strong base, or buffer solution. The separated ions, along with the salt-containing mobile phase, pass into a second high-capacity ion-exchange column, which reduces the acid, base, or buffer concentration to a low level. This permits detection of the trace amounts of separated ions by measurement of the electrical conductivity of the eluent from the second column. Sodium, potassium, and ammonium ions can be determined in milk by injection of a 100-μL sample onto a 0.6- $\times$ 25-cm column of pellicular strong-acid cation-exchange resin and elution with 0.005 M HCl at 2.3 mL/min. The separated ions are passed through a second column containing a high-capacity (nonpellicular) strongly basic anion-exchange resin. In this column all anions in the eluent are converted to OH$^-$. Thus the HCl eluent is converted to water, and the sodium and potassium salts in the sample to the corresponding hydroxides. The conductances of these hydroxides, now in a water background, can be readily measured. The ammonium ion is converted to the poorly conducting NH$_3$ species. A chromatogram of this separation is shown in Figure 19-4. Also shown is the separation of several anions in urine. Here the NaHCO$_3$–Na$_2$CO$_3$ eluent is

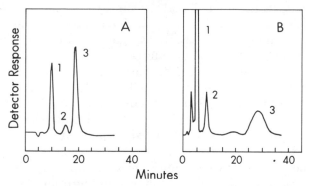

Figure 19-4. Chromatograms of the separation and determination of monovalent inorganic cations in milk, and of inorganic anions in urine, by ion chromatography. (A) 1, 6.1 ppm Na$^+$; 2, 12 ppm NH$_4^+$; 3, 17.3 ppm K$^+$. From Dionex Corp. (B) From *Clin. Chem.* 1976, 22, 1424. With permission.

converted to weakly ionized H_2CO_3 by the second column of strongly acidic cation resin. Since H_2CO_3 is only slightly conductive, it does not interfere in the detection of phosphate and sulfate in the sample.

19-2 ADSORPTION CHROMATOGRAPHY

Adsorption chromatography is a type of liquid-solid chromatography in which the stationary phase is a solid of large surface area. The most common stationary phase is hydrated silica, $SiO_2 \cdot xH_2O$, which has a surface area of about 500 m^2/g. The adsorption sites on silica are hydroxyl groups attached to silicon atoms; these sites interact with sample components primarily through hydrogen bonding, but also through attraction of the polar oxygen atoms for polar solutes. Another polar adsorbent is hydrated aluminum oxide that has been activated by heating to about 200°C to produce a material with a surface area of about 100 m^2/g. If heated to higher temperatures, the surface activity increases, and an undesirable base-catalyzed reaction of the alumina with certain solutes may occur. The nonpolar stationary phase, charcoal, has a surface area in the range of 1500 m^2/g. Another nonpolar adsorbent is a polymer of styrene and divinylbenzene synthesized in the form of beads consisting of fused microspheres. (An example is Amberlite XAD-2, manufactured by Rohm and Haas Company.)

In adsorption chromatography an adsorbent solid of uniform mesh size is packed into a column. The column is filled with mobile phase, and a sample placed at the inlet end of the column. Active sites on the solid may adsorb either solvent or solute molecules. which compete for adsorption sites. In adsorption chromatography, therefore, the mobile phase should be chosen so that interaction between it and the solute–stationary phase and mobile phase gives the best separations with reasonable partition ratios.

Solvents for polar adsorbents are described as weak or strong. Weak solvents such as pentane and carbon tetrachloride yield larger partition ratios;

strong solvents such as ethanol and dioxane yield smaller partition ratios. Tables of relative solvent strengths based on solvent polarities have been developed, and a number of empirical rules established, that aid in the prediction of the order of elution of solutes from a column and in the selection of suitable mobile phases. Relative solvent strengths tend to be similar on a variety of stationary phases and normally increase in the following order: pentane, carbon tetrachloride, benzene, chloroform, methylene chloride, ethyl acetate, acetone, dioxane, acetonitrile, ethanol, methanol, and water. A sequence of eluent solvents of increasing polarity is referred to as an *eluotropic series*.

In a complex mixture some solutes will be more tightly adsorbed than others. With pentane as the mobile fluid, solutes with low partition ratios will be eluted in a reasonable volume and time, while strongly adsorbed solutes, having high partition ratios, will not be. Stepwise, or gradient-elution, techniques are used for resolving such mixtures. *Gradient elution* involves changing solvent composition during elution to more strongly adsorbing solvent species so as to displace the more strongly held solutes. In this way components for which the adsorbent has a high affinity are displaced by solvents for which the adsorbent has a high affinity. The partition ratio is thereby reduced so that the solutes are eluted in acceptable times. Gradient elution can be programmed into modern liquid chromatographs by built-in microprocessors that control valves and pumps, so that two or even three solvents can be metered simultaneously in varying proportions to give the desired eluent properties.

The presence of a relatively few highly active adsorbent sites often gives rise to tailing. This can sometimes be reduced by operating at a higher temperature or by injecting samples sufficiently small that only the most active sites are employed. Tailing can also be reduced by deactivating, or blocking, the most active sites through a suitable chemical reaction or by the addition of traces of water or another polar solvent as part of the eluent.

Adsorption chromatography is used mainly for the separation of nonionic molecules with low vapor pressures. This category includes many important organic compounds.

EXAMPLE 19-1

Suppose that separation of a mixture of 1,2- and 1,4-dibromobenzene, along with quinoline and 1,2,3,4-dibenzanthracene, was attempted on an alumina column. What eluents might be recommended?

Beginning with pentane as the mobile phase, 1,2-dibromobenzene is eluted first, followed by the 1,4 isomer. The other two compounds are not eluted, being too strongly held by the alumina to be displaced. If pentane is replaced by a more polar solvent such as toluene, however, quinoline is displaced and is eluted, but dibenzoanthracene is not. A still stronger solvent such as chloroform is required. Thus all four components may be eluted separately in a reasonable time by resorting to successively more polar solvents (Figure 19-5).

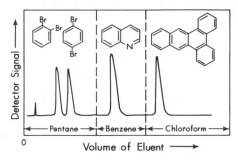

Figure 19-5. Schematic chromatogram for the separation of four aromatic compounds on alumina by successively more polar mobile phases.

19–3 LIQUID-LIQUID CHROMATOGRAPHY

In *liquid-liquid chromatography* one liquid serves as the mobile phase and a second, immiscible liquid supported on some inert material of large surface area as the stationary phase. The instrumentation and techniques in this branch of chromatography are similar to those in liquid-solid or adsorption chromatography. Compared with adsorption chromatography, it has both advantages and drawbacks. One advantage is that the affinity of the stationary phase for solutes is more uniform, leading to a partition ratio that is essentially independent of sample size. Tailing of bands and peaks is much less, so that the shapes tend to be more nearly Gaussian.

A disadvantage of adsorbed liquid stationary phases is that the liquid is always slightly soluble in the mobile phase. Consequently, as mobile phase passes through a column, it continually dissolves and removes some of the stationary liquid. One way to minimize this effect is to presaturate the mobile phase with stationary phase by passing it through a precolumn that contains the same stationary phase as the main column. Since the temperature coefficient of solubility of most substances is appreciable, this technique requires meticulous control of temperature. The most successful solution to this problem of loss of stationary phase is to chemically bond it to the surface of the solid support. For example, the hydroxyl groups on a silica support may be replaced through a series of chemical reactions with a carbon atom attached to a molecule of the desired stationary phase, after which the surface is composed of such groups as long-chain hydrocarbons (Figure 19-6). The stationary phase may then be considered to possess properties of both a solid and a liquid stationary phase. The mobile phase typically involves water and a miscible organic solvent such as methanol or acetonitrile.

19–4   EXCLUSION CHROMATOGRAPHY

Exclusion, or *gel, chromatography* is unique in that the mobile and stationary phases are chemically the same. In this type of chromatography,

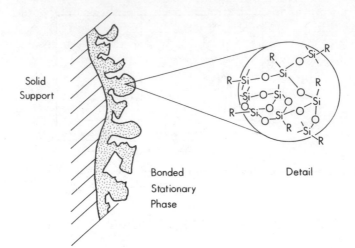

Solid
Support

Bonded
Stationary
Phase

Detail

Figure 19-6. A schematic cross section of a support with a chemically bonded organic stationary phase.

size rather than specific chemical properties forms the basis for separation. Solutes distribute themselves by normal diffusion between a liquid external to a porous packing and the same liquid within the packing.

The porous support and the liquid it contains are frequently referred to as the stationary phase. Molecules too large to enter the pores of the packing are eluted with a retention volume equal to the volume outside the packing. Molecules able to penetrate the pores are eluted in a larger volume. Large molecules therefore have small retention volumes, and small molecules have large retention volumes. The partition ratios can never be large; the maximum value is V_s/V_m, where V_s is the volume of the liquid within the pores and V_m the volume of the liquid outside the pores. When these volumes are comparable, the maximum partition ratio is about unity, and partition ratios range from zero (for the largest molecules) to about one. The mechanism of separation is shown schematically in Figure 19-7. The smallest molecules have all the pore spaces available to them, medium-sized ones can enter some of the pores, and the largest are excluded completely. Thus intermediate values of partition ratios are observed for molecules that are small enough to enter some, but not all, of the available pore spaces. Depending on uniformity of

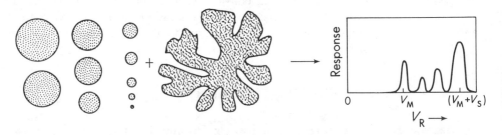

Figure 19-7. Schematic diagram illustrating separation of molecules on the basis of size.

pore size, a support may function as an efficient cutoff, separating molecules larger than a given size from those smaller; the more uniform the pores, the sharper is the size discrimination. In terms of resolution (Section 17-3) the partition-ratio factor is less favorable in exclusion chromatography than in conventional columns with the same number of theoretical plates but for which partition ratios greater than one are possible.

One rigid type of support consists of porous glass beads. These can be packed into columns uniformly and, because they are not compressible as are soft gels, give little resistance to the flow of mobile phase. The exclusion limit is controlled by selection of the appropriate glass porosity; a range of cutoff values are possible, from molecular weights as low as 300 to a million or more. Adsorption on the surface of the glass can have an undesirable influence. Active adsorption sites can be deactivated by chemical treatment of the surface with hexamethyldisilazane. Porous glasses are practicable for both aqueous and organic solvents.

Cross-linked polystyrene beads are regularly used in conjunction with organic solvents for the separation and characterization of polymers and large biological molecules. Provided that the polystyrene is properly prepared, pore sizes can be varied so that exclusion limits can range from molecular weights of 200 to several million. The molecular weight can be obtained from a calibration curve of log of the molecular weight as a function of retention volume.

Another molecular-exclusion medium, employed more in gas than in liquid chromatography, is a type of silicate called *molecular sieves*. These are crystalline aluminosilicates of uniform three-dimensional structure with pore volumes of about 0.5 mL/g. Although some of these materials, as zeolites, occur naturally, most are prepared synthetically. Linde 5A, for example, is a synthetic material with a highly uniform pore diameter of about 0.5 nm. *n*-Alkanes of low molecular weight can enter a cavity with an entrance 0.5 nm in diameter, whereas nonlinear, or branched-chain, hydrocarbons cannot. Accordingly, by this means separation of branched- from straight-chain hydrocarbons can be effected. Corresponding 4A and 3A materials have pore diameters of 0.4 and 0.3 nm. The 3A material is widely used as a desiccant because water and other small polar molecules are strongly adsorbed at room temperature. Adsorbed substances are removed by heating the sieves to 250°C in a vacuum for several hours.

19-5 HIGH-PRESSURE LIQUID CHROMATOGRAPHY

Because the diffusion coefficients encountered with liquid mobile phases are small, for difficult separations requiring optimum conditions, separation times of many hours or even days have been required. Several approaches to achieving fast yet effective separations have been developed. One such approach is to choose for the mobile phase a liquid of low viscosity such as hexane, acetonitrile, or methanol. With lower viscosity, diffusion coefficients in the liquid phase are somewhat more favorable, and at the same time higher flow rates can be achieved with a given inlet pressure.

When the term for resistance to mass transfer in the mobile phase (Section 17-2) is examined closely, it is found to include the. square of the particle diameter of the packing. Therefore a second effective way to achieve higher flow rates without increasing band broadening is to reduce the particle diameter. [2]   Since a reduced particle diameter increases flow resistance approximately as the inverse square of the particle diameter, higher inlet pressures are required for maintenance of the same flow rates. Still higher pressures are required for the increased flow rates that are desirable to speed elution without loss of resolution. Inlet pressures up to 14 megapascals (2000 psi) or even higher are currently being employed. Such *high-pressure liquid chromatography* requires special sample-injection systems, pumps, and columns to provide uniform flow rates. In addition, the stationary phase must be sufficiently rigid that it does not appreciably change dimensions with pressure.

To reduce the problem of resistance to mass transfer in the stationary phase, small solid silica beads of uniform diameter have been developed that possess a porous surface layer similar to the pellicular ion-exchanger packings described in Section 19-1. This layer serves as a stationary phase or as a substrate to which the stationary phase is bonded. With pellicular packings, resistance to mass transfer in the stationary phase is low, and flow resistance is decreased so that inlet pressures need not be so high. Disadvantages include low sample capacity and a relatively large dead-space volume.

The applications of liquid chromatography have grown by leaps and bounds with the development of instrumentation for high-pressure chromatography. The fundamental chromatographic processes and theory are not different from operations under low inlet pressures, and therefore ion exchange, adsorption, liquid-liquid or exclusion chromatographic processes may be involved. The use of either small-diameter particles or pellicular materials as stationary phases or supports has made possible higher flow rates that give better resolution than low-pressure methods in a fraction of the time. Improved sensitivity and detection limits permit the use of shorter columns of smaller diameter.

The principal remaining problem in high-pressure liquid chromatography is that of devising a sensitive universal detector for monitoring the eluate. The most broadly responsive detection system currently available measures the refractive index of the eluate, which changes when a sample component appears. The refractive-index detector is moderately responsive to small changes in solute concentrations but exceedingly sensitive to temperature variations; also, it cannot be used with solvent programming or gradient elution. Another popular detector measures ultraviolet absorption. Although only substances that absorb ultraviolet radiation can be sensed, present designs are highly sensitive and stable.

---

[2]Quantitatively, this means that for the product of the square of the particle diameter and the velocity to remain constant the velocity can be increased 16 times provided the particle diameter is decreased fourfold, say from 20 to 5 μm, but at the cost of higher pressure drop across the column. Separation time could be decreased by that same factor of 16 without loss of resolution.

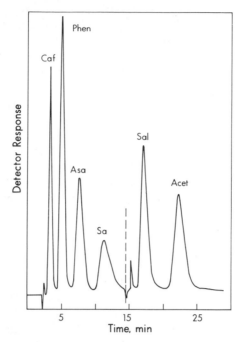

Figure 19-8. Chromatogram of a sample of a mixture of analgesics (caffeine, phenacetin, acetylsalicylic acid, salicylic acid, salicylamide and acetoaminophen). Chromatogram by F. Cantwell.

Qualitative and quantitative analysis by liquid chromatography involves the same principles as by gas chromatography (Section 18-3). Retention volumes provide a basis for identification, and peak areas (or peak heights under well controlled conditions) a basis for quantitative measurements. The combination of either gas or liquid chromatography with mass spectrometry is an important and powerful identification tool. New developments in high-pressure liquid chromatography, stationary phases, and instrumentation are proceeding rapidly. An example of the separation power of liquid chromatography is illustrated in Figure 19-8, which shows a chromatogram of a sample of a few microliters of a solution of analgesic drugs. The sample was injected directly into a chromatograph and the components separated and measured in a matter of minutes.

19-6 THIN-LAYER AND PAPER CHROMATOGRAPHY

In gas chromatography the stationary phase must be enclosed in a column to retain the mobile phase. In liquid chromatography the stationary phase is often packed into a column (a necessity with high-pressure techniques); under some circumstances, however, the stationary phase may simply be in the form of a thin layer. In *paper chromatography* a sheet of paper serves as the stationary phase or as a support for one. In *thin-layer chromatography* the stationary phase is spread as a thin film on a sheet of glass or plastic and the mobile phase is allowed to flow through it by capillary action. Since these types of chromatography are similar they are treated together.

A variety of stationary phases can be employed. In paper chromatography, the effective stationary phase is water associated with the cellulose fibers. Separation occurs through one or more of the following processes: (1) adsorption on active sites on the paper, (2) ion exchange with ionic sites on the paper, or (3) liquid-liquid distribution between the mobile phase and the aqueous stationary phase absorbed in the paper. Filter paper without special treatment is well suited to the separation of mixtures of polar or ionic compounds. Sometimes paper is treated to render it hydrophobic so that absorbed water is largely displaced by less polar materials such as glycol or glycerol to provide a paper with somewhat different characteristics. It can also be modified more drastically by impregnation with, for example, a silicone oil to produce a highly nonpolar surface layer. When the stationary phase is nonpolar, and a highly polar mobile phase such as water is used, the technique is sometimes referred to as *reversed-phase chromatography*.

In paper chromatography a moderate degree of mutual solubility of the mobile and stationary phases is tolerable, and therefore the range of possible mobile phases is wide. For the separation of ionic materials, mobile-phase mixtures containing water, an alcohol, and either ammonia or acetic acid are often effective. For less polar solutes, formamide in toluene or chloroform may be employed; for the least polar solutes, mixtures of kerosene and isopropanol are useful.

In thin-layer chromatography any of the stationary phases useful in liquid chromatography will generally work — ion-exchange resins, adsorptive solids, porous glass beads, or supported or bonded liquids. Similarly, any of a variety of mobile phases is possible. The most widely useful stationary phases are silica gel and alumina. These materials are blended into a slurry with a binder such as calcium sulfate and spread as a uniform layer about 0.25 mm thick on a glass or plastic sheet. The layer is dried and is then ready for use. Silica or alumina plates are usually pretreated by heating to a temperature sufficiently high to drive off water and adsorbed gases.

In either paper or thin-layer chromatography a few microliters of a solution of sample is placed near one end of the sheet and allowed to dry. Then the sheet is applied in a closed compartment (developing chamber), and the atmosphere allowed to become saturated with the solvent mixture. Next, the sheet is tilted so that the end nearest the sample spot is dipped in the solvent (called the developing solvent) in the chamber. Through capillary action the solvent rises up the stationary phase past the sample. As it moves upward, it carries the sample components at different rates according to their partition ratios. The process is stopped when the solvent reaches the other end of the thin-layer or paper strip.

Choice of mobile phase is critical for effective separation. A combination of two or three components, for example, a 1:2:1 mixture of water, pyridine, and butanol, is often employed. The optimum composition of solvent systems is best determined empirically.

Numerous variants and techniques of development are possible. Instead of ascending development, it can be descending, horizontal, or radial. The usual ascending technique reduces problems of channeling of the flow of mobile phase.

Sometimes a mobile phase has provided only partial separation of a mixture by the time it reaches the end of the strip. To separate the remaining components, a stepwise or multiple development may be employed. *Multiple development* involves vaporization of the solvent when it has reached the end and then repeating the development. *Stepwise* development is similar except that a second solvent is used. In *two-dimensional development* two solvents are used in succession, but the plate is turned 90° before immersion of the edge in the second solvent.

Special reagents may be incorporated into the stationary or mobile phase to assist in the separation of specific solutes. For example, silver ions interact particularly with double bonds in the sample components. The separating power of these techniques is high. TLC plates may provide separations equivalent to thousands of theoretical plates in a relatively short distance. Even molecules as similar as cis and trans isomers can be separated cleanly.

Sample components are not eluted from a column, but studied as a series of spots or bands on the thin layer or paper. The extent of solute movement is described by the R_f *value*, the ratio of the distance traveled by the band maximum to the distance travelled by the leading edge of the mobile phase.

Solutes and their locations on the chromatograms are easily identified when the components are colored. For normally colorless solutes identification of the position of the bands or spots is facilitated by chemical reactions or physical tests. Examination of a chromatogram under an ultraviolet lamp may reveal aromatic or other types of compounds that fluoresce as distinct spots. Quenching of fluorescence is another phenomenon that can be exploited. If a fluorescent material is mixed with the binder, upon exposure to ultraviolet radiation the whole plate fluoresces except for the spots where solute components cover the fluorescent substrate.

A thin-layer plate may be sprayed with a reagent that forms colored products with particular sample components. Thus, the reagent ninhydrin produces colored derivatives with α-amino acids, bromocresol green with carboxylic acids, and so on. When the stationary phase is chemically inert, a general reagent that can be used to reveal the positions of bands is sulfuric acid; upon heating, organic solutes are carbonized to give a series of dark spots.

Thin-layer and paper chromatography are important in the identification of tiny amounts of substances on the basis of their R_f values. Only semiquantitative data can be obtained from direct measurements of size and density of the spots. Indirectly, somewhat improved precision is possible by physical removal of a spot and extraction of the material within, followed by a spectral or other measurement (Chapter 14).

19-7 DETERMINATION OF TOTAL SALT IN A MIXTURE BY ION EXCHANGE

Background

In this experiment a cation-exchange resin in the hydrogen ion form is used to determine the total milliequivalents of salt in a sample. When a salt solution is passed through a column of this resin, cations in solution exchange with hydrogen ions on the resin to produce an amount of strong acid equivalent to the salt initially present. This acid is washed out of the column (eluted), and the effluent (eluate) is titrated with standard sodium hydroxide. The amount of base required is a direct measure of the amount of salt in the original solution. For titration of weak acids, solutions of sodium hydroxide must be kept free of carbon dioxide, or the carbonate formed will obscure indicator end points. Since in this experiment the acid being titrated is strong, and the indicator changes color in a low pH range, precautions to keep the sodium hydroxide free of carbonate are unnecessary.

Solutions of sodium hydroxide must be standardized after preparation. For this experiment potassium chloride is used as an acid-base primary standard in an unusual way. A sample is weighed and dissolved, the solution passed through an ion-exchange column, and the resulting hydrochloric acid titrated. Because the sample and standard are treated alike, errors in technique tend to cancel each other, particularly if alternation is used.

The resin must be converted initially to the acid form. This process, called *regeneration*, is brought about by passing hydrochloric acid through the column. The reaction is

$$NaR + H^+ \rightleftarrows HR + Na^+$$

Although the resin has a greater affinity for sodium than for hydrogen ions, use of a large volume of concentrated acid permits quantitative conversion of the resin to the acidic form. Before a sample is placed on a column, excess acid must be washed from the column to prevent its being titrated along with the sample and thereby producing high results.

Procedure (time 4 to 6 h)

Preparation of Column

Obtain two 25-cm ion-exchange columns (Figure 19-2). Half-fill the columns with water and insert a small portion of Pyrex glass wool.[3] Push the wool into a plug at the base of the column with a glass rod. Add 50- to 100-mesh, strongly acidic cation resin to a level of about 0.5 cm below the top of the narrow portion.

Backwash the columns by connecting a funnel to the column outlet with a 50-cm length of tubing and passing a flow of water up the column from the bottom. Alternatively, connect a wash bottle with an upward pointing tip to the column outlet, and use it to force water up the column. The resin should be raised

---

[3]Previously unused Pyrex glass-wool fiber often has a film on its surface that may be a source of contamination. Rinsing the wool with carbon tetrachloride before use eliminates this problem.

by the force of the water and then settle back to give a uniform bed. Take care not to lose resin during this operation. Next, to regenerate the resin, pass a 100-mL portion of 3 M HCl through each column at a flow rate of about 5 mL/min (1 drop/s). After regeneration, a 1.5- by 15-cm column has sufficient exchange capacity for about a dozen samples.

Try to avoid allowing the level of solution to fall below the surface of the resin. Should the liquid level accidentally drop below the top of the resin bed during use, upon the next addition of solution stir the top centimeter or two of resin with a stirring rod to dislodge any bubbles, which may cause channeling and inefficient contact between the liquid and resin phases. Rinse the rod on removal to minimize loss of sample.

Pass distilled water through the column until the effluent is neutral. To test for neutrality, collect about 30 mL of effluent and add 1 drop of methyl red indicator. At neutrality, 1 drop of 0.1 M NaOH is sufficient to change the indicator from red to the yellow of the alkaline form.

Standardization of 0.1 M NaOH Solution

Prepare about 1 L of 0.1 M NaOH solution. Weigh accurately 0.3-g portions of pure, dry KCl into 50-mL beakers. Dissolve each portion in 1 to 2 mL of distilled water.

Add one of the KCl solutions to the top of one column in as small a volume as possible. Allow the solution to drain to near the top of the resin, and then with 3- to 5-mL portions of wash water rinse the beaker and the sides of the column. Allow each portion to drain to the level of the resin before adding the next. Repeat several times. Continue washing with water at a flow rate of about 5 mL/min until the effluent is neutral; this will require 75 to 100 mL. Collect a 30-mL portion from the column. Add 1 drop of methyl red indicator. Fill the buret with NaOH titrant and record the initial buret reading. Add a drop of NaOH from the buret. If the indicator remains in the red form, the acid has not been completely eluted from the column and further washing is required. Quantitatively transfer all test portions to the receiving flask. When a 30-mL test portion shows the yellow alkaline form of the indicator upon addition of a drop of sodium hydroxide, indicating that the effluent is neutral, transfer that portion to the receiving flask also, add several drops of methyl red indicator to the total effluent, and titrate with 0.1 M sodium hydroxide solution.[4]

Calculate the molarity of the NaOH solution.

Analytical Procedure

Accurately weigh 0.20- to 0.30-g portions of dry sample. Dissolve each portion in water, pass through the ion-exchange column, and titrate with standard NaOH solution as before.[5] If the indicator fades, add more as needed. Report the salt

---

[4]It is important that all test portions be transferred quantitatively to the receiving flask for titration. Also important is that the volume of sodium hydroxide used in all the test portions be included in the volume of titrant recorded for the overall titration. To ensure that all has been included, record the buret reading prior to the first addition of sodium hydroxide to a test portion, then transfer all test portions containing sodium hydroxide titrant to the receiving flask, and record the final volume of sodium hydroxide upon completion of titration of the total contents of the receiving flask.

[5]Most errors in this experiment are caused by nonquantitative transfer of samples, incomplete elution of samples, and failure to use alternation. (Standards and samples should be alternated between the two columns so that each type is run on both columns.) Agreement among replicates should be excellent, two parts per thousand or better.

content of the sample as milliequivalents of salt per gram of sample.

Backwashing of the columns after the initial backwashing procedure upon obtaining the columns from the store room is not recommended; an exception would be if the resin is to be completely reconverted to the acid form by repetition of the regeneration procedure before addition of another sample. Do not backwash a column during a run or between samples, but only prior to regeneration.

Calculations

The equivalent weight of a substance in ion exchange is the number of grams of the substance that will replace one mole of H^+ or OH^- on a resin. For example, since one mole of pure KCl will displace one mole of H^+, the equivalent weight of KCl is the same as the molecular weight, 74.56, and the milliequivalent weight (meq wt) is 0.07456. But the milliequivalent weight of K_2SO_4 is mol wt/2000, or 0.08714 g/meq.

Milliequivalents of salt per gram of sample are reported here because the identity of the compound, and consequently the number of millimoles in the sample, are unknown.

19–8   ZINC AND NICKEL BY ION-EXCHANGE SEPARATION AND COMPLEXATION TITRATION

Background

This experiment illustrates an application of ion exchange to the separation of two components of a mixture by formation of a complex. In 2 M hydrochloric acid, zinc(II) forms the negatively charged complexes $ZnCl_3^-$ and $ZnCl_4^{2-}$, which can exchange with chloride ions on an ion-exchange resin in the chloride form. The equilibrium lies far on the side of the resin–zinc chloride complexes:

$$ZnCl_3^- + R^+Cl^- \rightleftarrows R^+ZnCl_3^- + Cl^-$$

$$ZnCl_4^{2-} + 2R^+Cl^- \rightleftarrows R_2^+ZnCl_4^{2-} + 2Cl^-$$

Nickel(II) does not form a stable complex with chloride under these conditions, and since cations do not interact with anion-exchange sites, nickel is not held by the resin. If anion-exchange resin in the chloride form is placed in a column and a solution containing zinc and nickel in 2 M HCl is passed through it, the zinc is held by the resin, but the nickel is not. Once the nickel has been eluted, 0.01 M HCl is passed through the column. This lowers the chloride ion concentration to a level where the anionic zinc–chloro complexes are no longer stable and permits the zinc to be eluted readily from the resin. The chloro complexes of the transition metals vary in stability over a wide range; by a systematic change in the concentration of HCl eluent a number of them can be separated by the procedure described here (Figure 19-9).

After separation the zinc and nickel may be determined in several ways.

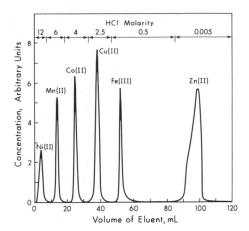

Figure 19-9. Separation of six transition metals on an anion-exchange resin by elution with successively lower concentrations of hydrochloric acid. Adapted from K. A. Kraus and G. E. Moore, *J. Amer. Chem. Soc.* 1953, 75,1460. By permission of the American Chemical Society.

One convenient procedure is by titration with EDTA. Zinc may be titrated in a solution buffered at about pH 8 with *tris*-hydroxymethylaminomethane (THAM, tris); murexide, the ammonium salt of purpuric acid, is a suitable indicator. Nickel may also be titrated with EDTA by use of murexide, in an ammonia buffer at pH 10. Murexide, being unstable in solution, is best added as a solid. Because so little is required for a titration, it is diluted with an inert material such as sodium chloride to permit closer control of the amount dispensed.

Procedure (time 9 to 15 h)

Preparation of 2 M THAM Buffer Solution

Dissolve 24 g of *tris*-hydroxymethylaminomethane in 100 mL of water.

Preparation of Standard 0.02 M EDTA

Use demineralized water throughout the experiment. Dissolve about 4 g of reagent-grade EDTA in 500 mL of water. Prepare a standard 0.06 M solution of $ZnCl_2$ by weighing, to the nearest 0.1 mg, 1 to 1.1 g of pure granulated zinc metal into a 250-mL volumetric flask.[6] Slowly add 5 mL of 12 M HCl. Warm on a hot plate until dissolution is complete; if the reaction stops, add 2 to 3 mL more HCl. When all the zinc is dissolved, rinse the sides of the flask with water, dilute to the mark, and mix. To standardize the solution of EDTA, pipet 10-mL portions of the zinc solution into 500-mL conical flasks. Add 150 mL of H_2O, 10 mL of THAM solution, and 0.2 g of 0.2% solid murexide indicator. Add 6 M HCl dropwise until the solution is bright yellow.[7] Titrate with 0.02 M EDTA solution until the solution is yellow-pink (peach). Next add another 5 mL of THAM solution (to

---

[6]If sufficient 0.025 M $ZnCl_2$ solution remains from the determination of aluminum (Section 16-4) it can be used here. As it is more dilute, 20-mL aliquots are necessary instead of the 10-mL portions specified here.

[7]The pH of the solution will be about 6 at this point. If too much acid is added the solution will turn orange again.

raise the pH of the solution to about 8), and titrate to the end point, which is indicated by a sharp change from the yellow of the zinc murexide complex to the pure pink of the free indicator.

Calculate the molarity of the EDTA solution.

Separation of Zinc and Nickel by Anion Exchange

Obtain a sample containing 3 to 6 mmol each of zinc and nickel, dissolve if necessary, and dilute to volume in a 100-mL volumetric flask, adding sufficient HCl that the final concentration of acid is about 2 M. Assume no HCl is present initially. Obtain two columns about 25 cm in length and 1.3 cm in diameter, and fill to about the 10-cm level with 50- to 100-mesh strong-base anion-exchange resin. Backwash the columns to remove air bubbles; break up any lumps of resin with a glass rod. Wash the columns with about 50 mL of 6 M NH$_3$ to remove any metals present as anionic species, followed by 100 mL of water and then 100 mL of 2 M HCl. The flow rate should be about 2 to 3 mL/min. Do not allow the liquid level to drop below the surface of the resin, or channeling and also entrapment of air bubbles may result.[8]

Once the columns are prepared, allow the liquid to fall to just above the resin, and then pipet a 10-mL portion of sample onto the top of each column.[9] Disturb the resin as little as possible during the addition of sample or eluent. Collect the effluent in a 200-mL conical flask. Wash the inner wall of the column above the resin with several 3- to 4-mL portions of 2 M HCl, permitting the liquid level to come to the surface of the resin bed each time before adding the next portion. Then pass about 50 mL of 2 M HCl through the column to elute the nickel.[10] After elution is complete, place the flask on a steam bath or hot plate in a hood and evaporate to dryness. Do not heat the residue too strongly, or the readily soluble nickel chloride may be converted to insoluble nickel oxide.

Place a clean 500-mL conical flask under the column outlet, and elute the zinc from the column by adding several 2 to 3-mL portions of 0.01 M HCl to the column, letting the liquid level drop to the top of the resin after each addition. Then pass 150-mL of 0.01 M HCl through the column at a flow rate of 1 to 2 mL/min (15 to 20 drops/min). After titration of the zinc the column will be tested to ensure complete removal of zinc as directed in the next paragraph. Do not regenerate the column until a test for absence of zinc has been made.

Titration of Zinc Samples with EDTA

To the flask containing the zinc eluate add 5 mL of THAM solution. Then add 0.2 g of solid 0.2% murexide indicator, mix, and add more THAM solution until the solution in the flask is bright yellow. Titrate with standard 0.02 M EDTA solution to the point where the solution is yellow-pink (orange or peach). Add another 5 mL of THAM solution, and continue the titration to a pure pink. At the

---

[8]Do not backwash the column to remove air bubbles once the sample is on the column. If removal of trapped air is required, clamp the column outlet and stir the top of the resin with a glass stirring rod.

[9]About 2½ h is required to elute completely the nickel and zinc from the column. Plan to complete the elutions in one laboratory period if possible.

[10]The nickel should be completely eluted after 50 mL. If in doubt, test for complete removal as follows. Collect 1 to 2 mL of the column effluent in a test tube, add sufficient NaOH solution to give a pH of about 7, and then add a few drops of 2% sodium dimethylglyoxime solution. A red precipitate indicates the presence of nickel. If nickel is present, add the test portion to the receiving flask, pass an additional 20 mL of 2 M HCl through the column, and retest for nickel.

end point record the volume of titrant, and pass 80 mL of 0.005 M NH_3 through the column at the same flow rate (15 to 20 drops/min), collecting the effluent in the titrated solution. If the yellow of the zinc–murexide complex returns, additional zinc has been removed from the column. In this case simply continue the titration to the pink end point. Record the new volume of titrant, and repeat the NH_3 washing and EDTA titration until all the zinc has been removed from the column, as shown by retention of the pink coloration on further washing. Record the total volume of EDTA solution required.

Before another sample is run through the column, the resin must be reconverted to the chloride form. To accomplish this, pass 50 mL of 0.1 M HCl and then 50 mL of 2 M HCl through the column at a flow rate of 2 to 3 mL/min.

Calculate and report the total grams of zinc present in the original sample.

Before returning the columns, pass 100 mL of 0.1 M HCl through each at a rate of 1 to 2 mL/min to convert the resin to the chloride form. Then wash with distilled water until the effluent is neutral. The resin is more stable when stored in the chloride form. Have your laboratory instructor test the pH of the contents of the column before returning it.

Titration of Nickel with EDTA

Dissolve the residue of nickel chloride in 25 mL of water, add 10 mL of pH 10 buffer (prepared by dissolution of 10 g of NH_4Cl in 150 mL of 6 M NH_3), and dilute the solution to about 100 mL with water. Add 0.2 g of solid 0.2% murexide indicator, and titrate immediately with standard 0.02 M EDTA solution until the yellow of the nickel–murexide complex changes to purple.[11] Record the volume of EDTA solution required.

Calculate and report the total grams of nickel in the original sample.

19–9   *ORTHO* AND *PARA* NITROANILINE IN A MIXTURE BY LIQUID CHROMATOGRAPHY

Background

Nitroanilines are important intermediates in the synthesis of dyes used as titration indicators and ball-point pen inks. The unknown sample in this determination is a mixture of the ortho and para isomers of nitroaniline dissolved in ethylene glycol, a commonly used solvent in these inks. Standards are supplied in the form of solutions of each of the pure isomers, also dissolved in ethylene glycol. Weighed portions of the two standards are combined in flasks and diluted with a methanol solution of an internal standard, 4-ethoxy-2-nitroaniline. A weighed portion of the sample mixture is also diluted with internal standard solution. Sample and standard solutions are injected onto a liquid-chromatographic column. Using peak-height ratios, a calibration curve of three points is plotted for each standard, and the weight of the isomers in the unknown is read from these curves.

The stationary phase in the liquid-chromatographic column is the nonpolar adsorbent Amberlite XAD-2, a highly cross-linked polystyrene resin. Methanol is the mobile phase. The detector consists of a Spectronic 20

---

[11]The purple is a combination of the pink free indicator plus the blue nickel–EDTA complex.

Table 19-3.   Weights to be Taken for Preparation of Standard and Sample Solutions

Flask	p-Nitro-aniline, g*	o-Nitro-aniline, g*	Sample, g*
Standard solution 1	5	15	0
Standard solution 2	10	10	0
Standard solution 3	15	5	0
Sample solution	0	0	20

*Weights should be within 0.5 g and weighed to the nearest milligram.

spectrophotometer set to 430 nm and fitted with a small-volume flow cell (cuvette) through which the eluted mobile phase is passed. A log-converter integrated circuit converts the percent-transmittance output signal from the Spectronic 20 to absorbance, which is fed to a strip-chart recorder; the peak heights thus are proportional to sample concentration in the eluate. Solutions are injected onto the chromatographic column with a 10-μL sample-injection valve.

Procedure (time 3 to 4 h)

Preparation of Internal Standard Solution

Transfer approximately 400 mg of 4-ethoxy-2-nitroaniline to a 100-mL volumetric flask, dissolve in methanol, dilute to volume with methanol, and mix well. This is the internal standard solution. Keep stoppered to minimize evaporation during use.

Preparation of Standard and Sample Solutions

Weigh four clean, dry 50-mL stoppered flasks to the nearest milligram. Using a disposable Pasteur pipet, deliver the specified weight of p-nitroaniline (Table 19-3) into the flasks. Reweigh each flask to the nearest milligram. Repeat this operation with o-nitroaniline, and then with the sample. (Each pipetful of ethylene glycol should weigh about 2 g.) Should you accidentally contaminate the rim or outside of any of the flasks with solution, wipe thoroughly with a lintless tissue moistened with methanol, and then dry with a second tissue. With a volumetric pipet, deliver 20 mL of internal standard solution into each of the four flasks, stopper, and mix well.

Chromatography Procedure [12]

Preliminary Operations. (The laboratory instructor may have performed the following operations. Check before proceeding.)

(1) Turn on the spectrophotometer and recorder power switches and allow 20 min

---

[12]This procedure is for a liquid chromatograph consisting of a pump capable of providing about 250-psi pressure, a 6.3-mm by 20-cm column of 12- to 44-μm XAD-2 resin, an 80-μL flow-through detector cell (178-QS, Hellma Corp.) in a Spectronic 20 spectrometer, and a Houston Omniscribe strip-chart recorder. The injection valve is a 10-μL model CSV slider valve (Laboratory Data Control).

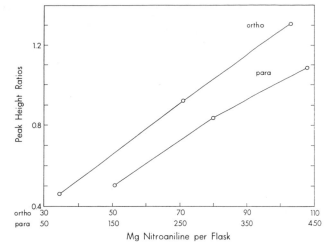

Figure 19-10. Calibration curves for *ortho*- and *para*- nitroaniline.

for warmup. (2) Set the pump for a methanol mobile-phase flow rate of 3 mL/min. (3) Adjust the recorder and spectrophotometer controls to obtain a suitable base-line recorder tracing. (4) Set the recorder speed to about ½ in./min, and turn on the chart-drive switch.

Sample Injection. Fill the injection-valve loop with standard solution, and introduce the sample onto the column. Immediately mark the recorder chart at the point of injection, and label it "standard number 1". While the chromatogram is developing, rinse the injection loop with methanol and fill it with standard solution 2. After the third peak of the first standard has been eluted, inject the second standard. Repeat for standard 3. Obtain each chromatogram in triplicate. It should be possible to make an injection every few minutes.

After the last chromatogram has been recorded, rinse the injection valve with methanol, turn off the recorder chart-drive switch, and inform the instructor. Do *not* shut off the pump, spectrophotometer, or recorder power switches unless instructed to do so.

Calculations

The three major peaks appearing in each chromatogram are, in order, p-nitroaniline, o-nitroaniline, and internal standard. The small peak before the p-nitroaniline peak marks the elution of ethylene glycol, an unretained component in this system. Draw a base line under the peaks, by connecting the base line before the first and after the last peak. With a ruler, measure the height of each peak as the vertical distance from the peak maximum to the base line. Estimate to the nearest 0.1 mm. See Section 18-3 for instruction on peak measurement.

Calculate the ratio of peak heights H_o/H_S and H_p/H_S for each chromatogram, where H_o, H_p, and H_S are the heights for the ortho and para isomers and the internal standard. Determine the average and median peak-height ratios from replicate injections for each standard solution. Plot two calibration curves, one of the average or median values of H_o/H_S against milligrams of o-nitroaniline in the standard-solution flask and the other of H_p/H_S against milligrams of p-nitroaniline. Because the spectral bandwidth of the light from the Spectronic 20 is wider than the natural bandwidth of the absorption spectra of the nitroanilines, the calibration curves will not be straight lines. To obtain a calibration curve, simply join each adjacent pair of points with straight lines (Figure 19-10).

Using the average or median peak-height ratio obtained for each in the sample solution, read the milligrams of each isomer in the sample flask from the calibration curve. From the weight of sample originally placed in the sample flask, calculate and report the milligrams of both o- and p-nitroaniline per gram of sample.

PROBLEMS

19-1.   List the major types of liquid chromatography. Why are there more types than of gas chromatography?

19-2.   Why are mobile-phase velocities normally held to low levels?

19-3.   Describe the chemical structure of typical anion- and cation-exchange resins.

19-4.   Define the distribution ratio for an ion-exchange process. Name the important factors that determine the value of the distribution ratio. How is it related to the partition ratio? (See Section 17-1.)

19-5.   What is meant by selectivity coefficient; by exchange capacity?

19-6.   Write the equations for the reactions involved in the removal of potassium chloride from water by cation- and anion-exchange resins.

19-7.   How can the distribution ratio be varied and controlled in ion-exchange chromatography?

19-8.   Rank the order in which chloride, bromide, and iodide ions would be expected to be eluted from an anion-exchange resin with 0.4 M sodium nitrate as the eluent. Explain your reasoning.

19-9.   A certain solute has a partition ratio of 5.0 on a 0.6- $\times$ 10-cm ion-exchange column where V_s is 2.0 mL and V_m is 1.5 mL. What is its retention volume?

19-10.   Solutions of the following mixtures were passed through a cation-exchange column in the hydrogen form: (a) $NaNO_3$ and $Ca(NO_3)_2$; (b) NaBr, KBr, and $CaBr_2$; (c) K_2SO_4, $MgSO_4$, and Ag_2SO_4. What is the expected order of appearance of ions in the effluent (Table 19-2) if the eluent was a dilute solution of hydrochloric acid?

19-11.   If the ion-exchange partition ratio for an ion on a 0.8- by 20-cm column of resin is 6.3, calculate the retention volume if the volume of the stationary phase is 4.8 mL and that of the mobile phase 4.2 mL. If the flow rate is 3 mL/min, what time is required to elute to the peak maximum?

19-12.   (a) Discuss the applications of ion-exchange chromatography to the removal of interferences; to trace analysis. (b) How is an ion-exchange column regenerated?

19-13.†   Trimethylamine (K_a of trimethylammonium ion is 6×10^{-5}) is eluted from a strong-acid cation-exchange resin in the sodium form. The mobile phase is an aqueous solution containing 0.1 M NaCl and 0.001 M buffer to control the pH. The pH of the mobile phase is 3.0. Indicate whether the retention volume of this compound will increase, decrease, or remain the same if the following changes are made: (a) the sodium chloride concentration is increased to 0.2 M; (b) the pH is increased to 5.0; (c) LiCl is substituted for NaCl; (d) the mobile-phase flow rate is increased.

19-14.   Describe the technique of ion chromatography. Why can the sensitivity with this technique be extremely favorable? Why is it highly desirable to use only dilute electrolyte as the mobile phase?

19-15.   What are the differences and similarities between ion exchange and adsorption chromatography?

19-16.   In adsorption chromatography, how is relative solvent strength described? Which of the following solvents would be described as the strongest: ether, hexane, methanol, chloroform, or benzene?

19-17.   (a) Name the most commonly used adsorbents in adsorption chromatography. (b) What are the main applications of adsorption chromatography?

19-18.   What is meant by gradient elution? How is it used?

19-19.   How can the problem of solubility of a liquid stationary phase in a liquid mobile phase be minimized or overcome?

19-20.   Explain the difference between a conventional liquid stationary phase and a chemically bonded stationary phase in liquid chromatography. Why are chemically bonded phases of lesser importance in gas chromatography?

19-21.   A mixture of organic dyes was separated with a 75-cm column containing a stationary phase of silicone oil chemically bonded to an inert support and a mobile phase of 75% methanol:25% water. All the dyes were eluted within a few minutes and resolution was poor. What recommendation would you make to the operator to improve the separation?

19-22.   What is the mechanism of separation in exclusion chromatography? What are molecular sieves?

19-23.   What are the advantages and disadvantages of high-pressure liquid chromatography? Why is it important to be able to reduce the packing particle diameter?

19-24.   In liquid chromatography, how is the problem of resistance to mass transfer in the stationary phase reduced? What is a pellicular resin?

19-25.   (a) List the advantages of thin-layer chromatography. (b) In what ways can colorless substances that have been separated on a paper or thin-layer chromatogram be detected?

19-26.   Show with the aid of a small diagram and an equation how the R_f value of a compound is measured in thin-layer chromatography.

19-27.   List the common stationary phases used in thin-layer chromatography. What is reversed-phase thin-layer chromatography? Discuss the various kinds of chromatogram development in thin-layer chromatography.

19-28.   What types of salts cannot be determined by the acid-base ion-exchange method for total salt (Section 19-7)? Which of sodium chloride, sodium borate, potassium nitrate, sodium hydroxide, and calcium nitrate can be determined by the procedure? Explain why the others cannot.

19-29.   A 0.2512-g sample of a mixture of sodium salts required 40.98 mL of 0.0972 M sodium hydroxide for titration after passage through an ion-exchange column in the hydrogen form. (a) How many milliequivalents were present in the sample? (b) How many milliequivalents were present per gram?

19-30.   A sample weighing 0.2386 g containing iron(II) sulfate and ammonium sulfate was dissolved in 10 mL of water and passed through a cation-exchange column in the hydrogen form. The effluent required 26.62 mL of 0.1120 M sodium hydroxide. Calculate the percentage of sulfate as SO_3 in the sample.

19-31.   A 0.2218-g mixture containing only sodium chloride and sodium nitrate required 36.71 mL of 0.0995 M sodium hydroxide for titration after treatment by the procedure of Section 19-7. What was the percentage of each salt in the mixture?

19-32.† Predict the number of milliequivalents per gram of each of the following that would be expected after passage through a cation-exchange column in the hydrogen form and titration of the effluent with sodium hydroxide to a methyl red end point: (a) KNO_3; (b) NaOH; (c) LiBr; (d) Na_3PO_4; (e) NaCN; (f) $NaC_2H_3O_2$; (g) Na_2SO_4; (h) $CaCl_2$.

19-33.   A sample weighing 0.6666 g and containing sodium, potassium, and ammonium sulfates was dissolved and passed through a cation-exchange column in the acid form. The effluent required 33.33 mL of 0.1234 M NaOH to a methyl red end point. What was the percentage of sulfate in the sample?

19-34.   In the procedure of Section 19-8 (a) why is the solution containing the separated nickel evaporated to dryness? (b) Why is the zinc retained on the ion-exchange column, whereas nickel is not? (c) Why is zinc strongly retained on an anion-exchange resin from 2 M HCl but not significantly from 0.01 M HCl?

19-35.   In the separation of zinc and nickel by ion exchange the ion-exchange columns are conditioned by passage of 50 mL of 6 M NH_3, followed by 100 mL of water and 100 mL of 2 M HCl. What is the purpose of the ammonia and 2 M HCl treatments?

19-36.   A 1.2437-g sample of an alloy was dissolved and analyzed for zinc and nickel by the procedure of Section 19-8. If 33.84 and 38.71 mL of 0.009728 M EDTA were required for titration of the zinc and the nickel eluted from the column, what was the percentage of each metal in the sample?

19–37.† The zinc and cadmium in a 0.2411-g sample are separated by cation-exchange chromatography. The effluent portions are titrated with 0.02467 M EDTA solution. The separation point occurs at 32.16 mL, and the total elution of both metals occurs at 41.67 mL. What is the percentage of each metal in the sample? (*Hint*: Predict which metal will be eluted first.)

19-38.   In the experiment of Section 19-9 must the weight of p-ethoxy–2-nitroaniline used to make up the internal standard solution be known accurately? Explain.

19-39.   In the experiment of Section 19-9 must the volume of internal standard solution delivered into each of the standard and sample flasks be known accurately? Must it be known precisely? Explain.

19-40.   In the experiment of Section 19-9, will the results be in error if methanol in the sample solution evaporates faster than methanol in the standard solution? Explain.

REFERENCES

F. Cantwell and D. Brown, *J. Chem. Educ.* In press. Provides additional details on the experiment of Section 19-9.

H. Determann, *Gel Chromatography*, Springer-Verlag, New York, 1968.

I. M. Hais and K. Macek, Eds., *Paper Chromatography*, Czechoslovak Academy of Sciences, Prague, 1963.

K. A. Kraus and G. E. Moore, *J. Amer. Chem. Soc.* **1953**, 75, 460. Description of the separation by anion-exchange chromatography of the transition metals as their chloro complexes.

W. M. MacNevin, M. G. Reilley, and T. R.Sweet, *J. Chem. Educ.* **1951**, 28, 389.

T. H. Maugh, *Science* **1980**, 208, 164. Describes recent developments leading to extremely favorable detection limits.

W. Rieman and H. F. Walton, *Ion Exchange in Analytical Chemistry*, Pergamon Press, New York, 1970.

R. H. Schuler, A. C. Boyd, Jr., and D. J.Kay, *J. Chem. Educ.* **1951**, 28, 192.

L. R. Snyder and J. J. Kirkland, Ed., *Introduction to Modern Liquid Chromatography*, 2nd ed., Wiley, New York, 1980.

E. Stahl, Ed., *Thin Layer Chromatography*, 2nd ed., Academic Press, New York, 1975.

20

AUTOMATION IN CHEMICAL ANALYSIS

The accomplishments resulting from automation are recognized even by laymen, who marvel at the knowledge gained by automated instruments in the analysis of the surfaces of the moon and of Mars.

I. M. Kolthoff and P. J. Elving

The increasing demand for chemical analyses is a result not only of the increasing need for quantitative chemical information but also of new instrumentation and methodology that make available measurements that were previously impossible. Important areas of application of automatic chemical methods include clinical chemistry, industrial process control, and environmental monitoring of chemicals in water, air, and soil. Analytical chemists, working with scientists and engineers from many areas, are incorporating into instruments more and more capability for automatic operation. The purposes of automation are several — to speed up analyses, to reduce the cost of manpower, and to improve reproducibility of operations. An additional benefit is relief from the tedium of repetitive, routine manual operations. Increasing government regulations and more tightly controlled plant operations (for efficiency and improved product quality) have increased the need for dedicated process- and product-measuring instruments. This whole field is rapidly expanding.

The electronics revolution has put at the analyst's disposal a powerful tool. Apparatus can now be built that will not only automatically perform individual operations such as weighing and pipetting but also add reagents, measure reaction rates or concentrations of products, and calculate analytical results. Further, instruments can be designed to carry out periodic checks on their own operation by running standards and making necessary adjustments.

The control for these operations is provided by computers of varying sophistication. A large number of analytical instruments today have small built-in computers, called *microprocessors*, that carry out internal programs to check on proper functioning. They can accept a limited range of commands, say to vary the solvent composition in gradient elution in liquid chromatography or to collect and process the absorbance data in an atomic-absorption spectrophotometer. Other instruments are connected, or interfaced, with separate minicomputers for control and data processing. The topics of computers, microprocessors, and their analytical applications cannot be treated here. For more information one of the advanced textbooks in the field should be consulted.

A distinction should be made between automatic and automated instruments. An *automatic* device performs an operation or sequence of operations in a repetitive way without human intervention. An automatic dishwasher, once started, can carry out washing, rinsing, and drying cycles without attention. Similarly, an automatic titrator might record the titration curve, or a spectrophotometer an absorption spectrum. In an *automated* device the instrument is controlled to some extent by input of prior information and to some extent by feedback of information from a sensing device. For example, a pH-stat will measure the pH of a solution and add acid or base as required to hold the pH at some preset value. An automated titrator may add an increment of titrant and, on the basis of the rate of change in the signal from the sensing electrodes, wait for the reaction to reach equilibrium before the next increment is added; or, on the basis of the net change in the signal from the sensors, it may add a larger or smaller increment of titrant the next time.

In this chapter we consider two automatic methods of analysis. The first employs instruments that perform a single operation, and the second, those that perform multistep analyses. Instruments in the second group can be divided further into two general types. In one type, a *continuous-flow analyzer*, samples are successively drawn into a portion of tubing, reagents added, and the mixture pumped through a measurement device (usually a flow-through photometer cell). In the second type, a *discrete analyzer*, all operations are performed on a sample in a single separate container.

The increasing reliability of sophisticated automated instruments, together with their speed and wide range of analytical applicability, has made them more and more important. The cost of incorporating these capabilities, though high, continues to drop, and the future for this field appears unlimited.

20–1 AUTOMATION OF SINGLE OPERATIONS

Early efforts at replacement of manual operations in the analytical laboratory were directed toward the measurement step. One was to follow the course of an acid-base titration by feeding the output from a pH meter to a strip-chart recorder. The titrant was delivered at a slow, constant rate so that the distance along the chart paper of the recorder was directly proportional to the volume of titrant; the operator then selected the end point of the titration from the curve. A sensing device was later added that compared the voltage difference between the pH and reference electrodes with the value corresponding to the selected end point. When the desired voltage was reached, an electrical signal activated a mechanical device to stop delivery of titrant. The operator then recorded the titrant volume at leisure. A similar arrangement, consisting of a lamp and photocell to detect the color transition of an indicator at the end point, was also devised and used for acid-base, oxidation-reduction, and complexation titrations.

Automatic samplers, which sample process or product streams at preset intervals, also have long been used. An example is gas-chromatographic sampling valves in industrial-process streams. These are activated by either small electric motors or by air pressure, and they deliver volumes of sample with high precision. Improvements in this area have been more mechanical and electrical than chemical. Automatic treatment of data has also become widespread through use of electronic calculators that provide rapid, inexpensive processing of output signals from measurement detectors (transducers). Inclusion of this capability in instruments saves both money and operator time and also leads to fewer calculation errors. However, the operator must possess a reasonable understanding of how the instrument functions and how the data are processed if he is to obtain maximum efficiency and productivity from it. Moreover, he must be able to recognize likely errors in both the input data provided to the processing unit and in the functioning of the unit itself.

Operations in advance of the measurement step, such as dissolution or separation, often require complex manipulations, and as a consequence they have not been so successfully converted to automation as other parts of the analytical process. Nevertheless, automation of the entire process, from sampling to measurement and recording of the result, is increasing.

20-2   CONTINUOUS ANALYSIS

Continuous-Flow Analyzers

Automatic analyzers are increasingly being employed in industrial and clinical laboratories where large numbers of samples similar in composition must be determined. Several types have been invented, and new designs and improvements appear regularly. One important system, especially in hospital and clinical laboratories, is the continuous-flow analyzer. In continuous-flow analysis all the individual steps, ordinarily performed manually as discrete operations, are performed by the instrument. A solvent is pumped at a constant rate through a series of components in which individual operations such as dialysis, reagent addition, and heating occur continuously. Samples are introduced into the solvent stream one after the other, each being subjected to all the operations in turn. The final solution then flows through a detector, often a colorimeter cell where absorption by the solution of visible radiation from a lamp is measured with a photocell, and the transmittance of the solution is recorded. Alternatively, an electrochemical sensor such as a selective electrode may constitute the measuring device. Many analyses have been automated by use of this type of apparatus, which can execute a range of operations such as sample dissolution, filtration, solvent extraction, dialysis, distillation, addition of color-producing reagents, incubation at elevated temperature, and photometric or potentiometric measurement.

By means of parallel multichannel systems as many as 20 components in a sample can be determined simultaneously at rates of up to 150 samples an

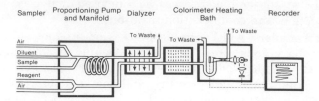

Figure 20-1. Components and flow scheme for Technicon single-channel analyzer.
From Technicon Operating Manual.

hour. [1] The results are often transmitted to an automatic printout device. Systems of this kind are used routinely for the determination in biological fluids of metal ions, enzymes, and other substances whose concentration levels provide diagnostic information.

One of the most widely used commercial continuous analyzers is manufactured by the Technicon Corporation.[2] Figure 20-1 is a schematic diagram of a single-channel Technicon AutoAnalyzer. Sample, reagent, diluent, and air are drawn into plastic tubing by a proportioning pump. The pump consists of a series of chain-drive rollers that press against a flat plate. The tubes containing reagent, sample, diluent, and air pass between the rollers and the plate, and their contents are squeezed ahead by positive displacement. The volume of each fluid delivered per unit time can be varied by changing the diameter of the individual tubes. The system is designed to introduce a series of air bubbles into the sample and reagent streams. These bubbles aid in reducing cross contamination between samples. They also help mix the contents of two streams such as sample and reagent after the streams have been combined. The pump draws sample or standard solution into the system through an inlet probe immersed in one of a sequence of small plastic cups on a turntable. In operation the inlet probe dips into the first cup and draws its contents into the tube for a period of time. The probe then moves out of the cup into a rinse solution of distilled water. The rinse helps separate individual samples and may provide a base-line check on the recorder output. While the probe is in the rinse solution, the turntable containing the sample cups advances another cup into position for the start of the next cycle. The rate at which the cups are sampled, along with the relative lengths of the sample and rinse periods, can be adjusted by changing a plastic cam on the sampler unit.

In analysis of real samples, substances that interfere with the measurement operation are the rule rather than the exception. Because methods of measurement possessing high specificity are rare, preliminary separations must generally be resorted to. A separation technique especially

---

[1]Technicon SMAC (sequential multiple analyzer, computer-controlled) has an effective rate of about 100 patient samples per hour, the other 50 runs being concerned with control and calibration samples and with rechecks of questionable values. The net result is about 2000 individual analyses per hour, a prodigious number for a single instrument.

[2]The AutoAnalyzer is based on early work and patents of a physician, L. Skeggs, who built the initial models in his basement during evenings and weekends. Technicon and AutoAnalyzer are trademarks of Technicon Instruments Corporation, Tarrytown, N.Y.

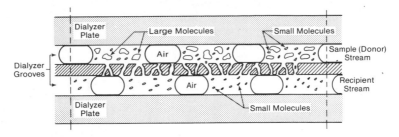

Figure 20-2. Schematic diagram of the process of dialysis in the Technicon system.
From Technicon Operating Manual.

useful in biological work is *dialysis*, which is the separation of solutes based on relative rates of diffusion through a porous membrane. A magnified cross section of a membrane would show a large number of small, tortuous pathways through which the solution components move by diffusion (Figure 20-2). Small molecules pass across the membrane readily; large ones pass only slowly, if at all. The driving force is provided by differences in concentration (Section 16-2) across the membrane. Dialysis separations are well suited to continuous-flow analyzers.

In the Technicon dialysis system the dialyzing membrane (typically cellophane) is sandwiched between two plates. A matching set of spiral grooves on the two plates is in contact with both sides of the membrane. As sample solution is pumped through the groove on one side of the membrane, the substance to be determined diffuses through the membrane to a solution of reagent being pumped along the corresponding channel on the other side. The sample is thereby separated from interfering substances. Since both samples and standards are subjected to the same conditions before and during dialysis, not all of a sample component need cross the membrane. If the quantity crossing is proportional to the amount present in the original sample, the results will be accurate.

Many reactions between sample and reagent are sufficiently rapid that the mixture from the dialysis operation can be pumped directly to the colorimeter. If the reaction is slow, additional time can be provided by inserting a long glass or plastic coil of tubing between the dialyzer and colorimeter to act as a delay. These delay coils are sometimes immersed in a heating bath to further accelerate reaction.

In the colorimeter the stream passes into a flow-through cell, where the percent transmittance is continually monitored by a phototube and displayed on a strip-chart recorder. As each sample passes through the cell, the recorder output rises to a plateau, or peak, value related to the amount of sample present. As the between-sample rinse solution passes, the output drops back to a base line corresponding to zero sample. This is followed by a new plateau value as the next sample moves through the colorimeter cell (Figure 20-3). If a sample is concentrated and the change in percent transmittance is therefore large, the recorder trace may not return completely to the base line between samples. This occurrence does not affect the results. After passage through the colorimeter the solutions are discarded.

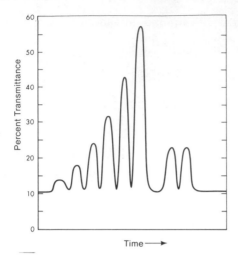

Figure 20-3. A strip-chart recording of a series of glucose standards on a Technicon AutoAnalyzer, followed by duplicate runs on a sample. In the method shown glucose reacts with ferricyanide; the resulting decrease in concentration of the ferricyanide reagent is measured. (See Section 20-3.)

Although analysis by the continuous-flow approach has been applied to many systems other than biological fluids, its foremost application is in clinical laboratories. Methods for over 30 different components in blood serum have been developed. A modification of the technique even allows white blood cells to be counted: whole blood is aspirated, cell concentrations are reduced by dilution, stains are added that selectively color different cell types, and the number of each cell type is counted by scattering and absorbance measurements at different wavelengths.

Replacement of photometers by other detectors such as flow-through selective electrodes has expanded the applicability of continuous-flow analysis to the determination of many cations, anions, and gases and to the indirect determination of many enzymes or substrates. For example, urea can be determined by measurement of the ammonia or carbon dioxide formed on hydrolysis catalyzed by the enzyme urease:

$$H_2O + (H_2N)_2C{=}O \rightarrow 2NH_3 + CO_2$$

Concentrations of either of the reaction products can be monitored with an appropriate gas-sensing electrode. The same reaction can be applied in the measurement of the activity of urease by providing an excess of the substrate urea.

Kinetic analyses can be performed in continuous-flow systems by mixing sample and reagents, taking a measurement, and then passing the mixture through a delay coil to a second detector. The difference in signal between the two detectors gives a measure of the rate of the reaction. In this way enzymes can be determined with relative freedom from the problems of single-point measurements, where competing nonspecific enzyme reactions or other conditions in the system might affect reaction rates.

In nonclinical analyses, dissolution of the sample is often the slow step. A system for handling solids such as drug tablets provides automatic introduction of the solid samples, one after another, into a blender where the sample is treated with an appropriate solvent and mixed for a prescribed time. The solution is then drawn into a sample tube for completion of the analysis. Continuous-flow routines have also been perfected for operations such as filtration, liquid-liquid extraction, ion-exchange separation, and distillation. Unstable reagents needed for an analysis can sometimes be synthesized on-line in a continuous-flow scheme and combined immediately with the sample.

Discrete Analyzers

Discrete analyzers take a measured portion of sample, add the necessary reagents, and execute the required steps such as stirring, filtration, incubation, and measurement. Discrete analysis has not received the widespread adoption of continuous-flow analysis because it is not so well suited to transport of sample from one step to the next. However, it is inherently more flexible in application, being more easily switched from one method to another and more useful for small numbers of samples of varying composition. Both the following examples of discrete analyzers are designed primarily for clinical application.

The DuPont ACA (Automatic Clinical Analyzer) system provides a separate plastic packet containing the reagents required for a single determination. The packet serves as the reaction chamber during the preliminary operations and as the cuvette for the photometric measurement. When a particular analysis is ordered, the appropriate analytical test packet is introduced into the instrument along with a portion of the sample. The instrument identifies the determination to be performed by reading a binary code on the packet with an optical scanner and programs itself to carry out the operations needed for that determination. The necessary amounts of sample and diluent are injected into the packet, mixed, incubated as needed, and the packet then is transported into a photometer compartment where air is injected into it to form a precise optical cell. After the photometer has read the transmittance, the packet is ejected, and a computer calculates the concentration value and prints the results on a report sheet.

For some determinations disposable chromatographic columns are built into the packet, and reagents in separate interior compartments can be introduced at different times by breaking internal seals. A variety of analyses can be performed on a sample by introduction of the necessary packets. Over 30 different methods are possible; a major advantage is that a variety of tests can be done quickly, in any order. Since the need for an individual packet for each analysis makes the cost per analysis generally greater than for other methods, the ACA system is most widely used in large hospitals for emergency analyses at night or on weekends.

Another discrete system is the centrifugal analyzer developed at Oak Ridge National Laboratory. Samples of a few microliters are pipetted into a ring of separate cups on a rotor. The necessary reagents are pipetted into another set of cups located in a second ring outside the first. The rotor is then spun at increasing speed until the sample is transferred by centrifugal force through a channel into the reagent cup, from which the mixture is transferred

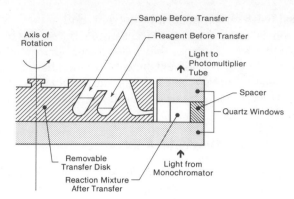

Figure 20-4. Cross section of the sample rotor and photometric analyzer section of the centrifugal analyzer. From R. T. Coleman, W. D. Schultz, M. T. Kelley, and J. A. Dean, *Amer. Lab.* 1971, *3*(7), 26.

into a photometer measuring cell (Figure 20-4). The photometer is synchronized to read the transmittance as a function of time for each of the cells in the perimeter of the instrument, and the data are stored in a computer. Transmittance readings are usually taken at a rotor speed of about 600 rpm, and so hundreds of readings can be taken and averaged by the computer in a short time. Since as many as 42 sample cups can be provided on a single rotor, samples and standards can be run at the same time. Determinations are possible on microliter quantities of sample, with results available within seconds.

The centrifugal analyzer does have some disadvantages. First, only a single determination can be performed on a sample per run, in contrast with the sequential multiple analyses possible in continuous-flow methods. Second, samples must be individually pipetted into the rotors (though automatic pipetting systems have been devised). The system, however, can be adapted to kinetic measurements if desired.

20–3 DETERMINATION OF GLUCOSE BY AUTOMATIC ANALYSIS

Background

The determination of glucose in blood or urine is a common clinical test. High values of glucose may be found, for example, in diabetes, nephritis, diseases of the liver and thyroid, and after administration of adrenalin and anaesthetics. Qualitative analysis for sugar in urine was one of the first routine medical tests.[3]

In this experiment the amount of glucose in a solid sample is determined on a continuous-flow analyzer by oxidation with potassium ferricyanide in alkaline solution. In solution glucose occurs primarily in two isomeric ring forms, α and β; these are in equilibrium with a small amount of an open-chain aldehyde form. Only this form reacts with ferricyanide; therefore, any increase in its concentration at equilibrium increases the rate of reaction.

---

[3]Early physicians tested urine samples for sugar by tasting.

Even though the reaction is slow and not stoichiometric, neither characteristic is a major drawback to its use in a continuous-analysis system provided samples and standards are run under the same conditions. Since both glucose and its oxidation products are colorless in the visible region of the spectrum, the reaction is followed by a decrease in intensity of the color of the ferricyanide solution due to reduction to ferrocyanide.

The glucose sample is dissolved and mixed with a diluent solution containing approximately 0.9% (0.15 M) sodium chloride. The salt is added to bring the sample solution to about the same ionic strength as the reagent solution on the other side of the dialyzer membrane. This salt concentration is also close to the level present in blood. Maintenance of the same ionic strength on both sides of the membrane minimizes solvent flow across the membrane by osmosis. Air bubbles are also introduced into the sample stream at this point. A wetting agent is added to the saline diluent and to the potassium ferricyanide oxidant to facilitate smooth flow of the solution through the tubing and to minimize holdup of portions of solution or of air bubbles in the system.

The sample stream is pumped into the dialysis cell on one side of the membrane. The glucose diffuses through the membrane, leaving behind the high-molecular-weight components of the sample. The recipient stream is an alkaline solution of potassium ferricyanide and sodium chloride. The glucose–ferricyanide mixture next passes through a glass coil in an oil bath at 95°C, where glucose reduces ferricyanide to ferrocyanide, and then flows to the debubbler and colorimeter cell.

The system is calibrated by running a series of standards containing varying known amounts of glucose, as shown in Figure 20-3. A calibration graph of change in absorbance against grams of glucose per liter is prepared and used to give the concentration of the unknown samples. Figure 20-5 displays an example of a calibration graph.

Figure 20-5. Example of a calibration graph for a set of glucose standards run on a Technicon AutoAnalyzer.

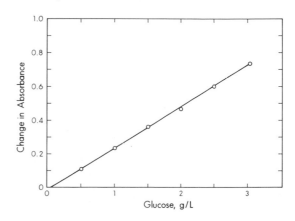

Procedure (time 3 h)

Preparation of Solutions

All solutions should be prepared well in advance of the time the analysis is to be run. The potassium ferricyanide and saline diluent solutions may be provided.

Stock Glucose Standard Solution. Dry about 2 g of glucose at 80°C for 1 to 2 h.[4] Weigh, to the nearest 0.1 mg, approximately 1.50 g of the dry glucose into a small beaker. Quantitatively transfer the glucose to a 250-mL volumetric flask, dilute to volume with distilled water, and mix.

Glucose Working-Standard Solutions. With a buret measure, to the nearest 0.01 mL, 8.0, 18.0, 25.0, 33.0, 42.0, and 50.0 mL of stock glucose standard solution into a set of six 100-mL volumetric flasks. Record the volumes. Dilute each to volume with distilled water and mix.

Glucose Sample Solution. Obtain a glucose sample. (Consult your instructor for details regarding sample preparation.) If it is a solid and the entire sample is to be taken, quantitatively transfer it to a 100-mL volumetric flask, using small portions of distilled water to rinse the last traces into the flask. Dilute the solution to volume with distilled water and mix. If only a portion of the original sample is to be taken, dry the sample at 80°C for 2 h and then weigh, to the nearest 0.1 mg, an amount sufficient to provide from 0.1 to 0.2 g of glucose per 100 mL of solution. Transfer the weighed sample quantitatively to a 100-mL volumetric flask, dilute the solution to volume with distilled water, and mix. If the sample is a solution, as one of blood serum, it may be ready for use directly or may require dilution to bring the glucose into the required concentration range of 0.1 to 0.2 g/100 mL.

Alkaline Ferricyanide Solution. Weigh, to the nearest 0.1 g, 4.5 g of sodium chloride, and quantitatively transfer it to a 500-mL volumetric flask. Also weigh 0.18 ± 0.01 g of potassium ferricyanide and 10 ± 0.1 g of anhydrous sodium carbonate, and add them to the flask. Dilute to volume with distilled water, and add 5 drops of Brij 35 wetting solution.[5] Mix well. Transfer to a 1-L amber plastic bottle. [6]

Saline Diluent (0.9% saline solution). In a 500-mL volumetric flask, dissolve 4.5 ± 0.1 g of sodium chloride in distilled water. Dilute to volume and mix. Then add 5 drops of Brij 35 wetting solution and mix again.

Measurement

The following procedure is written for a Technicon single-channel AutoAnalyzer. Operation of the instrument should be carried out under the supervision of the laboratory instructor.

With all the inlet lines in distilled water, lock the roller-head assembly of the pump into place and turn on the pump. (The on-off switch is located on the back of the proportioning pump module.) Turn on the colorimeter; a warm-up period of about 30 min is necessary to allow the mechanical and electronic systems to stabilize.

When ready to begin measurements, turn on the recorder main switch and chart drive. Keeping all the reagent lines and the sample probe in distilled water,

---

[4]Do not dry longer than 2 h, or some decomposition may occur.

[5]Addition of wetting agent before dilution to volume would affect adversely the shape of the meniscus. The volume of wetting agent is sufficiently small relative to 500 mL that it can be considered negligible.

[6]This reagent is best when fresh. If stored in an amber bottle, it may be used for up to 1 month after preparation.

with the 100%-T control on the colorimeter, adjust the recorder pen to between 96 and 99% transmittance on the recorder scale. Remove the cover from the colorimeter, insert the zero aperture (a solid strip of metal that blocks passage of light to the photodetector) on the sample side (where the flow cell is located), and adjust the recorder reading to zero with the zero control. Remove the zero aperture and replace the cover. Place the 0.090-in. reagent line (black and purple collars) in the saline diluent solution and the 0.100-in. reagent line (purple and orange collars) in the alkaline ferricyanide solution. When equilibrium is attained, the recorder should give a flat reagent base line of 13 ± 3% transmittance.[7]

Shake the glucose solutions thoroughly, and then rinse and fill six sample cups with the six glucose working-standard solutions. Similarly rinse and fill two sample cups with the sample solution. Fill all cups to the same height, not more than 1 to 2 mm from the top.[8] Place the cups in the sampler tray; the tray rotates clockwise. Follow this set of eight cups with two more sets of eight in the same order. Place a cup filled with distilled water between each set of eight. Install the dust cover and start the sampler motor. A set of peaks similar to those in Figure 20-3 should be obtained. After the last solution has been sampled and the probe placed in distilled water, turn off the sampler motor and remove the dust cover and sampler tray. Empty and wash the sample cups and invert to dry. Consult the laboratory instructor concerning shutdown of the instrument.

Calculations

If the recorder reads in "percent transmittance," convert the readings for the base line (alkaline potassium ferricyanide) and for each of the samples and working-standard peaks to absorbance. Subtract each peak absorbance from the base-line value to obtain the change in absorbance. Plot a graph of average absorbance change against milligrams of glucose per 100 mL of solution for the working standards, and read the concentration of the unknown sample from the plot for the standards. Report (1) the milligrams of glucose in the original sample if it was a solid and the whole sample was taken, (2) the percentage of glucose if only a weighed portion of solid sample was taken, or (3) the concentration of glucose in mg/100 mL if a liquid sample was provided.

PROBLEMS

20-1.   Define the following: continuous-flow analysis, discrete analysis, pH-stat, dialysis.

20-2.   What are the primary goals of automatic analysis? How are they achieved through modern instrumentation?

20-3.   Describe the differences between an instrument designed for monitoring of industrial quality control and one designed for an analytical research laboratory.

20-4.   Explain the difference between an automatic operation and an automated operation.

20-5.   Compare the advantages and disadvantages of automated analysis by continuous-flow and by discrete analyzers.

---

[7]If the base line is outside this transmittance range, the transmittance of the standard solution will not fall into the optimum range, and the accuracy of the results will be decreased.

[8]If the sample cups are so full that solution touches the dust cover, cross contamination may occur.

20-6.   Explain why in a typical hospital setting only about two thirds of the measurements on a Technicon SMAC continuous-flow analyzer are on patient samples.

20-7.   What are the differences in the isomeric forms of glucose? Why is the form in which glucose is present in solution important in its determination by the procedure of Section 20-3?

20-8.   Explain the function of each of the five major instrumental components of the Technicon AutoAnalyzer as used in the determination of glucose by ferricyanide oxidation.

20-9.   Although the reaction between glucose and ferricyanide is slow and nonstoichiometric, it can be used for the quantitative determination of glucose by continuous-flow analysis. Explain.

20-10.   Why is a wetting agent necessary in the solutions pumped through the Technicon AutoAnalyzer system?

20-11.   Explain the purpose of adding air bubbles to the streams of solution in the Technicon AutoAnalyzer system. Why must they be removed before the streams are passed through the colorimeter cell?

20-12.   In the glucose determination of Section 20-3, explain the purpose of (a) the dialysis step, (b) the saline diluent solution, and (c) the heating bath.

20-13.   For the determination of glucose in a sample of blood serum by the procedure of Section 20-3 the following data were recorded: weight of glucose standard taken, 1.5280 g; volumes of stock glucose standard solution taken for preparation of the working standards, 8.32, 17.05, 25.01, 33.06, 42.02, and 49.94 mL; % T for base line, 12.8; % T for working standards, 78.5, 57.8, 44.7, 33.9, 25.1, and 19.1; and % T for the sample of serum run directly, 47.9. Calculate the glucose in the serum in milligrams per 100 mL.

REFERENCES

K. S. Fletcher III and N. L. Alpert in *Instrumental Analysis*, H. H. Bauer, G. D. Christian, and J. E. O'Reilley, Eds., Allyn and Bacon, Boston, 1978, Chapter 24.

C. E. Klopfenstein and C. L. Wilkins, Eds., *Computers in Chemical and Biochemical Research*, Academic Press, New York, 1972, Vol. 1; 1974, Vol. 2.

S. P. Perone and D. O. Jones, *Digital Computers in Scientific Instrumentation: Applications to Chemistry*, McGraw-Hill, New York, 1973.

L. Snyder, J. Levine, R. Stoy, and A. Conetta, *Anal. Chem.* **1976**, *48*, 943A. Discussion of developments in automatic analysis, with emphasis on continuous-flow systems.

P. Wilks, *Anal. Chem.* **1980**, *52*, 222A. An assessment of the increasing demand for dedicated instruments in analytical process control.

21

SAMPLING.
PRELIMINARY STEPS IN ANALYSIS

The classic example of incorrect sampling procedure and its ridiculous consequences is given by the fable of the blind men and the elephant. The consequences are sometimes no less ridiculous for incorrect chemical sampling.

W. J. Blaedel and V. W. Meloche

As instruments become more sophisticated, with minicomputer or microprocessor control of many operations, the analytical process may appear to be approaching more and more the concept of a black box that requires only inserting a sample and pushing a button to produce an accurate result on a digital readout. But the reality is that an analytical black box, even though it can perform many operations more rapidly and reproducibly than heretofore possible, still requires an operator with analytical experience and competence. Nowhere is chemical knowledge and skill more critical than in ensuring that the sample is representative.

Sampling refers to the operation of obtaining a suitable amount of material for analysis that is representative of the bulk composition. The value of an analysis is often determined by the degree to which the sample analyzed truly represents the body of material under study. The aim in the sampling operation is to provide such a sample with the least effort and cost commensurate with the precision required. Because the process can be tedious and expensive, shortcuts may be tempting. Nevertheless, if analytical results are to be meaningful, proper attention to this step is essential.

Sampling requirements vary widely, depending on the nature of the bulk material. This material may be gaseous, liquid, or solid, heterogeneous or homogeneous, large or small. Some substances are readily sampled; the composition of a milliliter of blood taken from a patient's vein generally represents with reasonable accuracy the overall composition of the venous blood at that time. Others, such as an ore body to be analyzed for gold, a river sediment for methylmercury, or a warehouse of peanuts for aflatoxin, are not so easily sampled. Once a bulk sample has been obtained, numerous steps such as crushing, subdividing, and grinding may be needed to produce portions suitable in size and composition for analysis. Each of these steps can cause segregation and thereby bias the sample if precautions are not taken.

The sampling of materials that occur in discrete units (tablets in a pharmaceutical plant or tins of food in a cannery) presents special problems. Procedures have been worked out for materials composed of essentially uniform sampling units. Another special problem is presented by large volumes of liquids. Efficient sampling, especially of streams and lakes, requires knowledge of the degree of homogeneity of the liquid both vertically and horizontally. Natural mixing is generally much less than one might expect, and the amount of dissolved material may vary greatly from center to edge and from top to bottom.

21-1 OBTAINING REPRESENTATIVE SAMPLES

The first step in sampling is to identify the population from which the sample is to be obtained, that is, the total material to be sampled. Then a *gross sample* is taken, that is, a sample that includes representative amounts of all components in the bulk population in terms of both composition and particle size. As far as possible every member of the population should have an equal chance of being selected. The size of the gross sample is determined on the one hand by the heterogeneity of the bulk material and the precision desired in the analysis and on the other by the cost of the sampling operation. Gross samples of gases or solutions that are relatively homogeneous often need not be large. If possible, thorough mixing should precede sampling; where mixing is impossible, as with natural waters and many waste streams, samples should be taken from several locations at several different times. A convenient device for this purpose is a *sample thief*, a container that can be positioned where desired, opened, filled, and closed before removal.

Metals and alloys are typically sampled by drilling or sawing to obtain turnings or dust over a cross section of the material. A single piece of metal may vary considerably in composition across its width, especially in trace elements, and should be sampled with care.

Solids such as rock, soil, organic tissue, chemical products, and grains present a variety of challenges. A random gross sample is often most easily taken while the material is being loaded or unloaded. Heterogeneity is a serious problem with many solids; particles of completely different composition are frequently present, such as rock fragments in soil or coal. The problem of sampling particulate mixtures is considered in Section 21-2.

Reduction of the gross sample to a finely divided *laboratory sample* is often necessary. This is accomplished by carrying out a cycle of operations of crushing, sieving, mixing, and retaining a portion (often one half) several times until the final small amount of material is sufficiently finely ground to be suitable for analysis. As with the gross sample, the size of the laboratory sample depends on numerous factors, of which the principal ones are (1) the required precision and accuracy of the analysis, (2) the degree of heterogeneity in the original material, and (3) the level of concentration of the sought-for substance.

Particle size may be reduced by a variety of tools such as a jaw crusher or disk pulverizer for large lumps and samples, a ball mill for medium particles and samples, and an agate or mullite mortar and pestle for small samples. Since some components are reduced in size more easily than others, they may bias the sample if all the material is not ground. At this stage the procedure is to grind and sieve and then to regrind and resieve the material not passing through the first time. This sequence is continued until all the material has passed the sieve. Most solids for laboratory analysis should be ground to pass a 100-mesh sieve,[1] which generally ensures that all particles are sufficiently small. Sieves are commonly made of brass or stainless steel, though sometimes of plastic for materials that attack metals.

After the particles have been reduced to the required size, various methods are employed to divide the material. One is *coning and quartering*, in which opposite quarters of the sample are retained. Another is use of a *riffle*, a device consisting of a hopper that feeds the sample to a series of narrow, sloping chutes of equal width arranged so that adjacent ones pass portions of the sample in opposite directions. For samples of only a few grams the material can be poured onto a clean surface and portions for analysis taken from the pile.

A convenient method for mixing or blending a material during sample preparation is to fill a bottle or jar no more than two-thirds full and rotate it end over end for several hours. The best mixing device available for solids is a *V blender*, a rotating V formed from two intersecting cylinders, which splits and recombines the sample on each rotation. To mix small portions of a sample it may be poured onto a sheet of glazed paper and rolled back and forth by lifting alternate corners of the paper in succession.[2]

Segregation of the various sample components after mixing should be guarded against. Smaller particles and those of higher density tend to sift to the bottom of stored particulates. Homogeneity is obviously more likely to be maintained in a material of uniform density and particle size.

21–2 ERRORS IN SAMPLING

Sampling errors arise from both determinate and indeterminate (random) sources.

Determinate Error

Errors from determinate sources can be minimized by careful attention to the collection of a gross sample and its subsequent reduction to a laboratory sample. A significant source of determinate error in trace analysis is contamination. An apt example is the determination of lead in foods.

---

[1]The size of the openings is generally indicated on the sieve, along with the mesh number. In the U.S. Standard Series a 100-mesh sieve has 100 wires per inch, with openings of about 152 μm. Other examples are 35-mesh, 500 μm; 140 mesh, 104 μm; and 200 mesh, 76 μm.

[2]An impressive example of particulate material that cannot be mixed is the toy "Magic Window," manufactured by Wham-O. The light and dark components remain segregated after repeated attempts at mixing, apparently because of differences in density.

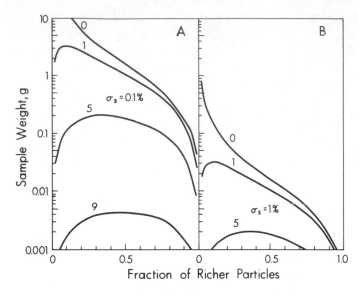

Figure 21-1. Relation between minimum sample size and fraction of the richer particles in a mixture of two types of spherical particles (diameter 0.1 mm and density 3) for sampling standard deviation of (A) 0.1% and (B) 1%. Richer particles contain 10% of substance of interest, and leaner ones contain 0, 1, 5, or 9%.

EXAMPLE 21-1

The average natural level of lead in fresh tuna muscle has been reported over the years by many investigators to be about 400 ng/g. Recent studies, however, indicate the level to be a thousand times less, on the order of 0.3 ng/g, the difference arising from contamination introduced during sampling operations and in handling of the samples prior to measurement. Much of the contamination appears to stem from industrial lead in brines used for refrigeration and thawing and from lead-containing dusts, lubricants, and alloys associated with canning operations. Special techniques in ultraclean laboratories may be necessary if contamination during sampling and analysis is to be avoided.[3]

Indeterminate Error

Even with a well mixed, homogeneous material, indeterminate error may result if too few particles are taken for analysis. The problem is particularly serious if the component being determined is present in only a small fraction of the particles; here the standard deviation in the sampling step may often be the largest single source of uncertainty in the entire analysis. The question to be decided initially is how much precision is demanded.

Figure 21-1 shows the relation between the minimum weight of sample that should be taken and the composition of a two-component mixture if the

[3]These techniques include use of pure-quartz dissolution vessels, polyethylene containers cleaned with 4 M HNO_3 to remove trace lead, and reproducible handling of samples and standards as to time, temperature, reagent lots, and so on. The final determination can be by measurements using stable-isotope dilution and a high-resolution mass spectrometer. (D. M. Settle and C. C. Patterson, *Science* **1980**, *207*, 1167.)

standard deviation in the sampling is to be held to a part per thousand (0.1%). The equation used to calculate the curves is obtained by calculation of the weight of sample required from the volume, density, and number of particles in the sample:

$$\text{wt sample} = (4/3)\pi r^3 dn$$

where r is the radius, d the average density, and n the number of spherical particles. The value of n may be calculated from the relation

$$n = p(1 - p)\left[\frac{d_1 d_2}{d^2}\right]^2 \left[\frac{100(P_1 - P_2)}{\sigma_s P_{av}}\right]^2 \qquad (21\text{-}1)$$

where p and $1 - p$ are fractions of the two kinds of particles in the materials being sampled, d_1 and d_2 are the densities of the two types of particles, P_1 and P_2 are the percentages of the component of interest in the two kinds of particles, P_{av} is the overall average composition of the sample, and σ_s is the percent relative standard deviation (or sampling error) of the sampling operation. If half the particles in a mixture contain 10% of the desired substance and the other half 5%, then from Figure 21-1 the sample weight should be at least 0.2 g if sampling error is to be held to a part per thousand. If the second half of the particles contain 0% of the substance of interest, a much larger sample of about 2 g would be required; if 9%, 0.004 g would be sufficient. The *relative* difference in particle composition is the significant factor. Exactly the same sample sizes would suffice if the composition of the two types of particles were 100% and 0, 50, or 90% or if they were 0.1% and 0, 0.05, and 0.09%. The same curves can be applied to any relative composition by substitution of $x\%$ for 10% and by substitution of $0.1x$, $0.5x$, and $0.9x\%$ for the curves corresponding to 1, 5, and 9% in Figure 21-1A. Also shown are the related curves for a standard deviation of 1% (Figure 21-1B). As expected, 100-times-smaller samples are adequate when a 10-times-larger standard deviation is acceptable.

A second point illustrated by the figure is that with mixtures of particles of similar composition (lower curve in Figure 21-1A) the sample weight required is greatest when an approximately 1:1 mixture of the two types of particles is taken and drops off at higher and lower ratios. But with mixtures of particles of widely differing composition (upper curves) the sample weight required increases markedly as the fraction of particles richer in the component of interest decreases. This conclusion is simply common sense. Say a ground sample of gold ore consists only of particles of worthless matrix minerals (gangue) and of pure gold. If the number of particles of gold in the ore is extremely small, an exceedingly large sample is necessary if it is to be representative of the composition with high probability. (Yet such an ore may be well worth mining.) Suppose that the spherical particles of 0.1-mm diameter and density 3 that we have been considering have an overall sample composition of 1% in the component of interest; a sample weight of 0.3 g is then required for a sampling standard deviation of 1%. For overall compositions of 0.1 and 0.01%, 3 and 30 g are required, while for 0.001%,

300 g is required! At much less than 0.01%, trace analyses of particulate materials are likely to provide only order-of-magnitude information solely because of the limitations imposed by indeterminate errors in sampling.

21-3 PRELIMINARY TREATMENT OF SAMPLES

The chemist faced with a new analytical problem involving a heterogeneous sample must first perform a preliminary evaluation. Perhaps the sample provided is a jar containing sludge suspended in a solvent, or a piece of coated wood, metal, or plastic, or gas in a bottle. If the analyst is a forensic chemist, the sample may be a bit of textile, a few shards of pottery or glass, or a stain on an object. In most instances the first question to be asked is, "What is the goal of the analysis?" The answer will frequently determine the next step, as in the following examples: "Is there an economic amount of phosphate present in this rock?" "What is the level of the pesticide DDT in this soil?" "Are alloying metals present in this steel in the right concentrations?" Thus the analyst must know (1) what substances are sought, (2) the approximate level of precision necessary in the analysis, and (3) the general nature of the interfering substances likely to be present.

The next question is the adequacy of the sample (Section 21-1). Once this has been assured, the most appropriate method of analysis must be selected. Criteria to apply to this decision include cost, time, availability of apparatus and instruments, and level of precision required. Having confirmed the adequacy of the sample and the aim of the analysis, the chemist selects the method to be applied and then prepares the sample for measurement. Preliminary operations common to most methods of analysis are drying before weighing, followed by dissolution in an appropriate solvent.

Drying of Samples

Most solid materials adsorb atmospheric water on their exposed surfaces. The amount adsorbed varies with the chemical nature of the solid, the amount of surface area exposed, and the relative humidity. Generally, this water should be removed before the samples are weighed, so that the result can be related to an accurately known quantity of material. The usual procedure (Section 8-7) is heating in an oven at a temperature sufficient to cause the water to be desorbed without sample decomposition, frequently 110°C for 2 h or overnight. If information on the amount of adsorbed water or other materials is of interest, analysis on an "as-received" basis may be called for, and one of the items of analytical information reported may be "percent weight loss on drying." Analysis "as-received" may be appropriate also in situations where heating may alter the compounds being determined.

For nonhygroscopic materials such as metals, alloys, and most ores the analysis can proceed without a drying step. For others such as potassium acid phthalate and silver nitrate, which adsorb water only under conditions of high relative humidity, drying may be dispensed with when the humidity in the laboratory is low. Certain materials should not be dried at elevated temperatures— ones that contain components of the atmosphere (such as oxygen, carbon dioxide, or water), ones that contain volatile components, and ones that may decompose at those temperatures.

Dissolution of Samples

Although a few analytical methods, such as emission spectroscopy in a dc arc, may require only grinding of a solid prior to packing into an electrode, most entail dissolution in an appropriate solvent. Mild solvents should ordinarily be tried first; these include water, dilute solutions of acids and bases, and certain organic solvents. In general, polar and ionic substances tend to dissolve more readily in water or polar organic solvents such as methanol and acetonitrile, and nonpolar substances in solvents such as hexane, toluene, and chloroform. Most salts of transition metals react with water to form hydroxy complexes with the metal ions. The extent to which complexation occurs varies with the metal; iron(III), cerium(IV), and thorium(IV) are particularly prone to hydrolysis. Mineral acids are often chosen as solvents for metal salts or oxides because hydrolysis is avoided and because most metal chlorides, sulfates, and nitrates are relatively soluble.

Dissolution of Metal Alloys and Other Inorganic Materials

Metal alloys also are usually dissolved in mineral acids. The choice of acid depends on the composition of the sample.

Almost all metals having negative standard electrode potentials dissolve in nonoxidizing acids with the formation of hydrogen gas. Exceptions are those that react only slowly, or become passive, in contact with certain acids. One example is the slow reaction of cadmium, cobalt, lead, and nickel with hydrochloric acid; another is that of chromium with nitric acid.

To dissolve metals having positive standard electrode potentials, an oxidizing acid such as nitric, hot concentrated sulfuric or perchloric, or hydrochloric plus bromine or a chlorate salt is necessary. Of these, nitric acid is the most versatile, dissolving all metals but aluminum and chromium, which become passive unless in alloys. Tin, antimony, and tungsten form insoluble hydrous oxides.

Recommended solvents for the more common metals are (1) hydrochloric acid for zinc, cadmium (slow), aluminum, titanium (slow, heating necessary), tin, chromium, iron, cobalt, and nickel; (2) nitric acid for copper, silver, zinc, cadmium, mercury, aluminum (slow), lead, bismuth, chromium (slow), molybdenum, iron, cobalt, and nickel; (3) sulfuric acid for zinc, cadmium (slow), aluminum (slow), tin (heating necessary), bismuth (heating necessary), chromium, iron, cobalt, and nickel; and (4) 3:1 hydrochloric acid–nitric acid (aqua regia) for gold and platinum.

Many minerals, particularly silicates and some oxides, are insoluble directly in mineral acids; they must first be heated to a high temperature (fused) in the presence of a basic compound such as sodium carbonate or lithium borate, called a *flux*. The fusion is frequently carried out in crucibles of nickel or platinum, though materials such as zirconium or gold are sometimes employed.

Another approach to the decomposition of insoluble silicates is treatment with a mixture of dilute hydrofluoric and sulfuric acids in a platinum dish. The silica is volatilized as SiF_4, and the mixture evaporated to fumes of sulfuric acid to expel the excess hydrofluoric acid. All the metals remain as the sulfate salts. Reactions with HF and other acids are sometimes carried out in a Teflon-lined container, which is tightly sealed and then heated to speed decomposition.

Ferrosilicon, tin(IV) oxide, metal borides or carbides, and many other materials require special treatment to put them into solution. Procedures for these uncommon materials may be found in books specializing in their properties and analysis.

Dissolution of Organic Materials

The initial treatment of organic materials depends on the information sought. If inorganic elements are to be determined, the organic matter is usually destroyed either by heat or by wet chemical oxidation.

In *dry ashing* the sample is heated in a furnace or over a flame in the presence of air or a stream of oxygen to convert the organic material to carbon dioxide, water, and other volatile substances. Sometimes a few drops of concentrated sulfuric acid are added initially to facilitate oxidation and to stabilize metals as their sulfates. The residue of metal oxides or sulfates can then be dissolved in water or an acid such as hydrochloric acid and the metals determined. Elements that may be lost, partly or entirely, by volatilization during this procedure include phosphorus, the halogens, arsenic, and selenium. Sometimes dry ashing is carried out at room temperature in a stream of ozone, generated from O_2 by a radio-frequency generator; microwave heating to drive off water is also sometimes applicable.

In *wet ashing* the sample is dissolved in a solution of a strong oxidant. This method is often preferred to dry ashing when loss of elements by volatilization must be avoided. Oxidizing acids, such as nitric, or acids that become strong oxidants when heated, such as sulfuric or perchloric, are useful here. Mixtures of acids or of acids plus oxidizing reagents may also be needed for some materials. Among the acids used for this purpose are concentrated sulfuric, to destroy organic material in the Kjeldahl procedure for nitrogen (Section 5-8), concentrated nitric, and concentrated perchloric. Of these a 1:1 mixture of nitric and perchloric acids is especially effective, provided suitable precautions are observed in the handling of the perchloric acid (Section 6-2). Carbon, hydrogen, and nitrogen volatilize as the corresponding oxides; sulfur is converted to sulfuric acid; arsenic, phosphorus, silica, and selenium are retained; and the metals remain in the highest stable oxidation states as the perchlorate salts.

When a mixture of concentrated nitric and perchloric acids is added to a sample at room temperature, the nitric acid immediately attacks the most readily oxidizable material. On gradual heating, oxidation is accelerated until at 100°C the full strength of nitric acid begins to act. This process continues to about 140 to 150°C, when the least resistant material has been destroyed and the nitric acid boils away. The more powerful oxidizing action of perchloric acid then begins to be exerted, increasing with temperatures between 150°C and the boiling point, 200°C. At 200°C few organic substances can resist the oxidizing power of perchloric acid (Teflon is one). For a few materials such as leather, certain synthetic fabrics, and some nitrogen heterocyclic compounds, addition of a trace of chromium or vanadium as catalyst is recommended. For safety, digestions involving perchloric acid must always be carried out under certain well defined conditions: (1) sufficient nitric acid must be provided initially to react with *all* the more easily oxidized material; (2) the increase in temperature must be gradual; and (3) meticulous precautions must be taken to minimize the condensation of perchloric acid vapors on any surface such as in a hood where contact with organic materials may occur.[4]

21-4   EXPERIMENTAL DETERMINATION OF SAMPLING ERROR

Background

Samples provided in introductory courses in chemical analysis are usually assumed to be sufficiently homogeneous that the data obtained can be used to evaluate the accuracy and · precision of the separation and measurement operations without concern for sampling error. However, in the analysis of real samples it is often difficult to sample in such a way that a portion taken for analysis is truly representative of the whole. Many sampling problems are caused by inhomogeneities in the material under study; they can occur in almost any mixture, including gases and liquids. For example, the level of nitrogen oxides in the atmosphere is greatly affected by distance from a highway.

For mixtures of solids of moderate particle size, the problem of sampling error can be particularly serious. Samples taken from the most carefully mixed materials may vary in composition, owing to statistical variations in the distribution of particles richer or leaner in the sought-after substance. For a two-component mixture Benedetti-Pichler related the relative standard deviation in sample composition σ_s to the total number n of the two types of uniformly sized particles in the sample. With some rearrangement of Equation (21-1), the equation for σ_s in parts per thousand is

$$\sigma_s = \sqrt{\frac{p(1-p)}{n}} \frac{d_1 d_2}{d^2} \frac{(P_1 - P_2)}{P_{av}} (1000) \qquad (21-2)$$

By judicious choice of parameters the theoretical sampling error σ_s can be restricted to any desired value. Experimentally, the overall variance in an analysis s_o^2 is the sum of the variances associated with the sampling procedure s_s^2 and of the subsequent analytical operations s_m^2:

---

[4]Where regular perchloric acid digestions are performed, a special hood is necessary. For occasional work a trap that draws the vapors into a water aspirator is acceptable. See Figure 4-1 in the reference on perchloric acid and its use at the end of the chapter.

$$s_0{}^2 = s_s{}^2 + s_m{}^2 \tag{21-3}$$

The standard deviation[5] of the sampling procedure is then

$$s_s = \sqrt{s_0{}^2 - s_m{}^2} \tag{21-4}$$

Hence, measurement as described later of s_0 and s_m permits direct estimation of the sampling standard deviation s_s.

In this experiment, attention is given to the uncertainty associated with the sampling operation. A two-component system consisting of potassium hydrogen phthalate (KHP) and an inert diluent of the same particle size and density is used. This mixture comprises two types of particles, one containing 100% [P_1 in Equation (21-2)] KHP and the other 0% [P_2 in (21-2)]. By precise titration of pure KHP with sodium hydroxide solution it is possible to determine the uncertainty associated with the measurement step $s_m{}^2$. This separate estimate of $s_m{}^2$, combined with knowledge of the overall uncertainty $s_0{}^2$, allows calculation by (21-4) of the uncertainty associated with the sampling step.

For a mixture of particles of KHP and inert diluent ($P_1 = 100$, $P_2 = 0$) the *absolute* standard deviation in the percentage of KHP in a sample, σP_{av}, depends on the relative number of particles of KHP and diluent in the initial mixture and on the total number of particles in the sample. If the particles are all of uniform size and density, as in this experiment, $d_1 = d_2 = d$, and (21-2) becomes

$$\sigma P_{av} = \sqrt{\frac{p(1-p)}{n}} \, (100 - 0) \tag{21-5}$$

This equation shows that the standard deviation is affected by the number of particles in the sample according to the square root of $1/n$.

The experiment consists of three steps. In the first, the standard deviation in the measurement operations is determined by titration of several portions of KHP with sodium hydroxide solution. In the second, a portion of KHP and diluent is thoroughly mixed, sampled, and the samples titrated. In the third, the percentage of KHP is calculated, along with the experimental value of s_s. The latter is then compared with theoretical values calculated for similar mixtures containing various numbers and sizes of particles.

Procedure (time 3 to 4 h)

Mixing of Sample

Obtain a sample of KHP and diluent. Place about 15 g of the mixture in a clean, dry weighing bottle, and store the uncapped bottle in a chamber of constant relative humidity (hygrostat or humidistat)[6] at a relative humidity of about 40%

---

[5]Recall that the quantity s_s is defined as the *estimate* of σ_s. The validity of the estimate will depend on the number of samples examined, the sampling procedure used, and the experimental proficiency with which it is carried out.

[6]Tobacconists use the word humidor.

for several hours or overnight.[7] Remove the bottle, stopper it, and mix the contents thoroughly by attaching the bottle to a device that rotates it slowly end over end for at least 2 h.

Sampling Operation

After the components have been thoroughly mixed, carefully open the container and pour the contents in a pile onto a clean sheet of glazed paper. Throughout sample removal avoid dislodging particles adhering to the walls or cap, as these are unlikely to be well mixed. Take 10 to 12 portions[8] in the range of 0.50 g ($\pm$ 0.02 g) from equally spaced points around the pile, and place them in separate 200-mL conical flasks, previously weighed to the nearest 0.1 mg. Reweigh each flask plus contents to obtain the sample weights to the nearest 0.1 mg.

Preparation of Standards

To assess the precision of the titration operation, weigh into 200-mL conical flasks, to the nearest 0.1 mg, six 0.2- to 0.3-g portions of pure KHP.

Titration Procedure

Dissolve each sample in 25 to 30 mL of distilled water, add 3 to 4 drops of 0.1% phenolphthalein indicator solution, and titrate with 0.05 M NaOH to the first permanent pink.[9] Back titrate immediately[10] by dropwise addition of dilute HCl (concentration adjusted so that 1 drop is equivalent to 0.01 mL of NaOH titrant) until the pink just disappears. Record the volume of NaOH required, subtracting from the final buret reading the number of drops of back titrant required (in units of 0.01 mL). This procedure yields precise end points.

Dissolve and titrate the portions taken from the mixtures in the same way as the samples of KHP.

· Calculations

1. From the titrations of pure KHP calculate the average molarity of the NaOH solution, along with the standard deviation and the relative standard deviation of the molarity s_m. Should you need to review the calculation of standard deviation, refer to Section 1-4.

2. Using the molarity of NaOH just calculated, calculate the percentage of KHP in each of the sample portions. Also calculate the standard deviation and the relative

---

[7]Buildup of static charge on the container and sample particles under conditions of low humidity, or clumping in high humidity, makes effective mixing and transfer difficult. Storage of the sample and mixing container for a few hours or overnight in an atmosphere of moderate humidity eliminates this problem. An effective hygrostat for this purpose is a chamber with a beaker containing a saturated aqueous solution of K_2CO_3, which provides a constant relative humidity of 43% at room temperature.

[8]A small sampling scoop designed to deliver reproducibly about 0.5 g is convenient here. The technique recommended for manipulating a sampling scoop is as follows: Push scoop into pile, working from edge toward center. Tap with finger to settle contents. Push scoop into pile again to fill to overflowing. Pass a spatula lightly over the top to remove excess. Tap again with finger to remove particles adhering to outside or handle. Transfer contents to previously weighed flask. Tap scoop on edge of flask if necessary to dislodge all particles.

[9]Use of a self-filling, self-zeroing buret is recommended where feasible to save time.

[10]The back titration must be immediate because the pink end point fades on standing, owing to absorption of carbon dioxide from the atmosphere.

standard deviation s_o of these values. Then calculate the relative standard deviation resulting from *sampling*, in parts per thousand, by the relation

$$s_s = \sqrt{(s_o^2 - s_m^2)}(1000)$$

The magnitude of s_m should be small relative to s_o if the experimental technique was proficient.

Report s_s in parts per thousand. Also report the average molarity of NaOH and the average percentage of KHP, along with the standard deviation and relative standard deviation of each.

3. As additional information, calculate the absolute and relative standard deviations in sampling predicted for the system you used if the particles studied had been spheres with diameters of (a) 0.8 mm, (b) 0.6 mm, or (c) 0.4 mm. Assume the density of both KHP and the diluent to be 1.63 g/mL. (*Hint*: From the density and weight of a sample, calculate its volume. Dividing this volume by the volume per particle gives n, which can be substituted in Equation (21-2).

4. From (21-2) calculate the size of n required if the value for the relative standard deviation of sampling were to be held to (a) 0.01 (1% relative); (b) 0.001 (0.1% relative). On this basis, and assuming a particle diameter of 0.6 mm, what weights of your mixture would be required to attain each of these relative standard deviations?

EXAMPLE 21-2

From the following data calculate the quantities s_o, s_m, and s_s, along with the percentage of KHP.

Values for the molarity of NaOH from the titrations of pure KHP are 0.03361, 0.03338, 0.03345, 0.03364, 0.03340, and 0.03322. The average is 0.03345; standard deviation, 0.00016; s_m, 4.7 ppt. The percentage of KHP found in sample portions is 17.79, 18.78, 18.56, 19.15, 18.96, 17.44, 18.75, 17.07, 18.95, 19.60, 19.27, 19.45. The average is 18.65%; standard deviation, 0.80; s_o, 43.1 ppt.

From Equation (21-4): $s_s = \sqrt{(43.1)^2 - (4.7)^2} = \sqrt{1835} = 43$ ppt.

PROBLEMS

21-1.   What is the objective of a sampling operation?

21-2.   Distinguish between a gross and a laboratory sample.

21-3.   What techniques are commonly used to reduce the amount of material in a sample?

21-4.   List the important factors that determine the weight of a sample for it to be representative of the whole.

21-5.   From Figure 21-1 estimate the sample weight required if the overall composition is 10% in the component of interest, the richer containing 50% and the leaner 5% (radius of particles 0.005 cm, density 3, and sampling variance 0.1%).

21-6.   What weight of sample is required if the particle radius is 0.005 cm, the density is 3, and 10^6 particles are required for a representative sample?

21-7.   For a sampling error of 0.1% relative standard deviation how many particles must a representative sample contain if it consists of two types of particles of density 3, one being 100% and the other 0% in the component of interest, and the overall composition is (1) 90%; (b) 10%; (c) 1%; (d) 0.1%; (e) 0.001%.

21-8.   In the preceding problem what would be the weight of sample in each instance if the particles were spheres of radius (a) 0.005 cm; (b) 0.001 cm; (c) 0.0005 cm?

21-9.   (a) If the overall relative standard deviation of an analysis, including measurement and sampling errors, is 0.43% and that of the measurement aspect alone is 0.32%, what is the relative standard deviation of the sampling error? (b) If the relative standard deviation of the measurement error alone were 0.12%, what would be the relative standard deviation of the sampling error?

21-10. The sampling error for the analysis of a mixture containing KHP was calculated according to the method outlined in Section 21-4. The concentration of sodium hydroxide from successive titrations of pure KHP was 0.04444, 0.04446, 0.04432, 0.04449, 0.04438, and 0.04441. The % KHP in successive sample portions was 22.22, 21.11, 20.62, 22.87, 21.52, 20.98, 20.22, 22.64, 20.29, 23.11, 22.62, and 21.14. Calculate the relative standard deviation of the measurement error and of the sampling error.

REFERENCES

A. A. Benedetti-Pichler, *Physical Methods of Chemical Analysis*, W. M. Bell, Ed., Academic Press, New York, 1956, Vol. 3, p 183. Gives original development of the theory of sampling of particulate mixtures.

W. E. Harris and B. Kratochvil, *Anal. Chem.* **1974**, *46*, 313. Discusses applications to practical systems of Benedetti-Pichler's approach.

B. Kratochvil, R. S. Reid, and W. E. Harris, *J. Chem. Educ.* **1980**, *57*, 518. Description of the sampling-error experiment of Section 21-4.

H. A. Laitinen and W. E. Harris, *Chemical Analysis*, 2nd Ed., McGraw-Hill, New York, 1975, Chapter 27. Concise discussion of sampling for analysis.

A. A. Schilt, *Perchloric Acid and Perchlorates*, G. F. Smith Chemical Co., Columbus, Ohio, 1979.

B. Williams, *A Sampler on Sampling*, Wiley, New York, 1978. A highly readable treatment of the concepts of sampling with a nonmathematical approach.

22

EVALUATION OF EXPERIMENTAL DATA

Coincidences, in general, are great stumbling blocks in the way of that class of thinkers who have been educated to know nothing of the theory of probabilities.

Edgar A. Poe

In Chapter 1 the subject of errors in chemical measurements was introduced. Emphasis in that section is on the identification of the types of errors that may arise in laboratory operations and on the ways of evaluating and reporting results based on small numbers of experiments to reflect the accuracy and precision of the analytical work. In this chapter the use of statistical methods to obtain additional information from the results of analyses is considered. For most statistical treatments the assumption of a Gaussian distribution of experimental observations is made because of the power of this approach. But it should be kept in mind that most chemical measurements contain bias, generally unidentified, that introduces skew into the results.

Statistics deals with groups of data. Statistics can supply estimates, but not absolute answers, at various levels of confidence. Usually, the broader the data base, the more reliable is the statistical evaluation. For unbiased systems, the more data that are provided, the greater is the level of confidence with which a result can be expressed, and the higher the probability that the result reflects accurately the true state of the system. In this chapter several procedures for the evaluation of data are presented.

22-1 INDETERMINATE (RANDOM) ERROR AND THE GAUSSIAN-DISTRIBUTION CURVE

Everybody believes in the exponential law of errors; the experimenters because they think it can be proved by mathematics; and the mathematicians because they believe it has been established by observation.

Lippman, quoted by Poincare

If a large number of measurements is made involving only indeterminate (random, or unbiased) errors and the frequency with which the various values occur is plotted against the values, a symmetrical curve like the one in Figure

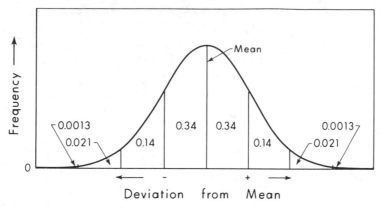

Figure 22-1. Gaussian-distribution curve. Vertical lines are shown at 0, 1, 2, and 3 standard deviations from the mean. Numbers refer to the fraction of the total area in each section.

22-1 is obtained. This curve, known as the *Gaussian-distribution, standard-error, normal-error, normal-distribution,* or *probability curve,* or sometimes as the *exponential law of errors,* shows the distribution of measurements expected for most analytical systems. The less precise the measurements, the broader will be the curve. The curve is defined by two parameters: the mean (average) $\bar{x}$, which locates the position of the curve on the horizontal axis, and the standard deviation s (Section 1-4), which denotes the spread.[1] For a set of measurements involving only random errors, 68% fall within one, 96% within two, and 99.7% within three standard-deviation units of the average.

Reliability of an Average

Because in real situations the true value is unknown, one of the best available approximations to it, the average, is generally used. From the standard deviation the distribution relation for the Gaussian curve can be applied to estimate the range of values within which the average for an infinite number of measurements lies with some level of probability. The average for an infinite number of measurements that are free of determinate error is the true value. This range around the measured average is expressed in terms of the *confidence limits,* or *confidence interval.* For unbiased, independent measurements having a Gaussian distribution, the equation expressing the confidence limits around the measured average within which the true value lies is

$$\text{confidence limits} = \bar{x} \pm ts/\sqrt{n} \tag{22-1}$$

---

[1]The symbol s is used here to denote standard deviation. When the number of unbiased measurements is exceedingly large (ideally, $n = $ infinity), the symbol σ is used. The accuracy of s improves with the number of measurements. In special cases, such as where a large number of unbiased values are available, the estimate s may be assumed to be σ. Consult the references at the end of the chapter for further information.

Table 22-1. *t* Values for Calculating Confidence Limits of Averages

No. of Measurements, n	Degrees of Freedom, $n-1$	t for 90% Probability	t for 95% Probability	t for 99% Probability
2	1	6.3	12.7	63.7
3	2	2.9	4.3	9.9
4	3	2.35	3.2	5.8
5	4	2.13	2.78	4.6
6	5	2.02	2.57	4.03
7	6	1.94	2.45	3.71
8	7	1.90	2.37	3.50
9	8	1.86	2.31	3.36
10	9	1.83	2.26	3.25
11	10	1.81	2.23	3.17
21	20	1.72	2.09	2.84
31	30	1.70	2.04	2.75
∞	∞	1.64	1.96	2.58

where t is the tabulated value for $n-1$ measurements at some desired probability level. Generally, the convenient values 90, 95, and 99% probability are employed (Table 22-1).

In Table 22-1 the value $n-1$ denotes the degrees of freedom in the system and is used in the calculation of confidence limits of the average for a set of numbers. This table confirms that the size of the factor t in Equation (22-1) depends on the number of measurements taken to obtain the standard deviation. Note that, for $n = 10$ or 20, t is already close to the limiting value for $n = \infty$.

EXAMPLE 22-1

For the measurements of density in Example 1-2, $\bar{x}$ is 1.015, s is 0.0055, and n is 6. The confidence interval is then $1.015 \pm t(0.0055/\sqrt{6})$, or $1.015 \pm t(0.0022)$. From Table 22-1, t for $n = 6$ is 2.02, 2.57, and 4.03 for 90, 95, and 99% probability. With 90% confidence the true value falls within the range 1.015 ± 0.0045; with 95% confidence, within 1.015 ± 0.006; and with 99% confidence, within 1.015 ± 0.009.

In this example we are 90% certain that the true value lies within 1.010 and 1.020, 95% certain that it lies between 1.009 and 1.021, and 99% certain that it lies between 1.006 and 1.024. Clearly, if we want to increase the certainty of the true value lying within some range of the average, the range must be increased. The choice of confidence level is governed by the level of certainty that is needed.[2] The 95% level, which incorporates about two standard-deviation units, is often used in estimates of the validity of analytical measurements. Remember that the use of confidence limits is valid only when determinate errors are negligible.

---

[2]Statisticians often approach the question of whether two values significantly differ by application of a *null hypothesis*. In this hypothesis the two values are assumed to be the same, and the probability that the differences present are the result of indeterminate (random) error is calculated. If the observed difference is greater than that calculated at, say, 95% confidence limits, the null hypothesis is considered to not apply, and the difference deemed significant.

Table 22-1 has several applications. In addition to providing confidence limits for a set of measurements, it may be employed in testing the accuracy of a method of measurement by comparing an experimentally observed average obtained by that method with the true value. This application is discussed in the next section.

22-2 TESTING WHETHER TWO MEANS DIFFER

An analyst must often decide whether the difference between the results of two sets of analyses is real or is caused by statistical variation. This situation may arise when a comparison is being made between two analytical procedures, or a procedure in two different laboratories, or a procedure in the same laboratory by two different analysts or with different apparatus. It is also encountered when a sample of known composition, such as a standard reference material, is analyzed by a laboratory and the results are compared with a pure compound or certified value. The decision can be aided by applying the *t test* in the following manner.

First, calculate the means $\bar{x}_1$ and $\bar{x}_2$ and the standard deviations s_1 and s_2 for the two sets of data. Second, calculate the value t from the expression

$$t = \frac{|\bar{x}_1 - \bar{x}_2|}{s_p}\left[\frac{n_1 n_2}{n_1 + n_2}\right]^{0.5} \tag{22-2}$$

Here n_1 and n_2 are the number of measurements in Sets 1 and 2, and s_p (called a pooled s by statisticians) is given by

$$s_p = \left[\frac{(n_1 - 1)(s_1^2) + (n_2 - 1)(s_2^2)}{(n_1 - 1) + (n_2 - 1)}\right]^{0.5} \tag{22-3}$$

Third, select the desired confidence level, and obtain from a t table such as Table 22-1 the value corresponding to that confidence level at $n_1 + n_2 - 2$ degrees of freedom. If the value in the table is greater than the value calculated from (22-2), $\bar{x}_1$ and $\bar{x}_2$ may be deemed to be from the same population and not to differ significantly at the chosen level of confidence. If the t value in the table is less than the calculated value, the two averages may be deemed to be different at the chosen level of confidence, and investigation of the source of the difference should be considered. Sometimes the calculated and tabular values for t fall close to each other. In such cases a definitive conclusion cannot be drawn, and collection of more data is advisable.

EXAMPLE 22-2
A second set of measurements of density was taken on the solution in Example 1-2 by a different technique, yielding values of 1.010, 1.004, 1.011, 1.003, 1.000, and 1.014. Do the two methods differ significantly at the 90% confidence level?

   For the first set of measurements, $\bar{x}_1 = 1.015$ and $s_1 = 0.0055$. For the second set, $\bar{x}_2 = 1.007$ and $s_2 = 0.0054$. Then

$$s_p = \left[\frac{(6-1)(0.0055)^2 + (6-1)(0.0054)^2}{(6-1) + (6-1)} \right]^{0.5} = 0.0055$$

$$t = \frac{1.015 - 1.007}{0.0055} \left[\frac{(6)(6)}{6+6} \right]^{0.5} = 2.52$$

From Table 22-1, for 10 degrees of freedom t at 90% probability is 1.81. Since the value for t calculated from (22-2), 2.52, is greater than the value from Table 22-1, 1.81, we may say with 90% confidence that the two methods are different. That is, 90 times out of 100 the difference will be the result of determinate error in one or both of the methods. Since we do not know the source of the difference, the discrepancy must be investigated by further experimental work.

When one value has been established with a high level of certainty through numerous analyses and crosschecks, and a second value is being compared with it, use of a pooled s is unnecessary. Instead, (22-2) may be simplified to

$$t = n^{0.5} \, |\bar{x} - \mu| \, /s \tag{22-4}$$

where μ is the true value and where $\bar{x}$ and s are obtained from n measurements on the sample by the method under study. In effect, what is done here is to estimate the confidence limits for the measured value $\bar{x}$ to see whether the true value μ falls within them.

EXAMPLE 22-3

A solution of 0.01000 M silver nitrate was carefully prepared from primary-standard silver nitrate and used to test the purity of a sample of sodium chloride. Four titrations yielded 99.94, 99.91, 99.96, and 99.92% NaCl. Can the average of this set be considered different from the expected value of 100% at the 95% confidence level?

Compare the absolute difference between the average for the titrations and the expected true value, $\bar{x} - \mu$, with the statistical variability in the measurements, $ts/\sqrt{n}$. If $\bar{x} - \mu$ is greater than $ts/\sqrt{n}$, the average can be considered different from the expected value at the 95% confidence level. If $\bar{x} - \mu$ is less than $ts/\sqrt{n}$, then the average and the expected value can be considered the same at the 95% confidence level. Here the average $\bar{x}$ is 99.93%, the standard deviation s is 0.022, and the expected true value μ is 100.00%. From Table 22-1, t is 3.2 for 3 degrees of freedom at the 95% confidence level. Then $|\bar{x} - \mu| = 0.07$, and $ts/\sqrt{n} = 0.035$. Since 0.07 is appreciably larger than 0.035, we may state that a difference exists at the 95% confidence level between the analytical results and the value expected for pure NaCl. The purity of the sample of sodium chloride is therefore suspect.

Should only the first three titrations be completed, will the results be the same? In this case $\bar{x} = 99.94$, $s = 0.025$, $t = 4.3$, $n = 3$, $|\bar{x} - \mu| = 0.06$, and $ts/\sqrt{n} = 0.062$. Because the values 0.06 and 0.062 are about the same, a definite statement cannot be made, and more titrations are necessary.

EXAMPLE 22-4

A new analytical method for the measurement of magnesium was evaluated by using it to determine the magnesium content of Standard Reference Material 929, magnesium gluconate dihydrate, a clinical-laboratory standard certified by the National Bureau of Standards, Washington, D.C., to contain 5.043 wt % magnesium. The proposed method gave results for magnesium of 4.995, 4.998, 5.032, 5.022, and 4.988. Does the proposed method give the correct magnesium content at the 95% confidence level?

For the new method, $\bar{x} = 5.007$ and $s = 0.019$.

$$t = (5)^{0.5}(5.043 - 5.007)/0.019 = 4.2$$

From Table 22-1, for four degrees of freedom $t = 3.2$. Therefore a 95% probability exists that the results by the new method differ significantly from the correct value. If the fourth result in the set had been 5.057 rather than 5.022, the value of t from (22-4) would be 2.24 (from $\bar{x} = 5.014$ and $s = 0.029$). In this case the calculated value of t would be less than the tabulated value, and the results obtained by the new method would be considered to be from the same population. Hence the new method could be considered satisfactory at the 95% confidence level. A less precise method with greater scatter might appear to give the correct value within 95% confidence limits but would be a poorer method because of the wide confidence limits.

22-3 TESTING WHETHER TWO STANDARD DEVIATIONS DIFFER

In the foregoing examples we assume that the standard deviations for the two sets of data under comparison are not statistically different. This assumption can be tested by a statistical method called the F test. This test can be applied to two sets of data even when their averages differ significantly. In the procedure the ratio of the larger variance to the smaller, s_1^2/s_2^2, is calculated and compared with values from a table such as Table 22-2 at the desired confidence level. If the calculated value is larger than the tabulated value, the standard deviations may be considered significantly different at that confidence level.

EXAMPLE 22-5

Are the two standard deviations from Example 22-2 different at the 95% confidence level?

From Example 22-2, $s_1 = 0.0055$ and $s_2 = 0.0054$ for the two sets of six measurements. Then $s_1^2/s_2^2 = (0.0055)^2/(0.0054)^2 = 1.04$. The values of $n_1 - 1$ and

Table 22-2. Values of F at the 95% Confidence Level

$n_2 - 1$			$n_1 - 1$			
	2	3	4	5	6	∞
2	19.00	19.16	19.25	19.30	19.33	19.50
3	9.55	9.28	9.12	9.01	8.94	8.53
4	6.94	6.59	6.39	6.26	6.16	5.63
5	5.79	5.41	5.19	5.05	4.95	4.36
6	5.14	4.76	4.53	4.39	4.28	3.67
∞	3.00	2.60	2.37	2.21	2.10	1.00

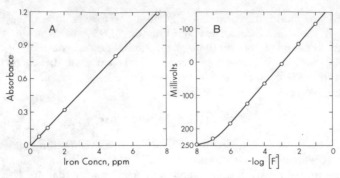

Figure 22-2. Examples of straight-line calibration plots for use in analysis. (*A*) Relation between absorbance and concentration of iron(II)–bipyridyl complex. (*B*) Relation between potential of a fluoride-selective electrode and the log of fluoride ion concentration.

$n_2 - 1$ are both 5. From Table 22-2 for five degrees of freedom for each set the tabuated value is 5.05. We conclude that the two standard deviations do not differ significantly at the 95% confidence level.

EXAMPLE 22-6

A set of five results for magnesium in the NBS standard magnesium gluconate of Example 22-4 from a colorimetry procedure gave a standard deviation of 0.038. The NBS reports the uncertainty in their value of magnesium to be 0.022. Assuming the uncertainty in this value to be based on a sufficient number of measurements that the degrees of freedom can be considered infinity (equivalent to saying that $s = \sigma$), do the two s values differ significantly at the 95% confidence level?

In this case $s_1^2/s_2^2 = (0.038)^2/(0.022)^2 = 2.98$. The F-table value for four degrees of freedom and ∞ is 2.37. We may state, then, that the two standard deviations differ significantly at the 95% confidence level.

22–4 FITTING STRAIGHT-LINE RELATIONS

In instrumental measurements for analytical work, a linear relation between the response signal of the instrument and the amount of sought-for substance is highly desirable because it allows the preparation of linear calibration plots. For example, linear plots are seen when the absorbance of a solute follows Beer's law over a range of iron concentrations (Figure 22-2A) or when the change in potential of a selective electrode varies with the log of the concentration of the sought-for species (Figure 22-2B). The slopes and intercepts of such plots can furnish useful information. Thus, from the expression $A = abc$, a plot of A against c should give a straight line with a slope of ab and a value of 0 for y when $x = 0$. Similarly, for $E = $ constant $- (0.059/n) \log [M]$ a plot of E against $-\log [M]$ should give a straight line with a slope of $0.059/n$.

Visual estimation of the best line through a set of points is subject to error, both in plotting the points and in fitting the line.[3] A preferable approach to

---

[3]Even though the eye tends to weight points near the ends of a set of data more heavily, visual plots of data often aid in identifying suspect points.

data in which only indeterminate errors are present, especially when considerable scatter exists, is to use the mathematical approach called the *method of least squares*. In the application of the method discussed here the values plotted on the x axis are assumed to be known with much higher precision and accuracy than those on the y axis, so that all the uncertainties in the points arise from uncertainty in the y values. This assumption is often justifiable. For example, in the determination of copper by atomic absorption (Section 15-8) the concentration of copper in the standard solution is known to within 0.1 % or so, and the volumes are measured to within about the same precision. The calculation then assumes that each value for y has an uncertainty associated with it that depends on the standard deviation in the measurement operations. The line representing the data, then, should be drawn so that the deviations of the y values from the line are slight. It can be shown mathematically that for random errors the best fit is obtained when the sum of the squares of the individual deviations is at a minimum.

In practice, for a set of data points — $x_1, y_1; x_2, y_2; ...; x_n, y_n$ — it is possible to calculate the values of a and b in the straight-line equation $y = a + bx$, where b is the slope and a the intercept on the y axis when $x = 0$. The best fit is assumed to occur when the sum of the squares of the deviations in y between actual points and points on the line (predicted by the equation for the line) is a minimum. The value of the slope b is obtained by the relation

$$b = \frac{\Sigma xy - [\Sigma x \, \Sigma y / n]}{\Sigma x^2 - [(\Sigma x)^2 / n]} \tag{22-5}$$

Once b is known, the intercept a can be learned from

$$a = (1/n) (\Sigma y - b\Sigma x) \tag{22-6}$$

In these equations, Σx and Σy are the sums of the n measurements of x and y, Σxy is the sum of the product of the individual xy pairs, and so on.[4]

EXAMPLE 22-7

Values of absorbance for the iron–bipyridyl complex as a function of iron concentration were recorded as follows: 0.500 ppm, 0.084; 1.000 ppm, 0.162; 2.000 ppm, 0.319; 5.000 ppm, 0.801; 7.500 ppm, 1.187. Calculate the equation for the best line through this set of points by the method of least squares. From the slope calculate the absorptivity of the iron–bipyridyl complex, assuming a 1-cm cell was used. From the intercept indicate whether the reagents contain sufficient iron to give a nonzero blank absorbance.

By substitution in Equation (22-5) a slope of 0.158 is obtained. Since the slope is equal to ab in Beer's law (Section 14-3), when b is 1.000 cm, a is equal to the absorptivity of the iron–bipyridyl complex. The units of absorptivity in this system, assuming 1 ppm to equal 1 mg/L, are liters per milligram-centimeter. The molar absorptivity is $(0.158)(55.85 \text{ g/mol})(10^3 \text{ mg/g}) = 8.82 \times 10^3$ L/mol-cm.

From Equation (22-6) the intercept is 0.005 absorbance units. Therefore a slight blank is indicated, corresponding to a concentration of iron on the order of

---

[4]With modern calculators capable of storing accumulated sums, calculations of quantities involving summations are no longer time-consuming.

$C = A/ab = (0.005)/(8820)(1) = 5 \times 10^{-7}$ M. However, see Example 22-9. A plot of this equation is shown in Figure 22-2A.

EXAMPLE 22-8

A study of the range of response of a solid-state fluoride-selective electrode was made by measurement of a series of solutions of known fluoride concentration. The values of $-\log [F^-]$ and millivolt readings were: 8.00, 247; 7.00, 229; 6.00, 184; 5.00, 124; 4.00, 64; 3.00, 4; 2.00, −56; and 1.00, − 116. Calculate the least-squares slope for this set of data. Does the value of the slope, which from the Nernst equation is $0.059/n$, correspond to an integral value of n? From the raw data, plotted as millivolts against $- \log [F^-]$, indicate a likely source of error in the least-squares values for the slope and intercept, and suggest a means of correcting for it.

For this set of data, from Equations (22-5) and (22-6) the slope is 54.4 mV and the intercept is − 160 mV. The value calculated for n is 1.1 electrons. On a plot of the data (Figure 22-2B) the last two points can be seen to curve away from a straight line. This curvature is the result of fluoride contribution to the solution from solubility of the lanthanum fluoride membrane at extremely low fluoride concentrations. Thus these points can legitimately be deleted from the graph. A second least-squares calculation without these points gives a slope of 60.0 mV per unit change in pF and an intercept of −176 mV. This slope corresponds to a value for n of 0.98, in excellent agreement with Nernstian behavior.

The foregoing example illustrates the important point that least squares should be used with caution when deviations from straight-line behavior are possible. A wise practice is to plot the data on a graph to see whether curvature is present at either end. If it is, a search for possible reasons should be made, with perhaps some additional measurements to confirm the trend.

Sometimes points in one region of a plot have higher precision than those in other regions. These points should be weighted more heavily in a least-squares calculation. The simplest approach is to enter them into the data set more than once, the number of multiple entries for a pair of values governed by the difference in precision between that pair and the others.[5]

The uncertainty in the values of the slope and intercept of a least-squares line can also be calculated. The expression for the standard deviation in the slope b is

$$s_b = (s_{y/x})/[\Sigma x^2 - (\Sigma x)^2/n]^{0.5} \qquad (22\text{-}7)$$

Here $s_{y/x}$ is given by

$$s_{y/x} = \left[\frac{1}{(n - 2)^{0.5}}\right] \left[\Sigma y^2 - \frac{(\Sigma y)^2}{n} - b^2\left[\Sigma x^2 - \frac{(\Sigma x)^2}{n}\right]\right]^{0.5} \qquad (22\text{-}8)$$

The standard deviation in the intercept a on the y axis where $x = 0$ is

$$s_a = s_{y/x} [\Sigma x^2/(nx^2 - (\Sigma x)^2)]^{0.5} \qquad (22\text{-}9)$$

---

[5]A common approach when a point on a graph is the average of several measurements is to weight the point by a value equal to the reciprocal of the standard deviation of the average.

The standard deviation in the slope is useful for estimating confidence limits for such values as the absorptivity in Example 22-7; that of the intercept is useful for estimating, for instance, whether a positive intercept is statistically significant, as in the same example. In the calculation of confidence limits the number of degrees of freedom is $n - 2$, where n is the number of points.

EXAMPLE 22-9

Calculate the standard deviation of the molar absorptivity in Example 22-7. What are the 90% confidence limits for the value? Also calculate the standard deviation and 95% confidence limits for the intercept on the y axis when $x = 0$. Is the positive value of the intercept significant at this confidence level?

From Equation (22-8), $s_{y/x} = 0.01673$, and from Equation (22-7), $s_b = 2.8 \times 10^{-3}$. From Table 22-1, for $n - 2$ degrees of freedom, where $n = 5$, the t value at 90% confidence limits is 2.35. The 90% confidence limits of the slope are then $0.158 \pm ts_b = 0.158 \pm 0.007$. Thus we can state that the molar absorptivity should fall within the range $8820 \pm 0.007(55.85)(10^3)$ or within 8450 to 9190, with 90% confidence.

From Equation 22-9, $s_a = 0.0117$. The 95% confidence limits for the intercept then are $0.005 \pm (2.35)(0.0117)$, or 0.005 ± 0.027. Thus at this level of confidence the difference between 0.005 and 0 cannot be considered statistically significant.

22-5 ESTIMATION OF UNCERTAINTY IN A RESULT FROM A CALIBRATION CURVE

Frequently in instrumental analysis a calibration curve is prepared, and values for unknown samples are read from it. If the calibration plot is a straight line, the standard deviation in a sample s_{unk} can be estimated from the measurements from which the straight line was obtained and the measurements made on the sample:

$$s_{unk} = \frac{s_{y/x}}{b} \left[(1/m + 1/n) + \frac{(\bar{y}_{unk} - \bar{y})^2}{b^2(\Sigma x^2 - (\Sigma x)^2/n)} \right]^{0.5} \qquad (22\text{-}10)$$

where $s_{y/x}$ is calculated from Equation (22-8), m is the number of replicate measurements of the unknown, and $\bar{y}_{unk}$ is the average of the m replicates. Note that s_{unk} is smallest when $\bar{y}_{unk} = \bar{y}$. This indicates that measurements of the unknown should be near the center of the calibration plot for minimum error; it also indicates that, since the uncertainty increases as $\bar{y}_{unk}$ moves away from $\bar{y}$, results obtained by extrapolation are prone to increased error. Although the confidence limits for the true value can be determined from Table 22-1, the degrees of freedom must be taken to be $n - 2$ (not m).

EXAMPLE 22-10

A calibration curve for the determination of part-per-million levels of nickel in copper metal by the liquid-liquid extraction and spectrophotometric measurement of nickel as the dimethylglyoxime complex (Section 16-3) gave the following absorbance values for the listed concentrations of standard nickel solution: 5.00 ppm, 0.125; 10.00 ppm, 0.240; 20.00 ppm, 0.495; 30.00 ppm, 0.720; and 35.00 ppm, 0.850. A 10.00-mL aliquot of the sample solution, treated in the same way as the standards, gave replicate absorbance readings of 0.424, 0.420, 0.422, and 0.430. Calculate the concentration of nickel in the sample solution. Then calculate the standard deviation in the value and the 95% confidence limits. What would be the standard deviation and confidence limits if only one measurement of 0.424 had been made on the sample solution?

First, calculate the coefficients of the equation for the calibration plot by the method of least squares. From Equation (22-5)

$$b = \frac{64.23 - (100)(2.43)/5}{2650 - (100)^2/5} = 0.0241$$

and from Equation (22-6)

$$a = (1/5) [2.43 - (0.0239)(100)] = 0.00369$$

Second, substitute the measured absorbance values for the unknown sample for y in the calibration equation, $y = 0.00369 + 0.0241x$, to calculate the corresponding values of x. The results are 17.43, 17.26, 17.35, and 17.68 ppm of nickel, with an average of 17.43.[6] Figure 22-3 illustrates this calibration plot. Third, calculate the value of s_{unk} as follows: From Equation (22-8)

$$s_{y/x} = [(0.37097 - (0.0238846)^2 \, 650)/(5 - 2)]^{0.5} = 0.00733$$

Substitution in Equation (22-10) gives

$$s_{unk} = (0.00733/0.024)[1/4 + 1/5 + (0.424 - 0.486)^2/(0.024)^2(2650 - (100)^2/5)]^{0.5}$$

$$s_{unk} = 0.412$$

---

[6]Values can also be obtained by plotting a calibration curve and reading the concentrations for the unknown sample from the plot. Advantages of this method are that a suspect point is more easily spotted, the presence of determinate errors may be compensated for more readily, and results are available more quickly.

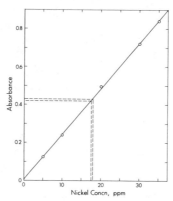

Figure 22-3. Calibration plot of nickel concentration against absorbance for data of Example 22-10. Readings shown for measurements of unknown sample at absorbances of 0.420 and 0.430.

Then $\bar{x} = 17.4 \pm 0.4$ ppm of nickel.

The 95% confidence limits are $\bar{x} \pm ts_{unk} = 17.4 \pm (3.2)(0.4) = 17.4 \pm$ 1.3 ppm, where t is the t-table value for $5 - 2$ degrees of freedom. If only one measurement of the unknown were made, with an absorbance value of 0.424, s_{unk} would be calculated as before, but with $m = 1$; a value of 0.669 would be obtained. The confidence limits in this instance would be $17.4 \pm (3.2)(0.34)$, or $17.4 \pm$ 2.1 ppm.

Note: For convenience of presentation many intermediate values in this example have been rounded. In practice, retain as many significant figures as possible throughout all calculations, rounding only at the end. Otherwise, appreciable error may be introduced from terms involving small differences between large numbers.

22-6 CORRELATION COEFFICIENTS

A problem often encountered in experimental work is to establish how analytical results are affected by variables such as temperature or time of reaction. To answer questions concerning the effect of one variable on another, Pearson introduced the concept of correlation coefficient. If a plot of a suspected dependent variable y against an independent variable x exhibits a trend rather than a random pattern, the two variables x and y are said to be *correlated*. In other words, given a value for x we can predict to some extent a value for y. When an increase in x is associated with an increase in y, the two variables are said to have a *positive correlation*; when an increase in x is associated with a decrease in y, the two are said to have a *negative correlation*. If a plot of x against y shows a random, shotgun pattern, correlation is considered low; if all the points fall near the trend line, correlation is high. The extent to which correlation occurs is measured by a quantity r, the *correlation coefficient*, defined by

$$r = \frac{s_{yx}}{s_x s_y} = \frac{n \, \Sigma xy - \Sigma x \Sigma y}{([n\Sigma x^2 - (\Sigma x)^2][n\Sigma y^2 - (\Sigma y)^2])^{0.5}} \qquad (22\text{-}11)$$

The quantity s_{yx} is called the *covariance* of y on x; s_x and s_y are the standard deviations for the sets of values of x and y.[7] The equation may be most easily recalled as

$$r = \text{covariance of } y \text{ on } x/s_x s_y$$

Although formidable in appearance, the form shown in Equation (22-11) is convenient for use in calculators. The value of r can range from $+1$ to -1. A value of $+1$ indicates perfect positive correlation between y and x, while a value of -1 also indicates perfect correlation, but in this case negative. A value of zero denotes absence of any trend in the relation between x and y.

---

[7]Covariance is similar to variance s^2 except that, instead of the sum of squares of deviations from the average, the sum of the product of the deviations of individual values of x and y from the averages $\bar{x}$ and $\bar{y}$ is taken. Thus for variance

$$s_x^2 = \Sigma(x - \bar{x})^2/(n - 1)$$

whereas for covariance

$$s_{yx} = \Sigma(x - \bar{x})(y - \bar{y})/(n - 1).$$

EXAMPLE 22-11

In a study of sampling error by the method of Section 21-4, the titration of potassium acid phthalate with standard NaOH yielded the following results, listed in the order of titration: 17.79, 18.78, 18.56, 19.15, 18.96, 17.44, 18.75, 17.07, 18.95, 19.60, 19.27, 19.45. Is there a correlation between the percentage of potassium acid phthalate found and the order of titration?

The raw data are $x = $ 1, 2, 3, ... for the corresponding values of $y = $ 17.79, 18.78, 18.56, $\cdots$. From (22-11)

$$r = \frac{12(1464.80) - (78)(223.37)}{([12(650) - (78)^2][12(4164.80) - (223.37)^2])^{0.5}} = 0.39$$

This value indicates moderate positive correlation between x and y.

It is important to be able to relate the value of r to the extent of interrelation between y and x. If 0 denotes no correlation, and 1 perfect correlation, a good fit of a set of data points to a line requires a value for r of close to 1. Of interest, therefore, is the uncertainty in a value of y obtained from a trend line, that is, a straight-line equation fitted to the data points.

An estimate of this uncertainty can be calculated from the correlation coefficient and the standard deviation of the values for y by the relation

$$\text{confidence limits} = ts_y (1 - r^2)^{0.5} \tag{22-12}$$

where the value for t is obtained from a t-table at $n - 2$ degrees of freedom.

EXAMPLE 22-12

In Example 22-11 what is the uncertainty in the value for the percentage of potassium acid phthalate at the 95% confidence level?

From Equation (22-12) the confidence limits $= (2.23)(0.80)(1 - (0.39)^2)^{0.5} = 1.65$, and so the correct value lies in the range 18.65 $\pm$ 1.65 with 95% confidence.

EXAMPLE 22-13

In Example 22-10 what is the correlation coefficient for the calibration line of Figure 22-3?

Substituting the five sets of data points of Example 22-10 in Equation (22-11), we obtain a value for r of 0.9998. A value of r extremely close to 1 is to be expected for a calibration curve of sufficient precision to be analytically reliable. If the absorbance of the data point at 20 ppm of nickel had been measured to be 0.520 instead of 0.495, a highly suspect value, the correlation coefficient would still be large, 0.9985.

In general, correlation coefficients are of most value in the determination of whether a suspected interaction between two variables is present, rather than in the estimation of the precision of a calibration curve for analytical use.[8]

---

[8]As Perrin states, "How close r must be to 1 depends upon context; however, in chemistry you run the risk of being laughed at if you publish with r less than 0.95." (*Mathematics for Chemists*, C. L. Perrin, Wiley, New York, 1970, p 162.)

22–7   EVALUATION OF PREJUDICE IN DATA

We have emphasized that statistical quantities such as means, standard deviations, and confidence limits have sensible meaning only when applied to data free of determinate error. As pointed out in Section 1-4, however, determinate error may arise in any experimental situation. Therefore, the experimentalist should ascertain whether bias is being introduced into his results by this type of error. One objective test suitable for examining small numbers of observations for determinate error is the Q test (Section 1-5); it is recommended for the evaluation of data in most chemical work.

The Chi-Square Test

When frequencies of occurrence of numbers can be expressed quantitatively in terms of probabilities, statistical tests may be applied. In this section such a test, the *chi-square test*, and its application to evaluation of the probability of prejudice are considered.

A determinate factor can be introduced into experimental observations through prejudice due to, say, a knowledge of preceding members of a set of replicates. In most chemical measurements in which a scale or meter is read, the terminal digits should be randomly distributed. Nevertheless, real observations rarely display a perfectly random distribution.

To evaluate the probability of prejudice, a chi-square value is calculated from the observed frequency f and the average (or expected) frequency F:

$$\text{chi square} = \Sigma[(F - f)^2]/F$$

EXAMPLE 22–14

Suppose that in a set of 61 observations the digits 0 through 9 occurred at the frequencies shown in the accompanying table. These might be the last digits in final buret readings or in analytical weighings. Is this the frequency distribution expected from chance, or does number prejudice (bias) exist? First, values of F and $(F - f)^2$ are tabulated as follows:

Terminal Digit	f	$F - f$	$(F - f)^2$
0	4	2.1	4.4
1	4	2.1	4.4
2	9	−2.9	8.4
3	9	−2.9	8.4
4	6	0.1	0.0
5	7	−0.9	0.8
6	4	2.1	4.4
7	6	0.1	0.0
8	11	−4.9	24.0
9	1	5.1	26.0
	Σf 61		$\Sigma[(F - f)^2]$ 80.9
	$F = \Sigma f/10 = 6.1$		

Table 22-3. Relation Between Chi Square and Probability
for Several Degrees of Freedom

Degrees of Freedom	Probability of a Greater Value								
	0.995	0.99	0.90	0.50	0.25	0.10	0.05	0.01	0.005
1	---	0.0002	0.02	0.45	1.32	2.71	3.84	6.6	7.8
2	0.01	0.02	0.21	1.39	2.77	4.61	5.99	9.2	10.6
3	0.07	0.11	0.58	2.37	4.11	6.25	7.81	11.3	12.8
4	0.21	0.30	1.06	3.36	5.39	7.78	9.49	13.3	14.9
5	0.41	0.55	1.61	4.35	6.63	9.24	11.1	15.1	16.7
9	1.73	2.09	4.17	8.34	11.4	14.7.	16.9	21.7	23.6

Then chi square is calculated to be $80.9/6.1 = 13.3$.

The average frequency is the sum of all the digits times the probability, or 61×0.1. Here the probability is 0.1 because the experimental observation can be any one of ten digits; it arises from a system that has nine degrees of freedom. The value of chi square, calculated by dividing the sum of $(F - f)^2$ by the average frequency F is in this case 13.3. The probability that a given chi-square value will be exceeded through pure chance can be determined by consulting a chi-square table under the listing for nine degrees of freedom (Table 22-3), or a figure such as Figure 22-4. More extensive tables can be found in handbooks of chemistry and in books on statistics. Interpolation in Table 22-3 indicates that the chi-square value 13.3 has a probability of about 0.15. When the probability is about 0.5, there is no evidence that number prejudice or other nonrandom factors have introduced a determinate bias into the observations. As the probability becomes appreciably greater or less than 0.5, we may conclude with. an increasing level of confidence that number prejudice exists. At 0.15 some prejudice is likely to be present.

The situation encountered most frequently in experimental work is prejudice for certain numbers and against others.

The numbers 0 and 5 often appear with greater frequency than would be due purely to chance. In such cases chi-square values greater than 8.3 are

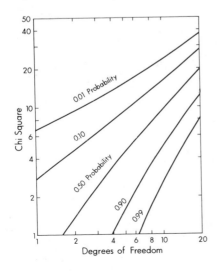

Figure 22-4. Graphs of chi-square. The 0.01 and 0.10 probability levels indicate suspiciously *poor* fit. The 0.90 and 0.99 levels indicate suspiciously *good* fit.

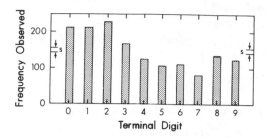

Figure 22-5. Frequency of observation of about 1500 randomly distributed terminal digits in buret readings. The space between the lines on the vertical axis denotes the 95% confidence limits. From W. E. Harris, *Talanta* 1978, *25*, 325.

found, corresponding to probability values less than 0.5. A strong suspicion of number prejudice is warranted for a small number of observations with probabilities in the range 0.2 to 0.005. A larger number of observations is needed to either confirm or discount the suspicion. At 0.05 the probability of prejudice is 95 out of 100, or 95%. At 0.001 the probability is 999 out of 1000 — a high level of confidence.

Chi-square values indicating probabilities of appreciably more than 0.5 are rare. A chi-square value of unity would mean that all numbers were observed with equal frequency, a highly improbable occurrence. Chi-square values much below 1 suggest that prejudice is being introduced by the observer, consciously or unconsciously, making number frequencies more even than they normally would be as a result of chance. At a chi square of 2.1 the chances are 99 out of 100 that the data are prejudiced.

EXAMPLE 22–15

A sample was obtained of the distribution of about 1500 terminal digits for buret readings, taken with care during the performance of analysis (Figure 22-5). Virtually all digits have an observed frequency falling well outside the expected standard deviation limits, and a chi-square calculation indicates number prejudice is almost certainly present. The bias is toward small numbers (0, 1, 2) and away from midrange (5, 6, 7) terminal digits and also toward even and against odd digits. There is additional bias toward negative errors and against positive errors, as shown by Figure 22-6.

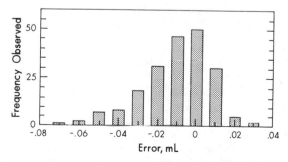

Figure 22-6. Distribution of absolute reading errors for 200 independent observations of the positions of buret meniscus. From W. E. Harris, *Talanta* 1978, *25*, 325.

PROBLEMS

22-1. Are the Gaussian-distribution curve and the conclusions that follow from it applicable to observations with determinate error? Explain.

22-2. Explain what is meant by a confidence limit. Give an example.

22-3. How do the standard deviation and the confidence limits indicate the reliabilities of different quantities? How are standard deviation and confidence limit related?

22-4. (a) If 11 measurements of a quantity have a mean of 11.23% with an absolute standard deviation of 0.13%, what are the 90% and 99% confidence limits? (b) Approximately how many additional measurements would be required to reduce the confidence limits to half the above values?

22-5. What are the 95% confidence limits for the amount of thorium-234 in Problem 1-16?

22-6. What are the 90% confidence limits for the average percentage of nickel in Problem 1-17?

22-7. Explain how to test experimentally whether a proposed new method is reliable.

22-8. Explain the difference in purpose of an F test and a t test.

22-9. A sample of nickel was analyzed for a trace of copper by atomic absorption with the following results: 0.343, 0.283, 0.300, 0.236, 0.272, 0.333, 0.322, and 0.308%. The sample was also analyzed coulometrically for copper with the results 0.333, 0.314, 0.303, 0.306, 0.324, 0.318, 0.326, and 0.330%. Do the results of the two methods differ significantly?

22-10. For the preceding problem are the standard deviations in the results from the two methods different at the 95% confidence level?

22-11. Do the results of the analyses in Problem 1-14 fall within the 95% confidence limits of the correct value?

22-12. Do the results of the analyses in Problem 1-18 fall within the 95% confidence limits of the correct value?

22-13. Do the weights of the coins in Problem 1-20 fall within the 95% confidence limits of the nominal value?

22-14. The method of least squares can be used to determine the slope and the intercept of a linear relation. What other method can be used? What are the relative merits and drawbacks of the two methods?

22-15. Six cobalt(II) chloride solutions were prepared and the absorbances measured at 510 nm. At molar concentrations of 0.0168, 0.0338, 0.0552, 0.0963, 0.103, and 0.124, absorbances of 0.097, 0.190, 0.311, 0.494, 0.591, and 0.724 were observed. (a) Calculate by the method of least squares the equation for the best straight line for these data. (b) Assuming the cell length was 1 cm, calculate the molar absorptivity from the slope. Is there a significant nonzero intercept?
    A sample solution gave absorbance readings of 0.381, 0.382, 0.395, and 0.386. Calculate the concentration of cobalt(II) chloride in the sample solution. (a) Calculate the standard deviation and the 95% confidence limits. (b) What would be the standard deviation and confidence limits if only one measurement of 0.381 had been made on the sample solution?

22-16. For measurement of the amino acid methionine, $CH_3SCH_2CH_2CH(NH_2)COOH$, an electrode was prepared by coating an ammonia-sensing electrode with the enzyme methionine lyase, which decomposes methionine into α-ketobutyrate, methyl mercaptan, and ammonia. A study of the response of the electrode in millivolts to varying concentrations of methionine produced the following results (−log [methionine] given first, followed by mV value): 1.70, 87; 2.37, 86; 2.82, 76; 3.37, 58; 3.82, 35; 4.37, 15; 4.82, 4. Calculate by the least-squares method the best equation for this data. Does the value of the slope, which is $0.059/n$ according to the Nernst equation, correspond to an integral value for n? Of what significance is the value of the intercept?

22-17. What is the significance of the use of a correlation coefficient in chemical measurements? What minimum value for the correlation coefficient would you consider acceptable for a calibration curve?

22-18. A set of titrations of an organic acid with standard sodium hydroxide was performed. The samples of acid were weighed and dissolved in distilled water and then titrated, each titration taking 5 min. The equivalent weights found in order of titration were 117.8, 118.8, 118.6, 119.2, 119.0, 117.4, 118.8, 117.1, 119.0, 119.6, 119.3, and 119.4. Later, slow hydrolysis of the acid was suspected to occur on standing in water, and the analyst became concerned as to whether this was affecting his results. Calculate the correlation coefficient between the values for equivalent weight and time, and indicate whether hydrolysis is indeed a possibility.

22-19. A student recorded the following frequencies for the last digit of the final buret readings for a series of titrations (last digit in parentheses): (0) 16, (1) 6, (2) 10, (3) 6, (4) 14, (5) 12, (6) 8, (7) 8, (8) 16, (9) 4. What is chi square and the probability of prejudice for this set of numbers? How would you interpret the results?

22-20. Use the data of the preceding problem to calculate chi square and the probability of prejudice involving even compared with odd numbers. What is chi square and the probability of prejudice for low numbers (0, 1, 2, 3, 4) compared with high numbers (5, 6, 7, 8, 9) in this set of data?

22-21. One of a pair of dice was suspected of being tampered with. The following frequencies were observed for the numbers 1 through 6 for a series of rolls. Die 1: (1) 8, (2) 18, (3) 11, (4) 19, (5) 9, (6) 25. Die 2: (1) 11, (2) 14, (3) 12, (4) 19, (5) 19, (6) 15. Calculate chi square and the probability of prejudice for each. What is the likelihood that one of the dice shows prejudice? How would you interpret the results?

22-22. Prepare a table of frequencies of the last digit of the *final* buret readings in your laboratory notebook for all titrations performed thus far in which indicators were used. For example, in the following readings the digit of interest is "6": final buret reading, 34.46; initial buret reading, 0.24; difference, 34.22. Calculate chi square for your table of digits, and determine the probability from Table 22-3. Is prejudice indicated by the results? If so, explain the type, extent, and probable cause.

22-23. Five determinations of the molarity of an NaCl solution by a new method under development gave values of 0.2033, 0.2035, 0.2030, 0.2036, and 0.2034 M. Another set of five determinations on the same solution by an old-reliable method gave 0.2030, 0.2028, 0.2029, 0.2026, and 0.2027 M. (a) For each method, calculate the mean and standard deviation. (b) Determine whether any of the results may be rejected on the basis of the Q test. (c) State whether you think the new method is statistically different from the old one, and give the reasons for your answer.

22-24. Lord Rayleigh prepared nitrogen samples by several different methods. The density of each sample was measured as the weight of gas required to fill a particular flask at a certain temperature and pressure. Weights of nitrogen samples prepared by decomposition of various nitrogen compounds were 2.29890, 2.29940, 2.29849, and 2.30054 g. Weights of "nitrogen" prepared by removing oxygen from air in various ways were 2.31001, 2.31163, and 2.31028 g. Is the density of nitrogen prepared from nitrogen compounds significantly different from that prepared from air? What are the chances of the conclusion being in error? (Study of this difference led to the discovery of the inert gases by Sir William Ramsey, Lord Rayleigh.)

22-25.   Final grades for a course in quantitative analysis for 15 students were as follows.

Final Grade, %	Practical Test Grade, out of 10	Overall Laboratory Grade, %
55	4	50
78	10	80
76	7	79
63	9	57
87	10	91
42	6	21
55	10	58
67	5	70
67	9	59
83	10	92
80	7	76
48	5	47
68	8	72
61	6	70
75	9	87
90	10	85

Calculate the correlation coefficient between the final grade and (a) the practical test grade and (b) the overall laboratory grade.

REFERENCES

W. J. Dixon and F. J. Massey, *Introduction to Statistical Analysis*, 3rd ed., McGraw-Hill, New York, 1969. Gives details of advanced statistical tests.

W. S. Gossett, *Biometrika* **1908**, *6*, 1. A famous paper that introduced the *t* test and *t* distribution. It was published under the pseudonym Student, and the test is known as Student's *t* distribution. The table of *t* values we use today was provided by R. A. Fisher (*Metron.* **1926**, *5*, 90).

H. A. Laitinen and W. E. Harris, *Chemical Analysis*, 2nd ed., McGraw-Hill, New York, 1974, Chapter 26. Concise discussion of useful statistics for analytical applications.

C. L. Perrin, *Mathematics for Chemists*, Wiley, New York, 1970. In addition to this book, an audio course by the same author comprising about seven hours of tapes plus a manual is produced by the American Chemical Society.

J. Topping, *Errors in Observation and Their Treatment*, Chapman and Hall, London, 1955. Straightforward information on least-squares calculations.

W. Weaver, *Lady Luck*, Doubleday (Anchor Books), New York, 1963. A delightful introduction to the ideas of statistics.

23

ANALYTICAL PROBLEM SOLVING. USE OF THE CHEMICAL LITERATURE

The more extensive a man's knowledge of what has been done, the greater will be his power of knowing what to do.

Disraeli

23-1 APPROACHING AN ANALYTICAL PROBLEM

One of the fundamental challenges of chemistry lies in investigating what and how much of a substance is present in a sample. Questions of this kind are encountered constantly by chemists in all fields — the organic or inorganic chemist who is characterizing new products, the physical chemist who is studying the kinetics of a system, and the industrial chemist who is trying to improve the yield and quality of a commercial product, as well as the analytical chemist who is developing and using new methods. Similar challenges are met by scientists in many other fields, including biology, medicine, engineering, and agriculture, who may seek information about pollution in the air, water, and land, about insecticide and herbicide concentrations in food, or about the composition of the moon.

Up to this point, involvement in basic principles and details of individual experiments may have obscured one of the basic purposes of analytical chemistry —the solution of chemical problems. An analytical chemist is a solver of problems. To give some perspective on applications of analytical thinking, a variety of analytical problems are presented at the end of the chapter. The selection of an appropriate method of analysis depends on a clear understanding and definition of the problem in terms of the nature of the substance to be determined, its approximate concentration, and the nature of other materials present.

EXAMPLE 23-1

In a mixture of hydrochloric and nitric acids, how can the amount of each be determined at the part-per-thousand level if no other solutes are present?

Several methods could be considered. The total amount of hydrochloric plus nitric acid could be obtained by titration with standard sodium hydroxide. The amount of hydrochloric acid could be obtained by precipitation and weighing of chloride as silver chloride or by titration with standard silver nitrate solution. The amount of nitric acid would then be obtained by difference.

23-2 THE LITERATURE OF ANALYTICAL CHEMISTRY

The majority of analytical methods in use today have been reported in scientific papers in research journals that specialize in analytical chemistry. Others, especially when fundamental in nature, may appear in more general journals. These articles constitute the *primary* literature sources. Books, reviews, abstracts of published papers, and other reference works constitute *secondary* sources that assist in tracking down what has been done. This section lists a selection of important primary and secondary sources, with brief comments.

Primary Literature: Journals

Analusis. France. Most widely read French journal in the field.

Analyst. Great Britain. Emphasizes methods for determinations of substances in complex samples.

Analytica Chimica Acta. The Netherlands. Most articles in English, but some in French or German. A separate section on computer techniques and optimization is published several times a year.

Analytical Biochemistry. U.S.A. Deals with applications of analytical chemistry to biological systems.

Analytical Chemistry. Published monthly by the American Chemical Society. The most widely read of all analytical journals. Contains, in addition to articles on original research, discussions of new methods and instruments by experts.

Analytical Letters. U.S.A. In two parts: A, chemical analysis and B, clinical and biochemical analysis. Designed for rapid communication of new findings.

Journal of the Association of Official Analytical Chemists. U.S.A. Reports mainly analytical methods pertaining to food and agriculture.

Microchemical Journal. U.S.A. This and the following journal emphasize analytical methods for small samples.

Mikrochimica Acta. Austria. Articles in English or German.

Talanta. Ireland. Most articles in English, with a few in German or French.

Zeitschrift für Analytische Chemie. Germany. The major German journal on analytical chemistry.

Zhurnal Analiticheskoi Khimii (Journal of Analytical Chemistry of the U.S.S.R.). Available in English translation from Consultant's Bureau, New York City.

Secondary Literature: Abstracts

Chemical Abstracts. Published every two weeks by the American Chemical Society. The most thorough coverage of the chemical literature available. Provides brief abstracts of papers dealing with chemistry of all types. Key-word indexes appear in each issue, and subject indexes each six months. The best place to learn what has been done, although the immense volume of material takes time to go through. Computer searches, which can be performed on authors' names, key words, or combinations of key words, may save much time if a specific item is sought.

Analytical Abstracts. Published monthly by the Society for Analytical Chemistry, London. Not so complete as the Analytical Chemistry section of *Chemical Abstracts*, but the narrower coverage makes it faster to use.

Chemical Titles. Published monthly by the American Chemical Society. Lists tables of contents of most chemical journals, thereby allowing rapid coverage of new titles.

Current Contents. Published weekly by the Institute for Scientific Information, Philadelphia. Similar to *Chemical Titles*, but also includes fields related to chemistry.

Science Citation Index. Published monthly by the Institute for Scientific Information, Philadelphia. Allows reader to identify recent citations to earlier papers and provides powerful "search-forward" capability. Useful also in providing the most recent references, before they have been published in *Chemical Abstracts*.

Collections of Methods and Data

American Society for Testing and Materials, *Annual Book of Standards*, Philadelphia, 1974. In 47 volumes. Covers many physical tests in addition to chemical methods.

Association of Official Analytical Chemists, *Association of Official Analytical Chemists, Methods*, 11th ed., Washington, D.C., 1970. Methods for the analysis of food and agricultural products.

W. G. Berl, Ed., *Physical Methods in Chemical Analysis*, 2nd ed., Academic Press, New York, 1960–61. In four volumes.

N. H. Furman and F. J. Welcher, Eds., *Scott's Standard Methods of Chemical Analysis*, 6th ed., D. Van Nostrand, New York, 1962–66. In five volumes. Emphasizes older, more classic methods.

N. W. Hanson, *Official Standardised and Recommended Methods of Analysis*, 2nd ed., Society for Analytical Chemistry, London, 1973.

I. M. Kolthoff and P. J. Elving, Eds., *Treatise on Analytical Chemistry*, Wiley, New York, 1959–. A multivolume series. Second edition begins in 1979. Part I covers theoretical aspects, Part II, methods for analysis of organic and inorganic compounds, and Part III, industrial analytical chemistry.

F. Korte, Ed., *Methodicum Chimicum*, Academic Press, New York, 1974. Volumes 1A and 1B cover analytical methods. A critical survey of procedures.

L. Meites, Ed., *Handbook of Analytical Chemistry*, McGraw-Hill, New York, 1963. A collection of useful tables of data and brief discussions of techniques.

C. N. Reilley, Ed., *Advances in Analytical Chemistry and Instrumentation*, Interscience, New York, 1960–. A multivolume series of monographs.

C. L. Wilson and D. W. Wilson, Eds., *Comprehensive Analytical Chemistry*, Elsevier, New York, 1959–64. A multivolume series.

Books and Reviews on Specific Topics

See also references to specific topics at the ends of appropriate chapters.

American Public Health Association, *Standard Methods for the Examination of Water and Wastewater*, 14th ed., New York, 1975.

American Society for Testing and Materials, *Chemical Analysis of Metals and Metal-Bearing Ores*, Philadelphia, 1974. Volume 12 of *The Annual Book of ASTM Standards*.

R. G. Bates, *Determination of pH: Theory and Practice*, 2nd ed., Wiley, New York, 1973. Clear, comprehensive treatment of pH measurement in water and nonaqueous solvents.

H. H. Bauer, G. D. Christian, and J. E. O'Reilly, Eds., *Instrumental Analysis*, Allyn and Bacon, Boston. 1978.

E. Bishop, Ed., *Indicators*, Pergamon Press, Oxford, 1972. Thorough coverage of indicators of all types.

J. S. Fritz, *Acid-Base Titrations in Nonaqueous Solvents*, Allyn and Bacon, Boston, 1973. Brief but highly practical treatment.

D. Glick, Ed., *Methods of Biochemical Analysis*, Interscience, New York, 1954–67. In 15 volumes. The most systematic available coverage of analytical methods in biochemistry.

R. L. Grob, Ed., *Modern Practice of Gas Chromatography*, Wiley-Interscience, New York, 1977.

F. Helfferich, *Ion Exchange*, McGraw-Hill, New York, 1962. An advanced, rather mathematical treatment, but the most thorough one available on fundamental properties of resins.

W. F. Hillebrand, G. E. F. Lundell, H. A. Bright, and J. I. Hoffman, *Applied Inorganic Analysis*, 2nd ed., Wiley, New York, 1953. The definitive work on rock and mineral analysis.

M. B. Jacobs, *The Chemical Analysis of Air Pollutants*, Interscience, New York, 1960.

B. L. Karger, L. R. Snyder, and C. Horvath, *An Introduction to Separation Science*, Wiley, New York, 1973.

I. M. Kolthoff and P. J. Elving, Eds., *Treatise on Analytical Chemistry*, Wiley, New York, 1963. Section D-2, Vol. 4, "Electrical Methods of Analysis."

I. M. Kolthoff, E. B. Sandell, E. J. Meehan, and S. Bruckenstein, *Quantitative Chemical Analysis*, 4th ed, New York, Macmillan, 1969. Thorough, detailed treatment of basic quantitative analysis.

H. A. Laitinen and W. E. Harris, *Chemical Analysis*, 2nd ed., McGraw-Hill, New York, 1975. Noninstrumental aspects of analytical chemistry at an advanced level.

J. Mitchell, Jr., I. M. Kolthoff, E. S. Proskauer, and A. Weissberger, Eds., *Organic Analysis*, Wiley, New York, 1953 —. A continuing series of volumes, each dealing with particular areas.

G. H. Morrison and H. Freiser, *Solvent Extraction in Analytical Chemistry*, Wiley, New York, 1957. Although dated, provides much basic information and many practical applications.

D. D. Perrin, *Masking and Demasking of Chemical Reactions*. Wiley, New York, 1970.

M. Reiner and D. Seligson, Eds., *Standard Methods of Clinical Chemistry*, Academic Press, New York, 1953–63. In four volumes.

W. Rieman and H. F. Walton, *Ion Exchange in Analytical Chemistry*, Pergamon, Oxford, 1970.

O. Samuelson, *Ion Exchange Separations in Analytical Chemistry*, Wiley, New York, 1963. A more practical, applications-oriented book than Helfferich.

G. Schwarzenbach and H. Flaschka, *Complexometric Titrations*, 2nd English ed., transl. by H. M. N. H. Irving, Methuen, London, 1969. The classic work on this subject.

S. Siggia, Ed., *Instrumental Methods of Organic Functional Group Analysis*, Wiley, New York, 1972.

S. Siggia, *Quantitative Organic Analysis Via Functional Groups*, 4th ed., Wiley, New York, 1979. Important source for critical evaluation of available methods for organic analysis. Includes many procedures.

D. A. Skoog and D. M. West, *Principles of Instrumental Analysis*, 2nd ed., Saunders, Holt, Rinehart, and Winston, New York, 1980.

L. R. Snyder and J. J. Kirkland, Ed., *Introduction to Modern Liquid Chromatography*, 2nd. ed., Wiley-Interscience, New York, 1979.

H. Strobel, *Chemical Instrumentation*, 2nd ed., Addison-Wesley, Boston, 1973.

A. Vogel, *Textbook of Quantitative Inorganic Analysis*, 4th ed., Longman, London, 1978.
   Provides procedures for most inorganic species.

23-3 CONSULTING THE CHEMICAL LITERATURE

Not the fact avails, but the use you make of it.

Ralph Waldo Emerson

Analytical chemistry is not so much performing analyses as it is solving chemical problems. Problems of analysis are solved with the assistance of two main resources. First and indispensable is knowledge of and practical experience with techniques of sample preparation, separations, and measurement. Second is the accumulated knowledge inherited from predecessors who have worked on similar problems. The purpose of the experimental work in this book is to build experience in the practice and applicability of a number of important methods of analysis. This experience, melded with pertinent information from the literature, should provide a solid foundation for attacking analytical problems.

Presently, about 10 000 papers in the area of analytical chemistry are published yearly. Consulting this store of information often reveals that a specific analysis can be performed by any of several methods; the question is then one of intelligent selection and, if necessary, adaptation. A specialist may incorrectly attempt to solve all the problems he encounters by the technique with which he is most familiar, be it mass spectrometry, polarography, chromatography, or any other. For one problem mass spectrometry might be better than polarography, for another the reverse might be true, and for still another, neither would be suitable. An intelligent decision as to the most appropriate must be based on sufficient knowledge of the whole arsenal of available methods. A published procedure can seldom be applied directly without some adaptation or development that also must be based on intelligent choices.

Searching the Literature

At the beginning of a search for the answer to a chemical problem an examination of the most recent issues of *Chemical Abstracts* will provide references to the original literature along with brief statements of the author's findings. It is advisable not to spend much time in the older abstracts. The recent original literature will include references leading to virtually all the pertinent preceding work and, in addition, provide critical comment that is usually lacking in abstracts. Although collections are likely to suggest only well tested methods, they tend to go out of date quickly and are often superseded by better methods reported in the current literature. Books are most useful in the later stages of decision making. They provide in-depth discussions of the theory and practice of selected techniques and are generally more up-to-date than collections. Scanning current issues of well chosen periodicals will often uncover the latest pertinent reports, but is time-consuming.

Fast, comprehensive searches of *Chemical Abstracts* are now possible by computer. These can be according to author or topic, within any time span. Particularly valuable is the option of selecting combinations of key words to provide more selective coverage. Because the number of articles abstracted is so large and the scope so diverse, careful choice of key-word combinations is necessary if coverage is to be complete without inclusion of large numbers of extraneous articles.

For information on the most recent reports in the current literature concerning problems in which you are interested, *Science Citation Index* is valuable. Once a recent paper has been found, earlier work can be traced from the references provided in the paper. This approach is a *retrospective search*. *Science Citation Index* can carry out the inverse process, providing *forward search* capability. Starting with a significant paper on the subject of interest, the searcher consults the index under the author's name to locate recent articles that refer to the paper in their lists of references.

23-4 A PROBLEM IN LITERATURE SEARCHING

Background

In this experiment an analytical problem is assigned to provide experience in using the chemical literature and in selecting the techniques most appropriate for a given situation. The suggested problems are intended to require consultation of the literature, and it is not presumed that the answer will be immediately obvious. The method you propose need not be limited to the ones introduced in this book.

Procedure (time 4 to 8 h)

Problems may be classified under areas such as agricultural, biological, chemical, clinical, environmental, industrial, geological, forensic, and nutritional. After you have chosen an area, a specific problem will be assigned. Guided by the literature and your experience, propose a solution to the problem.

Submit a summary of the proposed solution in 400 to 500 words. Include the following: (1) a clear statement of the problem, including one or two lines of background information; (2) the proposed solution, with (a) a brief summary of the method (*not* a detailed procedure); (b) any special precautions, limitations, or time requirements; (c) the precision and range, including upper and lower limits of the range, and (d) any instruments, techniques, and reagents that are expensive or uncommon; (3) the practical significance and utility of the analysis, including comparisons with other possible techniques if appropriate; (4) a list of references in correct format, identifying (a) the most recent significant reference and (b) the most helpful reference; (5) the search technique used, with a list of the topic headings surveyed.

When preparing the final copy, keep each section clearly separated and titled, following the above arrangement. It may help to think of the result as a report to an employer (another chemist) who wants to apply the best possible technique to the solution of a problem. In choosing a method, weigh all the factors in Items 2 and 3, as well as economic factors. Remember that the reader is influenced by clarity of expression, neatness, and your grasp of the problem as well as by your proposed solution.

An example of a report follows; examples of other literature-search problems are given at the end of the Problems section.

EXAMPLE 23-2

Question: How can the concentration in soil of the herbicide picloram be determined?

Problem: Picloram (4-amino-3,5,6-trichloropicolinic acid) is an effective herbicide for control of perennial weeds, poisonous plants, and shrubs. The anion of the acid is readily absorbed by organic and amorphous inorganic soil materials. Therefore a rapid method of determination or estimation of picloram residue in soil or plant tissue is desired.

Proposed Solution: Extraction and colorimetric determination of picloram in soil samples appears to be the most satisfactory method of analysis. Quantitative extraction of picloram can be rapidly performed on large numbers of samples. Radioisotope tracing, IR spectroscopy, and thin-layer, high-pressure liquid, or gas chromatography are also possible methods for picloram determination, but most entail lengthy procedures.

The colorimetric determination involves diazotization of the amino group on the picloram molecule with nitrite in sulfuric acid solution to form an azo dye. The absorbance of the stable bright-yellow product is then measured spectrophotometrically. The diazotization reaction is specific; no other soil substances or colored species interfere if proper blanks are utilized in the analysis.

Technique: The herbicide is extracted from the soil sample with either 1 M NH_4OAc or 2 M KCl or a 2 M KCl solution containing sufficient KOH to make the extract pH 7. After mixing and centrifuging, aliquots of the clear supernatant liquid are taken, followed by the addition of H_2SO_4 and then $NaNO_2$. After a 10-min period of color development in the dark, the absorbance is recorded and compared with standards.

Time Requirements: The soil suspension must be agitated mechanically with the extracting solution for 30 min and then either centrifuged or allowed to settle. Approximately 10 min is necessary for maximum color intensity to develop after addition of $NaNO_2$ to the acidified sample solution. The overall procedure requires about 1 to 1½ hours.

Instruments and Reagents: Reagents required include H_2SO_4; $NaNO_2$; 1 M NH_4OAc, 2 M KCl, or 2 M KCl-KOH; and standard picloram solutions for the calibration curve. Measurements may be made on any spectrophotometer. A centrifuge for soil separation is optional, but a mechanical mixer for the extraction step is strongly recommended. A room with low illumination is recommended for the diazotization step.

Precision and Limits: Picloram concentrations between 0.5 and 2.0 ppm (5-mL samples) exhibit a straight line when plotted against absorbance at 405 nm. The method is not sensitive enough to be useful at concentrations below 0.5 ppm.

Precautions: The developed color is unstable when exposed to sunlight or to incandescent or fluorescent lamps; a darkroom is therefore advisable. Because the choice of the extractant solution depends on the pH of the soil, it must be optimized for each soil.

Practical Applications: Picloram movement in soil columns under leaching conditions can be analyzed quickly by the proposed method without special instruments. The most used alternative, gas chromatography, is time-consuming because it requires the formation of carboxyl-group derivatives.

Search Technique: The topic of herbicide compounds and their identification was investigated, with "picloram" the primary word searched in *Chemical Abstracts*. The research summary by Cheng provided the most useful description of the analytical procedure, with Leahy and Taylor and also Grover contributing information on other methods such as gas chromatography.

References:

E. L. Bjerke, A. H. Kutschinski, and M. C. Ramsey, *Journal of Agricultural and Food Chemistry* **1967**, *15*, 469.

H. H. Cheng, *Journal of Agricultural and Food Chemistry* **1969**, *17*, 1174.

R. Grover, *Weed Research* **1967**, *7*, 61.

J. S. Leahy and T. Taylor, *Analyst* **1967**, *92*, 371. Data for this analysis by R. Hawkins.

PROBLEMS

Although in some instances other methods might be preferable, each of the first eleven problems can be solved by application of the methods treated in this book. Explain any necessary modifications.

23-1.   Suggest methods for determining the following, assuming each is present in concentrations of at least several percent: (a) the hydrochloric acid and sodium chloride in a mixture of the two; (b) the hydrochloric acid and sulfuric acid in a mixture of the two; (c) the nitric acid and sodium nitrate in a mixture of the two; (d) the bicarbonate in a sodium chloride brine; (e) the ethylene glycol in an automobile cooling system; (f) the iron(III) and iron(II) in a solution containing salts of these two ions; (g) the dichromate and chromium(III) in a solution containing chromium(III) sulfate, potassium dichromate, and potassium sulfate; (h) the copper and nickel in an alloy containing iron and tin; (i) the permanganate and periodate in a mixture of the two potassium salts.

23-2.   Devise a procedure for determining the amount of each component in the following mixtures, assuming each to contain a few percent of inert material: (a) hydrochloric acid and ammonium chloride; (b) calcium sulfate and sodium sulfate; (c) iron(III) nitrate and aluminum nitrate.

23-3.   Suggest a method for determining the amount of iron (a) in a rock consisting mostly of the mineral iron pyrite, FeS_2, and (b) in wine.

23-4.   How could the concentration of cyanide in an approximately 0.1 *M* solution of the sodium salt be determined?

23-5.   A worker in a chemical plant mistakenly added an unknown quantity of iron(III) chloride to a mixing chamber containing the chloride salts of aluminum, calcium, sodium, and iron(II). How could he determine how much iron(III) was present in the final mixture?

23-6.   An organic chemist needs to know whether a bottle of ethylene glycol he plans to use in a synthesis is contaminated with glycerol, $CH_2OHCHOHCH_2OH$. How could he determine to the nearest percent the amount of glycerol present?

23-7.   Maintenance of jet engines includes careful monitoring of bearings for excessive wear. An engineer wishes to take a sample of lubricating oil from a jet engine to determine whether the metal content of the oil may be used as a measure of the level of bearing wear. What methods could he use for rapid analysis of iron, copper, cobalt, and nickel at the 1% relative-error level in the oil?

23-8.   An organic research chemist in a pharmaceutical company has just isolated a new compound from juniper berries that he hopes will neutralize the physiological action of a hallucinogenic drug. The compound melts at 72°C, boils without decomposition at 180°C, and contains both carboxylic acid (–COOH) and primary amine (–NH$_2$) groups. How can the purity of the compound be determined?

23-9.   An entomologist working on mosquito breeding has set up an artificial stream in his laboratory. Unfortunately the larvae keep dying. Metal poisoning from plumbing in the circulating water system is suspected. Among the metals that may be present are copper, iron, and zinc. How could each of these metals be determined at the part-per-million level?

23-10.   Axel Grind was asked to devise a method for determining the ammonium ion in ammonium phosphate. He proposed an acid-base method as follows: "Treat a sample with excess sodium hydroxide and distill the released ammonia. Carefully collect the distillate in excess boric acid solution. Using methyl red indicator, titrate the ammonium borate–boric acid solution with a standard solution of sodium hydroxide." Is Axel's proposed method feasible?

23-11.   The bearings in a car motor have worn out prematurely. Leakage of ethylene glycol antifreeze into the crankcase is the suspected cause. Suggest a method for the determination of ethylene glycol in motor oil.

Each of the following problems is appropriate as a literature-search problem, as discussed in Section 23-4.

23-12.   Propose a method for the determination of the organo-phosphorus pesticides malathion and parathion in a cereal grain. (Agricultural)

23-13.   The concentration of sulfur dioxide in the atmosphere has been used as an indicator of urban air pollution. Recommend a simple, rapid method for the determination of sulfur dioxide in air at the level of 0.01 ppm. (Environmental)

23-14.   Among the chemicals considered hazardous to health in cigarette smoke is catechol. How can the amount of catechol in cigarette smoke be determined? (Biochemical)

23-15.   Some illnesses give rise to unusual proportions of alpha, beta, and gamma globulins in blood serum. How can the ratio of these proteins be determined? (Clinical)

23-16.   Insulating oils for electrical transformers must be kept as water-free as possible if their insulating capacity is to remain high. How can the water content of such an oil be determined? (Industrial)

23-17.   Molybdenum enhances the hardness, strength, and corrosion resistance of steels. How can the amount of molybdenum in a tool steel be determined? (Industrial)

23-18.   How can the amount of secondary butyl bromide in isobutyl bromide be measured? (Chemical)

23-19.   The molecules responsible for the transport of oxygen in the blood of crabs and other marine arthropods contain copper (unlike those of mammals, which contain iron). How can the amount of copper in the blood of a crab be determined? (Biological)

23-20.   How can the phosphorus content of milk be determined? (Nutritional)

23-21.   How can the polychlorobiphenyls in penguin eggs be determined? (Environmental)

23-22. How could oxygen at the level of a part per thousand be determined in a gaseous mixture of nitrogen and carbon dioxide? (Chemical)

23-23. How could the titanium in a sample of Australian beach sand containing rutile, zircon, garnet, and quartz be determined? (Geological)

23-24. How can the vinyl chloride in the atmosphere of a chemical plant be monitored? (Environmental)

23-25. Propose a method for the determination of Vitamin E in an infant formula. (Nutritional)

23-26. Suggest a procedure whereby traces of the lanthanide elements could be determined at the part-per-million level in geological material. (Geological)

23-27. Strychnine is used in low dosages as a stimulant of the central nervous system but in large doses may be fatal. How can the strychnine in a solution containing on the order of 100 ppm be determined in the presence of other basic organic compounds? (Clinical)

23-28. Estrogen concentration levels in urine are often important in assessing fetal growth and well-being. How can the total estrogens in a urine sample be determined? (Clinical)

23-29. Esters of phthalic acid are widely used as plasticizers. (Over 5×10^5 tonnes a year are produced for use in plastics.) Several of these esters are slow to degrade, and there is concern that they may be toxic to some aquatic organisms. Their presence has been reported in midocean water samples. How could the amount of dibutylphthalate or di-2-ethylhexylphthalate in a sample of sea water be measured? (Environmental)

23-30. N'-nitrosonornicotine, which causes tumors in the esophagus and lungs of laboratory animals, has been detected in tobacco in amounts as great as 90 ppm. How could a sample of tobacco be analyzed for this nitrosamine? (Agricultural)

23-31. Pulp mills often produce a mixture of wastes, known as black liquor, which contains alkylsulfides, polysulfides, and free sulfide. How could the free sulfide in a black liquor be determined? (Industrial)

23-32. The presence of morphine in postmortem specimens implies possible use of heroin. How could the amount of morphine in a sample of blood be measured? (Forensic)

23-33. Ozone in parts-per-billion concentrations in the air readily attacks a variety of organic materials, including rubber, microfilm, and human tissue. How can the levels of ozone concentration in air be determined? (Environmental)

23-34. The levels of oxygen are important in the strength and brittleness of metals. How can oxygen at the level of 100 ppm in copper metal be determined? (Industrial)

23-35. How can the sodium at a level of several percent be determined in a mixture of salts with part-per-thousand precision? (Chemical)

23-36. How can the amount of water in a sample of a strong-base ion-exchange resin be determined? (Chemical)

23-37. Coke for steel-making is prepared by heating the proper type of coal. Gases produced in the process often contain polynuclear aromatic hydrocarbons. Propose a method for the determination of benzo(a)pyrene in the effluent of a coke oven. (Environmental)

23-38. Propellants for rockets often contain plasticizers to convert propellant liquids to solids. Degradation of the material on storage leads to poor ballistics performance. Propose a method whereby the plasticizer triacetin could be determined in a propellant, along with the degradation product diacetin. (Industrial)

23-39. How could the amount of thorium at the 0.1 % level in a thorium ore be determined? (Geological)

23-40.   How could the concentration of tellurium(VI) be determined in the presence of tellurium(IV)? (Chemical)

23-41.   In Section 23-4 an example of a literature-search report is given. The report lists four references (Bjerke, Cheng, Grover, Leahy). Use *Science Citation Index* to obtain a listing of references to these publications since 1975. Should the report presented be modified in the light of the *Index* information? If so, how?

23-42.   For one of the problems listed below, proceed as follows: First examine the last *Decennial Index of Chemical Abstracts* to locate several pertinent references. With the information obtained from these references, examine *Science Citation Index* for the past three years to obtain a listing of up-to-date references. Then complete the assignment as directed in (a) Problem 23-12, (b) Problem 23-25, (c) Problem 23-30, (d) Problem 23-34.

23-43.   In Section 1-2 the changing detection limits for mercury with time are discussed. This has created an obstacle to understanding among members of the public. On the basis of a literature search, give the detection limits in 1950 and in 1980 for (a) sulfur dioxide in air, (b) PCB (polychlorinated biphenyls) in meat, (c) vinyl chloride in air, and (d) ozone in air.

REFERENCES

C. R. Burman, *How to Find Out in Chemistry*, Pergamon, Elmsford, New York, 1966. A useful guide to sources of information.

E. I. Crane, A. M. Patterson, and E. B. Marr, *A Guide to the Literature of Chemistry*, 2nd ed., Wiley, New York, 1957.

R. E. Maizell, *How to Find Chemical Information: A Guide for Practicing Chemists, Teachers, and Students*, Wiley, New York, 1979. Designed as a practical guide, especially for those just entering an industrial position.

M. G. Mellon, *Chemical Publications—Their Nature and Use*, 4th ed., McGraw-Hill, New York, 1965.

ANSWERS TO NUMERICAL PROBLEMS

(Answers prepared by C. Johnson, D. Brown, J. Nolan, and S. Reid.)

CHAPTER 1

1-3. 3, 4, 4, 3, 7, 3, 3; **1-4.** 3.30×10^5, 7.00×10^5; **1-5.** 43.64, 0.7778, 4.426×10^{-7}, 3.000×10^4; **1-6.** 231.735, 69.62, 258.32, 239.3, 18.0152; **1-7.** 0.802; **1-10.** (a) 1%, (b) 0.1%, (c) 0.05%; **1-11.** (a) 5, 0.5, 4 ppt; (b) 0.26%; (c) 3 ppt, 3 ppt; **1-12.** 1, 2, 4, 71 ppt; **1-13.** (a) av 40.065, av dev 0.06, rel av dev 2 ppt; (b) 0.002, 3 ppt; (c) 5, 4 ppt; **1-14.** av 33.31, av dev 0.15, std dev 0.20, rel std dev 6 ppt, median 33.30, prob dev 0.13, rel prob dev 4 ppt, accuracy 2 ppt; **1-15.** 0.26; **1-16.** av 3.49, median 3.54, rel std dev 34 ppt; **1-17.** av 0.300, median 0.304, rel std dev 117 ppt, rel prob dev 82 ppt; **1-18.** std dev 0.12, rel std dev 0.25%, accuracy 0.27%, rel accuracy = 0.57%; **1-20.** av 31.1925, std dev 0.0176 g = \$0.283/coin, av accuracy \$1.43/coin; **1-21.** av 10.07, median 10.05, report 10.07; **1-22.** av 10.34, median 10.18, report 10.18; **1-23.** av 39.73%, std dev 3 ppt, 8 ppt; **1-24.** median 40.07,, prob dev 0.06, reject 40.32, av 40.06, av dev 0.05, std dev 0.06; **1-25.** (a) median 12.37, prob dev 0.03, reject 12.45, av 12.36, av dev 0.02, std dev 0.02; (b) median 0.432, prob dev 0.002, av 0.432, av dev 0.002, std dev 0.003; (c) median 122.2, prob dev 0.4, reject 120.2, av 122.4, av dev 0.3, std dev 0.4; (d) median 87.40, prob dev 0.12, av 87.44, av dev 0.16, std dev 0.23.

CHAPTER 2

2-10. 20.0053 g, 19.9982 g, 20.0000 g, 20.0053 g; **2-11.** 1, 2, 4, 8, 16, 32, 64; 1, 2, 2, 5, 10, 20, 20, 50; **2-12.** 5 turns up; **2-13.** 5.0018 g; **2-15.** (a) 0.5007 g, (b) 2.9998 g; **2-16.** 60.4349 g, 17.6638 mL, 3.4175 g/mL; **2-17.** 24.936 g, 25.016 mL; **2-18.** 0.025 g/weighing; **2-19.** 0.33°C; **2-20.** 4.6 ppt; **2-22.** 0.002 mL; **2-23.** (a) 3 ppt, (b) 1 ppt; **2-24.** cum corr $+0.06$, $+0.12$, $+0.17$, $+0.29$, $+0.24$; **2-29.** cum corr $+0.03$, $+0.07$, -0.08, -0.04, $+0.01$; **2-30.** 0.00 mL; **2-31.** 21.38 mL; **2-33.** 10.02 mL, 0.6 cm; **2-34.** 999.1 mL; **2-36.** (a) 0.2001 M, (b) 0.1999 M, (c) 0.1996 M; 4°C; **2-39.** (a) 0.001138 mol, (b) 0.001134 mol, (c) 0.001136 mol; **2-40.** 0.2051 M; **2-41.** (a) 40.6058 g, (b) 24.1763 g, (c) 7.0430 g, (d) –0.0006 g.

CHAPTER 3

3-6. (a) 0.05, (b) 0.2, (c) 0.03, (d) 0.75, (e) 0.22, (f) 0.57, (g) 0.52, (h) 0.16; **3-7.** (a) 0.44, (b) 0.59,, (c) 0.44, (d) 0.017; **3-14.** 0.205 L of 85% H_3PO_4; **3-15.** (a) 0.0005 mol, (b) 0.01 mol, (c) 1.0 mol, (d) 0.013 mol; **3-16.** 9×10^{-10} M, 9×10^2 mol/km³; **3-17.** (a) 87.4 M, (b) 29.6 M, (c) 9.04 M, (d) 2.68 M, (e) 2.58 M, (f) 2.72 M; **3-18.** 7×10^{-6} M, 2×10^{-6} mol/kg; **3-19.** 25.6 M; **3-20.** 2×10^{-4} g, 6×10^{-7} mol; **3-21.** 2.26 mol, 9.4 L of 0.24 M NaOH, 0.12 L of 50% NaOH, 12.5 mol/kg; **3-22.** 0.03 L of 38% HCl, 0.07 L of 6 M HCl; **3-23.** (a) 0.005 M to 0.012 M, (b) 200 to 500 ppm; **3-24.** [Mg] 5.35×10^{-2} M, 5.22×10^{-2} mol/kg, 0.127%, 1.27 ppt; [Br] 8.3×10^{-4} M, 8.1×10^{-4} mol/kg, 0.0065%, 0.065 ppt; **3-25.** (a) 12.4 M, (b) 14.8%, (c) 17.5%; **3-26.** 2.5 g; **3-27** (a) 0.1656 g, (b) 1.095 g, (c) 0.8499 g; **3-28.** (a) 0.02 L, (b) 0.03 L, (c) 0.006 L.

CHAPTER 4

4-2. 47.3%; **4-3.** Cl 0.03 M, Br 0.01 M, I 0.008 M; **4-4.** 18.42 mL; **4-5.** 0.5821 g; **4-6.** 49 mL 4.00, 49.99 mL 4.98, 50.00 mL 5.00; **4-7.** (a) 5 mL 2.30, 10 mL 2.85, 15 mL 7.10, 25 mL 7.70; (b) 5.00; **4-8.** 8.85, 8.60, 8.12, 7.42, 7.25, 6.97, 6.13, 2.85, equiv pt 5.00; **4-9.** 11.09, 11.01, 10.62, 9.82, 8.80, 7.80, 6.05, 4.30, 3.30, 2.32; **4-10.** 0.41 to 0.44 g; **4-11.** 1.34, 1.48, 2.26, 2.99, 6.20, 9.39, 10.06; **4-12.** 1.55, 1.65, 2.40, 3.01, 5.70, 8.38, 9.13; **4-13.** 0.3 M; **4-14.** 0.0105 M; **4-15.** 257.5 g; **4-17.** 407 L; **4-18.** 0.12 M; **4-19.** 1.469%; **4-20.** 38.6% NaCl, 61.4% KBr; **4-21.** [Ag^+] = 3×10^{-5} M, [Cl^-] = 3×10^{-6} M; **4-22.** [Br^-] = 1×10^{-8} M, [I^-] = 3×10^{-12} M; **4-23.** pH = 8.9, pH_{max} = 10.7; **4-24.** 5 M; **4-25.** 10.67% NaOH, 53.59% NaCl, 35.74% H_2O; **4-28.** 2 M; **4-29.** 14.45%; **4-30.** 0.553 $_6$ mg/L; **4-31.** 0.1255 g; **4-32.** 3.98_6%; **4-33.** 56.44%; **4-34.** 16.83%.

CHAPTER 5

5-2. 3.64, 6.34, 1.02, 4.62, 8.49, 11.59; **5-3.** 6.9×10^{-3}, 4.2×10^{-6}, 7.1×10^{-8}, 8.1×10^{-9}, 7.6×10^{-12}, 2.1×10^{-5}, 1.8×10^{-10}, 5.9×10^{-10}, 1.4×10^{-4}, 4.2×10^{-7}, 1.3×10^{-2}; **5-9.** 3.30; **5-10.**

7.04, 7.00; **5-11**. 4.48; **5-12**. (a) 13.6, (b) 1.2, (c) 2.7, (d) 4.8, (e) 9.63, (f) 1.3, (g) 10.2;   **5-13**. (a)
α_{H2CO3} = 0.97, α_{HCO3-}= 0.034, α_{CO32-} = 1.7 × 10^{-7}, $[H_2CO_3]$ = 0.048 M, $[HCO_3^-]$ = 0.0017 M,
$[CO_3^{2-}]$= 8.5×$10^{-9}M$,(b) α_{H2CO3} = 1.9 × 10^{-4} M, α_{HCO3-}= 0.67, α_{CO32-} = 0.33, $[H_2CO_3]$ = 9.5 ×
10^{-6} M, $[HCO_3^-]$ = 3.3 × 10^{-2} M, $[CO_3^{2-}]$ = 1.7 × 10^{-2} M; **5-14**. (a) $[H_3PO_4]$ = 1.4 × 10^{-6} M,
$[H_2PO_4^-]$ = 0.031 M, $[HPO_4^{2-}]$ = 0.069 M, $[PO_4^{3-}]$ = 8.7 × 10^{-7} M; (b) 0.232; **5-15**. 4.7, 4.7;
5-18. 7 g; **5-19**. 0.8_6 mL of 1 M H_3PO_4; **5-20**. 700:1; **5-28**. 17 mL; **5-29**. pH 11.1, 9.26, 5.3, 1.7;
5-30. pH 11.4, 9.8, 5.6; **5-33**. 2.877%; **5-34**. 3.65%; **5-36**. 0.1853 M; **5-37**. 50.16%; **5-39**. 32.4%
Na $_2CO_3$, 58.25% NaOH; **5-40**. 0.96 at pH 6, 0.20 at pH 8; **5-41**. (a) 0.90, (b) 1.7 × 10^{-3} M, (c)
0.043; **5-43**. 0.43 g; **5-45**. (b) 4.12 × 10^{-3}%, 1.9%, 4.04 × 10^{-3}%; **5-46**. 3.65; **5-47**. (a) 1.63, (b)
2.15, (c) 4.65, (d) 7.15, (e) 9.78; **5-48**. 6.98; **5-49**. 0.46 M, 2.8%; **5-50**. 2.12 g; **5-51**. 2.4 g; **5-52**.
0.20 mol $CH_2ClCOOH$, 0.05 mol salt, 4:1.

CHAPTER 6

6-6. pK_s = 15.2, acid pH = 2, base pH = 13.2; **6-7**. pK_s = 33, acid pH = 2, base pH = 31;
6-8. pH = 6.2, buffer pH = 10.3; **6-10**. $10^{-10.7}$, 8.4; **6-11**. 9.6; **6-17**. formic acid 3.1, acetic acid
7.2, ammonia 14.9, acetonitrile 16; **6-18**. 0 mL, 17.9; 12.5 mL, 17.4; 24 mL, 16.2; 25 mL, 9.4; 26
mL, 2.7; 30 mL, 2.0; with H_2O 24 mL, 11.3; 26 mL, 2.7; **6-29**. 98.8%; **6-32**. 95.4%.

CHAPTER 7

7-9. 5.7 × 10^{-8}; **7-10**. 1 × 10^{-16}; **7-11**. α_{Y4-} = 0.16, α_{Cd2+} = 9 × 10^{-5}, $K'_{M'Y'}$ = 4 × 10^{11};
7-13. (a) 5.6 × 10^{11}, (b) 5.3 × 10^6; **7-14**. α_{Ag} = 3.3 × 10^{-19}, $[Ag^+]$ = 3.3 × 10^{-22} M; **7-15**. α_{Cu}
= 3.6 × 10^{-5}, $[Cu^{2+}]$ = 3.6 × 10^{-10} M; **7-16**. K_1 = 2 × 10^3, K_2 = 80, K_3 = 16, K_4 = 0.52;
7-17. (a) 1.4 × 10^{-11} M, (b) 1.7 × 10^{-6} M, (c) 1.4 × 10^{-2} M; **7-18**. (a) 1 × 10^{-2} M, (b) 1 × 10^{-2}
M; **7-19**. $[Ox^{2-}]$ = 9 × 10^{-4} M, $[Ca^{2+}]$ = 2 × 10^{-6} M; **7-20**. 0.2718 M; **7-22**. pH 5 α_{H2Y2-} = 0.9,
pH 9 α_{HY3-} = 0.9, pH 12 α_{Y4-} = 0.98; **7-23**. pH 7 1.9 × 10^{15}, pH 9 2.1 × 10^{17}, pH 10 1.4 × 10^{18},
pH 12 3.9 × 10^{18}; **7-25**. 6 × 10^{-7} M, 3 × 10^{-10} M; **7-27**. 0 mL, 2.00, 10 mL, 2.20, 20 mL, 2.41, 30
mL, 2.70, 40 mL, 3.23, 44 mL, 3.95, 44.9 mL, 4.95, 45 mL, 6.27, 45.1 mL, 7.59, 50 mL, 9.29; **7-28**. 0
mL, 2.00, 10 mL, 2.30, 20 mL, 2.70, 29 mL, 3.77, 29.9 mL, 4.78, 30 mL, 6.07, 30.1 mL, 7.37, 31 mL,
8.37, 40 mL, 9.37; **7-29**. 0 mL, 2.00, 10 mL, 2.30, 20 mL, 2.70, 29 mL, 3.77, 29.9 mL,4.78, 30 mL,
5.27, 30.1 mL, 5.76, 31 mL, 6.76, 40 mL, 7.76; **7-33**. 1.3 × 10^8; **7-35**. Zn, 4 to 9, Cd, 4 to 13, Cu, 3
to 14, Fe, 1 to 14, Al, 4 to 7; **7-41**. 0.01353 M; **7-42**. 0.8 mL; **7-43** 85.6_6% $CaCO_3$, 34.30% Ca;
7-44. 0.105 L; **7-45**. 74.7 mg/L of $CaCO_3$, 74.7 ppm of $CaCO_3$, 62.9 mg/L of $MgCO_3$.

CHAPTER 8

8-5. (a) 1.81 × 10^{-4}, (b) 2.52 × 10^{-15}; **8-6** PbOx = 5 × 10^{-6} M, CaF_2 = 2 × 10^{-4} M;   **8-7**.
Ag_2CrO_4 = 8.5×$10^{-5}M$,AgCl = 1.3 × 10^{-5} M; **8-8**. (a) 8.8 × 10^{14} M, (b) 7.5 × 10^{-3} M, (c) 1.3 ×
10^{-5} M; **8-11**. SO_4^{2-} = 3.0 × 10^{-4} mg; **8-15**. 5.0 × 10^2; **8-16**. $[SO_4^{2-}]$ = 3 × 10^{-8} M; **8-20**. 9.4;
8-25. 2 × 10^{-9} M; **8-26**. 0.12 L; **8-28**. (a) 0.06_8%, (b) 0.20_8%, (c) 0.3 L; **8-29**. 10 mL; **8-30**. (a)
35.453/143.32, (b) 107.868/143.32, (c) 137.33/233.40, (d) 58.70/288.93; **8-31**. 9.337%; **8-32** 10.94%;
8-33. (a) 0.1609, (b) 0.1938, (c) 0.3068, (d) 1.143; **8-34**. 2.384%; **8-35**. 58.98%; **8-36**. 42.3 mL;
8-37. 0.02021 M; **8-38**. 23.35%; **8-39**. 22.941 g/mol; **8-42**. 1.293%; **8-44**. 0.8 mL; **8-47**. 11.83%;
8-48 65.4_7%; **8-49** 77.4_9%; **8-50**. 2.4 g; **8-51**. 21.72%; **8-52**. 60 mL; **8-53**. 23.76% sulfur, 97.9_3
% $(NH_4)_2SO_4$; **8-54**. 83.9_9%; **8-55**. 21.93%; **8-56**. 50 mL, 0.53 g.

CHAPTER 9

9-11. 0° 0.0542, 100° 0.0740; **9-13**. (a) 0.55 V, (b) 0.50 V, (c) 0.79 V, (d) –0.29 V; **9-14**. 0.267
V; **9-15**. –0.121 V; **9-16**. – 0.122 V; **9-17**. –0.118 V; **9-18**. E_1 = 0.56 V, E_2 = 0.63 V; **9-19**. E_2 =
0.43 V, E_5 = 0.26 V; **9-20**. $E_{29°}$ = – 0.12 V, $E_{0°}$ = –0.11 V; **9-21**. E_1 = 0.335 V, E_2 = 0.332 V,
9-22. E = 0.35 V; **9-23**. E_{Cd} = –0.45 V; **9-24**. E = 0.15 V; **9-26**. E = 1.55 V; **9-27**. $[Fe^{2+}]$ = 2
× 10^{-28} M; **9-28**. –0.168 V; **9-29**. (a) 2 × 10^{14}, (b) 6 × 10^{-7}, (c) 3 × 10^{62}, (d) 4 × 10^{79}, (e) 3.2, (f)
7 × 10^{51}, (g) 0.01; **9-31**. 6 × 10^{47}; **9-32**. 2 × 10^{-8}; **9-33**. 4 × 10^{-12}; **9-34**. 8 × 10^{-3};   **9-35**. 7 ×
10^{-13}.

CHAPTER 10

10-3. 0.749 V; **10-5**. 3 × 10^{31}; **10-6**. 5 mL, –0.30 V; 15 mL, –0.26 V; 25 mL, –0.21 V; 29 mL,
–0.17 V; 30 mL, 0.68 V; 31 mL, 1.52 V; 40 mL, 1.58 V; **10-7**. 0.68 V, 3.6 × 10^{-18} M; **10-10**. (a)
0.92 V, (b) 0.35 V, (c) 0.98 V; **10-11**. K_{cond} = 7 × 10^{-14}; **10-12**. 6 × 10^{-17}; **10-13** 1 × 10^{-12} M;

10-14. (a) 0.48 V, (b) no reaction, (c) 0.50 V; **10-16.** (a) 3.1×10^3, (b) 3.1×10^{15}; **10-25.** 9.88_4 %; **10-26.** 26.0_2%; **10-27.** 0.0501 M; **10-28.** 1×10^{-6} M; **10-29.** 0.1693 M; **10-34.** 52.02%; **10-35.** 18 g; **10-37.** 67.1_1%; **10-38.** 19.10%; **10-40.** 0.01691 M; **10-41.** 0.01786 M; **10-42.** 49.1_2 %; **10-43.** 0.01870 M, 59.1_0%; **10-44.** 58.3_4%; **10-45.** 54.05%; **10-47.** 63.5_4%; **10-48.** 18.43%; **10-52.** 0.0622_4 M; **10-53.** 0.819_1 g; **10-54.** 3.4_6 g.

CHAPTER 11

11-4. 0.266 V; **11-7.** 0.272 V; **11-12.** (a) 0.163 V, (b) 0.0113 M, (c) 0.0117 M, (d) 3.5%, (e) $\pm$ 0.0003 V, (f) $\pm$ 0.00015 V; **11-16.** 0.965 V; **11-17.** (b) Ag 0.222 V, Cd –0.403 V; **11-33.** 6.08; **11-34.** 1.9×10^4; **11-35.** 1.6×10^{-3} M; **11-37.** 1 mL, –0.034 V; 10 mL, –0.021 V; 20 mL, 0.008 V; 25 mL, 0.188 V; 30 mL, 0.366 V; **11-38.** (a) 2.8×10^{-4}, (b) 1.1×10^{-4}, (c) –0.012 V; **11-39.** (a) 5.56, (b) 1.66, (c) 8.64, (d) 10.90; **11-40.** (a) 0.36 V vs S.H.E., 0.12 V vs S.C.E.; (b) 0.56 V vs S.H.E., 0.32 V vs S.C.E.; **11-41.** (a) –0.62 V, (b) 0.27 V, (c) 0.24 V vs S.H.E., 0.00 V vs S.C.E.; **11-42.** (b) 2.2×10^4; **11-46.** 2.373 meq, 1.613 meq; **11-49.** 14.10%; **11-50.** 14.86%; **11-51.** 0.17 V; **11-52.** (b) 0.355 V, (c) 0.98 V, (d) 1.61 V; **11-53.** E_{ref} – 0.44 V, K_{sp} = 1.1×10^{-5}; **11-54.** (a) 0.27 V, (b) 0.51 V.

CHAPTER 12

12-7. 0.00148 M; **12-8.** 0.002353 g; **12-9.** (a) 0.001893 M, (b) 3.783×10^{-5} mol, 3.333 mg; **12-11.** (b) 0.01620 g; **12-13.** (a) 1.18 V, (b) –0.45 V; **12-14.** (a) –0.74 V, (b) 10% –0.80 V, 1% – 0.86 V, 0.1% –0.92 V; **12-30.** 696 s; **12-34.** 5.90%; **12-35.** 2.67 $mg^{2/3}s^{-1/2}$; **12-43.** 26.4_8 mg.

CHAPTER 13

13-1. 4.9×10^{-5} M min^{-1}; **13-4.** 0.69 min^{-1}; **13-5.** 11.3 h; **13-6.** 5.0×10^3 s; **13-12.** 34.0 min; **13-13.** 103 mg; **13-14.** 6.84 min; **13-15.** 0.425 ± 0.009; **13-16.** 1.60×10^{-12} M; **13-18.** 0.569; **13-19.** 4.0×10^3 years; **13-22.** 3.65 pg; **13-23.** 9.71×10^{-16} g; **13-24.** 5.35×10^{-3}%; **13-26.** 0.00281; **13-27.** 43.5 days; **13-28.** 8.9×10^3 years; **13-33.** 6.8×10^2 years; **13-35.** 52 years; **13-36.** 0.86.

CHAPTER 14

14-1. 2.9970×10^{10} cm/s; **14-2.** 4.897 μm, 2.042×10^3 cm^{-1}, 8.44×10^{14} Hz; **14-7.** 39.5 cm^{-1} M^{-1}, 0.00128 M; **14-8** 30 $cm^{-1}M^{-1}$, 0.012 M; **14-9.** 3.5×10^{-4} M, 7.9×10^2 $cm^{-1}M^{-1}$; **14-10.** 0.664 mg/100 mL; **14-11.** 1.16×10^3 $cm^{-1}M^{-1}$; **14-12.** At 0.1 M, 650 nm: (a) 0.64, (b) 0.18, (c) 18, etc.; **14-16.** 2.4_0, 1.4_5, 1.1_0, 2.3_0; **14-18.** 0.070%; **14-19.** 0.32 mg/L; **14-21.** C′ = 5.00 μg/mL, c″ = $1.99 \times 10^2 \mu$g/mL; **14-24.** 0.24, 58%; **14-32.** 1:4 salt:acid; **14-34.** 3.2 μg/mL; **14-35.** 0.17 μg/mL; **14-36.** 2.10 μg/mL; **14-37.** 2.0 μg/g; **14-39.** 2.04 μg/100 mL; **14-42.** 19.6% Co, 20.2% Ni; **14-43.** w = 2.84×10^{-5} M; Q = 1.35×10^{-4} M; **14-45.** 1.66×10^{-4} M in Co, 4.33×10^{-5} M in Ni; **14-46.** 2.9×10^{25}.

CHAPTER 15

15-6. (a) 4.143×10^{-4} cm, 2.414×10^3 cm^{-1}; (b) 1.176×10^{-4} cm, 8.502×10^3 cm^{-1}; (c) 6.27 $\times 10^{-4}$ cm, 1.59×10^3 cm^{-1}; (d) 2.735×10^{-4} cm, 3.656×10^3 cm^{-1}; **15-7.** 1.56×10^3 cm^{-1}, 4.66 $\times 10^{13}$ Hz, 3.09×10^{-20} J; **15-8.** 2×10^{-4} mg/mL chloroform, 1.5×10^{-4} mg/mL methylene chloride, 4×10^{-5} mg/mL benzene; **15-9.** (a)$2._5 \times 10^{-4}$ mm, (b) $1._7 \times 10^{-4}$ mm; (c) $4._5 \times 10^{-4}$ mm; **15-16.** sample 1 0.15% Na_2O, 0.65% K_2O; sample 2 0.47% Na_2O, 0.47% K_2O; sample 3 0.30% Na_2O, 0.59% K_2O; **15-37.** 1.73×10^{-3} mg/mL; **15-38.** 0.155%; **15-40.** 0.33%; **15-42.** 0.14%; **15-43.** 4.3×10^{-4}%.

CHAPTER 16

16-12. 0.048; **16-13.** (a) 0.20, (b) 0.053; **16-14.** 4.5; **16-15.** 5.0; **16-16.** 4; **16-17.** 1.4 mg; **16-18.** 12; **16-20.** 1, $f_{x,w}$ = 0.01, $f_{x,org}$ = 0.99, $f_{y,w}$ = 0.98, $f_{y,org}$ = 0.02; 2, $f_{x,w}$ = 0.0001, $f_{y,w}$ = 0.96; 3, $f_{x,org}$ = 1.0, $f_{y,org}$ = 0.06; **16-26.** (a) for x (tubes 1, 2, 3, 4), equilibrium 1: 1.00, -, -, -; 2: 0.33, 0.67, -, -; 3: 0.12, 0.44, 0.44, -; 4: 0.03, 0.23, 0.44, 0.30. For y (tubes 1, 2, 3, 4), equilibrium 1: 1.00, -, -, -; 2: 0.83, 0.17, -, -; 3: 0.69, 0.28, 0.03, -; 4: 0.58, 0.35, 0.07, 0.00_5; (b) 1 equil, 1; 2 equil, 0.4, 3.9; 3 equil, 0.17, 1.6, 15; 4 equil,, 0.05, 0.66, 6.3, 60; **16-27.** (a) 0.71, (b) 0.0019; **16-28.** 1.58%; **16-35.** 3,5; **16-37.** 1.31 g; **16-38.** 1.8032 g; **16-39.** (a) 4.6_5, (b) 5.6_5; **16-40.** 19.72%; **16-41.** 0.681_3 g.

CHAPTER 17

17-4. 91%; **17-5**. 4.4 × 10^2; **17-6**. 3.33, 0.231 cm/s, 12.99 mL; **17-9**.25.7 mL ret vol, 23.3 mL net ret vol, 11 mL/g specific ret vol; **17-19**. (b) 1.6 for cyclohexane, 3.2 for benzene, 2.0; **17-20**. 12 mm base width; **17-25**. 0.55 mL; **17-26**. 7.6 s, 0.050 cm; **17-28** 0.0062 M; **17-29**. 23.2 mL, 25.4 mL, 0.412; **17-32**. 2.4; **17-33**. (a) increase by square root of 2, (b) decrease by 0.833:0.909, (c) increase by 0.167:0.091; **17-34**. 4.0, 1.5.

CHAPTER 18

18-5. 0.749; **18-11**. ∼ 870; **18-12**. 746; **18-17**. (a) ∼3%, (b) ∼5%; **18-23**. methylethyl ketone 2.2, MIBK 3.2; **18-28**. 53.50%; **18-29**. 0.35, 0.033.

CHAPTER 19

19-9. 9.0 mL; **19-11**. 31 mL, 10.2 min; **19-29**. 3.98 meq, 15.9 meq/g; **19-30**. 50.02%; **19-31**. 87.98% NaCl, 12.02% $NaNO_3$; **19-32**. (a) 9.9 meq/g, (b) 0, (c) 11.5 meq/g, (d) 6.1 meq/g, (e) 0, (f) 0, (g) 14.1 meq/g, (h) 18.0 meq/g; **19-33**. 29.63%; **19-36**. 1.730% Zn, 1.777% Ni; **19-37**. 21.51% Zn, 10.9_4% Cd.

CHAPTER 20

20-13. 1.396 mg/100 mL.

CHAPTER 21

21-5. 0.11, sample wt 3 g; **21-6**. 1.6 g; **21-7**. (a) 1.1 × 10^5, (b) 9 × 10^6, (c) 1 × 10^8, (d) 1 × 10^9, (e) 1 × 10^{11}; **21-8**. (0.005 cm, 0.001 cm, 0.0005 cm) (a) 0.17 g, 0.0014 g, 0.00017 g; (b) 14 g, 0.11 g, 0.014 g; (c) 1.6 × 10^2 g, 1.3 g, 0.16 g; (d) 1.6 × 10^3 g, 13 g, 1.6 g; (e) 1.6 × 10^5 g, 1.3 × 10^3 g, 1.6 × 10^2 g; **21-9**. 0.29%, 0.41%; **21-10**. 48 ppt, 48.

CHAPTER 22

22-4. (a) 90%, 11.23 ± 0.07; 99%, 11.23 + 0.12; (b) 40, 32; **22-5**. 3.48 ± 0.10; **22-6**. 0.300 ± 0.024; **22-9**. t = 1.5; **22-10**. 10; **22-11**. 33.31 ± 0.15; **22-12**. 47.29 ± 0.13; **22-13**. 31.1925 ± 0.0126; **22-15**. (a) b = 5.67, a = –0.0046, (b) 4.67 $M^{-1}cm^{-1}$, av = 0.0688 M $CoCl_2$, std dev – 0.0058, C.L. = 0.0066; for 1 meas, std dev = 0.0098, C.L. = 0.011; **22-16**. Y = –29.8x + 151, 2.0; **22-18**. 0.37, yes; **22-19**. x^2 = 16.8, 0.05; **22-20**. (a) x^2 = 7.84, 0.005; (b) x^2 = 0.16, 0.8; **22-21**. die 1: x^2 = 15.1, 0.01; die 2: x^2 = 3.87, 0.57; **22-23**. t = 4.74, yes; **22-24**. t = 16.79, yes. **22-25**. (a) 0.61, (b) 0.92.

APPENDIX: TABLES OF DATA

Ionization Constants

Acids

Acetic		1.9×10^{-5}
Benzoic		6.7×10^{-5}
Boric	K_1	5×10^{-10}
Carbonic	K_1	3.5×10^{-7}
	K_2	5×10^{-11}
Chloroacetic		1.5×10^{-3}
Chromic	K_2	1×10^{-7}
Citric	K_1	8.7×10^{-4}
	K_2	1.8×10^{-5}
	K_3	4×10^{-6}
Dichloroacetic		5×10^{-2}
EDTA	K_1	7×10^{-3}
	K_2	2×10^{-3}
	K_3	7×10^{-7}
	K_4	6×10^{-11}
Formic		2×10^{-4}
Glycine HCl	K_1	4.6×10^{-3}
	K_2	2.5×10^{-10}
Hydrocyanic		7×10^{-10}
Hydrofluoric		7×10^{-4}
Hypochlorous		3.7×10^{-8}
H_2S	K_1	9×10^{-8}
	K_2	1×10^{-15}
Oxalic	K_1	6×10^{-2}
	K_2	6×10^{-5}
Phenol		1×10^{-10}
Phosphoric	K_1	7×10^{-3}
	K_2	7×10^{-8}
	K_3	4×10^{-13}
Phthalic	K_2	4×10^{-6}
Silicic	K_1	2×10^{-10}
Succinic	K_1	6×10^{-5}
	K_2	2×10^{-6}
Sulfuric	K_2	1.2×10^{-2}
Sulfurous	K_1	2×10^{-2}
	K_2	6×10^{-8}
Trichloroacetic		2×10^{-1}

Bases

Ammonia	1.8×10^{-5}
2-Aminopyridine	5×10^{-8}
Aniline	4.2×10^{-10}
Hydrazine	3×10^{-6}
Hydroxylamine	1×10^{-8}
Methylamine	4×10^{-4}
Monoethanolamine	3×10^{-5}
Pyridine	1×10^{-9}
Trimethylamine	6.3×10^{-5}
Urea	1.5×10^{-14}

Ion product for water at 24°C
$$1 \times 10^{-14}$$

Solubility Products

Aluminum hydroxide	2×10^{-32}
Barium carbonate	5×10^{-9}
Barium chromate	1×10^{-10}
Barium oxalate	2×10^{-8}
Barium sulfate	1×10^{-10}
Cadmium sulfide	1×10^{-28}
Calcium carbonate	5×10^{-9}
Calcium fluoride	4×10^{-11}
Calcium oxalate	2×10^{-9}
Copper(II) hydroxide	2×10^{-20}
Copper(II) sulfide	1×10^{-36}
Iron(III) hydroxide	1×10^{-36}
Lead chromate	2×10^{-14}
Lead sulfate	2×10^{-8}
Lead sulfide	1×10^{-28}
Magnesium carbonate	1×10^{-5}
Magnesium hydroxide	1×10^{-11}
Magnesium ammonium phosphate	2×10^{-13}
Magnesium oxalate	9×10^{-5}
Manganese(II) sulfide	1×10^{-15}
Mercury(I) bromide	3×10^{-23}
Mercury(I) chloride	6×10^{-19}
Mercury(I) sulfide	1×10^{-52}
Potassium perchlorate	2×10^{-2}
Silver argenticyanide	4×10^{-12}
Silver bromide	8×10^{-13}
Silver carbonate	6×10^{-12}
Silver chloride	1×10^{-10}
Silver chromate	2×10^{-12}
Silver iodide	1×10^{-16}
Silver phosphate	1×10^{-19}
Silver sulfide	1×10^{-50}
Silver thiocyanate	1×10^{-12}
Strontium chromate	4×10^{-5}
Zinc hydroxide	5×10^{-18}
Zinc sulfide	1×10^{-24}

Acid-Base Buffer Systems

Buffer Couple	Useful pH Range
$CH_2ClCOOH$, $CH_2ClCOONa$	1.8 – 3.8
CH_3COOH, CH_3COONa	3.7 – 5.7
2-Aminopyridine(2-NH_2Py), 2-$NH_2Py \cdot HCl$	5.7 – 7.7
NaH_2PO_4, Na_2HPO_4	6.1 – 8.1
THAM, THAM·HCl	7.5 – 9.0
NH_3, NH_4^+	8.3 –10.3

Formation Constants[*]

	K_1	K_2	K_3	K_4	overall
$Ag(NH_3)_2^+$	2.0×10^3	8.1×10^3			1.6×10^7
$Cd(NH_3)_4^{2+}$	3.5×10^2	1.0×10^2	2.2×10^1	6.9	5.5×10^6
$Co(NH_3)_4^{2+}$	9.8×10^1	3.2×10^1	8.5	4.4	1.2×10^5
$Cu(NH_3)_4^{2+}$	1.1×10^4	2.7×10^3	1.3×10^2	3.0×10^1	5.6×10^{11}
$Ni(NH_3)_4^{2+}$	5.2×10^2	1.5×10^2	4.6×10^1	1.3×10^1	4.7×10^7
$Zn(NH_3)_4^{2+}$	1.6×10^2	2.0×10^2	2.3×10^2	1.1×10^2	7.8×10^8
$Fe(CNS)^{2+}$	1.0×10^3				
$Hg(CNS)_4^{2-}$	1.2×10^9	6.0×10^7	6.9×10	1.0×10^2	5×10^{21}
$Ag(CN)_2^-$		3.0×10^{20}			5×10^{20}
$Fe(CN)_6^{4-}$					2.5×10^{35}
$Fe(CN)_6^{3-}$					4.0×10^{43}
$Zn(CN)_4^{3-}$	2.0×10^5	5.9×10^5	9.5×10	3.7×10^3	4.2×10^{19}
$Cd(CN)_4^{2-}$	1.0×10^6	1.3×10^5	3.4×10^4	1.9×10^2	8.3×10^{17}
$Hg(CN)_4^{2-}$	1.0×10^{17}	5.6×10^{15}	3.6×10^3	4.6×10^2	9.3×10^{38}
$HgCl_4^{2-}$	5.5×10^6	3.0×10^6	7.6	1.0×10^1	1.3×10^{15}
HgI_4^{2-}	7.4×10^{12}	8.9×10^{10}	6.0×10^3	1.6×10^2	6.3×10^{29}
$Ag(S_2O_3)_2^{3-}$	6.6×10^8	7.1×10^4			4.7×10^{13}
$Ca(EDTA)^{2-}$	1.0×10^{11}				
$Mg(EDTA)^{2-}$	1.3×10^9				
$Al(OH)_4^-$	1.0×10^9	4.9×10^9			
			2.0×10^8	1.0×10^6	1.0×10^{33}
$Cr(OH)_2^+$	1.2×10^{10}	1.7×10^7			2×10^{17}
$Pb(OH)_3^-$	2.0×10^6	4.0×10^4	1.0×10^3		7.9×10^{13}
$Zn(OH)_4^{2-}$	1.0×10^5	1.3×10^6	3.2×10^2	1.6×10^1	6.3×10^{14}

[*]Values at 25° and zero ionic strength where available. From *Critical Stability Constants*, R. M. Smith and A. E. Martell, Plenum Press, New York, N.Y., Vol. 1, 1974, Vol. 4, 1976.

Aqueous Acid-Base Indicators at 25°C

Indicator	Transition Range	pK_{in}	Acid	Base
Thymol Blue	1.2- 2.8	1.6	Red	Yellow
Methyl yellow	2.9- 4.0	3.3	Red	Yellow
Methyl orange	3.1- 4.4	4.2	Red	Yellow
Bromocresol green	3.8- 5.4	4.7	Yellow	Blue
Methyl red	4.2- 6.2	5.0	Red	Yellow
Chlorophenol red	4.8- 6.4	6.0	Yellow	Red
Bromothymol blue	6.0- 7.6	7.1	Yellow	Blue
Phenol red	6.4- 8.0	7.4	Yellow	Red
Cresol purple	7.4- 9.0	8.3	Yellow	Purple
Thymol blue	8.0- 9.6	8.9	Yellow	Blue
Phenolphthalein	8.0- 9.8	9.7	Colorless	Red
Thymolphthalein	9.3-10.5	9.9	Colorless	Blue

Standard Electrode Potentials*

Half-reaction	E^0, V
$Na^+ + e \rightleftarrows Na$	**−2.713**
$H_2AlO_3^- + H_2O + 3e \rightleftarrows Al + 4OH^-$	−2.35
$Al^{3+} + 3e \rightleftarrows Al$	**−1.66**
$Sn(OH)_6^{2-} + 2e \rightleftarrows HSnO_2^- + H_2O + 3OH^-$	−0.90
$Zn^{++} + 2e \rightleftarrows Zn$	**−0.7628**
$AsO_4^{3-} + 3H_2O + 2e \rightleftarrows H_2AsO_3^- + 4OH^-$	−0.67
$U^{4+} + e \rightleftarrows U^{3+}$	−0.63
$Fe^{++} + 2e \rightleftarrows Fe$	**−0.44**
$Cr^{3+} + e \rightleftarrows Cr^{++}$	−0.38
$Cd^{++} + 2e \rightleftarrows Cd$	**−0.403**
$V^{3+} + e \rightleftarrows V^{++}$	−0.255
$Sn^{++} + 2e \rightleftarrows Sn$	**−0.14**
$Pb^{++} + 2e \rightleftarrows Pb$	**−0.126**
$2H^+ + 2e \rightleftarrows H_2$	**0.0000**
$TiO^{++} + 2H^+ + e \rightleftarrows Ti^{3+} + H_2O$	0.1
$S_4O_6^{2-} + 2e \rightleftarrows 2S_2O_3^{2-}$	0.09
$S + 2H^+ + 2e \rightleftarrows H_2S$	**0.14**
$Sn^{4+} + 2e \rightleftarrows Sn^{++}$	0.14
$SO_4^{2-} + 4H^+ + 2e \rightleftarrows H_2SO_3 + H_2O$	0.17
$AgCl + \rightleftarrows Ag + Cl^-$	**0.2223**
$BiO^+ + 2H^+ + 3e \rightleftarrows Bi + H_2O$	0.32
$UO_2^{++} + 4H^+ + 2e \rightleftarrows U^{4+} + 2H_2O$	**0.33**
$Cu^{++} + 2e \rightleftarrows Cu$	**0.34**
$VO^{++} + 2H^+ + e \rightleftarrows V^{3+} + H_2O$	0.34
$O_2 + 2H_2O + 4e \rightleftarrows 4OH^-$	0.401
$H_2SO_3 + 4H^+ + 4e \rightleftarrows S + 3H_2O$	0.45
$Cu^+ + e \rightleftarrows Cu$	**0.52**
$MnO_4^{2-} + 2H_2O + 2e \rightleftarrows MnO_2 + 4OH^-$	0.5
$I_3 + 2e \rightleftarrows 3I^-$	0.545
$H_3AsO_4 + 2H^+ + 2e \rightleftarrows HAsO_2 + 2H_2O$	0.56
$MnO_4^- + e \rightleftarrows MnO_4^{2-}$	**0.57**
$I_2 + 2e \rightleftarrows 2I^-$	**0.621**
$O_2 + 2H^+ + 2e \rightleftarrows H_2O_2$	**0.69**
$OBr^- + H_2O + 2e \rightleftarrows Br^- + 2OH^-$	0.76
$Fe^{3+} + e \rightleftarrows Fe^{++}$	**0.771**
$Hg_2^{++} + 2e \rightleftarrows 2Hg$	**0.792**
$Ag^+ + e \rightleftarrows Ag$	**0.7994**
$OCl^- + H_2O + 2e \rightleftarrows Cl^- + 2OH^-$	0.89
$2Hg^{++} + 2e \rightleftarrows Hg_2^{++}$	**0.907**
$VO_2^+ + 2H^+ + e \rightleftarrows VO^{++} + H_2O$	**0.999**

$2ICl_2^- + 2e \rightleftarrows I_2 + 4Cl^-$	1.06
$Br_2 + 2e \rightleftarrows 2Br^-$	1.08
$2IO_3^- + 12H^+ + 10e \rightleftarrows I_2 + 6H_2O$	**1.19**
$MnO_2 + 4H^+ + 2e \rightleftarrows Mn^{++} + 2H_2O$	**1.23**
$O_2 + 4H^+ + 4e \rightleftarrows 2H_2O$	1.229
$Cr_2O_7^{2-} + 14H^+ + 6e \rightleftarrows 2Cr^{3+} + 7H_2O$	**1.33**
$Cl_2 + 2e \rightleftarrows 2Cl^-$	**1.358**
$2HIO + 2H^+ + 2e \rightleftarrows I_2 + 2H_2O$	1.45
$PbO_2 + 4H^+ + 2e \rightleftarrows Pb^{++} + 2H_2O$	1.455
$Mn^{3+} + e \rightleftarrows Mn^{++}$	1.51
$MnO_4^- + 8H^+ + 5e \rightleftarrows Mn^{++} + 4H_2O$	**1.51**
$2BrO_3^- + 12H^+ + 10e \rightleftarrows Br_2 + 6H_2O$	1.5
$Bi_2O_4 + 4H^+ + 2e \rightleftarrows 2BiO^+ + 2H_2O$	1.59
$2HBrO + 2H^+ + 2e \rightleftarrows Br_2 + 2H_2O$	1.6
$H_5IO_6 + H^+ + 2e \rightleftarrows IO_3^- + 3H_2O$	$\simeq 1.6$
$Ce^{4+} + e \rightleftarrows Ce^{3+}$	1.61
$2HClO + 2H^+ + 2e \rightleftarrows Cl_2 + 2H_2O$	**1.63**
$MnO_4^- + 4H^+ + 3e \rightleftarrows MnO_2 + 2H_2O$	1.68
$H_2O_2 + 2H^+ + 2e \rightleftarrows 2H_2O$	**1.77**
$Ag^{++} + e \rightleftarrows Ag^+$	1.98
$S_2O_8^{2-} + e \rightleftarrows 2SO_4^{2-}$	2.0
$O_3 + 2H^+ + 2e \rightleftarrows O_2 + H_2O$	**2.07**
$F_2 + 2e \rightleftarrows 2F^-$	**2.87**

*Numbers in boldface are considered to be the most reliable by the Electrochemical Commission of the IUPAC. Values from W. M. Latimer, *Oxidation States of the Elements and Their Potentials in Aqueous Solutions*, 2nd ed., app.1, Prentice-Hall, Englewood Cliffs, N.J., 1952, and G. Charlot and others, *Selected Constants*, Supplement to *Pure Appl. Chem.*, IUPAC, Butterworth, London, 1971.

Concentrated Acids and Bases

	Molecular Weight	Density	Weight Percent	Molarity
CH_3COOH	60.05	1.05	99.5	17.4
H_2SO_4	98.07	1.83	94	17.6
HF	20.01	1.14	45	25.7
HCl	36.46	1.19	38	12.4
HBr	80.91	1.52	48	9.0
HNO_3	63.01	1.41	69	15.4
$HClO_4$	100.46	1.67	70	11.6
H_3PO_4	98.00	1.69	85	14.7
NaOH	40.00	1.53	50	19.1
NH_3	17.03	0.90	28	14.8

INDEX